TK6630　　　　　　　　　3233
G75　　　　GROB, BERNARD
1975　　　　BASIC
COPY 3　　　TELEVISION:
　　　　　　PRINCIPLES AND

BASIC TELEVISION
Principles and Servicing

Author of *Basic Electronics*
Coauthor of *Applications of Electronics*

Fourth Edition

BASIC TELEVISION
Principles and Servicing

Bernard Grob

Instructor, Technical Career Institutes, Inc.
(formerly RCA Institutes, Inc.)

McGraw-Hill Book Company

New York
St. Louis
Dallas
San Francisco
Auckland
Düsseldorf
Johannesburg
Kuala Lumpur
London
Mexico
Montreal
New Delhi
Panama
Paris
São Paulo
Singapore
Sydney
Tokyo
Toronto

Library of Congress Cataloging in Publication Data

Grob, Bernard
  Basic television, principles and servicing.

  Bibliography.
  Includes index.
  1. Television. I. Title.
  TK6630.G75   1975   621.388   75-1125
  ISBN 0-07-024927-X

Basic Television

Copyright © 1975, 1964 by McGraw-Hill, Inc. All rights reserved. Copyright 1954, 1949 by McGraw-Hill, Inc. All rights reserved. Printed in the United States of America. No part of this publication may be reproduced, stored in a retrieval system, or transmitted, in any form or by any means, electronic, mechanical, photocopying, recording, or otherwise, without the prior written permission of the publisher.

7890 DODO 832109

The editors for this book were Gordon Rockmaker and Zivile K. Khoury, the designer was Marsha Cohen, and the production supervisor was Patricia Ollague. It was set in Linofilm Helvetica by Textbook Services, Inc.

Dedicated to
Ruth and Harriet Deborah

Contents

PREFACE   xv

CHAPTER 1   APPLICATIONS OF TELEVISION   1

1-1 Television Broadcasting. 1-2 Television Broadcast Channels. 1-3 Color Television. 1-4 Cable Television (CATV). 1-5 Closed-circuit Television (CCTV). 1-6 Picture Phone. 1-7 Facsimile. 1-8 Satellites for Worldwide Television. 1-9 CRT Numerical Displays. 1-10 Video Recording. 1-11 Development of Television Broadcasting. Summary, Self-Examination, and Essay Questions.

CHAPTER 2   THE TELEVISION PICTURE   16

2-1 Picture Elements. 2-2 Horizontal and Vertical Scanning. 2-3 Motion Pictures. 2-4 Frame and Field Frequencies. 2-5 Horizontal and Vertical Scanning Frequencies. 2-6 Horizontal and Vertical Synchronization. 2-7 Horizontal and Vertical Blanking. 2-8 The 3.58-MHz Color Signal. 2-9 Picture Qualities. 2-10 The 6-MHz Television Broadcast Channel. 2-11 Standards of Transmission. Summary, Self-Examination, Essay Questions, and Problems.

CHAPTER 3   TELEVISION CAMERAS   31

3-1 Camera-tube Requirements. 3-2 Image Orthicon. 3-3 Vidicon. 3-4 Plimbicon. 3-5 Silicon Target Plate. 3-6 Solid-state Image Sensor. 3-7 Spectraflex Color Camera Tube. 3-8 Television Cameras. 3-9 Definitions of Light Units. Summary, Self-Examination, Essay Questions, and Problems.

CHAPTER 4   SCANNING AND SYNCHRONIZING   50

4-1 The Sawtooth Waveform for Linear Scanning. 4-2 Standard Scanning Pattern. 4-3 A Sample Frame of Scanning. 4-4 Flicker. 4-5 Raster Distortions. 4-6 The Synchronizing Pulses. 4-7 Scanning, Synchronizing, and Blanking Frequencies. Summary, Self-Examination, Essay Questions, and Problems.

CHAPTER 5 COMPOSITE VIDEO SIGNAL 65

5-1 Construction of the Composite Video Signal. 5-2 Horizontal Blanking Time. 5-3 Vertical Blanking Time. 5-4 Picture Information and Video Signal Amplitudes. 5-5 Oscilloscope Waveforms of Video Signal. 5-6 Picture Information and Video Signal Frequencies. 5-7 Maximum Number of Picture Elements. 5-8 Test Patterns. 5-9 DC Component of the Video Signal. 5-10 Gamma and Contrast in the Picture. 5-11 Color Information in the Video Signal. 5-12 Vertical Interval Test Signals (VITS).
Summary, Self-Examination, Essay Questions, and Problems.

CHAPTER 6 PICTURE CARRIER SIGNAL 88

6-1 Negative Transmission. 6-2 Vestigial-sideband Transmission 6-3 Television Broadcast Channels. 6-4 The Standard Channel. 6-5 Line-of-sight Transmission.
Summary, Self-Examination, Essay Questions, and Problems.

CHAPTER 7 TELEVISION RECEIVERS 103

7-1 Types of Television Receivers. 7-2 Receiver Circuits. 7-3 Signal Frequencies. 7-4 Intercarrier Sound. 7-5 Dividing the Receiver into Sections. 7-6 Receiver controls. 7-7 Receiver Tubes. 7-8 Solid-state Devices. 7-9 Special Components. 7-10 Printed-wiring Boards. 7-11 Localizing Receiver Troubles to One Section. 7-12 Multiple troubles. 7-13 Safety Features.
Summary, Self-Examination, Essay Questions, and Problems.

CHAPTER 8 COLOR TELEVISION 139

8-1 Red, Green, and Blue Video Signals. 8-2 Color Addition 8-3 Definitions of Color Television Terms. 8-4 Encoding the 3.58-MHz C Signal at the Transmitter. 8-5 Decoding the 3.58 MHz C Signal at the Receiver. 8-6 The Y Signal for Luminance. 8-7 Types of Color Video Signals. 8-8 The Color Sync Burst. 8-9 Hue Phase Angles. 8-10 The Colorplexed Composite Video Signal. 8-11 Desaturated Colors with White. 8-12 Color Resolution and Bandwidth. 8-13 Color Subcarrier Frequency. 8-14 Color Television Systems.
Summary, Self-Examination, Essay Questions, and Problems.

CHAPTER 9 COLOR TELEVISION RECEIVERS 167

9-1 Requirements for the Color Picture Tube. 9-2 Normal-Service Switch. 9-3 Monochrome Circuits in a Color Receiver. 9-4 Circuits for Color. 9-5 Manual Color Controls. 9-6 Functions of the Automatic Color Circuits. 9-7 $R-Y$, $B-Y$ Receivers. 9-8 X and Z Demodulators. 9-9 R, G, B Receivers. 9-10 How the Picture Tube Mixes Colors. 9-11 Localizing Color Troubles.
Summary, Self-Examination, Essay Questions, and Problems.

CHAPTER 10 PICTURE TUBES   185

10-1 Types of Picture Tubes. 10-2 Deflection, Focusing, and Centering. 10-3 Screen Phosphors. 10-4 The Electron Beam. 10-5 Electrostatic Focusing. 10-6 Magnetic Deflection. 10-7 Color Picture Tubes. 10-8 Picture Tubes with In-line Beams. 10-9 Grid-Cathode Voltage on the Picture Tube. 10-10 Picture Tube Precautions. 10-11 Picture Tube Troubles.
Summary, Self-Examination, Essay Questions, and Problems.

CHAPTER 11 ADJUSTMENTS FOR COLOR PICTURE TUBES   210

11-1 Color Purity. 11-2 Color Convergence. 11-3 Convergence Correction Waveshapes. 11-4 Screen-grid Adjustments. 11-5 Degaussing. 11-6 Pincushion Correction. 11-7 Overall Setup Adjustments.
Summary, Self-Examination, Essay Questions, and Problems.

CHAPTER 12 POWER SUPPLIES   226

12-1 Basic Functions of a Power Supply. 12-2 The B+ Supply Line. 12-3 Types of Rectifier Circuits. 12-4 Half-wave Rectifiers. 12-5 Low-voltage Supply with Half-wave Rectifier. 12-6 Full-wave Center-tap Rectifier. 12-7 Full-wave Bridge Rectifier. 12-8 Voltage Doublers. 12-9 Voltage Triplers and Quadruplers. 12-10 Heater Circuits. 12-11 Filter Circuits. 12-12 Voltage Regulators. 12-13 Low-voltage Supply with Series Regulator Circuit. 12-14 Troubles in the Low-voltage Supply. 12-15 Hum in the B+ Voltage. 12-16 Flyback High-voltage Supply. 12-17 Fuses and Circuit Breakers.
Summary, Self-Examination, Essay Questions, and Problems.

CHAPTER 13 VIDEO CIRCUITS   261

13-1 Requirements of the Video Amplifier. 13-2 Polarity of the Video Signal. 13-3 Amplifying the Video Signal. 13-4 Manual Contrast Control. 13-5 Video Frequencies. 13-6 Frequency Distortion. 13-7 Phase Distortion. 13-8 High-frequency Response of the Video Amplifier. 13-9 Low-frequency Response of the Video Amplifier. 13-10 Video Amplifier Circuits. 13-11 The Video Detector Stage. 13-12 Luminance Video Amplifier in Color Receivers. 13-13 Functions of the Composite Video Signal. 13-14 The 4.5-MHz Sound Trap.
Summary, Self-Examination, Essay Questions, and Problems.

CHAPTER 14 DC LEVEL OF THE VIDEO SIGNAL   296

14-1 Changes in Brightness. 14-2 Definitions of Terms for the DC Component. 14-3 How a Coupling Capacitor Blocks the Average DC Voltage. 14-4 DC Coupling and AC Coupling. 14-5 Average Value of the Video Signal. 14-6 DC Insertion. 14-7 Video Amplifier DC-coupled to Picture Tube.
Summary, Self-Examination, Essay Questions, and Problems.

CHAPTER 15 AGC CIRCUITS   309

15-1 Requirements of the AGC Circuit. 15-2 Airplane Flutter. 15-3 AGC Bias for Tubes. 15-4 AGC Bias for Transistors. 15-5 Keying or Gating Pulses for the AGC Rectifier. 15-6 AGC Circuits. 15-7 Twin Pentode for AGC and Sync Separator. 15-8 AGC Circuit in IC Chip. 15-9 Transistorized AGC Gate and Amplifier. 15-10 DC Voltages in the AGC Circuit. 15-11 AGC Adjustments. 15-12 AGC Troubles.
Summary, Self-Examination, Essay Questions, and Problems.

CHAPTER 16 SYNC CIRCUITS   329

16-1 Vertical Synchronization of the Picture. 16-2 Horizontal Synchronization of the Picture. 16-3 Separating the Sync from the Video Signal. 16-4 Integrating the Vertical Sync. 16-5 Noise in the Sync. 16-6 Horizontal AFC. 16-7 Sync Separator Circuits. 16-8 Phasing Between Horizontal Blanking and Flyback. 16-9 Sync and Blanking Bars on the Screen. 16-10 Sync Troubles.
Summary, Self-Examination, Essay Questions, and Problems.

CHAPTER 17 COLOR CIRCUITS   358

17-1 Signals for the Color Picture. 17-2 The Video Preamplifier Stage. 17-3 The Y-Channel Video Amplifier. 17-4 Chroma Bandpass Amplifiers. 17-5 The Burst Amplifier Stage. 17-6 Color Oscillator and Synchronization. 17-7 AFPC Alignment. 17-8 Chroma Demodulators. 17-9 Matrixing the Y Video and Color Video Signals. 17-10 Video Amplifiers for Color. 17-11 The Blanker Stage. 17-12 Troubles in the Color Circuits.
Summary, Self-Examination, and Essay Questions.

CHAPTER 18 AUTOMATIC COLOR CIRCUITS   394

18-1 Color Killer and Automatic Chroma Control (ACC). 18-2 Color Killer Circuit. 18-3 Color Killer Adjustment. 18-4 ACC Circuit. 18-5 Peak Chroma Control (PCC). 18-6 Automatic Tint Control (ATC). 18-7 Automatic Brightness Limiter (ABL). 18-8 One-Button Tuning.
Summary, Self-Examination, and Essay Questions.

CHAPTER 19 TROUBLES IN THE RASTER AND PICTURE   405

19-1 Raster Circuits. 19-2 Troubles in Height and Width. 19-3 No Brightness. 19-4 Picture Troubles. 19-5 Bars in the Picture. 19-6 Sound in the Picture. 19-7 Ghosts in the Picture. 19-8 Interference. 19-9 Hum Troubles.
Summary, Self-Examination, Essay Questions, and Problems.

CHAPTER 20 DEFLECTION OSCILLATORS 424

20-1 The Sawtooth Deflection Waveform. 20-2 Producing Sawtooth Voltage. 20-3 Blocking Oscillator and Discharge Tube. 20-4 Analysis of Blocking Oscillator Operation. 20-5 Blocking Oscillator Sawtooth Generator. 20-6 Transistorized Blocking Oscillator Circuit. 20-7 Frequency and Size Controls. 20-8 Synchronizing the Blocking Oscillator. 20-9 Types of Multivibrator Circuits. 20-10 Plate-coupled Multivibrator. 20-11 Cathode-coupled Multivibrator. 20-12 Multivibrator Sawtooth Generator. 20-13 Synchronizing the Multivibrator. 20-14 Frequency dividers. 20-15 Transistors in Multivibrator Circuits. 20-16 Sawtooth, Trapezoid, and Rectangular Voltages and Currents. 20-17 Incorrect Oscillator Frequency.
Summary, Self-Examination, Essay Questions, and Problems.

CHAPTER 21 VERTICAL DEFLECTION CIRCUITS 457

21-1 Triode Vertical Output Stage. 21-2 Transistorized Vertical Output Stage. 21-3 Vertical Output Transformers. 21-4 Vertical Linearity. 21-5 Internal Vertical Blanking. 21-6 Multivibrator for Combined Vertical Oscillator and Amplifier. 21-7 Transistorized Blocking Oscillator and Vertical Amplifier. 21-8 Miller Feedback Integrator Circuit. 21-9 Transistor Pair in Vertical Output Circuit. 21-10 Vertical Deflection Troubles.
Summary, Self-Examination, Essay Questions, and Problems.

CHAPTER 22 HORIZONTAL DEFLECTION CIRCUITS 477

22-1 Functions of the Horizontal Output Circuit. 22-2 The Horizontal Amplifier. 22-3 Damping in the Horizontal Output. 22-4 Horizontal Scanning and Damping. 22-5 Boosted B+ Voltage. 22-6 Flyback High Voltage. 22-7 Horizontal Deflection Controls. 22-8 Deflection Yokes. 22-9 Horizontal Output Transformers. 22-10 Analysis of Horizontal Output Circuit. 22-11 Complete Horizontal Deflection Circuit. 22-12 Protective Bias on Horizontal Output Tube. 22-13 Transistorized Horizontal Deflection. 22-14 SCR Horizontal Output Circuit. 22-15 High-voltage Limit Control. 22-16 Troubles in the Horizontal Deflection Circuits.
Summary, Self-Examination, Essay Questions, and Problems.

CHAPTER 23 THE PICTURE IF SECTION 513

23-1 Functions of the Picture IF Section. 23-2 Tuned Amplifiers. 23-3 Single-tuned Circuits. 23-4 Double-tuned Circuits. 23-5 Neutralization of Transistor IF Amplifiers. 23-6 The Picture IF Response Curve. 23-7 IF Wavetraps to Reject Interfering Frequencies. 23-8 Picture IF Amplifier Circuits. 23-9 IC Unit for the Picture IF Section. 23-10 Alignment of the Picture IF Amplifier. 23-11 Troubles in the Picture IF Amplifier.
Summary, Self-Examination, Essay Questions, and Problems.

xii CONTENTS

CHAPTER 24 THE RF TUNER 546

24-1 Tuner Operation. 24-2 Functions of the RF Amplifier. 24-3 RF Amplifier Circuits. 24-4 The Mixer Stage. 24-5 The Local Oscillator. 24-6 Automatic Fine Tuning (AFT). 24-7 Types of RF Tuners. 24-8 VHF Tuner Circuits. 24-9 UHF Tuner Circuits. 24-10 Cable Television Channels. 24-11 Varactors for Electronic Tuning. 24-12 RF Alignment. 24-13 Remote Tuning. 24-14 Tuner Troubles.
Summary, Self-Examination, Essay Questions, and Problems.

CHAPTER 25 ANTENNAS AND TRANSMISSION LINES 582

25-1 Resonant Length of an Antenna. 25-2 Definitions of Antenna Terms. 25-3 How Multipath Antenna Signals Produce Ghosts. 25-4 Straight Dipole. 25-5 Folded Dipole. 25-6 Broad-band Dipoles. 25-7 Long-wire Antennas. 25-8 Parasitic Arrays. 25-9 Multiband Antennas. 25-10 Stacked Antenna Arrays. 25-11 Transmission Lines. 25-12 Characteristic Impedance. 25-13 Transmission-line Sections as Resonant Circuits. 25-14 Impedance Matching. 25-15 Antenna Installation. 25-16 Multiple Installations. 25-17 Troubles in the Antenna System.
Summary, Self-Examination, Essay Questions, and Problems.

CHAPTER 26 CABLE DISTRIBUTION SYSTEMS 616

26-1 Head-end Equipment. 26-2 Distribution of the Signal. 26-3 Types of Distribution Losses. 26-4 Calculation of Total Losses. 26-5 System with Multitaps. 26-6 System with Single Taps at Wall Outlets. 26-7 Decibel (dB) Units. 26-8 Decibel Conversion Charts.
Summary, Self-Examination, Essay Questions, and Problems.

CHAPTER 27 THE FM SOUND SIGNAL 631

27-1 Frequency Changes in an FM Signal. 27-2 Audio Modulation in an FM Signal. 27-3 Definitions of FM Terms. 27-4 Pre-emphasis and De-emphasis. 27-5 Advantages and Disadvantages of FM. 27-6 Reactance Tube. 27-7 Receiver Requirements for the FM Sound Signal. 27-8 Slope Detection of an FM Signal. 27-9 Discriminator Operation. 27-10 Center-tuned Discriminator. 27-11 The Limiter. 27-12 Ratio Detector. 27-13 Quadrature Detector. 27-14 Sound IF Alignment. 27-15 The Audio Amplifier Section. 27-16 Complete Circuits for the Associated Sound Signal. 27-17 Intercarrier Buzz. 27-18 Multiplexed Stereo Sound.
Summary, Self-Examination, Essay Questions, and Problems.

CHAPTER 28 RECEIVER SERVICING   669

28-1 Troubleshooting Techniques. 28-2 Test Instruments. 28-3 DC Voltage Measurements. 28-4 Oscilloscope Measurements. 28-5 Alignment Curves. 28-6 Signal Injection. 28-7 Color-bar Generators. 28-8 Schematic Diagram of Monochrome Receiver with Tubes. 28-9 Schematic Diagram of Solid-state Color Receiver. 28-10 Transistor Troubles.
Summary, Self-Examination, Essay Questions, and Problems.

APPENDIXES   693

    A Television Channel Frequencies
    B FCC Frequency Allocations
    C Television Systems Around the World
    D RC Time Constant

BIBLIOGRAPHY   707

ANSWERS TO SELF-EXAMINATIONS   709

ANSWERS TO SELECTED PROBLEMS   715

INDEX   719

Preface

This book presents a comprehensive course in black-and-white and color television for electronic technicians and for television servicing. The practical explanations of television principles and receiver circuits are planned for the student just starting in television. The book is designed as a text for television courses that follow a course in radio or electronics fundamentals.

This edition has been completely revised to emphasize color television, solid-state circuits, and cable television. Principles of color television are integrated throughout the entire book since modern television is mainly in color. The first chapter, Applications of Television, describes the many uses besides broadcasting. Television receiver circuits are shown with vacuum tubes, transistors, and integrated circuits. However, the emphasis is on solid-state circuits in modern receivers. Chapter 7, Television Receivers, describes the main types of semiconductor devices.

On the practical side, the book explains troubleshooting principles that are helpful in repairing receivers. All the examples of circuits are from typical television receivers. Many photographs are used to show actual components. Common troubles are illustrated by photographs from the screen of the picture tube, including color photographs in color plates I to XV (between text pages 140 and 141). In addition, oscilloscope photographs show actual waveshapes.

Each chapter on receiver circuits ends with the main troubleshooting problems associated with that section. Furthermore, Chapter 19, Troubles in the Raster and Picture, describes common troubles for all the sections in the complete receiver. Troubleshooting techniques with details on the use of the oscilloscope, meters, and signal generators are explained in Chapter 28, Receiver Servicing. This chapter has a foldout page with complete circuit diagrams of a monochrome receiver with vacuum tubes and a solid-state color receiver, including oscilloscope waveforms and a pictorial diagram of the printed-wiring board.

Organization of contents. This fourth edition, like the first three, presents the basic material first, with continuity from topic to topic in explaining the television system. First is an introductory chapter on the applications of television. Then Chapters 2 to 6 describe the details of a television picture, how camera tubes convert the light image into an electrical signal, the scanning procedure, how the composite video signal is formed, and how the signal is transmitted.

After this analysis of the television system, the general requirements of television receivers are described, followed by a detailed description of picture tubes. Chapter 7 explains the basic circuits in television receivers, mainly for monochrome. Then Chapter 8 describes the principles of color television. After this, Chapter 9 explains color receivers, with emphasis on the added circuits compared with black-and-white receivers, especially for the color picture tube. Chapter 10 describes picture tubes for monochrome and color. Chapter 11 explains the special requirements for color picture tubes, particularly the purity and convergence adjustments.

The circuits for each receiver section are presented in detail in separate chapters. The sequence starts with power supplies, followed by the signal circuits, including video, AGC, and sync circuits. Then the fundamental receiver circuits for color are described, with a separate chapter on automatic color circuits, including one-button tuning.

At this point, the chapter on raster and picture troubles explains how the deflection circuits in the receiver produce the scanning raster. Both the vertical and horizontal scanning circuits are considered here, in terms of the raster on the screen of the picture tube. Then separate chapters explain the details of deflection oscillators, vertical deflection circuits, and horizontal deflection circuits, including the flyback high-voltage supply.

After the deflection circuits for the raster are discussed, there are chapters on the IF and rf circuits, including the FM sound signal. Antennas and transmission lines for the television receiver are explained, with a separate chapter on cable distribution systems.

The last chapter on receiver servicing describes troubleshooting techniques and test equipment, with details on color-bar generators. At the end are two complete receiver circuits to show how all the sections fit together.

New chapters and reorganization. The importance of color television can be seen from the fact that four chapters are devoted to it in this edition. Furthermore, the principles of color television are explained toward the beginning of the book instead of leaving this topic for last. After the principles are explained, the block diagram of a color television receiver is analyzed in Chapter 9. This is a new chapter with many important features of color receivers that do not depend on knowledge of color circuits. The details of color circuits are in Chapter 17, and an explanation of automatic color circuits is in Chapter 18, which is also a new chapter.

The new Chapter 19, Troubles in the Raster and Picture, has two purposes. It introduces general troubles for all the receiver sections without the need for details of the deflection circuits. Also, this chapter shows how the vertical and horizontal deflection circuits work together to form the raster.

The requirements of an installation with multiple receivers are explained in the new Chapter 26, Cable Distribution Systems. This chapter indicates the importance of cable TV in modern television. Last but not least, is the new Chapter 1, Applications of Television, which describes television broadcasting, color television systems, and additional ways of using television principles, including video tape recording.

Teaching aids. In this edition, each chapter has a short introduction that describes the main features, followed by a list of the topics explained in the chapter. As a result, the reader can visualize what material is to be learned in each step. The specific order usually helps in understanding the subject. Each chapter also has a summary at the end.

Because self-testing materials have proved successful, the chapters conclude with:

1. Self-examination questions, including multiple-choice, true-false, fill-in, and matching questions. These are based mainly on the summary. Answers are given at the back of the book.
2. Thorough and specific essay questions for review, including functions of components and troubleshooting problems. These questions generally require drawing schematic diagrams, referring to the diagrams in the text, reviewing definitions, and applying the text material.
3. A separate group of numerical problems for practice in quantitative work. Many of these problems review the fundamentals of electronic circuits applied to television receivers. Examples include *RC* time constant, decibels, induced voltages, and resonance. Answers to selected problems are given at the back of the book.

The self-testing exercises are helpful in organizing class work and reviewing the most important features of each chapter. In addition, the book can be used for self-study, since students can check their own progress. A bibliography at the end of the book lists helpful references for more information on television, especially color television, and further study on solid-state circuits.

Credits. The schematic diagrams and photographs have been made available by many companies in the television field, as noted in each illustration. This courtesy is gratefully acknowledged. I want to thank my colleague Gerald McGinty for his invaluable assistance in reviewing the entire manuscript.

For the final credit, it is a pleasure to thank my wife Ruth for her excellent work in typing the manuscript.

Bernard Grob

BASIC TELEVISION
Principles and Servicing

1 Applications of Television

Television means "to see at a distance." In our practical television system, the visual information in the scene is converted to an electric video signal for transmission to the receiver. Then the image is reassembled on the fluorescent screen of the picture tube (Fig. 1-1). In monochrome television, the picture is reproduced as shades of white, gray, and black. In color television, the main parts of the picture are reproduced in all their natural colors as combinations of red, green, and blue.

Originally, the techniques of television were developed for commercial broadcasting, which started in 1941. The ability to reproduce pictures electronically has proved so useful though that many more applications of television are used now for education, industry, business, and visual communications in general. The main applications are described briefly in the following topics:

- 1-1 Television broadcasting
- 1-2 Television broadcast channels
- 1-3 Color television
- 1-4 Cable television (CATV)
- 1-5 Closed-circuit television (CCTV)
- 1-6 Picture phone
- 1-7 Facsimile
- 1-8 Satellites for worldwide television
- 1-9 CRT numerical displays
- 1-10 Video recording
- 1-11 Development of television broadcasting

2 BASIC TELEVISION

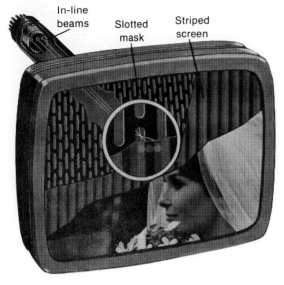

FIGURE 1-1 IMAGE ON FLUORESCENT SCREEN OF COLOR PICTURE TUBE IN TELEVISION RECEIVER. *(RCA)*

1-1 Television Broadcasting

The term "broadcasting" means to send out in all directions. As illustrated in Fig. 1-2, the transmitting antenna radiates electromagnetic radio waves that can be picked up by the receiving antenna. For commercial television broadcast stations, the service area is about 25 to 75 mi in all directions from the transmitter. The radiation is in the form of two rf carrier waves, modulated by the desired information. Amplitude modulation (AM) is used for the picture signal. However, frequency modulation (FM) is used for the sound signal.

Referring to Fig. 1-2, we see that the desired sound for the televised program is converted by the microphone to an audio signal, which is amplified for the sound-signal transmitter. For transmission of the picture, the camera

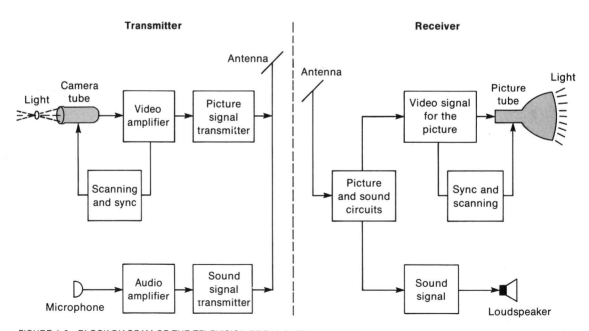

FIGURE 1-2 BLOCK DIAGRAM OF THE TELEVISION BROADCASTING SYSTEM.

tube converts the visual information into electrical signal variations. A camera tube is a cathode-ray tube with a photoelectric image plate. A common type is the vidicon camera tube shown in Fig. 1-3.

The electrical variations from the camera tube become the *video signal*,* which contains the desired picture information. The video signal is amplified and coupled to the picture-signal transmitter for broadcasting to receivers in the service area.

Separate carrier waves are used for the picture signal and sound signal, but they are radiated by one transmitting antenna. Furthermore, the picture and sound signals are included in the broadcast channel for each station. A television channel for a commercial broadcast station is made 6 MHz wide to include both the picture and sound. At the receiver also, one antenna is used for the picture and sound signals.

The receiving antenna intercepts the radiated picture and sound carrier signals, which are then amplified and detected in the receiver. The detector output includes the desired video signal containing the information needed to reproduce the picture. Then the recovered video signal is amplified and coupled to a pic-

*Video is Latin for "see"; audio means "hear."

FIGURE 1-4 MONOCHROME PICTURE TUBE TYPE NUMBER 18VAUP4. SCREEN DIAGONAL IS 18 IN. THE P4 INDICATES WHITE PHOSPHOR SCREEN. (*SYLVANIA ELECTRIC PRODUCTS*)

ture tube that converts the electric signal back into light.

Reproducing the picture. The picture tube in Fig. 1-4 is very similar to the cathode-ray tube used in the oscilloscope. The glass envelope contains an electron-gun structure that produces a beam of electrons aimed at the fluorescent screen. When the electron beam strikes the screen, it emits light.

When the signal voltage makes the control grid less negative, the beam current is increased, making the spot of light on the screen brighter. More negative grid voltage reduces the brightness. If the grid voltage is negative enough to cut off the electron-beam current at the picture tube, there will be no light. This state corresponds to black. A color picture tube has three electron guns for the tricolor screen.

The picture tube is also called a *kine-*

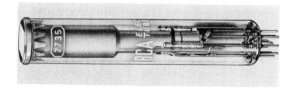

FIGURE 1-3 VIDICON CAMERA TUBE. LENGTH IS 6 IN. (*RCA*)

scope or a *CRT*. Its function is to convert the video signal into a picture.

Scanning and synchronizing. In order for the camera tube to convert the picture information into video signal, the image is dissected into a series of horizontal lines. Similarly, the picture tube reassembles the image line by line. These horizontal lines are produced by making the electron beam scan across the screen. There are 525 lines per picture frame. In addition to this horizontal scanning, vertical scanning is necessary to spread the lines from top to bottom of the screen. There are 30 complete picture frames per second.

Furthermore, the scanning at the camera tube and picture tube must be synchronized, or timed, with respect to the video signal. The synchronization is necessary to reassemble the picture information on the correct lines. These functions are provided by the block of scanning and synchronizing circuits shown in Fig. 1-2 for the transmitter and receiver. The term "synchronizing" is usually abbreviated *sync*.

Most programs are produced live in the studio but recorded on video tape at a convenient time, to be shown later. The quality is so good that the picture looks practically the same as a live program. The studio also has projectors to use 35-mm still slides, opaque slides, and motion-picture film, either 16 or 35 mm, as the program source.

For remote pickups, as in broadcasting a sports event, the signal is relayed to the studio for broadcasting in the assigned channel. When there is a national hookup for important programs, each station in the network receives video signal by intercity relay links, usually leased from the telephone company. A system for satellite relay stations covering the country is being developed for this nationwide television service.

1-2 Television Broadcast Channels

The band of frequencies assigned for transmission of the picture and sound signals is a television *channel*. Each television broadcast station is assigned by the Federal Communications Commission (FCC) a channel 6 MHz wide in one of the following frequency bands: 54 to 88 MHz, 174 to 216 MHz, and 470 to 890 MHz. The first and second bands are in that very high frequency (VHF) spectrum of 30 to 300 MHz originally used for television broadcasting. The band of 54 to 88 MHz includes channels 2 to 6, inclusive, which are the low-band VHF television channels; the band of 174 to 216 MHz includes channels 7 to 13, inclusive, which are the high-band VHF television channels. See Table 1-1.

The band of 470 to 890 MHz is in the ultrahigh-frequency (UHF) spectrum of 300 to 3,000 MHz. This band includes the UHF television channels 14 to 83, inclusive. These have been added to the previous VHF television channels to provide more channels for the expanding television broadcast service.

In all bands, each television broadcast channel is 6 MHz wide. The 6-MHz bandwidth is needed for the picture carrier signal, which is amplitude-modulated by the wide range of video signal frequencies up to 4.2 MHz. For color broadcasts, a 3.58-MHz color video signal is combined with the monochrome video signal so that both can be transmitted as one picture carrier signal. Also included in the 6-MHz channel is the associated sound carrier signal.

The following notes apply to Table 1-1. The band of 44 to 50 MHz was originally allocated as television channel 1, but it is now assigned to other services. The FM radio broadcast band of 88 to 108 MHz is just above channel 6, but this service is not related to television broadcasting. Between television channels 4

TABLE 1-1   TELEVISION CHANNELS

CHANNEL NUMBER	FREQUENCY BAND, MHz
1	not used
2	54–60
3	60–66
4	66–72
5	76–82
6	82–88
FM band	88–108
7	174–180
8	180–186
9	186–192
10	192–198
11	198–204
12	204–210
13	210–216
14–83	470–890

and 5 the frequencies skip from 72 to 76 MHz because this band of frequencies is assigned for other uses, including air navigation. More details on frequency allocations are listed in Appendix B at the back of the book.

1-3   Color Television

The block diagram in Fig. 1-2 illustrates the television broadcasting system for monochrome. In color television, a color camera is necessary at the transmitter and a color picture tube at the receiver. The color camera provides video signal for the red, green, and blue picture information. A color picture tube has red, green, and blue phosphors on the viewing screen to reproduce the picture in color. A typical color picture tube is shown in Fig. 1-5. This type has three electron guns for the tricolor screen. The phosphors can be dot trios of red, green, and blue, or vertical stripes of color as in Fig. 1-1. Then each gun produces an electron beam to illuminate the red, green, or blue phosphor dots on the fluorescent screen.

Although the camera and picture tube operate with red, green, and blue, all other colors including white can be reproduced by combinations of these three colors. Furthermore, in commercial television, the red, green, and blue signals are combined for broadcasting. The purpose is to transmit only a chrominance signal for color and a luminance signal that contains the monochrome information. It is necessary to transmit the luminance signal so that monochrome receivers can reproduce the picture in black and white. The chrominance signal or *chroma signal* has all the information needed to reproduce the picture in color.

The luminance signal is called the *Y* video signal. The chrominance signal can be called the *C* signal. Actually the *C* signal is a modulated subcarrier of 3.58 MHz. This 3.58-MHz *C* signal modulates the assigned picture carrier in the standard 6-MHz television broadcast channel. Furthermore, the 3.58-MHz chrominance signal itself is modulated by two color video sig-

FIGURE 1-5   COLOR PICTURE TUBE TYPE NUMBER 18VATP22. THE P22 INDICATES SCREEN WITH RED, GREEN, AND BLUE PHOSPHORS. (*ZENITH CORPORATION*)

nals. The process of interweaving the Y signal for luminance with the 3.58-MHz color subcarrier signal for color is called *multiplexing*. In terms of the modulated chrominance signal, 3.58 MHz is the frequency for color in the television broadcast system.

Almost all programs now are broadcast in color. However, monochrome receivers use only the luminance signal to reproduce the picture in black and white. Color receivers use both the chrominance and luminance signals. On the screen of the picture tube, the color information is superimposed on the black-and-white picture. It should be noted that the color receiver can also reproduce the picture in monochrome. In fact, if you turn down the color control, or if the 3.58-MHz chrominance signal is missing, you will see a normal black-and-white picture.

1-4 Cable Television (CATV)

When practical radio transmission started in the year 1901, the low radio frequencies of about 100 kHz were used for long distances of hundreds to thousands of miles. As radio developed, higher frequencies were used for services requiring more bandwidth. Now we have television broadcasting in the VHF band of 30 to 300 MHz and the UHF band of 300 to 3,000 MHz. However, the distance for wireless transmission becomes much shorter at these high frequencies. Broadcasting is practically limited to the line-of-sight distance between the transmitting and receiving antennas in the VHF and UHF bands. The useful service range is up to 75 mi for VHF stations and 25 to 35 mi for UHF stations.

Another problem in the VHF and UHF bands is that the wavelength is short enough to allow reflections of the signal by metal structures, such as bridges, steel buildings, and even airplanes. The result is multipath reception of the direct and reflected signals. In the picture, the multipath signals produce multiple images called *ghosts*.

In many cases, the problems of ghosts and weak signals are solved best by going back to the idea of sending signals by wire cable, as in a telephone system. A cable television system provides broadcasting service by a network of coaxial cables. The rf sound and picture carrier signals, including color, are distributed as standard 6-MHz channels to subscribers who pay for this service. Coaxial cable is used because its shielding eliminates pickup of interference or radiation by the lines. The cable supplies at least 1 mV of signal for a strong picture with no snow and no ghosts.

The block diagram in Fig. 1-6 illustrates the head end, trunk lines, and subscriber lines with the required amplifiers. The head end is the source of the signals for the trunk lines, which are the main distribution lines.

Cable television started as an aid for improving reception in two opposite types of locations. The signal is weak in remote areas far from the broadcast transmitter or in a valley blocked by mountains. In big cities, the transmitter may be close, but tall buildings cause multipath reflections. For both types of location, cable television has solved the problem. Because of its advantages, cable television has grown to the point where it supplies 50 percent of the homes in some cities and rural areas. In addition to its ability to supply good signal, cable television provides more channels for the subscriber and new services besides commercial broadcast programs. See Table 1-2. In a typical cable TV service, there are usually 24 channels, in two sets of 12. Typical costs for the subscriber are an installation fee of $10 to $25 and a monthly charge of $4 to $6 per month for each receiver outlet.

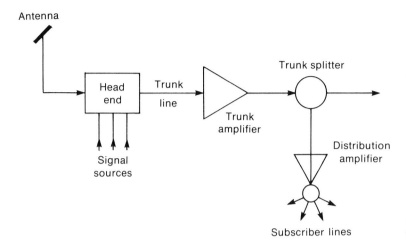

FIGURE 1-6 BLOCK DIAGRAM OF CABLE TV SYSTEM.

The methods of broadcasting television signals by cable apply to several other applications, which are also important but on a smaller scale. For instance, in a hotel, motel, or apartment house, a master antenna can supply signal for all the outlets in the building. This system is called *master-antenna television (MATV)*.

1-5 Closed-Circuit Television (CCTV)

In this system the video signal output from the camera is connected directly by cable to monitors at a remote position, where the picture is reproduced on the screen of the picture tube. A television monitor is a receiver without the rf and IF circuits for tuning. About a 1-V peak-to-peak video signal is required for the monitor. There are any number of possible uses for CCTV in education, industry, medicine, and the home, but just a few are listed here:

Education. One teacher for many classrooms; closeup views of experiments.

Industrial. Night watchman; remote inspection of materials; observe nuclear reactions.

TABLE 1-2 PROGRAM SOURCES FOR CABLE TELEVISION

TYPE	SOURCE
Commercial programs	Off the air from broadcast channels
Local community programs	Local CATV studio
Educational television	Direct wire from schools
Local sports	Remote pickup from mobile microwave equipment
Continuous time and weather	Local CATV studio
Continuous stock market quotations	Local CATV studio
Special events*—theatre and sports	Local CATV studio or direct wire from theatre

*May require extra subscription fee from subscriber.

Business. Train personnel; observe customers and salespeople.

Medicine. Show operation to students; observe patients in bed (Fig. 1-7).

Traffic Control. Observe both ends of a tunnel or bridge; control freight traffic in railroad yard.

Home. Door monitor; babysitter; observe person sick in bed.

Surveillance. Stores; banks; traffic control; crime control.

Since the video signal is not transmitted, CCTV equipment need not follow the standards of television broadcasting. The cameras are very compact, as shown in Fig. 1-8. CCTV equipment is available for monochrome and

FIGURE 1-8 APPLICATION OF CLOSED-CIRCUIT TELEVISION (CCTV) IN TEACHING.

color. The camera lenses are usually the same as for 16-mm film cameras. In some applications, the CCTV program is recorded, either on 16-mm film or on magnetic tape, to be played back at a later time. Furthermore the monitor display can be a large-screen projection unit, as shown in Fig. 1-9.

Theatre television. Special programs can be shown on a large screen by optical projection in a theatre, which usually presents important sports events that are not broadcast for the public. The theatre charges a box-office fee to see the show. Usually, the video signal is supplied by coaxial cable, as another application of closed-circuit television. For color, three projection picture tubes are used, one each for red, green, and blue. Then the three color pictures are superimposed in register on one screen.

Projection of the light image on the face of the picture tube is usually by the *Schmidt reflective system,* preferred because of its efficiency for maximum brightness. This method

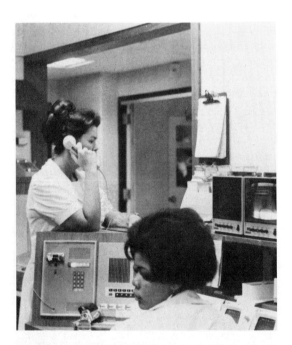

FIGURE 1-7 CLOSED-CIRCUIT TV IN HOSPITAL TO OBSERVE PATIENTS. (*COLUMBIA-PRESBYTERIAN MEDICAL CENTER*)

APPLICATIONS OF TELEVISION 9

service is also provided on an experimental basis in hotels and to homes in some areas. The features include first-run films, sports events, and cultural programs. Distribution of the signal can be by cable or over-the-air transmission in a particular channel. At the receiver, a special decoder or unscrambler is used that enables the receiver to produce the picture and to record the charge made for the program.

1-6 Picture Phone

This system adds television to the telephone service, so that we can see as well as hear each other. As shown in Fig. 1-10, a picture-phone installation includes a CRT display unit and small camera for the picture with a 12-button phone. The picture has 250 lines per frame, 30 frames

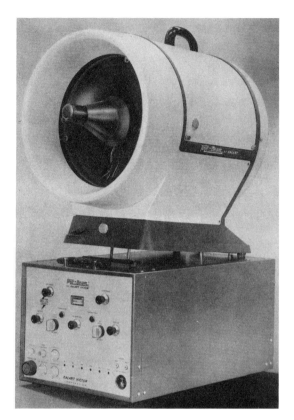

FIGURE 1-9 PROJECTION UNIT FOR TELEVISION PICTURE UP TO 9 × 12 FT. BASE OF PROJECTOR IS 13.5 IN. WIDE. (*KALART VICTOR CORPORATION*)

is adopted from the Schmidt astronomical camera, which uses a spherical mirror to reflect and enlarge the image. With 80 kV on the anode of the picture tube, there is enough light for a picture 20 × 15 ft, at a projection throw distance of 80 ft. The same idea of projection television can be used for a home video theatre with a screen 4 × 3 ft.

Pay television. The idea of charging a fee for special programs is pay TV, box-office TV, or subscription TV. Besides theatre television, this

FIGURE 1-10 PICTURE-PHONE INSTALLATION. (*BELL TELEPHONE LABORATORIES*)

per second, and video frequencies limited to 1 MHz. However, more scanning lines and a wider video-frequency bandwidth may be used for a picture with better resolution. Picture-phone service is also useful for showing still pictures of drawings, photographs, and equipment.

1-7 Facsimile

This application is the electronic transmission of visual information, usually a still picture, over telephone lines. Facsimile is also called *slow-scan television*. Since there is no motion in the scene, the scanning for facsimile can be relatively slow. A typical rate is 360 scanning lines per minute. One important use is the facsimile mail service being used between New York City and Washington, D.C. by the U.S. Post Office. A typical facsimile unit is shown in Fig. 1-11.

1-8 Satellites for Worldwide Television

Transmission in the VHF and UHF bands is limited to the line-of-sight distance to the horizon. Therefore, satellite stations circling the globe are necessary for television transmission over long distances. A satellite orbit is hundreds to thousands of miles above the earth. Satellites for worldwide communications are now an established part of television and telephone service, operated in the United States by Communications Satellite Corporation (COMSAT) in cooperation with the 82-nation International Telecommunications Satellite (*Intelsat*) Consortium. There are satellites over the Atlantic, Pacific, and Indian oceans operating as relay stations between 40 ground stations around the world. The first satellite used for television was Telstar in 1962.

A further step for satellites is a domestic system in each country. In Canada, Anik 1 and Anik 2 are the first domestic satellites to provide communications between the east and west coasts for Canada and the United States, including Alaska. For television in the United States, work is being done on satellites to broadcast directly to receivers throughout the country, using the standard 6-MHz channels.

The Intelsat IV in Fig. 1-12 is a relay station for international communications, providing either 5,000 two-way phone circuits or 12 simultaneous color TV broadcasts. Typical transmission frequencies are 6,000 MHz for the up path to the satellite and 4,000 MHz for the down path to the ground station.

The ground station converts the satellite signal to local standards for transmission on commercial TV broadcast channels. With different television standards in Europe and the United States, a converter at the ground station changes the video signal to the required form. The scanning standards are 525 lines per frame and 30 frames per second in the United States, Canada, South America, and Japan. However, most of Europe uses 625 lines per frame and 25 frames per second. Furthermore, our color system uses the National Television Systems Committee (NTSC) method. In Europe, the Phase-Alternate-Line (PAL) system is generally used for

FIGURE 1-11 FACSIMILE UNIT FOR SENDING COPIES OF DOCUMENTS OVER TELEPHONE LINES. (*TELAUTOGRAPH CORPORATION*)

APPLICATIONS OF TELEVISION 11

FIGURE 1-12 INTELSAT IV SATELLITE FOR INTERNATIONAL COMMUNICATIONS. (*HUGHES AIRCRAFT COMPANY*)

FIGURE 1-13 CRT NUMERICAL DISPLAY FOR STOCK QUOTATION SERVICE. (*BUNKER-RAMO*)

color television. The converter at the ground station can change between scanning standards: from PAL to NTSC color or from NTSC to PAL color. More details on television standards throughout the world are listed in Appendix C.

1-9 CRT Numerical Displays

The cathode-ray tube used for reproducing the picture in television is basically an electronic display that can be used for many kinds of visual information. An example is the numerical display, shown in Fig. 1-13, for quoting stock prices as a financial service. Although it is a numerical display, the picture tube still requires video signal, scanning, and synchronizing. The video signal varies the intensity of the electron beam to illuminate the characters. The synchronizing signal regulates the scanning in accordance with a preestablished pattern to determine the shape for each digit or letter.

1-10 Video Recording

The video signal can be recorded on magnetic tape for picture reproduction, similar to audio tape recording for the reproduction of sound. Figure 1-14 illustrates the elements of a magnetic tape recorder. The tape consists of very fine particles of magnetic iron oxide coated on a plastic base. The transport mechanism pulls the tape past the record-playback head. Since the head contains many turns of fine wire, its magnetic field reacts with the magnetic tape. On the recording position, signal current magnetizes the tape, with the same variations as the signal.

12 BASIC TELEVISION

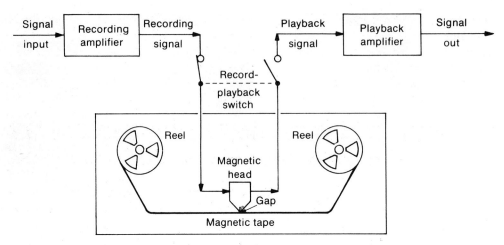

FIGURE 1-14 ESSENTIAL FUNCTIONS OF A REEL-TO-REEL TAPE RECORDER. THE TAPE CAN ALSO BE IN A CASSETTE OR CARTRIDGE.

On playback, the moving magnetic tape induces signal current in the head. The record-playback switch determines whether the machine records or plays back. Although a reel-to-reel recorder is shown here, the same functions apply to cassettes or cartridges that hold the tape.

The problem in video recording is the extreme ratio of highest to lowest signal frequencies. Compared with 20 to 20,000 Hz for audio signal, the video-frequency range is from 30 Hz to 4 MHz, approximately. High-frequency response is limited by the size of the gap in the head and by the speed of the tape. The nonmagnetic gap is usually silicon dioxide, which is similar to glass, with a width only 850×10^{-9} cm.

To provide an effective increase in tape speed, rotating heads can be used. The relative head-to-tape speed is the *writing speed*. Another technique is helical scan, where the head records diagonally across the width of the tape. Finally, the video-frequency bandwidth is reduced to about 2.5 MHz, for portable recorders.

In all cases, the video is recorded as an FM signal on a carrier of approximately 5 MHz. The main reason for the FM recording is to reduce amplitude distortion on playback that can be caused by mistracking of the rotating heads. The tape speed is generally $7\frac{1}{2}$ in./s.

Almost all television studio programs are taped, but the picture quality is so good that it looks practically the same as a live program. Professional video tape recorders generally use rotating heads with full video bandwidth of 4 MHz on 1-in. tape for monochrome or color. However, smaller and lower-priced video tape recorders using $\frac{1}{2}$-in. tape are made for closed-circuit TV or for use in the home. They can record and play back programs on a television monitor or a receiver, in monochrome and color. In addition, small portable cameras are available for a complete television system with the recorder (Fig. 1-15). These portable systems are also used for taping television programs from a remote location away from the broadcast studio. A reel-to-reel recorder is shown in Fig. 1-15, but video cassettes are also used.

In addition to video tape recording (VTR), playback units also use video disk recording (VDR). The disks are similar to phonograph records, but a much greater frequency response is needed for video signal. As one example, an 8-in. disk rotating at 1,500 rpm has a playback time of 10 min. Either a mechanical stylus can be used or the disk is scanned optically by a laser light beam. The video disks are much more economical than magnetic tapes. However, the disks are generally used for playback only, while video tape is for recording and playback.

1-11 Development of Television Broadcasting

The start can be taken from the year 1945 when the Federal Communications Commission (FCC) assigned the VHF channels 2 to 13 now used for commercial television. Originally, channel 1 at 44 to 50 MHz was also allocated, but these frequencies were assigned in 1948 to mobile radio services because of interference problems.

The first popular television receiver, RCA model 630 TS, was marketed in 1946. This was a 30-tube receiver, with a 10-in. round screen. Two important circuits for television first used in this receiver are as follows:

1. *Flyback high-voltage supply.* This circuit produces anode voltage for the picture tube, which is needed for brightness, from the sharp change in current during horizontal retrace. Unless the horizontal scanning circuits are operating, there is no high voltage for the picture tube and no light on the screen.
2. *Horizontal automatic frequency control (AFC).* This circuit improves the horizontal synchronization. With horizontal AFC, the picture will break into diagonal segments if it is out of sync.

These circuits are still used in practically all television receivers, for color or monochrome, that use tubes or transistors.

In 1947, R. B. Dome of General Electric proposed the method of intercarrier sound for television receivers. In this system, the sound signal in the receiver is detected as the 4.5-MHz difference frequency between the rf picture and sound carrier frequencies. The advantage lies in easier tuning of the sound signal, especially on UHF channels. All television receivers now use intercarrier sound, with a common IF amplifier for both the picture and sound signals. As a result, there will be no picture and no sound if the picture IF amplifier is not operating.

Color television broadcasting developed from earlier systems that suffered from mechanical problems, incompatible scanning standards, and excessive bandwidth. In 1949, experimental systems were demonstrated to the FCC by Columbia Broadcasting System (CBS) and by Radio Corporation of America (RCA). The CBS method used a color wheel rotating in front of a black-and-white picture tube. The color information was produced in a sequential pattern that required scanning frequencies different from the monochrome standards for a 6-MHz broadcast channel. The RCA method was all electronic and compatible with monochrome

FIGURE 1-15 VIDEO TAPE RECORDER WITH HAND-HELD CAMERA 15 IN. LONG. (*SONY CORPORATION OF AMERICA*)

broadcasting. The FCC approved the CBS color system for commercial broadcasting because of excellent color reproduction. However, this approval was soon withdrawn in favor of the color system developed by the National Television Systems Committee (NTSC). This committee, as part of the Electronic Industries Association EIA, also developed the standards used for black-and-white television. The NTSC system adopted by the FCC in 1953 is the method used now for all commercial color television in the United States and in most of the countries of North and South America. The details of the NTSC system with its 3.58-MHz chrominance signal are described in Chap. 8 on Color Television.

In 1952, the UHF channels 14 to 83 were assigned by the FCC to provide for more television broadcast stations. Some of these channels are reserved for educational television operated by schools.

In 1962, worldwide television transmission was made possible by the use of satellites circling the earth. The Telstar project of American Telephone and Telegraph Company was the start. Now there are several satellites for worldwide communications, operated by COMSAT in the United States and by Intelsat internationally.

In 1964 a federal law was passed that all television receivers shipped in interstate commerce must have rf tuners for both the VHF and UHF channels. In 1972, it was required that the UHF tuning must be as accurate as the VHF tuning. These laws were passed to increase the use of the UHF television broadcast channels. What was the channel 1 position on the VHF tuner is now used for switching on the UHF tuner.

SUMMARY

1. A commercial television broadcast station uses a 6-MHz channel for the AM picture carrier signal and FM sound carrier signal. Channels 2 to 13 are in the VHF band; channels 14 to 83 are in the UHF band.
2. The camera tube converts the light image into video signal voltage. The picture tube converts the video signal back into light to reproduce the picture.
3. In color television, 3.58 MHz is the frequency used for the chrominance signal. A color picture tube has red, green, and blue phosphors to reproduce the picture in all its natural colors.
4. In cable television (CATV), the rf picture and sound signals are distributed by a network of coaxial cable to subscribers who pay for this service.
5. In closed-circuit television (CCTV), the video and audio signals are connected directly by coaxial cable to monitors or receivers at a remote location.
6. Facsimile, or slow-scan television, is a system for transmitting still pictures over telephone lines.
7. Satellites are used as relay stations to provide worldwide television broadcasting.
8. In video tape recording, the video signal is recorded on magnetic tape, similar to the process of tape recording for audio signal.

Self-Examination (Answers at back of book)

Answer True or False.
1. The picture carrier signal is AM, but the sound carrier signal is FM. T
2. Channel 6 is 66 to 72 MHz. F
3. In color television, 3.58 MHz is the frequency of the chrominance signal. T
4. A picture tube converts video signal to audio signal. F
5. A camera tube converts a light image into video signal. T
6. With a flyback high-voltage supply, there cannot be any light on the screen of the picture tube unless the horizontal scanning circuits are operating. T
7. The range of video signal frequencies is much greater than for audio frequencies. F
8. For intercarrier sound receivers, the frequency for the sound signal is 4.5 MHz. T
9. Facsimile is for still pictures, but television can show motion in the scene. T
10. A television picture frame consists of 525 horizontal scanning lines, repeated every $1/30$ s. T

Essay Questions

1. Define the following abbreviations: AM, FM, FCC, NTSC, EIA, VTR, CATV, CCTV, and AFC.
2. Give the frequency bands for the following channels: 2, 6, 7, 13, 14, and 83.
3. Give two uses for closed-circuit television.
4. Describe briefly the functions of the camera tube and picture tube.
5. Define intercarrier sound.
6. Describe briefly four applications of the principles of a television picture.

2 The Television Picture

Television is basically a system for reproducing a still picture such as a snapshot. However, the pictures are shown one over the other fast enough to give the illusion of motion. One picture frame by itself is just a group of small areas of light and shade. This structure can be seen in Fig. 2-1b, which is a magnified view to show the details of the still picture in a. All the details with varying light and dark spots provide the video signal for the picture information.

We consider black-and-white or monochrome pictures first because these requirements apply for color also. A color television picture is just a monochrome picture with color added in the main areas of the scene. More details are described in the following topics:

- 2-1 Picture elements
- 2-2 Horizontal and vertical scanning
- 2-3 Motion pictures
- 2-4 Frame and field frequencies
- 2-5 Horizontal and vertical scanning frequencies
- 2-6 Horizontal and vertical synchronization
- 2-7 Horizontal and vertical blanking
- 2-8 The 3.58-MHz color signal
- 2-9 Picture qualities
- 2-10 The 6-MHz television broadcast channel
- 2-11 Standards of transmission

THE TELEVISION PICTURE 17

(a)

(b)

FIGURE 2-1 (a) A STILL PICTURE. (b) MAGNIFIED VIEW TO SHOW PICTURE ELEMENTS.

2-1 Picture Elements

A still picture is fundamentally an arrangement of many small dark and light areas. In a photographic print, fine grains of silver provide the differences in light and shade needed to reproduce the image. When a picture is printed from a photoengraving, there are many small black printed dots which form the image. Looking at the magnified view in Fig. 2-1b, we can see that the printed picture is composed of small elementary areas of black and white. This basic structure of a picture is evident in newspaper photographs. If they are examined closely, the dots will be seen because the picture elements are relatively large.

Each small area of light or shade is a *picture element* or *picture detail*. All the elements contain the visual information in the scene. If they are transmitted and reproduced in the same degree of light or shade as the original and in proper position, the picture will be reproduced.

As an example, suppose that we want to transmit an image of the black cross on a white background, shown at the left in Fig. 2-2, to the right side of the figure. The picture is divided into the elementary areas of black and white shown. Picture elements in the background are white, while the elements forming the cross are black. When each picture element is transmitted

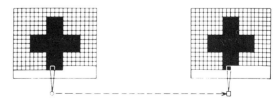

FIGURE 2-2 REPRODUCING A PICTURE BY DUPLICATING ITS PICTURE ELEMENTS.

to the right side of the figure and reproduced in the original position with its shade of black or white, the image is duplicated.

2-2 Horizontal and Vertical Scanning

The television picture is scanned in a sequential series of horizontal lines, one under the other, as shown in Fig. 2-3. This scanning makes it possible for one video signal to include all the elements for the entire picture. At one instant of time, the video signal can show only one variation. In order to have video signal for all the variations of light and shade, all the picture details are scanned in a sequential order of time.

The scanning makes a television picture reproduction different from a photographic print. In a photo, the entire picture is reproduced at one time. In television, the picture is reassembled line after line and frame over frame. This factor of time is the reason why a television picture can appear with the line structure torn apart in diagonal segments and the frames rolling up or down the screen.

The scanning is done in the same way you read to cover all the words in one line and all the lines on the page. Starting at the top left in Fig. 2-3, all the picture elements are scanned in successive order, from left to right and from top to bottom, one line at a time. This method, called *horizontal linear scanning*, is used in the camera tube at the transmitter to divide the image into picture elements and in the picture tube at the receiver to reassemble the reproduced image.

The sequence for scanning all the picture elements is as follows:

1. The electron beam sweeps across one horizontal line, covering all the picture elements in that line.

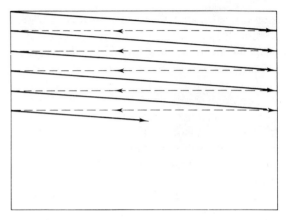

FIGURE 2-3 HORIZONTAL LINEAR SCANNING. SOLID LINES SHOW TRACE FROM LEFT TO RIGHT; DASHED LINES FOR RETRACE FROM RIGHT TO LEFT.

2. At the end of each line, the beam is returned very quickly to the left side to begin scanning the next horizontal line. The return time is called *retrace* or *flyback*. No picture information is scanned during retrace because both the camera tube and picture tube are blanked out for this period. The retraces must be very rapid, therefore, since they are wasted time in terms of picture information.
3. When the beam is returned to the left side, its vertical position is lowered so that the beam will scan the next lower line and not repeat over the same line. This is accomplished by the vertical scanning motion of the beam, which is provided in addition to horizontal scanning.

Lines per frame. The number of scanning lines for one complete picture should be large in order to include the highest possible number of picture elements and, therefore, more details. However, other factors limit the choice, and it has been standardized at a total of 525 scanning lines for one complete picture or frame. This number is the optimum number of scanning lines per frame for the standard 6-MHz bandwidth of the television broadcast channels.

Frames per second. Note that the beam moves slowly downward as it scans horizontally. This vertical scanning motion is necessary so that the lines will not be scanned one over the other. The horizontal scanning produces the lines left to right, while the vertical scanning spreads the lines to fill the frame top to bottom.

The vertical scanning is at the rate of 30 Hz for the frame frequency of 30 frames per second. This value is exactly one-half the ac power-line frequency of 60 Hz. The frame rate of 30 per second means that 525 lines, for one complete frame, are scanned in $1/30$ s.

2-3 Motion Pictures

With all the picture elements in the frame televised by means of the scanning process, it is also necessary to present the picture to the eye in such a way that any motion in the scene appears on the screen as a smooth and continuous change. In this respect the television system is very similar to motion-picture practice.

Figure 2-4 shows a strip of motion-picture film. Note that it consists of a series of still pictures with each picture frame differing slightly from the preceding one. Each frame is projected individually as a still picture; but they are shown one after the other in rapid succession to produce the illusion of continuous motion.

In standard commercial motion-picture practice, 24 frames are shown on the screen for every second during which the film is projected. A shutter in the projector rotates in front of the light source. The shutter allows the light to be projected on the screen when the film frame is still, but blanks out any light during the time when the next film frame is being moved into position. As a result, a rapid succession of still-film frames is seen on the screen.

Persistence of vision. The impression made by any light seen by the eye persists for a small fraction of a second after the light source is removed. Therefore, if many views are presented to the eye during this interval of persistence of vision, the eye will integrate them and give the impression of seeing all the images at the same time. It is this persistence effect that makes pos-

FIGURE 2-4 A STRIP OF MOTION-PICTURE FILM. *(EASTMAN-KODAK CO.)*

sible the televising of one basic element of a picture at a time. When the elements are scanned rapidly enough, they appear to the eye as a complete picture.

In addition, to create the illusion of motion, enough complete pictures must be shown during each second. This effect can be produced by having a picture repetition rate greater than 16 per second. The repetition rate of 24 pictures per second used in motion-picture practice is satisfactory and produces the illusion of motion on the screen.

Flicker in motion pictures. The rate of 24 frames per second, however, is not rapid enough to allow the brightness of one picture to blend smoothly into the next during the time when the screen is black between frames. The result is a definite flicker of light as the screen is made alternately bright and dark. This flicker is worse at higher illumination levels. In motion-picture films, the problem of flicker is solved by running the film through the projector at a rate of 24 frames per second, but showing each frame twice so that 48 pictures are flashed on the screen during each second. A shutter is used to blank out light from the screen not only during the time when each frame is being changed, but once between. Then each frame is projected twice on the screen. There are 48 views of the scene during each second, and the screen is blanked out 48 times per second, although there are still the same 24 picture frames per second. As a result of the increased blanking rate, flicker is eliminated.

2-4 Frame and Field Frequencies

A similar process is used in television to reproduce motion in the scene. Not only is each picture broken down into its many individual picture elements, but the scene is scanned rapidly enough to provide sufficient complete pictures or frames per second to give the illusion of motion. Instead of the 24 rate in commercial motion-picture practice, however, the frame repetition rate is 30 per second in the television system. This repetition rate provides the required continuity of motion.

The picture repetition rate of 30 per second is still not rapid enough to overcome the problem of flicker at the light levels encountered on the picture tube screen. Again the solution is similar to motion-picture practice. Each frame is divided into two parts, so that 60 views of the scene are presented to the eye during each second. However, the division of a frame into two parts cannot be accomplished by the simple method of the shutter used with motion-picture film, because the picture is reproduced one element at a time in the television system. Instead, the same effect is obtained by interlacing the horizontal scanning lines in two groups, one with the odd-numbered lines and the other with the even-numbered lines. A group of odd or even lines is called a *field*.

The repetition rate of the fields is 60 per second, as two fields are scanned during one frame period of $1/30$ s. In this way, 60 views of the picture are shown during 1 s. This repetition rate is fast enough to eliminate flicker.

The frame repetition rate of 30 is chosen in television, rather than the 24 of commercial motion pictures, because most homes in the United States are supplied with 60-Hz ac power. Having the frame rate of 30 per second makes the field rate exactly equal to the power-line frequency of 60 Hz. In countries where the ac power-line frequency is 50 Hz, the frame rate is 25 Hz, making the field frequency 50 Hz. Television standards for the United States and other countries are compared in Appendix C at the back of the book.

2-5 Horizontal and Vertical Scanning Frequencies

The field rate of 60 Hz is the vertical scanning frequency. This is the rate at which the electron beam completes its cycles of vertical motion, from top to bottom and back to top again, ready to start the next vertical scan. Therefore, vertical deflection circuits for either the camera tube or picture tube operate at 60 Hz. The time of each vertical scanning cycle for one field is $1/60$ s.

The number of horizontal scanning lines in a field is one-half the total 525 lines for a complete frame, since one field contains every other line. This results in $262\frac{1}{2}$ horizontal lines for each vertical field. Since the time for a field is $1/60$ s and since it contains $262\frac{1}{2}$ lines, the number of lines per second is

$$262\frac{1}{2} \times 60 = 15{,}750$$

Or, considering 525 lines for a successive pair of fields, which is a frame, we can multiply the frame rate of 30 by 525, which equals the same 15,750 lines scanned in 1 s.

This frequency of 15,750 Hz is the rate at which the electron beam completes its cycles of horizontal motion, from left to right and back to left again, ready to start the next horizontal scan. Therefore, horizontal deflection circuits for either the camera tube or picture tube operate at 15,750 Hz.

The time for each horizontal scanning line is $1/15{,}750$ s. In terms of microseconds,

$$H \text{ time} = \frac{1{,}000{,}000}{15{,}750} \mu s = 63.5\ \mu s \quad \text{approx}$$

This time in microseconds indicates that the video signal for picture elements within a horizontal line can have high frequencies in the order of megahertz. If there were more lines, the scanning time would be shorter, resulting in higher video frequencies. Actually, in our 525-line system, the highest video frequency is limited to approximately 4 MHz because of the restriction of 6 MHz for the commercial television broadcast channels.

2-6 Horizontal and Vertical Synchronization

Time in scanning corresponds to distance in the image. As the electron beam in the camera tube scans the image, the beam covers different elements of the image and provides the corresponding picture information. Therefore, when the electron beam scans the screen of the picture tube at the receiver, the scanning must be exactly timed to assemble the picture information in the correct position. Otherwise, the electron beam in the picture tube can be scanning the part of the screen where a man's mouth should be while at that time the picture information being received corresponds to his nose. To keep the transmitter and receiver scanning in step with each other, special synchronizing signals must be transmitted with the picture information for the receiver. These timing signals are rectangular pulses used to control both transmitter and receiver scanning.

The synchronizing pulses are transmitted as a part of the complete picture signal for the receiver, but they occur during the blanking time when no picture information is transmitted. The picture is blanked out for this period while the electron beam retraces. A horizontal synchronizing pulse at the end of each horizontal line begins the horizontal retrace time, and a vertical synchronizing pulse at the end of each field begins the vertical retrace time. As a result, the receiver and transmitter scanning are synchronized.

Without the vertical field synchronization, the reproduced picture at the receiver does not hold vertically, and it rolls up or down on the picture tube screen. If the scanning lines are not synchronized, the picture will not hold horizontally, as it slips to the left or right and then tears apart into diagonal segments.

In summary, then, the horizontal line-scanning frequency is 15,750 Hz, and the frequency of the horizontal synchronizing pulses is also 15,750 Hz. The frame repetition rate is 30 per second, but the vertical field-scanning frequency is 60 Hz, and the frequency of the vertical synchronizing pulses is also 60 Hz.

It should be noted that the scanning frequencies of 15,750 and 60 Hz are exact for monochrome but only approximate for color television. In color broadcasting, the horizontal line-scanning frequency is exactly 15,734.26 Hz, with 59.94 Hz for the vertical field frequency. These exact scanning frequencies are used to minimize interference between the color subcarrier at exactly 3.579545 MHz and the intercarrier sound signal at exactly 4.5 MHz. However, the horizontal and vertical scanning frequencies can be considered generally as 15,750 and 60 Hz. The reason is that the deflection circuits are automatically synchronized at the required scanning frequencies for both monochrome and color broadcasting.

2-7 Horizontal and Vertical Blanking

In television, "blanking" means going to black. As part of the video signal, blanking voltage is at the black level. Video voltage at the black level cuts off beam current in the picture tube to blank out light from the screen. The purpose of the blanking pulses is to make invisible the retraces required in scanning. Horizontal blanking pulses at 15,750 Hz blank out the retrace from right to left for each line. Vertical blanking pulses at 60 Hz blank out the retrace from bottom to top for each field.

The period of time for horizontal blanking is approximately 16 percent of each H line. The total time for H is 63.5 μs, including trace and retrace. The blanking time for each line then is $63.5 \times 0.16 = 10.2$ μs. This H blanking time means that the retrace from right to left must be completed within 10.2 μs, before the start of visible picture information during the scan from right to left.

The time for vertical blanking is approximately 8 percent of each V field. The total time for V is $1/60$ s, including the downward trace and upward retrace. The blanking time for each field, then, is $1/60 \times 0.08 = 0.0013$ s. This V blanking time means that, within 0.0013 s, the vertical retrace must be completed from bottom to top of the picture.

The retraces occur during blanking time because of synchronization of the scanning. The synchronizing pulses coincide with the start of the retraces. Each horizontal synchronizing pulse is inserted in the video signal within the time of the horizontal blanking pulse. In summary, a blanking pulse comes first to put the video signal at black level; then a synchronizing signal occurs to start the retrace in scanning. This sequence applies to blanking horizontal and vertical retraces.

2-8 The 3.58-MHz Color Signal

The system for color television is the same as for monochrome, but in addition the color information in the scene is used. This is accomplished by considering the picture information in terms of red, green, and blue. When the image is scanned at the camera tube, separate video signals are produced for the red, green, and blue picture information. Optical color filters separate the colors for the camera. For broadcasting in the standard 6-MHz television channel, however, the red, green, and blue color video signals are combined to form two equivalent signals, one for brightness and the other for color. Specifically, the two transmitted signals are as follows:

1. *Luminance Signal.* Contains only brightness variations of the picture information, including fine details, as in a monochrome signal. The luminance signal is used to reproduce the picture in black and white. This signal is generally labeled Y signal (not for yellow).
2. *Chrominance Signal.* Contains the color information. This signal is transmitted as the modulation on a subcarrier at 3.58 MHz for all stations. Therefore 3.58 MHz is the frequency for color. This is generally labeled C signal for chrominance, chroma, or color.

In a color television receiver, color signal is combined with luminance signal to recover the original red, green, and blue video signals. These are then used for reproducing the picture in color on the screen of a color picture tube. The color screen has phosphors that produce red, green, and blue fluorescence. All colors can be produced as mixtures of red, green, and blue. A typical color television picture is shown in color plate I.

In monochrome receivers, the Y signal reproduces the picture in black and white. The 3.58-MHz color signal is just not used. In this case, 3.58 Mhz is filtered out of the video signal, to prevent interference with the monochrome picture.

As a result, the color and monochrome systems are completely compatible. When a program is televised in color, the picture is reproduced in color by color receivers, while monochrome receivers show the picture in black and white. Furthermore, programs televised in monochrome are reproduced in black and white by both monochrome and color receivers. The tricolor picture tube can also reproduce white by combining red, green, and blue.

2-9 Picture Qualities

Assuming it is synchronized to stay still, the reproduced picture should also have high brightness, strong contrast, sharp detail, and the correct proportions of height and width. These requirements apply for monochrome and color. In addition, the color picture should have strong color or saturation, with the correct tints or hues.

Brightness. This is the overall or average intensity of illumination, which determines the background level in the reproduced picture. Individual picture elements can then vary above and below this average brightness level. Brightness on the screen depends on the amount of high voltage for the picture tube and its dc bias in the grid-cathode circuit. In television receivers, the brightness control varies the picture tube dc bias.

It should be noted that the fluorescent screen of the picture tube is illuminated on only one small spot at a time. Therefore, the brightness of the complete picture is much less than the actual spot illumination. The bigger the picture, the more light needed from the spot to produce enough brightness.

Contrast. By "contrast" is meant the difference in intensity between black-and-white parts of the reproduced picture, as distinguished from brightness, which is average intensity. The contrast range should be great enough to produce a strong picture, with bright white and dark black for the extreme intensity values. The amount of ac video signal determines the contrast of the reproduced picture. It is the ac signal amplitude that determines how intense the white will be, compared with black parts of the signal. In television receivers, the contrast control varies the peak-to-peak amplitude of the ac video signal coupled to the picture-tube grid-cathode circuit.

Keep in mind the fact that black in the picture is the same light level you see on the picture tube screen when the set is shut off. With a picture, this level looks black in contrast to the white fluorescence. However, the black cannot appear any darker than the room lighting reflected from the picture tube screen. The surrounding illumination must be low enough, therefore, to make black look dark. At the opposite extreme, the picture appears washed out, with little contrast, when viewed in direct sunlight because so much reflected light from the screen makes it impossible to have dark black.

Detail. The quality of detail, which is also called *resolution* or *definition,* depends on the number of picture elements that can be reproduced. With many small picture elements, the fine detail of the image is evident. Therefore, as many picture elements as possible should be reproduced to have a picture with good definition. This technique makes the picture clearer. Small details can be seen, and objects in the image are outlined sharply. Good definition also gives apparent depth to the picture by bringing in details of the background. The improved quality of a picture with more detail can be seen in Fig. 2-5, which shows how more picture elements increase definition.

In our commercial television broadcasting

(a)

(b)

FIGURE 2-5 PICTURE QUALITY IMPROVES WITH MORE DETAIL. (a) COARSE STRUCTURE WITH FEW DETAILS AND POOR DEFINITION. (b) FINE DETAIL AND GOOD DEFINITION.

system, the picture reproduced on the picture tube screen is limited to a maximum of 150,000 picture elements, approximately, counting all details horizontally and vertically. Such definition allows about the same detail as in 16-mm film. This maximum applies to any size frame, from a small picture 4 × 3 in. to a projected image 20 × 15 ft. The reason is that the maximum definition in a television picture depends on the number of scanning lines and on the bandwidth of the transmission channel.

Color level. In effect, the color information is superimposed on a monochrome picture. How much color is added depends on the amplitude of the 3.58-MHz chrominance signal. The amount of color, or color level, is varied by controlling the gain or level for the C signal. In color television receivers this control is called *color, chroma,* or *saturation.* The color control should vary the picture from no color, to pale and medium colors, up to vivid, intense colors.

Hue. What we generally call the color of an object is more specifically its hue, or *tint.* For instance, grass has a green hue. In the color television picture, the hue, or tint, depends on the phase angle of the 3.58-MHz chrominance signal. This phase with respect to a color synchronizing signal is varied by the *hue, or tint, control.* You can set this control for the correct hue of any known color in the scene, such as blue sky, green grass, or pink flesh tones. Then all other hues are correct, as the color synchronization holds the hues in their proper phase.

Aspect ratio. This is the ratio of width to height of the picture frame. Standardized at 4:3, this aspect ratio makes the picture wider than its height by the factor 1.33. Approximately the same aspect ratio is used for the frames in conventional motion-picture film. Making the frame wider than the height allows for motion in the scene, which is usually in the horizontal direction.

Only the proportions are set by the aspect ratio. The actual frame can be any size from a few square inches to 20 × 15 ft as long as the correct aspect ratio of 4:3 is maintained. If the picture tube does not reproduce the picture with this proportion of width to height, people in the scene look too thin or too wide.

The picture tube rectangular screen has the proportions of 4:3, approximately, for width to height. Therefore, when the horizontal scanning amplitude just fills the width of the screen and the vertical scanning just fills the height, the reproduced picture has the correct aspect ratio.

Viewing distance. Close to the screen, we see all the details. However, the individual scanning lines are visible. Also, we may see the fine grain of the picture reproduction. In television, the grain consists of small white speckles, called *snow,* produced by noise in the video signal. The best viewing distance is a compromise, therefore, about 4 to 8 times the picture height.

2-10 The 6-MHz Television Broadcast Channel

The group of frequencies assigned by the FCC to a broadcast station for transmission of their signals is called a *channel.* Each television station has a 6-MHz channel within one of the following bands allocated for commercial television broadcasting:

54 to 88 MHz for low-band VHF channels 2 to 6
174 to 216 MHz for high-band VHF channels 7 to 13
470 to 890 MHz for UHF channels 14 to 83

In all the bands, each TV channel is 6 MHz wide. As an example, channel 3 is 60 to 66 MHz. The specific frequencies for all the TV channels are listed in Table 6-2 in Chap. 6.

Video modulation. The 6-MHz bandwidth is needed mainly for the picture carrier signal. This carrier is amplitude-modulated by the video signal with a wide range of video frequencies up to approximately 4 MHz. The highest video modulating frequencies of 2 to 4 MHz correspond to the smallest horizontal details in the picture.

Chrominance modulation. For color broadcasts, the 3.58-MHz chrominance signal has the color information. This color signal is combined with the luminance signal to form one video signal that modulates the picture carrier wave for transmission to the receiver.

The FM sound. Also included in the 6-MHz channel is the sound carrier signal for the picture, which is called the *associated sound*. The sound carrier is an FM signal modulated by audio frequencies in the range of 50 to 15,000 Hz. This audio frequency range is the same as for stations in the commercial FM broadcast band of 88 to 108 MHz. In the TV sound signal, the maximum frequency swing of the carrier is ±25 kHz for 100 percent modulation. This swing is less than the ±75 kHz for 100 percent modulation in the commercial FM broadcast band. However, the television sound has all the advantages of FM compared with AM, including less noise and interference.

It should be noted that AM is better for the picture signal because the ghosts resulting from multipath reception are less obvious. With AM, the ghosts stay still, but with FM the ghosts would flutter in the picture.

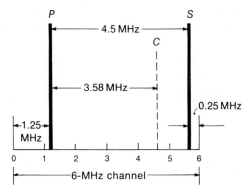

FIGURE 2-6 THE 6-MHZ TELEVISION BROADCAST CHANNEL. *P* IS PICTURE CARRIER; *S* IS SOUND CARRIER; *C* IS COLOR SUBCARRIER.

Carrier frequencies. Figure 2-6 shows how the different carrier signals fit into the standard 6-MHz channel. The picture carrier frequency labeled *P* is always 1.25 MHz above the low end of the channel. At the opposite end, the sound carrier frequency labeled *S* is 0.25 MHz below the high end. This spacing of the carrier frequencies applies for all TV channels in the VHF and UHF bands, whether the broadcast is in color or monochrome.

To apply the standard spacing to actual rf carriers, consider channel 3 as an example. This channel is 60 to 66 MHz, which is a band of 6 MHz. The picture carrier frequency is $60 + 1.25 = 61.25$ MHz. The sound carrier frequency is $66 - 0.25 = 65.75$ MHz.

Intercarrier sound. The rf sound carrier can also be figured as 4.5 MHz above the picture carrier because these two frequencies are always separated by exactly 4.5 MHz. This difference is important because all television receivers use 4.5 MHz as the frequency for the sound IF signal. The 4.5-MHz signal is called the *intercarrier sound* signal. The sound signal is made to beat with the picture carrier, to produce the difference frequency always equal to exactly 4.5 MHz. The intercarrier-sound method makes it much easier for the receiver to tune in the sound associated with the picture, especially for the UHF channels. It should be noted that the 4.5-MHz sound is still an FM signal with its original audio modulation.

2-11 Standards of Transmission

Mainly because scanning must be synchronized, the receiver depends on the transmitter for proper operation. This setup makes it necessary to establish standards for the transmitter, so that a receiver will work equally well for all stations. These have been specified as a list of transmission standards* by the Federal Communications Commission. Several points in the standards are listed here to summarize briefly the main requirements of the television system:

*Communications Commission Rules Governing Radio Broadcast Services, Part 3, Subpart E, Rules Governing Television Broadcast Stations. This also gives channel assignments by states and cities.

1. It is standard to scan at uniform velocity in horizontal lines from left to right, progressing from top to bottom of the image, when viewing the scene from the camera position.
2. The number of scanning lines per frame period is 525.
3. The frame repetition rate is 30 per second, and the field repetition rate is 60 per second.
4. The width of the channel assigned to a television broadcast station is 6 MHz. This bandwidth applies to VHF channels and UHF channels, either for monochrome or for color.
5. The associated sound is transmitted as an FM signal. Maximum frequency swing is ±25 kHz for 100 percent audio modulation. The sound carrier signal is included in the 6-MHz television channel.
6. The picture carrier is amplitude-modulated by both picture and synchronizing signals. The two signals have different amplitudes on the AM picture carrier.
7. The color subcarrier has the exact frequency of 3.579545 MHz. (This is rounded off to 3.58 MHz.)

SUMMARY

1. The smallest area of light or shade in the image is a picture element.
2. Picture elements are converted to electric signal by a camera tube at the studio. This signal becomes the video signal to be broadcast to receivers. The picture tube in the receiver converts the video signal back into visual information.
3. The electron beam scans all the picture elements from left to right in one horizontal line and all the lines in succession from top to bottom. There are 525 lines per picture frame.
4. The complete picture frame is scanned 30 times per second.
5. Blanking means going to black so that retraces cannot be seen.

6. For vertical scanning, the 525 lines in each frame are divided into two fields, each with $262 1/2$ lines. The odd lines are scanned separately; then the even lines are scanned. This procedure is interlaced scanning.
7. The vertical scanning frequency is the field rate of 60 Hz.
8. The horizontal scanning frequency is 15,750 Hz.
9. Synchronization is necessary to time the scanning with respect to picture information. The synchronizing pulse frequencies are 15,750 and 60 Hz, respectively, the same as horizontal and vertical scanning frequencies.
10. "Brightness" is the average or overall illumination. On the picture tube screen, brightness depends on high voltage and dc grid bias for the picture tube.
11. "Contrast" is the difference in intensity between black-and-white parts of the picture. The peak-to-peak ac video signal amplitude determines contrast.
12. Detail, resolution, or definition is a measure of how many picture elements can be reproduced. With many fine details, the picture looks sharp and clear.
13. The aspect ratio specifies 4:3 for the ratio of width to height of the frame.
14. A standard commercial television broadcast channel is 6 MHz wide. This includes the AM picture carrier signal 1.25 MHz above the low end of the channel and the FM sound carrier signal 0.25 MHz below the high end. The two carrier frequencies are separated by 4.5 MHz.
15. In color television broadcasting, red, green, and blue video signals corresponding to the picture information are converted into luminance and chrominance signals for transmission in the standard 6-MHz broadcast channel. The luminance signal has the black-and-white picture information; the chrominance signal provides the color.
16. The color subcarrier frequency is 3.58 MHz.
17. The amount of color in the picture, or color intensity, is the color level, chroma level, or saturation. This depends on the amplitude of the modulated chrominance signal.
18. The tint of the color is its hue. The hue depends on the phase angle of the chrominance signal. See Table 2-1.

TABLE 2-1 PICTURE QUALITIES

QUALITY	PICTURE	SIGNAL
Contrast	Range between black and white	Amplitude of ac video signal
Brightness	Background illumination	Dc bias on picture tube
Resolution	Sharpness of details	Frequency response of video signal
Color saturation	Intensity or level of color	Amplitude of 3.58-MHz chroma signal
Hue	Tint of color	Phase angle of 3.58-MHz chroma signal

Self-Examination (Answers at back of book)

Fill in the missing word or number in the following statements:

1. Picture frames are repeated at the rate of ___30___ per second.
2. The number of scanning lines is ___525___ per frame.
3. The number of fields is ___2___ per frame.
4. The number of scanning lines is ___262½___ per field.
5. The number of scanning lines is ___15,750___ per second.
6. The horizontal line-scanning frequency is ___15,750___ Hz.
7. The vertical field-scanning frequency is ___60___ Hz.
8. Video signal amplitude determines the picture quality called ___Contrast___.
9. Light is converted to video signal by the ___Camera___ tube.
10. Video signal is converted to light by the ___CR___ tube.
11. The bandwidth of a television channel is ___6___ MHz.
12. The type of modulation on the picture carrier signal is ___AM___.
13. The type of modulation on the sound carrier signal is ___FM___.
14. The assigned band for channel 3 is ___60-66___ MHz.
15. The difference between the picture and sound carrier frequencies for channel 3 is ___4.5___ MHz.
16. Scanning in the receiver is timed correctly by ___Sync___ pulses.
17. Retraces are not visible because of ___Blanking___ pulses.
18. Black on the picture tube screen results from ___0___ beam current.
19. The color subcarrier frequency is approximately ___3.58___ MHz.
20. The amount of color saturation in the picture depends on the amount of ___Chrominance___ signal.

Essay Questions

1. Why is the television system of transmitting and receiving the picture information called a sequential method?

2. Why is vertical scanning necessary in addition to the horizontal line scanning?
3. Define aspect ratio, contrast, brightness, and resolution.
4. Name the two signals transmitted in color television.
5. How is flicker eliminated by using interlaced scanning?
6. How would the reproduced picture look if it were transmitted with the correct aspect ratio of 4:3, but if on the picture tube screen at the receiver the frame was square?
7. What is the difference between color level and hue?
8. Give two ways in which color and monochrome television broadcasting are compatible.

Problems (Answers to selected problems at back of book)

1. Give the frequencies included in channels 2, 6, 7, 13, and 14.
2. Calculate the time of one horizontal line for the following examples: (a) frames repeated at 60 Hz with 525 lines per frame (for progressive scanning without interlacing); (b) frames repeated at 25 Hz with 625 lines interlaced per frame (for European standards).
3. A picture has 400 picture elements horizontally and 300 details vertically. What is the total number of details?
4. How long does it take to scan across two elements if 400 are scanned in 50 μs?
5. Show calculations for the horizontal blanking time of 10.2 μs as 16 percent of H.

3 Television Cameras

The video signal for the picture begins at the camera tube, inside the camera head (Fig. 3-1). The input is light from the scene to be televised, and the output is electric signal corresponding to the picture information.

Operation of a camera tube is illustrated in Fig. 3-2. This applies the same way to monochrome or the red, green, and blue camera tubes for color television. Light from the scene is focused by the lens onto the photoelectric image plate. If you could look in, you would see the optical image. The photoelectric properties of the image plate then convert the different light intensities into corresponding electrical variations. With an electron beam scanning across the image plate, line by line, and field by field, the camera signal is produced for the entire picture area. After the camera signal is processed, with sync and blanking added, the result is composite video signal that can be transmitted to the receiver. The main types of camera tubes, in their order of development, are the image orthicon shown in Fig. 3-8, the vidicon in Fig. 3-9, and the plumbicon in Fig. 3-12. More details are described in the following topics:

- 3-1 Camera-tube requirements
- 3-2 Image orthicon
- 3-3 Vidicon
- 3-4 Plumbicon
- 3-5 Silicon target plate
- 3-6 Solid-state image sensor
- 3-7 Spectraflex color camera tube
- 3-8 Television cameras
- 3-9 Definitions of light units

FIGURE 3-1 MONOCHROME TELEVISION CAMERA WITH LENSES OF DIFFERENT FOCAL LENGTHS ON REVOLVING TURRET. (*RCA*)

3-1 Camera-Tube Requirements

First, the image plate must be photoelectric to convert variations of light intensity into electrical variations. In photoemission, electrons are emitted; more light produces more electrons. In photoconduction, the conductance or resistance is changed; more light decreases the resistance. The image orthicon operates by photoemission, while the vidicon and plumbicon depend on photoconduction to produce the required camera signal. A third possibility is the photovoltaic effect, where light on a semiconductor junction can generate a potential difference.

In addition to the photoelectric conversion, scanning of the image plate is necessary to provide signal variations in a successive order, from left to right and top to bottom. The scanning process dissects the image into its basic picture elements. Although the entire image plate is photoelectric, its construction isolates the picture elements so that each discrete small area can produce its own signal variation.

Photoemission. Certain metals emit electrons when light strikes the surface. These emitted electrons are called *photoelectrons*, and the emitting surface is a *photocathode*. Especially sensitive to light are the elements cesium, silver, sodium, potassium, and lithium, in the group of alkali metals. Cesium oxide is often used because its photoemission is sensitive to incandescent light. The photoelectric effect is illustrated in Fig. 3-3.

The explanation of the photoelectric effect is that light consists of small bundles of energy called *photons*. When photons collide with electrons at or near the surface of the photocathode, they give off energy. The energy is used in forcing electrons to escape. The amount of emitted electrons depends on the light intensity. The maximum velocity of the photoelectrons depends only on the wavelength of light, however, which is its color. This factor is why a camera tube has different sensitivities for colors in the light spectrum.

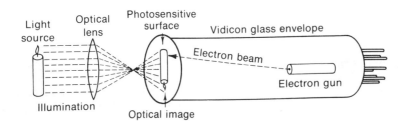

FIGURE 3-2 TELEVISING AN IMAGE WITH VIDICON CAMERA TUBE. EXTERNAL COILS FOR FOCUSING AND DEFLECTING THE ELECTRON SCANNING BEAM ARE NOT SHOWN.

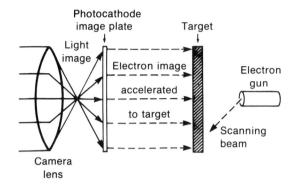

FIGURE 3-3 PHOTOEMISSION PRODUCING AN ELECTRON IMAGE IN A CAMERA TUBE.

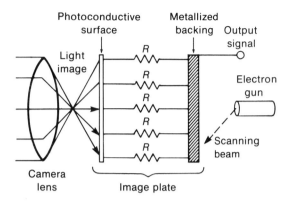

FIGURE 3-4 PHOTOCONDUCTIVITY OF IMAGE PLATE IN A CAMERA TUBE.

Photoconductivity. This photoelectric effect is a decrease in resistance with more light. In general, the semiconductor metals including selenium, tellurium, and lead, with their oxides, have this property. The vidicon and plumbicon camera tubes use an image plate that is a thin photoconductive layer. This function is illustrated in Fig. 3-4. For the vidicon, as an example, the resistance of the photoconductive image plate can decrease from 20 MΩ for black to 2 MΩ for white. A disadvantage, compared with photoemission, is that photoconducting materials generally have a slight lag in the buildup and decay of resistance when the light intensity changes.

The electron scanning beam. An electron gun produces a narrow beam of electrons for scanning. In camera tubes, the electron beam scans the image plate or target plate; in picture tubes, the electron beam scans the fluorescent screen. Enclosed in a vacuumed glass envelope, the gun assembly has a heater, cathode, control grid, and one or more accelerating grids (Fig. 3-5). Since the electrons emitted from the cathode must be concentrated in a beam, the

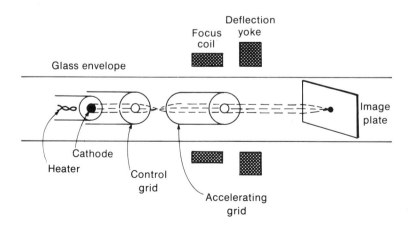

FIGURE 3-5 ELEMENTS OF AN ELECTRON GUN, ILLUSTRATED FOR VIDICON CAMERA TUBE.

34   BASIC TELEVISION

grid structures are in the form of metal cylinders with a small pinhole, or aperture. The control grid voltage determines the amount of beam current, while the accelerating grid attracts the electron beam toward the surface to be scanned.

The gun supplies the electron beam, but the scanning is produced by horizontal and vertical coils in the deflection yoke around the glass neck. In addition, focusing of the electron beam into a small, sharp spot is accomplished by an external focusing coil. This is magnetic focusing, but electrostatic focusing can be used instead, by varying the voltage on an accelerating grid. Camera tubes generally use magnetic focusing while picture tubes often use electrostatic focusing. Magnetic deflection with an external deflection yoke is used for both camera tubes and picture tubes.

Optical lenses. With the smaller vidicon and plumbicon camera tubes having a diameter of 1.25 in., the camera can use the same lenses as in 16-mm photography. The focal lengths are generally 4 to 8 in. for wide-angle views and closeup shots. Shorter lenses have a wider angle of view. A typical lens opening is $f/5.6$ for normal light levels. Lens shades are generally used to reduce reflections (Fig. 3-1). In addition to the different lenses, several cameras can be used for different angles in televising the scene.

The optical lens inverts the image for the camera tube, as shown in Fig. 3-6. *Inversion* means the image is reversed left to right and top to bottom. Then the image is scanned starting at the bottom right corner of the image plate, corresponding to top left in the scene.

Light splitters. An electron scanning beam is the same in any camera tube or any picture tube. What happens in the color camera is that the red, green, and blue components of the incident light are separated by optical prisms or by dichroic mirrors. Then each camera tube has light input proportional to the intensity of each primary color in the scene. For instance, the red camera tube has light only for the red parts of the picture. Then its output is red video signal for this component of the color information. Similarly, the blue and green camera tubes produce blue and green video. Each color video signal is just a sequence of electrical variations, but they represent the intensities for that particular color.

Figure 3-7 illustrates the principles of a four-way beam splitter. The achromatic mirrors shown reflect light of all colors. The *dichroic* mirrors are coated to transmit light of only one color and reflect the other colors. In Fig. 3-7, light from the lens on the turret is split into two separate beams. One of these beams is transmitted into the Y, or monochrome, channel. The

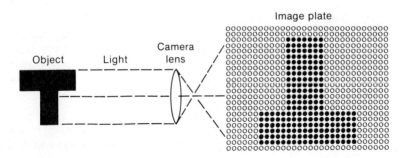

FIGURE 3-6 INVERSION OF PICTURE ON IMAGE PLATE OF CAMERA TUBE.

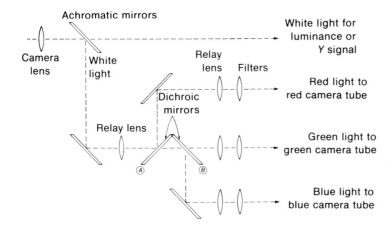

FIGURE 3-7 BEAM SPLITTER FOR THE LIGHT INTO A COLOR CAMERA.

other beam is reflected into a color-beam splitter. In this arrangement, the white-light beam is reflected through a relay lens onto a set of dichroic mirrors, A and B. These mirrors are mounted on opposing 45° angles. Each mirror is coated to reflect light of one particular color while allowing light of all other colors to pass straight through. In the figure, mirror A reflects red light into the red channel. However, green and blue light pass through to mirror B. Here the blue light is reflected, and the green light passes through. In this way, each camera tube has light input only for its respective color.

It should be noted that a three-way beam splitter is used in color cameras with three camera tubes. In this case, the Y luminance signal is derived as a combination of the red, green, and blue video signals. There are also color cameras with two camera tubes or even one camera tube. A single color tube must have a special image plate for the colors, or a motor-driven color wheel rotates in front of a conventional camera tube. This is a sequential arrangement that must be converted to the standard NTSC color signal for broadcasting.

Development of camera tubes. After the mechanical scanning disks were discarded, television broadcasting used the image dissector as the first camera tube, invented by P. T. Farnsworth. The next successful camera tube was the iconoscope, invented by V. K. Zworykin. This was the first camera tube to use the principle of allowing light to accumulate charge on the image plate. The result is the equivalent of *light storage* to increase camera sensitivity. The *light sensitivity* is the ratio of signal output to the incident illumination. High camera sensitivity is necessary to televise scenes at low light levels.

Next to be developed was the image orthicon (I.O.) camera tube. The I.O. has high sensitivity but is relatively large. The simplest and smallest camera tube is the vidicon. Similar to the vidicon is the plumbicon camera tube. This camera tube uses a different image plate made of lead monoxide (PbO). In later types of vidicons the target plate is an array of many tiny silicon photodiodes.

Another type of camera pickup is the *flying-spot scanner*. The spot of light from a cathode-ray tube scans a film image, and the light variations are picked up by a photoelectric tube on the opposite side. This method is seldom used anymore because of its size and complexity.

3-2 Image Orthicon

As shown in Fig. 3-8, the image orthicon (I.O.) is constructed in three main sections: the image section, scanning section, and electron multiplier. Light from the scene to be televised is focused onto the photocathode in the image section. This action produces a photoelectric image, which is then converted to an electrical charge image on the target plate. One side of the target plate receives the electrons emitted from the photocathode, while the opposite side of the target is scanned by the electron beam from the scanning section. As a result, signal current for the entire image is produced by the scanning beam. The signal current is then amplified in the electron-multiplier section, which provides the desired camera output signal.

Producing the camera signal output. The signal action in the image orthicon can be summarized briefly as follows:

1. Light from the televised scene is focused onto the photocathode, where the light

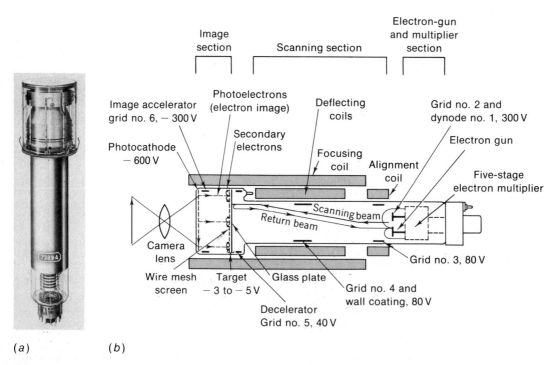

FIGURE 3-8 IMAGE ORTHICON CAMERA TUBE. LENGTH IS 15 IN. (a) PHOTO. (b) CONSTRUCTION AND OPERATING VOLTAGES. (*RCA*)

image produces an electron image corresponding to the picture elements.
2. The electron image is accelerated to the target to produce secondary emission from the glass plate.
3. The secondary emission produces on the target a pattern of positive charges corresponding to the picture elements in the scene. White is most positive.
4. The low-velocity scanning beam from the electron gun provides electrons that land on the target to neutralize the positive charges. Scanning-beam electrons in excess of the amount needed to neutralize the positive charges turn back from the target and go toward the electron gun.
5. As the beam scans the target, therefore, the electrons turned back from the glass plate provide a signal current that varies in amplitude in accordance with the charge pattern and the picture information. The signal current is maximum for black.
6. The returning signal current enters the electron multiplier, where the current is amplified. The amplified current flowing through the load resistor in the multiplier's anode circuit produces the camera signal output voltage.

With a signal current of 5 μA from the highlights in the scene and a 20-kΩ R_L as typical values, the camera signal output is 100,000 μV, or 0.1 V. A typical dark signal current is 30 μA for 0.6 V across R_L. The peak-to-peak camera signal voltage, then, is $0.6 - 0.1 = 0.5$ V.

Sticking picture. The sticking picture is an image of the televised scene with reversed black and white, which remains after the camera has been focused on a stationary bright image for several minutes, especially if the image orthicon is operated without sufficient warmup. The sticking picture can usually be erased, however, by focusing on a clear white screen or wall.

3-3 Vidicon

As shown in Fig. 3-9, the vidicon is a very small camera tube of relatively simple construction. Figure 3-10 shows the circuit for camera signal output. The vidicon has just a photoconductive target plate and electron gun. With the optical image focused on the target, it produces a charge image that is scanned by the electron beam from the gun. Vidicons are 5 to 8 in. long, with a diameter of 0.58 to 1.6 in. The $^3/_4$-in. vidicon is commonly used for closed-circuit television. Average illumination required is about 150 footcandles in the scene or 1 to 10 footcandles on the target plate.

Target. The target has two layers. One is a transparent film of conducting material, coated directly on the inside surface of the glass faceplate. This conductor is the signal-plate electrode for camera output signal. Light passes through to the second layer, which is an extremely thin coating of photoconductive material. Either selenium or antimony compounds are used. The photoconductive property means that its resistance decreases with the amount of incident light.

Charge image. The photolayer is an insulator with a resistance of approximately 20 MΩ for the 0.00003-in. thickness, in the dark. Incident light can reduce the resistance to 2 MΩ, as indicated in Fig. 3-10. Note that the image side of the photolayer contacts the signal plate at a +40 V potential with respect to the cathode. The opposite side returns to cathode through the electron beam, which has an approximate resistance of 90 MΩ. With an image on the target, the potential of each point on the gun side of

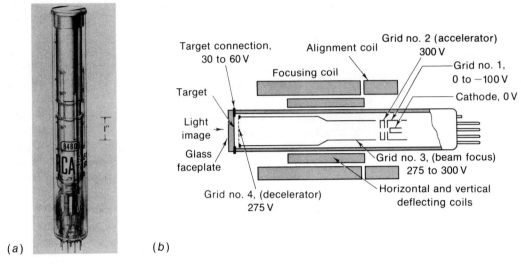

FIGURE 3-9 VIDICON CAMERA TUBE. (a) PHOTO. (b) CONSTRUCTION AND OPERATING VOLTAGES. (RCA)

the photolayer depends on its resistance to the signal plate at +40 V. As examples: a white area with low resistance can be close to the signal-plate voltage, approximately at +39.5 V; the potential of a dark area with high resistance is lower at +35 V. The result then is a pattern of positive potentials on the gun side of the photolayer, producing a charge image corresponding to the optical image. Maximum white in the picture is most positive in the charge pattern. The charge image on the target plate is scanned by the electron beam from the gun.

Electron gun. As shown in Fig. 3-9, the gun includes a heated cathode, control grid (no. 1), accelerating grid (no. 2), and focusing grid (no. 3). The electrostatic field of grid 3 and the magnetic field of an external-focus coil are both used to focus the electron beam on the target plate. Deflection of the beam for scanning is produced by horizontal and vertical deflection coils in an external deflection yoke.

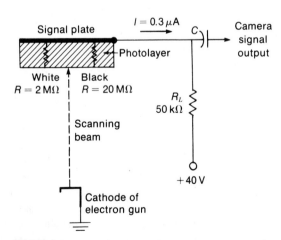

FIGURE 3-10 CIRCUIT FOR CAMERA SIGNAL OUTPUT OF VIDICON.

Note that grids 3 and 4 are connected internally. Grid 4 is a wire mesh to provide a uniform field near the target plate. Since the target is at a lower potential of 30 to 60 V, compared with 275 V on grid 4, electrons in the beam are decelerated just before they reach the target plate. The lower potential is used to slow down the electrons so that the low-velocity beam can deposit electrons on the charge image without producing secondary emission from the photolayer.

Signal current. Each point in the charge image has a different positive potential on the side of the photolayer toward the electron gun. Electrons in the beam are then deposited on the photolayer surface, reducing the positive potential toward the cathode voltage of zero. Excess electrons not deposited on the target are turned back, but this return beam is not used in the vidicon.

Consider a white picture element in the charge image on the photolayer. Its positive potential is close to the +40 V on the signal plate just before the electron beam strikes. Then, as electrons are deposited, the potential drops toward zero. This change in potential causes signal current to flow in the signal-plate circuit, producing output voltage across R_L. For black in the picture, where the photolayer is less positive than white areas, the deposited electrons cause a smaller change in signal current.

The signal current results from the changes in potential difference between the two surfaces of the photolayer. The path for signal current can be considered a capacitive circuit provided by 5 pF capacitance of the signal plate to the electron gun. Considering polarity, the output signal produces the least positive or most negative voltage output across R_L for white highlights in the image. With 1-footcandle illumination on the target, white highlights can produce 0.3-μA signal current for a voltage drop of 0.015 V across the 0.05-MΩ load resistor.

Light-transfer characteristics. Three curves of vidicon output current are shown in Fig. 3-11. Each is for a specific value of dark current, which is the output with no light, corresponding to black in the picture. The dark current is set by adjusting target voltage. Sensitivity and dark current both increase as the target voltage is increased. Typical output for the vidicon is 0.4 μA for white highlights, with a dark current of 0.02 μA.

Lag. This term refers to the time lag of the photoconductive layer. The lag can cause smear, with a tail or comet following fast-moving objects in the scene. Also, an x-ray appearance, as you seem to be able to see through objects, can be caused by the lag. The photoconductive lag increases at high target voltages, where the vidicon has its best sensitivity. The lag is most

Illumination: Uniform over photoconductive layer; scanned area of photoconductive layer $1/2'' \times 3/8''$; faceplate temperature = 30°C approx.

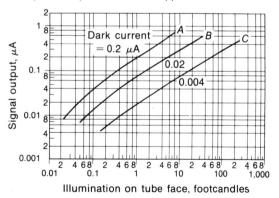

FIGURE 3-11 LIGHT-TRANSFER CHARACTERISTICS OF VIDICON. (RCA)

troublesome, at low light levels, therefore, when the target voltage is increased for more signal current.

Special types for low lag are:

Lead oxide vidicon. This tube has a target plate made of PbO, similar to the plumbicon described in Sec. 3-4.

Silicon-diode vidicon. This tube has a target plate made of silicon photodiodes, as described in Sec. 3-5.

3-4 Plumbicon

As shown in Fig. 3-12, this is a small camera tube like the vidicon. The electron gun is essentially the same as for the vidicon shown in Fig. 3-9. However, the plumbicon has a new type of target using lead* monoxide (PbO) for the photoconductive plate. The plumbicon target operates effectively as a PIN semiconductor diode coating on the inner surface of the glass faceplate. P-type semiconductor is doped to have an excess of positive charges; N-type has an excess of electrons as negative charges. Intrinsic semiconductor, or I-type, is pure to be neutral without doping.

In the manufacturing process for the target, a thin transparent conductive film of tin oxide (SnO_2) is deposited directly on the inside surface of the glass faceplate. This layer of conductor is the signal plate. Then a layer of pure lead monoxide (PbO) is deposited over the tin oxide film. Finally, the scanning surface of the pure PbO layer is doped to complete the semiconductor requirements of the target. In these layers, the SnO_2 signal plate is N-type. The layer of pure PbO in the middle is intrinsic or I-type. The layer of doped PbO on the scanned side of

---

*Lead (Pb) is a semiconductor element with the valence of 4. Since the valence of oxygen is 2, PbO is lead monoxide; lead oxide is PbO_2. The element tin (Sn) is also a semiconductor element with the valence of 4.

(a)

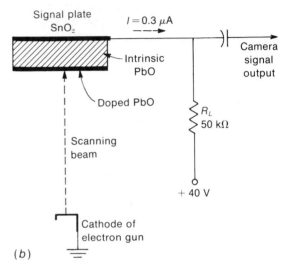

(b)

FIGURE 3-12 PLUMBICON CAMERA TUBE. LENGTH IS 8 IN. (*NORTH AMERICAN PHILIPS COMPANY INC.*)

the target is P-type. As a result, the sandwich of layers has the properties of a PIN semiconductor diode. The overall thickness of the target is 15×10^{-6} m. The PbO layer is granular in structure, with individual particles of 1×10^{-6} m.

Scanning the target. In the signal circuit, the conductive film of tin oxide (SnO_2) is connected to the target supply of 40 V through an external load resistor R_L to develop the camera output signal voltage (Fig. 3-12). As the electron beam scans the target, the signal current varies with the amount of light for each picture element.

The photoelectric conversion is similar to the vidicon, except for the method of discharging each storage element. In the standard vidicon, each element acts as a leaky capacitor, with the leakage resistance decreasing with more light. In the plumbicon, however, each element serves as a capacitor in series with a light-controlled diode. Without light, the diode is reverse-biased to prevent conduction, and there is little or no output. Typical values of this dark current are 4×10^{-9} A. With light, the diode is forward-biased for minimum resistance and maximum current. A typical signal current for highlights is 0.3×10^{-6} A or 0.3 μA.

The forward bias on each diode results from photoexcitation of the semiconductor junction between the pure PbO and the doped layer. Furthermore, the layers can be modified to increase sensitivity to red, green, and blue for color cameras.

General characteristics. Typical operation is illustrated by the following specifications for the plumbicon types: CCTV 111 for monochrome; CCTV 111 B, CCTV 111 G, and CCTV 111 R for blue, green, and red.

OPTICAL	ELECTRICAL
Size of image: 0.5 × 0.375 in.	Heater: 6.3 V at 95 mA
	Cathode: 0 V
Illumination: 0.8 footcandle for 0.3-μA signal current	Signal electrode: 25 to 45 V
	Grid No. 2: 300 V
Maximum illumination: 50 footcandles	Grid No. 3: 600 V
	Grid No. 4: 675 V
Maximum resolution: 600 lines	Beam current: 0.2 to 0.4 μA for highlights
Lag: 2 percent residual signal after dark pulse of 50 ms	

Spectral response. Figure 3-13 compares the sensitivity of the image orthicon, vidicon, and plumbicon for light of different colors. The baseline of the graph is in angstrom (Å) units for the wavelength, which determines the hue. The angstrom unit is 10^{-10} m. Also, 10 Å equals 1 millimicron (10^{-9}) unit.

The response of the human eye is centered at about 5500 Å, which is the wavelength for green. In the graph, the plumbicon response is closest to this response. However, this response is weaker toward the lower wavelengths for red.

3-5 Silicon Target Plate

This type of target is not deposited on the glass faceplate of the camera tube. Instead, the target is prepared from a thin N-type silicon wafer. The final result is an array of silicon photodiodes for the target plate. Then the silicon-diode target plate is mounted in a vidicon-type camera tube. The advantages of the silicon-diode array are resistance to burns from excessive light, low lag time, and high sensitivity for visible light, which can be extended to the infrared region.

The silicon target plate is typically 0.001 in. thick and about $1/2$ in. square. One side is oxidized to form a film of silicon dioxide, which is an insulator. Then by photolithographic processes similar to those used in making miniature integrated circuits, an array of openings is produced in the film. This layer with its holes is used as a diffusion mask for producing the individual photodiodes. The doping element boron (B) is vaporized through the array of holes, forming islands of P-type silicon on one side of the N-type silicon substrate. An overlay of gold is centered on each P-type area (Fig. 3-14). The resulting PN photodiodes have a diameter of 0.0003 in. A typical array has 540 × 540 diodes.

In typical operation, the N-type silicon substrate or platform has +10 V applied. Positive polarity on the N-side reverse-biases the photodiodes. The electron beam in the camera

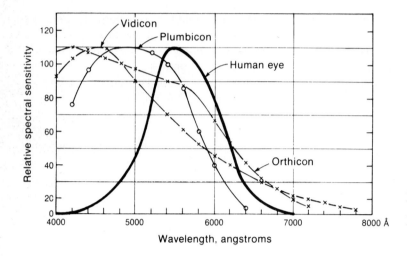

FIGURE 3-13 COMPARISON OF SENSITIVITY TO LIGHT OF DIFFERENT WAVELENGTHS FOR ORTHICON, VIDICON, AND PLUMBICON CAMERA TUBES. (*NORTH AMERICAN PHILIPS CO. INC.*)

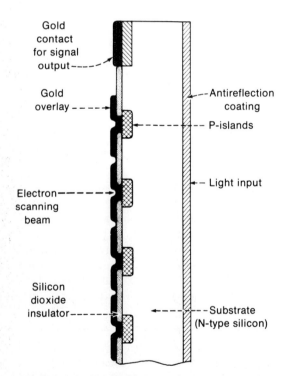

FIGURE 3-14 CONSTRUCTION OF TARGET PLATE OF SILICON PHOTODIODES. (*BELL TELEPHONE LABORATORIES*)

tube scans the photodiode side, depositing electrons on the gold. From the opposite side, light on the target plate penetrates the silicon substrate to make it less negative. As a result, the reverse bias is reduced on the N-side of the photodiode. Furthermore, the scanning beam reduces the reverse bias on the P-side. The signal current then consists of increases of target current for more light as each element is scanned by the electron beam. A typical value of peak target current is 0.7 μA for white highlights of 1 footcandle. A camera tube using a silicon-diode array tor the target plate is shown in Fig. 3-15. This type has the general name of *silicon diode–array vidicon.*

FIGURE 3-15 EPICON CAMERA TUBE USING SILICON-DIODE ARRAY FOR TARGET PLATE. TUBE DIAMETER IS 1 IN. (*GENERAL ELECTRIC*)

TELEVISION CAMERAS 43

FIGURE 3-16 SOLID-STATE IMAGE SENSOR. SIZE IS $3/4 \times 1/2$ IN., WITH 120,000 ELEMENTS. (*RCA*)

3-6 Solid-State Image Sensor

This new device consists of a flat silicon chip with an array of metal electrodes. Figure 3-16 shows such an image sensor $3/4 \times 1/2$ in., containing 120,000 elements. It does not need an electron gun, scanning beam, high voltage, or vacuum envelope of a conventional camera tube, since the entire image-sensor assembly is contained in the one chip of solid-state semiconductor.

The silicon chip is oxidized on one side with a linear array of electrodes deposited on this surface. The electrodes operate in groups of three, with every third electrode connected to a common conductor. The spot under the center of each triplet serves as one light-sensitive element, or resolution cell.

In operation, the image is focused onto the silicon chip. The light causes electrons to be produced within the silicon. More charges are generated with more light. In each trio, the center electrode is the most positive. The charges generated by that element collect at the surface of the silicon under this center electrode. As a result, the pattern of collected charges represents the image.

The charge at one element is transferred along the surface of the silicon chip by applying a more positive voltage to the adjacent electrode, while reducing the voltage on the electrode over the charge packet. The change is applied by voltage pulses in successive order to all the elements. This method of dissecting the image is equivalent to scanning. When the charge packets reach the output electrode, they are collected to form the signal current. The potential required to move the charges is only 5 to 10 V. Transferring the charges from one element to the next is called *charge coupling*. This type of image sensor, then, is a charge-coupled device (CCD).

3-7 Spectraflex Color Camera Tube

The construction is similar to a vidicon, but the faceplate has a dichroic stripe filter to separate colors for the target. As illustrated in Fig. 3-17, the dichroic filter results in vertical stripes of

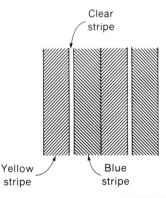

FIGURE 3-17 COLOR STRIPES FOR SPECTRAFLEX CAMERA TUBE. (*RCA*)

44 BASIC TELEVISION

yellow and cyan, which is blue-green. The clear stripes between are for monochrome.

When the electron scanning beam sweeps across the target plate, the stripe pattern generates subcarrier frequencies for two color signals and the luminance signal. A third color signal can be derived by mixing the two color signals in the proper proportions. As a result, one camera tube can supply all the information needed to reproduce a color picture.

3-8 Television Cameras

Figure 3-18 shows a color camera that can be used either for studio work or for remote pickup assignments in the field. This camera head has three plumbicons, with built-in color filters and a beam splitter. It uses just one lens, which is the motorized zoom type for a wide range of focal lengths. The viewfinder shows the cameraman a reproduction of the scene, on a monochrome picture tube with an 8-in. screen. Note the cue light at the top. This light shows when the camera is connected to be "on the air." The camera can be operated at light levels as low as 5 foot-

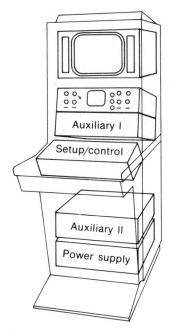

FIGURE 3-19 CONTROL CONSOLE FOR CAMERA IN FIG. 3-18. (*RCA*)

candles in the scene. However, typical operation is with an illumination of 125 footcandles, with an *f*/4 lens opening.

The camera signal output is 1 V peak to peak for the coaxial cable input to the control room. As shown in Fig. 3-19, the control console includes a monitor picture tube and oscilloscope monitor below it. The control desk switches cameras, provides special effects, and sets black level. The auxiliary panels include the generators for synchronizing pulses and the 3.58-MHz color subcarrier. A camera with its control equipment is a *camera chain*.

Televising motion-picture films and still slides. A separate studio is used, containing projectors for 35- and 16-mm film. In addition, slides are used for station identification and titles. The projector throws the light image di-

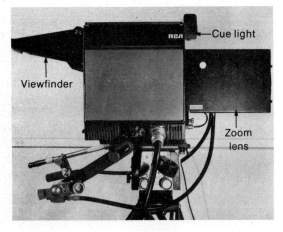

FIGURE 3-18 STUDIO-FIELD COLOR CAMERA. (*RCA*)

the time for five television frames is $5/30$ s, which is also $1/6$ s. As a result, the time for scanning 30 television frames matches the 24 film frames.

Video tape. Most studio programs are recorded on video tape, for broadcasting at a convenient time. Also, the tapes are available for rebroadcasting by stations in different areas. Figure 3-21 shows a video tape machine for broadcast use in color television. There are four record-playback heads, mounted on a wheel rotating at 14,400 rpm. The heads are phased 90° apart so that the four gaps contact the tape in sequential order. As a result, the video signal is recorded as a series of transverse tracks across the tape. The sound signal is recorded on a separate track.

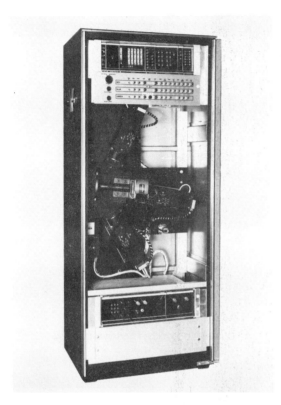

FIGURE 3-20 COLOR FILM CAMERA, WITH THREE VIDICONS. *(RCA)*

rectly onto the image plate of the camera tubes in the film camera (Fig. 3-20). A mirror triplexer enables one film camera to be used with three projectors.

When commercial motion-picture films are televised, a special projector is necessary to convert from 24 to 30 frames per second. The film moves at 24 frames per second to keep the sound track normal, but an intermittent shutter projects 60 images per second. Specifically, one film frame is projected for two television fields ($2/60$ s), but the next frame is scanned with three fields ($3/60$ s). After four film frames, the two extra fields make five television frames. The time for four film frames is $4/24$ or $1/6$ s. Similarly

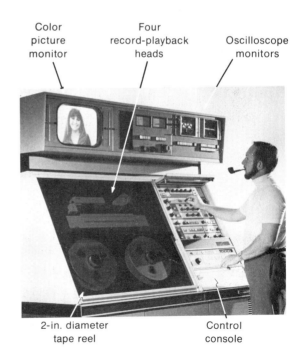

FIGURE 3-21 VIDEO TAPE MACHINE FOR COLOR TELEVISION BROADCASTING. *(AMPEX CORPORATION)*

3-9 Definitions of Light Units

Optical lenses are made of glass or plastic, with the function of concentrating light. A converging lens can bend parallel rays of light inward to converge at a point, as shown in Fig. 3-22. The focal point f is where the light rays converge. Past this crossover point the image is inverted, left to right and top to bottom.

Focal length. This length is the distance from the lens to point f. Typical values are 5 to 20 in. A short focal length is a wide-angle lens, as a wide field of view is focused at f. A long focal length is a telephoto lens allowing a closeup image for the same distance of lens to subject.

f-number. This number is the focal length divided by the diameter of the lens. For a focal length of 12 in. and lens diameter of 1.5 in., the f-number is 12 in./1.5 in. = 8. This is an $f/8$ lens, therefore. The smaller the f-number, the more light the lens takes in, which is a fast lens. The lens is rated at its maximum f-number, but it can be stopped down by an iris diaphragm for smaller openings.

The usual f stops are $f/1.4$, $f/2$, $f/2.8$, $f/5.6$, $f/8$, $f/11$, $f/16$, and $f/32$. These values are chosen

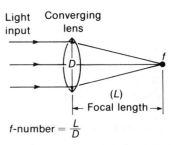

FIGURE 3-22 FOCAL LENGTH AND f-NUMBER OF AN OPTICAL LENS. D IS LENS DIAMETER.

TABLE 3-1 TYPES OF CAMERA TUBES

TYPE	SIZE	IMAGE PLATE	TYPICAL SIGNAL CURRENT, μA DARK	WHITE	NOTES	LIFE, hr
Image orthicon	Length 15–20 in.; diameter 3–4 in.	Photocathode	30	6	High quality; high sensitivity; signal current maximum for black	1,500–6,000
Vidicon	Length 5–8 in.; diameter 0.6–1.6 in.	Selenium photoconductor	0.02	0.4	Simple construction; used for film pickup; has photoconductive lag for low light levels	5,000–20,000
Plumbicon	Length 8 in.; diameter 1.2 in.	PbO photoconductor	0.004	0.3	Simple construction; low lag; sensitivity low for red light	2,000–3,000
Silicon diode–array vidicon	Length 6 in.; diameter 1 in.	Silicon-diode array	0.01	0.6	Low lag; sensitive to red and infrared light	2,000–3,000

for each higher f stop to allow exactly one-half the light of the previous stop. Smaller f stops are used when maximum light is necessary, but a higher f stop allows better depth of focus. This means that subjects a little closer or farther from the best distance for focus will still be sharp.

Illumination. Originally, the light from a standard candle was used as a reference, the intensity of this source being defined as 1 candlepower or simply 1 candle (cd). When the source is at the center of a hollow sphere with a radius of 1 ft, the amount of luminous flux on 1 ft² equals *1 lumen* (lm). Since the surface area of the sphere is 4π ft², and 1 lm is for 1 ft², a 1-cd source provides 4π lm of light radiation. As an example, a 25-W bulb as a light source has an illuminating power of 20.7 cd or 260 lm. These units correspond to power, as they define the rate at which light energy is being radiated.

The illumination on a surface, such as the image plate, is the amount of light energy received from a source per unit area of the illuminated surface. One *footcandle* is the illumination for 1 ft² at a distance 1 ft away from a source of 1 cd. Therefore, 1 footcandle and 1 lm/ft² represent the same amount of surface illumination. For example, the desired illumination for reading is about 10 footcandles, or 10 lm/ft².

Wavelength. Different wavelengths of light have different hues, such as red, green, and blue. The unit is the angstrom (Å) equal to 10^{-10} m, the micron of 10^{-6} m, or the millimicron of 10^{-9} m. The visible spectrum of colors extends from about 4500 Å for blue, through 5500 Å for green, and 6500 Å for red. These hues are shown in color plate VIII. Longer wavelengths for infrared are not visible to the human eye. Also, shorter wavelengths below 4500 Å for ultraviolet are not visible.

FIGURE 3-23 SMALL MONOCHROME CAMERA WITH VIDICON FOR CLOSED-CIRCUIT TV. LENGTH OF THE CASE IS 11 IN. (*JERROLD ELECTRONICS CORPORATION*)

SUMMARY

The characteristics of the image orthicon (I.O.), vidicon, plumbicon, and silicon diode–array vidicon are summarized in Table 3-1. The vidicon and plumbicon are the camera tubes generally used in television broadcasting. This includes live scenes in the studio, film cameras, and portable TV cameras, either in color or monochrome.

48 BASIC TELEVISION

Also included are applications in closed-circuit TV for education, surveillance, and medical electronics. Because of its small size and low cost, the vidicon is often used in compact monochrome television cameras (Fig. 3-23). The silicon diode-array vidicons are also used for closed-circuit TV, but are not considered to have the quality for broadcast use.

Self-Examination (Answers at back of book)

Answer True or False.
1. A photocathode emits more electrons with more light.
2. A photoconductor has less resistance with more light.
3. The millimicron is a unit of wavelength equal to 10^{-9} meter.
4. In the image orthicon, maximum white on the photocathode produces maximum positive charge on the glass target plate.
5. Camera signal output of the image orthicon is taken from the photocathode in the image section.
6. The image orthicon has high sensitivity because of charge storage by the target plate and the use of an electron multiplier.
7. Dichroic mirrors can split white light into colors.
8. In a camera with four camera tubes, one is for the Y luminance signal.
9. In the vidicon and plumbicon, the signal output is taken from the target plate.
10. In the image orthicon, the return beam provides the variations in camera signal.
11. The image orthicon uses silicon photodiodes for the target plate.
12. The plumbicon uses lead monoxide (PbO) for the target plate.
13. The amount of target voltage in the vidicon and plumbicon determines the amount of dark current.
14. At low values of dark current in the vidicon, lag in the photoconductive layer can produce a tail that follows fast-moving objects in the scene.
15. Typical output current for the vidicon and plumbicon is approximately 140 mA.
16. The solid-state image sensor does not use an electron scanning beam.
17. Without light on a silicon diode-array vidicon, the diodes have maximum reverse bias to limit conduction in the forward direction.
18. The cue light on the camera head shows when the camera is "on the air."
19. A converging lens inverts the optical image.
20. A typical lens opening for television cameras is $f/50$.

Essay Questions

1. Define the following terms: photocathode, photoconductor, photoelectrons, photodiode, I.O., and C.C.D.

2. List the types of image plate for the light image in the image orthicon, standard vidicon, plumbicon, and silicon vidicon camera tubes.
3. Give the function for each of the three sections in the image orthicon.
4. What is the difference between reverse bias and forward bias for a semiconductor diode?
5. List four semiconductor elements having a valence of 4.
6. Make a drawing to illustrate construction and operation of a standard vidicon. Explain briefly how camera output signal is produced.
7. What is meant by "dark current" in the vidicon?
8. Explain briefly how four camera tubes are used in a color camera. Do the same for a color camera with three camera tubes.
9. List the main items of equipment in a camera chain.
10. Give one application for each of the camera tubes listed in Table 3-1.

Problems (Answers to selected problems at back of book)

1. With 10-μA peak-to-peak output from an image orthicon, how much is the camera signal voltage across a 20-kΩ R_L?
2. With 0.42 μA from a vidicon for white highlights and a dark current of 0.02 μA, how much is the peak-to-peak camera signal voltage across a 60-kΩ R_L?
3. Refer to curve C of the vidicon light-transfer characteristics in Fig. 3-11, with 0.004-μA dark current. Give the output current produced by white highlights of 100 footcandles, as provided by film projection directly on the target.

4 Scanning and Synchronizing

The rectangular area scanned by the electron beam as it is deflected horizontally and vertically is called the *raster*. Figure 4-1 shows the scanning raster on the picture tube screen, without any picture information. With video signal, the picture tube reproduces the picture on the raster. In addition, the deflection must be synchronized with the picture. To time the horizontal and vertical scanning correctly, synchronizing pulses are transmitted as part of the signal. The scanning and synchronizing are explained in the following topics:

4-1 The sawtooth waveform for linear scanning
4-2 Standard scanning pattern
4-3 A sample frame of scanning
4-4 Flicker
4-5 Raster distortions
4-6 The synchronizing pulses
4-7 Scanning, synchronizing, and blanking frequencies

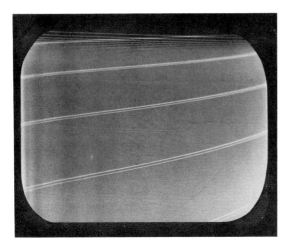

FIGURE 4-1 SCANNING RASTER ON SCREEN OF PICTURE TUBE (RETRACES ARE NORMALLY BLANKED OUT). THIS RASTER IS NOT INTERLACED BECAUSE THERE IS NO VERTICAL SYNCHRONIZATION.

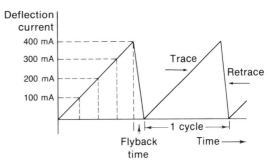

FIGURE 4-2 SAWTOOTH SCANNING WAVEFORM.

4-1 The Sawtooth Waveform for Linear Scanning

As an example of linear scanning, consider the sawtooth waveshape in Fig. 4-2 as scanning current for an electromagnetic tube. This current flows through the deflection coils in the yoke on the neck of the picture tube. Let the peak value be 400 mA. If 100 mA is needed to produce a deflection of 5 in., then 400 mA will deflect the beam 20 in. Furthermore, the linear rise on the sawtooth wave provides equal increases of 100 mA for each of the four equal periods of time shown. Each additional 100 mA deflects the beam another 5 in.

Horizontal scanning. This linear rise of current in the horizontal deflection coils deflects the beam across the screen with a continuous, uniform motion for the trace from left to right. At the peak of the rise, the sawtooth wave reverses direction and decreases rapidly to its initial value. This fast reversal produces the retrace or flyback.

The start of horizontal trace is at the left edge of the raster. The finish is at the right edge, where the flyback produces retrace back to the left edge. See Fig. 4-3a. Note that "up" on the sawtooth wave corresponds to horizontal deflection to the right.

Vertical scanning. This sawtooth current in the vertical deflection coils moves the electron beam from top to bottom of the raster. While the electron beam is being deflected horizontally,

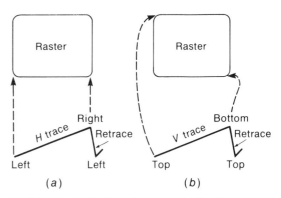

FIGURE 4-3 DIRECTIONS FOR TRACE AND RETRACE: (a) HORIZONTAL; (b) VERTICAL.

the vertical sawtooth deflection waveshape moves the beam downward with uniform speed. Then the beam produces complete horizontal lines, one under the other.

The trace part of the sawtooth wave for vertical scanning deflects the beam to the bottom of the raster. Then the rapid vertical retrace returns the beam to the top. See Fig. 4-3b. Note that up on the sawtooth wave for vertical deflection shows increasing current to deflect the beam downward.

Scanning frequencies. Both trace and retrace are included in one cycle of the sawtooth wave. Since the number of complete horizontal lines scanned in 1 s equals 15,750, for horizontal deflection the frequency of the sawtooth waves is 15,750 Hz. For vertical deflection, the frequency of the sawtooth waves equals the field-scanning rate of 60 Hz. The vertical scanning motion at 60 Hz is much slower than the horizontal sweep rate of 15,750 Hz. As a result, many horizontal lines are scanned during one cycle of vertical scanning. We can consider that the vertical deflection makes the horizontal lines fill the raster from top to bottom.

Retrace time. During flyback time, both horizontal and vertical, all picture information is blanked out. Therefore, the retrace part of the sawtooth wave is made as short as possible, since retrace is wasted time in terms of picture information. For horizontal scanning, retrace time is approximately 10 percent of the total line period. With 63.5 μs for a complete line, 10 percent equals 6.35 μs for horizontal flyback time. Practical limitations in the circuits producing the sawtooth waveform make it difficult to produce a faster flyback.

The lower-frequency vertical sawtooth waves usually have a flyback time less than 5 percent of one complete cycle. A vertical retrace 3 percent of $1/_{60}$ s, as an example, equals 0.0005 s, or 500 μs. Although vertical retrace is fast compared with vertical trace, note that 500 μs is much longer than a complete horizontal line, which takes 63.5 μs. Actually, the vertical retrace time of 500 μs includes approximately eight lines.

4-2 Standard Scanning Pattern

The scanning procedure that has been universally adopted employs horizontal linear scanning in an odd-line interlaced pattern. The FCC scanning specifications for television broadcasting in the United States provide a standard scanning pattern that includes a total of 525 horizontal scanning lines in a rectangular frame having a 4.3 aspect ratio. The frames are repeated at a rate of 30 per second with two fields interlaced in each frame.

Interlacing procedure. Interlaced scanning can be compared with reading the interlaced lines written in Fig. 4-4. Here the information on the page is continuous if you read all the odd lines from top to bottom and then go back to the top to read all the even lines down to the bottom. If the whole page were written and read in this interlaced pattern, the same amount of information would be available as though it were written

The horizontal scanning lines are interlaced in the odd lines are scanned, omitting the even lines. the television system in order to provide two Then the even lines are scanned to complete the views of the image for each picture frame. All whole frame without losing any picture information.

FIGURE 4-4 INTERLACED LINES. READ THE FIRST AND ODD LINES AND THEN THE SECOND AND EVEN LINES.

in the usual way with all the lines in progressive order.

For interlaced scanning, therefore, all the odd lines from top to bottom of the frame are scanned first, skipping over the even lines. After this vertical scanning cycle, a rapid vertical retrace moves the electron scanning beam back to the top of the frame. Then all the even lines which were omitted in the previous scanning run are scanned from top to bottom.

Each frame is therefore divided into two fields. The first and all odd fields contain the odd lines in the frame, while the second and all even fields include the even scanning lines. With two fields per frame and 30 complete frames scanned per second the field repetition rate is 60 per second and the vertical scanning frequency is 60 Hz. In fact, it is the doubling of the vertical scanning frequency from the 30-Hz frame rate to the 60-Hz field rate that makes the beam scan every other line in the frame.

Odd-line interlacing. The geometry of the standard odd-line interlaced scanning pattern is illustrated in Fig. 4-5. Actually, the electron gun aims the beam at the center, which is where the scanning starts from. For convenience, however, we can follow the motion starting at the upper left corner of the frame at point A. For this line 1, the beam sweeps across the frame with uniform velocity to cover all the picture elements in one horizontal line. At the end of this trace the beam then retraces rapidly to the left side of the frame, as shown by the dashed line in the illustration, to begin the next horizontal line.

Note that the horizontal lines slope downward in the direction of scanning because the vertical deflecting signal simultaneously produces a vertical scanning motion, which is very slow compared with horizontal scanning. Also note that the slope of the horizontal trace from left to right is greater than during retrace from right to left. The reason is that the faster retrace does not allow the beam so much time to be deflected vertically.

After line 1, the beam is at the left side ready to scan line 3, omitting the second line. This skipping of lines is accomplished by dou-

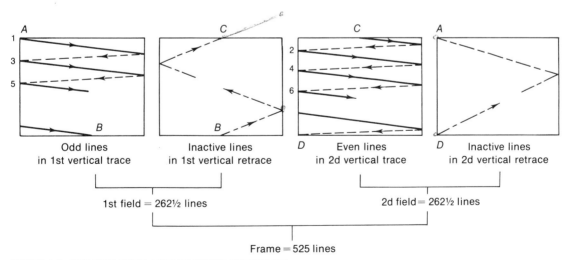

FIGURE 4-5 ODD-LINE INTERLACED SCANNING PROCEDURE.

bling the vertical scanning frequency from the frame repetition rate of 30 to the field frequency of 60 Hz. Deflecting the beam vertically at twice the speed necessary to scan 525 lines produces a complete vertical scanning period for only $262\frac{1}{2}$ lines, with alternate lines left blank. The electron beam scans all the odd lines, then, finally reaching a position such as B in the figure at the bottom of the frame.

At time B the vertical retrace begins because of flyback on the vertical sawtooth deflecting signal. Then the beam is brought back to the top of the frame to begin the second, or even, field. As shown in Fig. 4-5, the beam moves from point B up to C, traversing a whole number of horizontal lines.

This vertical retrace time is long enough for the beam to scan several horizontal lines. We can call these *vertical retrace lines,* meaning complete horizontal lines scanned during vertical flyback. Note that the vertical retrace lines slope upward, as the beam is moving up while it scans horizontally. The upward slope of vertical retrace lines is greater than the downward slope of lines scanned during vertical trace because the flyback upward is much faster than the trace downward. Any lines scanned during vertical retrace are not visible, though, because the electron beam is cut off by blanking voltage during vertical flyback time. The vertical retrace lines are inactive because they are blanked out.

Horizontal scanning during the second field begins with the beam at point C in Fig. 4-5. This point is at the middle of a horizontal line because the first field contains 262 lines plus one-half a line. After scanning a half line from point C, the beam scans line 2 in the second field. Then the beam scans between the odd lines to produce the even lines that were omitted during the scanning of the first field. The vertical scanning motion is exactly the same as in the previous field, giving all the horizontal lines the same slope downward in the direction of scanning. As a result, all the even lines in the second field are scanned down to point D. Points D and B are a half line away from each other because the second field started with a half line.

The vertical retrace in the second field starts at point D in Fig. 4-5. From here, vertical flyback returns the beam to the top. With a whole number of vertical retrace lines, the beam finishes the second vertical retrace at A. The beam will always finish the second vertical retrace where the first trace started because the number of vertical retrace lines is the same in both fields. At point A, then, the scanning beam has just completed two fields or one frame and is ready to start the third field to repeat the scanning pattern.

All odd fields begin at point A and are the same. All even fields begin at point C and are the same. Since the beginning of the even-field scanning at C is on the same horizontal level as A with a separation of one-half line, and since the slope of all the lines is the same, the even lines in the even fields fall exactly between the odd lines in the odd field. The essential requirement for this odd-line interlace is that the starting points at the top of the frame be separated by exactly one-half line between even and odd fields.

4-3 A Sample Frame of Scanning

A complete scanning pattern is shown in Fig. 4-6 with the corresponding horizontal and vertical sawtooth waveforms to illustrate odd-line interlacing. A total of 21 lines in the frame is used for simplicity, instead of 525. The 21 lines are interlaced in two fields per frame. One-half the 21-line total, or $10\frac{1}{2}$ lines, are in each field. Of the $10\frac{1}{2}$ lines in a field, we can assume 1 line is scanned during vertical retrace for a convenient vertical flyback time. Then $9\frac{1}{2}$ lines are

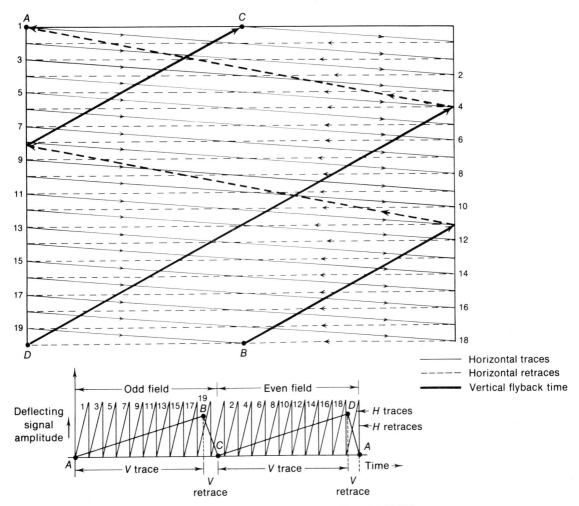

FIGURE 4-6 A SAMPLE SCANNING PATTERN FOR 21 INTERLACED LINES, WITH CORRESPONDING SAWTOOTH DEFLECTION WAVEFORMS. BEGINNING AT POINT A, THE SCANNING MOTION CONTINUES THROUGH B, C, D, AND BACK TO A AGAIN.

scanned during vertical trace in each field.

The entire frame has $2 \times 9\frac{1}{2}$, or 19, lines scanned during vertical trace, plus 2 vertical retrace lines.

Starting in the upper left corner at A in Fig. 4-6, the beam scans the first line from left to right and retraces to the left for the beginning of the third line in the frame. Then the beam scans the third and succeeding odd lines down to the bottom of the frame. After scanning $9\frac{1}{2}$ lines, the beam is at point B at the bottom when vertical flyback begins. Notice that this vertical retrace starts in the middle of a horizontal line. One line is scanned during vertical retrace, consisting of two half lines in this illustration, sloping upward in the direction of scanning. During this vertical retrace the scanning beam is brought up to point C, separated from point A by exactly a half line, to start scanning the second field.

Because of this half-line separation between points A and C, the lines scanned in the even field fall exactly between the odd lines in the previous field. The beam then scans $9\frac{1}{2}$ even lines from point C to D where the vertical retrace begins for the even field. This vertical retrace starts at the beginning of a horizontal line. Vertical retrace time is the same for both fields. Therefore, the one vertical retrace line in the second field returns the beam from D at the bottom to A at the top left corner of the frame where another odd field begins.

It should be noted that the points at which vertical retrace and the downward scan begin need not be exactly as shown in Fig. 4-6. These points could all be shifted by any fraction of a horizontal line without loss of interlace if the half-line difference were maintained.

The half-line spacing between the starting points in alternate fields is automatically produced in the sawtooth deflecting signals and the scanning motion because there is an odd number of lines for an even number of fields. Proper interlacing is assured, therefore, when the required frequencies of the horizontal and vertical sawtooth scanning signals are maintained precisely and the flyback time on the vertical sawtooth wave is constant for all fields.

4-4 Flicker

Interlaced scanning is used because the flicker effect is negligible with 60 views of the picture presented each second. Although the frame repetition rate is still 30 per second, the picture is blanked out during each vertical retrace 60 times per second. Then the change from black between pictures to the white picture is too rapid to be noticeable. If progressive scanning were used instead of interlacing, with all the lines in the frame simply scanned in progressive order from top to bottom, there would be only 30 blankouts per second and objectionable flicker would result. Scanning 60 complete frames per second in a progressive pattern would also eliminate flicker in the picture, but the horizontal scanning speed would be doubled, which would double the video frequencies corresponding to the picture elements in a line.

Although the increased blanking rate with interlaced scanning largely eliminates the effect of flicker in the image as a whole, the fact that individual lines are interlaced can cause flicker in small areas of the picture. Any one line in the image is illuminated 30 times per second, reducing the flicker rate of a single line to one-half the flicker rate for the interlaced image as a whole. The lower flicker rate for individual lines may cause two effects in the picture called *interline flicker* and *line crawl*. The interline flicker is sometimes evident as a blinking of thin horizontal objects in the picture, such as the roof line of a house. Line crawl is an apparent movement of the scanning lines upward or downward through the picture, due to the successive illumination of adjacent lines. These effects may be noticed sometimes in bright parts of the picture because the eye perceives flicker more easily at high brightness levels.

4-5 Raster Distortions

Since the picture information is reproduced on the scanning lines, distortions of the raster are in the picture. A rectangular shape for the raster, the correct proportions of width to height, and, finally, uniform deflection are required in order for the edges not to be distorted with respect to the center.

Incorrect aspect ratio. Two cases are illustrated in Fig. 4-7. In *a* the raster on the picture tube screen is not wide enough for its height, compared with the 4:3 aspect ratio used in the camera tube. Then people in the picture look too tall and thin, with the same geometrical distortion as the raster. This raster needs more width. In *b*, the raster is not high enough for its width, and people in the picture will look too short. This raster needs more height. For both *a* and *b*, these troubles are generally caused by insufficient output from the horizontal and vertical deflection circuits.

Pincushion and barrel distortion. If deflection is not uniform at the edges of the raster, compared with the center, the raster will not have straight edges. For scanning lines bowed inward as in Fig. 4-8*a*, this effect is *pincushion distortion*. Barrel distortion is shown in *b*.

Pincushion distortion is a problem with large-screen picture tubes. Since the faceplate is almost flat, the distance is longer from the point of deflection to the corners of the screen. The electron beam is deflected more than at the center, resulting in a raster with the corners stretched out. The pincushion distortion can be corrected, however, by a compensating magnetic field. Small permanent magnets for pincushion correction are mounted on the deflection yoke for monochrome picture tubes. With color picture tubes, the deflection current in the yoke is modified by pincushion correction circuits.

Trapezoidal distortion. In Fig. 4-9*a*, the scanning lines are wider at the top than at the bottom. This raster has the shape of a keystone or a trapezoid. The geometrical form of a trapezoid

FIGURE 4-7 INCORRECT ASPECT RATIO IN RASTER. BLACK SHOWS AREAS OF SCREEN NOT COVERED BY SCANNING. (*a*) INSUFFICIENT WIDTH. (*b*) INSUFFICIENT HEIGHT.

FIGURE 4-8 PINCUSHION DISTORTION OF RASTER IN (*a*) AND BARREL DISTORTION IN (*b*).

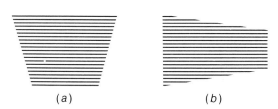
FIGURE 4-9 TRAPEZOIDAL RASTER. (*a*) KEYSTONED SIDES CAUSED BY UNSYMMETRICAL HORIZONTAL SCANNING. (*b*) KEYSTONED AT TOP AND BOTTOM BECAUSE OF UNSYMMETRICAL VERTICAL SCANNING.

has straight edges that are not parallel. This effect in the raster is called *keystoning,* which is a form of trapezoidal distortion. The cause is unsymmetrical deflection, either left to right as in Fig. 4-9*a* or top to bottom as in *b*. For picture tubes, the symmetry in scanning is provided by the balanced coils in the deflection yoke.

Nonlinear scanning. The sawtooth waveform

with a linear rise for trace time produces linear scanning, as the beam is made to move with constant speed. With nonlinear scanning, however, the motion of the beam is too slow or too fast. If the scanning spot at the receiver moves too slowly, compared with scanning in the camera tube at the transmitter, picture information will be crowded together. Or, if scanning, is too fast, the reproduced picture information is spread out. Usually the nonlinear scanning causes both effects at opposite ends of the raster. This is illustrated in Fig. 4-10a for a horizontal line with picture elements spread out at the left and crowded at the right. When the same effect occurs for all the horizontal lines in the raster, the entire picture is spread out at the left side and crowded at the right side. With people in the picture, a person at the left appears too wide while someone at the right looks too thin.

The vertical scanning motion must also be uniform; otherwise the horizontal lines will be bunched at the top or bottom of the raster and spread out at the opposite end. This effect is illustrated in Fig. 4-10b for spreading at the top and crowding at the bottom. Then a person in the picture will appear distorted with a long head and short legs. For both a and b, scanning nonlinearity generally is caused by amplitude distortion in the deflection amplifier circuits.

Poor interlacing. In each field the vertical trace from the top must start exactly a half line from the start of the previous field for odd-line interlacing. If the downward motion is slightly displaced from this correct position, the spot starts scanning too close to one of the lines in the previous field, instead of scanning exactly between lines. This incorrect start produces a vertical displacement between odd and even lines that is carried through the entire frame. As a result, pairs of lines are too close, with extra

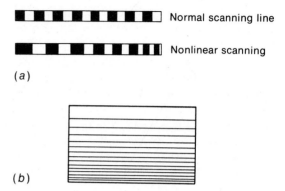

FIGURE 4-10 EFFECTS OF NONLINEAR SCANNING. (a) PICTURE ELEMENTS SPREAD OUT AT LEFT AND CROWDED AT RIGHT, CAUSED BY NONLINEAR HORIZONTAL SCANNING. (b) SCANNING LINES SPREAD OUT AT TOP AND CROWDED AT BOTTOM, CAUSED BY NONLINEAR VERTICAL SCANNING.

space between pairs. Then you can see too much black between the white scanning lines. This defect in the interlaced scanning is called *line pairing*. For the extreme case, lines in each successive field may be scanned exactly on the previous field lines. Then the raster contains only one-half the usual number of horizontal lines.

When the picture has diagonal lines as part of the image, poor interlacing makes them appear to be interwoven in the moiré effect shown in Fig. 4-11. This effect, also called *fishtailing*, is more evident in diagonal picture information, when the interlacing varies in successive frames.

Poor interlacing is caused by inaccurate vertical synchronization. Although the period of a field is $1/60$ s, which is a relatively long time, vertical scanning in every field must be timed much more accurately for good interlace. If the vertical timing is off by $1/4$ μs in one field compared with the next, the interlaced fields are shifted the distance of one picture element.

FIGURE 4-11 FAULTY INTERLACING. LINE DIVISIONS IN HORIZONTAL WEDGES ARE INTERWOVEN IN A MOIRÉ EFFECT.

4-6 The Synchronizing Pulses

At the receiver the picture tube scanning beam must reassemble picture elements on each horizontal line with the same left-right position as the image at the camera tube. Also, as the beam scans vertically the successive scanning lines on the picture tube screen must present the same picture elements as corresponding lines at the camera tube. Therefore, a horizontal synchronizing pulse is transmitted for each horizontal line to keep the horizontal scanning synchronized, and a vertical synchronizing pulse is transmitted for each field to synchronize the vertical scanning motion. The horizontal synchronizing pulses then have a frequency of 15,750 Hz, and the frequency of the vertical synchronizing pulses is 60 Hz.

The synchronizing pulses are transmitted as part of the picture signal but are sent during the blanking period when no picture information is transmitted. This is possible because the synchronizing pulse begins the retrace, either horizontal or vertical, and therefore occurs during retrace time. The synchronizing signals are combined with the picture signal in such a way that part of the modulated picture signal amplitude is used for the synchronizing pulses and the remainder for the camera signal. The term *sync* is often used for brevity to indicate the synchronizing pulses.

The form of the synchronizing pulses is illustrated in Fig. 4-12. Note that all pulses have the same amplitude but differ in pulse width or waveform. The synchronizing pulses shown include from left to right three horizontal pulses, a group of six equalizing pulses, a serrated vertical pulse, and six additional equalizing pulses which are followed by three more horizontal pulses. There are many additional horizontal pulses after the last one shown, following each other at the horizontal line frequency until the equalizing pulses occur again for the beginning of the next field. For every field there must be one wide vertical pulse, which is actually composed of six individual pulses separated by the five serrations.

Each vertical synchronizing pulse extends over a period equal to six half lines or three complete horizontal lines, making it much wider than a horizontal pulse. This is done to give the vertical pulses an entirely different form from the horizontal pulses. Then they can be completely separated from each other at the receiver, one furnishing horizontal synchronizing signals alone while the other provides only vertical synchronization.

The five serrations are inserted in the vertical pulse at half-line intervals. The equalizing pulses are also spaced at half-line intervals. These half-line pulses can serve for horizontal synchronization, alternate pulses being used for

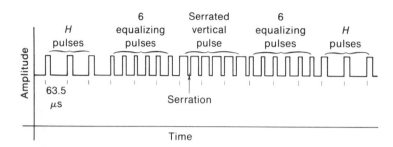

FIGURE 4-12 THE SYNCHRONIZING PULSES.

even and odd fields. The reason for using equalizing pulses, however, is related to vertical synchronization. Their effect is to provide identical waveshapes in the separated vertical synchronizing signal for even and odd fields so that constant timing can be obtained for good interlace. Since the equalizing pulses are repeated at half-line intervals, their repetition rate is twice 15,750, or 31,500 Hz.

The synchronizing signals do not produce scanning. Sawtooth generator circuits are needed to provide the deflection of the electron beam that produces the scanning raster. However, the sync enables the picture information reproduced on the raster to hold still in the correct position. Without vertical sync, the picture which is reproduced appears to roll up or down the raster; without horizontal sync the picture drifts to the left or right, and then the line structure breaks into diagonal segments. These effects are shown in Fig. 4-13.

4-7 Scanning, Synchronizing, and Blanking Frequencies

The sync and blanking pulses always have the same rate as the scanning frequency. These values are summarized in Table 4-1. For vertical

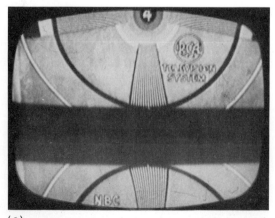

(a) (b)

FIGURE 4-13 EFFECTS OF NO SYNC. (a) PICTURE ROLLING UP OR DOWN WITHOUT VERTICAL SYNC. (b) PICTURE TORN INTO DIAGONAL SEGMENTS WITHOUT HORIZONTAL SYNC.

TABLE 4-1 SCANNING, SYNCHRONIZING, AND BLANKING FREQUENCIES*

FREQUENCY, Hz	APPLICATION
60	V sync to time V field scanning
60	V scanning to make lines fill raster
60	V blanking to blank out V retraces
15,750	H sync to time H scanning
15,750	H scanning to produce lines
15,750	H blanking to blank out H retraces
31,500	Equalizing pulses

*These are values for monochrome and only nominal values for color television broadcasting, where V is exactly 59.94 Hz and H is exactly 15,734.26 Hz.

deflection, the sawtooth scanning waveform has the frequency of 60 Hz because this is determined by the V sync pulses repeated every $1/60$ s. The vertical retrace is blanked out because the flyback is triggered by the V sync pulse during the time of the V blanking pulse. The relation of V sync within V blanking can be seen in Fig. 5-5 in Chap. 5 on the Composite Video Signal.

Similarly, for horizontal deflection, the sawtooth scanning waveform has the frequency of 15,750 Hz because this is the repetition rate of the H sync pulses. H flyback is blanked out because H retrace is within the time of H blanking.

The groups of equalizing pulses are repeated every $1/60$ s but each of the pulses is spaced at half-line intervals with the frequency of 31,500 Hz. The function of the equalizing pulses is to equalize the vertical synchronization in even and odd fields for good interlacing. However, there is no scanning or blanking at the equalizing-pulse rate of 31,500 Hz.

It may be of interest to note that for color television broadcasting the V field frequency is exactly 59.94 Hz and the H line frequency is 15,734.26 Hz. These values are derived as submultiples of the exact color subcarrier frequency of 3.579545 MHz. However, these frequencies are so close to 60 and 15,750 Hz that the deflection circuits can easily lock in for synchronized V and H scanning. The main factor is that when the V and H scanning frequencies are shifted slightly for color television, both the sync and blanking pulses are also changed to the new frequencies. A synchronized scanning circuit automatically locks into the sync frequency.

SUMMARY

1. The sawtooth waveform for deflection provides linear scanning. The linear rise on the sawtooth is the trace part; the sharp drop in amplitude is for the retrace or flyback. Both trace and retrace are included in one cycle.
2. The frequency of the sawtooth waveform for horizontal deflection is the horizontal line rate of 15,750 Hz.
3. The frequency of the sawtooth waveform for vertical deflection is the field rate of 60 Hz. Vertical flyback time 5 percent or less of $1/60$ s is long enough to include several complete lines. These horizontal lines scanned during vertical retrace are vertical flyback lines.
4. In odd-line interlacing, an odd number of lines (525) is used with an even number of fields (60), so that each field has a whole number of lines plus one-half. Then

successive fields start scanning a half line away from the previous field, interlacing odd and even lines in the frame.
5. Interlaced scanning eliminates flicker because of the 60-cycle vertical blanking rate, while maintaining the 30-cycle rate for complete picture frames.
6. Distortions of the scanning raster include keystone, trapezoid, pincushion, and barrel effects.
7. Incorrect aspect ratio can make people in the picture look too tall or too short, because of incorrect height or width of the raster.
8. Nonlinear scanning spreads or crowds picture information at one end of the raster compared with the opposite end. This effect also distorts the shape of people in the picture.
9. The synchronizing pulses time the scanning with respect to the position of picture information on the raster. Horizontal sync pulses time every line at 15,750 Hz; vertical sync pulses time every field at 60 Hz. All the sync pulses have the same amplitude, but a much wider pulse is used for vertical sync. The equalizing pulses and the serrations in the vertical pulse occur at half-line intervals with the frequency of 31,500 Hz.

Self-Examination (Answers at back of book)

Choose (a), (b), (c), or (d).

1. In the sawtooth waveform for linear scanning the (a) linear rise is for flyback; (b) complete cycle includes trace and retrace; (c) sharp reversal in amplitude produces trace; (d) beam moves faster during trace than retrace.
2. With a vertical retrace time of 635 μs, the number of complete horizontal lines scanned during vertical flyback is (a) 10; (b) 20; (c) 30; (d) 63.
3. One-half line spacing between the start positions for scanning even and odd fields produces (a) linear scanning; (b) line pairing; (c) fishtailing; (d) exact interlacing.
4. The number of lines scanned per frame in the raster on the picture tube screen is (a) 525; (b) 262$\frac{1}{2}$; (c) 20; (d) 10.
5. In the interlaced frame, alternate lines are skipped during vertical scanning because the (a) trace is slower than retrace; (b) vertical scanning frequency is doubled from the 30-Hz frame rate to the 60-Hz field rate; (c) horizontal scanning is slower than vertical scanning; (d) frame has the aspect ratio of 4:3.
6. With 10 percent for horizontal flyback, this time equals (a) 10 μs; (b) 56 μs; (c) 6.4 μs; (d) 83 μsec.
7. Which of the following is *not* true? (a) Line pairing indicates poor interlacing. (b) People will look too tall and thin on a square raster on the picture tube screen. (c) A person can appear to have one shoulder wider than the other because of nonlinear horizontal scanning. (d) The keystone effect produces a square raster.

SCANNING AND SYNCHRONIZING   63

8. The width of a vertical sync pulse with its serrations includes the time of (a) six half lines or three lines; (b) five lines; (c) three half lines; (d) five half lines.
9. Sawtooth generator circuits produce the scanning raster, but the sync pulses are needed for (a) linearity; (b) timing; (c) keystoning; (d) line pairing.
10. Which of the following frequencies is wrong? (a) 15,750 Hz for horizontal sync and scanning; (b) 60 Hz for vertical sync and scanning; (c) 31,500 Hz for equalizing pulses and serrations in the vertical sync pulse; (d) 31,500 Hz for the vertical scanning frequency.

Essay Questions

1. Draw the interlaced scanning pattern for a total of 25 lines per frame, interlaced in two fields. Also show the corresponding sawtooth waveforms for horizontal and vertical scanning, as in Fig. 4-6. Assume one line scanned during each vertical flyback.
2. Define the following terms: (a) scanning raster; (b) pincushion effect; (c) line pairing; (d) interline flicker; (e) moiré effect.
3. Why are the lines scanned during vertical trace much closer together than lines scanned during vertical flyback?
4. Suppose the sawtooth waveform for vertical scanning has a trace that rises too fast at the start and flattens at the top. Will the scanning lines be crowded at the top or bottom of the picture tube raster? How will people look in the picture?
5. Draw two cycles of the 15,750-Hz sawtooth waveform, showing retrace equal to 0.08H to exact scale. Label trace, retrace, and time of one cycle in microseconds.
6. Draw two cycles of the 60-Hz sawtooth waveform, showing retrace equal to 0.04V to exact scale. Label trace, retrace, and time of one cycle in microseconds.
7. Where is the electron scanning beam at the time of: (a) start of linear rise in H sawtooth; (b) start of H flyback; (c) start of linear rise in V sawtooth; (d) start of V flyback?

Problems   (Answers to selected problems at back of book)

1. How many flyback lines are produced during vertical retrace for each field and each frame for retrace time equal to (a) 0.02V; (b) 0.08V?
2. Compare the time in microseconds for horizontal flyback equal to 0.08H and vertical flyback of 0.04V.
3. (a) How much time elapses between the start of one horizontal sync pulse and the

next? (b) Between one vertical pulse in an odd field and the next in an even field?
4. What frequencies correspond to the following periods of time for one cycle: (a) $1/60$ s; (b) 63.5 μs; (c) 53.3 μs.
5. Refer to the sawtooth waveforms in Fig. 4-14 below. For the opposite polarities in a and b, indicate trace and retrace on each H sawtooth, also, the corresponding left and right edges in the raster. For the opposite polarities in c and d indicate trace and retrace on each V sawtooth, also, the corresponding top and bottom edges in the raster.

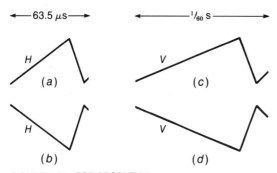

FIGURE 4-14 FOR PROBLEM 5

5 Composite Video Signal

"Composite" means that the video signal includes separate parts. These parts are (1) camera signal corresponding to the desired picture information, (2) synchronizing pulses to synchronize the transmitter and receiver scanning, and (3) blanking pulses to make the retraces invisible. How these three components are added to produce the composite video signal is illustrated in Fig. 5-1. The camera signal in *a* is combined with the blanking pulse in *b*, and then the sync pulse to produce the composite video signal in *c*. The result shown here is composite video signal for one horizontal scanning line. In addition, for color the 3.58-MHz chrominance signal and color sync burst are included. With signal for all the lines, the composite video contains the information needed to reproduce the complete picture. More details are explained in the following topics:

 5-1 Construction of the composite video signal
 5-2 Horizontal blanking time
 5-3 Vertical blanking time
 5-4 Picture information and video signal amplitudes
 5-5 Oscilloscope waveforms of video signal
 5-6 Picture information and video signal frequencies
 5-7 Maximum number of picture elements
 5-8 Test patterns
 5-9 Dc component of the video signal
 5-10 Gamma and contrast in the picture
 5-11 Color information in the video signal
 5-12 Vertical interval test signals (VITS)

66 BASIC TELEVISION

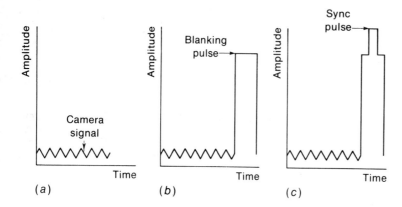

FIGURE 5-1 THE THREE COMPONENTS OF COMPOSITE VIDEO SIGNAL. (a) CAMERA SIGNAL FOR ONE HORIZONTAL LINE. (b) HORIZONTAL BLANKING PULSE ADDED TO CAMERA SIGNAL. (c) SYNC PULSE ADDED TO BLANKING PULSE.

5-1 Construction of the Composite Video Signal

In Fig. 5-2, successive values of voltage or current amplitude are shown for the scanning of three horizontal lines in the image. Note that the amplitude of the video signal is divided into two sections. The lower 75 percent is used for camera signal, with the upper 25 percent for synchronizing pulses. In the camera signal, the lowest amplitudes correspond to the whitest parts of the picture while the darker parts of the picture have higher amplitudes. This is the way the signal is transmitted, using a standard negative polarity of transmission. "Negative transmission" means that white parts of the picture are represented by low amplitudes in the transmitted picture carrier signal. Higher amplitudes correspond to progressively darker picture information until the black level is reached.

The composite video signal and scanning. Referring again to Fig. 5-2, consider the amplitude variations shown as the desired video signal obtained in scanning three horizontal lines at the top of the image. Starting at the extreme left in the figure at zero time, the signal is

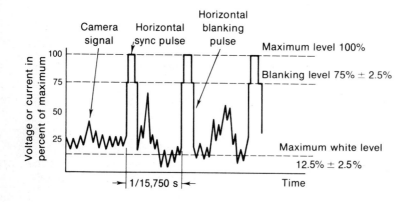

FIGURE 5-2 COMPOSITE VIDEO SIGNAL FOR THREE CONSECUTIVE HORIZONTAL LINES.

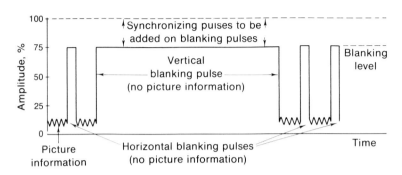

FIGURE 5-3 HORIZONTAL AND VERTICAL BLANKING PULSES IN VIDEO SIGNAL. SYNC PULSES ARE NOT SHOWN.

at a white level and the scanning beam is at the left side of the image. As the first line is scanned from left to right, camera signal variations are obtained with various amplitudes that correspond to the required picture information. After horizontal trace produces the desired camera signal for one line, the scanning beam is at the right side of the image. The blanking pulse is then inserted to bring the video signal amplitude up to black level so that retrace can be blanked out.

After a blanking time long enough to include retrace, the blanking voltage is removed. The scanning beam is then at the left side ready to scan the next line. Each horizontal line is scanned successively in this way. Notice that the second line shows dark picture information near the black level. The third line has gray values with medium amplitudes of 40 to 50 percent.

With respect to time, the signal amplitudes just after blanking in Fig. 5-2 indicate information corresponding to the left side at the start of a scanning line. Just before blanking, the signal variations correspond to the right side. Information exactly in the center of a scanning line occurs at a time halfway between blanking pulses.

The blanking pulses. The composite video signal contains blanking pulses to make the retrace lines invisible by raising the signal amplitude to black level during the time the scanning circuits produce retraces. All picture information is cut off during blanking time because of the black level. The retraces are normally produced within the time of blanking.

As illustrated in Fig. 5-3, there are horizontal and vertical blanking pulses in the composite video signal. The horizontal blanking pulses are included to blank out the retrace from right to left in each horizontal scanning line. The repetition rate of horizontal blanking pulses, therefore, is the line-scanning frequency of 15,750 Hz. The vertical blanking pulses have the function of blanking out the scanning lines produced when the electron beam retraces vertically from bottom to top in each field. Therefore, the frequency of vertical blanking pulses is 60 Hz for every field.

5-2 Horizontal Blanking Time

Details in the horizontal blanking period are illustrated in Fig. 5-4. The interval between horizontal scanning lines is indicated by H. This time for scanning one complete line, including trace and retrace, equals $1/15{,}750$ s, or 63.5 μs. However, the horizontal blanking pulse has a width only $0.14H$ to $0.18H$. We can consider the average of 16 percent of the line period as a typi-

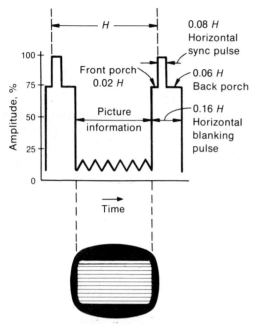

FIGURE 5-4 DETAILS OF HORIZONTAL BLANKING AND SYNC PULSES. H EQUALS 63.5 μs.

porch is 3 times longer than the front porch. These values are summarized in Table 5-1, with the required tolerances. The minimum time of 3.81 μs is specified for the back porch in order to provide time for a burst of 3.58-MHz color synchronizing signal. This is shown in Fig. 5-16b.

TABLE 5-1 DETAILS OF HORIZONTAL BLANKING

PERIOD	TIME, μs
Total line (H)	63.5
H blanking	9.5–11.5
H sync pulse	4.75 ± 0.5
Front porch	1.27 (minimum)
Back porch	3.81 (minimum)
Visible line time	52–54

cal value. Then horizontal blanking time is 0.16×63.5 μs for H, which equals 10 μs, approximately. Subtracting from 63.5 μs, we have a remainder of 53.5 μs as the time for visible scanning, without blanking, in each line. The 10 μs for blanking allows time for retrace.

Superimposed on the blanking pulses are the narrower sync pulses. As noted in Fig. 5-4, each horizontal sync pulse is 0.08H, or one-half the average width for the blanking pulse. This sync time equals $^{10}/_2$ or 5 μs.

For the remaining half of blanking time, which is also 5 μs, the signal is at the blanking level. The part just before the sync pulse is called the *front porch*, and the *back porch* follows the sync pulse. The front porch is 0.02H and the back porch 0.06H. Note that the back

Blanking time is slightly longer than typical values of retrace time, which depend on the horizontal deflection circuits in the receiver. As a result, a small part of the trace usually is blanked out at the start and end of every scanning line. This effect of horizontal blanking is illustrated by the black bars at the left and right sides of the raster in Fig. 5-4. The black at the right edge corresponds to the front porch of horizontal blanking, before retrace starts. Generally, horizontal retrace starts at the leading edge of the sync pulse. Just before retrace, when the scanning beam is completing its trace to the right, therefore, the blanking level of the front porch makes the right edge black. With a small part of every line blanked this way, a black bar is formed at the right edge. This black bar at the right can be considered as a reproduction of the front-porch part of horizontal blanking.

After the front porch of blanking, horizontal retrace is produced when the sync pulse starts. The flyback is definitely blanked out because the sync level is blacker than black. Although retrace starts with the sync pulse, how much time is needed to complete the flyback depends on the scanning circuits. A typical horizontal flyback time is 7 μs. Blanking time after the front porch is longer, however, and approximately equal to 9 μs. Therefore, 2 μs of blanking remains after retrace is completed to the left edge. Although the blanking is still on, the sawtooth deflection waveform makes the scanning beam start its trace after flyback. As a result, the first part of trace at the left is blanked. After 2 μs of blanked trace time at the left edge, the blanking pulse is removed. Then video signal reproduces picture information as the scanning beam continues its trace for 53.5 μs of visible trace time. However, the small part of every line blanked at the start of trace forms the black bar at the left edge of the raster. This black edge at the left represents part of every back porch after horizontal sync.

The blanking bars at the sides have no effect on the picture other than decreasing its width slightly, compared with unblanked raster. However, the amplitude of horizontal scanning can be increased to provide the desired width.

5-3 Vertical Blanking Time

The vertical blanking pulses raise the video signal amplitude to black level so that the scanning beam is blanked out during vertical retraces. The width of the vertical blanking pulse is $0.05V$ to $0.08V$, where V equals $1/60$ s. If we take 8 percent for the maximum, vertical blanking time is $0.08 \times 1/60$ s, which equals 1,333 μs. Note that this time is long enough to include many complete horizontal scanning lines. If we divide 1,333-μs vertical blanking time by the total line period of 63.5 μs, the answer is 21. Therefore, 21 lines are blanked out in each field, or 42 lines in the frame. The total number of blanked lines in the frame can also be calculated as $0.08 \times 525 = 42$. This relatively long time blanks not only vertical retrace lines but also a small part of vertical trace at the bottom and top. The maximum blanking time of $0.08V$ is generally used to allow time for the vertical interval test signals described in Sec. 5-12.

The sync pulses inserted in the composite video signal during the wide vertical blanking pulse are shown in Fig. 5-5. These include equalizing pulses, vertical sync pulses, and some horizontal sync pulses. The signals are shown for the time intervals between the end of one field and the next, to illustrate what happens during vertical blanking time. The two signals shown one above the other are the same except for the half-line displacement between successive fields necessary for odd-line interlacing.

Starting at the left in Fig. 5-5, the last four horizontal scanning lines at the bottom of the raster are shown with the required horizontal blanking and sync pulses. Immediately following the last visible line, the video signal is brought up to black level by the vertical blanking pulse in preparation for vertical retrace. The vertical blanking period begins with a group of six equalizing pulses, which are spaced at half-line intervals. Next is the serrated vertical sync pulse that actually produces vertical flyback in the scanning circuits. The serrations also occur at half-line intervals. Therefore, the complete vertical sync pulse is three lines wide. Following the vertical sync is another group of six equalizing pulses and a train of horizontal pulses. During this entire vertical blanking period no picture information is produced, as the signal level is black or blacker than black so that vertical retrace can be blanked out. The details of all the pulses in the vertical blanking interval are summarized in Table 5-2.

Notice the position of the first equalizing pulse at the start of vertical blanking in Fig. 5-5. In the signal at the top, the first pulse is a full line

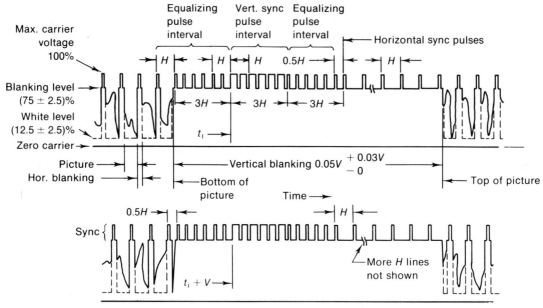

FIGURE 5-5 SYNC AND BLANKING PULSES FOR SUCCESSIVE FIELDS. V EQUALS $1/60$ s.

away from the previous horizontal sync pulse; in the signal below for the next field, the first pulse is one-half line away. This half-line difference in time between even and odd fields continues through all the following pulses, so that the vertical sync pulses for successive fields have the timing required for odd-line interlacing.

The serrated vertical sync pulse forces the vertical deflection circuits to start the flyback. However, the flyback generally does not begin with the start of vertical sync because the sync pulse must build up charge in a capacitor to trigger the scanning circuits. If we assume vertical flyback starts with the leading edge of the third serration, the time of one line passes during vertical sync before vertical flyback starts. Also, six equalizing pulses equal to three lines occur before vertical sync. Then 3 + 1, or 4, lines are blanked at the bottom of the picture, just before vertical retrace starts.

How long the flyback is depends on the scanning circuits, but a typical vertical retrace time is five lines. As the scanning beam retraces from bottom to top of the raster, then, five complete horizontal lines are produced. This vertical retrace is easily fast enough to be completed within vertical blanking time.

With four lines blanked at the bottom before flyback and five lines during flyback, 12

TABLE 5-2 DETAILS OF VERTICAL BLANKING

PERIOD	TIME
Total field (V)	$1/60 = 0.0167$ s
V blanking	0.05–0.08 V, or 0.0008–0.0013 s
Each V sync pulse	$H/2 = 31.75$ μs
Total of 6 V sync pulses	$3H = 190.5$ μs
Each E pulse	$0.04H = 2.54$ μs
Each serration	$0.07H = 4.4$ μs
Visible field time	0.92–0.95 V, or 0.015–0.016 s

lines remain of the total 21 during vertical blanking. These 12 blanked lines are at the top of the raster at the start of the vertical trace downward.

In summary, 4 lines are blanked at the bottom and 12 lines at the top in each field. In the total frame of two fields, these numbers are doubled. The scanning lines that are produced during vertical trace, but made black by vertical blanking, form the black bars at top and bottom of the raster in Fig. 5-4. The result is a slight reduction in height of the picture with blanking, compared with the unblanked raster. However, the height is easily corrected by increasing the amplitude of the sawtooth waveform for vertical scanning.

5-4 Picture Information and Video Signal Amplitudes

Two examples are shown in Fig. 5-6 to illustrate how the composite video signal corresponds to visual information. In a, the video signal corresponds to one horizontal line in scanning an image with a black vertical bar down the center of a white frame. In b, the black and white values in the picture are reversed from a.

Starting at the left in Fig. 5-6a, the camera signal obtained in active scanning of the image is initially at the white level corresponding to the white background. The scanning beam continues its forward motion across the white background of the frame, and the signal continues at the same white level until the middle of the picture is reached. When the black bar is scanned, the video signal rises to the black level and remains there while the entire width of the black bar is scanned. Then the signal amplitude drops to white level corresponding to the white background and continues at that level while the forward scanning motion is completed to the right side of the image.

At the end of the visible trace the horizontal blanking pulse raises the video signal amplitude to black level in preparation for horizontal retrace. After retrace, the forward scanning motion begins again to scan the next horizontal line. Each successive horizontal line in the even and odd fields is scanned in this way. As a result, the corresponding composite video signal for the entire picture contains a succession of signals with a waveform identical with that shown in Fig. 5-6a for each active horizontal scanning line. For the image in b the idea is the

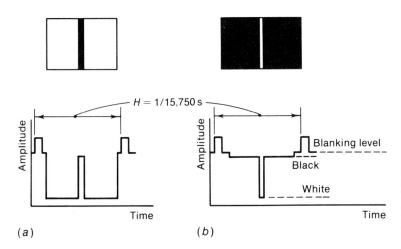

FIGURE 5-6 COMPOSITE VIDEO SIGNAL AND ITS CORRESPONDING PICTURE INFORMATION. (a) IMAGE WITH BLACK VERTICAL LINE ON WHITE BACKGROUND. (b) WHITE LINE ON BLACK BACKGROUND.

72 BASIC TELEVISION

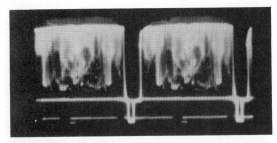

(a)

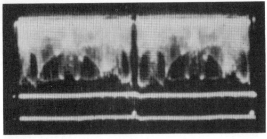

(b)

FIGURE 5-7 OSCILLOSCOPE PHOTOGRAPHS OF COMPOSITE VIDEO SIGNAL, SHOWN WITH SYNC POLARITY DOWNWARD. (a) TWO LINES OF HORIZONTAL PICTURE INFORMATION BETWEEN HORIZONTAL BLANKING AND SYNC PULSES. OSCILLOSCOPE SWEEP AT 7,875 HZ. (b) TWO FIELDS OF VERTICAL PICTURE INFORMATION BETWEEN VERTICAL BLANKING AND SYNC PULSES. OSCILLOSCOPE SWEEP AT 30 HZ.

same, but the camera signal corresponds to a white vertical bar down the center of a black frame.

These are simple types of images, but the correlation can be carried over to an image with any distribution of light and shade. If the pattern contains five vertical black bars against a white background, the composite video signal for each horizontal line will include five rapid variations in amplitude from white to the black level.

As another example, suppose the pattern consists of a horizontal black bar across the center of a white frame. Then most of the horizontal lines will contain white picture information for the entire trace period, with the camera signal amplitude remaining at white level except for the blanking intervals. However, those horizontal lines that scan across the black bar will produce camera signal at black level for the complete active scanning time.

Black setup. In Fig. 5-6, notice that the 75 percent blanking level is offset from black peaks of camera signal. This difference in black amplitudes is called *black setup*. Standard practice in broadcasting is to use 5 to 10 percent black setup to make sure that black picture information cannot extend into the sync amplitudes.

Typical video signal voltages. An actual picture consists of elements having different amounts of light and shade with a nonuniform distribution in the horizontal lines and through the vertical fields. Especially with motion in the scene, the video signal contains a succession of continuously changing voltages. Within each line there are variations in camera signal amplitude for different picture elements. Furthermore, the waveforms of camera signal for the lines change within the field. The resulting waveforms are shown by the oscilloscope photos of typical video signal in Fig. 5-7. This signal is at the control grid of the picture tube, with an amplitude of 100 V peak to peak.

5-5 Oscilloscope Waveforms of Video Signal

When you look at oscilloscope patterns, the camera signal variations are usually hazy as they change with motion in the scene. However, the oscilloscope trace locks in for the horizontal blanking and sync pulses at the steady repetition rate of 15,750 Hz or for the vertical pulses at 60 Hz. The internal horizontal sweep frequency of the oscilloscope is preferably set at one-half these frequencies to show video signal either for two lines as in Fig. 5-7a or for two fields as in b. Then each cycle is shown as wide as possible and with continuity through blanking time.

Line rate. With the oscilloscope sweep at 15,750/2 or 7,875 Hz, you see two lines of video signal (Fig. 5-7a). Any horizontal motion in the scene shows as horizontal motion in the signal variations between the sync pulses.

Field rate. With the oscilloscope sweep at $^{60}/_2$ or 30 Hz, you see two fields of video signal (Fig. 5-7b). Any vertical motion in the scene shows as motion in the camera signal variations across the trace between the sync pulses. The lines extending across the top and bottom of the vertical sync are caused by the horizontal sync.

You do not see the equalizing pulses in this pattern because the scope is locked at the vertical scanning frequency. In order to see the equalizing pulses and the serrations in the vertical pulses, you must set the scope internal sweep to 31,500 Hz or a submultiple. Also, the scope's horizontal sweep usually must be expanded.

Sync polarity. You can observe the oscilloscope patterns of video signal either with positive sync amplitudes upward or with negative sync down as in Fig. 5-7. In terms of the picture tube, negative sync is the correct polarity at the control grid to cut off beam current for black level. At the cathode, positive sync polarity is required, since positive voltage at the cathode corresponds to negative voltage at the control grid. With the blanking voltage driving the picture tube to cutoff for black, then white in the video signal produces maximum beam current for the highlights in the picture. In any case, remember that the white amplitudes are always

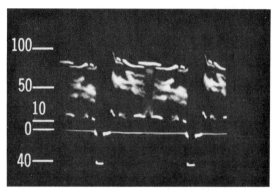

FIGURE 5-8 VIDEO SIGNAL ON STANDARD IRE SCALE OF 140 WITH 10 UNITS FOR BLACK SETUP.

opposite the sync pulses, away from the blanking level.

IRE amplitude scale. On oscilloscope monitors at the studio, the video signal is checked with negative sync polarity down, to fit the IRE scale shown in Fig. 5-8. The *IRE* stands for Institute of Radio Engineers, now called the Institute of Electrical and Electronic Engineers (IEEE). Of the 140 IRE units, 40 are for sync amplitudes and 10 for black setup. The 140 units correspond to 1-V peak-to-peak video signal to feed 75-Ω coaxial cable between amplifier units.

5-6 Picture Information and Video Signal Frequencies

In general, signal variations within any line have high frequencies because the horizontal scanning is fast. In the vertical direction, the signal amplitude variations are much lower in frequency because of the slower speed of vertical scanning. As a result, the video signal for a complete picture has high-frequency variations for the horizontal details and low-frequency variations for the vertical information.

Video frequencies associated with horizontal scanning. Referring to the checkerboard pattern in Fig. 5-9, the square-wave signal shown at the top represents the camera signal variations of the composite video signal obtained in scanning one horizontal line. It is desired to find the frequency of this square wave. The frequency of the camera signal variations is very important in determining whether or not the television system can transmit and reproduce the corresponding picture information.

In determining the frequency of any signal variation, the time for one complete cycle must be known. A cycle includes the time from one point on the signal waveform to the next succeeding point which has the same magnitude and direction. Frequency can then be found as the reciprocal of the period for a cycle. As an example, the period of one horizontal scanning line is 1/15,750 s and the line-scanning frequency is 15,750 Hz. The camera signal variations within one horizontal line, however, necessarily have a shorter period and a higher frequency.

Note that one complete cycle of camera signal in Fig. 5-9 includes the information in two adjacent picture elements, one white and the other black. Only after scanning the second square does the camera signal have the same magnitude and direction as at the start of the first square. Therefore, to find the frequency of the camera signal variations it is necessary to determine how long it takes to scan across two adjacent squares. This time is the period for one cycle of the resultant camera signal.

Now the period of one complete cycle of the square-wave camera signal variations in Fig. 5-9 can be calculated. The horizontal line period is 1/15,750 s, or 63.5 μs, including trace and retrace. With a horizontal blanking time of 10.2 μs, the time remaining for visible trace equals 53.3 μs. This is the time to scan across all the picture elements in a line. For 12 squares across one line in 53.3 μs, the beam scans two squares in $2/12$ or $1/6$ of 53.3 μs. Then 53.3/6, or 8.9 μs, is the time to scan two squares.

This time of 8.9 μs is the period of one complete cycle of the square-wave signal. The reciprocal equal to 1/8.9 μs is the frequency, therefore, which is 0.11×10^6 Hz, or 0.11 MHz. This is the frequency of the square-wave camera signal variations in Fig. 5-9.

When a typical picture is scanned, the scattered areas of light and shade do not

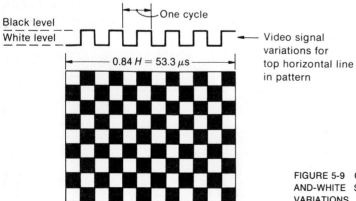

FIGURE 5-9 CHECKERBOARD PATTERN OF 12 BLACK-AND-WHITE SQUARES WITH CORRESPONDING SIGNAL VARIATIONS.

produce symmetrical square-wave signal. However, the differences of light and shade correspond to changes of camera signal amplitude in the same way. The frequency of the resultant camera signal variations always depends on the time to scan adjacent areas with different light values. When large objects with a constant white, gray, or black level are scanned, the corresponding camera signal variations have low frequencies because of the comparatively long time between changes in level. Smaller areas of light and shade in the image produce higher video frequencies. The highest signal frequencies correspond to variations between very small picture elements in a horizontal line, especially the vertical edge between a white area and black area.

Video frequencies associated with vertical scanning. At the opposite extreme, signal variations which correspond to picture elements adjacent in the vertical direction have low frequencies because the vertical scanning is comparatively slow. Variations between one line and the next correspond to a frequency of approximately 10 kHz. Slower changes over larger distances in vertical scanning produce lower frequencies. The very low frequency of 30 Hz corresponds to a variation in light level between two successive fields.

Video frequencies and picture information. Figure 5-10 shows how the size of the picture information can be considered in terms of video frequencies. The main body of the image in *a* is shown separately in *b* with only the large areas of white and black. These video frequencies extend up to 100 kHz. However, the detail with sharp edges and outlines is filled in by the high video frequencies from 0.1 to 4 MHz shown in *c*. Notice that the canopy of the building is reproduced in *b*, but its stripes and the small lettering need the high-frequency reproduction in *c*.

5-7 Maximum Number of Picture Elements

If we consider a checkerboard pattern such as Fig. 5-9 with many more squares, the maximum possible number of picture elements can be calculated where each square is one element. The total elements in the area equal the maximum details in a line horizontally, multiplied by the details in a vertical row. However, horizontal detail and vertical detail must be considered separately in a television picture because of the scanning process. For horizontal detail, the problem is to determine how many elements correspond to the high-frequency limit of 4-MHz video signal. The vertical detail is a question of how many elements can be resolved by the scanning lines.

Maximum horizontal detail. Proceeding in the same manner as in the previous section, the number of elements corresponding to 4 MHz can be determined to show the maximum number of picture elements in a horizontal line and the size of the smallest possible horizontal detail. The period of one complete cycle for a 4-MHz signal variation is $1/(4 \times 10^6)$ s, or 0.25 μs. This is the time required to scan two adjacent picture elements. With two elements scanned in 0.25 or $1/4$ μs, then eight elements are scanned in 1 μs. Finally, 8×53.3, or 426, picture elements can be scanned during the entire active line period of 53.3 μs. If there were 426 squares in the horizontal direction in the checkerboard pattern in Fig. 5-9, therefore, the resultant camera signal variations would produce a 4-MHz signal.

Utilization ratio and vertical detail. Each scanning line can represent at best only one detail in the vertical direction. However, a scanning line may represent no vertical detail at all, by missing a vertical detail com-

(a)

(b)

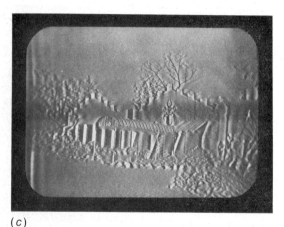

(c)

FIGURE 5-10 EFFECT OF VIDEO FREQUENCIES ON PICTURE REPRODUCTION. (a) NORMAL PICTURE REPRODUCTION. (b) ONLY LARGE AREAS IN PICTURE REPRODUCED BY LOW VIDEO FREQUENCIES UP TO 0.1 MHZ. (c) ONLY HORIZONTAL EDGES AND OUTLINES REPRODUCED BY HIGH VIDEO FREQUENCIES BETWEEN 0.1 AND 4 MHZ.

pletely. Also, two lines may straddle one picture element. The problem in establishing the useful vertical detail, then, is determining how many picture elements can be reproduced along a vertical line by a given number of scanning lines. The ratio of the number of scanning lines useful in representing the vertical detail to the total number of visible scanning lines is called the *utilization ratio*. Theoretical calculations and experimental tests show that the utilization ratio ranges from 0.6 to 0.8 for different images with typical picture content. We can use 0.7 as an average.

Now the maximum possible number of vertical elements can be determined. The number of visible lines equals 525 minus those scanned during vertical blanking. With a vertical blanking time of 8 percent, the number of lines blanked out for the entire frame is 0.06×525, or approximately 42 lines. Some of these lines occur during vertical retrace, and others are scanned at the top or bottom of the frame, but all are blanked out. When we subtract 42 from 525, then 483 visible lines remain. The number of lines useful in showing vertical detail is 483×0.7, since this is the utilization ratio. Then there are 338 effective lines. Therefore, the maximum number of vertical details that can be reproduced with 525 total and 483 visible scanning lines is about 338, the exact value depending upon the utilization ratio.

Total number of picture elements. On the basis of the previous calculations, the maximum number of picture elements possible for the entire image is 426 × 338, or about 144,000. This number is independent of picture size.

Since the total number of picture elements can be regarded as the figure of merit, it may be compared with motion-picture reproduction. A single frame of 35-mm motion-picture film has about 500,000 picture elements. The smaller 16-mm frame contains one-fourth as many, or about 125,000. The televised reproduction, therefore, can have about the same amount of detail as 16-mm motion pictures.

5-8 Test Patterns

In order to adjust a television system and compare performances, a standard picture is desirable. This picture is usually in the form of a test pattern. For the EIA pattern in Fig. 5-11, the horizontal and vertical bars form a large square within the white circle. Each bar is numbered from 1 to 10 as a gray scale with ten logarithmic steps between black and maximum white. The contrast range is approximately 30 to 1. With the correct aspect ratio, these four bars form a perfect square. The wedges at the center should be of equal length for good linearity of the horizontal and vertical scanning. The bottom wedge is also marked in the corresponding video-frequency response for horizontal resolution. For instance, 300-line resolution in the horizontal direction corresponds to a little under 4-MHz

video. In addition to the wedges, vertical and horizontal groups of parallel lines are placed around the edges of the picture to check scanning linearity. The lines should be equally spaced. The black bars at the top and bottom of the white circle are used to check streaking, which indicates phase distortion for low video frequencies.

Resolution. The picture detail or resolution is measured on the test pattern in number of lines. If the vertical resolution is 300 lines in the reproduced picture, this means that it is possible to see 300 individual horizontal lines consisting of 150 black lines separated by 150 white lines. For equal resolution in the horizontal direction and an aspect ratio of 4:3, 300 × $^4/_3$, or 400, vertical lines can be resolved in the picture. These include 200 black lines separated by 200 white lines. However, this is still considered 300-line resolution because the resolution is measured in terms of the picture height when indicating either horizontal or vertical detail. The purpose is to provide a common basis for comparison.

The line divisions in the side wedges measure vertical resolution. They also indicate good interlacing when there is little moiré effect in the diagonal lines. The line divisions in the top and bottom wedges measure horizontal resolution.

More specific test signals are illustrated in Fig. 5-12. The *window signal* in *a* provides maximum white and black in large areas with a sharp transition to check edge distortion, streaking, and smearing. In *b* the uniformly spaced black-and-white bars can be used to check scanning linearity. The bars may be vertical, or horizontal, or both may be used in a *crosshatch* pattern. The *staircase signal* in *c* illustrates uniform changes in signal amplitude from white, progressing through gray values in equal steps to black. This is called a *gray scale*.

78 BASIC TELEVISION

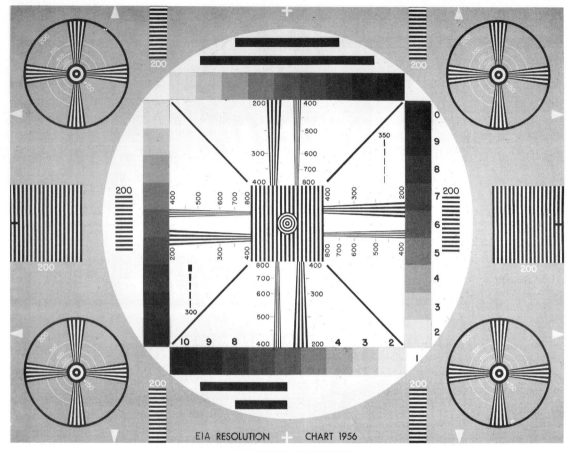

FIGURE 5-11 EIA TEST PATTERN. (*ELECTRONIC INDUSTRIES ASSOCIATION*)

5-9 DC Component of the Video Signal

In addition to continuous amplitude variations for individual picture elements, the video signal must have an average value corresponding to the average brightness in the scene. Otherwise the receiver cannot follow changes in brightness. As an example of the importance of brightness level, the ac camera signal for a gray picture element on a black background will be the same as signal for white on a gray background, if there is no average-brightness information to indicate the change in background.

The average level of a signal is the arithmetic mean of all the instantaneous values measured from the zero axis. In Fig. 5-13a the average level is higher than in b because the camera signal variations have higher amplitudes. Now it is important to remember that for any signal variation, its average value for a complete cycle is its dc component. Therefore,

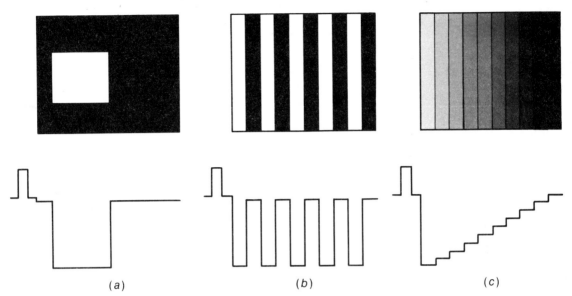

FIGURE 5-12 PATTERNS FOR TEST PURPOSES, WITH CORRESPONDING VIDEO SIGNAL FOR ONE HORIZONTAL LINE. (a) WINDOW SIGNAL. (b) BAR PATTERN. (c) STAIRCASE SIGNAL.

the dc component in *a* is closer to the black level than in *b*. Although illustrated here for one scanning line, for convenience, the required dc component of the video signal is its average value for complete frames, since the background information of the frame indicates the brightness of the scene.

When the average value or dc component of the video signal is close to the black level, as in Fig. 5-13a, the average brightness is dark.

The same ac signal variations in *b* have a lighter background because the dc axis is farther from the black level.

The method of using the black level for reference voltage can be followed by starting with the camera signal. The camera output voltage is amplified in several stages before being coupled to a control amplifier, where sync and blanking are added. At this point the camera signal has no dc component, since the dc level is

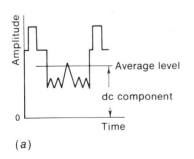

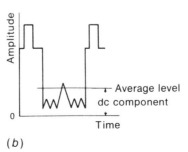

FIGURE 5-13 VIDEO SIGNALS WITH SAME AC VARIATIONS BUT DIFFERENT AVERAGE-BRIGHTNESS VALUES. ONLY ONE LINE OF FRAME ILLUSTRATED. (a) DARK SCENE WITH AVERAGE VALUE CLOSE TO BLACK LEVEL. (b) LIGHT SCENE WITH AVERAGE VALUE FARTHER FROM BLACK LEVEL.

Note that clipping lower on the blanking pulses makes the brightness darker. For the opposite case, a higher clipping level means the signal will have a lighter background.

5-10 Gamma and Contrast in the Picture

Gamma is a numerical factor used in television and film reproduction for indicating how light values are expanded or compressed. Referring to Fig. 5-15, the exponent of the equations for the curves shown is called *gamma* (γ). The numerical value of gamma is equal to the slope of the straight-line part of the curve where it rises most sharply. A curve with a gamma of less than 1 is bowed downward as in *a*, with the greatest slope at the start and the relatively flat part at the end. When the gamma is more than 1, the curve is bowed upward as in *b*, making the start comparatively flat while the sharp slope is at the end. With a gamma of 1 the result is a straight line as in *c*, where the slope is constant.

A gamma value of 1 means a linear characteristic that does not exaggerate any light values. When gamma is greater than 1 for the white parts of the image, the reproduced picture looks "contrasty" because the increases in white level are expanded by the sharp slope, to emphasize the white parts of the picture. Commercial motion pictures shown in a darkened theater have this high-contrast appearance. Gamma values

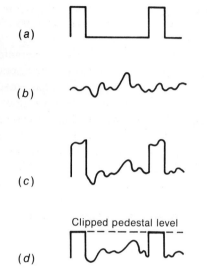

FIGURE 5-14 FORMATION OF PEDESTALS BY CLIPPING TOP OF BLANKING PULSES. (*a*) BLANKING PULSES ALONE. (*b*) CAMERA SIGNAL WITHOUT BLANKING. (*c*) BLANKING PULSES ADDED TO CAMERA SIGNAL. (*d*) PULSES CLIPPED TO PROVIDE PEDESTAL LEVEL FOR SYNC PULSES.

blocked by capacitive coupling in either the camera tube or the amplifier stages.

To produce composite video signal, the sync pulses are superimposed on the blanking pulses. Before sync is added, though, the tops of blanking pulses are cut off by a clipper stage in the control amplifier (see Fig. 5-14). The level at which the pulses are clipped becomes the blanking level.

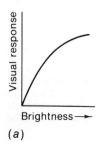
(*a*)

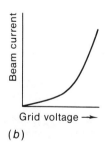
(*b*)

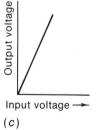
(*c*)

FIGURE 5-15 GAMMA CHARACTERISTICS. (*a*) VISUAL RESPONSE OF THE EYE; GAMMA LESS THAN 1. (*b*) CONTROL-GRID CHARACTERISTIC OF PICTURE TUBE; GAMMA EQUALS 2.5. (*c*) LINEAR CHARACTERISTIC OF AN AMPLIFIER; GAMMA EQUALS 1.

of less than 1 for the white parts of the image compress the changes in white levels to make the picture appear softer, with the gradations in gray level more evident.

Any component in the television system can be assigned a value of gamma to describe the shape of its response curve and contrast characteristics. As a typical example, picture tubes have the control-characteristic curve illustrated in b of Fig. 5-15. The video signal voltage is always impressed on the control grid of the picture tube with the polarity required to make the signal variations for the white parts of the picture fall on that part of the response curve with the steep slope. As a result, a variation in video signal amplitude at the white level produces a greater change in beam current and screen brightness than it would at a darker level. Therefore, picture tubes emphasize the white parts of the picture, with typical gamma values of 2.5 to 3.5. Commercial film also has a gamma greater than 1, an average value being 1.5. In general, monochrome pictures are reproduced with high gamma to make up for the loss of color contrasts.

Amplifiers have a gamma characteristic that is very nearly unity, using linear operation (see Fig. 5-15c). The straight-line response shows that output signal voltage is proportional to input voltage without emphasizing any signal level. If desired, however, an amplifier can be made to operate over the curved portion of its transfer-characteristic curve by shifting the operating bias. The nonlinear amplifier can be used as a *gamma-control* stage, to compensate for the camera tube and picture tube.

5-11 Color Information in the Video Signal

For color television, the composite video includes the 3.58-MHz chrominance signal. As a comparison, Fig. 5-16 shows video signal with and without color. The polarity is shown with

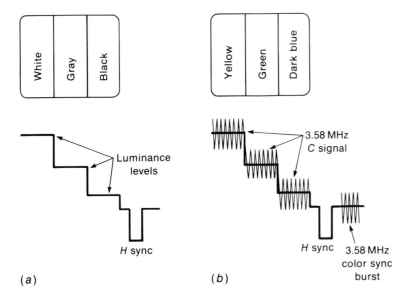

FIGURE 5-16 VIDEO SIGNAL WITH AND WITHOUT COLOR INFORMATION. (a) MONOCHROME SIGNAL ALONE. (b) COMBINED WITH 3.58-MHZ CHROMINANCE SIGNAL.

sync and black down while white is upward. The relative amplitudes in a drop from white for the first bar at the left, to gray level, and then close to black level. These levels correspond to the relative brightness or luminance values for the monochrome information. In b, the video signal has the 3.58-MHz chrominance signal added for the color information in the yellow, green, and blue bars. The specific colors in the C signal are not evident because the relative phase angles and amplitudes are not shown. The main point here is that the difference between monochrome and color television is the 3.58-MHz chrominance signal for color. More details of the 3.58-MHz C signal are explained in Chap. 8 on Color Television.

It is important to note that the luminance levels in Fig. 5-16a are shown the same as the average levels for the signal variations in b. This factor means that without the C signal the color bars in b would be reproduced in monochrome as the white and gray bars in a.

Note also that the color signal in Fig. 5-16b has on the back porch of horizontal sync a color sync burst. This burst consists of 8 to 11 cycles of the 3.58-MHz color subcarrier. Its purpose is to synchronize the 3.58-MHz color oscillator in the receiver. The burst and C signal are both 3.58 MHz, but the burst has no picture information, as it is present only during blanking time.

5-12 Vertical Interval Test Signals (VITS)

The vertical blanking interval represents a valuable period of time, of relatively long duration, that can be used for test signals. The purpose of the test signals is to provide reference values of color and luminance for broadcast stations, or different programs from the same station. In addition, video test frequencies are transmitted to check bandwidth. The test signals include:

1. Multiburst at the frequencies of 0.5, 1.25, 2.0, 3.0, 3.58, and 4.2 MHz (see Fig. 5-17).
2. Vertical interval reference signal (VIRS) for checking chrominance and luminance (Fig. 5-18). The phase of the test chrominance signal is the same as burst phase.

The multiburst test signal of Fig. 5-17 is inserted on line 18 of odd fields. Line 18 of even fields has a color-bar test signal. The VIR signal of Fig. 5-18 is inserted on line 20 of every field. Line 1 in odd fields is counted from the first to the third equalizing pulse. Remember these pulses occur at one-half the horizontal line rate. In even fields, line 1 starts from the second equalizing pulse.

On the screen of the picture tube, the vertical sync and VITS appear as in Fig. 5-19. You can see this vertical blanking bar by rolling the picture out of synchronization with the hold

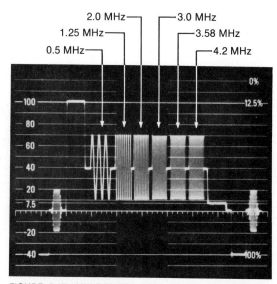

FIGURE 5-17 MULTIBURST TEST SIGNAL TRANSMITTED DURING LINE 18 OF VERTICAL BLANKING IN ODD FIELDS. (TEKTRONIX INC.)

COMPOSITE VIDEO SIGNAL   83

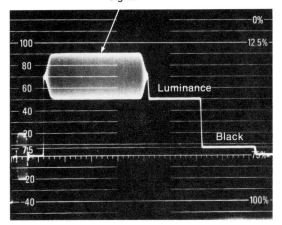

FIGURE 5-18 VERTICAL INTERVAL REFERENCE (VIR) SIGNAL TRANSMITTED DURING LINE 20 OF VERTICAL BLANKING IN BOTH FIELDS. (*TEKTRONIX INC.*)

control and turning up the brightness. The black *hammerhead* pattern is formed by the equalizing pulses just before and after vertical sync.

Time and channel-number encoding. A system has been proposed to use line 21 of the vertical blanking interval for encoding additional information. The code would include the exact time from the National Bureau of Standards, a precise 1-MHz reference signal, 64 characters of text, and a signal identifying the channel number. The text is intended for deaf viewers, emergency codes, and network communications. The channel identification can be used to reproduce the channel number for a short time on the screen of the picture tube, as shown in Fig. 5-20.

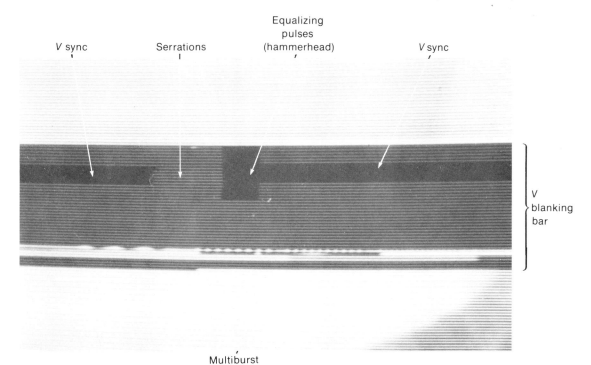

FIGURE 5-19 VERTICAL INTERVAL TEST SIGNALS AND HAMMERHEAD PATTERN IN VERTICAL BLANKING BAR. YOU CAN ROLL PICTURE OUT OF *V* SYNC TO SEE THIS.

FIGURE 5-20 CHANNEL NUMBER 16, WITH EXACT TIME, REPRODUCED ON SCREEN OF PICTURE TUBE FROM ENCODED INFORMATION IN LINE 21 OF THE VERTICAL BLANKING INTERVAL. (*HEATH COMPANY*)

SUMMARY

1. The composite video signal includes camera signal with picture information, synchronizing pulses to time scanning, and blanking pulses.
2. The signal is transmitted with 25 percent amplitude for sync, and the remaining 75 percent for camera signal. Specifically, the peak amplitude is tip of sync; 75 percent level is blanking level; maximum white is 12.5 ± 2.5 percent amplitude. Black setup is usually 7 percent from the blanking level.
3. Horizontal blanking pulses at 15,750 Hz blank out retrace for every line by raising the signal to the 75 percent blanking level. Average pulse width for horizontal blanking is 0.16H, or 10 μs. The width for the horizontal sync pulses is approximately 5 μs.
4. Vertical blanking pulses at 60 Hz blank vertical retrace for every field by raising the signal to the 75 percent blanking level. Much longer than horizontal blanking time, a typical value is 0.08V for vertical blanking. This equals the time for 21 lines blanked in every field.
5. The camera signal variations in the video signal correspond to the picture information in the image. Amplitude changes between the 12.5 and 68 percent levels indicate variations in light level between white and black, with 7 percent black setup.
6. The high video frequencies in the camera signal correspond to horizontal detail. Approximately 4.2 MHz is the highest video frequency that can be broadcast in the 6-MHz transmission channel.

7. The average utilization ratio of 0.7 means 70 percent of the visible scanning lines are useful in showing details in the vertical direction.
8. A test pattern usually includes black, white, and gray lines and areas to check picture reproduction. Vertical wedges indicate horizontal resolution as the beam scans across individual line divisions. The ability to resolve divisions in the side wedges indicates vertical resolution. The side wedges also show poor interlacing by moiré effect in the diagonal lines. In addition, equal lengths for the horizontal wedges show linear horizontal scanning; equal lengths for the vertical wedges show linear vertical scanning.
9. The dc component of any signal is its average-value axis.
10. Gamma is a numerical factor indicating how contrast is expanded or compressed. Picture tubes have a characteristic curve with gamma more than 1, which emphasizes white signal voltages.
11. For color television, the composite video includes the 3.58-MHz chrominance signal, as shown in Fig. 5-16.
12. VITS indicates the vertical interval test signals transmitted during vertical blanking time. VIRS indicates the vertical interval reference signal. (See Figs. 5-17 and 5-18.)

Self-Examination   (Answers at back of book)

Answer true or false.
1. The three components of composite video signal are camera signal, blanking pulses, and sync pulses.
2. Sync pulses transmitted during vertical blanking time include equalizing pulses, the serrated vertical sync pulse, and horizontal sync pulses.
3. During the front-porch time before a horizontal sync pulse the scanning beam is at the left edge of the raster.
4. The 10 percent amplitude in composite video signal corresponds to maximum white picture information.
5. The video signal is at the 75 percent black level during horizontal blanking but not during vertical blanking.
6. The visible trace time for one horizontal line is approximately 10 μs.
7. When the vertical blanking pulse starts, the scanning beam position is at the top of the raster.
8. The equalizing pulses and serrations in the vertical sync pulse are spaced at half-line intervals.
9. The horizontal blanking pulses can produce vertical black bars at the sides of the raster.
10. The vertical sync pulse for one field starts a half line away from its timing in the previous field.

11. Camera signal variations between successive horizontal blanking pulses correspond to information from left to right in the picture.
12. The picture has the left half white and the right half black. The corresponding signal frequency in scanning across one line is approximately 19 kHz.
13. The picture has the top half white and the bottom half black. The corresponding signal frequency in scanning vertically through one frame then must be slightly more than 200 Hz.
14. The high video frequencies correspond to horizontal detail in the picture.
15. Ability to resolve individual lines in the top and bottom wedges of the test pattern indicates horizontal resolution.
16. Ability to resolve individual lines in the side wedges of the test pattern indicates vertical resolution.
17. Picture tubes have a gamma value greater than 1, emphasizing white to increase contrast in the reproduced picture.
18. Average brightness of the reproduced picture depends on the dc bias of the picture tube.
19. The chrominance signal and sync burst shown in Fig. 5-16b have the same frequency of 3.58 MHz.
20. The VIR test signal in Fig. 5-18 is transmitted during horizontal blanking time.

Essay Questions

1. Show the picture and draw the composite video signal of two consecutive lines in scanning across the following patterns: (a) all-white frame; (b) two vertical white bars and two black bars equally spaced; (c) 10 pairs of vertical bars. Why does this signal have a higher frequency than in b?
2. Why are the synchronizing pulses inserted during blanking time?
3. What is the function of the horizontal blanking pulses? The vertical blanking pulses?
4. Why are the horizontal blanking pulses wider than the horizontal sync pulses?
5. Trace the motion of the scanning beam from the beginning to end of vertical blanking.
6. Define: white level, black setup, gamma, utilization ratio, resolution, crosshatch pattern, and staircase signal.
7. Why do thin vertical lines produce higher video signal frequencies than wide vertical bars?

Problems (Answers to selected problems at back of book)

1. In the checkerboard pattern of Fig. 5-9, if there are 300 squares in a line, what is the frequency of the corresponding signal variations? Use 53.3 μs for visible trace time.
2. With a utilization ratio of 0.7, what would be the maximum vertical detail for a vertical blanking time of 0.08V?
3. Assume a facsimile reproduction with specifications of 200 lines per frame, progressive scanning, and 5 frames per second. Calculate the following: (a) time to scan one line, including trace and retrace; (b) visible trace time for one line with 4 percent blanking; (c) video frequency corresponding to 100 total black-and-white elements in a line.
4. Calculate the frequency of video signal produced in horizontal scanning of the window signal in Fig. 5-12. Assume 53.3 μs visible trace time.
5. Referring to the Table of Television Standards for foreign countries in Appendix C, calculate the time to scan one line, including trace and retrace for the systems in: (a) United States; (b) England; (c) Western Europe; (d) France.
6. Show the pictures corresponding to the video signals in Fig. 5-21. Assume that all the lines of video signal are the same as the one shown.

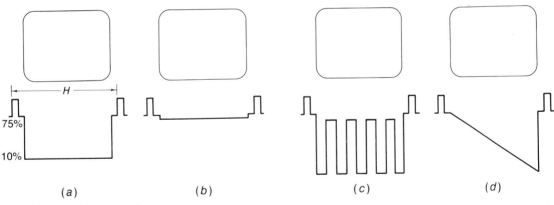

FIGURE 5-21 FOR PROBLEM 6.

6 Picture Carrier Signal

The method of transmitting the amplitude-modulated (AM) picture signal is similar to the more familiar system of AM radio broadcasting. In both cases, the amplitude of an rf carrier wave is made to vary with the modulating voltage. Modulation is necessary so that each broadcast station can have its own rf carrier frequency. Then the rf section of the receiver can be tuned to different stations. In television broadcasting, the composite video signal modulates a higher-frequency carrier wave to produce the AM picture signal* illustrated in Fig. 6-1. The amplitudes in the transmitted carrier wave vary with the video modulating signal. More details of the AM picture signal are explained in the following topics:

6-1 Negative transmission
6-2 Vestigial-sideband transmission
6-3 Television broadcast channels
6-4 The standard channel
6-5 Line-of-sight transmission

*The term "picture signal" is used here for the modulated rf carrier wave, while "video" represents the signal that can be used directly to reproduce the desired visual information when applied to a picture tube, corresponding to "audio" in a sound system.

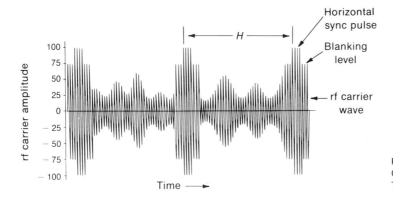

FIGURE 6-1 TRANSMITTED PICTURE CARRIER WAVE, AMPLITUDE-MODULATED BY COMPOSITE VIDEO SIGNAL.

6-1 Negative Transmission

The transmitted picture carrier signal in Fig. 6-1 is shown with the negative polarity of modulation that is an FCC standard. Negative transmission means that changes toward white in the picture decrease the amplitude of the AM picture carrier signal.

In Fig. 6-1 the tips of sync voltage produce the peaks of rf amplitude in the AM carrier wave. This peak carrier amplitude is its 100 percent level. The blanking level in the composite video signal is transmitted at a constant level of 75 percent of peak carrier amplitude. Smaller amplitudes in the modulated rf carrier correspond to picture information that varies between black and maximum white. The whitest parts of the picture produce a carrier amplitude 10 to 15 percent of the peak value. All these relative amplitudes are the same for the top or bottom of the envelope because the modulated rf carrier wave has a symmetrical envelope of amplitude variations.

The *negative transmission* refers to the polarity of video modulating voltage, not individual cycles of the rf carrier, both being positive and negative. The result of white video signal amplitudes having negative polarity at the grid of the modulator is negative polarity of modulation. Then the carrier amplitude decreases for white video modulating voltage. If the video modulating signal had opposite polarity, the transmitted carrier would have positive polarity of transmission.

An advantage of negative transmission is that noise pulses in the transmitted rf signal increase the carrier amplitude toward black, which makes noise less obvious in the picture. Also, the transmitter uses less power with lower carrier amplitudes for pictures that are mostly white.

6-2 Vestigial-Sideband Transmission

The AM picture signal is not transmitted as a normal double-sideband signal. Instead, some of the sideband frequencies are filtered out before transmission in order to reduce the bandwidth of the channel needed for the modulated picture signal. To see how this vestigial-sideband transmission is accomplished, we can consider first the idea of how amplitude modulation produces sideband frequencies.

Amplitude modulation. In Fig. 6-2 an rf carrier wave is amplitude-modulated by a sine-wave

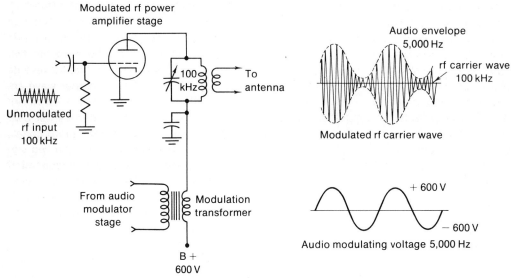

FIGURE 6-2 PLATE-MODULATION CIRCUIT. THE 100-kHz RF CARRIER IS AMPLITUDE-MODULATED BY THE 5,000-Hz AUDIO VOLTAGE.

audio signal in a plate-modulation arrangement. For simplicity the rf carrier frequency is taken as 100 kHz and the audio as 5,000 Hz. The B+ voltage for the rf power amplifier is assumed to be 600 V, and the peak value of the audio sine-wave modulating voltage is also 600 V, allowing 100 percent modulation.

Note that the audio voltage across the secondary of the modulation transformer is in series with the B+ supply and the rf amplifier plate circuit. Therefore, the audio modulating voltage varies the plate voltage of the rf amplifier at the audio rate (see Table 6-1).

The varying amplitudes of the rf carrier wave provide an envelope that corresponds to the audio modulating voltage. Both the positive and negative peaks of the rf carrier wave are symmetrical above and below the center axis and have exactly the same amplitude variations. The envelope is symmetrical because the

TABLE 6-1 MODULATION VALUES FOR FIG. 6-2

AUDIO VOLTAGE	B+ VOLTAGE	RF AMPLIFIER PLATE VOLTAGE	MODULATED RF SIGNAL AMPLITUDE
0	600	600	Carrier level
+600	600	1,200	Double carrier level
0	600	600	Carrier level
−600	600	0	Zero
0	600	600	Carrier level

changes in amplitude of the negative and positive half cycles of the rf signal are equal, as the carrier amplitude is varied at the audio rate, which is much slower than the rf variations. Any point on the audio waveform includes many cycles of the rf carrier. The result of the modulation in this case, is to produce an rf carrier wave at a frequency of 100 kHz with an amplitude that varies at the audio rate of 5,000 Hz. Either the top or bottom envelope of the amplitude-modulated carrier wave corresponds to the 5,000-Hz audio modulating signal.

Side-carrier frequencies. Referring now to Fig. 6-3, the AM wave is shown to be equal to the sum of the unmodulated rf carrier and two side-carrier frequencies. Notice that the carrier and its equivalent side frequencies all have a constant level. Also, the amplitude of the side carriers equals one-half the unmodulated carrier level, for 100 percent modulation. Each side frequency differs from the carrier by the audio modulating frequency. The upper side frequency is 105 kHz, and the lower side frequency 95 kHz in this illustration.

The question as to whether the transmitted signal is a carrier with varying amplitudes or a constant-amplitude carrier with its two side carriers is without meaning, because the two concepts are the same. The constant-level side carriers plus the unmodulated carrier wave are equal to the AM carrier signal. Or, the AM carrier wave is equal to the unmodulated carrier plus two side carriers of proper amplitude, phase, and frequency. The equivalence of the two signals is due to the fact that the modulated rf carrier wave is distorted slightly from true sine-wave form by the audio amplitude variations, producing new frequency components, which are the side frequencies.

The envelope is not a side frequency. The rf side-carrier frequencies should not be confused with the audio envelope. The envelope is an audio frequency. The side carriers are radio frequencies close to the carrier frequency. For the case here, the envelope is the audio modulating signal of 5,000 Hz, while the rf side frequencies are 105 and 95 kHz. If the audio modulating frequency is 1,000 Hz, the rf side

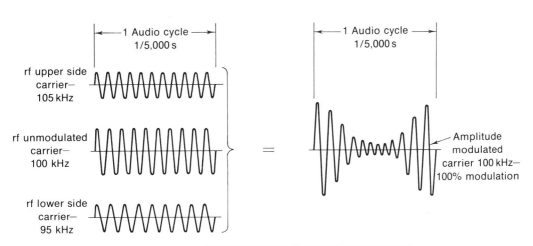

FIGURE 6-3 EQUIVALENCE OF AM WAVE TO THE CARRIER PLUS TWO RF SIDE CARRIERS PRODUCED BY MODULATION.

frequencies will be 101 and 99 kHz. Then the modulated carrier will have an audio envelope of 1,000 Hz.

Side bands. When the carrier is modulated by a voltage that includes many frequency components, each audio modulating frequency produces a pair of rf side frequencies. In each pair, one side frequency is higher than the carrier frequency and one is lower. All the higher side frequencies can be considered as the *upper side band* of the carrier with all the lower side frequencies the *lower side band*. This idea is illustrated in Fig. 6-4 for the case of modulation with a continuous band of audio frequencies from 0 to 5,000 Hz. The graph here indicates the corresponding side frequencies for a 100-kHz carrier. The upper side band includes all side frequencies from the carrier frequency of 100 up to 105 kHz; the lower side band includes the side frequencies down to 95 kHz. The bandwidth required for the two side bands in this case is ±5 kHz centered around the carrier frequency of 100 kHz, or a total bandwidth of 10 kHz. Note that the required bandwidth for the AM carrier with two side bands is double the highest modulating frequency.

The fact that different audio modulating frequencies produce different side frequencies in AM should not make it be confused with frequency modulation. In FM the rf carrier frequency varies in accordance with the amount of audio modulating *voltage,* but in AM the side frequencies depend on the audio modulating *frequency.*

The baseband. This is the band of modulating frequencies for either AM or FM. In this example of audio modulating frequencies up to 5,000 Hz, the baseband is 0 to 5 kHz. In television broadcasting the baseband is 50 to 15,000 Hz for the FM sound signal and 0 to 4 MHz, approximately, for the AM picture signal.

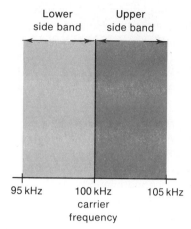

FIGURE 6-4 DOUBLE SIDE BANDS RESULTING FROM AMPLITUDE MODULATION OF 100-kHz CARRIER WITH ALL MODULATING FREQUENCIES UP TO 5 kHz.

Vestigial side bands. Note that the information of the modulating signal is in the side bands of the amplitude-modulated rf carrier. Frequency of the modulation is indicated by how much the side frequencies differ from the carrier frequency; modulating voltage is indicated by the amplitude of the two side carriers. For the case of 100 percent modulation, each side carrier has one-half the unmodulated carrier amplitude. Furthermore, the upper and lower side frequencies have the same information, since they are of equal amplitude and each differs from the carrier frequency by the same amount. The desired modulating signal can be transmitted by one side band, therefore, and it does not matter whether the upper or lower side band is used. With only one side band, the transmitted signal has the advantage of only one-half the bandwidth of two side bands. The amplitude modulation normally produces double side bands, but one side band can be filtered out if desired.

Figure 6-5 illustrates just one side frequency transmitted with the carrier. Notice that the resultant modulated wave has amplitude

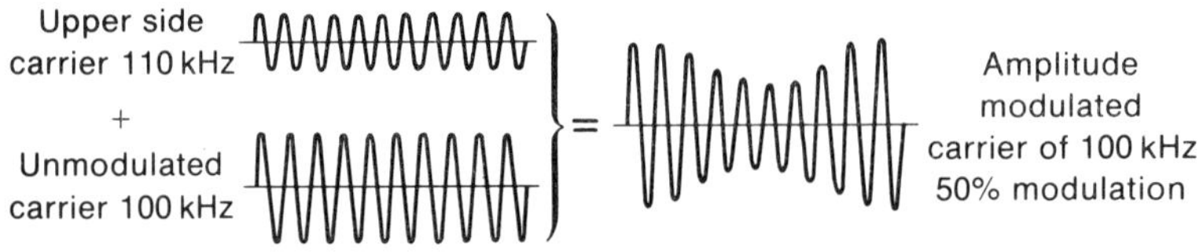

FIGURE 6-5 EQUIVALENCE OF CARRIER AND ONE SIDE CARRIER TO AM WAVE.

variations for only 50 percent modulation, instead of the 100 percent modulation produced with both side bands. In Fig. 6-5, the modulated wave varies in amplitude 50 percent above and below the unmodulated carrier amplitude, but with 100 percent modulation the peak carrier amplitude doubles the unmodulated level and decreases to zero. Therefore, a signal transmitted with one side band has effectively one-half the percent modulation, compared with double-sideband transmission. The same reduction factor of $1/2$ applies when the varying modulating voltage produces different amounts of modulation less than 100 percent.

Except for the amount of amplitude swing, the envelope of the carrier plus one side band can be considered essentially the same as with double-sideband transmission, although there is slight distortion for high percentages of modulation. Notice that the envelope in Fig. 6-5 still corresponds to the audio modulating voltage. Furthermore, the envelope is not cut off at the top or bottom of the AM wave. Remember that one rf side frequency is filtered out, but not the audio envelope. In order to remove one part of the envelope it would be necessary to rectify the modulated carrier signal.

The method just described can be considered single-sideband transmission with the carrier. In many applications of single-sideband transmission, however, the carrier itself is not transmitted in order to save power by suppressing the carrier. Then only one side band is transmitted. In this situation, the carrier must be reinserted at the receiver to detect the signal.

A compromise between double-sideband transmission and the single-sideband method is used for broadcasting the AM picture carrier signal. In this system, called *vestigial-sideband transmission*, all of one side band is transmitted but only a part, or vestige, of the other side band. The picture carrier itself is transmitted. More specifically, all the upper side band of the AM picture signal is transmitted, to include all video modulating frequencies up to 4 MHz. The lower side band, however, includes only video modulating frequencies up to 0.75 MHz approximately, to conserve bandwidth in the broadcast channel. The vestigial-sideband transmission used in television is designated by the FCC as emission type A5C.

6-3 Television Broadcast Channels

Each television broadcast station is assigned a 6-MHz channel for transmission of the AM picture signal and the FM sound signal. Vestigial-sideband transmission is used for the picture signal, so that video modulation frequencies up to 4 MHz can be broadcast in the 6-MHz channel.

Assigned channels. Since the picture carrier frequency must be much higher than the highest video modulating frequency of 4 MHz, the television channels are assigned in the VHF band of 30 to 300 MHz and the UHF band of 300 to 3,000 MHz. Table 6-2 lists the channels and frequen-

TABLE 6-2 TELEVISION CHANNEL ALLOCATIONS

CHANNEL NUMBER	FREQUENCY BAND, MHz	CHANNEL NUMBER	FREQUENCY BAND, MHz
1*	—	42	638–644
2	54–60	43	644–650
3	60–66	44	650–656
4	66–72	45	656–662
5	76–82	46	662–668
6	82–88	47	668–674
7	174–180	48	674–680
8	180–186	49	680–686
9	186–192	50	686–692
10	192–198	51	692–698
11	198–204	52	698–704
12	204–210	53	704–710
13	210–216	54	710–716
		55	716–722
14	470–476	56	722–728
15	476–482	57	728–734
16	482–488	58	734–740
17	488–494	59	740–746
18	494–500	60	746–752
19	500–506	61	752–758
20	506–512	62	758–764
21	512–518	63	764–770
22	518–524	64	770–776
23	524–530	65	776–782
24	530–536	66	782–788
25	536–542	67	788–794
26	542–548	68	794–800
27	548–554	69	800–806
28	554–560	70†	806–812
29	560–566	71	812–818
30	566–572	72	818–824
31	572–578	73	824–830
32	578–584	74	830–836
33	584–590	75	836–842
34	590–596	76	842–848
35	596–602	77	848–854
36	602–608	78	854–860
37	608–614	79	860–866
38	614–620	80	866–872
39	620–626	81	872–878
40	626–632	82	878–884
41	632–638	83	884–890

*The 44- to 50-MHz band was television channel 1 but is now assigned to other services.

†Channels 70 to 83 are also allocated for land mobile radio services. For television, these UHF channels are used mainly by translator stations (see Sec. 6-5).

cies assigned by the FCC for commercial television broadcast stations in the United States. The television channel frequencies can be considered in three groups: the five low-band channels 2 to 6 in the VHF range, seven high-band channels 7 to 13 also in the VHF range, and the 70 UHF channels 14 to 83. Frequencies between these television broadcasting bands are used by other radio services.

The number of channels available for television broadcast stations in any one locality depends upon its population, varying from one channel in a smaller city to twelve stations for New York City, including VHF and UHF channels. Most cities have at least one channel reserved for a noncommercial educational television broadcast station.

One channel can be used by many broadcast stations, but they must be far enough apart to minimize interference between them. Such stations using the same channel are *cochannel stations*. They must be separated by 170 to 220 mi for VHF channels or 155 to 205 mi for UHF channels. Stations that use channels adjacent in frequency, like channels 3 and 4, are *adjacent-channel stations*. To minimize interference between them, adjacent-channel stations are separated by 60 mi for VHF channels or 55 mi for UHF channels. However, channels consecutive in number but not adjacent in frequencies, such as channels 4 and 5, channels 6 and 7, or channels 13 and 14, can be assigned in one area.

6-4 The Standard Channel

The structure of a standard television channel is illustrated in Fig. 6-6a. The width of the channel is 6 MHz, including the picture and sound carriers with their sideband frequencies. The picture carrier is spaced 1.25 MHz from the lower edge of the channel, and the sound carrier is 0.25 MHz below the upper edge of the channel. As a result, there is always a fixed spacing of 4.5 MHz between the picture and sound carrier frequencies. The specific picture and sound carrier frequencies for all television broadcast channels are listed in Appendix A.

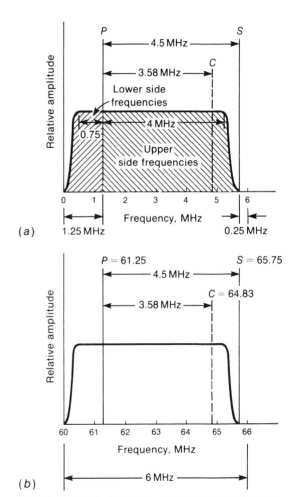

FIGURE 6-6 THE STANDARD TELEVISION BROADCAST CHANNEL. P IS PICTURE CARRIER, S IS SOUND CARRIER, AND C IS COLOR SUBCARRIER. (a) FREQUENCY SEPARATIONS FOR ANY CHANNEL. (b) SPECIFIC FREQUENCIES FOR CHANNEL 3, 60 TO 66 MHz.

The standard channel characteristics shown in Fig. 6-6 should not be interpreted as an illustration of the picture signal. The graph merely defines the signal frequencies that can be transmitted in the television channel, with their relative amplitudes. The picture carrier is shown with twice the amplitude of the sideband frequencies, which are their relative amplitudes for 100 percent modulation.

Since the sound signal is frequency-modulated, its sideband frequencies do not have the same type of amplitude characteristic as in the picture signal, and these are not shown. The sound carrier signal is a conventional FM signal, with a bandwidth of approximately 50 kHz for a frequency swing of ±25kHz with 100 percent modulation.

In the picture signal, all upper sideband frequencies up to approximately 4-MHz video modulation are transmitted with their normal amplitude. The color subcarrier at 3.58 MHz can be considered as just another high video-frequency signal that modulates the picture carrier with the desired color information. For monochrome information, the high video frequencies represent only edge details. For a color program, however, 3.58 MHz represents the color signal needed for all the color information.

The lower side-carrier frequencies that differ from the picture carrier by more than 0.75 MHz but less than 1.25 MHz are gradually attenuated. Lower side-carrier frequencies below the picture carrier by 1.25 MHz or more are outside the channel. These frequencies must be completely filtered out at the transmitter so that they will not be radiated to interfere with the lower adjacent channel. Note that upper side-carrier frequencies more than 4 MHz above the picture carrier frequency are also attenuated to eliminate interference with the associated sound signal.

Examples of rf channel frequencies. Numerical values for channel 3 as a typical television channel are shown in Fig. 6-6b. The channel has a bandwidth of 6 MHz, from 60 to 66 MHz. The picture carrier is 1.25 MHz above the lower edge of the channel, which is 61.25 MHz for this channel. The sound carrier at 65.75 MHz is 4.5MHz above the picture carrier frequency.

With vestigial-sideband transmission, all the upper sideband frequencies up to 65.25 MHz are transmitted. However, only that part of the lower sideband frequencies down to 60.5 MHz, approximately, is transmitted without attenuation. As an example, when the video modulating voltage has a frequency of 0.75 MHz, both the upper and lower side frequencies of 62 and 60.5 MHz are transmitted. For this case, the picture carrier is a normal double-sideband signal. The same is true for any video modulating signal having a frequency less than 0.75 MHz.

However, for components of video modulating signal with a frequency higher than 0.75 MHz, only the upper side carrier is transmitted with normal amplitude. For 2-MHz video modulation, as an example, the upper side frequency of 63.25 MHz is in the channel. The lower side frequency is 59.25 MHz, which is outside channel 3 and must be filtered out at the transmitter. In this case, then, only the upper side frequency is transmitted with the picture carrier, resulting in single-sideband transmission. The result is a vestigial-sideband transmission system because double-sideband transmission is used for video modulating frequencies lower than 0.75 MHz, but single-sideband transmission is used for higher video modulating frequencies up to 4 MHz, approximately.

The color video signal of 3.58 MHz is at 64.83 MHz as an upper side frequency of the modulated picture carrier in channel 3. This frequency is calculated as 61.25 + 3.58 = 64.83 MHz. If the receiver does not pass 64.83 MHz in the antenna and rf circuits for channel 3, there will be no color.

Advantage of vestigial-sideband transmission. This advantage can be seen from the fact that the picture carrier is 1.25 MHz from the end of the channel, allowing video modulating frequencies up to 4 MHz to be transmitted in the 6-MHz channel. A video-frequency limit of about 2.5 MHz would be necessary if double-sideband transmission were used with the picture carrier at the center of the channel. This would represent a serious loss in horizontal detail, since the high-frequency components of the video modulation determine the amount of horizontal detail in the picture.

It might seem desirable to place the picture carrier at the lower edge of the channel and use single-sideband transmission completely. This method would allow the use of video modulating frequencies higher than 5 MHz and increased horizontal detail. However, this is not practicable. The elimination of undesired side-carrier frequencies is accomplished by a filter circuit at the transmitter, which cannot have ideal cutoff characteristics. Therefore, it would not be possible to remove side carriers that are too close to the carrier frequency without introducing objectionable phase distortion for the lower video signal frequencies, which causes smear in the picture. The low video frequencies have the most important information for the picture.

The practical compromise of vestigial-sideband transmission that is used provides for complete removal of the lower side band only where the side-carrier frequencies are sufficiently removed from the picture carrier to avoid phase distortion. The picture carrier itself and the side frequencies close to the carrier are not attenuated. The net result is normal double-sideband transmission for the lower video frequencies corresponding to the main body of picture information for large areas in the picture. The single-sideband transmission is used only for the higher video frequencies that represent details of edges or outlines in the picture.

Compensating for the vestigial-sideband transmission. It should be noted that the picture signal is distorted in terms of relative amplitude for different frequencies. Remember that a signal transmitted with only a single side band and the carrier represents 50 percent modulation in comparison with a normal double-sideband signal with 100 percent modulation. Therefore, the higher video frequencies provide signals with one-half the effective carrier modulation, compared with the lower video frequencies that are transmitted with both side bands. This is in effect a low-frequency boost in the video signal. However, it is corrected by deemphasizing the low video frequencies to the same extent in the IF amplifier of the television receiver.

6-5 Line-of-Sight Transmission

Propagation of radio waves in the VHF and UHF bands is produced mainly by ground-wave effects, rather than sky waves from the ionized atmosphere. The ground wave is that part of the radiated signal affected by the presence of the earth, and it can be considered as being propagated along the surface of the earth from the transmitting antenna. Since the television broadcast channels are in the VHF and UHF bands, transmission of the picture and sound carrier signals is determined primarily by ground-wave propagation.

Horizon distance. The transmission distance that can be obtained for the ground-wave signal is limited by the distance along the surface of the earth to the horizon, as viewed from the transmitting antenna. This is called *line-of-sight* transmission; the line-of-sight distance to the horizon is the *horizon distance* (see Fig. 6-7). The horizon distance for the transmitted radio wave, however, is about 15 percent longer than the optical horizon distance because the path of the ground wave curves slightly in the same direction as the earth's curvature. This bending of the radio waves by the earth's atmosphere is called *refraction*. The graph in Fig. 6-8 shows the radio horizon distance directly for any antenna height up to 10,000 ft.

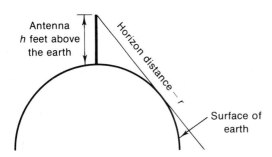

FIGURE 6-7 HORIZON DISTANCE r DEPENDS ON ANTENNA HEIGHT h.

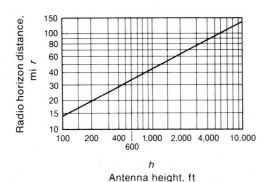

FIGURE 6-8 GRAPH SHOWING HOW RADIO HORIZON DISTANCE r INCREASES WITH ANTENNA HEIGHT h.

In considering the line of sight from transmitter to receiver, the horizon distance of each antenna must be added. As an example, a transmitting antenna at a height of 1,100 ft has the horizon distance of 55 mi. For an antenna height of 150 ft at the receiver, this horizon distance is 17 mi. Therefore, the line-of-sight distance between these two antennas equals $55 + 17 = 72$ mi.

Television transmitters. The peak rf power output of a typical picture or sound signal VHF transmitter is 1 to 50 kW. However, the effective radiated power can be higher because it includes the gain of the transmitting antenna. The minimum effective radiated power specified by the FCC for a population of 1 million or more is 50 kW, with a transmitting antenna height of 500 ft. For areas with a population under 50,000 the minimum effective radiated power is 1 kW with an antenna height of 300 ft.

The radiated power of the sound carrier signal is not less than 50 percent or more than 150 percent of the radiated power for the picture carrier signal, for monochrome transmission. In color transmission, the sound power is limited to 50 to 70 percent of the picture power for minimum sound interference in the picture.

The frequency tolerance for the picture or sound carrier is ±0.002 percent. However, the exact carrier frequencies for different stations on the same channel are offset from each other by +10 kHz or −10 kHz, in order to reduce interference between cochannel stations. This system is called *offset carrier operation*.

Reflections. As the ground wave travels along the surface of the earth, the radio signal encounters buildings, towers, bridges, hills, and other obstructions. When the intervening object is a good conductor and its size is an appreciable part of the radio signal's wavelength, the obstruction will reflect the radio wave, similar to the reflection of light from a mirror or other reflecting surface. What happens is that the conductor intercepts the radio wave, current flows as in an antenna, and the conductor reradiates the signal. Reflection of radio waves can occur at any frequency but more easily at higher frequencies because of the shorter wavelengths.

For the television channel frequencies between 54 and 890 MHz the wavelengths are between 17 and 1 ft, depending on the frequency. Objects of comparable size, or bigger, can reflect the television carrier waves. When the reflected picture carrier signal arrives at the receiving antenna in addition to the direct wave or along with other reflections, the multipath signals produce multiple images called *ghosts* in the reproduced picture.

Shadow areas. Where an object in the path of the ground wave reflects the radio signal, the area behind the obstruction is shadowed and therefore has reduced signal strength. The shadowing effect is more definite at higher frequencies because of the shorter wavelengths, just like reflection of the radio waves. Reception of television signal in shadow areas behind an obstruction like a tall building is often accomplished by utilizing waves reflected from other buildings nearby.

Booster stations. Some areas are either shadowed by mountains or too far from the nearest transmitter for satisfactory television broadcast service. In this case, a booster station can be used. The station is in a suitable location for reception and rebroadcasts the program to receivers in the local area. Some booster stations convert the VHF channel frequencies for rebroadcasting on an unused UHF channel, to minimize interference problems. These are *translator* stations.

SUMMARY

1. The amplitude-modulated picture carrier signal has a symmetrical envelope, which is the composite video signal used to modulate the carrier wave. The two main features of the transmitted picture carrier signal are negative polarity of modulation and vestigial-sideband transmission.
2. Negative transmission means that the video modulating signal has the polarity required to reduce the carrier amplitude for white camera signal. Darker picture information raises the carrier amplitude. Tip of sync produces maximum carrier amplitude for the 100 percent level.
3. Vestigial-sideband transmission means that all the upper side frequencies but only some of the lower side frequencies are transmitted in the 6-MHz channel for the modulated picture carrier signal. The band of upper side frequencies approximately 4 MHz above the carrier frequency is transmitted with full amplitude. Similarly, lower side frequencies separated by 0.75 MHz, or less, from the carrier frequency are also transmitted. However, the lower side frequencies below the picture carrier enough to be outside the assigned channel are not transmitted.

4. Channels 2 to 6 are low-band VHF channels between 54 and 88 MHz; channels 7 to 13 are high-band VHF channels from 174 to 216 MHz; channels 14 to 83 are UHF channels from 470 to 890 MHz. In all bands the station broadcasts in a standard channel 6 MHz wide.
5. The standard 6-MHz channel includes the AM picture carrier 1.25 MHz above the low end, and the FM sound carrier 0.25 MHz below the high end, with 4.5 MHz between the picture and sound carrier frequencies. The color subcarrier signal is transmitted as a side frequency 3.58 MHz above the picture carrier.

Self-Examination (Answers at back of book)

Choose (a), (b), (c), or (d).

1. The modulated picture carrier wave includes the composite video signal as the (a) average carrier level; (b) symmetrical envelope of amplitude variations; (c) lower side band without the upper side band; (d) upper envelope without the lower envelope.
2. Which of the following statements is true? (a) Negative transmission means the carrier amplitude decreases for black. (b) Negative transmission means the carrier amplitude decreases for white. (c) Vestigal-sideband transmission means both upper and lower side bands are transmitted for all modulating frequencies. (d) Vestigial-sideband transmission means the modulated picture carrier signal has only the upper envelope.
3. With 2-MHz video signal modulating the picture carrier for channel 4 (66 to 72 MHz), which of the following frequencies are transmitted? (a) 66-MHz carrier and 68-MHz upper side frequency; (b) 71.75-MHz carrier, with 69- and 73-MHz side frequencies; (c) 67.25-MHz carrier, with 65.25- and 69.25-MHz side frequencies; (d) 67.25-MHz carrier and 69.25 MHz upper side frequency.
4. With 0.5-MHz video signal modulating the picture carrier, (a) both upper and lower side frequencies are transmitted; (b) only the upper side frequency is transmitted; (c) only the lower side frequency is transmitted; (d) no side frequencies are transmitted.
5. In all standard television broadcast channels the difference between picture and sound carrier frequencies is (a) 0.25 MHz, (b) 1.25 MHz, (c) 4.5 MHz, (d) 6 MHz.
6. The difference between the sound carrier frequencies in two adjacent channels equals (a) 0.25 MHz; (b) 1.25 MHz; (c) 4.5 MHz; (d) 6 MHz.
7. With 7 percent black setup, maximum black in the picture corresponds to what percent amplitude in the modulated carrier signal? (a) 5; (b) 68; (c) 75; (d) 95.
8. Line-of-sight transmission is a characteristic of propagation for the (a) VHF and UHF bands; (b) VHF band but not the UHF band; (c) low radio frequencies below 1 MHz; (d) AM picture signal but not the FM sound signal.

9. In channel 14 (470 to 476 MHz) the 3.58-MHz color signal is transmitted at the frequency of (a) 471.25 MHz; (b) 473.25 MHz; (c) 474.83 MHz; (d) 475.25 MHz.
10. The difference between the sound carrier and color subcarrier frequencies is (a) 4.5 MHz; (b) 1.25 MHz; (c) 0.92 MHz; (d) 0.25 MHz.

Essay Questions

1. Define negative transmission.
2. Give one advantage and one disadvantage of vestigial-sideband transmission.
3. For each of the following channels, list the picture carrier and sound carrier frequencies with their frequency separation: channels 2, 5, 7, 13, 14, and 83.
4. Define the following terms: (a) offset carrier operation; (b) ghost in the picture; (c) shadow area.
5. Which television channel numbers are in the band of 30 to 300 MHz? Which are in the band of 300 to 3,000 MHz?
6. Draw a graph similar to Fig. 6-6 showing frequencies transmitted in channel 8, indicating picture carrier, sound carrier, and color subcarrier with their frequencies.
7. Why is reflection of the transmitted carrier wave a common problem with the picture signal in television, but not in the radio broadcast band of 535 to 1,605 kHz?
8. What is the effect in the reproduced picture of multipath signals caused by reflections?

Problems (Answers to selected problems at back of book)

1. List the rf side frequencies transmitted with the picture carrier for the following modulation examples: (a) 0.25-MHz video modulation of channel 2 carrier; (b) 3-MHz video modulation of channel 5 carrier; (c) 0.5-MHz video modulation of channel 14 carrier; (d) 4-MHz video modulation of channel 14 carrier.
2. In negative transmission, what is the relative amplitude of the modulated picture carrier wave for (a) maximum white; (b) tip of sync; (c) blanking level; (d) black in picture; (e) medium gray?
3. List the picture and sound carrier frequencies for channels 7, 8, and 9.
4. Give the frequency separation for the following combinations: (a) picture and associated sound carriers; (b) picture carrier and lower adjacent-channel sound carrier; (c) picture carriers in two adjacent channels; (d) sound carriers in two adjacent channels.

5. Give the frequency separation for the combinations of (a) picture carrier and color subcarrier; (b) sound carrier and color subcarrier.
6. List the picture and sound carrier frequencies for all channels from 2 to 14, inclusive.
7. Give the exact picture carrier frequency for (a) channel 2 offset −10 kHz; (b) channel 5 offset +10 kHz.
8. Refer to Fig. 6-9 below. Indicate the picture and sound carrier frequencies for three VHF channels adjacent in frequency, other than 7, 8, and 9.

FIGURE 6-9 FOR PROBLEM 8.

7 Television Receivers

The receiver circuits use three signals, including color. One is the FM sound carrier, modulated with audio signal for the loudspeaker. The second is the AM picture carrier, modulated with video signal for the picture tube. In color receivers, the third signal is the 3.58-MHz colorplexed subcarrier. This is part of the picture carrier signal. The three signals enable us to see the picture and hear the sound. We consider circuits only for monochrome receivers here, as the principles of color television and color receivers are explained in more detail in the next two chapters. It is important to realize, though, that all the circuits for a black-and-white picture are also needed for color. The color television picture is just a monochrome picture with color added in the main areas of picture information.

In addition to the signal circuits, the television receiver has vertical and horizontal deflection circuits for scanning the raster. Finally, with the picture on the raster, synchronization is necessary. The deflection sync includes vertical and horizontal pulses that time the scanning circuits to hold the picture steady on the raster. The functions of the receiver circuits and controls are described in the following topics:

- 7-1 Types of television receivers
- 7-2 Receiver circuits
- 7-3 Signal frequencies
- 7-4 Intercarrier sound
- 7-5 Dividing the receiver into sections
- 7-6 Receiver controls
- 7-7 Receiver tubes
- 7-8 Solid-state devices
- 7-9 Special components
- 7-10 Printed-wiring boards
- 7-11 Localizing receiver troubles to one section
- 7-12 Multiple troubles
- 7-13 Safety features

7-1 Types of Television Receivers

For either monochrome or color, the receiver may use tubes for all stages, have all solid-state transistors and integrated circuits, or combine tubes and transistors as a *hybrid receiver*. A typical hybrid chassis for a monochrome receiver is shown in Fig. 7-1.

All-tube receivers. This type applies mainly to monochrome receivers and older color receivers. All the functions are provided by about 12 tubes for monochrome and 18 tubes for color receivers. Included are multipurpose tubes with two or three stages in one envelope. The B+ for plate and screen voltage is 140 V or 280 V.

Solid-state receivers. In this type, all the stages except the picture tube use semiconductor diodes, transistors, and integrated circuits. The dc supply voltage, then, is about 12 to 90 V, for collector voltage. The only heater is for the picture tube. A separate filament transformer is used, or the filament can be heated with direct current from the low-voltage power supply.

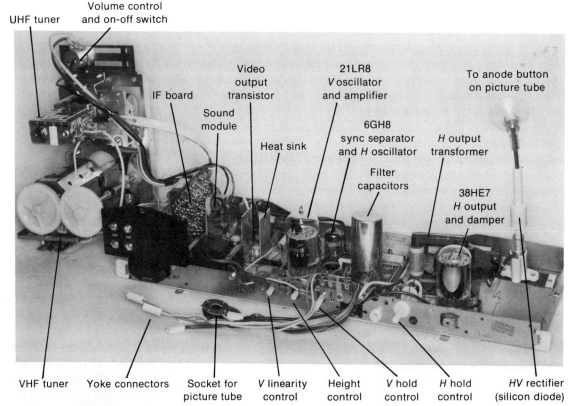

FIGURE 7-1 CHASSIS OF HYBRID RECEIVER WITH TUBES AND TRANSISTORS. (*MAGNAVOX CORPORATION*)

Hybrid receivers. In this type, the deflection circuits generally use power tubes, while the signal circuits use transistors and integrated circuits. These receivers usually have an ac-dc power supply, with series heaters for the tubes. The collector voltage of 12 to 28 V for the transistors can be taken from the dc bias voltage of the horizontal deflection amplifier.

7-2 Receiver Circuits

See the block diagram in Fig. 7-2, with typical waveshapes. These circuits are essentially the same for monochrome and color receivers. For a color broadcast, the chrominance signal is part of the video signal. In monochrome receivers, however, this 3.58-MHz C signal is just not used, as the video amplifier attenuates frequencies above 3.2 MHz. The type number 18VAUP4 for the monochrome picture tube indicates that 18 in. is the screen diagonal and P4 is the phosphor for a white screen.

Antenna input. Starting with the antenna signal, the picture and sound rf carrier signals are intercepted by a common receiving antenna. The transmission line connects the antenna to the receiver input terminals for the rf tuner. Twin lead is generally used. This type is a balanced line, without a ground, and unshielded. The characteristic impedance for rf is 300 Ω.

When there is a problem of interference, it may be better to use shielded coaxial cable. This line has high attenuation of the signal, however, especially for the UHF channel frequencies. Coaxial cable has a characteristic impedance of 75 Ω.

Two ungrounded screw terminals on the receiver are connections for 300-Ω twin lead. A grounded jack is for coaxial cable. To convert one type of line to the other type of connection, a balancing transformer (balun) is used. Most receivers now have an antenna jack for coaxial cable, which is generally used for master antenna and cable-television installations.

There are two tuners, one for VHF channels 2 to 13 and the other for UHF channels 14 to 83, each with its own antenna input terminals. When the antenna and transmission line are the same for VHF and UHF, then a signal splitter at the receiver input is used to separate the signal for the two tuners.

VHF tuner. The antenna input provides rf picture and sound signals for the rf amplifier stage. The amplified rf output is then coupled into the mixer stage. Also coupled into the mixer is the output of the local oscillator to heterodyne with the rf picture and sound signals. When the oscillator frequency is set for the channel to be tuned in, the picture and sound signals of the selected station are heterodyned down to the lower intermediate frequencies of the receiver.

The oscillator beating with the two rf carrier signals produces two IF carrier signals. One is the picture IF signal from the rf picture signal, and the other is the sound IF signal from the sound rf signal. The original modulating information of the rf carrier signals is present in the IF signals out of the mixer, for both the AM picture signal and the FM sound signal. Furthermore, the 4.5-MHz separation between the rf carrier frequencies is maintained in the IF carrier frequencies. The intermediate frequencies in the output of the mixer stage are 45.75 MHz for the IF picture carrier and 41.25 MHz for the IF sound carrier. Note the 4.5-MHz difference between 45.75 and 41.25 MHz.

The rf amplifier, mixer, and local oscillator stages are on an individual subchassis, called the *front end,* or *rf tuner.* Either tubes or transistors can be used. With tubes, the local oscillator and mixer functions are usually combined in one stage, called the *frequency converter.*

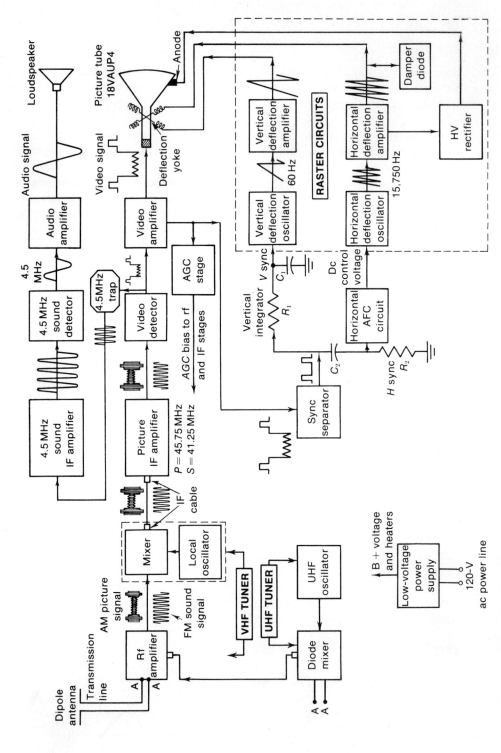

FIGURE 7-2 FUNCTIONS OF CIRCUITS IN TELEVISION RECEIVER. P IS PICTURE CARRIER SIGNAL; S IS SOUND CARRIER SIGNAL. WAVESHAPES SHOWN ARE NOT TO SCALE.

The tuner selects the channel to be received by converting its picture and sound rf carrier frequencies to the intermediate frequencies of the receiver. Then the selected signals can be amplified in the IF stages. The station selector is a ganged switch that changes the tuned circuits for the rf amplifier, mixer, and oscillator. The fine tuning control sets the oscillator frequency exactly for the best picture. It is important to note that the oscillator frequency determines which channel can be amplified by the IF section. Any problem of receiving the wrong channel is an oscillator trouble.

UHF tuner. When the VHF channel selector is set to its UHF position, the UHF tuner operates. The antenna input is obtained from a separate UHF antenna or the splitter from a combination VHF-UHF antenna. The UHF tuner is a separate unit, including transistor oscillator and crystal diode mixer. These two stages serve as the frequency converter to heterodyne the UHF channels down to the intermediate frequencies of the receiver. The UHF oscillator is tuned for the desired UHF station.

On the UHF position of the VHF tuner, the VHF oscillator is disabled. Then the rf amplifier and mixer on the VHF tuner are tuned to the IF values of 45.75 MHz for the picture signal and 41.25 MHz for the sound signal. A short length of coaxial cable plugs into both tuners, connecting the IF output of the UHF tuner to the rf amplifier of the VHF tuner, now serving as an IF amplifier.

Picture IF signal. The picture IF amplifier includes two to four tuned stages using tubes, transistors, or an integrated circuit. The bandwidth is enough for the IF picture signal with its side frequencies and for the IF sound signal. The main function here is amplifying the picture IF signal from the mixer to provide several volts for the video detector. This section is also called the *video IF amplifier* in schematic diagrams. The gain of the IF amplifier is controlled automatically by the AGC bias, according to the strength of the signal. A typical IF module board containing the complete picture IF amplifier section and video detector is shown in Fig. 7-3. The IF amplifier is usually connected to the mixer output on the rf tuner by a short length of coaxial cable, with a plug on both ends.

The IF value standardized by the Electronic Industries Association for the picture carrier frequency is 45.75 MHz in all receivers. Then the sound IF carrier is automatically 41.25 MHz, separated by 4.5 MHz from the picture carrier. The chrominance signal in the IF amplifier has the frequency of 42.17 MHz, which is 3.58 MHz from the picture carrier at 45.75. The sound and color IF values are below the picture carrier frequency because the rf oscillator beats above the rf signal frequencies in the frequency conversion by the rf tuner.

Video detector. The modulated IF picture signal is rectified and filtered here to recover its AM envelope, which is the composite video signal needed for the picture tube. The main purpose of the video detector is to provide for the video amplifier the composite video with its camera signal, sync, and blanking.

For color broadcasts, the video detector output also includes the 3.58-MHz chrominance signal. This signal is used in the color amplifier in color receivers.

Intercarrier sound. In addition to the video signal, the output of the video detector in Fig. 7-2 includes the 4.5-MHz intercarrier sound signal. All television receivers, monochrome or color, use the intercarrier method of demodulating the sound as a 4.5-MHz signal for all channels, VHF or UHF. However, color receivers have a separate 4.5-MHz sound converter, instead of using the video detector for this function. The

FIGURE 7-3 MODULE BOARD WITH PICTURE IF AMPLIFIER, VIDEO DETECTOR AND VIDEO PREAMPLIFIER CIRCUITS IN COLOR RECEIVER. AFT IS AUTOMATIC FINE TUNING. THE LENGTH OF ENTIRE BOARD IS 7 IN. *(RCA)*

reason is to minimize interference between the sound and color signals. The advantage of the intercarrier system is that the 4.5-MHz sound signal is automatically tuned in with the picture.

Video amplifier. Consisting of one or more stages, this section amplifies the composite video signal enough to drive the grid-cathode circuit of the picture tube. The camera signal variations change the instantaneous grid-cathode voltage, modulating the intensity of beam current. Then the variations of light intensity, as the spot scans the screen, enable the picture to be reproduced on the raster.

The amount of composite video signal required for the picture tube is about 100 V peak to peak for strong contrast. Cathode drive is generally used. This requires video signal with positive sync polarity and negative white amplitudes at the picture tube cathode, as shown in Fig. 7-2. Remember that negative cathode voltage corresponds to positive voltage at the control grid, to increase the beam current for white signal amplitudes.

The contrast control in the video amplifier varies the amount of video signal for the picture tube. More video signal means more contrast.

The blanking pulses in the composite video signal drive the picture tube grid-cathode voltage to cutoff, blanking out retraces. Although the sync is included in the video signal for the picture tube, the only effect of sync voltage here is to drive the grid more negative than cutoff. Note that the composite video signal is also coupled to the sync circuits (where the synchronizing pulses are separated for use in timing the receiver scanning) and to the AGC stage that produces AGC bias for the rf and IF amplifiers.

Automatic gain control. This automatic-gain-control (AGC) circuit is similar to the automatic-volume-control (AVC) system in AM radios. The stronger the picture carrier is, the greater the AGC bias voltage produced and the less the gain of the receiver. The result is relatively constant video signal amplitude for different carrier-signal strengths. Therefore, AGC for the picture signal is useful as an automatic control of contrast in the reproduced picture. However, the AGC circuit affects both the picture and sound, since it controls the gain of the rf and IF stages.

Synchronizing circuits. The video detector output includes the synchronizing pulses as part of the composite video signal. This signal, therefore, is also used in the sync circuits. The sync provides the timing pulses needed for controlling the frequency of the vertical and horizontal deflection oscillators.

A sync separator is a clipper stage that can separate the synchronizing pulse amplitude from the camera signal in the composite video. Remember that the top 25 percent of the video signal amplitude is used for sync. When the tips of the video signal are clipped off and amplified, the output consists only of sync pulses.

Since there are synchronizing pulses for both horizontal and vertical scanning, Fig. 7-2 shows the output of the sync separator divided into two parts. The integrator is a low-pass *RC* filter circuit that filters out all but the vertical pulses from the total separated sync voltage. Then the vertical synchronizing signals can lock in the vertical deflection oscillator at 60 Hz. For horizontal synchronization, an automatic-frequency-control (AFC) circuit is used to lock in the horizontal deflection oscillator at 15,750 Hz.

Deflection circuits. As shown in Fig. 7-2, these include the vertical oscillator and amplifier for vertical scanning at 60 Hz, with the horizontal oscillator and amplifier for scanning at 15,750 Hz. For either vertical or horizontal scanning, the oscillator stage generates deflection voltage to drive the amplifier at the required frequency. The deflection amplifiers are power output stages to provide enough scanning current in the deflection yoke for a full-sized raster.

The horizontal output circuit also includes the damper diode and high-voltage rectifier. The damper has the function of reducing sine-wave oscillations in the horizontal sawtooth scanning current, which occur immediately after flyback. The high-voltage rectifier produces dc anode voltage for the picture tube.

The deflection circuits produce the required scanning current and the resultant raster with or without the synchronizing signals. Since the deflection oscillators are free-running, they do not require any external signal for operation. However, the sync is needed to hold the deflection oscillators at exactly the right frequency so that the picture information is reproduced in the correct position on the raster. Without sync, the deflection circuits scan the raster, but the picture will not hold still.

Low-voltage supply. Two power supplies are shown in Fig. 7-2. One is the usual B+ supply for dc operating voltages on the tubes or transistors. This is the low-voltage supply because its output voltage is low compared with the high-voltage supply for anode voltage on the picture tube. For sufficient brightness, the anode voltage for black-and-white picture tubes is 9 to 20 kV, while color tubes use about 18 to 25 kV.

The dc output of 140 to 280 V supplies plate voltage for vacuum tube amplifiers. Silicon diodes are generally used as the rectifiers. For transistors and integrated circuits, the required dc supply is about 12 to 90 V, in either positive or negative polarity.

When tubes are used, the heaters can be in parallel with a 6-V winding on the power transformer, or in series for ac-dc sets. With series heaters, all the heater voltages add to equal the power-line voltage of 120 V. An open in any one heater opens the entire series string. The heater current for a series string usually is either 450 or 600 mA.

In tube sets having "instant-on" operation, the heaters have about one-half their normal voltage with the power switch off. Full power is applied when the receiver is turned on, and the tubes are on almost immediately. This feature has been developed to be similar to solid-state devices, which are on immediately because they have no heater.

High-voltage supply. The high-voltage rectifier obtains its ac input from the horizontal amplifier. This arrangement is called a *flyback supply* because the high voltage is generated as an induced voltage during the fast horizontal retrace. The resultant voltage is stepped up by the horizontal output transformer for the required amount of high voltage. The rectified output is the dc anode voltage needed by the picture tube to produce brightness on the phosphor screen. Because the anode voltage depends on the horizontal output with a flyback circuit, there cannot be any brightness on the picture tube screen if the horizontal scanning circuits are not operating.

Signal voltages. For a picture without snow, the required antenna signal is about 1,000 μV or 1 mV. The signal voltage from the tuner into the IF amplifier is about 10 mV. For linear operation of the video detector diode, about 1 to 3 V of IF signal is needed. These are all rms values for the modulated picture carrier signal.

In the output of the video detector, a typical value of composite video signal for tubes is about 3 V, peak to peak, with an average dc level of about 2 V. For transistors in the video amplifier, one-half these voltages are typical values. The peak-to-peak signal out of the video amplifier to drive the picture tube is 80 to 120 V for good contrast. The required dc level or bias for the correct brightness is about 40 V, negative at the control grid or positive at the cathode of the picture tube.

7-3 Signal Frequencies

The operation of the receiver can be examined in more detail by considering the frequencies in different sections of the receiver for picture, sound, and color.

Rf and IF values. As an example, let the VHF tuner in Fig. 7-2 be set for tuning in channel 3 at 60 to 66 MHz. Then the rf amplifier and mixer input are tuned to this band of frequencies for the picture carrier (P) at 61.25 MHz and the sound carrier (S) at 65.75 MHz. The local oscillator is at 107 MHz, which is 45.75 MHz above P and 41.25 MHz above S.

Similarly, for UHF channel 14 at 470 to 476 MHz, P is 471.25 MHz and S is 475.75 MHz. The UHF oscillator is at 517 MHz. Then UHF channel 14 is converted to the IF picture carrier signal at 45.75 MHz and the IF sound signal at 41.25 MHz.

In all cases the difference between P and S is exactly 4.5 MHz, for VHF or UHF channels, for color or monochrome, for rf and IF signals. This frequency of 4.5 MHz is the intercarrier sound signal for the sound section.

Inversion of the intermediate frequencies. It should be noted that the oscillator beats above the rf signal frequencies. As a result, the IF signal frequencies become inverted. In the rf signals, S is higher than P, but in the IF signals S at 41.25 MHz is lower than P at 45.75 MHz. The reason is just that, in the rf signal, the frequency for S is closer to the oscillator frequency than P is. Then the difference is less for S.

Video frequencies. When the picture carrier signal is rectified in the video detector, the carrier frequency is filtered out. The envelope is the composite video signal with camera signal, sync, and blanking. The frequencies in the camera signal variation include high video frequencies up to 4 MHz, approximately, for horizontal details. At the low end, frequencies down to 30 Hz show shades of light and dark in the vertical direction. In addition, the dc level of the signal indicates the average brightness level.

Double superheterodyne. The feature of a superheterodyne circuit is converting the rf signal frequencies for different stations to the intermediate frequencies of the receiver. This way the rf carrier signals are all shifted in frequency to make them fit the receiver's IF amplifier, instead of changing all the tuned circuits in the receiver for each rf signal. Practically all the selectivity and all the gain of the receiver result from the IF amplification.

In a double superheterodyne circuit, the signal frequency is beat down twice for a second IF value lower than the first. The television receiver is actually a double superheterodyne for the 4.5-MHz sound signal and for the 3.58-MHz chrominance signal. These IF values are listed in Table 7-1 to illustrate the sequence of frequencies.

TABLE 7-1 THE IF SIGNAL FREQUENCIES

Picture IF value	= 45.75 MHz
First chrominance IF value	= 42.17 MHz
First sound IF value	= 41.25 MHz
Second sound IF value	= 4.5 MHz
Second chrominance IF value	= 3.58 MHz

7-4 Intercarrier Sound

Note the sequence of stages for sound in Fig. 7-2. After being amplified in the rf tuner with the picture signal, the sound signal passes through the picture IF amplifier so that both the sound and picture IF signals are coupled into the video detector. This IF amplification for both signals is accomplished by having a little more bandwidth in the picture IF amplifier to provide some gain for the sound IF signal. However, there is much more gain for the picture signal. The gain for the picture IF signal at 45.75 MHz is about 10 times more than for the sound IF signal at 41.25 MHz. As a result, the video detector can produce the 4.5-MHz sound signal as the difference frequencies between the strong IF picture carrier beating with the weaker IF sound signal. This action corresponds to the operation of the frequency converter in the rf tuner, where the relatively strong output of the local oscillator beats with the weak rf signal.

The 4.5-MHz sound section. The 4.5-MHz wavetrap in the output of the video detector in Fig. 7-2 is the *sound takeoff* circuit. This resonant tuned circuit filters out the 4.5-MHz sound signal from the video frequencies. Note that the takeoff circuit, the sound IF amplifier, and the FM sound detector are always tuned to 4.5 MHz. This combination is the mark of intercarrier sound receivers.

The sound IF stage amplifies the 4.5-MHz signal enough to drive the FM detector. This de-

112 BASIC TELEVISION

tector is necessary because the 4.5-MHz intercarrier sound is not an audio signal. The frequency modulation in the 4.5-MHz signal is the same as in the rf and first IF sound signals. After this FM signal is detected at 4.5 MHz, the detector output is the desired audio signal. Then the audio signal is amplified in a conventional audio amplifier to drive the loudspeaker.

Buzz in the sound. The only problem with intercarrier sound is that the vertical blanking pulses in the picture signal can interfere with the 4.5-MHz sound signal. The result is buzz at 60 Hz. It sounds like 60-Hz hum but is rougher because the buzz is produced by square waves. The buzz is eliminated by keeping the amplitude low for the first IF sound signal and using AM rejection circuits for the 4.5-MHz FM sound signal. Many receivers have a *buzz control* on the back of the chassis, which is adjusted for minimum buzz.

The intercarrier buzz may increase when the picture goes to white, as the picture carrier amplitude then decreases. If the transmitter modulates white too close to zero amplitude, there will not be enough picture carrier, causing the intercarrier sound signal to be interrupted.

Split-sound receivers. Practically all television receivers use intercarrier sound, compared with the old method of split-sound. These old receivers amplified and detected the sound at its first IF value, with a separate IF amplifier tuned to 21.25 MHz. The old standard IF values were 21.25 MHz for the sound signal and 25.75 MHz for the picture signal. Split-sound receivers became obsolete when television broadcasting started to use the UHF channels from 470 to 890 MHz. At these frequencies, it is too difficult for the local oscillator in the rf tuner to tune in the sound as a separate signal.

7-5 Dividing the Receiver into Sections

The picture is reproduced on the screen as the combined result of the raster, video signal, and sync. These functions are summarized in Tables 7-2 and 7-3. In addition, Fig. 7-4 illustrates the successive steps in forming the raster and superimposing the picture on the raster.

Illumination. Just the spot of light on the screen in Fig. 7-4a shows that the picture tube and high-voltage supply are operating. However, it should be noted that this illustration was produced with an external high-voltage source. The flyback supply in the receiver cannot produce high voltage without horizontal output.

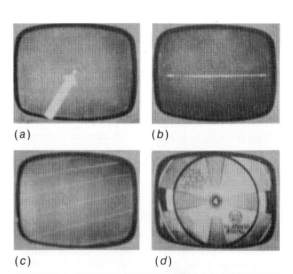

FIGURE 7-4 HOW THE RECEIVER PUTS THE PICTURE ON THE RASTER. (a) ILLUMINATION ON SCREEN. (b) ILLUMINATION PLUS HORIZONTAL SCANNING. (c) HORIZONTAL AND VERTICAL SCANNING TO PRODUCE THE WHITE RASTER. (d) VIDEO SIGNAL PRODUCING A PICTURE ON THE RASTER.

Horizontal scanning. The single horizontal line in Fig. 7-4*b* shows illumination and horizontal scanning. The horizontal deflection circuits, including the horizontal oscillator, amplifier, and damper stages, produce the horizontal scanning.

Vertical scanning. The vertical oscillator and amplifier stages produce vertical scanning. Then the horizontal scanning lines fill the entire screen area from top to bottom to form the scanning raster. The white raster in Fig. 7-4*c* shows that the vertical and horizontal deflection circuits, the picture tube, and the high-voltage supply are operating.

Picture. Figure 7-4*d* shows a picture reproduced on the raster. The circuits for picture signal, from the antenna input to the picture tube grid, provide video signal for the picture information. Then the video signal voltage varies the intensity of the electron beam, while the deflection circuits produce scanning, to reproduce the picture as shades of white, gray, and black on the raster.

Deflection sync. Vertical synchronization allows successive frames to be superimposed exactly over each other so that the picture will not appear to be rolling up or down the screen. Horizontal synchronization prevents the line structure of the picture from tearing apart into diagonal segments. The synchronizing circuits in the receiver provide the horizontal and vertical sync for the deflection oscillators to hold the picture steady.

Circuits for the raster. In Table 7-2 only the raster circuits are listed with the requirements for illumination. Assuming that the picture tube is operating with high voltage to produce light, the horizontal and vertical deflection circuits then can produce the raster.

Circuits for signal. Table 7-3 lists these circuits as separate groups, based on the receiver block diagram in Fig. 7-2. With intercarrier sound, only the 4.5-MHz circuits and audio amplifier in the second column are for sound alone. Almost all the signal circuits are for both

TABLE 7-2. CIRCUITS FOR THE RASTER (for block diagram in Fig. 7-2)

ILLUMINATION	SCANNING	
	HORIZONTAL	VERTICAL

Picture tube
  Heater voltage
  Bias voltage
  Screen-grid voltage
	H oscillator
	H amplifier
	H damper
		V oscillator
		V amplifier

High-voltage supply	Deflection yoke	

TABLE 7-3 CIRCUITS FOR THE SIGNAL (for block diagram in Fig. 7-2)

PICTURE AND SOUND	SOUND ONLY	PICTURE ONLY
Rf tuner	4.5-MHz IF amplifier	Video amplifier
Picture IF and detector	4.5-MHz FM detector	Sync separator
AGC circuit	Audio amplifier	Horizontal AFC

picture and sound. As listed in the first column, the rf tuner, picture IF amplifier, and video detector are common to the picture and sound signals. The AGC circuit controls the gain of these rf and IF amplifiers.

Only the video amplifier is listed in the third column for picture alone, with the sync circuits to hold the picture steady. Even the video amplifier can be common to the sound, however, in receivers where the 4.5-MHz signal is taken from the video amplifier output circuit.

7-6 Receiver Controls

These can be considered in two groups: the setup adjustments mainly for the raster and the operating controls in the signal circuits. The setup adjustments for the scanning raster are usually on the rear apron of the chassis. The operating controls are in the front panel or at the side of cabinet.

Raster adjustments. The vertical height and linearity controls are adjusted to fill the screen top to bottom, with the scanning lines equally spaced for good linearity. If there is a width control, it is adjusted to make the raster just cover the left and right edges of the screen. It should be noted that the raster is the same for all stations and is present with or without signal. However, the raster with signal can be a little smaller as blanking crops the edges slightly.

If the raster is off-center, there are usually magnet rings on the neck of the picture tube that can be rotated to shift the raster vertically and horizontally. It should be noted that the horizontal hold control can shift the picture slightly with respect to the raster.

Channel selector. The VHF channel switch tunes in the desired station for channels 2 to 13. On the UHF position of this switch, dc voltage is supplied to the UHF oscillator to operate the UHF tuner. Then the UHF channel control can be used to select any UHF channel from 14 to 83.

Fine tuning. This control provides more exact setting of the frequency for the VHF oscillator. With intercarrier sound, the fine tuning can be set for the best picture, independently of the sound. The best picture with good contrast and fine detail results by setting the fine tuning just off the point where you see wide horizontal sound bars that move with the voice modulation.

Volume. This is a typical audio level control, usually a potentiometer to vary the signal voltage input to the first audio amplifier. Some receivers may also have a tone control to adjust the response for high audio frequencies.

Buzz. Adjust this control, if necessary, for minimum 60-Hz buzz in the sound.

Contrast. Since the receiver generally has AGC to vary the gain of the rf and IF amplifiers, the contrast control is in the video amplifier circuit to adjust the amplitude of video signal voltage for the picture tube.

Video peaking. Some receivers have this control to vary the high-frequency response of the video amplifier for sharper outlines in the picture. This control is also called *fidelity,* or *sharpness.* However, the picture may look better with reduced bandwidth and less sharpness if there is noise or interference in the signal.

AGC level. For proper range of the contrast control, the AGC level setup adjustment at the back of the chassis must be set properly. Adjust the AGC level for full contrast on the strongest station with the contrast control at maximum.

Keep the AGC level below the point of overload distortion, however, where black and white are reversed and the picture is out of sync with buzz in the sound.

Brightness. This control varies the dc bias for the grid-cathode circuit of the picture tube. Adjust for the desired overall illumination on the screen. If the picture goes completely out of focus at high brightness levels, the trouble may be insufficient high voltage or an old picture tube that probably has weak emission.

Vertical hold. This control adjusts the frequency of the vertical deflection oscillator close enough to 60 Hz to allow the sync to lock in the vertical scanning. When the picture rolls up or down, the vertical hold control is varied to make the picture stay still.

Horizontal hold. This control adjusts the horizontal deflection oscillator. When the picture shifts horizontally and tears apart into diagonal segments, the horizontal hold control is varied to establish horizontal synchronization and provide a complete picture.

It is interesting to note that many of the controls are similar in their function of varying an ac voltage level. Turning up the volume control increases the amount of audio signal for more volume. Similarly, the contrast control increases the amount of video signal for more contrast. Also, the color control increases the amount of 3.58-MHz chroma signal for more color in the picture. In addition, the height control increases the amount of vertical sawtooth scanning current for more height in the raster. A width control increases the amount of horizontal sawtooth scanning current.

7-7 Receiver Tubes

Where tubes are used instead of transistors to amplify the sound and picture signals, these stages are usually the miniature glass types shown in Fig. 7-5. The heater pins are 3 and 4 for the seven-pin base in *a* or 4 and 5 for the nine-pin *novar* base in *b*. The wider spacing between end pins is the guide for inserting the tube in its socket. The nine-pin tubes come in several different sizes, with a slightly larger base and larger pins for power tubes.

The 6-V tubes such as 6GH8A are for parallel heater circuits in receivers with an ac power supply using a power transformer, which includes a 6-V winding. Heater ratings such as 3BZ6, 8FQ7, and 17D4 are for series heater cir-

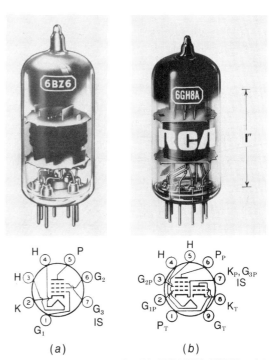

FIGURE 7-5 MINIATURE GLASS AMPLIFIER TUBES. (*a*) SEVEN-PIN BASE. (*b*) NINE-PIN BASE. *IS* IS INTERNAL SHIELD.

116 BASIC TELEVISION

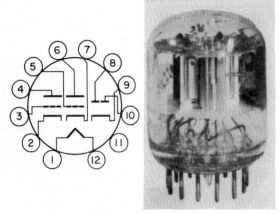

FIGURE 7-6 COMPACTRON MULTIFUNCTION TUBE 6B10, COMBINING TWO DIODES AND TWO TRIODES. HEIGHT IS 1½ IN. (GE)

A tube for the damper must have a high inverse-voltage rating between cathode and heater, to prevent arcing to chassis ground. A typical maximum rating is −5,500 V. The damper takes a long time to heat up because of the heavy cathode insulation.

The damper stage conducts immediately after horizontal flyback to damp the output circuit. Without damping, there are vertical, thin, white lines at the left edge of the raster, produced by oscillations in the horizontal scanning current. In addition, the rectified deflection voltage from the damper supplies boosted B+ voltage for the plate of the horizontal amplifier tube. Finally, the damper current, while the diode is

cuits in those receivers that do not have a power transformer. The 3BZ6 or 6BZ6 is a pentode often used in the picture IF amplifier.

The *compactron* tube in Fig. 7-6 combines three stages in one envelope. Note the *duodecar* twelve-pin base used for this type and for power tubes in general. The heaters are the end pins 1 and 12. At the opposite extreme, the *nuvistor* is a miniaturized tube not much larger than a transistor, often used as the triode rf amplifier in VHF tuners.

Power tubes are needed for the deflection circuits, especially in the horizontal output stage, which takes the most power from the low-voltage supply. The 6JE6 in Fig. 7-7 has a maximum power rating of 30 W plate dissipation. Average cathode current is about 350 mA. Note the plate connection to the top cap.

The damper (Fig. 7-8) in the horizontal output stage is a diode power rectifier. The 6CT3 damper tube in *a* has a maximum rating of 1.2 A for peak plate current. Peak ratings for the semiconductor diode damper in *b* are 10 A for forward current and 320 V for reverse voltage.

FIGURE 7-7 HORIZONTAL POWER OUTPUT TUBE 6JE6 WITH NOVAR BASE. TOP CAP IS PLATE CONNECTION. HEIGHT IS 4 IN. (RAYTHEON COMPANY)

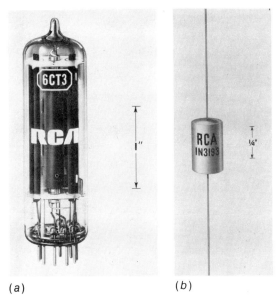

FIGURE 7-8 DAMPER DIODE FOR HORIZONTAL OUTPUT CIRCUIT. (a) TUBE. (b) SILICON DIODE. (RCA)

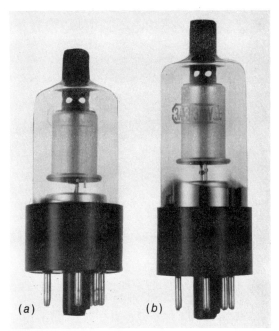

FIGURE 7-9 HIGH-VOLTAGE DIODE RECTIFIER TUBES. (a) 1G3/1B3 FOR MONOCHROME RECEIVERS. (b) 3A3 FOR COLOR RECEIVERS. HEIGHT IS 4 IN. (RAYTHEON COMPANY)

conducting, provides horizontal scanning for the left side of the raster. For this reason, the horizontal damper is also called an *efficiency diode*. It should be noted that when boosted B+ voltage is used for the horizontal amplifier, this stage cannot operate without the damper.

The high-voltage rectifiers in Fig. 7-9 feature very high peak inverse-voltage rating, in a flyback supply. This rating is 36 kV for the 1G3 in a. The heater power of 1.25 V at 0.2 A is taken from the horizontal output circuit, as is the high-voltage ac input of 18 kV. This tube is for a black-and-white picture tube where beam current is less than 1 mA for the rectifier's load current. For three-gun color picture tubes, the 3A3 in b has a current rating of 2 mA, with dc high-voltage output of 25 kV. Both of these tubes have an octal socket and a top cap for the plate connection. The silicon unit in Fig. 7-10 is rated at 18 kV, but ratings up to 35 kV are available.

7-8 Solid-State Devices

These include transistor amplifiers, diode rectifiers, and the integrated circuit (IC). Typical transistors and IC units are shown in Figs. 7-11 and 7-12.

The semiconductor elements germanium (Ge) and silicon (Si) are generally used. Silicon is a dark, crystalline solid that is brittle, like

FIGURE 7-10 SILICON-DIODE STICK FOR HIGH-VOLTAGE RECTIFIER. LENGTH IS 3 IN. CATHODE END MARKED + FOR DC OUTPUT.

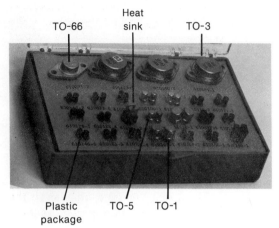

FIGURE 7-11 TRANSISTOR KIT FOR SERVICING TELEVISION RECEIVERS. "TO" IS TRANSISTOR OUTLINE NUMBER. DIAMETER OF METAL TO-3 CASE IS 1.2 IN. (*MAGNAVOX CORPORATION*)

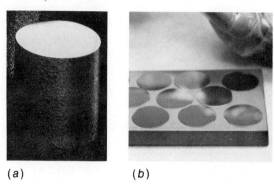

FIGURE 7-13 FORMS OF PURE SILICON. (a) SOLID BAR. (b) WAFER DISKS. DIAMETER IS $1\frac{1}{2}$ IN. (*DOW CORNING*)

glass. A bar of pure silicon and wafer disks cut from the bar are shown in Fig. 7-13. After the silicon is processed to alter its electrical characteristics, each disk can be cut into more than 100 sections for transistors and IC chips. An IC chip, typically 0.05 in.^{2}, combines transistors, diodes, and resistors and capacitors to form complete circuits. A transistor chip is even smaller.

FIGURE 7-12 TYPICAL *IC* UNITS FOR TELEVISION RECEIVERS. (a) DUAL IN-LINE PACKAGE (DIP). LENGTH IS $\frac{3}{4}$ IN. NOTCH AT END INDICATES PIN 1. POWER RATING IS 850 mW. (b) TO-5 PACKAGE MOUNTED ON NINE-PIN TUBE SOCKET. (*SPRAGUE ELECTRIC COMPANY*)

The transistor was invented in 1948, using germanium. Now almost all solid-state devices use Si because it can operate at higher temperatures and has less leakage current, compared with Ge. The name "transistor" is derived from "transfer resistor." The advantages of a transistor are its small size and efficient operation. Semiconductor devices use the electric charges in the solid material. There is no heater or cathode. Without a heater, there is no warmup time, allowing instant operation when the electrode voltages are applied. Normal life of a semiconductor device, without any troubles of opens or shorts, is probably about 20 years. The disadvantage of semiconductors is that their characteristics are sensitive to changes in temperature and that they are easily damaged by excessive heat.

Doping. In their pure form, semiconductors* have few free electrons in their atoms. However, impurity elements can be added to increase the number of free charges by a factor of 10 to 50 times. This doping process alters the distribu-

---

*More details of semiconductors can be found in transistor books and in the authors' "Basic Electronics," 3d ed. McGraw-Hill Book Company. See Bibliography at back of book.

tion of the valence electrons in the semiconductor atoms. When electrons are added, the semiconductor is N-type. The P-type has a deficiency of electrons. The positive charge resulting from a vacancy of one electron in the valence band is called a *hole charge.* Current in a semiconductor can be either electron flow or a drift of positive hole charges.

Hole current. The positive hole has exactly the same amount of charge as the negative electron but of opposite polarity. A hole is not a proton but a new type of positive charge produced only in P-type semiconductors. When hole charges move, the hole current is in the same direction as conventional current, opposite to electron flow. All arrow symbols for semiconductor devices show the direction of hole current. It should be noted, though, that hole charges are moving only in the P semiconductor. Externally, electrons are moving in the opposite direction.

The PN junction. When P-type and N-type materials are joined in a continuous crystalline structure, the edge where the opposite types meet is a PN junction. The junction prevents the hole charges and electrons from neutralizing each other. The reason is a small barrier potential across the junction. Its polarity repels holes back into the P material and electrons into the N material. The barrier voltage results from ions of the impurity elements used in the doping. The amount of this barrier potential is 0.3 V for Ge junctions and 0.7 V for Si junctions. These voltages apply to all Ge and Si junctions, large or small, as the value is a characteristic of the element.

The junction voltage is the basis of operation for all semiconductor diodes and junction transistors. A diode rectifier is just a PN junction. Also, PNP and NPN transistors consist of two junctions with opposite polarities. The application of the junction is to use an external voltage to aid or oppose the barrier voltage, in order to control current through the junction. Actually, the P and N bulk materials only provide electrodes to connect the junction to an external circuit.

Forward and reverse voltages. When external voltage is applied to cancel all or part of the internal barrier voltage, this polarity is called *forward voltage* (V_F). As shown at the emitter-base junctions in Fig. 7-14, V_F is positive at the P side and negative at the N side to produce *forward current* through the junction. In short, the polarity for forward voltage is the same as the P and N electrodes.

For reverse voltage (V_R) the connections are the opposite. In Fig. 7-14, V_R for the collector-base junction has the positive side to the N electrode and the negative side to P. Reverse voltage prevents any forward current from flowing across the junction.

PNP and NPN junction transistors. As shown in Fig. 7-14, three sections or electrodes are used to form two junctions. Either a P is between two N sections, or the N is between two P sections. With two junctions one can control the current through the other to provide amplification of the input signal. This junction type is also called a *bipolar transistor* because it uses both electrons and hole charges. The three electrodes are called *emitter, base,* and *collector.* The base section is in the middle.

The function of the emitter is to supply charges, either electrons or holes, through its junction with the base. As the source of charges, the emitter function is like the cathode of a vacuum tube. The collector receives these charges through its junction with the base. This function is similar to the plate or anode of a tube, but the collector is positive for NPN transistors or negative for PNP transistors. The base is part of both junctions, to control how many

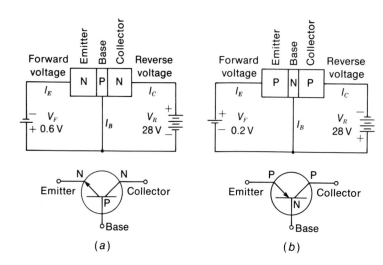

FIGURE 7-14 ELECTRODES AND SYMBOLS FOR JUNCTION TRANSISTORS. (a) NPN, USING Si WITH FORWARD BIAS OF 0.6 V. (b) PNP, USING Ge WITH FORWARD BIAS OF 0.2 V.

charges can move from the emitter, through the base, to the collector. The function of the base is similar to the control grid in a tube.

Referring to Fig. 7-14, the requirements for electrode voltages in a junction transistor can be summarized as follows:

1. The emitter-base junction must have forward voltage to bias the transistor into conduction. For a class A amplifier, the forward bias is approximately 0.2 V for all Ge transistors or 0.6 V for all Si transistors. Zero voltage from emitter to base means no forward bias. Then the transistor is cut off, without any output current in the collector circuit.
2. The collector-base junction always has reverse voltage. This polarity is necessary to prevent easy current in the direction from collector to base. Actually, though, the reverse voltage at the collector is the right polarity to attract charges in the base supplied by the emitter. The amount of reverse collector voltage is about 3 to 20 V for small, signal transistors and up to 200 V for power transistors.

In comparing the operation of tubes with transistors, a tube conducts with plate voltage applied when the negative control grid bias is less than cutoff. However, a transistor must be made to conduct by applying forward bias. This corresponds to heating the cathode in order to produce the charges needed for conduction.

Transistor circuits. With three electrodes for a PNP or NPN transistor there are three possible circuit configurations. These depend on which electrode is used as a common terminal for the input and output signals. The circuits are:

1. Common-emitter (CE): input to the base and output from the collector (Fig. 7-15a)
2. Common-base (CB): input to the emitter and output from the collector (Fig. 7-15b)
3. Common-collector (CC): input to the base and output from the emitter (Fig. 7-15c)

Note that in all the amplifiers, the electrode for signal output has the load resistance R_L. Also, the common terminal is shown grounded, but this can be just a common return

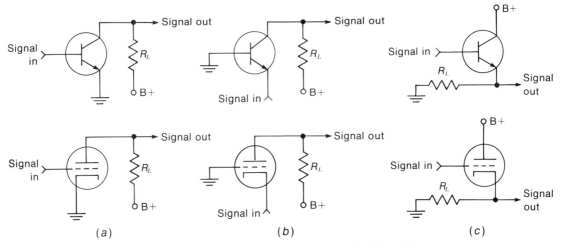

FIGURE 7-15 TRANSISTOR AMPLIFIERS AND THEIR CORRESPONDING VACUUM TUBE CIRCUITS. NPN TRANSISTORS SHOWN. (a) COMMON-EMITTER (CE) AND GROUNDED CATHODE. (b) COMMON-BASE (CB) AND GROUNDED GRID. (c) COMMON-COLLECTOR (CC) OR EMITTER-FOLLOWER AND CATHODE-FOLLOWER.

connection without chassis ground. Furthermore, the ground return need not be a dc connection. A bypass capacitor can provide an ac ground for signal, even when the electrode has a dc potential. The circuit is named for the common electrode, which is the one that does not have any input or output signal voltage.

The common-emitter circuit is the one used generally for amplifiers. It has the best combination of voltage gain and current gain.

The common-collector circuit is called an *emitter-follower*, corresponding to the cathode-follower circuit with vacuum tubes. This circuit is used for impedance matching because of its high input resistance and low output resistance. The voltage gain is less than 1, but the stage has current gain. Two emitter-followers in cascade are called a *Darlington pair*. The two stages are usually dc-coupled and packaged in one unit.

Figure 7-16 shows more details of the CE circuit, using a silicon NPN transistor. The positive 28-V supply is used for reverse voltage on the N collector. R_L is the collector load. The collector voltage V_C is 12 V because the $I_C R_L$ voltage drop is 16 V. In the base circuit, R_1 and R_2 form a voltage divider to supply positive forward bias. In the emitter leg, R_E has a voltage drop of 1 V produced by I_E. The net base-emitter bias is

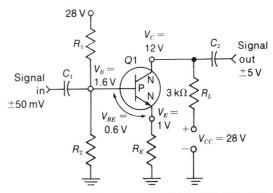

FIGURE 7-16 TYPICAL COMMON-EMITTER (CE) AMPLIFIER CIRCUIT.

1.6 − 1.0 = 0.6 V for the Si transistor, positive at the P base.

The emitter resistance R_E is used to stabilize the CE amplifier against *thermal runaway*, where I_C can become excessive if the forward bias is increased because of reverse leakage current. Note that the common terminal for input and output signals is the emitter, although it is not directly grounded.

The signal input is applied to the base in the CE stage, with amplified voltage output from the collector. With an R_L of 3 kΩ, a typical voltage gain is 100. The input resistance of the CE stage is 1 kΩ with an output resistance of 50 kΩ, approximately. The amplified output signal voltage has inverted polarity, compared with the input. The CB and CC circuits do not invert the polarity.

Field-effect transistor (FET). Field-effect transistors are used as amplifiers, like junction transistors, but the FET construction is different. The purpose is to eliminate the problems of low input resistance with forward bias and leakage current from reverse collector voltage. The FET provides very high input resistance, in megohms. Also, the FET can take input signal in the order of volts, compared with less than 0.1 V for junction transistors.

In Fig. 7-17, the bulk or substrate material is neutral or lightly doped. This part only serves as a platform on which the channel and gate electrodes are diffused. The source and drain are just ohmic connections at opposite ends of the channel. When voltage is applied between the drain and source, the current through the channel is controlled by the gate electrode. The channel can be either N-type or P-type, but current flows only from source to drain. The FET is a *unipolar* device, as the charge carriers in the conducting path through the channel have only one polarity. In summary, the FET electrodes are as follows:

Source. The source is the terminal where charge carriers enter the channel bar to provide current. The source corresponds to the emitter. The substrate is usually connected internally to the source.

Drain. This is the terminal where current leaves the channel. The drain corresponds to the collector.

Gate. This electrode controls the conductance of the channel to vary the current from source to drain. Input signal voltage is usually applied to the gate. This corresponds to the base.

In a FET, gate voltage controls the electric field in the channel, while base current controls the collector current in a bipolar transistor. Actually, the gate function is more similar to the control grid in a

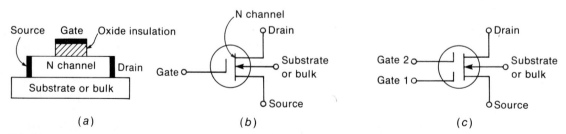

FIGURE 7-17 FIELD-EFFECT TRANSISTOR (FET) WITH INSULATED GATE. (a) CONSTRUCTION. (b) SYMBOL. (c) DOUBLE-GATE.

TABLE 7-4 COMPARISON OF TRANSISTORS AND VACUUM TUBES

↓	VACUUM TUBE	JUNCTION TRANSISTOR	FIELD-EFFECT TRANSISTOR
Source of charges	Cathode	Emitter	Source
Control of charges	Control grid	Base	Gate
Collector of charges	Plate or anode	Collector	Drain
Amplifier type	Grounded cathode	Common emitter	Common source
Amplifier type	Grounded grid	Common base	Common gate
Follower circuit	Cathode-follower	Emitter-follower	Source-follower

vacuum tube. These comparisons of junction transistors, field-effect transistors, and vacuum tube amplifiers are listed in Table 7-4.

The FET in Fig. 7-17 is the insulated-gate (IGFET) type, with one gate in *b* or dual gates in *c*. This is also called a metal-oxide semiconductor FET, or *MOSFET*. When the channel can conduct with zero gate bias, the FET is the *depletion* type. The *enhancement* type requires a gate bias of about 3 V. In schematic symbols, the depletion type has a solid channel bar, as in Fig. 7-17. For the enhancement type the channel line is shown broken by the *D, B,* and *S* electrodes. Another type is the JFET, which uses a PN junction with reverse bias for the gate electrode. The insulated-gate types of FET require a metal shorting ring for the gate while it is out of the circuit, in order to prevent accumulation of static charge. Or, they may have internal protective diodes.

Transistor characteristics. For junction transistors, the ratio I_C/I_B is called beta (β). As an example, with an I_C of 50 mA and I_B of 1 mA, the beta is $50/1 = 50$. Typical values of β are 20 to 200. The beta value is the current amplification factor in the common-emitter circuit.

For the common-base circuit, the ratio I_C/I_E is used, which is *alpha* (α). As an example with an I_C of 50 mA and an I_E of 51 mA, the alpha is $50/51 = 0.98$. Alpha is always less than 1 because I_C must be less than I_E.

For field-effect transistors, g_m is the mutual transconductance between gate and drain, defined as I_D/E_G. As an example, with a drain current of 6 mA and gate voltage of 1 V, the g_m is 6 V/1 mA = 0.006 mho or 6,000 μmhos.

These values of β, α, and g_m are static or dc values. When small changes in i or e are considered for signal variations, the characteristics are dynamic or ac values.

Type numbers. PNP and NPN triode transistors are numbered 2N, as in 2N1613. The "N" is for a semiconductor, and the prefix number 2 is the number of junctions used. Semiconductor diodes are numbered 1N, as in 1N4785. Field-effect transistors are numbered 3N, as in 3N128. This system is for semiconductors registered with the EIA. Foreign transistors are often num-

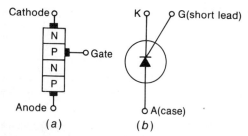

FIGURE 7-18 SILICON CONTROLLED RECTIFIER OR SCR. (a) CONSTRUCTION. (b) SYMBOL.

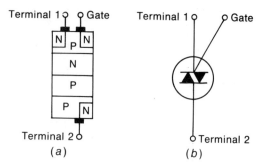

FIGURE 7-19 TRIAC. (a) CONSTRUCTION. (b) SYMBOL.

bered 2SA for PNP or 2SC for NPN types. A rectifier stack with diodes in series may have the prefix "CR" as in CR4.

Small-signal transistors are rated at 150 mW, with collector current of 1 mA, approximately. These are practically all silicon, NPN transistors, in a plastic case. Power transistors, rated at 1, 5, 25 W, or more, are often silicon NPN or germanium PNP. These are in a metal case, or the plastic case has a metal *heat sink* connected to the collector to radiate heat. With a metal case, the collector is usually connected internally to the case to dissipate heat. When the collector is not at chassis ground potential, a mica insulating washer must be used. This washer is coated with a thermal grease for better heat transfer.

Remember that all Si junction transistors, large or small, take a typical forward bias of 0.6 V with 0.2 V used for Ge. The ac signal swing is ±0.1 V or less. Small transistors are often soldered into the socket instead of using a socket. The reason is that oxidation at the socket connections can affect the very small base-emitter voltage.

Silicon controlled rectifier (SCR). As shown in Fig. 7-18, the SCR is a four-layer device used as a silicon rectifier. However, the start of conduction is controlled by a gate electrode. When forward voltage is applied between cathode and anode, in series with an external load, no appreciable load current can flow until the barrier voltage at the gate-cathode junction is overcome. This internal reverse bias at the gate is about 0.7 V.

Thyristors. This is the general name for semiconductors that are controlled rectifiers like the SCR. The characteristics are similar to thyratron gas tubes. The main types of thyristors are the SCR, triac, diac, and unijunction transistor (UJT). The SCR and triac are specific types in the general class of four-layer devices that can serve as gate-controlled switches (GCS).

As shown in Fig. 7-19, the *triac* construction enables the thyristor to conduct for either polarity of load voltage in the main circuit. Also, the gate can trigger the triac for positive or negative voltage. The diac in Fig. 7-20 is also

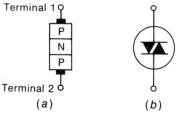

FIGURE 7-20 DIAC. (a) CONSTRUCTION. (b) SYMBOL.

FIGURE 7-21 UNIJUNCTION TRANSISTOR (UJT). (a) CONSTRUCTION. (b) SYMBOL.

bidirectional, but it does not have a gate. In many applications the diac is used to provide pulses for triggering a triac. The unijunction transistor in Fig. 7-21 is a bipolar PNP transistor but with two connections to the base. One of the base connections is used as a gate trigger.

Semiconductor diode rectifiers. A diode is just a PN junction. Since it conducts only in one direction, the diode is a simple, efficient rectifier. Generally Si is used for power-supply rectifiers, with Ge for signal detectors. A detector is just a low-power rectifier circuit for a few volts of ac signal. Additional types of solid-state diodes include copper oxide and selenium used as *metallic rectifiers*.

The standard symbol for a semiconductor diode is a bar and arrow for the PN junction, showing the direction of hole current (Fig. 7-22). This symbol is often marked on the diode.

If not, a dot, band, or any mark at one end indicates the cathode side.

Zener* diodes. These are silicon diodes designed for a specific value of reverse breakdown voltage, which is generally 3.9 V to 27 V. Heavy doping is used to allow appreciable reverse current. For the zener breakdown voltage and a wide range of higher voltages, the reverse voltage across the diode remains relatively constant at the zener value. This characteristic makes the diode useful as a voltage regulator. Zener diodes are also called *voltage-reference diodes* and *avalanche diodes,* as the relatively large reverse current can be considered an avalanche current. The diode is not damaged, however, assuming the reverse current is within the power rating.

Since the diode is a PN junction, reverse voltage is applied as positive voltage at the N-cathode side. The reverse characteristic is used because higher voltages can be applied, compared with forward voltage.

Figure 7-23 shows a voltage regulator circuit with the zener diode to maintain a constant 12-V output across the load R_L. The dc input volt-

*Named after C. A. Zener, a physicist who analyzed voltage breakdown of insulators.

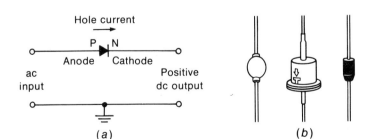

FIGURE 7-22 SEMICONDUCTOR DIODE RECTIFIER. (a) CIRCUIT. (b) TYPICAL DIODES, ACTUAL SIZE. PELLET AT LEFT, HIGH-HAT AND TUBULAR TYPES. RATED 3A AND PIV OF 1,000 V.

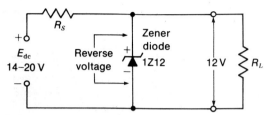

FIGURE 7-23 CIRCUIT WITH ZENER DIODE TO STABILIZE OUTPUT VOLTAGE ACROSS R_L.

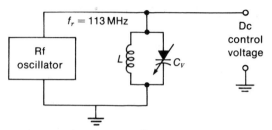

FIGURE 7-24 CIRCUIT WITH VARACTOR DIODE TO TUNE RF OSCILLATOR TO 113 MHz.

age of 14 to 20 V is enough to keep the zener diode in conduction. The series resistance R_S limits the current to values within the power rating. Note that the polarity of the input is positive at the N side of the diode for reverse voltage. Since the diode is in parallel with R_L, the voltage across both is constant at 12 V.

Varactor diodes. A PN junction with reverse bias is actually a capacitor. The P and N electrodes are the two conductor plates, on both sides of the depletion zone within the junction. This area serves as an insulator because it has no free charges. C may be 20 pF with reverse bias of 4 V, for typical values. Most important, C changes with the amount of reverse bias. Semiconductor diodes made for this application of a variable capacitance controlled by a dc voltage are *varactors* or *varicaps*. The varactor capacitance decreases with more reverse voltage.

The block diagram in Fig. 7-24 illustrates how the varactor is used for electronic tuning. As part of the LC circuit, the varactor capacitance C_V tunes the rf oscillator to resonance at 113 MHz. However, when C_V is varied by controlling its reverse voltage, the oscillator frequency changes.

Integrated circuit (IC). Transistor chips are small, but their use as discrete components with separate resistors, capacitors, and inductors makes the complete circuit relatively large. As an IC example, Fig. 7-25 shows a transistor, resistor, and capacitor on a single wafer. Inductors are not integrated because they take too much space. Transistors and diodes are the easiest components to integrate. Resistors are not precise, but the circuits are designed to depend on ratios of resistances rather than absolute values. Field-effect transistors also are easily integrated.

In Fig. 7-25a, the layers at the left form an NPN transistor. Note that the collector has a reverse NP junction with the wafer to isolate the transistor. In the middle section, the resistance between the two ohmic contacts of the P material is determined by the length and width of the

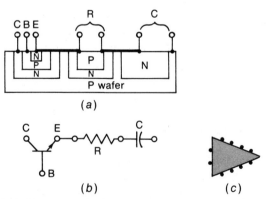

FIGURE 7-25 INTEGRATED CIRCUIT. (a) CONSTRUCTION. (b) SCHEMATIC. (c) SYMBOL FOR *IC* AMPLIFIER UNIT.

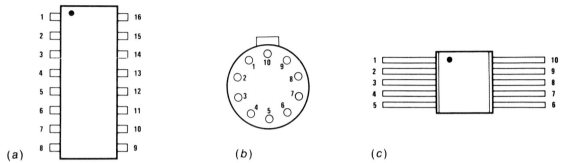

FIGURE 7-26 PIN CONNECTIONS FOR IC PACKAGES. DOT AT PIN 1. (a) 16-PIN DIP. THIS IS TO-116 CASE. (b) 10-PIN TO-5. (c) 10-PIN CERAMIC FLAT PACK. THIS IS TO-91 CASE.

strip. C at the right is the capacitance across the reverse-biased junction of the N material to the P wafer. Some areas may need very heavy doping, indicated as N+ or P+. The schematic diagram of this integrated circuit is shown in b, while the symbol in c is the general form for an IC amplifier.

Three popular types of IC packages are shown in Fig. 7-26. Common applications in the television receiver include the picture IF amplifier section with video detector, 4.5-MHz sound IF amplifier with FM audio detector, and 3.58-MHz chroma amplifier with color demodulators.

Special diodes. Schematic symbols are shown in Fig. 7-27 for additional types of semiconductor diodes. In a is the *tunnel diode,* or Esaki diode, named after the man who discovered that heavy doping can cause a tunneling effect of charge carriers through the depletion zone at the junction. Because of its negative resistance characteristic, the tunnel diode can be used as an amplifier or oscillator at microwave frequencies.

The light-emitting diode (LED) in b radiates light when forward voltage is applied. For greater efficiency, special materials such as gallium arsenide are used. Since the diodes are very small, they can be arranged to display numbers and letters. The experimental flat "wall screen," instead of a picture tube, is made of light-emitter diodes.

The photoconductive diode in c is made of a photosensitive material, such as cadmium sulfide. Its resistance decreases with more light. The applications include many light-control devices, including automatic control of brightness in the television picture according to the ambient light in the room.

Two additional diodes without special symbols are the *hot-carrier diode* and *compensating diodes*. These have compensating temperature characteristics for bias stabilization in transistor circuits. The *hot-carrier* diode uses a metal-to-semiconductor junction, such as cop-

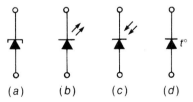

FIGURE 7-27 SCHEMATIC SYMBOLS FOR (a) TUNNEL DIODE; (b) LIGHT-EMITTING DIODE (LED); (c) PHOTOSENSITIVE DIODE; (d) TEMPERATURE-SENSITIVE DIODE.

per to N-silicon. This is used as a signal diode rectifier with better high-frequency response than conventional diodes. One possible application is the video detector.

Protective devices. These include varistors and thermistors. A *thermistor* (Fig. 7-28a) decreases its resistance with temperature. A *varistor* (Fig. 7-28b and c) decreases its resistance with higher voltage. The varistor is also called a *voltage-dependent resistor* (VDR).

The collector often has a varistor to protect against voltage pulses that have too much amplitude. As an example, a varistor rated for 70-V peak value can clamp voltage spikes at this value, in either polarity. Also, a diode rectifier can be connected in the collector circuit to isolate the wrong polarity of collector voltage caused by transient voltage surges.

7-9 Special Components

The following list includes special types of capacitors and inductors often used in television receivers.

Ferrite beads. As shown in Fig. 7-29, a bare wire is used as a string for one or more ferrite beads. The wire may be a connection between terminals or one lead from a component. A fer-

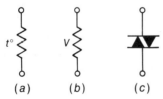

FIGURE 7-28 SCHEMATIC SYMBOLS FOR (a) THERMISTOR; (b) SILICON CARBIDE VARISTOR; (c) DUAL-DIODE SILICON VARISTOR.

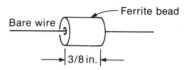

FIGURE 7-29 FERRITE BEAD USED AS RF CHOKE.

rite is a ceramic material that is magnetic like iron, but the ferrites are insulators. They are generally used for the core in IF and rf coils because the high resistance reduces eddy currents. A ferrite bead on a wire serves as a simple, economical rf choke. The dc resistance is zero because there is no coil. Also, there is no radiation. The ferrite beads are also considered as noise suppressors or inductance multipliers. They are commonly used in the heater line with tubes, or in the rf, IF, and video circuits for decoupling, shielding, and suppression of parasitic oscillations. As a typical value, a single bead on a 1-in. length of No. 22 wire has an inductance of 300 μH at 100 MHz.

Spark-gap capacitors. These combine the functions of a bypass capacitor and spark gap. They are generally used at the electrodes of a color picture tube, especially in transistorized receivers. The purpose of the spark gap is to bypass any arc produced in the picture tube, to prevent damage to the transistor amplifiers by voltage surges. The *gap capacitor* or *snap capacitor* is often constructed as a flat ceramic with a slit at the top for the spark gap.

Neon bulbs. As a gas tube, a neon bulb has high resistance until the striking voltage of about 90 V ionizes the bulb to make it a short circuit. Therefore, a neon bulb can also be used as a spark-gap protective device for the picture tube electrodes.

Bipolar electrolytic capacitors. This type consists of two electrolytics connected internally in series-opposition. Then the bipolar capacitor can be used in either polarity. However, the capacitance is cut one-half.

Electrolytic filter capacitors. For transistors, the filter capacitance is usually very high, with a low dc working voltage, compared with tube-type receivers. Typical values are 500 to 2,000 μF, at 50 V.

Electrolytic coupling capacitors. In audio and video circuits, for good low-frequency response, transistor amplifiers use 5- to 10-μF electrolytic capacitors. Large values of capacitors are needed with low resistance.

7-10 Printed-Wiring Boards

Whether tubes or transistors are used, all receivers now have printed wiring, instead of the old hand-wired method with individual solder connections. The components are inserted by machine into small metal holes or "eyelets," and all connections are made at once by dip-soldering. Figure 7-30 shows variable resistors with leads that fit into the eyelet connections.

Some components have metal prongs about 1 in. long, which are used for wire wraparound connections. This arrangement is called a *stake*. It provides a convenient point for connecting test equipment to the board.

Smaller boards for individual circuits are called *modules*. These have pin connections that plug into the main chassis, or spring clips, to hold the board in place. The board must be tight. Sometimes oxidation of the contacts causes intermittent connections. If it is necessary, the entire board can easily be replaced.

The following techniques can be helpful when working on a printed board:

1. Use a small soldering iron, of about 25 W, to prevent heat from lifting the printed wiring off the board or loosening the eyelet connections.
2. With IC units, a very thin tip is needed because of the close spacing of pins. You can wrap a piece of No. 12 wire around the tip of the iron and use the wire tip.
3. When soldering a semiconductor component, hold the lead with long-nose pliers, or connect an alligator clip. This conducts heat away to prevent damage to the junction.
4. When soldering field-effect transistors, the leads and the soldering iron should be grounded. The reason is that the gate electrode can accumulate static charge.
5. Resistors and capacitors can often be replaced without disturbing the printed wiring. Just break the old component in the middle, with diagonal-cutting pliers. Then solder the new component to the old leads.
6. A crack in the printed wiring can be repaired by soldering a piece of bare wire over the open.

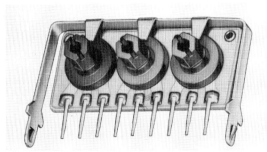

FIGURE 7-30 SET OF THREE VARIABLE RESISTORS FOR SETUP CONTROLS ON PRINTED-WIRING BOARD. HEIGHT IS $3/4$ IN. (*CENTRALAB*)

7. If a larger section of printed wiring is damaged, you can bridge the open with a length of hookup wire soldered at two convenient end terminals.
8. You can see the printed wiring more clearly by placing a bright light at one side of the board and looking through the opposite side.

To remove components from the eyelet connections, either of the following methods is generally used, working from the printed side of the board: (1) Use a "solder sucker" tool, which acts as a vacuum to pull up the solder after it is melted. (2) Use a piece of wire braid, such as a grounding strap, to pull up the molten solder by capillary action. Some solder paste on the braid helps attract the solder.

The desoldered eyelet connections must be cleaned of all solder so that the component practically falls out. Otherwise, the printed wiring can be damaged in removing the component.

When any semiconductor board, module, IC, or transistor is moved in or out of its socket, the power should be off. Also, discharge the dc voltage supply. These precautions are necessary because a transient pulse of voltage produced by the circuit change can damage the semiconductor. A typical module board is shown in Fig. 7-31.

FIGURE 7-31 MODULE BOARD FOR COLOR DEMODULATOR CIRCUITS. WIDTH IS 3 IN. (*ZENITH RADIO CORPORATION*)

7-11 Localizing Receiver Troubles to One Section

We can use four indicators: the sound, raster, picture, and color in the picture for color receivers. Several examples are now analyzed to illustrate how the receiver itself indicates where the trouble is. These troubles are based on the block diagram in Fig. 7-2 for a monochrome receiver.

No raster, with normal sound. Since the sound is normal, the receiver has ac power input, and the low-voltage supply is operating. Assuming that the heater of the picture tube is lit, the trouble of no brightness is usually the result of no anode voltage from the high-voltage supply. Remember that the horizontal deflection circuits must be operating to produce flyback high voltage.

No picture and no sound, with normal raster. This trouble is in the signal circuits, before the sound takeoff point, because both the picture and sound are affected. The circuits common to the picture and sound signals are the rf section, IF amplifier, second detector, and AGC circuit.

It is useful to see if there is snow in the raster. *Snow* is receiver noise generated in the mixer stage. No picture with a snowy raster indicates the trouble is in the rf amplifier or antenna circuits, as the snow from the mixer stage is coming through.

No picture, with normal raster and normal sound. In the signal circuits, all the stages operating on the sound signal must be normal. The one section in Fig. 7-2 operating only on signal for the picture tube is the video amplifier. Therefore, the trouble must be in this stage.

No sound, with normal raster and normal picture. The trouble must be in the sound circuits, after the sound takeoff point, because only the sound is affected. This includes the 4.5-MHz sound takeoff circuit, the 4.5-MHz sound IF amplifier, the 4.5-MHz FM detector, the audio amplifier, and the loudspeaker.

Only a horizontal line on the screen. The horizontal deflection circuits are producing the horizontal line on the picture tube screen and high voltage for brightness. Only vertical deflection is missing. Therefore, the trouble must be in the vertical deflection section of the raster circuits, which includes the vertical deflection oscillator and the vertical output stage.

No raster and no sound. The screen is completely black, without illumination, and there is no sound. This trouble means the raster circuits and signal circuits are not operating. The defect is likely to be in the low-voltage power supply, since this section affects both the raster and the signal. A common trouble is an open heater in a series string.

7-12 Multiple Troubles

Usually only one defect occurs at a time, but the circuit arrangement can cause multiple effects. The most common examples are series heaters, multiple tubes, flyback high voltage, a common voltage supply, and the AGC circuit.

Series heaters. When the tube heaters are in a series string, an open in any one heater means that none of the tubes in the string can light. If the receiver has all tubes in one string, the receiver will be completely dead, without any raster, picture, or sound because all the tubes are cold, including the picture tube.

Multiple tubes. As an example, the pentode section of the 6AN8 may be a common IF amplifier, while the triode is used for the vertical oscillator. Then an open heater in this 6AN8 results in no vertical deflection and no picture or sound.

Common dc voltage lines. In many receivers, the picture IF amplifier tubes obtain plate supply voltage through the audio output stage. In effect, the IF section is in series with the audio amplifier for dc supply voltage. As a result, the audio output stage affects the IF amplifier. If the audio tube does not conduct, there will be no supply voltage for the IF section, resulting in no picture and no sound.

Another example of stages related for dc voltage often occurs in hybrid receivers. In these circuits, the transistor amplifiers can obtain the collector supply voltage of about 12 V from the control grid or cathode of the horizontal output tube.

AGC troubles. The AGC circuit affects both picture and sound by controlling the bias of the rf and picture IF amplifiers. When an AGC amplifier tube is used, the AGC trouble can produce a reversed picture, out of sync, with buzz in the sound, caused by overload distortion.

7-13 Safety Features

In all television receivers the ac line cord is disconnected when the back cover is removed. To operate the receiver with the back cover off, a "cheater cord" is used to fit the male socket on the chassis. The ac-dc type of receiver has a polarized line plug, where the large prong connects to the grounded side of the ac power line. This requires the polarized type of a cheater cord.

To check which side of the outlet is grounded, you can use a neon-bulb tester or ac voltmeter. From the low side of the outlet to any metal path to earth ground, the ac voltage should be zero, and the neon bulb does not light.

The capacitor of about 0.04 μF as an rf filter across the ac power line is a special nonshorting type. This capacitor should not be replaced with a conventional capacitor.

In addition to the ac interlock for the back cover, the B supply is usually opened when the yoke plug or convergence plug is disconnected.

When servicing a receiver that is not isolated from the power line, it is important to use an isolation transformer. This safety feature prevents a possible short circuit when using line-connected test equipment on the receiver.

The fuses in a television receiver are designed to prevent fire hazard. Do not use larger values and never jump a fuse. Fusible resistors are designed to open with excessive current. These should not be replaced by a conventional resistor. When you are changing wire-fuse links, the replacement must be the exact gage number and length or else the wire fuse will not have the same current rating. A wire-fuse link is often in a sleeve of insulating tubing to catch any drippings, should the fuse melt with an overload. For wired-in fuses, do not mount clips on top for a replacement fuse, as the added weight can move the connections to cause a short circuit.

The high-voltage rectifier for anode voltage is usually in a metal cage as protection against shock hazard and x-radiation. The cover should always be replaced after working in the cage. In some receivers the high-voltage connection to the top cap of the rectifier is open unless the shield cover is in place.

All the metal shields or fish-paper insulators should be in place. The metal shields reduce radiation of signal frequencies between circuits in the receiver, to prevent interference in the picture. Besides insulation, the fish-paper separators help reduce x-radiation.

The lead dress should be kept the same for several reasons. High-voltage leads must be placed to prevent arcing. Leads with high-frequency signals can produce feedback or crosstalk that causes interference in the picture. In some cases the leads to the picture tube are dressed with specific spacings to serve as spark-gap protection. Always make sure that no wire is touching a hot component, such as a power tube, where the lead may become hot enough to burn.

HEW* requires that the receiver not be able to produce a viewable picture when the anode voltage for the picture tube may produce x-radiation. In some receivers, the horizontal oscillator is disabled, resulting in no brightness, when the high voltage exceeds the limiting value of approximately 25 kV. Or, the horizontal synchronization may be removed to produce a picture that is torn apart in diagonal segments.

For ac-dc receivers without a power transformer, the following procedures are recommended to test for leakage current that can cause a shock hazard at exposed metal parts of the receiver.

Leakage current cold check. Disconnect plug from outlet and place a jumper across the two prongs. Turn the receiver switch on. Use an ohmmeter to check from the shorted plug to exposed metal parts such as the antenna handle, control shafts, and metal overlays and mounting screws. Any exposed metal parts that have a return path to chassis should read 2.2 to 3.3 MΩ. Those parts without a return path to

*U.S. Department of Health, Education, and Welfare.

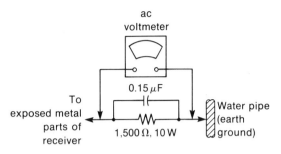

FIGURE 7-32 CONNECTIONS FOR TESTING LEAKAGE CURRENT FROM AC LINE ON RECEIVER WITHOUT A POWER TRANSFORMER.

chassis should read infinite ohms for an open circuit.

Leakage current hot check. Plug line cord into ac outlet and turn receiver on. Connect a 1,500-Ω resistor and 0.15-μF capacitor in parallel, as shown in Fig. 7-32. Using long clip leads, put this combination in series between earth ground and exposed metal parts of the receiver. Measure the ac voltage across the RC combination. Do this with the ac plug connected in both polarities, if the plug is not polarized. The ac voltage measured across R and C must not exceed 0.35 V, rms value. Use a portable voltmeter, not plugged into the power line.

SUMMARY

The functions for all stages in the block diagram of Fig. 7-2 are summarized in Table 7-5. The signal stages are listed first, followed by the raster circuits and power supplies. The double lines across the table indicate separate sections. Note the following abbreviations: P for picture signal, S for sound signal, and IC for integrated circuit. Although 6-V tubes are listed, different heater ratings are used for series strings. The color section is not included, as these circuits are explained in Chap. 9.

TABLE 7-5 FUNCTIONS OF STAGES IN THE TELEVISION RECEIVER

STAGE	INPUT SIGNAL	OUTPUT SIGNAL	COMMON TUBES	NOTES
VHF rf amplifier	Rf P and S from antenna	Amplified rf P and S for VHF mixer	6CW4 and 6DV4 nuvistor triodes, 6CY5, transistor	Low noise for no snow
VHF rf local oscillator	None	Beats with rf P and S in VHF mixer	6EA8 and 6KZ8 triode-pentodes, transistor	Combined with VHF mixer; tunes 45.75 MHz above P
VHF mixer	Rf P and S plus VHF oscillator output; or IF P and S from UHF mixer	IF P and S to common IF amplifier	6EA8 and 6KZ8 triode-pentodes, transistor	Converter or first detector on VHF; IF amplifier on UHF
UHF rf local oscillator	None	Beats with rf P and S in UHF mixer	Transistor	Tunes 45.75 MHz above P
UHF mixer	Rf P and S from antenna for UHF, plus UHF oscillator output	45.75-MHz P and 41.25-MHz S to VHF tuner as IF amplifier	Crystal diode for low noise	Feeds VHF tuner; may have UHF rf amplifier
Picture IF or video IF section	IF P and S from VHF mixer	Amplified IF P and S for video detector	6BZ6, 6EH7, 6JH6 pentodes, transistors, or IC	Two to four stages tuned for 45.75-MHz P and 41.25-MHz S
Video detector	Amplified IF P and S from IF amplifier	Composite video signal, 3.58-MHz C, 4.5-MHz S	Crystal diode	Picture detector or second detector
Sound IF amplifier	4.5-MHz second sound IF signal	Amplified 4.5-MHz S for FM detector	6EH7 or IC	Always tuned to 4.5 MHz
FM sound detector	Amplified 4.5-MHz intercarrier S	Audio signal to audio amplifier	6DT6, crystal diodes, or IC	Quadrature-grid tube or ratio detector with two diodes
Audio section	Audio signal from FM detector	Audio power output for loudspeaker	6BQ5, transistors	One or two stages
Video amplifier	Composite video signal from video detector	Amplified video signal for picture tube	6CL6, transistors	Also Y video amplifier in color receivers

TABLE 7-5 FUNCTIONS OF STAGES IN THE TELEVISION RECEIVER (Continued)

STAGE	INPUT SIGNAL	OUTPUT SIGNAL	COMMON TUBES	NOTES
Sync separator	Composite video from video amplifier	H and V sync for deflection circuits	6BY6, transistor	Clips and amplifies deflection sync pulses
Vertical oscillator	60-Hz V sync pulses	60-Hz deflection voltage to drive V amplifier	6DR7 double-triode, or transistors	Output with or without sync input
Vertical amplifier	60-Hz deflection voltage from V oscillator	60-Hz sawtooth current in V coils of yoke	6DR7 double-triode, or transistors	Often combined with V oscillator
Horizontal AFC	15,750-Hz H sync pulses	Dc control voltage for H oscillator	Two diodes as phase detector	Holds H scanning frequency
Horizontal oscillator	Dc control voltage from AFC circuit	15,750-Hz deflection voltage to drive H amplifier	6CG7, 6GH8, or transistors	Output with or without dc control voltage
Horizontal amplifier	15,750-Hz deflection voltage from H oscillator	15,750-Hz sawtooth current in H coils of deflection yoke	6DQ6, 6JE6, power transistor, or SCR	Also supplies ac input to HV rectifier and damper
Damper	H deflection voltage at 15,750 Hz	Rectified deflection voltage	6AU4, 6W4, 6BH3, and 6CT3 diodes, or silicon diode	Dc output added for boosted B+ voltage
High-voltage rectifier	H flyback pulses at 15,750 Hz stepped up to 15 to 25 kV.	Dc high voltage for anode of picture tube	1B3, 1K3/1J3, 1X2, 2BJ2, 3A3, and 3CA3 tubes, or silicon diode	Needs H output for operation
Low-voltage supply	Ac power from 120-V main line	Plate voltage and heater power for tubes; collector or drain voltage for transistors	Silicon diodes	May be ac supply or transformerless type; heaters may be parallel or series

Self-Examination (Answers at back of book)

Part A Match the functions listed at the left with the circuits at the right.

1. Anode voltage for CRT
2. Height of raster
3. Width of raster
4. Contrast of picture
5. Brightness of picture

(a) Horizontal amplifier
(b) Video amplifier
(c) Dc bias on picture tube
(d) High-voltage rectifier
(e) Vertical amplifier

Part B Match the troubles listed at the left with the circuits at the right. Assume single troubles without multiple effects, for the receiver in Fig. 7-2.

1. No picture, no sound, no raster
2. One bright line across center of screen
3. Snowy picture, with good antenna signal
4. No sound, but picture is normal
5. No brightness, but sound is normal
6. Picture in diagonal bars, out of sync
7. No picture, but raster and sound are normal
8. No picture, no sound, but raster is normal

(a) Vertical oscillator
(b) Rf amplifier
(c) 4.5-MHz IF amplifier
(d) Common IF amplifier
(e) Horizontal AFC
(f) Low-voltage rectifier
(g) Horizontal amplifier
(h) Video amplifier

Part C Match the controls at the left with the functions at the right.

1. Fine tuning
2. Volume
3. Contrast
4. Brightness
5. Station selector

(a) Tunes rf, oscillator, and mixer stages
(b) Varies frequency of oscillator in tuner
(c) Gain or level for audio signal
(d) Gain or level for video signal
(e) Dc bias for picture tube

Part D For semiconductor devices, answer true or false for the following statements.

1. Silicon NPN transistors are generally used for rf and IF amplifiers.
2. Any silicon junction transistor has a forward bias of approximately 0.6 V as a class A amplifier.

3. The collector always has reverse voltage.
4. An NPN transistor takes negative collector voltage for reverse bias.
5. A junction transistor with zero forward bias will not conduct collector current.
6. The common-collector (CC) circuit is similar to a cathode-follower.
7. The field-effect transistor has very low input resistance, compared with junction transistors.
8. Zener diodes are used for voltage regulation.
9. A varactor is a capacitive diode.
10. The drain electrode in the FET corresponds to the plate in a vacuum tube.

Essay Questions

1. Define the following: flyback high voltage, intercarrier sound, monochrome signal, and chrominance signal.
2. State the use for each of the following frequencies: 67.25, 71.75, 45.75, 42.17, 41.25, 4.5, and 3.58 MHz.
3. Give the function of the (a) antenna, (b) transmission line.
4. State the channel numbers tuned in by the (a) VHF tuner, (b) UHF tuner.
5. Give three functions of the composite video signal.
6. What is the advantage of intercarrier sound, compared with split sound?
7. In Fig. 7-2, why is the sync section included in the signal circuits and not in the raster circuits?
8. Using all the stages listed in Table 7-5, classify the stages under the following headings: picture and sound, picture alone, sound alone, synchronization, illuminated raster.
9. List the circuits or stages for the following controls: station selector, fine tuning, contrast, brightness, video peaking, V hold, H hold, AGC level, height, V linearity, width, volume, tone, and buzz.
10. Where would you connect an oscilloscope to see the voltage waveshapes in Fig. 7-33a, b, and c?
11. Give two features of printed-circuit boards.
12. List three safety features of television receivers.
13. Name five types of semiconductor devices, and give their schematic symbols.
14. Define the following: NPN transistor, FET, JFET, IGFET, SCR, triac, zener diode, and varactor.
15. (a) Name the three electrodes in PNP or NPN transistors corresponding to cathode, grid, and plate in a triode tube. (b) Do the same as in a for a field-effect transistor.
16. Draw the circuit of a common-source amplifier with a FET, corresponding to the CE amplifier in Fig. 7-16.

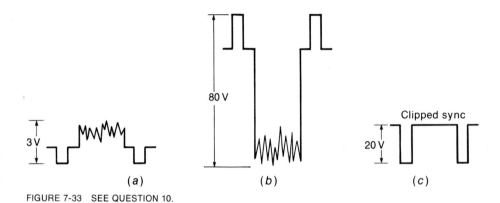

FIGURE 7-33 SEE QUESTION 10.

Problems

1. Calculate the oscillator frequencies for tuning in (a) channel 6, (b) channel 7, (c) channel 13, (d) channel 14, (e) channel 83.
2. Calculate the difference frequency in each of the following pairs of frequencies: (a) 83.25 and 87.75 MHz; (b) 175.25 and 179.75 MHz; (c) 45.75 and 41.25 MHz; (d) 83.25 and 86.83 MHz; (e) 45.75 and 42.17 MHz.
3. Draw a block diagram of the monochrome receiver in Fig. 28-14 on a foldout page in Chap. 28, without the rf tuner unit. Give tube type numbers for each stage.

8 Color Television

A color picture is actually a monochrome picture on a white raster but with colors added for the main parts of the scene. The required color information is in the 3.58-MHz chrominance (C) signal broadcast with the monochrome signal in the standard 6-MHz television broadcast channel. To illustrate this idea of the color in a separate signal, you can turn down the color control at the receiver to eliminate the color signal, and the result is a black-and-white picture. With the C signal, however, the picture is reproduced in natural colors. Practically all colors can be produced as combinations of red, green, and blue. A typical picture reproduced on the screen of a tricolor picture tube is shown in color plate I. More details are explained in the following topics:

8-1 Red, green, and blue video signals
8-2 Color addition
8-3 Definitions of color television terms
8-4 Encoding the 3.58-MHz C signal at the transmitter
8-5 Decoding the 3.58-MHz C signal at the receiver
8-6 The Y signal for luminance
8-7 Types of color video signals
8-8 The color sync burst
8-9 Hue phase angles
8-10 The colorplexed composite video signal
8-11 Desaturated colors with white
8-12 Color resolution and bandwidth
8-13 Color subcarrier frequency
8-14 Color television systems

8-1 Red, Green, and Blue Video Signals

The color television system begins and ends with red, green, and blue color video signals corresponding to the color information in the scene. It is interesting to consider that, in television, voltages correspond to colors. At the transmitter, light of different colors is converted to different video signal voltages. The picture tube in the receiver converts the color video voltages to their respective colors.

A color camera has different camera tubes for red, green, and blue. The screen of the color picture tube has red, green, and blue phosphors to reproduce these colors from the corresponding video signals. Furthermore, video signal voltages can be combined to provide the same effect as mixing colors. As a result, red, green, and blue with all their color combinations, including white, are shown on the screen of the color picture tube.

Color voltages. When the image is scanned in the color camera, separate camera tubes are used for each color, as illustrated in Fig. 8-1. The red, green, and blue in the scene are separated for the camera tubes by optical color filters. As a result, the output from camera tube 1 is red (R) video signal with information for only the red parts of the scene. Similarly, tubes 2 and 3 produce green (G) video signal and blue (B) video signal.

In Fig. 8-2, the picture tube has three electron guns for the red, green, and blue phosphor dots on the screen. Each gun has the usual function of producing a beam of electrons, but the beam excites only one color. The reason is that the shadow mask has tiny holes aligned with the dot trios. When the beams converge at the proper angles, the electrons pass through the mask and excite the color dots. The red gun

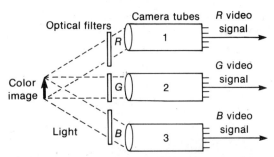

FIGURE 8-1 TELEVISING A SCENE TO OBTAIN R, G, AND B VIDEO SIGNALS FOR RED, BLUE, AND GREEN COLORS IN THE SCENE.

produces a red raster and picture on the screen; the green gun and blue gun do the same for their colors. If only one gun is working, you see just one color. With all three guns operating, the screen reproduces red, green, and blue and their color mixtures. In fact, the white raster is actually a combination of red, green, and blue. Color dots are shown here, but the screen can have vertical stripes of red, green, and blue phosphors.

Encoding and decoding. If we had closed-circuit television, the red, green, and blue video signals would be the only information needed to reproduce the picture. For broadcasting, however, these color signals are not compatible with the monochrome television system for black-and-white receivers. Therefore, the color video signals are encoded by combining them in specific proportions, to provide the same video information in a different form. The end result of the encoding is the formation of two separate signals: the C chrominance or chroma signal for color and the Y luminance or brightness signal for black-and-white information.

At the receiver, the color picture tube still needs R, G, and B video signals, corresponding to the color phosphors on the screen. However,

I NORMAL COLOR PICTURE. (*HEATH COMPANY*)

II COLOR FRINGING CAUSED BY POOR CONVERGENCE.

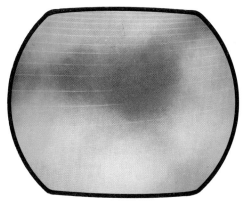
III INCORRECT ADJUSTMENT OF PURITY MAGNET. (*HEATH COMPANY*)

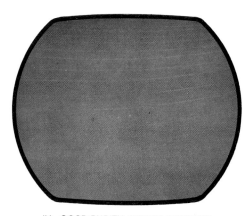
IV GOOD PURITY. (*HEATH COMPANY*)

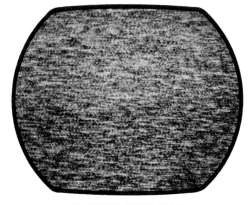
V COLOR SNOW. (*RCA*)

VI NO COLOR SYNC. (*RCA*)

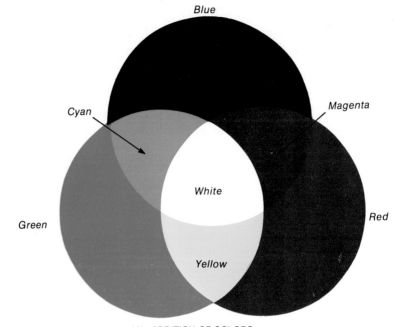

VII ADDITION OF COLORS.

(a)

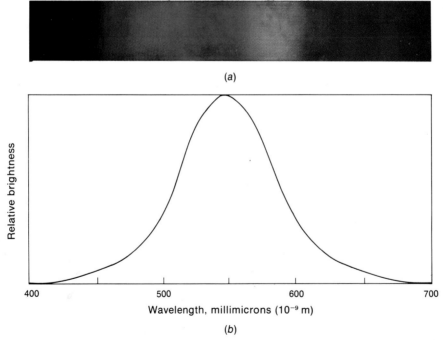

(b)

VIII RELATIVE BRIGHTNESS RESPONSE OF THE EYE. (a) HUES OF DIFFERENT WAVELENGTHS. (b) LUMINANCE OF SIGNAL-VOLTAGE RESPONSE.

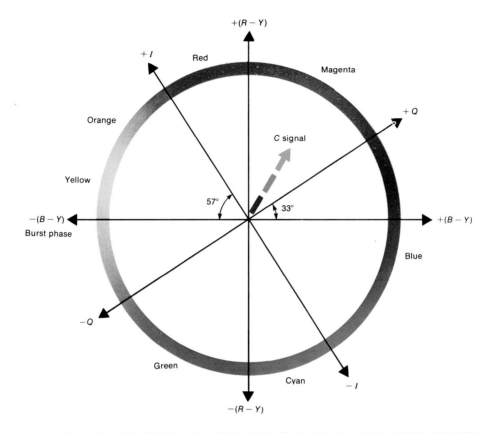

IX COLOR CIRCLE DIAGRAM, SHOWING DIFFERENT HUES FOR PHASE ANGLES OF CHROMINANCE SIGNAL.

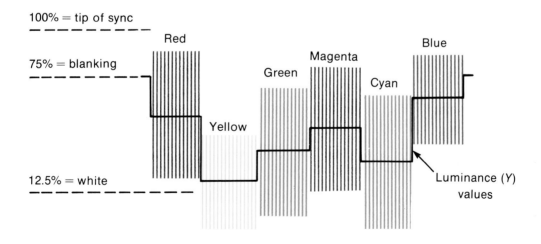

X AMPLITUDE VALUES OF CHROMINANCE AND LUMINANCE SIGNALS FOR COLOR BARS. (*RCA*)

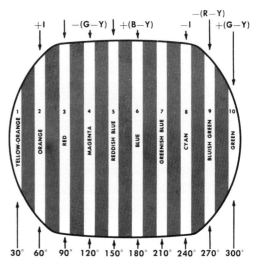

XI COLOR BAR PATTERN CORRESPONDING TO HUE PHASE ANGLES IN PLATE XII.

XII PHASE ANGLES FOR COLOR BAR PATTERN IN PLATE XI. THESE ANGLES ARE CLOCKWISE FROM BURST PHASE.

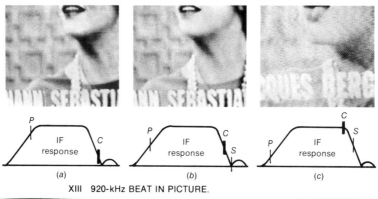

XIII 920-kHz BEAT IN PICTURE.

XIV SILVERY EFFECT IN HIGHLIGHTS OF PICTURE BECAUSE OF LOW EMISSION. ILLUSTRATED FOR RED GUN ALONE. (*RCA*)

XV COLOR HUM BARS. (*RCA*)

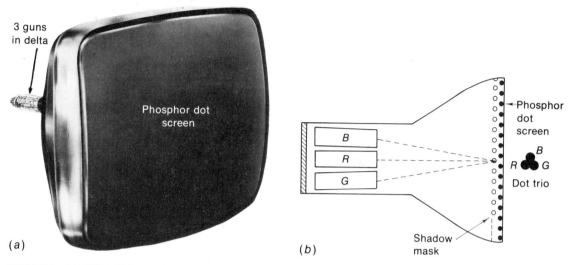

FIGURE 8-2 (a) PHOTO OF THREE-GUN TRICOLOR PICTURE TUBE, TYPE NUMBER 25AB22. SCREEN DIAGONAL IS 25 IN.; COLOR PHOSPHOR NUMBER IS P22. (b) INTERNAL CONSTRUCTION WITH THREE ELECTRON GUNS, SHADOW MASK, AND DOT TRIOS OF RED, GREEN, AND BLUE PHOSPORS.

the C signal is decoded by demodulation. This detected output is then combined with the luminance signal to recover the original red, green, and blue video signals for the color picture tube. More details of the R, G, and B video signals are illustrated in Figs. 8-3 to 8-5.

Color video amplitudes. In Fig. 8-3, the separate R, G, and B video signals are shown for a horizontal line scanned across the image with vertical red, green, and blue bars. The bars just represent picture information of their particular color. The R signal voltage has its full amplitude while the red bar is scanned. However, there is no R video signal for the green or blue information. Similarly, the G video voltage is produced only when green picture information is scanned, and the B video voltage indicates blue information.

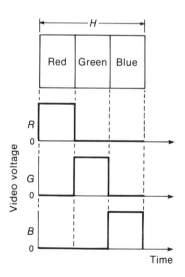

FIGURE 8-3 R, G, AND B VIDEO SIGNALS FOR RED, GREEN, AND BLUE COLOR BARS. H INDICATES ONE HORIZONTAL SCANNING LINE FOR WIDTH OF PICTURE.

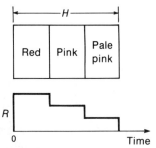

FIGURE 8-4 DECREASING AMPLITUDES OF R VIDEO SIGNAL FOR RED, PINK, AND PALE PINK BARS, INDICATING WEAKER COLORS.

Different values of color voltage are illustrated in Fig. 8-4. Here the red, pink, and pale pink bars have decreasing values of color intensity. Therefore, the corresponding video voltages have decreasing amplitudes. We can say, then, that R, G, or B video voltage indicates information of that color, while the relative amplitude depends on the color intensity.

Color video frequencies. In Fig. 8-5, all the color bars are red, but the bars just became narrower. This case is just a question of less scanning time across smaller details of picture information. The result is higher video frequencies. For either chrominance or luminance information, the high-frequency components of the video signal determine the amount of horizontal detail that can be reproduced in the picture.

8-2 Color Addition

Almost any color can be produced by adding red, green, and blue in different proportions. The additive effect is obtained by superimposing the individual colors. In a tricolor picture tube, the color addition results as the red, green, and blue information on the screen is integrated by the eye to provide the color mixtures in the actual scene.

Additive color mixtures. The idea of adding colors is shown in color plate VII. There are three circles in red, green, and blue, which partially overlap. Where the circles are superimposed, the color shown is the mixture produced by adding the primary colors. At the center, all three color circles overlap, resulting in white.

Where only green and blue add, the resultant color is a greenish-blue mixture called *cyan*. The red-purple color shown by the addition of red and blue is called *magenta* or violet. More blue with less red produces purple. Yellow is an additive color mixture of approximately the same amount of red and green. More red with less green produces orange. Similarly, practically all natural colors can be produced as additive mixtures of red, green, and blue.

Primary colors. These are combined to form different color mixtures. The only requirement is that no primary can be matched by any mixture of the other primaries. Red, green, and blue are the primary colors used in television because they produce a wide range of color mixtures when added. Therefore, red, green, and blue are *additive primaries*.

Complementary colors. The color that produces white light when added to a primary is its

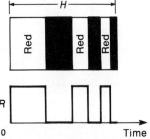

FIGURE 8-5 INCREASING FREQUENCIES OF R VIDEO SIGNAL FOR NARROWER RED BARS, INDICATING SMALLER DETAILS OF COLOR INFORMATION.

complement. For instance, yellow added to blue produces white light. Therefore, yellow is the complement of the blue primary. The fact that yellow plus blue equals white follows from the fact that yellow is a mixture of red and green. Therefore, the combination of yellow and blue actually includes all three additive primaries. Similarly, magenta is the complement of green, while cyan is the complement of red. Sometimes the complementary colors cyan, magenta, and yellow are indicated as minus-red, minus-green, and minus-blue, respectively. The reason is that each can be produced as white light minus the corresponding primary.

The complementary colors are also known as the *subtractive primaries*. In a reproduction process such as color photography, color mixtures are obtained by subtracting individual colors from white light with color filters. Then cyan, magenta, and yellow are the subtractive primary colors used to filter out red, green, and blue.

A primary and its complement can be considered opposite colors. The reason is that for any primary its complement contains the other two primaries. This idea is illustrated by the color circle in Fig. 8-6, with the dashed lines between each primary and its opposite complementary color.

The hue of the complementary colors can be seen in color plate IX. Cyan is a greenish blue and magenta is a purplish red. When considered as primary colors in a subtractive system, these colors are often labeled simply blue, red, and yellow. However, this blue is really cyan with green and blue; the red is magenta combining blue with red; the yellow combines green and blue.

Adding for colors. It can be useful to consider how mixtures of red, green, and blue video signal voltages are the same as combinations of the

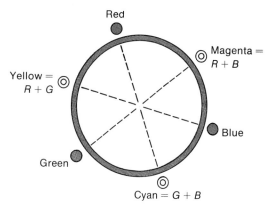

FIGURE 8-6 COLOR WHEEL SHOWING PRIMARY COLORS RED, GREEN, AND BLUE WITH THEIR OPPOSITE COMPLEMENTARY COLORS CYAN, MAGENTA, AND YELLOW.

primary colors and their complements. The reason for the similarity of the voltages and colors is the fact that a tricolor picture tube really operates as a color mixer. What you see on the screen is the superimposed combination of red, green, and blue. This effect is most obvious when looking at a raster alone, without any picture. If the blue gun is cut off by reducing either the bias or screen-grid voltage, the electron beams of the red and green guns can produce a yellow raster. If there is more red and less green, you will see the raster color become orange. Similarly, the blue and green guns operating without the red gun produce cyan; or the red and blue guns produce magenta. With all three guns operating to reproduce red, green, and blue in the correct proportions, the raster is white. For the opposite case, all three guns are cut off to reproduce black, which is just the absence of light.

Complementary voltage polarities. Another important feature of color video voltages is the fact that opposite polarities correspond to complementary colors. To illustrate this idea, assume the blue video voltage has the polarity that increases beam current of the blue gun. More blue video voltage produces more beam current from the blue gun to reproduce more blue in the raster and picture. Although all three guns are operating, we are looking only at the effect of blue for this example. It should be noted now that the opposite polarity of blue video voltage decreases the beam current. Then less beam current from the blue gun reduces the amount of blue on the screen. The effect is the same as increasing the red and green, which is a yellow combination. Furthermore, yellow is the complement of blue. The result, then, is that reversing the polarity of a color video voltage is a change to its complementary color.

8-3 Definitions of Color Television Terms

Any color has three characteristics needed to specify the visual information. First is its hue or tint, which is what we generally call the color; second is its saturation; and third is the luminance. Saturation indicates how concentrated, vivid, or intense the color is. Luminance indicates the brightness, or what shade of gray the color would be in black-and-white. These qualities of colors and other important terms are defined now in order to analyze the special features of color television broadcasting.

White. Actually, white light can be considered as a mixture of the red, green, and blue primary colors in the proper proportions. A glass prism can produce the colors of the rainbow from white light. For the opposite effect, red, green, and blue can be added to produce white. The reference white for color television is a mixture of 30 percent red, 59 percent green, and 11 percent blue.

These percentages for the luminance signal are based on the sensitivity of the eye to different colors. As shown in color plate VIII, green has the greatest relative brightness.

The degree of white is generally specified as a color temperature of 6,500 K. This is a bluish white, like daylight. The symbol "K" indicates degrees Kelvin on the absolute temperature scale, which is 273° below °C.

Hue. The color itself is its hue or tint. Green leaves have a green hue; a red apple has a red hue; the color of any object is distinguished primarily by its hue. Different hues result from different wavelengths of the light producing the visual sensation in the eye.

Saturation. Saturated colors are vivid, intense, deep, or strong. Pale or weak colors have little saturation. The saturation indicates how little the color is diluted by white. As an example, vivid red is fully saturated. When the red is diluted by white, the result is pink, which is really a desaturated red. It is important to note that a fully saturated color has no white. Then the color has only its own hue, without any other components that could be added by the red, green, and blue of white.

Chrominance. This term is used to indicate both hue and saturation of a color. In color television, the 3.58-MHz color signal more specifically is the chrominance signal. In short, the chrominance includes all the color information, without the brightness. The chrominance and brightness together specify the picture information completely. Chrominance is also called *chroma*.

In color television, we can reserve the term "chrominance" or "chroma" for the 3.58-MHz modulated subcarrier. This C signal contains the hue and saturation for all the colors. Its frequency is 3.58 MHz. However, before modulation and after demodulation, the color information is in the red, green, and blue color video signals. The range of these modulation frequencies, or the baseband for color, can be considered practically as 0 to 0.5 MHz.

Luminance. Luminance indicates the amount of light intensity, which is perceived by the eye as brightness. In a black-and-white picture, the lighter parts have more luminance than the dark areas. Different colors also have shades of luminance, however, as some colors appear brighter than others. This idea is illustrated by the relative luminosity curve in color plate VIII. The curve shows that the green hues between cyan and orange have maximum brightness.

The luminance really indicates how the color will look in a black-and-white reproduction. Consider a scene being either photographed on black-and-white film or televised in monochrome. The picture includes a colorful costume with a dark red skirt, yellow blouse, and a light blue hat. For the same illumination, these different hues will have different brightness values and will therefore be reproduced in different shades of monochrome. As shown by the graph in color plate VIII for relative brightness values of different hues, dark red has low brightness, yellow has high brightness, and blue is medium in relative brightness. Therefore, the monochrome picture reproduction will show a white blouse (for yellow) with a black shirt (for dark red) and a gray hat (for light blue). It is the relative brightness variations for different hues that make it possible to reproduce scenes that are naturally in color as similar pictures in black and white.

In color television, the luminance information is in the luminance or Y signal. This abbreviation should not be confused with yellow, as the luminance signal contains the brightness variations for all the information in the picture.

Compatibility. Color television is compatible with black-and-white television because essentially the same scanning standards are used and the luminance signal enables a monochrome receiver to reproduce in black and white a picture televised in color. In addition, color television receivers can use a monochrome signal to reproduce the picture in black and white. Color television broadcasting uses the same 6-MHz broadcast channels as in monochrome transmission. Also, the same picture carrier frequency is used.

Subcarrier. This is a carrier with modulation that modulates another carrier wave of higher frequency. In color television, the color information modulates the 3.58-MHz color subcarrier, which modulates the main picture carrier in the standard broadcast channel.

Multiplexing. This is the name for the technique of using one carrier for two separate signals. In color television, the 3.58-MHz C signal is multiplexed with the Y signal as both modulate the main picture carrier wave. Another example of multiplexing is stereo broadcasting in the commercial FM broadcast band, to transmit left and right audio signals on one rf carrier.

8-4 Encoding the 3.58-MHz C Signal at the Transmitter

Now we can consider more details of how the chrominance signal is produced for transmission to the receiver. First, the R, G, and B video voltages provide the color information. Then these primary signals are encoded to form separate chrominance and luminance signals. The luminance signal is formed directly from the R, G, and B video signals. However, the chrominance signal is formed from two color-mixture video signals, instead of from the three primary signals. The purpose is to have two quadrature signals 90° out of phase with each other. Furthermore, these two quadrature signals have all the chrominance information because the color mixtures include red, green, and blue.

The primary color video signals. The camera receives red, green, and blue light corresponding to the color information in the scene, to produce the primary color video signals in Fig. 8-7. These waveforms illustrate the voltages obtained in scanning one horizontal line across the color bars indicated. The red camera tube produces full output for red, but no output for green and blue. Similarly, the green or blue camera tube has output only for that color.

Notice the values shown for yellow, as an example of a complementary color. Since yellow includes red and green, video voltage is produced for both these primary colors, as the red and green camera tubes have light input through their color filters. However, there is no blue in yellow. This is why the blue video voltage is at zero for the yellow bar.

The last bar at the right is white, as an example that includes all three primaries. Then all three camera tubes have light input, and there are color video signals for red, green, and blue.

The red, green, and blue video voltages are then combined in order to encode the primary color voltages as brightness and chrominance signals for transmission to the receiver. This process is illustrated by the block diagram in Fig. 8-8.

Matrix section. A matrix circuit forms new output voltages from the signal input. The matrix at the transmitter combines the R, G, and B voltages in specific proportions to form three video signals that are better for broadcasting. One signal has the brightness information. The other two signals have the chrominance information. Specifically, these three signals from the matrix are as follows:

1. Luminance signal, designated the Y signal. This combination of R, G, and B contains the brightness variations, corresponding to a monochrome video signal. The Y signal is formed by taking 30 percent of R video, 59 percent of G video, and 11 percent of B video.
2. A color mixture, designated the I signal. Positive polarity of I signal is orange; negative polarity is cyan. These colors result from the fact that $I = 0.60R - 0.28R - 32B$.
3. A color mixture, designated the Q signal. Positive Q signal is purple; negative polarity is yellow-green. These colors result from the fact that $Q = 0.21R - 0.52G + 0.31B$.

The letter "Q" here is for *quadrature*, as the Q signal modulates the color subcarrier 90° out of phase with the I signal. The letter "I" indicates *in phase*. Note that the negative signs for subtracting R, G, or B signals mean adding

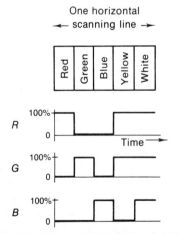

FIGURE 8-7 R, G, AND B VIDEO SIGNAL VOLTAGES FOR COLOR-BAR PATTERN.

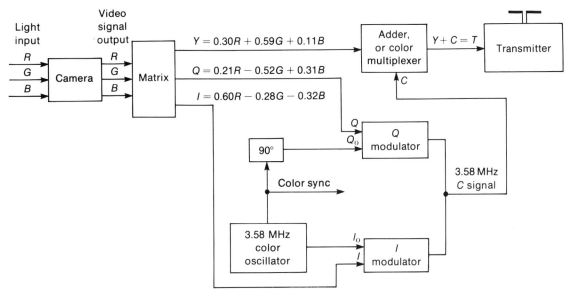

FIGURE 8-8 FUNCTIONS OF THE TRANSMITTER IN ENCODING THE PICTURE INFORMATION FOR COLOR TELEVISION BROADCASTING. SEE TEXT FOR DETAILS.

the video voltages in negative polarity. The orange and cyan of the I signal are chosen as the best colors for very small details. Automatically, then, the Q signal colors are magenta and yellow-green because this color axis is 90° off the I color axis.

The I and Q signals are specified by the FCC for modulation at the transmitter. The I signal has more bandwidth, equal to 1.5 MHz, compared with 0.5 MHz for the Q signal. The purpose of extra bandwidth in the I signal is to allow more color detail at the receiver. However, receivers practically never use the added bandwidth of the I signal, because the circuits are much simpler when all the color video signals have the same bandwidth of 0.5 MHz.

Specifically, the receivers generally use $R-Y$ and $B-Y$ demodulators. These two color video signals are also in quadrature with each other, but with a different phase angle than the I and Q signals. Actually, the receiver can detect the hue information of the C signal in different ways, as explained later with the hue-phase diagram in Fig. 8-16. Eventually, though, for any method of detection, the color information at the receiver must be converted to red, green, and blue video for the color picture tube.

Color subcarrier. The I and Q signals are transmitted to the receiver as the modulation side bands of a 3.58-MHz subcarrier, which in turn, modulates the main carrier wave. As an example, the picture carrier wave at 67.25 MHz for channel 4 can be modulated by the 3.58-MHz video-frequency color subcarrier. Note that the color subcarrier has the same video frequency of 3.58 MHz for all stations, although the assigned picture frequency is different for each channel. The value of 3.58 MHz is chosen as a high video frequency to separate the chrominance signal from low video frequencies in the luminance signal.

Chrominance modulation. Referring to Fig. 8-8, the output from the 3.58-MHz color subcarrier oscillator is coupled to the I and Q modulators. They also have I and Q video signal input from the matrix. Each circuit produces amplitude modulation of the 3.58-MHz subcarrier. Notice the separate inputs for I and Q but the common output combines the I and Q modulation. This combined output is the 3.58-MHz modulated chrominance (C) signal.

The 3.58-MHz oscillator input to the Q modulator, labeled Q_0, is shifted by 90°, with respect to I_0. Modulating the subcarrier in two different phases keeps the I and Q signals separate from each other. The 90° angle provides maximum separation in phase between two signals.

Suppressing the subcarrier. Using only the modulation side bands, without the carrier itself, is called *suppressed-carrier transmission*. The purpose of suppressing the subcarrier is to reduce interference at 3.58 MHz, which can produce a fine dot pattern on the screen.

Color sync burst. With suppressed-carrier transmission, the receiver must have a 3.58-MHz oscillator circuit that generates the subcarrier, in order to detect the chrominance signal. Furthermore, a sample of the 3.58-MHz subcarrier is transmitted with the C signal as a phase reference for the color oscillator at the receiver. In color television, phase angle is hue.

The color synchronization for correct hues in the picture is accomplished by a burst of 8 to 11 cycles of the 3.58-MHz subcarrier on the back porch of each horizontal blanking pulse. This color sync burst controls the frequency and phase of the 3.58-MHz oscillator at the receiver.

Total colorplexed video signal. The C signal with the color information and the Y luminance signal are both coupled to the adder section or colorplexer. This stage combines the Y signal with the 3.58-MHz C signal to form the total colorplexed video signal marked T in Fig. 8-8.

This signal is transmitted to the receiver by amplitude modulation of the picture carrier wave in the station's assigned 6-MHz channel. The T modulation is a composite video signal, including deflection sync and blanking pulses. Negative polarity of transmission, 4.5-MHz separation from the sound carrier, and vestigial side bands are standard for the modulated picture carrier wave, as in monochrome broadcasting.

The oscilloscope waveshape for colorplexed video signal is shown in Fig. 8-9. This waveform illustrates how the signal includes all the information needed for the picture. The shaded areas are 3.58-MHz C signal, corresponding to color bars. Peak-to-peak amplitude

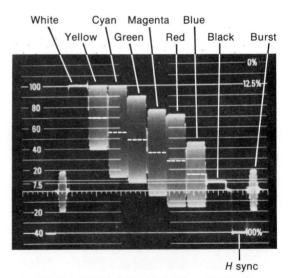

FIGURE 8-9 OSCILLOGRAM OF COLORPLEXED COMPOSITE VIDEO WITH C SIGNAL, COLOR SYNC BURST, Y SIGNAL, AND H DEFLECTION SYNC. THE WHITE DOTTED LINES SHOW LEVELS OF Y SIGNAL. (*TEKTRONIX INC.*)

of the C signal depends on the saturation or color intensity. The phase angles of the C signal for different hues cannot be seen because individual cycles are not shown.

In addition to the peak-to-peak amplitudes for the color bars, note that the average level is different for each bar. Specifically, the distance from the blanking level to the average level of the C signal is a measure of how dark or light the information is. These changing luminance levels are the variations in the Y luminance signal. If the 3.58-MHz C signal would be filtered out, the luminance levels would still remain to indicate the relative brightness values. In Fig. 8-9, the average luminance levels form a staircase of voltages from white at the left to black at the right.

Hue and saturation in the C signal. The two-phase modulation of the 3.58-MHz color subcarrier has the effect of concentrating all the color information into one chrominance signal. Consider the example of a strong I signal, with little Q signal. Then the resultant C signal has a phase angle close to the orange hue of I signal. For the opposite case, with a strong Q signal and little I signal, the modulated C signal has a phase angle close to the purple hue of Q signal. With equal amplitudes of I and Q modulating voltages, the phase of the C signal is between the I and Q phase angles, corresponding to a hue between orange and purple. The result, then, is that the instantaneous phase angle of the 3.58-MHz modulated C signal indicates the hue of the color information.

Furthermore, the amplitude variations of the modulated C signal indicate the strength or intensity of the color information. This variation corresponds to how saturated the color is. As a result, the hue is in the phase angle of the C signal, while its amplitude determines the saturation.

8-5 Decoding the 3.58-MHz C Signal at the Receiver

Most receivers decode the C signal into $B - Y$ and $R - Y$ color video signals. The $B - Y$ video is a color mixture that is close to blue, as it consists of blue video and $-Y$ video. $B - Y$ hue has the phase angle in the C signal exactly 180° opposite from the phase of the color sync burst. The phase in quadrature is $R - Y$ video. This hue is close to red.

Furthermore, $B - Y$ video and $R - Y$ video can be combined to supply $G - Y$ video, as the Y signal has green in it. The bandwidth of all three of these color video signals is 0 to 0.5 MHz.

Separating the C signal. Starting with the receiving antenna, the modulated picture carrier of the selected channel is amplified in the rf and IF stages to be rectified in the video detector. The detector output is the total colorplexed video signal, including the Y and C components. After the video detector, the video circuits divide into two paths, as shown in Fig. 8-10. One path is for the Y luminance signal, while the other is for the 3.58-MHz C signal.

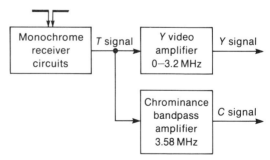

FIGURE 8-10 SEPARATING THE Y LUMINANCE SIGNAL AND THE 3.58-MHz MODULATED CHROMA SIGNAL AT THE RECEIVER.

The output of the Y video amplifier is Y signal without the 3.58-MHz C signal. The reason is that the amplifier response is limited to frequencies below 3.2 MHz, approximately. The output of the chrominance bandpass amplifier is C signal. The reason is that this stage is tuned to 3.58 MHz, with a bandpass of ±0.5 MHz, usually. In all color receivers, the chrominance or color amplifier is resonant at 3.58 MHz, for any channel. We can consider 3.58 MHz as the color intermediate frequency for the receiver.

The amplified 3.58-MHz C signal includes the chrominance modulation and color sync for the color section of the receiver. The original color information is in the modulated C signal, but it must be demodulated. The color sync controls the 3.58-MHz color oscillator used for the demodulator circuits.

Synchronous demodulation. When a modulated signal is transmitted without the carrier, or subcarrier, the original carrier wave must be reinserted at the receiver to detect the modulation. As shown in Fig. 8-11, the 3.58-MHz color oscillator supplies the subcarrier which is coupled to the demodulators with the C signal. Furthermore, it is important to note that this type of demodulator or detector has maximum output for the phase of the modulated signal that is the same as the oscillator input.

This circuit is a synchronous demodulator because it detects the modulation information that is synchronous with the reinserted carrier. There is practically no output for signal in quadrature with the oscillator voltage. For this reason, two demodulators are needed to detect two color signals. Also, the two demodulators are generally 90° out of phase, for minimum interference between the two detected signals.

In Fig. 8-11 the demodulators produce $B - Y$ and $R - Y$ video. Then these two color mixtures are combined to provide $G - Y$. This addition is possible because green is in the Y component.

The picture tube as a matrix. When the receiver uses $B - Y$, $R - Y$, and $G - Y$ video signals, the picture tube can serve as the matrix to produce the primary colors. As shown in Fig. 8-12, the Y signal is coupled to the cathode for all three guns. Actually $-Y$ signal is used because this polarity at the cathode cancels the $-Y$ component of the color voltages at each control grid. The results are the same as adding the components algebraically. For red (R), as an example,

$$(R - Y) - (-Y) = R - Y + Y = R$$

The same results for the three primary colors mean that the screen reproduces the color information as the required mixtures of red, green, and blue. As a summary of the different signals, Table 8-1 lists their sequence from camera tube to antenna at the transmitter, and from antenna to picture tube at the receiver.

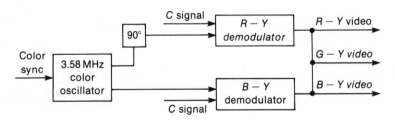

FIGURE 8-11 DETECTING THE MODULATED 3.58-MHz C SIGNAL WITH SYNCHRONOUS DEMODULATORS 90° OUT OF PHASE TO OBTAIN $R - Y$ AND $B - Y$ VIDEO SIGNALS. THESE TWO VOLTAGES ARE COMBINED TO OBTAIN THE $G - Y$ SIGNAL.

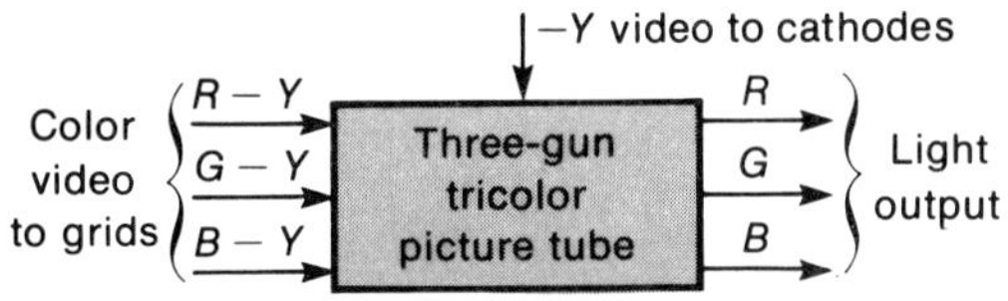

FIGURE 8-12 PICTURE TUBE AS A MATRIX THAT COMBINES Y SIGNAL WITH COLOR-DIFFERENCE VOLTAGES TO PRODUCE RED, GREEN, AND BLUE.

8-6 The Y Signal for Luminance

Now we can consider more details of the luminance signal, which contains the brightness variations of the picture information. The Y signal is formed by adding the primary red, green, and blue video signals in the proportions:

$$Y = 0.30R + 0.59G + 0.11B \qquad (8\text{-}1)$$

These percentages correspond to the relative brightness of the three primary colors. Therefore, a scene reproduced in black and white by the Y signal looks the same as when it is televised in monochrome.

Voltage values for Y signal. Figure 8-13 illustrates how the Y signal voltage in d is formed from the specified proportions of R, G, and B voltages for the color-bar pattern. Notice that the bars include the primary colors R, G, and B, their complementary mixtures of two primaries, and white for all three primaries.

The Y signal has its maximum relative amplitude of unity, 1.0 or 100 percent for white, because it includes R, G, and B. This value for white is calculated as

$$Y = 0.30 + 0.59 + 0.11 = 1.00$$

As another example, the cyan color bar includes G and B but no R. Then the Y value for cyan is calculated as

$$Y = 0 + 0.59 + 0.11 = 0.70$$

All the voltage values of the Y signal can be calculated in this way. The resulting voltages are the relative luminance values for each of the color bars. If the Y signal alone were used to reproduce this pattern, it would appear as monochrome bars shading off from white at the

TABLE 8-1 SEQUENCE OF COLOR SIGNALS

TRANSMITTER	RECEIVER*
1. R, G, and B video from camera.	1. Antenna signal is rf picture carrier modulated by colorplexed T signal.
2. Y, I, and Q video from matrix.	2. Modulated picture carrier is rectified in video detector.
3. I and Q modulate 3.58-MHz chrominance signal.	3. Synchronous demodulators for 3.58-MHz C signal provide $B - Y$ and $R - Y$ video, which are combined for $G - Y$.
4. Colorplexed T signal with Y and 3.58-MHz C signals.	4. $B - Y$, $R - Y$, and $G - Y$ video to picture tube, with $-Y$ video to cathodes.
5. Antenna signal is rf picture carrier modulated by colorplexed T signal.	5. Red, green, and blue, with their color mixtures, on screen of picture tube.

*For receivers using $B - Y$ and $R - Y$ demodulators and picture tube as matrix.

152 BASIC TELEVISION

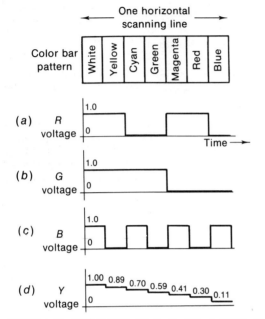

FIGURE 8-13 WAVEFORMS OF R, G, AND B VIDEO TO FORM THE Y SIGNAL, FOR COLOR-BAR PATTERN.

left to gray in the center and black at the right. These light values correspond to the staircase of Y video voltages in Fig. 8-13d, for the decreasing relative brightness of these color bars.

Bandwidth of Y signal. This signal is transmitted with the full video-frequency bandwidth of 0 to 4 MHz, the same as in monochrome broadcasting. However, most receivers cut off the response for video frequencies at 3.2 MHz, approximately. The purpose is to minimize interference with the 3.58-MHz C signal. In monochrome receivers, the IF bandwidth is generally limited to 3 MHz.

Matrix for Y signal. A matrix has the function of adding several input voltages in the desired proportions to form new combinations of output voltage. An example is illustrated in Fig. 8-14 for forming the Y signal. This circuit consists of three voltage dividers with a common resistor R_4. Each divider proportions the R, G, or B input signals in accordance with the resistance of R_1, R_2, or R_3, compared with R_4. The result is Y signal across R_4 as the common load resistor for the R, G, and B video voltages. Similarly, different resistance values can be used for separate voltage dividers to proportion the R, G, or B voltages in the percentages necessary to form I and Q signals at the transmitter.

8-7 Types of Color Video Signals

The main types are for the primary colors, as the system starts with R, G, and B voltages for the camera tube and finishes with R, G, and B at the picture tube. However, color mixtures are used for encoding and decoding. The reason is that two color-mixture signals can have all the color information of the three primary colors, allowing the third signal to be Y signal for luminance.

I signal. This color video voltage is produced in the transmitter matrix as the following com-

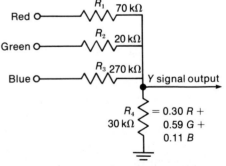

FIGURE 8-14 RESISTIVE VOLTAGE DIVIDER USED AS MATRIX TO FORM Y SIGNAL AT TRANSMITTER.

bination of the red, green, and blue primary colors:

$$I = 0.60R - 0.28G - 0.32B \qquad (8\text{-}2)$$

The minus indicates the addition of video voltage of negative polarity. For instance, $-0.32B$ means 32 percent of the total blue video signal but inverted from the polarity that reproduces blue. With $+I$ polarity, the signal includes red and minus blue, or yellow, which add to produce orange. For $-I$ signal, the polarity is reversed for all the primary components. Then the combination includes green and blue for cyan, with minus-red, which is cyan. As a result, opposite polarities of the I video signal represent the complementary colors orange and cyan, approximately. These hues are in color plate IX, which shows the main color video voltages.

Note that the negative components of $-0.28G$ and $-0.32B$ total -0.60 to equal the positive value of $0.60R$. These values are chosen to make the amplitude of the I video signal become zero for white.

Q signal. The primary color voltages are combined in the transmitter matrix in the following proportions for the Q signal:

$$Q = 0.21R - 0.52G + 0.31B \qquad (8\text{-}3)$$

With $+Q$ polarity, this signal includes minus-green, or magenta, with red and blue, which combine to form purple hues. For $-Q$ signal, this polarity includes mainly green with minus-blue, or yellow, for a combination of yellow-green. As a result, opposite polarities of Q signal represent the complementary colors purple and yellow-green.

Note that the positive components of $0.21R$ and $0.31B$ total 0.52 to equal the negative component of $-0.52G$. These values are chosen to make the amplitude of Q signal zero for white. Both the I and Q signals are zero for white, as there is no chrominance information in white. The luminance information for shades of white is in the Y signal.

B − Y signal. The hue of this signal is mainly blue, but it is a color mixture because of the $-Y$ component. When we combine 100 percent blue with the primary components of the Y signal, the result is

$$B - Y = 1.00B - (0.30R + 0.59G + 0.11B)$$
or
$$B - Y = -0.30R - 0.59G + 0.89B \qquad (8\text{-}4)$$

Note that $-R$ and $-G$ combined equal the complement of yellow, which is blue. However, a little more minus-green shifts the hue toward magenta, resulting in a purplish blue. When the $B - Y$ signal is combined with Y signal in the picture tube, it reproduces the blue information.

R − Y signal. The hue of $R - Y$ is a purplish red. Combining red with the primary components of the Y signal results in

$$R - Y = 1.00R - (0.30R + 0.59G + 0.11B)$$
or
$$R - Y = 0.70R - 0.59G - 0.11B \qquad (8\text{-}5)$$

The minus-green is magenta, which is combined with red to produce a purple-red for positive polarity of $R - Y$ signal. The opposite polarity of $R - Y$ signal has the hue of cyan-blue. When $R - Y$ signal is combined with Y signal in the picture tube, it reproduces the red information.

154 BASIC TELEVISION

G − Y signal. Combining −Y signal and 100 percent G results in

$$G - Y = 1.00G - (0.30R + 0.59G + 0.11B)$$
or
$$G - Y = -0.30R + 0.41G - 0.11B \qquad (8\text{-}6)$$

The hue of $G - Y$ signal is a bluish green. The opposite polarity is a purplish red.

Then, with $G - Y$ signal added to the Y signal in the color picture tube, the green information is reproduced.

In the receiver, $G - Y$ video is obtained by combining $R - Y$ and $B - Y$ in the following proportions:

$$G - Y = -0.51(R - Y) - 0.19(B - Y) \qquad (8\text{-}7)$$

Summary of color video signals. All the color-mixture voltages are related since each is a combination of R, G, and B. As additional examples, the I and Q signals can be specified in terms of the color difference signals as follows:

$$I = -0.27(B - Y) + 0.74(R - Y) \qquad (8\text{-}8)$$
$$Q = 0.41(B - Y) + 0.48(R - Y) \qquad (8\text{-}9)$$

All these video signals are color mixtures, combining R, G, and B so that two signals contain all the color information of the three primaries.

It is important to note the difference between these color video voltages, without modulation, and the 3.58-MHz modulated C signal. There is only one C signal, always at 3.58 MHz. This signal is encoded with the chrominance information as hue and saturation, corresponding to the phase and amplitude of the modulation on the 3.58-MHz color subcarrier.

However, the different color video signals exist before modulation of the 3.58-MHz color subcarrier at the transmitter and after demodulation at the receiver. The color video signals and their main features are summarized in Table 8-2. The colors listed are for positive signal voltage. The opposite polarity for each signal has the opposite hue. You can see these hues in the color circle diagram in color plate IX. The bandwidth is 0.5 MHz for all except the I signal.

Relative gain values. The amplitudes of the color video signals are modified in transmission to prevent modulation past the maximum white and black levels. For instance, yellow with high luminance can overmodulate white; blue with low luminance can overmodulate black. As a result, the receiver must compensate with the following proportions of gain:

$$B - Y \text{ gain is } 2.03 = \frac{1}{49\%}$$

TABLE 8-2 TYPES OF COLOR VIDEO SIGNALS

NAMES	HUES	BANDWIDTH, MHz	NOTES
B − Y	Blue-purple	0–0.5	Opposite phase from color sync
R − Y	Red-purple	0–0.5	In quadrature with B − Y
G − Y	Green-yellow	0–0.5	Combines B − Y and R − Y
Q	Purple	0–0.5	In quadrature with I
I	Orange	0–1.5	Maximum color bandwidth

$R - Y$ gain is $1.14 = \dfrac{1}{87.7\%}$

$G - Y$ gain is $0.70 = \dfrac{1}{142.3\%}$

As an example, the receiver gain for $B - Y$ signal is made almost double the gain for $R - Y$ signal. The reason is that, in modulation at the transmitter, the $B - Y$ component is reduced to 49 percent of its normal level.

8-8 The Color Sync Burst

Figure 8-15 shows the details of the 3.58-MHz color sync burst transmitted as part of the total composite video signal. The color burst synchronizes the phase of the 3.58-MHz color oscillator. This stage reinserts the 3.58-MHz color subcarrier in the synchronous demodulators to detect the chrominance signal. The phase of the reinserted oscillator voltage determines the hues in the detector output. Therefore, the color sync is necessary to establish the correct hues for the demodulators. Then the color AFC, which provides automatic frequency control of the color oscillator, can hold the hue values steady. Without color synchronization, the picture has drifting color bars.

The color sync burst is transmitted only for a color broadcast. The presence or absence of the burst determines how a color receiver recognizes whether a program is in color or monochrome.

The burst is 8 to 11 cycles of the 3.58-MHz subcarrier transmitted on the back porch of each horizontal blanking pulse. Peak value of the burst is one-half the sync pulse amplitude. However, the average value of the burst coincides with the blanking level. This value corresponds to zero for deflection sync. As a result, the color burst does not interfere with synchronization of the deflection oscillators.

The burst and C signal are both 3.58 MHz. However, the burst is on during blanking time only, when there is no picture information. The C signal is on during visible trace time for the color information in the picture. This comparison is illustrated in Fig. 8-15b.

8-9 Hue Phase Angles

Figure 8-16 illustrates how the hues of the modulated C signal are determined by its varying phase angle with respect to the constant

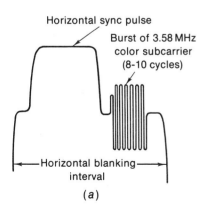

(a)

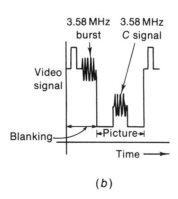

(b)

FIGURE 8-15 (a) COLOR SYNC BURST ON BACK PORCH OF EACH H SYNC PULSE. (b) COMPARISON OF BURST AND C SIGNAL.

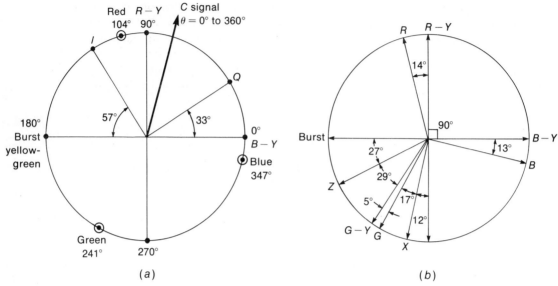

FIGURE 8-16 PHASE ANGLES OF DIFFERENT HUES. RELATIVE AMPLITUDES NOT TO SCALE. (a) C SIGNAL WITH EQUAL I AND Q MODULATION. (b) COLOR AXES OFTEN USED FOR DETECTION IN RECEIVER.

phase angle of the color sync burst. Note that the hue of the color sync burst corresponds to yellow-green. When picture information of this hue is being scanned at the transmitter, the phase angle of the chrominance signal is made the same as the phase of the color sync burst. For other hues, the C signal has different phase angles. How much the phase angle differs from the sync burst phase determines how the hue differs from yellow-green. In Fig. 8-16a, the phase angle of the C signal indicates the hue of purple-red between the angles for blue and red. This phase angle results from equal amounts of I and Q modulation. You can see all the hues and their phase angles in color plate IX. Opposite hues 180° apart are on a straight line called a color axis.

It should be noted that hue phase angles are indicated two different ways. Standard measure for angles counts counterclockwise as the positive direction from zero, as in Fig. 8-16a.

Then $B-Y$ is at 0°, and the color sync burst is at 180°. However, since burst phase is the reference, the hue phase angles are often counted clockwise from burst. Then $B-Y$ phase is at 180°. For the receiver demodulation axes in Fig. 18-16b, the angles are indicated by how much they are off the horizontal and vertical axes.

I and Q axes. These are the color video signals used to modulate the 3.58-MHz subcarrier for broadcasting. As shown in Fig. 8-16a, the I axis is 57° off the phase of color sync burst. The Q axis is in quadrature. These angles put the phase of $B-Y$ exactly opposite burst phase.

B − Y and R − Y axes. The receiver can recover these hues in demodulating the C signal by reinserting the 3.58-MHz color subcarrier at these phase angles. As shown in Fig. 8-16b, $B-Y$ phase is 180° from burst phase, and $R-Y$

phase is in quadrature. Two synchronous demodulators are needed to detect hues on the $R-Y$ and $B-Y$ axes. Then the $R-Y$ and $B-Y$ color video signals are added to form $G-Y$. Counting clockwise from burst, the phase angles are 90° for $R-Y$, 180° for $B-Y$, and 304° for $G-Y$.

X and Z axes. As shown in Fig. 8-16b, the X axis is close to $-(R-Y)$; the Z axis is close to $-(B-Y)$ or burst phase. Counting clockwise from burst, the X axis is at 282° and the Z axis at 333°. Approximately 51° apart, the X and Z axes can be used in the receiver demodulators even though they are not in quadrature. The advantage is that $R-Y$, $G-Y$, and $B-Y$ can be formed in three amplifier stages that are balanced, to reduce the possibility of color drift.

R, G, and B axes. Three demodulators can be used to recover red, green, and blue directly. These three angles are almost equally spaced around the color circle, close to the phases of $R-Y$, $G-Y$, and $B-Y$.

8-10 The Colorplexed Composite Video Signal

Formation of the total video signal combining luminance and chrominance is illustrated in Fig. 8-17 in successive steps. Starting with the primary colors, the R, G, and B video voltages in Fig. 8-17a, b, and c are shown for the time of scanning one horizontal line across the color bars. The colors are fully saturated, without any white. This means the relative voltage value for R, G, and B is at 100 percent, or 1.0. Also, the saturated complementary colors yellow, cyan, and magenta have only two primary colors, since there is no white to add the third primary.

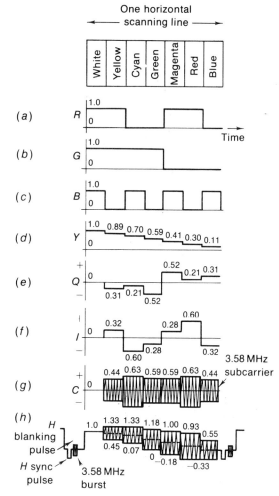

FIGURE 8-17 CONSTRUCTION OF COLORPLEXED COMPOSITE VIDEO SIGNAL FROM Y, I, AND Q VOLTAGES. WAVEFORM IN H AT BOTTOM SHOWS MULTIPLEXED Y AND C SIGNALS.

Amplitudes of Y signal. The luminance signal in Fig. 8-17d shows the brightness component for each bar. The relative values for Y are calculated with Eq. (8-1). As an example, for magenta, combining red and blue without green;

$Y = 0.30R + 0 + 0.11B$
$Y = 0.41$

Amplitudes of I and Q signals. The I and Q waveforms in e and f have the relative voltages indicated according to their proportions of primary colors. These values are calculated with Eqs. (8-2) and (8-3). As examples, for yellow with red and green but no blue;

$$I = 0.60R - 0.28G - 0.00B = 0.32$$
$$Q = 0.21R - 0.52G + 0.00B = -0.31$$

Note that the I and Q voltages can have positive or negative polarity because their components include positive and negative primary colors.

Phasor addition for C signal. The next waveform in g shows the 3.58-MHz color subcarrier modulated by the I and Q signals in quadrature. The two-phase amplitude modulation results in varying amplitudes and phase angles for the C signal. The phase angles cannot be shown, but the varying amplitudes can be calculated. The method of phasor* addition for the I and Q signals in quadrature is the same as combining two ac voltages 90° out of phase in series ac circuits, written as

$$C = \sqrt{I^2 + Q^2}$$

As an example, for yellow with values of 0.32 for I and -0.31 for Q,

$$C = \sqrt{(0.32)^2 + (-0.31)^2} = \sqrt{0.102 + 0.096}$$
$$C = \sqrt{0.1980} = 0.44$$

This method can be used to calculate all the C values for the color bars in Fig. 8-17, by phasor addition of the I and Q amplitudes.

There is no polarity for the C signal because it is a carrier wave with both positive and negative half-cycles. Note that the peak amplitude of 0.44 for blue or yellow C signal means that it varies 0.44 units above and below the zero axis of this modulated ac waveform. Yellow and blue have the same amplitudes but opposite phase angles because they are complementary colors.

Phase angle of C signal. If we want to know the phase angle θ for the hue, this angle has the tangent equal to Q/I. As an example, for red with the Q of 0.21 and I of 0.60, then $0.21/0.60 = 0.35$ for $\tan \theta$. This value of 0.35 for $\tan \theta$ defines the angle θ of 19°. This angle is 19° from I, toward Q.

Adding the Y and C signals. For the total video signal waveform in h, the Y amplitudes for luminance are combined with the C signal. The result is to shift the C signal variations to the axis of the Y luminance level, instead of its zero axis. As an example, blue has the level of 0.11 in the Y signal and the peak amplitude of 0.44 in the C signal. When we combine these Y and C signals, the result in the colorplexed video signal for blue is that the positive peak goes up to $0.44 + 0.11 = 0.55$. The negative peak goes down to $-0.44 + 0.11 = -0.33$. These maximum and minimum values are still ± 0.44, but around the average axis of 0.11 for the Y signal. The same idea applies to all the color bars shown. You can see how the color bars match the combined Y and C signals in color plate X.

It is important to realize that the Y signal for luminance information is inserted as the average level of the C signal variations for color information. If the C signal is removed from the colorplexed S signal in h, the result will be the same staircase of Y signal variations shown in d. In monochrome receivers, 3.58 MHz is filtered out to remove the color information, but the Y signal remains to provide the luminance variations.

---

*Components with different angles in time are phasors; vectors have different angles in space.

8-11 Desaturated Colors with White

The relative voltage values shown in Fig. 8-17 are for vivid colors that are 100 percent saturated. In this case, there is no primary color video for hues not included in the color. As examples, saturated R has zero B and G video voltage; saturated yellow (red-green) has zero B video voltage. This follows from the fact that with zero light input to a given color camera there is no signal output.

In natural scenes, however, most colors are not 100 percent saturated. Then any color diluted by white light has all three primaries. The following example illustrates how to take into account the amount of desaturation for weaker colors. Assume 80 percent saturation for yellow. Now this color has two components: 80 percent saturated yellow and 20 percent white. First consider the primary color video signals produced by each camera tube:

80 percent yellow (red-green) produces	$0.80R$	$0.80G$	$0.00B$
20 percent white (red-green-blue) produces	$0.20R$	$0.20G$	$0.20B$
Total camera output is	$1.00R$	$1.00G$	$0.20B$

These percentages of primary color video voltages can then be used for calculating relative amplitudes of the Y signal and color video signals for 80 percent saturated yellow. As examples, this desaturated color has the Y value of 0.912, Q value of -0.248, and I value of 0.256. Compare these with the values of 0.89, -0.31, and 0.32 shown in Fig. 8-17 for 100 percent saturated yellow. Note that the addition of white to desaturate a color increases the luminance value and decreases the chrominance value, compared with 100 percent saturation.

8-12 Color Resolution and Bandwidth

The Y signal is transmitted with the full video-frequency bandwidth of 4 MHz for maximum horizontal detail in monochrome. However, this bandwidth is not necessary for the color video signals. The reason is that for very small details the eye can perceive only the brightness, rather than the color. Therefore, the color information can be transmitted with a restricted bandwidth much less than 4 MHz. This feature allows the narrowband chrominance signal to be multiplexed with the wideband luminance signal in the standard 6-MHz broadcast channel. All the color video signals have the bandwidth of 0 to 0.5 MHz, except the I signal with a bandwidth of 0 to 1.5 MHz. These values are in the third column of Table 8-2 on page 154.

The I signal for orange and cyan has more bandwidth because smaller details can be resolved for these colors. However, the I bandwidth of 0 to 1.5 MHz must be considered in two parts, in terms of its modulation side bands above and below the 3.58-MHz color subcarrier. Frequencies of 0 to 0.5 MHz in the I signal are transmitted with double side bands, using both the upper and lower side frequencies produced by modulation. However, for frequencies between 0.5 and 1.5 MHz, only the lower side bands are transmitted. This method of vestigial-sideband transmission on the 3.58-MHz color subcarrier is used to provide maximum bandwidth for the I signal without extending into the frequencies of the sound carrier signal 4.5 MHz from the picture carrier signal. The bandwidths for the Y, I, and Q signals are illustrated by the graphs in Fig. 8-18.

The extra bandwidth of the I signal is seldom used in color receivers. The reason is that the color circuits are much simpler when all the color video signals have the same bandwidth of 0.5 MHz, the practical baseband for color.

As a result, we can consider the video frequencies of 0 to 0.5 MHz as the practical bandwidth for color. How this information is reproduced is illustrated by the pattern in Fig. 8-19. The squares and bars are drawn to scale for a screen width of 20 in. Keep in mind the fact that the video frequency of 0.5 MHz corresponds to horizontal details $1/50$ of the width of this picture. This value of 50 details is calculated as $1/8$ of 400 approximately, just as 0.5 MHz is $1/8$ of 4 MHz. All the squares are in color because their width is more than $1/50$ of 20 in., equal to 0.4 in. The vertical bar at the left is in color, as its width is 0.4 in. The vertical bar at the left is in color, as its width is 0.4 in. The entire background is in color as these large areas represent low video frequencies. Also, any horizontal bars are in color because vertical scanning represents low video frequencies.

The vertical bar at the right is not in color because it is less than 0.4 in. wide. The vertical edges between the pattern and the background also are in monochrome because these narrow details correspond to video frequencies higher than 0.5 MHz. The only parts of the picture not in color are vertical edges and bars less than 0.4 in. wide, corresponding to video frequencies higher than 0.5 MHz.

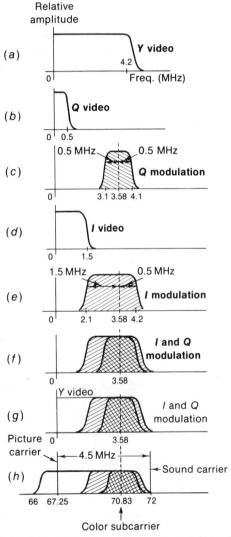

FIGURE 8-18 BANDWIDTH FOR Y AND C SIGNAL FREQUENCIES. GRAPH IN h AT BOTTOM IS FOR COLORPLEXED COMPOSITE VIDEO SIGNAL MODULATING CHANNEL 4 PICTURE CARRIER AT 67.25 MHz.

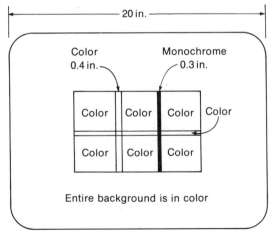

FIGURE 8-19 AREAS OF COLOR IN PICTURE WITH VIDEO BANDWIDTH OF 0 TO 0.5 MHz.

8-13 Color Subcarrier Frequency

This must be a high video frequency, in the range between 2 and 4 MHz approximately. If the color subcarrier frequency is too low, it can produce excessive interference with the luminance signal. At the opposite extreme, the chrominance signal can interfere with the sound signal. The choice of approximately 3.58 MHz for the color subcarrier is a compromise that allows 0.5-MHz side bands for chrominance information below and above the subcarrier frequency. Also, there is room for the extra 1 MHz of lower side frequencies of the *I* signal. Most important for compatibility, 3.58 MHz is a video frequency high enough to have little response in monochrome receivers. These sets use the luminance signal alone, with practically no effect from the 3.58-MHz chrominance signal.

The exact frequency of the color subcarrier is based on the following additional factors:

1. The transmitted picture carrier and sound carrier frequencies cannot be changed, in order to preserve the 4.5-MHz beat for intercarrier sound receivers.
2. There will be an interfering beat frequency of approximately 0.92 MHz or 920 kHz between the color subcarrier frequencies near 3.58 MHz and the intercarrier sound at 4.5 MHz.
3. There will be interfering beat frequencies between the chrominance signal and higher video frequencies of the luminance signal.

In order to minimize these interference effects, the color subcarrier frequency is made exactly 3.579545 MHz. This frequency is determined by harmonic relations for the color subcarrier, the horizontal line-scanning frequency, and the 4.5-MHz intercarrier beat.

Horizontal frequency. Specifically, the sound frequency of 4.5 MHz is made to be the 286th harmonic of the horizontal line frequency. Therefore,

$$f_H = \frac{4.5 \text{ MHz}}{286} = 15{,}734.26 \text{ Hz}$$

where f_H is the horizontal line-scanning frequency for color television broadcasting. Notice that 286 is the even number that will make f_H closest to the value of 15,750 Hz used for horizontal scanning in monochrome television. The slight difference has practically no effect on horizontal scanning and sync in the receiver because of the horizontal AFC circuit.

Vertical frequency. The vertical scanning frequency is also changed slightly since there must be 262.5 lines per field. Then the vertical field-scanning frequency is

$$f_V = \frac{15{,}734.26}{262.5} \text{ Hz} = 59.94 \text{ Hz}$$

The slight difference of 0.06 Hz below 60 Hz has practically no effect on vertical scanning and sync in the receiver, because an oscillator that can be triggered by 60-cycle pulses can also be synchronized by 59.94-Hz pulses. It should be noted that when the scanning frequencies are shifted slightly for color television, the transmitted sync is also changed to the new frequencies for f_V and f_H.

Color frequency. With the horizontal line-scanning frequency chosen, now the color subcarrier can be determined. This value is made to be the 455th harmonic of $f_H/2$:

$$C = 455 \times \frac{15{,}734.26 \text{ Hz}}{2} = 3.579545 \text{ MHz}$$

To obtain this exact frequency, the color oscillator is crystal-controlled. A typical quartz crystal unit is shown in Fig. 8-20. The multiple of 455 is chosen as an odd number that makes C close to 3.58 MHz.

Frequency interlace. Because of the odd-line interlaced scanning pattern, picture information for video frequencies that are odd multiples of $f_H/2$ tends to cancel in the effect on the screen of the picture tube. The cancellation results because these frequencies have opposite voltage polarities for the picture information on even and odd scanning lines. You can see this canceling effect by coupling from a signal generator output voltage at 2 to 4 MHz into the video amplifier to produce a diagonal-bar interference pattern on the picture tube screen. By adjusting the generator frequency carefully and watching the screen pattern closely, at certain frequencies the interference pattern will disappear. These frequencies are odd multiples of one-half the horizontal line-scanning frequency.

This technique of interlacing odd and even harmonic components of two different signals in order to minimize interference between them is *frequency interlace*. As a result, the chrominance signal can be transmitted in the same 6-MHz channel as the luminance and sound signals with practically no interference.

8-14 Color Television Systems

The standards described here are for the NTSC system, an abbreviation of National Television System Committee. This group formed by the Electronic Industries Association also prepared the standards for monochrome television in the United States. The Federal Communications Commission (FCC) approved the monochrome standards in 1941. The NTSC color television system was adopted in 1953.

Historically, color television broadcasting began experimentally about 1949 with two competing systems by RCA and CBS. The CBS system used a mechanical color wheel with red, green, and blue in sequential fields. This method used scanning frequencies that were not compatible with monochrome broadcasting. The RCA system used compatible scanning standards. The CBS system was adopted for a short time in 1951 but used very little. Then the NTSC prepared new standards based on the RCA system. After field trials, the NTSC system was adopted by the FCC. The NTSC color television system is standard in the United States, Canada, Japan, and most countries in the Western Hemisphere.

In Europe, the main color television systems are PAL and SECAM.* The PAL (for Phase Alternation by Line) system is similar to the NTSC system, but for each successive line one component of the chrominance signal is reversed in polarity. The purpose is to average out any errors in hue phase. This system is used in

FIGURE 8-20 CRYSTAL FOR COLOR OSCILLATOR IN RECEIVER. HEIGHT IS 1 IN. FREQUENCY IS EXACTLY 3,579.545 kHz OR 3.579545 MHz.

*For more details on PAL and SECAM, see C. R. G. Reed, "Principles of Colour Television Systems," Isaac Pitman and Sons Ltd., London, 1969.

Germany and most European countries except France. SECAM is a French system with a sequential technique and memory storage. In this method, two chrominance signals are transmitted, one at a time, for successive lines.

It should be noted that other countries may have scanning standards and a channel width different from the United States. Also, the color subcarrier frequency of 3.58 MHz is essentially for a 6-MHz broadcast channel. Standards for principal television systems of the world are listed in Appendix C at the back of the book.

SUMMARY

The following definitions, in alphabetical order, summarize the main features of color television.

B − Y signal. Color mixture close to blue. Phase is 180° opposite from color sync burst. Bandwidth is 0 to 0.5 MHz.

Burst. Color sync. Is 8 to 11 cycles of 3.58-MHz color subcarrier transmitted on back porch of every horizontal pulse. Needed to synchronize phase of 3.58-MHz oscillator in receiver for correct hues in C signal. The hue of color sync phase is yellow-green.

Chrominance signal. Also called *chroma* signal or C signal. Is 3.58-MHz color subcarrier with quadrature modulation by I and Q color video signals. Amplitude of C signal is saturation; phase angle is hue.

Colorplexer. Also called *multiplexer*. Combines C signal and Y luminance signal. Result is total colorplexed composite video signal transmitted to receiver as amplitude modulation of picture carrier.

Compatibility. Ability of monochrome receiver to use Y signal for picture in black and white. Also allows color receiver to reproduce monochrome picture. Compatibility results from transmission of Y signal for luminance and use of practically the same scanning standards for color and monochrome.

Complementary color. Opposite hue and phase angle of primary color. Cyan, magenta, and yellow are complements of red, green, and blue, respectively.

Decoding. Converting hue and saturation in the C signal to R, G, and B primary color video signals for the tricolor picture tube.

Encoding. Converting the R, G, and B primary color video signals to hue and saturation in the C signal.

Frequency interlace. Placing harmonic frequencies of C signal midway between harmonics of horizontal scanning frequency f_H. Accomplished by making color subcarrier frequency exactly 3.579545 MHz.

G − Y signal. Color mixture close to green. Bandwidth is 0 to 0.5 MHz. Usually formed by combining B − Y and R − Y video signals.

Hue. Also called *tint*. Wavelength of light for the color. The varying phase angles in the 3.58-MHz C signal indicate the different hues in the picture information.

I signal. Color video signal transmitted as amplitude modulation of the 3.58-MHz C signal. Hue axis is orange and cyan. This is the only color video signal with bandwidth of 0 to 1.5 MHz.

Luminance. Also brightness, for either color or monochrome information. Luminance information is in the Y signal.

Matrix. Combines signals in specific proportions. Transmitter matrix forms Y, I, and Q video signals in the output for R, G, and B input. At the receiver, the three-gun picture tube is often the matrix for input of $R-Y$, $B-Y$, $G-Y$, and Y signals to produce red, green, and blue light from the screen.

Monochrome. In black and white. Just luminance or brightness without color. Also called *achromatic.* The Y signal is a monochrome signal.

Multiplexing. Combining two signals on one carrier.

NTSC. National Television System Committee. The name for the standard color television system adopted by the FCC for use in the United States.

Primary colors. Red, green, and blue. Opposite voltage polarities are the complementary colors cyan, magenta, and yellow.

Q signal. Color video signal that modulates 3.58-MHz C signal in quadrature with I signal. Hues are green and magenta. Bandwidth is 0 to 0.5 MHz.

$R-Y$ signal. Color mixture close to red. Phase is in quadrature with $B-Y$. Bandwidth is 0 to 0.5 MHz.

Saturation. Intensity of color. Full saturation means no dilution by white. Different saturation values are varying peak-to-peak amplitudes in the 3.58-MHz modulated C signal.

Subcarrier. A carrier that modulates another carrier wave of higher frequency. In color television the 3.58-MHz color subcarrier modulates the rf picture carrier wave in the standard broadcast channel.

Suppressed subcarrier. Transmission of the modulation side bands without the subcarrier itself. Requires reinsertion of the subcarrier at the receiver for detecting the modulation.

Synchronous demodulator. Detector circuit for a specific phase of the modulated signal.

White. Contains red, green, and blue in the proportions $Y = 0.30R + 0.59G + 0.11B$.

Self-Examination (Answers at back of book)

Choose (a), (b), (c), or (d).

1. Brightness variations of the picture information are in which signal? (a) I; (b) Q; (c) Y; (d) $R-Y$.
2. The hue 180° out of phase with red is (a) cyan; (b) yellow; (c) green; (d) blue.
3. Greater peak-to-peak amplitude of the 3.58-MHz chrominance signal indicates more (a) white; (b) yellow; (c) hue; (d) saturation.

4. The interfering beat frequency of 920 kHz is between the 3.58-MHz color subcarrier and (a) 4.5-MHz intercarrier sound; (b) picture carrier; (c) lower adjacent sound; (d) upper adjacent picture.
5. The hue of color sync phase is (a) red; (b) cyan; (c) blue; (d) yellow-green.
6. Which signal has color information for 1.5-MHz bandwidth? (a) I; (b) Y; (c) $R - Y$; (d) $B - Y$.
7. Which of the following is false? (a) I video hues are orange or cyan; (b) the transmitter matrix output includes Y, I, and Q video; (c) a three-gun picture tube can serve as a matrix; (d) a fully saturated color is mostly white.
8. The color with the most luminance is (a) red; (b) yellow; (c) green; (d) blue.
9. What is the hue of a color 90° leading sync burst phase? (a) yellow; (b) cyan; (c) blue; (d) orange.
10. The average voltage value of the 3.58-MHz modulated chrominance signal is (a) zero for most colors; (b) close to black for yellow; (c) the brightness of the color; (d) the saturation of the color.
11. The IF value for color in receivers, for any station, is (a) 0.5 MHz; (b) 1.5 MHz; (c) 3.58 MHz; (d) 4.5 MHz.
12. If the 3.58-MHz C amplifier in the receiver does not operate, the result will be (a) no color; (b) no red; (c) too much blue; (d) too much yellow.

Essay Questions

1. What is meant by color addition? Name the three additive primary colors.
2. What color corresponds to white light minus red? White minus blue? White minus green?
3. Why are the primary color video voltages converted to Y and C signals for broadcasting?
4. Define hue, saturation, luminance, and chrominance.
5. What is the video-frequency bandwidth of the Y signal?
6. What parts of the picture are reproduced in black and white by the Y signal? What parts are reproduced in full color as mixtures of red, green, and blue?
7. What hues correspond to the following: $+I$, $-I$, $+Q$, $-Q$, $R - Y$, $-(R - Y)$, $(B - Y)$, and color sync burst?
8. How is the 3.58-MHz modulated chrominance signal transmitted to the receiver? Why is the 3.58-MHz signal called a subcarrier?
9. Describe the color sync burst signal and give its purpose.
10. How does the colorplexed video signal indicate hue, saturation, and luminance of the picture information?
11. Why is the chrominance signal transmitted with the subcarrier suppressed?
12. Why is the color subcarrier frequency made exactly 3.579545 MHz?

13. A scene displays a wide yellow vertical bar against a black background. How will this picture appear in a monochrome reproduction?
14. What is the effect of the chrominance signal in a monochrome receiver with video-frequency response up to 3.2 MHz?

Problems (Answers to selected problems at back of book)

1. Show the calculations for Y luminance values of blue, red, green, yellow, and white.
2. Prove that if $G - Y$ is $[-0.51(R - Y) - 0.19(B - Y)]$, then this equals $-0.30R + 0.41G - 0.11B$, by substituting the R, G, and B values for Y.
3. (a) Calculate the value of C voltage when $I = 0.4$ and $Q = 0.3$. (b) What is the approximate hue of this color?
4. A signal voltage varies between the peak of 0.79 V and minimum of 0.11 V. Calculate (a) peak-to-peak value and (b) average level.

9 Color Television Receivers

Figure 9-1 shows a common chassis layout with the setup adjustments at the back. Most of the receiver is used for a monochrome picture on a white raster. This part includes the circuits for picture and sound signals, the raster, and the power supplies. The 3.58-MHz color circuits are on a separate board at the rear of the chassis. Block diagrams for all these circuits are shown in two separate parts in Figs. 9-2 and 9-3. In Fig. 9-2 the monochrome part illustrates how a black-and-white picture is reproduced on the raster. Figure 9-3 shows details of the 3.58-MHz color circuits. Actually, a white raster is necessary, as a mixture of the red, green, and blue phosphors on the screen of the color picture tube. Furthermore, the Y luminance signal produces a monochrome picture. Then the color circuits can add color to the picture by providing red, green, and blue video signals for the picture tube. More details are explained in the following topics:

9-1 Requirements for the color picture tube
9-2 Normal-service switch
9-3 Monochrome circuits in a color receiver
9-4 Circuits for color
9-5 Manual color controls
9-6 Functions of the automatic color circuits
9-7 $R-Y, B-Y$ receivers
9-8 X and Z demodulators
9-9 R, G, B receivers
9-10 How the picture tube mixes colors
9-11 Localizing color troubles

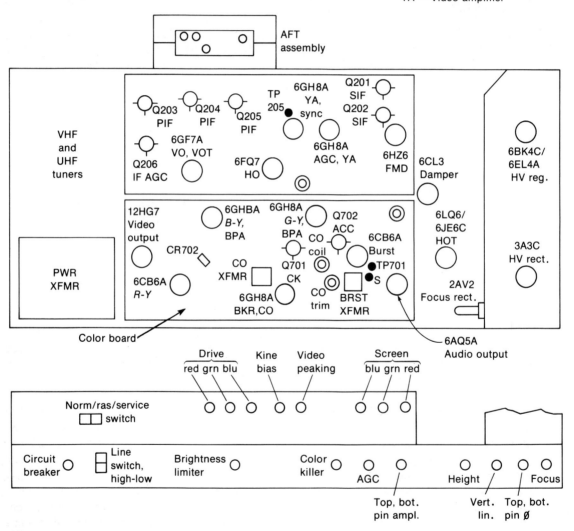

FIGURE 9-1 REAR VIEW OF CHASSIS LAYOUT WITH SETUP CONTROLS FOR COLOR TELEVISION RECEIVER. (*RCA CTC39 CHASSIS*)

9-1 Requirements for the Color Picture Tube

The details of picture tubes are described in Chaps. 10 and 11, but the requirements for the color tube in the block diagram of Fig. 9-2 can be stated briefly. The tube has three electron guns, one each for red, green, and blue. Also, the screen has either dots or vertical stripes of red, green, and blue phosphors that are excited by the three guns. All colors, including white, are produced as a mixture of red, green, and blue. The type number 21VAZP22 indicates this tube has a P22 phosphor for red, green, and blue, while the screen size is 21 in. diagonally.

The components mounted on the neck of the picture tube include the deflection yoke, convergence yoke, and purity ring magnet. *Convergence* means making each of the three beams go through an aperture mask at the proper angle to excite its respective color phosphor on the screen. In brief, the functions are as follows:

1. *Deflection yoke.* Deflects all three beams horizontally and vertically to produce the scanning raster.
2. *Convergence yoke.* Has individual adjustments for each beam. Converges the beams to produce white as a combination of red, green, and blue, without color fringing at the edges of picture information.
3. *Purity magnet.* Positions all three beams to produce pure red, green, or blue, without any effect from the other colors. The mounting for the purity ring usually includes another small magnet called the *blue lateral adjustment,* for converging the blue beam horizontally.

Screen controls. These adjustments on the back of the receiver chassis vary the screen-grid voltage for each of the electron guns. The purpose is to make them all have the same cutoff characteristic. The screen controls are adjusted to produce a white raster, at a very low light level.

Drive controls. These adjustments on the back of the receiver chassis vary the amount of Y video signal for each of the electron guns. The purpose is to balance the ac driving signals to compensate for different phosphor efficiencies, as the red gun may require more signal. The drive controls are adjusted to produce white highlights in a monochrome picture.

9-2 Normal-Service Switch

Most color receivers have this switch at the back of the receiver chassis to provide a raster alone, or just a horizontal line. The purpose is to help in making adjustments for the color picture tube. There are usually three settings:

1. *Raster position.* The AGC circuit cuts off the IF amplifier to provide a clean raster without any snow. This position may also be labeled "purity" because the raster is used when making purity adjustments for the picture tube.
2. *Service or line.* In this position of the switch, the vertical deflection is disabled to collapse the raster into one horizontal line across the screen. The purpose is to concentrate the illumination to help determine visual cutoff when adjusting the screen-grid voltages of the picture tube. The line may be misconverged, with separate colors, because the convergence correction signals are missing, but this does not matter for the screen adjustments.

3. *Normal.* This is the operating position of the switch, with a normal picture on the raster.

Notice that setting up the color picture tube is mainly a problem of obtaining good white. More details of these setup adjustments are explained in Chap. 11.

9-3 Monochrome Circuits in a Color Receiver

See Fig. 9-2. The only difference in the rf and IF circuits is the importance of bandwidth for color receivers. Remember that video frequencies around 3.58 MHz just show fine detail in monochrome, but these frequencies are essential for color information. Without the 3.58-MHz C signal, there is no color—just a monochrome picture. For this reason, the fine tuning control on the rf tuner must be tuned exactly for color in the picture, without the 920-kHz beat interference from the sound signal. Most color receivers have automatic fine tuning (AFT) which controls the frequency of the local oscillator on the rf tuner.

Intercarrier sound is used, but the sound takeoff circuit is before the video detector. The sound converter at the top of Fig. 9-2 is a diode rectifier. Input consists of P at 45.75 MHz and S at 41.25 MHz to produce the difference frequency of 4.5 MHz in the rectified output for the sound IF amplifier. A separate converter is used, instead of taking the 4.5-MHz sound from the video detector. The reason is to minimize the 920-kHz beat interference between the 4.5-MHz sound signal and the 3.58-MHz color signal.

The video detector output for a color broadcast includes the 3.58-MHz C signal multiplexed with the Y luminance signal. The waveform shown out of the video detector includes the following three components:

1. C signal for color information. This is shown for one vertical color bar down the center of the picture.
2. Luminance level corresponding to gray, midway between white and black.
3. Color sync burst of 3.58 MHz on the back porch of each horizontal blanking pulse.

The luminance information is indicated by its level between white and black. For the 3.58-MHz C signal, its average value is the luminance level.

The video amplifier in a color receiver serves as the Y amplifier, cutting off at 3.2 MHz to amplify the luminance signal without the chrominance signal. Then the video output, consisting of Y signal, is applied to each cathode of the three electron guns in the picture tube. The red, green, and blue drive controls proportion the amount of Y signal. The combined result is a good monochrome picture, without color in the white highlights.

The video amplifier in a color receiver usually includes a preamplifier, which is usually an emitter-follower in transistor circuits. The output of the video preamplifier supplies composite video signal to several stages for different functions, as follows:

1. AGC stage for AGC bias
2. Sync separator for vertical and horizontal synchronization
3. C signal for the color section
4. Y signal for video output to the picture tube

The Y signal for the video amplifier requires a *delay line.* This unit delays the Y signal with respect to the C signal. Both the Y and C signals are applied to the grid-cathode

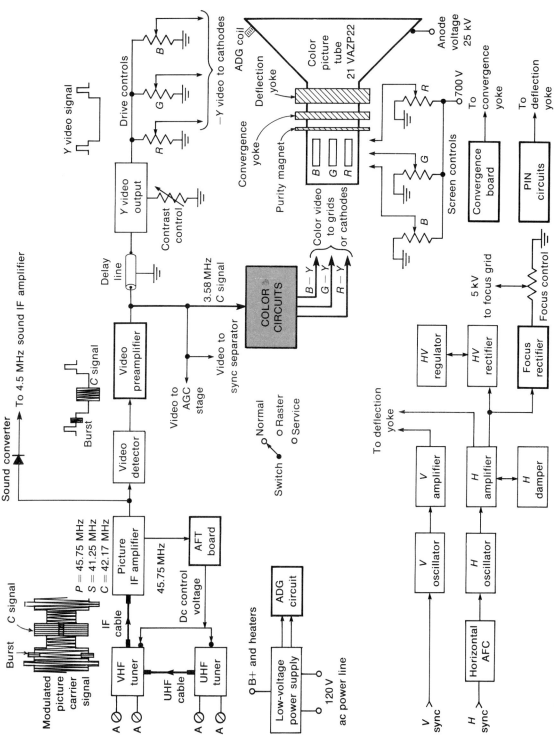

FIGURE 9-2 BLOCK DIAGRAM OF COLOR TELEVISION RECEIVER. CIRCUITS FOR COLOR SECTION ARE IN FIG. 9-3. SIGNAL WAVEFORM SHOWN FOR ONE VERTICAL COLOR BAR. AFT IS AUTOMATIC FINE TUNING. AFPC IS AUTOMATIC FREQUENCY AND PHASE CONTROL. ADG IS AUTOMATIC DEGAUSSING.

circuit of the picture tube but the Y signal would be there earlier in time. The reason is that the Y video amplifier is more resistive, with more bandwidth. The time delay inserted is 0.8 μs. Without this delay for the Y signal, the color information would be out of register, about 0.3 in. to the right on a screen 20 in. wide.

The raster circuits for a color receiver are essentially the same as for monochrome, but more power is needed with a bigger yoke for the color picture tube. Flyback high voltage is used for an anode voltage of 25 to 30 kV for a 19- to 25-in. picture tube. A separate rectifier for focus voltage on the color picture tube is shown in Fig. 9-2, but this can also be obtained with a voltage divider on the high-voltage supply. In addition, the high-voltage power supply usually has a regulator to maintain constant anode voltage with different load currents. The voltage regulation is important in maintaining focus and good convergence at high brightness levels.

In the low-voltage power supply, the ac voltage for the power line is also used for the automatic degaussing (ADG) circuit. "Degaussing" means demagnetizing. The ADG coil is around the rim at the front of the color picture tube. This coil has 60-Hz alternating current to demagnetize the tube each time the receiver is turned on or off. The degaussing is necessary to improve purity of the colors.

The convergence board shown in Fig. 9-2 supplies correction current for the convergence yoke on the color picture tube. The section labeled "PIN" includes pincushion correction circuits. These are adjusted for a rectangular raster that is not bowed in at the outside edges. Convergence and pincushion adjustments are explained in more detail in Chap. 11.

9-4 Circuits for Color

More details of the color section of the receiver are shown by the block diagram in Fig. 9-3. Color receivers are generally classified according to how the luminance and color signals are combined to provide red, green, and blue for the picture tube. This addition of the Y and C signals is *matrixing* in the receiver. The matrixing can be done by the picture tube itself or in the circuits before the picture tube.

For matrixing in the picture tube, the $R-Y$, $B-Y$, and $G-Y$ signals are fed to each of the guns, with the Y signal to all three. This method is generally used with vacuum tubes in the color circuits. It is classified as an $R-Y$, $B-Y$ receiver, which assumes there is $G-Y$ signal also. Figure 9-3 illustrates this type.

With transistors in the color circuits, the $R-Y$, $B-Y$, and $G-Y$ color signals are usually mixed with the Y signal before the picture tube. Then each of the three guns has one video signal that combines red, green, or blue with the luminance information. This type is classified as an R, G, B receiver.

Input signal to the color section. We can start with the C signal from the video preamplifier in Fig. 9-3. This signal is taken off before the delay line for the Y video amplifier. The waveshape shown here includes the 3.58-MHz C signal and the color sync burst. We cannot see the hue of the C signal because the phase angle with respect to burst phase is not evident. If the C phase is the same as burst phase, the hue of the color in the C signal will be yellow-green. The amplitude of the C signal indicates the color saturation.

It should be noted that the burst occurs during blanking time when there is no picture. However, the C signal has the color information during active trace time for the picture. Although the burst and C signal are both 3.58 MHz, they are not present at the same time.

Chrominance bandpass amplifier. This section includes one or two stages, fixed-tuned to 3.58 MHz as an IF amplifier, with a bandpass of ±0.5 MHz for the modulated C signal. Although the input includes luminance and chrominance,

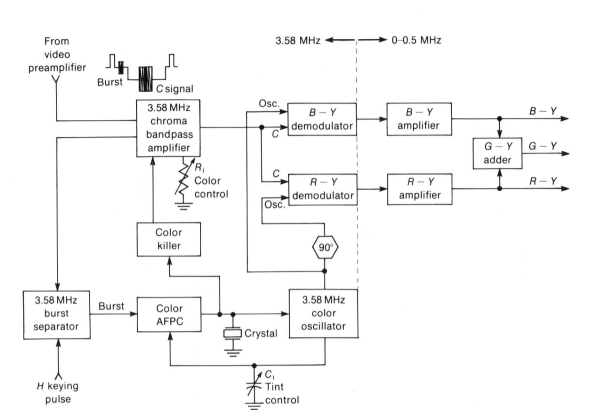

FIGURE 9-3 COLOR CIRCUITS FOR $R-Y$, $B-Y$ RECEIVER.

the output is 3.58-MHz C signal because this stage is tuned to 3.58 MHz. The output includes the 3.58-MHz C signal and the 3.58-MHz burst for color sync. The 3.58-MHz output of the color amplifier is then coupled to the color demodulators to detect the color information and to the burst separator to remove the color sync burst. The 3.58-MHz bandpass amplifier is the color or chroma amplifier, as its gain determines the amount of color in the picture.

The C signal has the frequency band of 3.58 MHz $\pm$ 0.5 MHz, which equals 3.08 to 4.08 MHz. A basic problem here is that the Y luminance signal also has frequency components in this band. The result can be interference between the color signal and high-frequency components of the Y signal. This effect causes sparkle or glitter at the edges of picture information with high-frequency components.

Color demodulators. The demodulators in Fig. 9-3 detect $R-Y$ and $B-Y$ color video signals because these two phases are provided by the output of the 3.58-MHz color oscillator. With C signal input and $B-Y$ phase of oscillator voltage into the $B-Y$ demodulator, the detected output is $B-Y$ video signal. Similarly, with C signal input to the second demodulator and $R-Y$ phase of oscillator voltage shifted 90° from $B-Y$ phase, the detected output is $R-Y$ video signal.

The $B-Y$ and $R-Y$ video signals are combined in the proper proportions in the $G-Y$ adder stage to produce the $G-Y$ video signal. Then the three color-difference voltages, $B-Y$, $R-Y$, and $G-Y$, are available for the picture tube. These three color signals go to each control grid or cathode of the three-gun picture tube. The Y signal also is applied to all three cathodes. This method is equivalent to using the picture tube as a matrix to produce red, green, and blue information in the reproduced picture.

Which color video signals are detected depends on the phase of 3.58-MHz reinserted oscillator voltage. For this reason, the stages are called *synchronous demodulators*. Maximum output is produced for the phase of the oscillator input and 180° phase. The output is minimum for the quadrature phase. With two demodulators, two color-difference voltages are detected. The receiver can also use three demodulators for red, green, and blue.

Color oscillator. This stage is a crystal-controlled oscillator, tuned to 3.579545 MHz. Its function is to generate the color subcarrier suppressed in transmission. The output is a carrier wave (cw), meaning it has no modulation.

Burst separator. The 3.58-MHz color sync burst is on the back porch of every horizontal blanking pulse. To obtain the color sync alone, the C signal is coupled into the burst separator. This stage is a 3.58-MHz amplifier, tuned to the color subcarrier frequency, but keyed on for horizontal flyback time only. During trace time, the burst separator is held cut off. Therefore, the output is produced for 3.58-MHz signal only for the color sync burst during flyback time. This stage amplifies the color burst alone, without the chrominance information.

Although the burst separator is a tuned amplifier for 3.58 MHz like the chrominance amplifier, they have opposite times for conduction. The burst separator is on during flyback time and off during trace time. The chrominance amplifier is on during trace time to supply the chrominance signal needed for the color demodulators.

Color synchronization. The amplified output of the burst separator supplies the required color synchronizing voltage for the 3.58-MHz color subcarrier oscillator. Although the oscillator has a 3.579545-MHz crystal for frequency stability, its phase must be accurately controlled for correct hues in the picture. Therefore, the color sync voltage goes to an AFPC circuit. This abbreviation is for automatic frequency and phase control. Its function is to control the oscillator phase. The color AFPC circuit holds the hue of the reproduced colors at the correct values, after the manual *hue* or *tint* control has been set.

Color killer stage. This name describes its function, which is to cut off the color amplifier for programs broadcast in monochrome. The purpose is to prevent Y signal components in the 3.08- to 4.08-MHz range from being amplified in the color circuits. The result would be color noise, called *confetti*, in the picture, which looks like snow but with larger spots in color.

The receiver automatically recognizes a color or monochrome signal by the presence or absence of the 3.58-MHz color sync burst. This voltage is rectified to provide a dc bias that cuts off the color killer. When the killer is off, the bandpass amplifier is on for a color program.

Most receivers have a color killer adjustment on the back of the chassis. Set the control at the point where it just barely cuts off the color for a monochrome program. In some circuits,

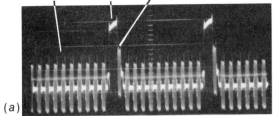

(a)

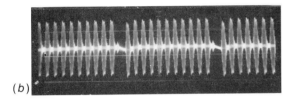

(b)

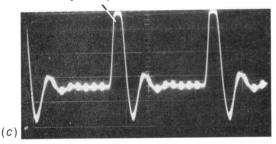

(c)

(d)

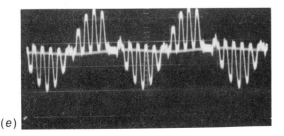

(e)

the best point is where the color just comes in. For both cases, check that color is normal on a color program.

Because of the color killer, it is important to note that the trouble of no color sync burst causes no color, instead of no color synchronization. The reason is that no burst corresponds to a monochrome signal. This makes the color killer cut off the color amplifier.

Tuning in the color. For a program that is being broadcast in color, the fine tuning control must be set to provide chrominance signal for the receiver. Without the rf side bands corresponding to the 3.58-MHz color signal, the picture is reproduced in black and white by the luminance signal. With the fine tuning control adjusted, the color is maximum just off the point where the 920-kHz beat interferes in the picture. This is the beat frequency between the 4.5-MHz sound and 3.58-MHz color signal. The result is about 60 diagonal bars with FM "wiggles" drifting through the picture. See color plate XIII.

Oscilloscope waveforms for color signals. In order to provide red, green, and blue video signals for the picture tube, the main requirements of the color section are indicated by the oscilloscope waveforms in Fig. 9-4. For transistor circuits, it is preferable to check voltages at the collector output, rather than the base input. The base-emitter junction with forward bias is actually a nonlinear diode that can distort the oscilloscope waveforms.

FIGURE 9-4 OSCILLOSCOPE WAVEFORMS IN COLOR CIRCUITS, WITH COLOR-BAR GENERATOR. (a) VIDEO DETECTOR OUTPUT. (b) 3.58-MHz C SIGNAL OF BANDPASS AMPLIFIER. (c) INPUT TO BURST AMPLIFIER. (d) 3.58-MHz OSCILLATOR CW OUTPUT. (e) R − Y VIDEO SIGNAL FROM DEMODULATOR.

It should be noted that the oscilloscope need not have response to 3.58 MHz in order to see the color video signals at 0 to 0.5 MHz. Furthermore, high-frequency response in the oscilloscope is used only to see individual cycles of the 3.58-MHz signal. Usually, it is just a question of seeing the amplitude of the signal. The main requirements are as follows:

1. There must be 3.58-MHz C signal into and out of the chrominance bandpass amplifier. See Fig. 9-4a and b. Without the C signal there is no color.
2. There must be color burst to synchronize the color oscillator. The burst separator must have the horizontal keying pulse shown in Fig. 9-4c to produce the separated burst.
3. There must be 3.58-MHz cw output from the oscillator, in order for the demodulator to detect the color information in the C signal. See Fig. 9-4d. Without the oscillator output there is no color.
4. Finally, each color video amplifier must supply detected signal to drive the three guns of the picture tube. See Fig. 9-4e. Here there can be a problem of individual colors missing, if one amplifier or one gun is not operating.

The waveforms in Fig. 9-4 are taken with a signal from a color-bar generator. A typical dot-bar generator is shown in Fig. 11-3, while color plate XII shows the rainbow of color bars for every 30° of hue phase.

9-5 Manual Color Controls

The only two additional operating controls for color receivers are the color level control and tint control. These are on the front panel of the receiver.

The color control varies the amplitude of 3.58-MHz chrominance signal voltage, usually adjusting the gain of the chrominance bandpass amplifier. As a result, the intensity or amount of color in the picture is varied. The color control is also called *saturation* or *chroma*. This is R_1 in Fig. 9-3.

The tint or hue control varies the phase of the 3.58-MHz oscillator cw output with respect to the color sync burst. The control can be in either the oscillator or its AFPC circuit. This is C_1 in Fig. 9-3.

9-6 Functions of the Automatic Color Circuits

The color receiver usually has these circuits to produce the best picture without manual adjustments.

Automatic color control (ACC). This circuit controls the gain of the chrominance bandpass amplifier for a constant color level in the picture. The idea is the same as automatic gain control (AGC) for the 3.58-MHz color IF amplifier, to prevent overload distortion. The strength of the color signal is indicated by the amplitude of burst, which is rectified to provide the dc bias for the ACC circuit.

Automatic fine tuning (AFT). This circuit controls the frequency of the local oscillator in the rf tuner to keep it tuned for the best color. There is an on-off switch to disable the AFT for cases where manual fine tuning is needed to reduce interference.

Automatic tint control (ATC). This circuit shifts the hue axis for more red to emphasize flesh tones. There is usually an on-off switch to disable the ATC if necessary.

One-button tuning. This switch turns on the AFT and ATC, and it sets the color to a preset level.

AFPC. This circuit controls the frequency and phase of the 3.58-MHz color oscillator, by comparing its output to burst phase. The AFPC is for color lock, to hold the hues at their correct values.

Automatic degaussing (ADG). This circuit demagnetizes the color picture tube each time the receiver is turned on or off, for good purity.

Automatic brightness limiter (ABL). This circuit keeps the brightness from becoming too high, to prevent blooming in the picture.

9-7 R − Y, B − Y Receivers

In this system, the luminance and video signals are matrixed in the picture tube (Fig. 9-5). It serves as the matrix by combining the beam currents from the red, green, and blue guns. Assume that $R - Y$, $B - Y$, and $G - Y$ color video signals are coupled to each of the control grids. Also, $-Y$ video signal is coupled to the three cathodes. This signal has positive sync polarity for the composite video and negative-going white amplitudes for cathode drive on the picture tube. Remember, though, that an increase in negative voltage at the cathode increases the beam current, equivalent to a positive change in grid voltage.

With $-Y$ signal at the cathodes and color signals at each grid, they are combined as follows:

$B - Y - (-Y) = B - Y + Y =$ blue
$R - Y - (-Y) = R - Y + Y =$ red
$G - Y - (-Y) = G - Y + Y =$ green

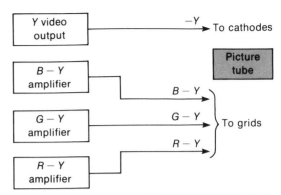

FIGURE 9-5 MATRIXING IN PICTURE TUBE WITH $R - Y$, $B - Y$ DEMODULATORS.

The $-Y$ signal is subtracted because the effect on beam current is opposite for cathode voltage, compared with control-grid voltage.

The matrixing to provide red, green, and blue applies only to the 0- to 0.5-MHz band for color. Actually, for fine detail in a black-and-white picture, the Y signal can have frequencies up to 4 MHz. The red, green, and blue information provides a color overlay for larger areas in the picture, with video frequencies up to 0.5 MHz for the color baseband.

9-8 X and Z Demodulators

Figure 9-6 illustrates this variation in $R - Y$, $B - Y$ receivers, using the X and Z hue axes for detecting the color signal. For the X demodulator, the phase of its oscillator cw input is 282°, clockwise from burst, only 12° past $- (R - Y)$ at 270°. For the Z demodulator the phase angle of its hue axis is 333°, which is close to $- (B - Y)$ at 360°. The X and Z axes, then, are 51° apart. Although not in quadrature, the X and Z demodulators provide the advantage of balanced amplifiers for the detected color video

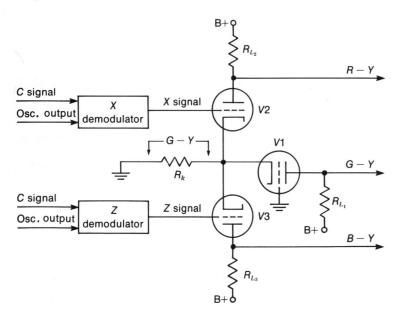

FIGURE 9-6 X AND Z DEMODULATORS USED IN $R-Y$, $B-Y$ RECEIVER.

signals. Then any drift with aging components is the same for all the colors.

The balance in Fig. 9-6 results from using identical values of load resistors R_L. Furthermore, the cathode resistance R_k is common for all three amplifiers. This means the signal voltage across R_k for V1 is combined with the grid signal for V2 and V3.

We can start with the output for $G-Y$ color video signal from V1. By the choice of component values, $G-Y$ is the color video signal developed across R_k. This is the signal input to the grounded-grid amplifier V1. There is no other input signal because the grid is grounded. Since the input signal is at the cathode, there is no phase inversion, resulting in amplified output of $G-Y$ signal from the plate of V1.

The demodulated input signal for V2 is X signal, from grid to ground. However, it has a large component of $-(R-Y)$ because this hue axis and the X axis are so close in phase. There is also some $G-Y$ signal in the grid voltage.

However, R_k in the cathode circuit has $G-Y$ signal. This cancels the $G-Y$ signal at the grid. The combined result of the two signals between grid and cathode is $-(R-Y)$ signal. This is amplified and inverted to produce $R-Y$ output from V2.

The demodulated input signal for V3 is Z signal from grid to ground. This has a large component of $-(B-Y)$ signal because this hue axis and the Z axis are close in phase. There is also some $G-Y$ signal at the grid. With $G-Y$ signal across R_k, the combined result of the two signals between grid and cathode is $-(B-Y)$ signal. This is amplified and inverted to produce $B-Y$ output from V3.

In summary, the functions of the three amplifier stages are as follows:

1. The $G-Y$ amplifier V1 has $G-Y$ input at the cathode and $G-Y$ output from the plate.
2. The $R-Y$ amplifier V2 has $G-Y$ input at the cathode and X input at the grid. The

phase of the X demodulator is 282° to make the X signal have the required components of $-(R-Y)$ signal to be amplified in V2 and $G-Y$ to be canceled by the signal voltage across R_k.

3. The $B-Y$ amplifier V3 has $G-Y$ input at the cathode and Z input at the grid. The phase of the Z demodulator is 333° to make the Z signal have the required components of $-(B-Y)$ signal to amplify in V3 and $G-Y$ to be canceled by the signal voltage across R_k.

It should be noted, though, that the color video signals for the picture tube are still $R-Y$, $B-Y$, and $G-Y$. These are matrixed with the Y signal in the picture tube, to provide red, green, and blue for the picture reproduction.

9-9 R, G, B Receivers

In solid-state color circuits, the matrix method in Fig. 9-7 is often used. Here, the Y signal is combined with the color video signals before the picture tube. As a result, the cathode of each gun has video signal consisting of both the luminance and the R, G, or B component.

One feature of the R, G, B method is that it can provide better tracking of color wth luminance for each gun. The reason is that each video output stage supplies one video signal combining the color and luminance information. In addition, dc coupling to the picture tube provides the correct dc level for both color and luminance. However, gray-scale adjustments are easier with an $R-Y$, $B-Y$ receiver, as the screen and drive controls are adjusted for a good monochrome picture. In the R, G, B receiver, the brightness and contrast controls are in the video preamplifier, before the matrixing of the Y and color signals.

An important factor is arcing or flashover to the control grid in the picture tube. With tran-

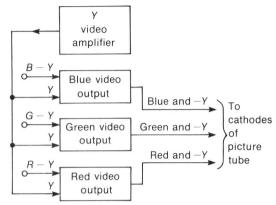

FIGURE 9-7 Y SIGNAL ADDED TO COLOR VIDEO BEFORE PICTURE TUBE IN R, G, B RECEIVER.

sistors, a voltage surge caused by the arc can damage the amplifier feeding signal to the control grid. In the R, G, B method, usually no video signal is applied to the control grids.

The R, G, B receiver requires three video output stages. Each has the 4-MHz bandwidth of the Y signal, instead of the 0.5-MHz bandwidth of each color video signal. Bandwidth is not a big problem with transistor circuits, however, because of their low impedance.

The general idea of combining the Y and color video signals is illustrated in Fig. 9-8. Across R_1 in the input, the signal is Y video alone. In addition, the coupling transformer T_1 is used to couple $B-Y$ video for this example. The same method can be used with $R-Y$ or $G-Y$ video. With the two input signals, the output across R_2 includes $B-Y$ plus Y video. They combine to equal blue video alone for color.

It should be noted that the cancellation of the Y component applies only to the 0- to 0.5-MHz band of frequencies in the color video. For high video frequencies from 0.5 up to 4 MHz, the Y signal is not canceled. The Y signal alone has the details of picture information in monochrome.

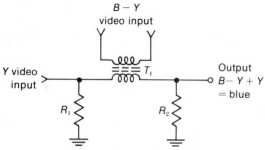

FIGURE 9-8 GENERAL METHOD OF COMBINING Y VIDEO WITH COLOR-DIFFERENCE SIGNAL TO PRODUCE R, G, OR B VIDEO.

9-10 How the Picture Tube Mixes Colors

You do not see the individual dots or stripes of phosphor colors on the screen. Therefore, the eye integrates the effects of red, green, and blue light emitted by the phosphors. The intensity of each color increases with more beam current. Furthermore, the amount of beam current is controlled by the cathode-to-grid signal voltage.

To make a white raster on the screen, the beam currents for the red, green, and blue guns are proportioned to produce white. The dc voltages for the electron guns are set by the screen adjustments for a white raster at a low light level. For a lighter white in the picture information, the red, green, and blue beam currents are increased in proportion by video signal for the three guns. Black in the picture is just the absence of light, as a result of cutting off beam current.

Suppose the picture information is blue. Then the blue gun operates, without any beam current from the red and green guns. Similarly, red or green can be produced by beam current from each gun, without any contribution from the other colors.

Suppose the picture information is yellow. Then the red and green guns operate without blue. For orange, more red can be added. Similarly, the green and blue guns produce beam current for the combination of cyan. Also, red and blue are added for magenta or violet.

This addition of colors by the picture tube illustrates why the inverted polarity of a color video voltage is its complementary color. Consider white information, combining red, green, and blue. Suppose that the blue component of video voltages is inverted in polarity. In terms of the picture tube, we can consider that positive blue video grid voltage increases the beam current for the blue gun. When the polarity is inverted to negative blue video voltage, now this video signal decreases the beam current. However, the red and green components stay the same. The yellow combining red and green is stronger, therefore, with less blue. If the blue is removed completely, the result will be yellow alone. The result, then, is that negative blue corresponds to yellow. Similarly, negative red corresponds to cyan with blue and green. Also negative green corresponds to magenta with red and blue. In each case, the negative polarity refers to inverting the effect on the amount of beam current in the picture tube.

9-11 Localizing Color Troubles

It can be useful to consider that the color is superimposed on a monochrome picture on a white raster. White in the raster is set by the combination of R, G, and B screen-grid controls for the picture tube. The monochrome picture is produced by the combination of R, G, and B drive controls for Y video signal to the three guns. Color is added by the 3.58-MHz chrominance signal, demodulated into separate color video signals. The 3.58-MHz C signal is for all the colors, but the demodulated video signals are for individual red, green, or blue.

Another important factor in separate colors is the picture tube with its three electron guns. When one gun does not operate, that color is missing. Typical troubles in one gun are weak emission or a cathode-heater short. The short circuit to the grounded heater kills video signal coupled to the cathode.

Raster not white. This can be caused by the wrong dc bias voltages at the cathode of each gun, or by incorrect settings of the *R, G,* and *B* screen controls on the color picture tubes. Then the raster has too much of one color, or not enough, resulting in a raster that has color instead of being white.

The trouble of an individual color in the entire raster can result from a defective picture tube. Weak emission from the cathode and a short circuit at the cathode or control grid are typical troubles for one gun alone.

It should also be noted that the color video amplifiers are often dc-coupled to the picture tube. Then a trouble that changes the dc voltages in the amplifier stage will affect the bias voltage for each gun. The dc coupling is illustrated in Fig. 9-9. Here the bias is $200 - 140 = 60$ V, negative at the control grid of the picture tube.

Monochrome picture not white. Color in the raster will also cause color in a monochrome picture. However, the raster can be white while the monochrome picture has color when the *Y* video signals are not balanced for the three guns. Check the *R, G,* and *B* drives for *Y* signal to each gun. Also, a cathode-to-heater short in one gun means this component of the *Y* signal will be missing.

No color. Assuming a normal raster and a good monochrome picture, no color at all means there is no 3.58-MHz chrominance signal or no

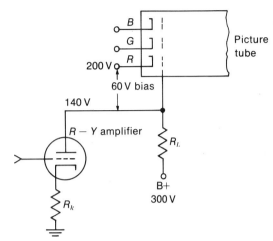

FIGURE 9-9 DC COUPLING TO PICTURE TUBE DETERMINES ITS BIAS.

3.58-MHz oscillator cw output. These stages are necessary to all the color video signals. Remember that the color killer cuts off the bandpass amplifier if there is no burst signal.

Weak color. This means weak 3.58-MHz color signal. Check the chroma bandpass amplifier. However, it must have normal color signal from the rf, IF, and video circuits.

One color missing. Check if the monochrome picture is white. If not, one gun may be shorted for *Y* signal. If white is normal, then one of the $R - Y, B - Y$, or $G - Y$ color video signals may be missing. Where the picture tube is used as the matrix, one gun can be shorted for *Y* video at the cathode but not for color video at the grid.

Wrong hues. Check that the raster and monochrome picture are white. If they are, then check color oscillator phase. Correct alignment of the AFPC circuit and chroma bandpass amplifier is necessary for correct hues.

Color bars in picture. See color plate VI. These bars indicate no color sync. Check the AFPC circuit.

Changing hues in picture. This effect indicates weak color sync. Check the AFPC circuit.

TABLE 9-1 FUNCTIONS OF THE COLOR STAGES (for block diagram in Fig. 9-2)

CIRCUIT	INPUT	OUTPUT	FREQUENCIES	FUNCTION	NOTES
Chrominance bandpass amplifier	C signal from first video	C signal for color demodulators	3.58 ± 0.5 MHz	Amplifies C signal	Tuned amplifier. Has manual color control.
Color oscillator	Dc control voltage from AFPC circuit	Cw to color demodulators	3.58 MHz	Reinserts color subcarrier	Uses 3.579545-MHz crystal. Has manual tint control.
B − Y demodulator	C signal plus oscillator cw	B − Y video	Input is 3.58 MHz. Output is 0 to 0.5 MHz.	Detects color video	Oscillator cw is at 0° phase angle.
R − Y demodulator	C signal plus oscillator cw	R − Y video	Input is 3.58 MHz. Output is 0 to 0.5 MHz.	Detects color video	Oscillator cw is at 90° phase angle.
B − Y amplifier	B − Y video	B − Y video	0–0.5 MHz	Amplifies color video	Drives picture tube cathode or grid.
R − Y amplifier	R − Y video	R − Y video	0–0.5 MHz	Amplifies color video	Drives picture tube cathode or grid.
G − Y adder	B − Y and R − Y video	G − Y video	0–0.5 MHz	Amplifies color video	Drives picture tube cathode or grid.
Burst separator	C signal and H flyback pulses	Color sync burst	3.58 MHz	Separates burst from C signal	Conducts only during H flyback time.
AFPC	Burst from separator and cw from oscillator	Dc control voltage for color oscillator	Input is 3.58 MHz. Output is dc control voltage.	Controls frequency and phase of color oscillator	Rectified burst can also be used for ACC and color killer.
Color killer	Dc bias from AFPC	Dc bias for bandpass amplifier	Dc voltage	Cuts off color amplifier	For monochrome picture

SUMMARY

The stages in the color section of the receiver are summarized in Table 9-1, with their input and output signals, frequencies, and functions. Notice there is only one chroma signal at 3.58 MHz before the demodulators, but after detection the color information is in separate color video signals with a bandwidth of 0.5 MHz.

Self-Examination (Answers at back of book)

Answer True or False.
1. The color saturation control varies the amount of C signal.
2. The tint control varies the phase of the color oscillator output with respect to burst.
3. The chrominance bandpass amplifier is tuned to 3.58 MHz.
4. The $G - Y$ adder stage is tuned to 3.58 MHz.
5. The R, G, B drive controls vary the amount of Y signal.
6. The R, G, B screen controls vary the amount of C signal.
7. The color killer is cut off on a color program.
8. The $R - Y$ demodulator is in quadrature with the $B - Y$ demodulator.
9. $B - Y$ phase is 180° opposite from burst phase.
10. The burst amplifier separates the color sync burst from the C signal.
11. The 3.579545-MHz crystal oscillator is needed to reinsert the color subcarrier.
12. Color bars in the picture indicate no color AFPC.
13. When $R - Y$, $B - Y$, and $G - Y$ color video signals are coupled with Y signal to the picture tube, it serves as the matrix.
14. The normal-service switch is used for purity and screen adjustments on the picture tube.
15. Color receivers generally use the video detector to produce the 4.5-MHz intercarrier sound signal.
16. A 920-kHz beat pattern in the picture is interference between the color and sound signals.
17. When blue video signal is inverted in polarity, the hue becomes yellow.
18. If one gun does not operate, the monochrome picture will not be white.
19. If the color oscillator does not operate, there will be no color in the picture.
20. A white raster depends on the gain of the $R - Y$, $B - Y$, and $G - Y$ amplifiers.

Essay Questions

1. Give five differences between color and monochrome receivers, not counting the circuits in the color section.

2. Give the frequencies and functions for the following stages: (a) chrominance bandpass amplifier; (b) color oscillator; (c) $B - Y$ demodulator; (d) $B - Y$ amplifier; (e) burst separator.
3. Describe briefly how the color control and tint control function.
4. Describe briefly how to set the fine tuning control manually for color in the picture.
5. What controls are adjusted for a white raster?
6. What controls are adjusted for a white picture?
7. Give the two input voltages and one output for the burst separator.
8. Give the two input voltages and one output for the $R - Y$ demodulator.
9. For the following abbreviations of automatic circuits, give the meaning and the circuit each controls: AFT, ACC, ATC, AFPC, ADG, and ABL.
10. Give the functions for the three positions of the normal-service switch.
11. Compare the methods of matrixing for an $R - Y$, $B - Y$ receiver and R, G, B receiver.
12. How do X and Z demodulators differ from $R - Y$ and $B - Y$ demodulators?
13. Give the function of the color killer stage.
14. Give two color troubles in the picture with a possible cause for each.
15. Why is the color level control in the chroma amplifier, rather than the $B - Y$, $R - Y$, or $G - Y$ amplifier?
16. Explain briefly why the brightness control in the Y video amplifier can vary the bias for all three electron guns in the picture tube.
17. Why is the burst separator on during horizontal retrace time?
18. Why is the chroma amplifier on during horizontal trace time?

Problems

1. For channel 3, 60 to 66 MHz, give the following frequencies in the receiver: (a) rf local oscillator in the VHF tuner; (b) picture, color, and sound carriers in the rf amplifier; (c) picture, color, and sound carriers in the picture IF amplifier; (d) color signal out of the video detector.
2. Referring to Fig. 9-9, if the $R - Y$ amplifier does not conduct any plate current, calculate the (a) plate voltage; (b) grid bias for picture tube.

10 Picture Tubes

As shown in Fig. 10-1, the picture tube is a cathode-ray tube (CRT) consisting of an electron gun and phosphor screen inside the evacuated glass envelope. The narrow neck contains the electron gun to produce a beam of electrons. The beam is accelerated to the screen by positive anode voltage. The anode is a conductive coating on the inside surface of the wide glass bell. To form the screen, the inside of the faceplate is coated with a luminescent material that produces light when excited by electrons in the beam. A monochrome picture tube has one electron gun and a continuous phosphor coating that produces a picture in black and white. For color picture tubes, the screen is formed with dot trios or vertical lines of red, green, and blue phosphors. There are three electron beams, one for each color phosphor. More details are described in the following topics:

    10-1   Types of picture tubes
    10-2   Deflection, focusing, and centering
    10-3   Screen phosphors
    10-4   The electron beam
    10-5   Electrostatic focusing
    10-6   Magnetic deflection
    10-7   Color picture tubes
    10-8   Picture tubes with in-line beams
    10-9   Grid-cathode voltage on the picture tube
    10-10   Picture tube precautions
    10-11   Picture tube troubles

186 BASIC TELEVISION

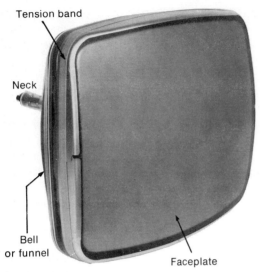

FIGURE 10-1 COLOR PICTURE TUBE, TYPE NUMBER 25UP22. (*RCA*)

10-1 Types of Picture Tubes

If we take the type number 25UP22 as an example, 25 in. is the diagonal of its rectangular screen, within $1/2$ in. If the dimension falls exactly on 0.5 in., the next larger number is used. The letters in the middle of the type designation are assigned alphabetically, in order of registration with the EIA.

The P-number at the end of the type designation specifies the phosphor screen. This is P4 for all black-and-white picture tubes. For color, the number is P22 for all tubes with red, green, and blue phosphors.

Many picture tubes have the letter "V" just after the screen size, as in 18VBKP22. The V is for viewing dimensions. Without a V, the screen size is the outside glass diagonal. With a V, the diagonal indicates the minimum viewing area of the screen itself. In this case, the glass faceplate may be little larger than the nominal screen size.

Heater voltage. This voltage is not specified by the type number, but the heater rating is usually 6.3 V. A few picture tubes for battery operation use 2 to 4 V. The heater current is generally 450 or 600 mA for monochrome tubes with one electron gun. For color tubes, typical ratings are 800 to 1,800 mA at 6.3 V. Although there are three electron guns, the heaters are parallel internally to provide just one pair of pins for the heater voltage. The two end pins on the base connections are usually for the heater.

Anode voltage. The anode is a conductive graphite coating, generally called *aquadag*, on the inside glass wall. This coating on the wide bell extends from the faceplate about halfway into the narrow neck. As a result, the electric field of the accelerating potential is symmetrical around the electron beam. Typical values of anode voltage are 18 kV for a 19-in. monochrome tube and 25 kV for a color tube. More high voltage is needed for the same brightness with color tubes because of losses through the aperture mesh used with a three-color screen. The anode current, which is the load for the high-voltage power supply, is typically 0.6 mA for a monochrome picture tube and 1.8 mA for a color tube with three guns.

A separate anode connection is provided at the top or side of the glass bell. The connection is through a recessed cavity of about $1/4$ in. diameter. Into this fits a metal ball or spring clip with a wire from the high-voltage rectifier to contact the anode coating on the inside wall. When installing a picture tube, put the anode connection at the same side as the high-voltage cage on the chassis.

External wall coating. Picture tubes also have a graphite coating on the outside surface of the glass bell. The external coating must be connected to chassis ground. Usually, a grounded wire spring on the mounting for the picture tube brushes against the outside coating.

High-voltage filter capacitance. The grounded coating on the picture tube provides a filter capacitance for the high voltage on the anode. The external coating is one conductor; the anode coating is the other conductor, with the glass bulb serving as the insulator between the two. This filter capacitance is about 2,000 pF. The capacitance can hold its charge for a long time after the anode voltage is turned off. Before handling a picture tube, therefore, make sure the high voltage is discharged by shorting the anode button to the grounded wall coating.

Input capacitance. This is C_{in} for video signal applied to the electron gun, either at the control grid or at the cathode. Usually, the video signal is applied to the cathode, with about 5 pF for C_{in}.

Faceplate. Approximately $1/2$-in. thickness provides the strength required for the large faceplate to withstand the air pressure on the vacuumed glass envelope. There is also a safety-glass window. This window was separate in older receivers, but most picture tubes now have a protective glass panel sealed to the faceplate. The panel is usually made of a dark-tint glass to improve picture contrast.

Deflection angle. This is the maximum angle the beam can be deflected without striking the side of the bulb. Typical values of deflection angle are 70, 90, 110, and 114°. The deflection angle is the total angle. For instance, a deflection angle of 110° means the electron beam can be deflected 55° from center. The angle for the picture tube specifies the deflection angle for the deflection yoke.

The advantage of a larger deflection angle is that the picture tube is shorter. Then it can be installed in a smaller cabinet. However, a larger deflection angle requires more power from the deflection circuits. For this reason, these tubes have a narrow neck to put the deflection yoke closer to the electron beam. A deflection yoke for a 110° tube has a hole diameter of $1 1/8$ in., compared with $1 7/16$ in. for tubes with a smaller deflection angle.

Different screen sizes can be filled with the same deflection angle. For instance, a 90° yoke will fill the screen of 17-, 19-, or 21-in. picture tubes if they all have the same deflection angle of 90°. The reason is that bigger tubes with the same deflection angle are longer.

10-2 Deflection, Focusing, and Centering

Either electrostatic or electromagnetic deflection can be used for a cathode-ray tube. For electrostatic deflection, two pairs of metal deflection plates are attached to the gun structure within the tube. Sawtooth voltage applied across each pair of plates provides an electric field to deflect the beam. This method is generally used in oscilloscope tubes, with anode voltage of 5 kV or less. However, picture tubes with anode voltage of 15 kV or more would require too much sawtooth deflection voltage. In addition, the relatively small angle for electrostatic deflection would result in a picture tube that is too long.

Picture tubes use magnetic deflection, therefore, with two pairs of deflection coils in the yoke housing on the neck of the picture tube. The associated magnetic field of sawtooth current in the deflection coils moves the electron beam horizontally and vertically to scan the raster.

Deflection yoke adjustment. In Fig. 10-2, note that the yoke is placed forward on the neck of the picture tube, against the wide bell of the

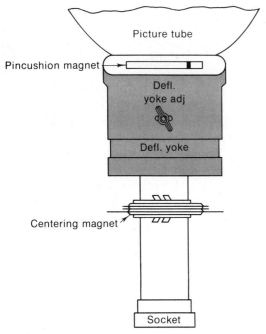

FIGURE 10-2 PLACEMENT OF DEFLECTION YOKE AND CENTERING MAGNETS ON MONOCHROME PICTURE TUBE.

envelope. If the yoke is too far back, the beam will hit the sides of the envelope for large deflection angles. The result, then, is a dark or shadowed corner on the screen. All four corners may be shadowed, especially with 110° yokes, producing a round raster.

The entire yoke can be turned in its housing, by loosening the wing nut at the top. Turning the yoke left or right tilts the raster. Adjust for a raster with edges parallel to the screen, or else the picture will be tilted. For color tubes with three electron guns, one yoke is used to deflect all three beams.

Focus adjustment. The electron beam must be focused to a small spot of light on the screen. Usually, focus is sharpest in the center area. Older picture tubes used magnetic focusing, with a focus magnet on the neck behind the deflection yoke. Now, though, practically all tubes use electrostatic focus. In this method, voltage applied to the focusing electrode of the electron gun controls beam focus.

For monochrome picture tubes, the focusing grid has about 0 to 300 V. There is usually no focus control, as this focusing voltage is not critical. However, there may be a terminal board with connections for different focusing voltages.

In color picture tubes the focus grids of the three guns are connected internally, for one focusing voltage. This is usually one-fifth the anode voltage. For instance, with 25 kV for the anode, the focusing voltage would be 5 kV. This focusing voltage is usually variable. Adjust the focus control for sharp scanning lines in the raster and fine details in the picture.

Centering adjustments. Electrical centering can be done by supplying direct current through the horizontal and vertical deflection coils. Then two rheostats on the chassis serve as horizontal and vertical centering controls. This method is not used often, however, because of the added current drain on the low-voltage power supply.

Monochrome picture tubes usually have a pair of permanent magnet rings for centering, as shown in Fig. 10-2. When the tabs are together, the magnetic poles are opposing. Then there is no field to displace the beam. Spreading the tabs apart increases the field strength. Then rotating the assembly of both rings can move the beam horizontally, vertically, or at an angle, by the amount needed for centering.

Most color receivers do not have any centering adjustments. Actually, the purity ring magnet does essentially the same thing, as the three beams of the trigun picture tube are centered properly when the purity or beam-landing adjustment is correct.

Beam-bender magnet. See Fig. 10-3. This small permanent magnet is clamped around the neck on older monochrome picture tubes, near the base. The magnet is used with tubes that have a bent gun, which aims the electron beam to the side. Then the magnet is positioned to deflect the beam back to the screen. Adjust exactly for maximum brightness, or there will be no raster at all. The purpose of this arrangement is to prevent ions in the electron beam from striking the center of the screen where they can produce a brown burned area called *ion spot*. That is why the beam bender is also called an *ion-trap magnet*. However, all picture tubes now have an aluminized screen, which prevents ions from reaching the screen. As a result, the ion-trap magnet is not used anymore.

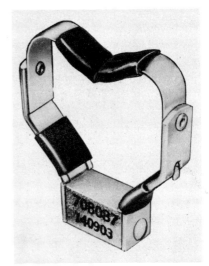

FIGURE 10-3 ION-TRAP MAGNET, USED ON OLDER MONOCHROME PICTURE TUBES.

10-3 Screen Phosphors

Most common are the P1 green phosphor for oscilloscope tubes, the P4 white phosphor for monochrome picture tubes, and the P22 phosphor for color tubes. These are listed in Table 10-1, along with other phosphors for cathode-ray tubes. The missing P-numbers have similar uses, but some types are obsolete.

The phosphor chemicals are generally light metals such as zinc and cadmium in the form of sulfide, sulfate, and phosphate compounds. For the green P1 phosphor, a form of zinc silicate called *willemite* is generally used. The P4 white phosphor usually combines com-

TABLE 10-1 TYPICAL SCREEN PHOSPHORS FOR CATHODE-RAY TUBES

PHOSPHOR NUMBER	COLOR	PERSISTENCE	USES
P1	Green	Medium	Oscilloscopes
P4	White	Medium-short	Monochrome picture tubes
P7	White, yellow	Short, long	Two-layer screen
P14	Blue, orange	Short, long	Two-layer screen
P15	Green-ultraviolet	Very short	Flying-spot scanner
P22	Red, green, and blue	Medium	Tricolor picture tubes

pounds of zinc sulfide, cadmium sulfide, or zinc silicate. This phosphor is actually a combination of yellow and blue, as no one phosphor can produce white. For color screens, the P22 phosphor includes zinc sulfide for blue, zinc silicate for green, and special rare-earth elements such as europium and yttrium for red.

The phosphor material is processed to produce very fine particles which are then applied to the inside of the glass faceplate. This very thin coating to form the screen is a uniform layer for monochrome tubes. For color tubes, though, the phosphor is deposited in dots or vertical lines for each color. You can see the individual color dots or lines with a small, portable microscope of 50× power against the screen while the picture is on. The spaces around the color dots or lines are usually made black to improve contrast.

In terms of molecular structure, the phosphors are crystals where an activator material such as manganese or silver can be added to distort the crystal lattice. Then high-velocity electrons excite the phosphor to emit light. Electrons within atoms of the phosphor are forced to move to a higher energy level. As these electrons fall back to a lower level, energy is radiated. Radiation of light from the screen as it is excited by the electron beam is *luminescence*. When the light is extinguished after excitation, the screen is *fluorescent*. *Phosphorescence* is continued emission of light after excitation.

Screen persistence. This can be defined as the time for light emitted from the screen to decay to 1 percent of its maximum value. Medium persistence is desirable to increase the average brightness and to reduce flicker. However, the persistence must be less than $1/30$ s for picture tubes so that one frame does not persist into the next and cause blurring of objects in motion.

The decay time for picture tubes is approximately 0.005 s, or 5 ms, which is a medium-short persistence. The P1 green phosphor for oscilloscope tubes has a longer persistence of 0.05 s.

Aluminized screen. Practically all picture tubes now have a very thin coating of aluminum on the back surface of the screen phosphor. With anode voltage of 10 kV or more, the electrons in the beam have enough velocity to pass through the aluminum and excite the phosphor. There are several advantages. First, the metal backing reflects light from the screen out through the faceplate. Also, negative ions in the beam cannot penetrate the aluminum coating because the heavy ions do not have enough velocity. An ion-trap magnet is not necessary, therefore, with an aluminized screen. Finally, the aluminum coating is connected to the anode wall coating inside the tube. As a result, the full anode voltage is applied to the screen phosphor.

The first step in applying the aluminum coating is to spray a thin plastic film over the entire screen area. This film prevents aluminum molecules from penetrating the phosphor material. After the plastic film has dried thoroughly, a rod of pure aluminum is evaporated near the screen in a vacuum. The aluminum vapor then condenses on the screen surface to form the aluminized coating. The plastic film is removed later in a baking-out process.

10-4 The Electron Beam

The electron gun in Fig. 10-4 includes a heater, cathode, control grid G1, screen grid or accelerator G2, and focusing grid G3. Each grid structure is a metal cylinder with a small aperture or hole in the center. Figure 10-5 shows a control-grid cylinder and disk cover with an aperture of 0.04 in. The electrodes are generally made of

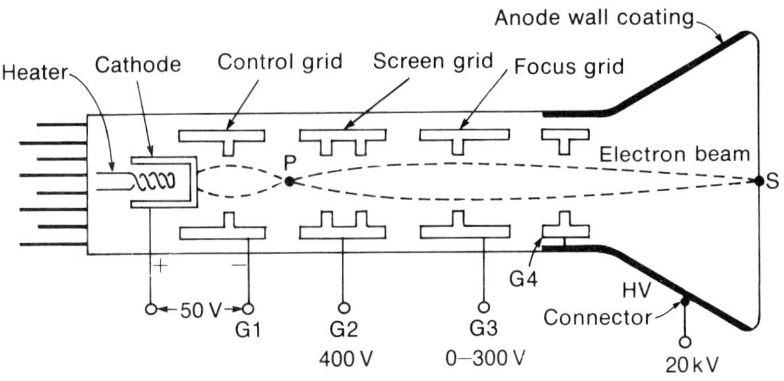

FIGURE 10-4 ELEMENTS OF AN ELECTRON GUN WITH LOW-VOLTAGE ELECTROSTATIC FOCUSING.

nickel or a nickel alloy, mounted on ceramic insulator supports inside the glass neck of the cathode-ray tube.

The control grid G1 has a negative bias with respect to cathode, in order to control the space charge of electrons from the heated cathode. The succeeding grids have positive potentials, with the anode at the highest voltage to accelerate the electron beam to the screen.

Note that the accelerating grid G4 is connected internally to the anode. This is accomplished by spring clips that contact the inside wall coating. Most of the electrons go through the apertures and are not collected by the positive electrodes because their circular structure provides a symmetrical accelerating field around all sides of the beam.

The electron beam has a complete circuit for current because of secondary emission from the screen. These secondary electrons are collected by the aluminum backing, which is connected to the anode. The path for electron flow is from cathode to the screen to the anode and returning to cathode through the high-voltage supply. A typical value of beam current for one gun is 0.6 mA with 20-kV anode voltage.

10-5 Electrostatic Focusing

Electrons emitted from the cathode tend to diverge because they repel each other. However, the electrons can be forced to converge to a point by an electric or magnetic field. This action is similar to focusing a beam of light by optical lenses. Therefore, the term "focusing" is used for producing a narrow beam, while the

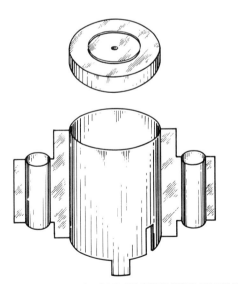

FIGURE 10-5 CONTROL-GRID CYLINDER WITH APERTURE DISK. DIAMETER IS $3/4$ IN.

focusing system is an *electron lens*. Two electron lenses are used. The first is the electrostatic* field between cathode and control grid produced by their difference in potential. This voltage focuses the beam to a spot called the *crossover point*, just beyond the control grid. The second lens may be either an electrostatic field or a magnetic field, to focus the beam just before deflection. As a result of the two electron lenses, the beam is focused to a small sharp spot of light on the screen.

Crossover point. Details of the first electron lens formed by the electrostatic field between cathode and control grid are illustrated in Fig. 10-6. The lines of force in *a* are shown tending to repel electrons back to the cathode because the control grid is negative. The lines are straight where the cathode and grid are parallel. Such straight lines indicate a uniform change of potential in the space between grid and cathode. However, where the grid does not have uniform distances to the cathode, the lines of force curve. Notice that the curved lines of force have the direction that repels electrons toward the center axis. Electrons diverging the most have the greatest force toward the center.

Remember now that the positive G2 voltage and anode voltage provide a forward accelerating force. The net result is that diverging paths are bent so that the electrons go through the grid aperture (Fig. 10-6*b*). The diverging beam then is focused at point *P* just beyond the control grid. Note that electrons emitted in the direction *KA* are made to follow the curved path *KDP*. Similarly, electrons from path *KB* are forced into path *KEP*. Electrons in a straight path along the center axis stay in this line.

---

*Any voltage has an associated electric field, just as any current has an associated magnetic field. When the voltage has a steady value, its field is electrostatic, meaning that it does not vary with respect to time.

The focal point *P* is the crossover point produced by the first electron lens. *P* serves as a point source of electrons to be imaged onto the screen by the second electron lens for a sharp spot. Fine focus can be produced this way because the crossover point is much smaller than the cathode area supplying electrons for the beam.

Low-voltage focusing. The focusing grid, which is usually G3, has 0 to 300 V. This is shown in Fig. 10-4. The electron beam is focused because of deceleration when the G3 voltage is less than the G2 voltage. Most monochrome picture tubes use this method.

High-voltage focusing. The voltage for the focusing grid is generally about one-fifth the anode voltage. For instance, the focusing is 5 kV with 25 kV on the anode. In this type of electron gun, the anode voltage is connected internally to an accelerating grid before the focusing grid. Then the electron beam is decelerated in passing through the electrostatic field between the two grids. Color picture tubes use this method. Generally, there is a control to vary the focus voltage for the sharpest scanning lines and small details in the picture.

10-6 Magnetic Deflection

Two pairs of deflection coils are used, as illustrated in Fig. 10-7, mounted externally around the neck of the tube just before the bell. The pair of coils above and below the beam axis produce horizontal deflection; the coils left and right of the beam deflect it vertically. This perpendicular displacement results because current in each coil has a magnetic field that reacts with the magnetic field of the electron beam to produce a force that deflects the electrons at right angles to both the beam axis and the deflection field.

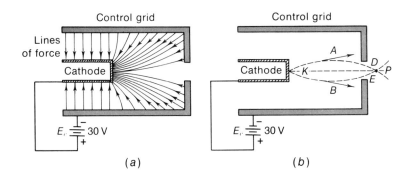

FIGURE 10-6 ELECTROSTATIC FOCUSING BY DECELERATION OF ELECTRONS IN THE BEAM. (a) ELECTRIC LINES OF FORCE BETWEEN CONTROL GRID AND CATHODE. ELECTRON BEAM NOT SHOWN. (b) DIVERGING BEAM FROM K FOCUSED AT CROSSOVER POINT P. LINES OF FORCE NOT SHOWN.

To analyze the deflecting action, remember that the reaction between two parallel fields always exerts a force toward the weaker field. Consider the horizontal deflection coils first in Fig. 10-7. The windings are in a horizontal plane above and below the beam axis. Using the left-hand rule, the thumb points in the direction of the field inside a coil when the fingers curve in the direction of the electron flow around the coil. Therefore, the deflection field for the horizontal windings is downward. When the direction of the electron beam is into the paper, as indicated by the cross in the center, its magnetic field has lines of force counterclockwise around the beam in the plane of the paper. To the left of the beam axis, the magnetic field of the electron beam is down in the same direction as the deflecting field, while the fields are opposing on the right. The electron beam is deflected to the right, therefore, as the resultant force moves the beam toward the weaker field. In a similar manner, the vertical deflection coils deflect the electron beam downward. Deflecting currents for both sets of coils are applied simultaneously, deflecting the beam to the lower right corner of the screen.

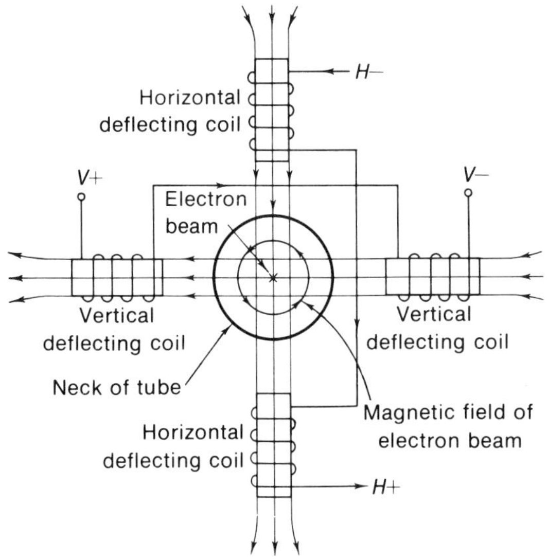

10-7 Color Picture Tubes

The screen has red, green, and blue phosphors, with three electron beams for these primary colors. See Fig. 10-8a and the aperture mask in b. This metal plate has very small holes that allow electrons to go through and excite the phosphor dots. The mask is made of steel 0.006 in. thick. Electrons that do not have the required angle are blocked. For this reason, it is generally

FIGURE 10-7 MAGNETIC DEFLECTION. ELECTRON BEAM WILL MOVE DOWN AND TO THE RIGHT FOR ELECTRON FLOW SHOWN IN SCANNING COILS.

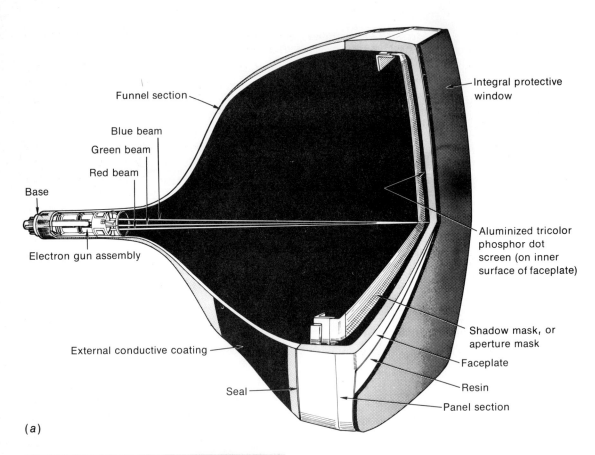

FIGURE 10-8 (a) STRUCTURE OF A COLOR PICTURE TUBE, WITH THREE ELECTRON BEAMS AND PHOSPHOR DOT TRIOS. (*ZENITH RADIO CORPORATION*) (b) PARTIAL VIEW OF APERTURE MASK ALONE TO ILLUSTRATE ITS TRANSPARENCY.

called a *shadow mask*. When the three beams converge at the correct angle, each beam excites its respective color, without straddling the other two colors in the dot trios. You do not see the individual dots because a 21-in. screen has over 450,000 trios. Similarly, the screen can have trios of red, green, and blue vertical stripes.

The screen phosphors produce the colors. All electrons are the same, and any electron beam has no color. However, each color phosphor has its own electron beam. If you operate the red gun alone, by cutting off the other two guns you will see a red raster without video signal or a red picture with signal. The color intensity increases with more beam current. Similarly, the screen can be made all green or blue by operating only one gun.

In normal operation, though, the three guns excite all three color phosphors to produce red, green, and blue on the screen. What you see is the picture in all its natural colors as the red, green, and blue components are superimposed. Just about any color can be reproduced by a mixture of these primary colors. For instance, yellow is a combination of red and green. Furthermore, white is produced by the proper combination of red, green, and blue. Black in the picture is just the absence of excitation when all three beams are cut off.

Figure 10-9 shows the schematic diagram of a color picture tube with three electron guns. There are three separate cathodes and three separate control grids. Separation is necessary so that individual color signals can be coupled to the grid-cathode circuit to modulate the beam intensity. In many circuits, the Y signal goes to the three cathodes, with each color video signal to its control grid. Or, the Y signal with each color video signal goes to its cathode. In this case, the three control grids can have one common connection to chassis ground.

Each gun has a separate screen grid. This voltage is adjusted for the desired cutoff characteristic on each gun.

The three guns have separate focus grids, but these are connected internally to one pin for a common focusing voltage. In addition, the three accelerating grids are connected internally to the anode voltage.

Also connected internally to the anode is the convergence electrode for all three guns. This assembly has pole pieces to concentrate the magnetic field for current in the external convergence yoke. More details of the three-gun assembly can be seen in Fig. 10-10.

Serving for all three guns are the deflection yoke, convergence yoke, and purity magnet mounted on the neck of the picture tube. Their functions are as follows:

1. *Deflection yoke.* Its vertical and horizontal coils deflect the three beams to form the scanning raster.
2. *Convergence yoke.* This yoke has individual adjustments for the red, green, and blue beams to make them converge through the openings in the aperture mask. For each beam, there is a small permanent magnet and coil. The three assemblies for red, green, and blue are symmetrical around the yoke frame, usually with blue at the top. The magnets are adjusted for *static convergence* in the center area of the screen. The coils have correction current for *dynamic convergence* at the top, bottom, left, and right edges of the screen.
3. *Blue lateral magnet.* It moves only the blue beam left or right to help in making the convergence adjustments. This permanent magnet is usually on the purity-ring assembly.
4. *Purity magnet.* Adjusts the three beams to produce pure red, green, and blue without any effect from the other colors. The purity

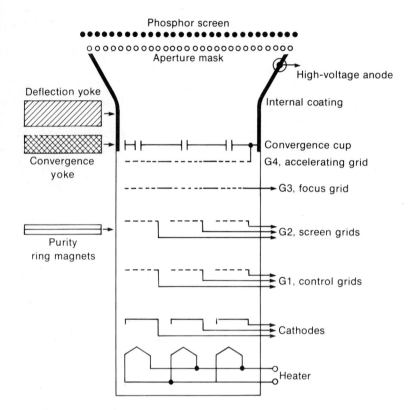

FIGURE 10-9 SCHEMATIC DIAGRAM OF COLOR PICTURE TUBE WITH THREE ELECTRON GUNS. DEFLECTION YOKE, CONVERGENCE YOKE, AND PURITY MAGNETS ARE AROUND THE NECK OF TUBE.

adjustment is also called *beam landing*. The purity magnet consists of two rings, like a centering magnet, for all three guns.

How these components are mounted on the neck of the color picture tube is shown in Fig. 10-11. In addition, note the metal shield over the front section of the tube. The shield minimizes the effect of the earth's magnetic field on the electron beams in the tube.

The details of how to make the convergence and purity adjustments are explained in the next chapter. In brief, though, the test of good convergence is a black-and-white picture without color fringing around edges of objects. This is very noticeable with numbers and letters.

An example of poor convergence is shown in color plate II. The test of good purity is a solid red, green, or blue raster. These can then be combined for a white raster without patches of color. The lack of purity shows as the wrong colors in some parts of the picture. An example of poor purity in the raster is shown in color plate III.

Electron-gun arrangements. See Fig. 10-12. In *a*, the three guns are 120° apart in a circle. This form is called a *delta-gun assembly*. Because of the equal spacing, each gun can have the maximum possible size in the tube neck. However, the convergence adjustments must correct for the guns in different planes.

PICTURE TUBES 197

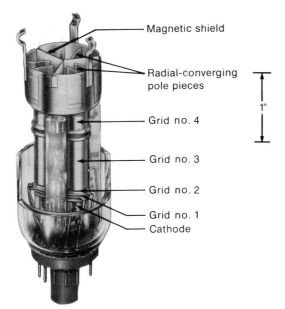

FIGURE 10-10   THREE ELECTRON GUNS IN DELTA ASSEMBLY. (*RCA*)

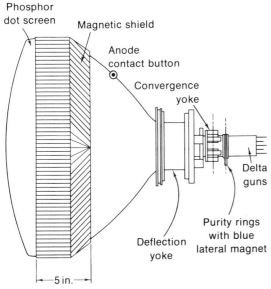

FIGURE 10-11   MAGNETIC COMPONENTS ON NECK OF COLOR PICTURE TUBE.

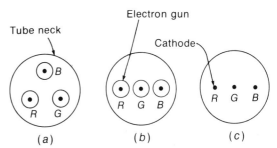

FIGURE 10-12   METHODS OF HAVING THREE ELECTRON BEAMS IN ONE PICTURE TUBE. (*a*) DELTA GUNS. (*b*) IN-LINE GUNS. (*c*) ONE GUN WITH THREE IN-LINE CATHODES.

Usually, the guns are angled in toward the center to help the convergence. In *b*, the three guns are in one horizontal plane, also angled in toward the center. Convergence is easier with this *in-line* type. Each gun must be smaller, compared with the delta guns. With a smaller gun, it is more difficult to obtain a high-intensity spot with sharp focus. In *c*, there are three cathodes to produce three electron beams, but all three beams are focused and accelerated as one common electron gun.

Types of color screens. When the red, green, and blue phosphors are in color dots, they form trios or triads. This color screen is often used with delta guns (Fig. 10-13*a*). However, the color phosphors can also be vertical stripes of red, green, and blue, with a slotted mask (Fig. 10-13*b*). Green is usually at the center. The striped screen is generally used with in-line guns.

With color dots, the shadow mask has holes opposite the triads. With stripes, the mask has vertical slots. In both cases, the shadow mask blocks an appreciable amount of beam current. Otherwise, correct convergence and color purity would be impossible. Typically, less than 20 percent of the beam current excites the phosphor screen. This factor is why color pic-

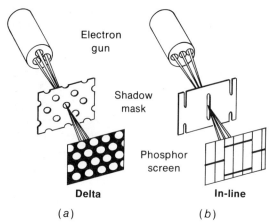

FIGURE 10-13 TWO BASIC TYPES OF COLOR PICTURE TUBES. (a) DELTA GUNS WITH PHOSPHOR-DOT TRIOS. (b) IN-LINE BEAMS WITH SLOTTED MASK FOR VERTICAL PHOSPHOR STRIPES.

ture tubes need higher values of anode voltage, beam current, cathode current, and heater power, compared with monochrome tubes.

10-8 Picture Tubes with In-Line Beams

Two examples are the Sony Trinitron and RCA in-line types. These use in-line beams with vertical phosphor stripes. The GE Portacolor tube uses in-line beams with dot trios. Type numbers for these picture tubes, with their characteristics, are listed in Table 10-3 in the summary at the end of this chapter.

The main feature of in-line guns with vertical phosphor stripes is that convergence and purity are simplified. First, all the beams are in the same plane. Then it is only necessary to adjust the outside beams left and right with respect to the center beam. Also, the vertical stripes make this type less sensitive to beam-landing errors caused by the earth's magnetic field.

Sony Trinitron picture tube. See Fig. 10-14. There is no convergence yoke because electrostatic convergence is used. The phosphor screen of vertical stripes is illustrated in Fig. 10-15. The aperture grille is similar to a shadow mask, but instead of holes the grille has one vertical slot for each set of red, green, and blue stripes. Green is in the center directly behind each slot. The aperture grille increases the transparency for beam current, compared to a shadow mask, allowing more brightness.

The electron-gun assembly of the Trinitron is illustrated in Fig. 10-16. There is actually just one electron gun, but with three cathodes for the red, green, and blue beams. The control grid (G1) is a single cup with three apertures for the three beams. Video signal must be coupled to the cathodes, therefore, because they are the only separate electrodes for each color. The screen grid (G2) is common, and one focus grid is used for all three beams.

The electrostatic convergence is accomplished by the four convergence plates shown in Fig. 10-15. The green beam passes through the center plates, while the other two beams pass between the outer plates. Both center plates are connected internally to the

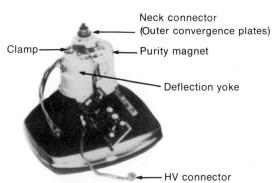

FIGURE 10-14 TRINITRON COLOR PICTURE TUBE. SCREEN DIAMETER IS 12 IN. (SONY CORPORATION)

PICTURE TUBES 199

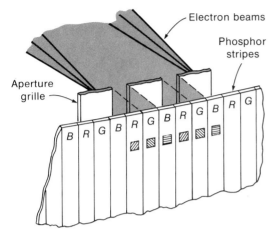

FIGURE 10-15 VERTICAL PHOSPHOR STRIPES ON SCREEN OF TRINITRON COLOR PICTURE TUBE.

RCA in-line picture tube. The construction is illustrated in Fig. 10-17. Continuous vertical phosphor strips are used with in-line beams, but the aperture mask is slotted to provide a spherical shape. The few components mounted on the neck are shown in Fig. 10-18. These include the deflection yoke and a magnet assembly for purity and static convergence. However, these are cemented to the tube, because their position is critical. This arrangement allows permanent factory adjustments, so that very little setup is required for a new receiver or a new picture tube. For replacement, the entire assembly of picture tube, yoke, and magnets is changed as one unit. The magnet assembly on the neck is for static convergence, as there are no adjustments for dynamic convergence.

The deflection yoke uses toroidal windings instead of a saddle yoke. A toroidal coil is made in the form of a ring magnet, to provide a strong magnetic field inside the ring.

The characteristics of type number 15VADP22, as an example, are listed in Table 10-3. This tube has a deflection angle of 90° and anode voltage of 25 kV for the 15-in. screen. There are three separate cathodes, but the control grid is common. Also, the screen grid (G2) and focusing grid (G3) serve for all three guns.

anode. Between the plates, though, the potential difference is zero. As a result, the green beam is not affected by convergence voltage applied to the outer plates. The red and blue beams are moved horizontally for the convergence adjustments. Dc voltage is applied for static convergence. Also, ac correction voltage can be adjusted for dynamic convergence. The voltage for the outer convergence plates is about 450 V less than the anode voltage.

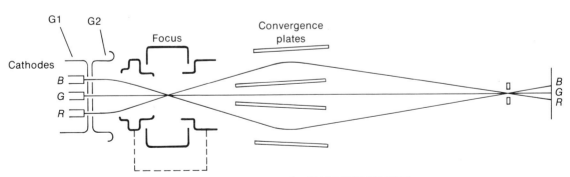

FIGURE 10-16 ELEMENTS IN THE ELECTRON GUN OF TRINITRON COLOR PICTURE TUBE.

10-9 Grid-Cathode Voltage on the Picture Tube

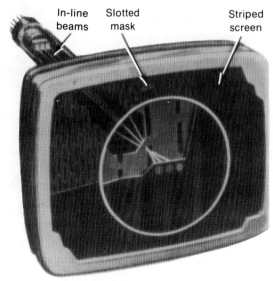

FIGURE 10-17 CONSTRUCTION OF COLOR PICTURE TUBE WITH IN-LINE GUNS AND VERTICAL PHOSPHOR STRIPES. (*RCA*)

For either monochrome or color, and for delta guns or in-line guns, the dc bias voltage between control grid and cathode of the electron gun determines the average beam current or anode current. The effect of two different bias voltages is shown in Fig. 10-19. Note that the bias value A at -60 V allows the anode current of approximately 0.1 mA. This beam current is close to cutoff, resulting in very low brightness on the screen. When the dc voltage is shifted to the bias B at -40 V, the smaller negative bias allows more anode current. This higher beam current is approximately 0.4 mA, producing higher screen brightness. In summary, then, the screen is brighter with less negative grid bias. Methods of supplying the dc bias to control the brightness are shown in Fig. 10-20. The brightness control adjusts the brightness of the raster and the picture on the raster.

Negative dc bias at the control grid. In Fig. 10-20*a*, the bias of -40 V is provided for the control grid, with respect to the grounded cathode.

FIGURE 10-18 DEFLECTION YOKE, CONVERGENCE, AND PURITY MAGNETS ON NECK OF PICTURE TUBE SHOWN IN FIG. 10-17.

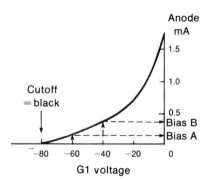

FIGURE 10-19 THE DC BIAS B AT -40 V PRODUCES MORE BEAM CURRENT AND HIGHER BRIGHTNESS THAN BIAS A AT -60 V.

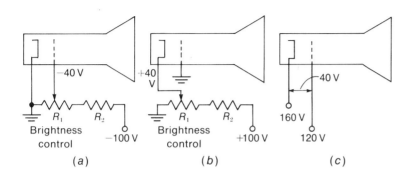

FIGURE 10-20 METHODS OF DC BIAS FOR GRID-CATHODE CIRCUIT OF PICTURE TUBE. (a) NEGATIVE GRID BIAS OF −40 V. (b) POSITIVE CATHODE BIAS OF 40 V. (c) POTENTIAL DIFFERENCE OF 40 V.

The brightness control R_1 varies the bias. Smaller negative values allow more brightness. Higher negative bias can cut off the beam to extinguish the raster.

Positive dc bias at the cathode. In Fig. 10-20b, the bias of 40 V is positive at the cathode, with respect to the grounded grid. The brightness control R_1 operates the same way as in a but with cathode bias. Less positive cathode bias allows more beam current for higher brightness. In the opposite direction, when the positive cathode bias is increased to cutoff, there is no beam current and the screen is black. The raster can be extinguished by cutoff bias whether the cathode is positive or the grid is negative. Actually, cathode bias is generally used for picture tubes, rather than grid bias.

It is the difference in potential between cathode and grid that determines the beam current. An example is illustrated in Fig. 10-20c. Here, both the cathode and grid are positive. The cathode is at 160 V and the grid at 120 V, but these voltages are with respect to chassis ground. The grid is less positive, which makes it negative with respect to cathode. Specifically, for the voltages shown, the potential difference is 160 − 120 = 40 V at the cathode with respect to the grid.

Brightness control for the picture tube. In monochrome receivers, the brightness usually is set by varying the cathode bias on the picture tube, as in Fig. 10-20b. For color tubes with three guns, however, this method would require three controls. Therefore, in color receivers the brightness control varies the dc bias on the Y video amplifier. This stage is dc coupled to the three guns of the picture tube.

How the ac video signal varies the beam intensity. The grid-cathode bias is a dc voltage that sets the brightness of the entire screen area. Then the ac video signal voltage varies the instantaneous values of beam intensity to reproduce the details of picture information. Figure 10-21 illustrates this idea with video signal for one horizontal line. Grid drive is shown in a, with cathode drive in b. For either method, the effect is the same as any ac signal at the control grid of a vacuum tube having a steady dc bias voltage. The bias sets the operating point, as the ac signal varies the beam current around this average level. This procedure of varying the beam current is called *intensity modulation* or *Z axis modulation*.

The peak-to-peak amplitude of the ac video signal determines the contrast in the picture, between peak white with maximum beam

202 BASIC TELEVISION

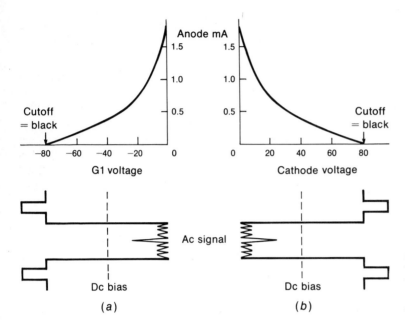

FIGURE 10-21 HOW AC VIDEO SIGNAL VARIES BEAM CURRENT TO REPRODUCE THE PICTURE INFORMATION. (a) NEGATIVE SYNC AND POSITIVE WHITE AT CONTROL GRID. (b) POSITIVE SYNC AND NEGATIVE WHITE AT CATHODE.

current and black at cutoff. The contrast control is in the video amplifier to vary its gain, which determines the amount of video signal for the picture tube. In color receivers, the red, green, and blue drive controls vary the proportion of Y luminance signal for each gun.

Grid drive. In Fig. 10-21a, the video signal for the control grid is shown with negative sync polarity. This means the blanking level drives the grid voltage more negative than the bias, to cut off the beam current for black. Specifically, the grid cutoff voltage here is −80 V. The white peaks in the video signal drive the grid voltage less negative than the bias for maximum beam current.

Cathode drive. In Fig. 10-21b, the video signal for the cathode is shown with positive sync polarity. This means the blanking level drives the cathode voltage more positive than the cathode bias to cut off the beam current. Specifically the cathode cutoff voltage here is +80 V. The white peaks in video signal drive the cathode voltage less positive than the bias for maximum beam current. It should be noted that cathode drive is generally used for the picture tube, because it provides more contrast than grid drive for the same amount of video signal. The transfer-characteristic curves in Fig. 10-22 illustrate how the beam current varies with video signal, for grid drive and cathode drive.

10-10 Picture Tube Precautions

Since the anode voltage is usually 15 kV or more, high-voltage safety precautions should be followed to avoid electrical shock. Turn off the power before touching the picture tube. The anode capacitance should be discharged with a

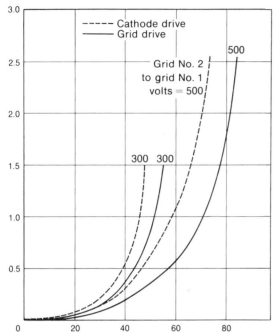

FIGURE 10-22 GRID-CATHODE DRIVE CHARACTERISTICS FOR 23CP4 PICTURE TUBE. (RCA)

jumper lead to chassis ground. Connect the ground end first.

Because of its large surface area, a picture tube has a very strong force of air pressure on the glass envelope. Normal air pressure on the outside is approximately 15 lb/in.^{2}, while the inside is a vacuum. For a 20-in. tube the surface area is about 1,000 in.^{2} and total force on the bulb then is $15 \times 1,000 = 15,000$ lb. The envelope is strong enough to withstand this force if there are no flaws and if the tube is handled gently. Do not carry the tube by its neck. If there is a crack in the glass, the external force can produce an *implosion* inward before the tube explodes outward. It is recommended that shatterproof safety glasses be worn when handling picture tubes for removal and installation.

X-radiation. X-rays are invisible radiation with wavelengths much shorter than visible light. Prolonged exposure to x-rays can be harmful. The x-rays are produced when a metal anode is bombarded by high-velocity electrons, generally with an anode voltage above 16 kV. Color receivers with an anode voltage of 18 to 25 kV can produce soft x-rays. This radiation is easier to shield than hard x-rays produced with much higher accelerating voltages up to 100 kV. Lead and leaded glass are used for shielding against x-ray penetration in general. For soft x-rays, some attenuation is provided by wood, cardboard, pressed paper, metals, and glass.

The main sources of x-radiation in a television receiver are the picture tube, especially from the metal shadow mask, the high-voltage rectifier tube, and the high-voltage regulator tube. However, when solid-state devices are used for the high-voltage supply and its regulation, they do not produce x-rays. Picture tubes with an improved faceplate for x-ray shielding may have the letter "V" in the type designation, or the prefix letters "XR."

Television receivers are designed to limit the level of x-radiation below the value set by the U.S. Department of Health, Education, and Welfare. The limit is 0.5 milliroentgen per hour (mR/hr) as measured at a point 2 in. from any surface of the receiver. This dosage is extremely small, as it probably is less than the radiation from the face of a luminous clock.

For color picture tubes, the x-radiation is mainly a question of not allowing the high voltage to exceed the recommended value. It is required that the receiver not be able to produce a viewable picture when the high voltage exceeds a specified limit. Two methods often used are to cut off the horizontal scanning or to remove horizontal synchronization when the anode voltage gets too high. This system is called a *horizontal disabling circuit*.

Spark-gap protection. Because of close spacing between electrodes and the high voltages, arcing or flashover can occur in the electron gun, especially to the control grid. The arcing causes voltage surges that damage transistors. For protection of the receiver circuits during arcing, spark gaps are used to provide a shunt path for the arc. The idea is to have the arc dissipated across the spark gap, instead of in the receiver circuits. Typical values for the gap breakdown voltage are 2 to 4 kV. Spark gaps are provided for the three control grids, the three screen grids, and possibly the focusing voltage.

Spark-gap capacitors are often used. One type looks like a flat ceramic disk capacitor but with an open slit at the top. The arc-voltage rating for these capacitors is 1.5 or 2.5 kV. This type is sometimes called a *snap capacitor*. At the picture tube socket, a 1,000-Ω resistor may be inserted in series with each lead, except for the heater. In newer picture tubes, spark gaps are incorporated into the base and socket connections.

Dc biasing of the heater. To prevent arcing from the cathode, the heater in a color picture tube usually has a dc bias, in addition to its ac power for heating. The dc offset is about 125 V to chassis ground. This voltage raises the dc potential of the heater close to the cathode voltage. With less potential difference, the danger of arcing is reduced.

10-11 Picture Tube Troubles

Typical troubles are weak emission from the cathode, an open heater, or an internal short circuit, usually from grid to cathode or cathode to

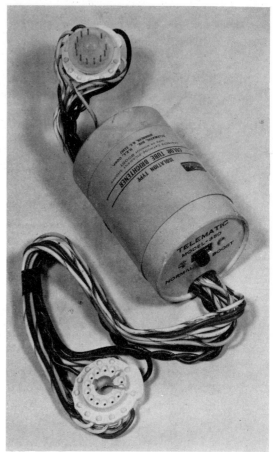

FIGURE 10-23 BRIGHTENER OR BOOSTER FOR HEATER OF PICTURE TUBE.

heater. An open heater means no emission and no brightness. You can look through the glass neck to see if the heater is lit. It is important to note, though, that the most common cause of no brightness is no anode voltage from the flyback high-voltage supply.

With weak emission, the raster and picture cannot be made bright enough. Insufficient brightness can also result from voltage troubles. However, weak emission results in a characteristic silvery effect in white areas of the picture when you turn up the brightness. The cause is saturation limiting of the beam current emitted by the cathode. Then the details are lost in white highlights. This effect also occurs with color tubes. See color plate XIV. When only one gun is weak, it may help to rearrange the cathode leads for Y signal, to supply more drive to the weak gun.

Weak emission in old picture tubes can often be improved temporarily by using a *tube brightener* or *filament booster* (Fig. 10-23). This is just a small filament transformer to step up the heater voltage, usually from 6.3 to 7.8 V. The connections are simple, as the picture tube plug goes on the brightener and its plug on the tube. The brightener must have the same type of socket as the picture tube, and provide for either series or parallel heater connections. Brighteners are available for monochrome or color tubes.

When the brightener uses a transformer with a separate secondary winding, it isolates the heater of the picture tube from chassis ground. This isolation can be used to repair the trouble of a short from cathode to heater in the picture tube. The short is not eliminated, but without the heater connection to ground, there is no effect on the cathode signal.

When a picture tube is replaced, either a new or rebuilt tube can be used. Rebuilt picture tubes cost less because they use the old envelope, or "dud," but the internal parts are all new.

Continuing spot. This luminous spot remains at the center of the screen for a few seconds after the receiver is turned off. The *afterglow* results because anode voltage remains on the filter capacitance. This spot really does not damage the screen. However, one method of eliminating the afterglow is to turn the brightness up for maximum beam current just before you turn off the receiver. Then the high voltage can discharge quickly.

Screen burn. If a spot or line is on the screen, instead of the complete raster, the brightness should be turned down to avoid burning the screen. The burn causes a dark brown area of the phosphor that cannot produce any light. The higher the anode voltage, the greater is the danger of the screen burn. However, it should be noted that when the receiver has a service

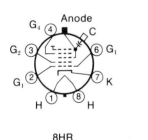

8HR

Anode = $G_3 + G_5$ + CL
Focusing electrode = G_4

(a)

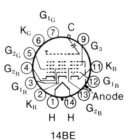

14BE

Anode = $G_4 + G_5$ + CL
Focusing electrode = G_3

(b)

Pin 1: Heater
Pin 2: Cathode of red gun
Pin 3: Grid No. 1 of red gun
Pin 4: Grid No. 2 of red gun
Pin 5: Grid No. 2 of green gun
Pin 6: Cathode of green gun
Pin 7: Grid No. 1 of green gun
Pin 9: Grid No. 3
Pin 11: Cathode of blue gun
Pin 12: Grid No. 1 of blue gun
Pin 13: Grid No. 2 of blue gun
Pin 14: Heater
C: External conductive coating
CL: Collector (anode wall coating)
Anode cap: Anode (Grid No. 4, screen, collector)

(c)

FIGURE 10-24 TYPICAL BASE-PIN CONNECTIONS FOR PICTURE TUBES. (a) 8-PIN BASE 8HR FOR MONOCHROME TUBES. (b) 14-PIN BASE 14BE FOR COLOR TUBES WITH DELTA GUNS. (c) LISTING OF ELECTRODES FOR BASE 14BE.

switch to collapse the raster into a horizontal line, the brightness is automatically reduced by the switch.

Blooming. This term describes a picture that increases in size and goes completely out of focus when the brightness is increased. The blooming results from a sharp drop in the amount of high voltage with the increase in beam current. Usually, the trouble is a weak rectifier having excessive internal resistance, or no regulation in the high-voltage supply.

Breathing. This term describes blooming that occurs at a slow, regular rate. Usually, the trouble is a weak regulator, often used in the high-voltage supply for color picture tubes. Both blooming and breathing are high-voltage troubles, and they do not indicate any defect in the picture tube.

SUMMARY

Typical monochrome picture tubes are summarized in Table 10-2, while the characteristics of some common color picture tubes are listed in Table 10-3. In addition, Fig. 10-24 illustrates base-pin diagrams. The 8-pin base 8HR in *a* is common for black-and-white picture tubes. For the three guns in a color picture tube, the 14-pin base 14BE in *b* is typical. The 70° color tubes are an older type with a round screen and plastic base. The 90° color tubes have a rectangular screen. The neck is narrower with a glass base that has thin pins. In both diagrams, the end pins are the heater connections.

TABLE 10-2 TYPICAL BLACK-AND-WHITE PICTURE TUBES

TYPE NUMBER	DEFLECTION ANGLE	HEATER, V/mA	MAXIMUM ANODE, kV	TYPICAL G2 VOLTAGE	NOTES*
8YP4	110°	6.3/600	22.0	400	Test picture tube *E* focus
11CP4	110°	6.3/450	15.0	400	*E* focus
16RP4	70°	6.3/600	17.5	300	*M* focus; needs *IM*
17DSP4	110°	6.3/600	20.0	400	*E* focus
19EZP4	114°	6.3/450	19.8	45	*E* focus
21AMP4	90°	6.3/600	20.0	300	*M* focus
21FVP4	114°	6.3/600	20.0	400	*E* focus
23CQP4	114°	6.3/600	23.5	500	*E* focus

*IM is ion-trap magnet; *E* is electrostatic focusing; *M* is magnetic focusing.

TABLE 10-3 TYPICAL COLOR PICTURE TUBES

TYPE NUMBER	DEFLECTION ANGLE	HEATER V/mA	ANODE, kV	G2, V	FOCUS* VOLTAGE	NOTES
10VADP22	70°	6.3/900	18	450	17–20%	GE in-line gun; phosphor stripes
15LP22	90°	6.3/900	20	150–390	−75–400 V	Delta gun
15VADP22	90°	6.3/900	25	335–670	17–20%	RCA in-line gun; phosphor stripes
18VAHP22	90°	6.3/900	25	285–685	17–20%	Delta gun
19EYP22	90°	6.3/900	25	285–685	17–20%	Delta gun
19VBLP22	110°	6.3/900	25	265–665	17–20%	Delta gun
19VDCP22	90°	6.3/900	25	170–430	17–20%	GE in-line gun; phosphor-dot trios
21FJP22	70°	6.3/1,800	25	310–690	17–20%	Delta gun
23EGP22	92°	6.3/1,350	25	320–625	17–22%	Delta gun
25YP22	90°	6.3/900	25	285–685	17–20%	Delta gun
1830P22	90°	6.3/900	25	285–685	17–20%	Delta gun; test tube; 19-in. screen
470DLB22	114°	6.3/570	25	100–700	1,100 V	Sony Trinitron; 17-in. screen; phosphor stripes

*Focus voltage given in percent of anode voltage for high voltage focusing, or in volts.

Self-Examination (Answers at back of book)

Answer True or False.

1. The 18VBKP22 is a color picture tube with an 18-in. screen.
2. The 12VAQP4 is a monochrome picture tube with a 12-V heater.
3. When the raster is tilted, it can be straightened by adjusting the deflection yoke.
4. An aluminized screen does not need an ion-trap magnet.
5. A picture tube with a deflection angle of 110° is longer than a 90° tube for the same screen size.
6. The heater connections are usually the end pins on the base of the picture tube.
7. The outside wall coating on a picture tube forms part of the anode capacitance.
8. The video signal is generally coupled to the cathode of the picture tube.
9. The brightness control sets the dc bias for the picture tube.

10. The bias of 60 V at the cathode is equivalent to −60 V for the control grid, when the grid is grounded.
11. Poor color purity causes patches of color in a white raster.
12. Poor convergence causes color fringing in the picture but not in the raster.
13. The anode voltage for a 20-in. color picture tube is about 10 kV.
14. A brightener for the picture tube boosts the heater voltage about 15 percent.
15. With cathode drive for the video signal at the picture tube, the sync pulses have positive polarity.
16. Black in the picture results from cutoff voltage in the grid-cathode circuit of the picture tube.
17. The Trinitron color picture tube uses electrostatic convergence.
18. The 25YP22 color tube listed in Table 10-3 uses a focusing voltage of about 5 kV.
19. Lead and leaded glass are used for x-ray shielding.
20. The spot of afterglow on the screen results when high voltage remains on the anode a short time after the receiver is turned off.

Essay Questions

1. Give the function of the following parts of a picture tube: electron gun, control grid (G1), accelerating grid or screen grid (G2), anode wall coating, phosphor screen, and external conductive coating.
2. Define: deflection angle, crossover point, pincushion magnet, blue lateral magnet, electrostatic focusing, and magnetic deflection.
3. What phosphor number is used for monochrome picture tubes? For color picture tubes? For oscilloscope tubes?
4. Give two advantages of an aluminized screen.
5. Cite two precautions to remember when installing a picture tube.
6. List the voltages required for a picture tube to produce screen brightness.
7. What causes afterglow, or a continuing spot on the screen of the picture tube?
8. Why can the anode retain its high voltage for a short time after the receiver is turned off?
9. How is the high voltage usually connected to the anode?
10. What is a brightener, or booster, for the picture tube?
11. What is the difference between delta guns and in-line guns for color picture tubes?
12. List the components mounted on the neck of a color picture tube with delta guns.
13. Give two types of color picture tubes that do not use delta guns.
14. Draw the circuit of a manual brightness control for a monochrome picture tube from (a) +100-V supply; (b) −100-V supply.

Problems (Answers to selected problems at back of book)

1. The raster is tilted. How would you straighten it?
2. The raster is off to one side, with a wide black bar at the left edge. What would you adjust, for a monochrome tube?
3. Give two causes of no brightness on the screen of the picture tube.
4. Give a possible cause of the following troubles: (a) picture tube six years old has little brightness and weak contrast with a silvery effect in white highlights; (b) picture blooms out of focus and screen becomes black when brightness control is turned up to maximum.
5. Referring to the characteristic curves in Fig. 10-22, give the value of anode mA with grid-cathode voltage 40 V from cutoff, for each of the four curves shown.
6. For the values of grid voltage and beam current listed below, draw the corresponding transfer-characteristic curve.

Grid, V	Beam, µA
−80	0
−60	100
−40	200
−20	500
0	1,000

11 Adjustments for Color Picture Tubes

Figure 11-1 shows a typical mounting assembly for a color picture tube in the receiver cabinet. This is a delta-gun tube, using a shadow mask with a phosphor-dot screen. The magnetic components on the neck include the deflection yoke against the wide bell, the convergence magnet assembly for red, green, or blue, and the purity magnet assembly, which also has the blue lateral magnet near the tube base. The convergence assembly, over the internal convergence cup, usually is about $1/2$ in. from the deflection yoke, and the purity rings are 1 in. behind, at the space between grids 1 and 2. These magnets are all adjusted for correct purity and convergence on the phosphor-dot screen. Also shown is the external magnetic shield, with a wraparound degaussing coil for demagnetizing the picture tube. More details of the setup adjustments for this type of color picture tube are explained in the following topics:

- 11-1 Color purity
- 11-2 Color convergence
- 11-3 Convergence correction waveshapes
- 11-4 Screen-grid adjustments
- 11-5 Degaussing
- 11-6 Pincushion correction
- 11-7 Overall setup adjustments

ADJUSTMENTS FOR COLOR PICTURE TUBES 211

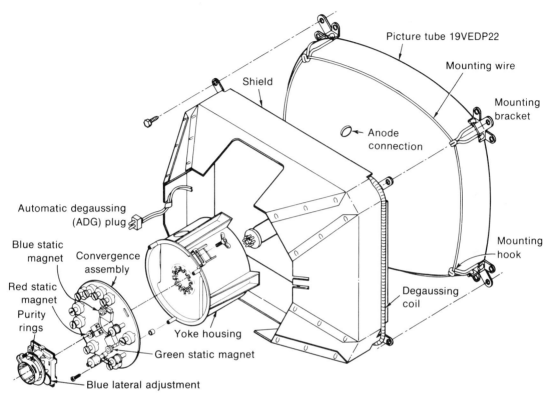

FIGURE 11-1 EXPLODED VIEW OF MOUNTING ASSEMBLY FOR COLOR PICTURE TUBE WITH DELTA GUNS, SHADOW MASK, AND PHOSPHOR-DOT TRIOS. (*RCA CTC71 CHASSIS*)

11-1 Color Purity

Each beam should strike the center of each color dot. Then the light is a pure color, without any mixing from the other two. Obtaining good purity, therefore, is also called the *beam-landing* adjustment. The test of good purity is a white raster that does not have any areas of color. The white raster means that pure red, green, and blue are produced by the color phosphors, forming a pure white mixture.

An example of poor purity is shown in color plate III, with patches of other colors in a red raster. With only one beam operating, the raster should be a uniform red, green, or blue.

The purity is usually adjusted for a solid red raster. Red is preferred because this phosphor requires the most beam current. Then any misadjustment of beam landing has the greatest effect on purity.

The position of the purity magnet near the base of the picture tube, between grid 1 and grid 2, is illustrated in Fig. 11-1, while Fig. 11-2 shows a closer view of the ring magnets. Although the purity assembly usually has a blue lateral magnet for convergence, this is not used for the purity adjustments.

The purity assembly serves as a centering magnet for all three beams. The two rings have

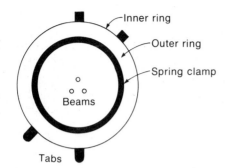

FIGURE 11-2 CONSTRUCTION OF PURITY MAGNET.

tabs, one square and the other rounded, to mark the opposite ends. When the same tabs are together, the ring magnets are bucking and there is no net magnetic field to shift the position of the beams. Spreading the tabs produces a stronger field for a greater effect in positioning. Rotating both rings together changes the direction of the shift in centering. You can move the beams up or down or at a diagonal.

The purpose of this positioning is to make the beams begin deflection from the correct point at the center of the area surrounded by the deflection yoke. Otherwise, the beam landing cannot be correct when the beams are deflected to form the raster. The purity magnet is adjusted for a pure red area in the center of the raster, with only the red gun operating.

No signal is used when checking purity because this test is for the raster, not for the picture. In receivers with a service switch, it is set to the raster position for purity adjustments.

Correct adjustment of purity is critical for proper colors and good convergence. Do not try to adjust a tube just brought in from the cold. The tube should be operating at a high brightness level for at least 10 min. Difficulty in adjusting purity may mean that manual degaussing is necessary. (See Sec. 11-5.) Also, the convergence assembly must be in its correct position for static convergence at the center.

With only the red gun operating for a red raster, the following procedure can be used to adjust purity:

1. Loosen the deflection yoke and move it back toward the convergence assembly. In many cases, there are wing nuts at the sides of a plastic housing for the yoke.
2. Rotate purity ring and separate the tabs, if necessary, to produce a pure red area at the center of the screen. Start with similar tabs together and rotate both rings. If the red is centered, this is all that is needed; if not, separate the tabs and rotate the rings again.
3. After the pure red is centered, push the deflection yoke forward to produce a uniform red raster over the entire screen area. Make sure the raster is not tilted. In most cases, the yoke will be up against the bell of the tube, but use the position that produces the best purity at the edges.

After a pure red raster has been obtained, operate the green and blue guns alone to check the green raster and blue raster. Usually, when the red raster has good purity, the green and blue rasters look even better. With the correct purity, all three guns can operate to produce a solid white raster.

11-2 Color Convergence

The process of adjusting each beam separately to make all three beams hit the same spot is *convergence*. Otherwise, the result is color fringes around the edges of objects in the picture. This effect of a color outline is only seen with a picture, especially in black and white. The raster can be perfect because it has no picture information with edges. An example of color fringing because of poor convergence is shown in color plate II.

Dot-bar generator. See Fig. 11-3. This type of signal generator is usually needed for convergence adjustments. It produces rows of dots or a crosshatch pattern of lines, all in black and white for observing the convergence, plus color bars for testing the color circuits. The dot pattern is shown in Fig. 11-4a with the crosshatch in b. These patterns are used for the picture because you need vertical and horizontal edges that stay still in order to see color fringing. When the dots and lines are white without any colors at the edges, this result shows convergence of red, green, and blue.

The dot-bar generator in Fig. 11-3 also has a superpulse setting to produce one white rectangle in the center of the picture. The output is available on either channel 3 or channel 9 for the receiver antenna terminals, at IF signal frequencies, or as video signal into the video amplifier. The dot pattern is generally used for adjustments at the center, which is *static convergence*. The adjustments for *dynamic convergence* at the borders of the screen usually require the crosshatch pattern.

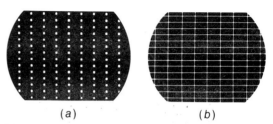

FIGURE 11-4 BLACK-AND-WHITE PATTERNS USED FOR CONVERGENCE ADJUSTMENTS. (a) DOTS. (b) CROSSHATCH.

Convergence yoke. See Fig. 11-5. The entire assembly is mounted over the internal pole pieces of the three guns. Then the magnetic flux provided by each section is confined to affect only the one beam. Adjusting the strength of the flux from the convergence magnets moves each beam slightly. Because of the delta arrangement with the guns 120° apart, the red and green beams move diagonally, when you watch the dot pattern on the screen. However, the blue beam moves vertically. For this reason, the blue gun also has the lateral adjustment magnet to move the blue beam left or right. The blue lateral magnet is usually part of the purity ring assembly (Fig. 11-1). It should be noted that the picture tube can be designed to move any one beam vertically and horizontally, while the other two move diagonally for convergence, but blue is generally used for the perpendicular displacements.

The net result of the four possible displacements allows convergence of the three beams to produce white, as shown in Fig. 11-6. Only one dot at the center is shown. Assume that there is no convergence to start. Adjust the red magnet to move the red dot diagonally about $1/4$ in. up to the right. Then use the green magnet to move the green dot up a little to the left to form a yellow dot. Next you can move the blue dot down with the blue magnet to be

FIGURE 11-3 DOT-BAR SIGNAL GENERATOR. WIDTH IS 10 IN. (*RCA*)

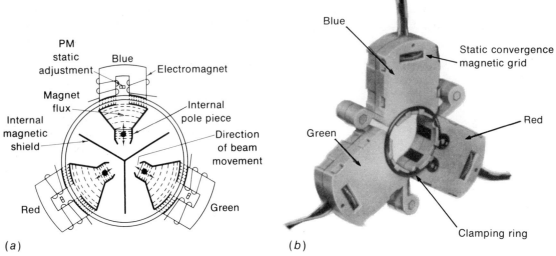

(a)

(b)

FIGURE 11-5 RADIAL CONVERGENCE MAGNET ASSEMBLY. (a) CONSTRUCTION. (b) PHOTO.

next to the yellow dot. Finally, the blue lateral magnet will move the blue beam to the left to converge all three beams for a white dot.

Each convergence magnet consists of a permanent magnet and an electromagnet. The permanent magnets are usually adjusted by turning small plastic thumbwheels. Blue is generally at the top and center of the delta. The magnets can move each color dot about $\pm 1/2$ in., but the required adjustment is usually much less. Adjusting the permanent magnets is static convergence. In addition, there is a coil magnet for each gun. The coil magnets have correction current from the deflection circuits for dynamic convergence at the top, bottom, left, and right edges of the screen.

Static convergence adjustments. Make sure the raster controls for height, width, focus, and high voltage are set properly before convergence is adjusted. It is usually easy to adjust the static magnets for white dots over about 50 percent of the screen area at the center. First, move red and green in their opposite diagonals to merge into yellow. Then adjust the blue vertically and laterally to produce white dots. With a crosshatch pattern, the lines at the center will be white without color fringing. However, the

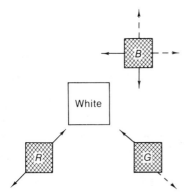

FIGURE 11-6 EFFECT OF STATIC CONVERGENCE MAGNETS IN MOVING COLOR DOTS TO PRODUCE WHITE.

ADJUSTMENTS FOR COLOR PICTURE TUBES 215

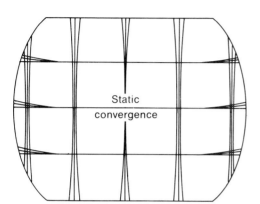

FIGURE 11-7 CROSSHATCH PATTERN WITH GOOD STATIC CONVERGENCE AT CENTER BUT POOR DYNAMIC CONVERGENCE AT EDGES OF SCREEN.

top of the cabinet so that you can vary the controls while watching the screen from the front. There are about 12 adjustments to converge the crosshatch pattern for white lines at the top, bottom, left, and right edges. The controls include adjustable coils to vary the phase of the convergence correction current and variable resistors to adjust the amplitude.

Usually, red and green are adjusted together, separate from blue. If necessary, the blue beam can be killed temporarily to watch only red and green. Or, blue can be moved out for separate lines by adjusting the blue lateral magnet. After red and green are converged for yellow lines throughout the picture, the blue can be moved in by the static magnets for a white crosshatch pattern. When the manufacturer's service notes are available, follow this procedure to save time for dynamic convergence.

Do not be afraid to separate the colors, especially blue, if this helps in making the dynamic convergence. Normally, it is easy to merge all the colors again with the static magnets. Watch for fringing that is the same at the edges and the center. For instance, suppose that all the bars or dots across the screen have

edges may need converging, as shown in Fig. 11-7. The static convergence at the center must be correct before doing the dynamic convergence for the edges.

Dynamic convergence adjustments. The adjustments may be on the convergence yoke, as in Fig 11-1, or on a separate convergence board, as in Fig. 11-8. The board usually mounts at the

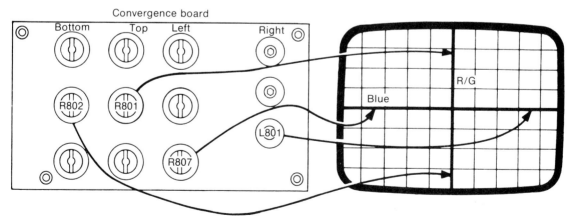

FIGURE 11-8 CONVERGENCE-BOARD ADJUSTMENTS WITH EFFECTS ON CROSSHATCH PATTERN. (*RCA CTC46 CHASSIS*)

blue fringing, always to the right. This can easily be corrected by a slight readjustment of the blue lateral magnet. In general, the static convergence should be checked periodically while adjusting the dynamic convergence. The convergence at the center should be perfect, while the edges are usually about 80 to 90 percent converged.

Once the convergence has been done, the adjustments are stable. However, do not make any big changes in focus, height, or width of the raster. Receivers delivered from the manufacturer are converged. The only time the complete adjustments may be necessary is when the picture tube is changed.

11-3 Convergence Correction Waveshapes

Basically, a parabola is needed for the waveshape of correction current in the coils of the convergence yoke. As shown at the top of Fig. 11-9, the parabola has symmetrical values at its two peaks, with little amplitude in the middle. In a parabola for vertical deflection at 60 Hz, the start and finish correspond to the top and bottom of the raster. A parabola for horizontal deflection at 15,750 Hz corresponds to the left and right sides. As a result, parabolic waveforms are used for dynamic convergence at the edges of the screen, with little effect on the static convergence at the center.

The parabolic waveform can be obtained by charging a capacitor with sawtooth current. This method is used for both vertical and horizontal parabolas. Convergence fields in synchronism with the scanning are the result of using output from the vertical and horizontal deflection circuits.

In addition to the parabola, a sawtooth component is added to tilt the waveform, as

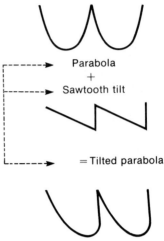

FIGURE 11-9 WAVEFORMS FOR DYNAMIC CONVERGENCE CORRECTION SIGNALS. TWO CYCLES SHOWN.

shown in the parabola at the bottom of Fig. 11-9. The tilt is necessary to favor one end with respect to the other. Both the vertical and horizontal correction parabolas need the tilt because of the position of the guns in a delta. For instance, the electron beam from the blue gun at the top must travel further when deflected to the bottom of the screen, compared with the top. Also, the red and green guns are mounted below center. Therefore, these beams travel further to the top. In addition, the red and green are off center horizontally. This requires tilt for horizontal dynamic convergence. Actually, the dynamic convergence adjustments for red and green are combined to reduce the number of controls.

The tilted parabola can correct for unequal distances to the edges of the screen, for each beam in a delta gun. Opposite polarities for the sawtooth component produce opposite directions of tilt. For vertical scanning, the correction can be increased either at the top or bottom. Similarly, the convergence correction can be increased left or right by reversing the tilt of the parabola for horizontal correction. For both cases, though, the parabolic correction does not affect the center, where the permanent magnets provide static convergence.

To help in reducing the effect of dynamic adjustments on the center convergence, the correction circuits generally use clamping diodes. A diode clamp circuit keeps the middle of the parabola at a fixed voltage value. The purpose is to maintain the center value while adjusting the parabola tilt and amplitude. A typical convergence diode assembly is shown in Fig. 11-10.

The parabolic correction signals are connected to the coils in the convergence yoke, as shown in Fig. 11-11, *not* to the deflection yoke. Note that separate coils are used for the vertical and horizontal correction signals. The electromagnet is illustrated for the blue gun, but the construction is the same for the red and green magnets.

FIGURE 11-10 CONVERGENCE RECTIFIER ASSEMBLY WITH FOUR DIODES. LENGTH IS 1 IN. (*INTERNATIONAL RECTIFIER*)

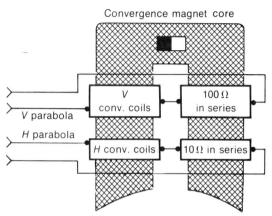

FIGURE 11-11 ILLUSTRATING HOW BOTH THE VERTICAL AND HORIZONTAL CORRECTION SIGNALS ARE CONNECTED TO THE CONVERGENCE MAGNET. THIS CONSTRUCTION IS THE SAME FOR THE BLUE, GREEN, AND RED MAGNETS.

It should be noted, though, that the red and green magnets are connected in series with each other. As a result, they both have the same vertical and horizontal deflection signals. Also, only one set of controls is used for both red and green convergence. This convenience is possible because the red and green guns require similar dynamic correction. Both guns are in the same horizontal plane, and they are displaced by the same amount off-center.

11-4 Screen-Grid Adjustments

These controls on the back of the receiver chassis adjust the screen-grid dc voltage for each gun. They are usually called *red screen, green screen,* and *blue screen.* The screen adjustment for each gun determines how much voltage is needed at the control grid to produce cutoff for black and the maximum beam current for white highlights.

The screen controls are dc adjustments for the raster, without signal. To see this effect, remove the picture to view the raster alone. If you turn up the red screen, with the other two controls down, you will see a red raster. Similarly, green or blue can be turned up, with the others down, to see just the one color. If you turn up the red screen and green screen, without blue, you will see a yellow raster. Add blue, and the raster becomes white.

The screen adjustments determine how the three beam currents will be balanced to make white. The purpose is to obtain a bright picture, while still maintaining proper white and gray, without color, for all settings of the brightness control. For this reason, the adjustment is often called *black-and-white setup, gray-scale tracking,* or *white color temperature*. The degree of white is specified in terms of temperature, usually at 9000 K. This white has a little more blue than the reference white of 6500 K, but people seem to prefer a bluish white on the television screen.

Actually, most adjustments for a color picture tube are made on white because this is the first requirement for having the correct colors. In brief, the screen-grid adjustments are set to produce a white raster. In a receiver with a normal-service switch and separate screen controls, the procedure can be as follows:

1. Preliminary. Brightness, contrast, and color controls at their normal positions.
2. Set the normal-service switch for just a horizontal line on the screen.
3. Turn all the screen adjustments down to cut off the line on the screen.
4. Turn the red screen up to the point where the red line is just barely visible. Then turn up the green screen for a faint yellow line and the blue screen for a white line. In each case, adjust to the point where the line is barely visible or just cut off.
5. Set the normal-service switch to the position that produces a raster, which should now be white. If the raster is not white, the screen controls can be readjusted slightly. Vary the brightness control to see that the raster is white for different brightness settings.
6. In the "normal" position of the service switch, check for a good black-and-white picture.

If any screen control does not produce a line, readjust the brightness control to make the line visible. Also, if the raster is not bright enough after the screen adjustments have been made, try readjusting them at a lower setting of brightness. Then the brightness control can be turned up for a lighter raster. Usually, at least one of the screen controls should be at its maximum setting.

If the raster tracks from light to dark but there is some color in white highlights of the picture, try readjusting the red, green, and blue drive controls. These adjust the amount of Y luminance signal for the three guns. In some cases, it may be necessary to rearrange the cathode leads for drive to the three guns. These wires are yellow with red, green, or blue tracers.

11-5 Degaussing

This term refers to demagnetizing iron and steel parts of the picture tube. Specifically, the steel shadow mask with its frame inside and the metal hood around the tube can become magnetized by a steady magnetic field. The main effect is from the earth's magnetic field (terrestrial magnetism). This magnetization affects the beam register on each color phosphor, causing poor purity, especially at the edges. When perfect purity is a problem, degaussing may be necessary.

In general, the principle of demagnetizing is to use the varying magnetic field of an alternating current and slowly reduce it to zero. The 60-Hz current from the ac power line is used.

Automatic degaussing (ADG). Receivers now have a coil around the front rim of the picture tube for automatic degaussing. In Fig. 11-1, the coil is wrapped around the metal hood made of 0.5-mm silicon steel. Current for the coil is taken from the 60-Hz ac power line. When the set is turned on, a high value of alternating current flows, which is then reduced to zero by the ADG circuit. In some receivers, the automatic degaussing operates when the set is turned off. In either case, the ADG circuit automatically degausses the picture tube each time the receiver is turned on or off. Because of the automatic degaussing, the receiver does not have to be facing in any particular direction with respect to the earth's magnetic field.

ADG circuit. A typical circuit for automatic degaussing is shown in Fig. 11-12. When the receiver is turned on, the ac voltage drop across the thermistor is about 70 V. This voltage is applied to the ADG coils in series with the varistor. As the thermistor heats up with current, its resistance decreases. At the same time, the varistor resistance increases. Then the voltage across the ADG coils decreases, resulting in less current for degaussing. As a result, the current in the degaussing coils drops from about 2 A to practically 0 in less than 1 s. The receiver must be off for about 30 min for the thermistors to

FIGURE 11-13 DEGAUSSING COIL WITH SWITCH FOR POWER-LINE CORD. (*WALSCO ELECTRONICS*)

cool down to room temperature, or else the ADG circuit will not demagnetize the picture tube.

Manual degaussing. In some cases, it may be necessary to demagnetize the picture tube with a separate degaussing coil (Fig. 11-13). Connect the coil to the 60-Hz ac power line. Hold it up to the screen, and move the coil around the top, bottom, and sides of the picture tube. After about 5 s, move the coil slowly at least 6 ft away, put the coil on the floor, and shut off the current. The purpose of moving the coil away and laying it flat is to prevent its magnetic field from affecting the picture tube when the demagnetizing field collapses. Otherwise, the transient drop of current in midcycle can magnetize the tube.

The requirements for degaussing are typically 2-A current through 450 turns of No. 20 wire, with a coil diameter of 12 to 13 in. This arrangement produces the magnetomotive force of 900 ampere-turns.

11-6 Pincushion Correction

The pincushion shape of the raster is shown in Fig. 11-14. The reason for the distortion is that the corners of a flat screen are further from the

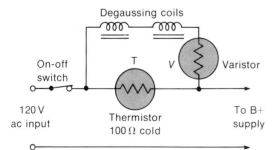

FIGURE 11-12 CIRCUIT FOR AUTOMATIC DEGAUSSING (ADG).

220 BASIC TELEVISION

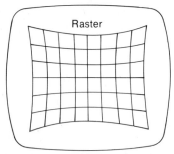

FIGURE 11-14 PINCUSHION DISTORTION OF RASTER.

point of deflection, compared with the center. Then the electron beam is moved more horizontally and vertically at the extreme angles of deflection. The pincushion problem is more severe with wide-angle tubes of 90° or more.

For monochrome picture tubes, the pincushion distortion is corrected by permanent magnets mounted on the housing of the deflection yoke. These pincushion magnets cannot be used with color picture tubes, however, because they would affect the three beams by different amounts, resulting in more problems with color purity and convergence. Therefore, dynamic pincushion correction is used with color picture tubes.

The pincushion correction signals are applied to the deflection yoke. The horizontal scanning coils have pincushion correction current to straighten the sides of the raster. The vertical scanning coils have pincushion correction current to straighten the top and bottom of the raster.

Horizontal pincushion correction. The problem here is to push out the sides of the raster, at the center. Thus more horizontal scanning current is needed for the horizontal lines produced midway between the top and bottom. To accomplish the side correction, the horizontal deflection current at 15,750 Hz is modulated at the vertical scanning rate at 60 Hz. The vertical modulation is a 60-Hz parabolic waveform, which stretches the width of the horizontal lines in the middle of the vertical scan but not at the start and finish.

The circuit generally used for horizontal pincushion correction is illustrated in Fig. 11-15. The vertical modulation is taken from the vertical output circuit. This sawtooth component is shaped into a parabola by C_1 and R_1 in parallel. Then the $B+$ voltage for the horizontal amplifier is in series with the 60-Hz parabolic voltage. As a result, the amplitude of individual cycles of the 15,750-Hz horizontal output varies in step with the 60-Hz component in the supply voltage. The modulated waveform of current for the horizontal deflection coils shows that the horizontal amplitude is increased for more width for those scanning lines in the middle of the raster. This pincushion correction signal, therefore, has the effect of pulling out the center to straighten the sides.

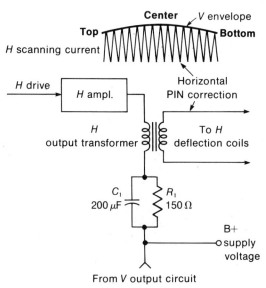

FIGURE 11-15 HORIZONTAL PINCUSHION CORRECTION FOR SIDES OF RASTER.

ADJUSTMENTS FOR COLOR PICTURE TUBES

Vertical pincushion correction. Also called top-and-bottom (TB) correction, vertical pincushion correction requires current in the vertical scanning coils with three features: (1) more vertical deflection in the middle of the horizontal lines; (2) correction only at the top and bottom of the raster; (3) opposite corrections at the top and bottom. Then the vertical pincushion correction current will stretch the raster upward at the top and pull the raster down toward the bottom, at the center of the raster. The correction should be zero at the vertical center.

The waveshapes required for the TB correction are illustrated in Fig. 11-16a. First is the 60-Hz sawtooth scanning current, without pincushion correction. Note that this waveform (1) is a negative-going sawtooth. More amplitude in the positive direction at the start will stretch the top of the raster. More negative amplitude at the finish will stretch the bottom. Next is shown the waveshape (2) of 15,750-Hz correction current. This waveform (2) can include sine waves, clipped sine waves, or parabolic waves derived from the horizontal output circuit. These amplitudes decrease to zero at the center of the vertical scan and then increase with opposite polarity. The combined waveform (3) for vertical pincushion correction then has the butterfly component shown in darker shading. This correction has the following effects:

1. At the top of the raster, the vertical scanning current is increased in the direction of more height at the top just for the center of the horizontal lines. The PIN phase control adjusts the phasing of the horizontal component to put the correction in the middle of the lines.
2. At the vertical center there is no TB correction.
3. At the bottom of the raster, the vertical scanning current is increased in the direction of more height at the bottom, just for the center of the horizontal lines.

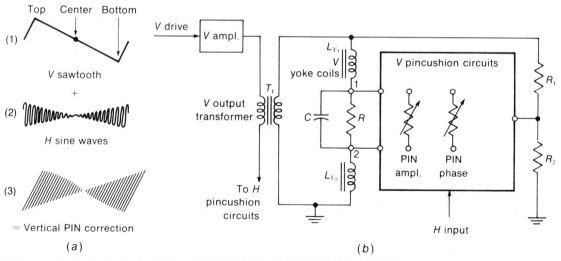

FIGURE 11-16 VERTICAL PINCUSHION CORRECTION FOR TOP AND BOTTOM OF RASTER. (a) WAVEFORMS. (b) CONNECTIONS TO VERTICAL COILS IN DEFLECTION YOKE.

The circuit arrangement for supplying the vertical pincushion correction current for the vertical scanning coils in the deflection yoke is illustrated in Fig. 11-16b. The dark outline indicates the board that has the PIN correction circuits. For vertical correction, the combined waveshape (3) in Fig. 11-16a is available at points 1 and 2 between the vertical coils L_{y_1} and L_{y_2}. The RC network here provides a low impedance for connection to the PIN board. Actually, the entire vertical PIN circuit is a low impedance between the center of the yoke coils and the center tap of the damping resistors R_1 and R_2.

A common method of obtaining waveform (2) in Fig. 11-16a uses a *saturable reactor*. This reactor is a special transformer with a closed iron core that can be easily saturated. One winding serves as a control winding to saturate the core.

Pincushion adjustments. There usually are no adjustments for side pincushion correction. The vertical TB controls are factory-set and seldom need adjustment. If necessary, the adjustments are made on a crosshatch pattern. The PIN phase control is adjusted to move any curvature of the horizontal lines to the middle. Then adjust the PIN amplitude control for straight scanning lines at the top and bottom of the raster. In some receivers, there may be a permanent magnet adjustment on the saturable reactor to adjust the crossover point between top-and-bottom correction.

11-7 Overall Setup Adjustments

When a color picture tube is installed, it may be necessary to do a complete job on purity, convergence, and gray-scale tracking. First, make sure all the simpler adjustments not related to color are set properly. These adjustments include height and width of the raster, with good focus and the required amount of high voltage. Any changes can affect the purity and convergence. Pincushion correction of the raster usually need not be readjusted, but, if necessary, this should be done now. Then start on the adjustments for the color picture tube. The following steps illustrate the usual sequence.

Step 1: Degaussing. Although the receiver has ADG, it usually helps in adjusting purity and convergence to do manual degaussing first. Have the receiver in its normal viewing position. Then the degaussing field can neutralize the effect of the earth's magnetic field.

Step 2: Purity. Adjust the purity magnet for the red "fireball" at the center of the raster, with only the red gun operating. Set the deflection yoke approximately for good red around the edges.

Step 3: Static convergence. Turn on all three guns for white. Connect the generator for a dot pattern. Adjust the permanent magnets on the convergence yoke to converge red, green, and blue for white dots at the center.

Step 4: Recheck purity. If the static convergence required much adjustment, the purity magnet should be checked again. Now you can tighten the deflection yoke in its final position for good purity at the edges. Be sure the raster is not tilted. Check the purity for red, green, and blue. Then combine the colors for a white raster.

Step 5: Dynamic convergence. Switch the generator to the crosshatch pattern. The manufacturer's instructions should be followed for dynamic convergence. However, the general procedure is to converge all the lines in the crosshatch, as follows:

Vertical center line, for *R/G*
Vertical lines at left and right, for *R/G*
Horizontal lines at top and bottom, for *R/G*
Horizontal center line, for *R/G*
Horizontal center line, for *B*
Horizontal lines at top and bottom, for *B*

The convergence board often has a drawing that shows the effect of each adjustment on the crosshatch pattern. It may be helpful to turn off the blue gun while making the *R/G* adjustments. Or, move the blue lines out temporarily with the static magnets. As the convergence improves, you can touch up previous adjustments for white lines in the crosshatch and white dots in the dot pattern. The convergence should be very good for almost the entire screen area, up to about $1\frac{1}{2}$ in. from the edges, on a 21-in. tube.

Step 6: Gray-scale tracking. This adjustment depends on the screen-grid voltage adjustments. The purpose is to obtain the brightest monochrome picture possible, without blooming. This is also the black-and-white setup, or color temperature, adjustment. Too much blue can cause a loss of red in the picture. Too much red gives the picture a sepia appearance. In receivers with separate screen controls and a normal-service switch, the usual best procedure is to extinguish the horizontal line in red, green, and blue. This method balances the cutoff characteristics of the three guns. Then turn on the white raster and set the switch to its "normal" position to check a black-and-white picture.

It should be noted that for some picture tubes with in-line guns, the deflection yoke, convergence yoke, and purity magnet are factory-set and cemented into position. This feature eliminates the need for purity and convergence adjustments.

SUMMARY

1. The purity or beam-landing adjustment is made by adjusting the purity magnet for pure red at the center of the raster and positioning the deflection yoke for pure red at the edges.
2. Convergence adjustments are made with a dot-bar signal generator producing a white dot or a crosshatch pattern on the screen of the picture tube.
3. The convergence-yoke assembly consists of a permanent magnet for static adjustments and an electromagnet for dynamic adjustments, for the blue, green, and red guns. The blue beam moves vertically; the red and green beams move diagonally. In addition, the blue lateral magnet moves the blue beam horizontally. This magnet is usually on the purity ring assembly.
4. Static convergence consists of adjusting the permanent magnets for white dots at the center of the pattern.
5. Dynamic convergence consists of adjusting the convergence controls for straight white lines in the crosshatch pattern.

6. The screen controls are adjusted to the point of visual cutoff for the red, green, and blue beams. Then the three guns are balanced for a white raster and monochrome picture. This procedure is gray-scale tracking or white color–temperature adjustment.
7. Degaussing means demagnetizing the iron and steel parts of the color picture tube, especially the shadow mask. The 60-Hz ac line current is used for the demagnetizing field. The automatic degaussing (ADG) circuit demagnetizes the picture tube when the receiver is turned on or off.
8. The pincushion (PIN) correction circuits straighten the sides of raster and the top and bottom (TB) edges.

Self-Examination (Answers at back of book)

Answer True or False.
1. The purity magnet is usually mounted against the wide bell of the picture tube.
2. The blue lateral magnet moves the blue beam left and right.
3. The green static magnet moves the green beam diagonally.
4. The static convergence magnets straighten the horizontal lines in the crosshatch pattern at the top and bottom edges.
5. Dynamic convergence is obtained by parabola correction current in the vertical and horizontal coils of the convergence yoke.
6. The pincushion correction current flows in the vertical and horizontal coils of the convergence yoke.
7. A degaussing coil uses 15,750-Hz current from the horizontal output circuit.
8. The screen controls adjust the cutoff voltage for the red, green, and blue guns.
9. A dot-bar generator can produce color bars, white dots, or a white crosshatch pattern.
10. The position of the deflection yoke affects purity at the edges of the raster.

Essay Questions

1. List the magnetic components mounted on the neck of a color picture tube using delta guns, with their functions.
2. Describe briefly how to make the purity adjustment.
3. What is the difference between static and dynamic convergence?
4. Give three functions of the normal-service switch.
5. Describe briefly how to adjust the red, green, and blue screen controls.

6. Give the functions for the PIN phase and amplitude controls.
7. Describe briefly the procedure for manual degaussing.
8. In dynamic convergence, why are the red and green beams adjusted together?

Problems (Answers at back of book)

1. The raster is tilted. What would you adjust?
2. The raster is too green. What would you adjust?
3. The raster is bowed in at the top and bottom. What would you adjust?
4. White lettering in the picture has blue fringes, always to the right. What would you adjust?
5. Label the components marked *a* to *g* in the drawing below for a color picture tube using delta guns and a phosphor-dot screen.

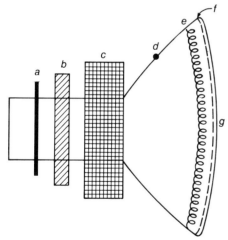

FIGURE 11-17 FOR PROBLEM 5.

12 Power Supplies

A power supply converts the ac input of the 60-Hz power line to dc output voltage. The television receiver has two power supplies. One is the low-voltage supply for B+ voltage and heater power for tubes. The other is a high-voltage supply for the anode voltage of the picture tube. The total amount of power used from the 120-V ac line is about 150 W in monochrome receivers and 250 W in color receivers. For vacuum tube amplifiers, the B+ voltage generally is about 140 to 280 V for plate and screen voltages. For transistors, the required collector voltage is 15 to 100 V. In the high-voltage supply, the required anode voltage is 10 to 20 kV for monochrome or 15 to 30 kV for color picture tubes. The rectifier that converts the ac input to dc output is generally a silicon diode (Fig. 12-1). Methods of using diode rectifiers in different types of power supplies are explained in the following topics:

12-1 Basic functions of a power supply
12-2 The B+ supply line
12-3 Types of rectifier circuits
12-4 Half-wave rectifiers
12-5 Low-voltage supply with half-wave rectifier
12-6 Full-wave center-tap rectifier
12-7 Full-wave bridge rectifier
12-8 Voltage doublers
12-9 Voltage triplers and quadruplers
12-10 Heater circuits
12-11 Filter circuits
12-12 Voltage regulators
12-13 Low-voltage supply with series regulator circuit
12-14 Troubles in the low-voltage supply
12-15 Hum in the B+ voltage
12-16 Flyback high-voltage supply
12-17 Fuses and circuit breakers

POWER SUPPLIES

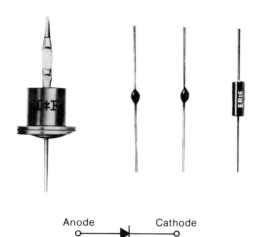

FIGURE 12-1 SILICON-DIODE RECTIFIERS. ACTUAL SIZE, WITH SCHEMATIC SYMBOL BELOW. FROM LEFT TO RIGHT, HIGH-HAT, PELLET, AND STICK TYPES.

12-1 Basic Functions of a Power Supply

Basically, all that the power supply needs is a rectifier to change the ac input to dc output. However, filter capacitors generally are used to remove the pulsating variations from the dc output voltage of the rectifier. In addition, a power transformer is often needed to step up or step down the ac input voltage. Finally, a voltage regulator may be used, especially with transistor circuits. A regulator keeps the dc output voltage constant when the dc load current changes. These basic functions are illustrated by the block diagram in Fig. 12-2. The supply is shown with positive dc output for B+ voltage. Positive polarity is needed for plate and screen voltages on tubes and collector voltage on NPN transistors.

The rectifier. Because it conducts in one direction, a rectifier converts its ac input to dc output. The current flows only for the polarity of input voltage that makes the anode positive. Or, negative cathode voltage makes the diode conduct. A single diode is a half-wave rectifier, as it can conduct only for a half-cycle of the ac input.

Power supplies generally use silicon-diode rectifiers (Fig. 12-1). The current rating is 500 mA for the smallest diodes up to 10 A or more. Larger units are stud-mounted. The advantage of a silicon diode is its very low internal voltage drop of 1 V or less, compared with 18 V, approximately, for a vacuum tube diode. Silicon diodes are generally labeled *D*, *SD*, *X*, *Y*, or *CR* for crystal rectifier.

Note the schematic symbol in Fig. 12-1 for a semiconductor diode. The arrowhead is the anode, while the bar is the cathode, corresponding to plate and cathode in a vacuum tube.

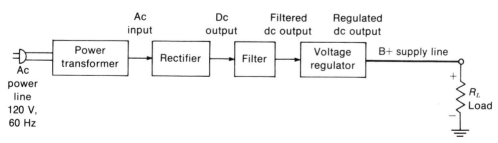

FIGURE 12-2 BASIC FUNCTIONS OF A POWER SUPPLY, WITH POSITIVE DC OUTPUT VOLTAGE.

On the diode unit, a bar, circle, or dot indicates the cathode terminal. The direction of the arrow shows hole current, as the forward current, from anode to cathode. Remember that hole current is a motion of positive charges, however, in the opposite direction from electron flow.

Peak reverse voltage (PRV). This is the maximum voltage that can be applied in the nonconducting direction, with the anode negative. Too much reverse voltage across the rectifier can produce breakdown at the junction, resulting in a shorted diode.

The actual amount of reverse voltage across a diode rectifier is about twice the dc output voltage, as shown in Fig. 12-3. In addition to the negative ac input at the anode, the dc voltage at the cathode is series-aiding, when you consider the potential difference across the diode. Typical PRV ratings for small silicon diodes are 400 to 1,000 V.

Checking the rectifier. A semiconductor diode can be easily checked with an ohmmeter to see if it is open or shorted. Out of the circuit, just measure the diode resistance in both directions by reversing the ohmmeter leads. The diode resistance should be very low in the forward direction and at least 100 times higher in the reverse direction. If the resistance is very low in both directions, the diode must be shorted. If the resistance is very high in both directions, the diode must be open. When an ohmmeter is used to check the diode in the circuit, make sure that the power is off, the filter capacitors are discharged, and one side of the diode is disconnected.

Filter. Although the dc output of the rectifier has only one polarity, the rectified voltage still includes the amplitude variations of the ac input. This ac component is called *ripple* in the dc out-

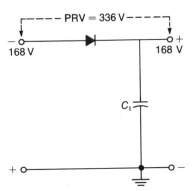

FIGURE 12-3 PEAK REVERSE VOLTAGE (PRV) ACROSS RECTIFIER DIODE.

put voltage. A filter can eliminate practically all the ripple.

Voltage regulator. Voltage regulation means maintaining a constant dc voltage output, even when the ac input voltage increases or decreases, or when the dc load current changes. A voltage regulator is often included in the low-voltage supply to maintain constant height and width in the scanning raster. For the high-voltage supply in color receivers, a regulator is used to prevent changes in focus and convergence for the color picture tube.

Power transformer. The ac voltage needed may be more or less than the 120 V of the 60-Hz power line. Then a power transformer is used (Fig. 12-4). By mutual induction between L_P and L_S, the ac secondary voltage is stepped up or down in direct proportion to the turns ratio. The isolated secondary in a has no connection to the primary, as two separate windings are used. In an autotransformer, such as in b, one winding is used for both L_S and L_P. The tap makes L_P just a part of the entire coil with L_S. Therefore, L_S is not isolated from L_P. Where the coil is tapped determines the turns ratio.

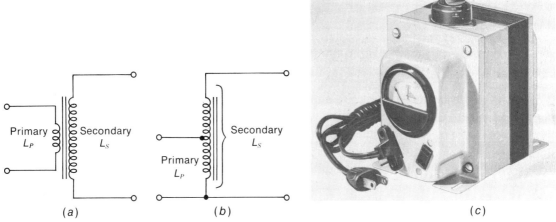

FIGURE 12-4 POWER TRANSFORMERS. (a) SCHEMATIC WITH ISOLATED SECONDARY. (b) AUTOTRANSFORMER, WITH SECONDARY NOT ISOLATED. (c) ISOLATION TRANSFORMER WITH VARIABLE OUTPUT VOLTAGE AND METER FOR TESTING RECEIVERS. (TRIAD-UTRAD)

With either type of transformer, the power supply can operate only with ac input. This type is called an *ac power supply*, which only means that it has a transformer requiring ac voltage. A supply without a transformer is connected directly to the power line for ac input voltage. This type can operate on either ac input or dc input of the correct polarity that allows the rectifier to conduct. Actually, the dc input does not have much meaning because very few homes have a dc power line. However, the name "ac-dc power supply" is used to indicate there is no power transformer.

Isolation. A supply that has a power transformer with a separate secondary winding is isolated from the ac power line. The primary is connected to the ac line, but the rectifier circuit with dc output is in the secondary. The receiver chassis, which is usually B— for the dc output, then, is a "cold" chassis because it is isolated. This type is called a *line-isolated* supply.

With an autotransformer or no power transformer at all, however, the receiver chassis can be "hot," meaning it is connected to the high side of the ac power line. Then there is danger of electric shock when touching the receiver chassis or exposed metal parts. This type is called a *line-connected* supply.

Portable receivers generally use a line-connected supply. The purpose is to eliminate the size and weight of the power transformer.

Remember that the ac power line has one side connected to earth ground, while the other side is the live connection. If the ac plug is inserted in the polarity that connects the receiver chassis to the live side of the ac power, then touching the chassis is the same as touching the 120-V power line.

To avoid this danger in a power supply which is not isolated, plug the receiver into the ac outlet with the grounded side connected to the receiver chassis. Many receivers now have a plug with one wide tip. It can be plugged in only the one way that grounds the chassis. However, a polarized type of socket is necessary to take this plug.

Isolation transformer. This power transformer has a 1:1 turns ratio with an isolated secondary winding. It is a good safety practice to use the isolation transformer when testing a receiver that has a line-connected power supply. The wattage rating should be 250 W or more, for color receivers. The transformer can also be made with a variable turns ratio. Its purpose is to test the receiver with voltages higher or lower than normal. See Fig. 12-4c.

12-2 The B+ Supply Line

Although tubes and transistors generally amplify ac signals, they must have dc electrode voltages in order to conduct any current at all. These operating voltages are provided by the B+ voltage from the low-voltage power supply. For tubes, they use the B+ for plate and screen voltage. Each stage has a separate path returning to the B+ supply line. These paths are in parallel with each other. For transistors, the branches for collector supply voltages are in parallel.

The details of a B+ supply line are shown in Fig. 12-5 for transistors. Only two transistors Q1 and Q2 are indicated here, but the same idea applies to all transistor stages operating from this low-voltage supply. For the example of Q1, its collector returns through R_1 to the 28-V supply. R_1 is the collector load resistor. The collector current I_C assumed here for Q1 is 10 mA.

It is important to note that the base voltage for Q1 is also taken from the B+ supply. The required forward bias between base and emitter is only a few tenths of a volt, but the R_3R_4 voltage divider provides the voltage across R_4 needed for the base. A bleeder current of 1 mA is assumed through the R_3R_4 branch.

Similarly, R_2 is in a parallel path for the collector of Q2. Also, the R_5R_6 provides base voltage for Q2. The bleeder current in this branch is 3 mA.

The total current in the four parallel paths adds up to $10 + 1 + 14 + 3 = 28$ mA. This load current of 28 mA at 28 V is equivalent to a load resistance of 28 V/0.028 A, or 1 kΩ, as shown at the right in Fig. 12-5. If there are 20 transistors instead of 2, and if the total load current is 280 mA, the equivalent load resistance will be 100 Ω for the low-voltage supply.

12-3 Types of Rectifier Circuits

Figure 12-6 shows the four most common types: half-wave, half-wave doubler, full-wave center-tap, and full-wave bridge. All the circuits are shown with an isolating power transformer, but they can be line-connected, except for the full-

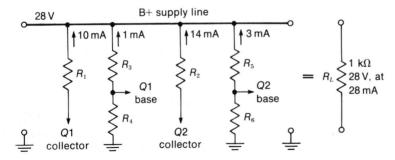

FIGURE 12-5 DETAILS OF THE B+ SUPPLY LINE FOR TRANSISTOR AMPLIFIERS WITH BRANCH CURRENTS. DIRECTION SHOWN FOR ELECTRON FLOW.

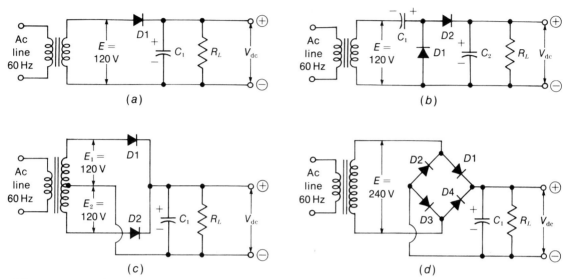

FIGURE 12-6 THE FOUR MOST COMMON TYPES OF RECTIFIER CIRCUITS. (a) HALF-WAVE. (b) HALF-WAVE VOLTAGE DOUBLER. (c) FULL-WAVE, CENTER-TAP. (d) FULL-WAVE BRIDGE.

wave rectifier in c, which needs a power transformer for the center tap. The polarity is positive for dc output here, but it can be inverted by reversing the diodes.

The approximate amount of dc output voltage can be calculated, if we assume a silicon diode, the capacitor-input filter C_1, and a typical load current of about 300 mA. The formula is

$$V_{dc} = 1.2 E_{rms} \qquad (12\text{-}1)$$

The factor 1.2 is a constant for the specified conditions. E_{rms} is the ac input, which equals 120 V.

Then the dc output voltage for the half-wave rectifier in a is 1.2 × 120 = 144 V.

For the voltage doubler in b, the dc output is 2 × 144 = 288 V.

For the full-wave center-tap rectifier in c, the dc output is 144 V. We assume the total ac secondary voltage is 240 V, with 120 V for E_1 and E_2 to each diode.

In d, the full-wave bridge rectifier uses the entire ac secondary voltage for dc output voltage. Then the dc output here would be 288 V, with 240 V as ac input to the bridge.

The ripple frequency is 60 Hz for the half-wave rectifier circuits in a and b, as one-half of each cycle of the 60-Hz ac input is used to produce dc output. For the full-wave rectifiers in c and d, the ripple frequency is 120 Hz because both halves of the cycle are used.

12-4 Half-Wave Rectifiers

The details of this basic circuit are illustrated in Figs. 12-7 to 12-11. The half-wave rectifier is fundamental to all types of power supplies because any one diode by itself can be only a half-wave rectifier. The other circuits are just combinations of half-wave diodes. A full-wave rectifier is really two half-wave diodes back to back, to conduct on opposite polarities of the ac input. A voltage doubler can be considered as two half-wave rectifier circuits that are effectively in series for dc output voltage.

Voltage polarities. See Fig. 12-7a and b for the opposite diode connections that make the dc output voltage either positive or negative. No filter capacitor is used here in order to show the basic waveforms of half-wave rectification. Note the ac input is shaded to mark only the half-cycle that allows the diode to conduct. The diode in a conducts on the positive half-cycle, without any output for the negative half-cycle. The dotted half-cycle in the waveform of voltage across R_L indicates the part of the ac input that is missing in the dc output. In b, the diode connections are reversed to allow conduction on the negative half-cycle of the ac input, without any output for the positive half-cycle.

Positive dc output. In Fig. 12-7a, D1 conducts when the ac input makes the diode anode positive. When the diode conducts, it connects the high side of R_L to the positive side of the ac source. The secondary winding L_S on the power transformer is the source supplying ac input to this rectifier circuit. Electrons flow from the grounded side of L_S up through R_L and from cathode to anode in D1, returning to the positive terminal of the ac source at the top of L_S. As a result, the top of R_L is the positive side of the voltage drop V_{R_L}. The basic reason is that when the diode conducts, it connects the high side of R_L to the high side of L_S only when this point has positive ac input voltage.

Negative dc output. In Fig. 12-7b, D1 conducts when the ac input makes the diode cathode negative. Note that the ac input voltage is reversed now. Making the cathode negative is equivalent to making the anode positive.

With the diode conducting in Fig. 12-7b, we can consider electron flow starting from the top of L_S, from cathode to anode in D1, through R_L, and returning to the grounded end of L_S. As

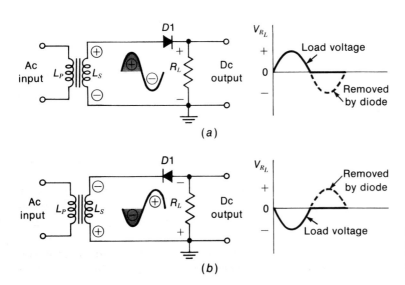

FIGURE 12-7 HALF-WAVE RECTIFIER CIRCUITS WITHOUT FILTER. (a) POSITIVE DC OUTPUT. (b) DIODE REVERSED FOR NEGATIVE DC OUTPUT.

a result, the top of R_L is the negative side of the voltage drop V_{R_L}. A fundamental way to consider the negative dc output voltage in b is the fact that when the diode conducts, it connects the high side of R_L to the high side of L_S only when the ac input is negative.

The negative dc output voltage in Fig. 12-7b is negative with respect to chassis ground. There is no law that says chassis ground must be the most negative point. The polarities are determined by the source supplying the ac input. In this circuit with negative dc output voltage, chassis ground is actually positive because the ac input is positive at the grounded end when the diode conducts.

The two polarities for dc output voltage can be summarized briefly as follows:

1. Connect the ac input to the anode of the diode for positive dc voltage at the load in the cathode circuit.
2. Connect the ac input to cathode of the diode for negative dc voltage at the load in the anode circuit.

Line-connected half-wave rectifier. No power transformer is used in Fig. 12-8 for the typical case of a half-wave power supply that is not isolated from the power line. C_1 is the filter capacitor to remove the 60-Hz ac ripple from the dc output voltage. The capacitor input filter is necessary to provide dc output voltage during the time when the diode is not conducting.

R_1 in series with $D1$ is used as a surge-protection resistor. It limits the peak current through the diode when C_1 is charging. The resistance is generally 3 to 7 Ω, with a power rating of 5 W. When a power transformer is used, R_1 is not needed because the secondary winding has the required series resistance.

This circuit has positive dc output at the cathode for B+ voltage. The ac input is applied

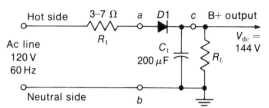

FIGURE 12-8 LINE-CONNECTED HALF-WAVE RECTIFIER, WITH FILTER.

to the anode. Note that the B− side of the power supply is connected to the grounded, or neutral, side of the ac input from the power line, to reduce shock hazard.

Effects of the filter capacitor. In Fig. 12-8, C_1 charges fast through the very low resistance of the rectifier diode when it conducts, but discharges slowly through the load resistance. As a result, C_1 maintains the dc output voltage during the half-cycle when the diode is not conducting. In this process of fast charge and slow discharge, the filter capacitor has the following effects:

1. The 60-Hz ac ripple is practically eliminated from the dc output.
2. The dc output voltage is maintained during the entire cycle of ac input.
3. The dc voltage across C_1 puts reverse bias on the diode so that it can conduct only at the peak of the ac input. As a result, the diode is a peak rectifier.

The waveforms in Fig. 12-9 illustrate the diode operation with the input filter capacitor. In a, the sine-wave voltage is shown for the ac input across terminals a and b in Fig. 12-8. Note that 120 V is the rms value of the sine wave, while the peak value is $1.4 \times 120 = 168$ V.

The waveform in Fig. 12-9b shows the dc output voltage across C_1 and R_L in parallel. This

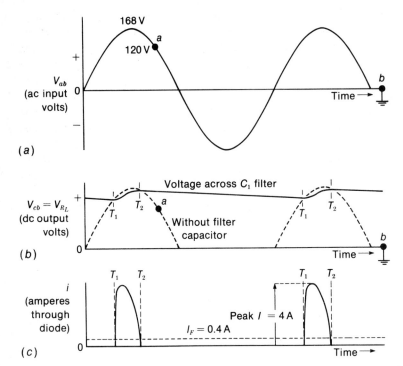

FIGURE 12-9 WAVEFORMS OF HALF-WAVE RECTIFIER. (a) AC INPUT VOLTAGE V_{ab}. (b) DC OUTPUT VOLTAGE V_{cb} OR VOLTAGE ACROSS R_L. (c) CURRENT THROUGH DIODE.

voltage is labeled V_{cb} between points c and b in Fig. 12-8. Assume the circuit has been operating for a few cycles of the ac input to charge C_1. Then the diode is forward-biased just for the time from T_1 to T_2. Only during this period does the ac input voltage make the diode anode more positive than the dc output voltage at the cathode. Remember that positive voltage at the diode cathode is reverse voltage that prevents conduction.

During the short conduction time from T_1 to T_2, a large pulse of current flows through the diode and the ac input circuit to charge C_1. As a result, the voltage across the capacitor rises rapidly. The time constant is short with the very low resistance of the forward-biased diode. The peak capacitor voltage is practically equal to the peak value of the ac input voltage, minus the 1-V drop across the silicon diode.

After the time T_2, the ac input voltage is not positive enough to forward-bias the diode, and it stops conducting. The diode is then an open circuit, and it cannot charge C_1. Then C_1 discharges slowly from time T_2 to T_1 through the relatively high resistance of the load R_L.

This same sequence of fast charge and slow discharge for C_1 is repeated every cycle. As a result, V_{cb} across C_1 and R_L in parallel is a relatively steady dc output voltage, with a very small component of 60-Hz ac ripple.

Charge and discharge current. The peaks of current that flow through the diode to charge C_1 are shown in Fig. 12-9c. A peak value of 4 A is shown here. This figure is 10 times the typical forward current (I_F) of 400 mA or 0.4 A assumed for the load.

The path for charging current is shown in Fig. 12-10a. Here electrons flow from point b of the ac input to charge the low end of C_1 negative. Electrons repelled from the positive side of C_1 flow from cathode to anode in the diode and return through R_1 to point a at the source. It is important to note that the higher the capacitance of the filter C_1 is, the higher is the peak value of the charging current through the diode.

In Fig. 12-10b, C_1 discharges through R_L when the diode is not conducting. Now the circuit consists only of C_1 and R_L. The diode is an open circuit, and it effectively disconnects the ac input. This discharge occurs between times T_2 and T_1 in Fig. 12-9. Actually, the filter capacitor supplies the current to the load for about 90 percent of the time, while the diode is not conducting. During the 10 percent of the cycle when the diode is on, it allows the ac input voltage to charge C_1 while supplying the load current.

12-5 Low-Voltage Supply with Half-Wave Rectifier

These circuits are usually line-connected ac-dc supplies, without a power transformer. All the heaters, including the picture tube, are in a series string with voltage ratings that add up to the 120 V of the ac power line.

In Fig. 12-11, CR1 is the silicon half-wave diode rectifier. The diode is connected to the ac power line through the 3.6-Ω surge-protective resistor R_1, a 2-A fuse, and the switch S_1. With ac input to the anode of CR1, the positive half-cycles are rectified to produce positive dc output at the cathode. The B+ voltage at the input filter capacitor C_{3A} is 150 V. A two-section filter is used with the 0.3-H filter choke L_2 and 1.5-kΩ filter resistor R_2. The output filter capacitor is the 100-μF C_{3C}, while the 300-μF C_{3B} is the midsection filter capacitor. All three filters are electrolytic capacitors in one can with the half-circle, square, and triangle marks to indicate each section.

In the dc output voltages, note that 140 V is less than the unfiltered B+ voltage of 150 V because of the 10-V drop across the dc resistance of L_2. Also, the voltage drop of 20 V across R_2 reduces the filtered output voltage from 140 to 120 V.

Quick-on operation. Many receivers with tubes now have *instant-on, instant-view,* or *quick-on operation* for the heaters. This feature allows the heaters to be on with half power, when the receiver is off. Then, when the receiver is turned on, full power is applied to the heaters. The picture and sound come on almost instantly because the heaters are not cold.

Referring to Fig. 12-11, the switches S_1 and S_2 apply power to the receiver. S_1 is the "vacation switch." When the receiver is not used for a long period, S_1 is opened to prevent

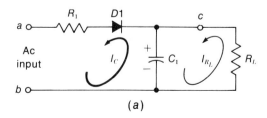
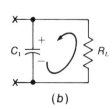

FIGURE 12-10 CHARGE AND DISCHARGE OF THE FILTER CAPACITOR C_1 IN HALF-WAVE RECTIFIER. (a) DIODE CHARGES C_1 BETWEEN TIMES T_1 AND T_2 ON WAVEFORMS IN FIG. 12-9. (b) C_1 DISCHARGES THROUGH THE LOAD BETWEEN T_2 AND T_1 WHEN DIODE IS NOT CONDUCTING.

236 BASIC TELEVISION

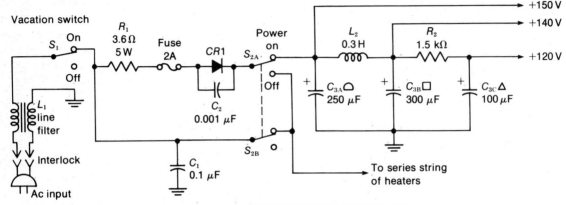

FIGURE 12-11 TYPICAL LOW-VOLTAGE POWER SUPPLY WITH LINE-CONNECTED HALF-WAVE RECTIFIER AND SERIES HEATERS FOR VACUUM TUBES. *(GE CHASSIS D2)*

any power being applied. Normally, though, S_1 is closed. Then the ac input is connected to the power supply, but the on-off switch S_2 determines how. With S_{2A} open, the filtering circuit is open for no B+ voltage. However, then CR1 is connected by S_{2A} to the series string of heaters. As a result, the positive half-cycles of the ac input are used to operate the heaters at half power. Then the heaters are warm and the tubes are in a ready state to be turned on quickly.

When the power switch is closed to turn the set on, S_{2A} completes the rectifier circuit for B+ voltage. Also, the lower section S_{2B} connects the heater string directly to the ac line for full power. The quick-on circuit shortens the turn-on time of the receiver from 30 s or more to 2 s or less.

Hash filter. Interference from motors, neon signs, or any sparking contacts is called *hash* because it does not have a specific resonant frequency. In order to filter any hash interference, either into or out of the power line, the line filter L_1 is used with C_1 as in Fig. 12-11. L_1 is a balanced choke with series inductance in both sides of the line. A typical value is 1 mH.

Silicon hum bar. When the silicon diode conducts, a large pulse of charging current flows in the ac input circuit. You can see this pulse with an oscilloscope across the surge-limiting resistor. The repetition rate of the pulses is 60 Hz. Because of its fast rise time, the pulse can radiate interference into the rf, IF, or video circuits. The result then is a black or white horizontal bar across the picture. This bar can be shifted between top and bottom of the picture by reversing the ac plug. To minimize the interference, a silicon-diode rectifier usually has a 0.001-μF shunt capacitor such as C_2 in Fig. 12-11. In addition, a large value can be used for the hash filter capacitor C_1.

Ac power interlock plug. When the back cover of the receiver is removed, the ac power cord is disconnected. In order to operate the receiver with the cover off, a "cheater cord" is used, which is a substitute line with its own plug.

Some interlocks have one pin bigger than the other, for a polarized plug. The wider pin is the ground side.

12-6 Full-Wave Center-Tap Rectifier

This circuit uses two diodes to rectify both half-cycles of the ac input. Each diode is a half-wave rectifier, but they have a common load R_L. See Fig. 12-12. Only one-half the total secondary voltage is used for each diode, from each end of L_S to the center. A power transformer is needed to provide the center-tap.

It is important to note that opposite ends of a coil always have opposite polarities with respect to the center. When the top end of L_S is positive, the bottom end is negative, as shown in Fig. 12-12a. For this polarity of ac input, the anode of D1 is positive, and this diode conducts. On the next half-cycle, though, all the polarities are opposite. Then the bottom of L_S is positive to make the D2 anode positive, and the bottom diode conducts.

When diode 1 conducts, we can trace the electron flow of I_1 from the center-tap, through R_L, from cathode to anode in D1 and returning to the top of L_S. Therefore, the cathode side of R_L at the right is the positive side of V_{R_L} with respect to the grounded center-tap, which is B−. See the top waveshape in Fig. 12-12b.

On the next half-cycle, diode 2 conducts. The electron flow of I_2 is from the center-tap, through R_L, from cathode to anode in D2 and returning to the bottom of L_S, which is now positive. Again, the cathode end of R_L at the right is the positive side of V_{R_L}. See the middle waveshape in b.

Note that both I_1 and I_2 are in the same direction through R_L, as the load resistance is common to both diodes. The combined result of V_{R_L} for both diodes is shown in the bottom waveshape. The ripple frequency is 120 Hz because the dc output voltage varies for each half-cycle.

In Fig. 12-13, the full-wave circuit is shown with a capacitor input filter. Each diode conducts only at the peak of the positive ac input

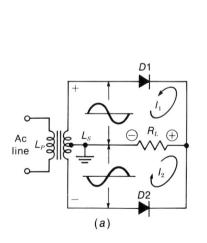

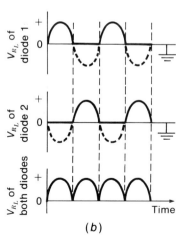

FIGURE 12-12 FULL-WAVE, CENTER-TAP RECTIFIER. (a) CIRCUIT. I_1 AND I_2 SHOWN IN DIRECTION OF ELECTRON FLOW. (b) WAVESHAPES WITHOUT FILTER.

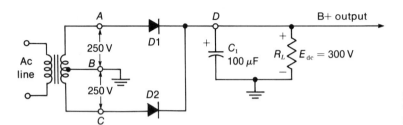

FIGURE 12-13 FULL-WAVE, CENTER-TAP RECTIFIER CIRCUIT WITH FILTER CAPACITOR C_1.

to charge C_1. The waveshape of the peak charging current for each diode is the same as shown in Fig. 12-9 for a half-wave rectifier. Since both diodes charge C_1, however, a smaller filter capacitor can be used in a full-wave rectifier.

The dc output voltage in Fig. 12-13 is calculated as $1.2 \times 250 = 300$ V. This is a typical B+ voltage for vacuum tube amplifiers in an ac receiver with a center-tapped power transformer. This circuit was commonly used with a dual-diode vacuum tube rectifier such as the 5U4. With solid-state rectifiers now, though, it is preferable to use circuits that do not need a center-tap. For vacuum tube rectifiers, bridge circuits and voltage doublers have the complications of extra heater connections, but this is eliminated with solid-state diodes.

12-7 Full-Wave Bridge Rectifier

This circuit is a full-wave rectifier that does not need a center-tapped power transformer. As shown in Fig. 12-14, the entire ac input is used for dc output voltage. In a, T_1 is used as an isolation transformer. The turns ratio is 1:1, to supply the same 120-V ac input as the line-connected supply in b. If necessary, though, the power transformer can step up or step down the ac voltage. Note that the surge-limiting resistor R_1 in b takes the place of the transformer resistance in a.

The advantage of a bridge rectifier is that both half-cycles of the ac input are rectified, without a center-tap to provide opposite polarities. The disadvantage is that four diodes are necessary. This factor was inconvenient with vacuum tube rectifiers because of the heater connections. However, semiconductors do not

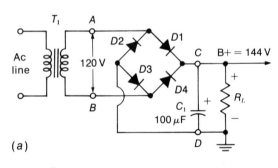

(a)

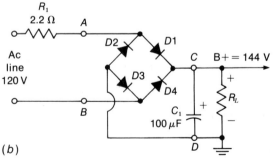

(b)

FIGURE 12-14 FULL-WAVE BRIDGE RECTIFIERS. (a) LINE-ISOLATED WITH POWER TRANSFORMER. (b) NO ISOLATION FROM POWER LINE.

have heaters. A small bridge rectifier unit with four diodes is shown in Fig. 12-15.

The details of conduction by the four diodes in the bridge are shown in Fig. 12-16. In brief, the operation of the bridge allows two diodes to conduct in series with the load R_L on each half-cycle. The equivalent circuit in a shows that R_L is the center arm of the bridge. The right side at point C is at a rectifier cathode, either for D1 or D4, for positive dc voltage output. The opposite end of R_L at point D is chassis ground. The result is B+ voltage at the cathode side of R_L, with respect to ground.

When point B is negative in Fig. 12-16b, the cathode is negative and anode positive for D3 and D1, making these diodes conduct. The electron flow I_{31} is from the bottom of L_S, from cathode to anode in D3, through R_L, from cathode to anode in D1, and returning to point A at the top of L_S. D2 and D4 are not shown in this diagram. They are effectively out of the circuit because of reverse bias with positive cathode voltage on D2 and negative voltage for the anode of D4.

On the next half-cycle when point A is negative in Fig. 12-16c, D2 and D4 conduct. The electron flow I_{24} is from the top of L_S, from cathode to anode in D2, through R_L, from cathode to anode in D4, and returning to point B at the bottom of L_S. Now D1 and D3 are out of

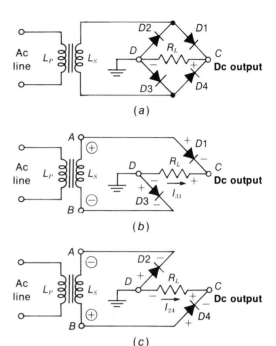

FIGURE 12-16 HOW THE BRIDGE RECTIFIER CONDUCTS ON OPPOSITE HALF-CYCLES OF THE AC INPUT. (a) THE LOAD R_L IS IN THE CENTER ARM OF THE BRIDGE. (b) WHEN POINT A IS POSITIVE, D1 AND D3 CONDUCT THROUGH R_L. (c) WHEN POINT B IS POSITIVE, D4 AND D2 CONDUCT THROUGH R_L.

the circuit because they are biased off with reverse voltage.

Note that I is in the same direction through R_L for both half-cycles. The result is B+ voltage at the cathode side of R_L, with respect to ground. If all four diodes were reversed, the dc output voltage would be negative.

The ripple frequency of the full-wave bridge is 120 Hz because both half-cycles of the ac input produce dc output. Note that the filter capacitor C_1 in Fig. 12-14 has the same value of 100 µF as C_1 for the full-wave center-tap rectifier in Fig. 12-13. The dc output voltage in Fig. 12-14 is calculated as $1.2 \times 120 = 144$ V.

FIGURE 12-15 BRIDGE RECTIFIER WITH FOUR DIODES IN ONE UNIT. SIZE IS 1/2-IN. SQUARE. (ERIE TECHNOLOGICAL PRODUCTS INC.)

12-8 Voltage Doublers

By series combinations of half-wave diodes and their filter capacitors, the amount of dc output voltage can be doubled, tripled, or quadrupled. Typical circuits are shown in Figs. 12-17 to 12-21. Even higher multiplication is possible, but the voltage regulation becomes worse and very large filter capacitors are necessary. In practical applications, the voltage doubler is often used for the B+ supply, especially the half-wave line-connected supply shown in Fig. 12-17. Voltage triplers and quadruplers can be used in the high-voltage supply for anode voltage on color picture tubes.

Half-wave voltage doubler. See Fig. 12-17. This circuit is also called a *cascade doubler* because the rectified output of $D1$ is used for input to $D2$. The two diodes conduct on opposite half-cycles. We can start with the negative half-cycle of the ac input that drives the cathode of $D1$ negative. Then this diode conducts to charge C_1. Note the polarity of the dc voltage across C_1. On the next half-cycle, when point A is positive, the C_1 voltage is connected series-aiding with the ac input voltage. Then the input for $D2$ is doubled, with the anode positive for conduction. As $D2$ charges C_2, the dc output voltage across C_2 is doubled. R_1 is the surge-limiting resistor for both diodes.

This circuit is a half-wave rectifier because dc output across C_2 is produced only on the positive half-cycle of the ac input when $D2$ conducts. The ripple frequency, then, is 60 Hz.

An important feature of this doubler is that one side of the ac input and dc output are common. For this reason, the half-wave doubler is generally used as a line-connected supply, without a power transformer. As shown in Fig. 12-17, the grounded side of the dc output is common to the neutral or grounded side of the ac input. The common connection reduces the problem of stray pickup of 60-Hz hum in the receiver.

Because the half-wave doubler is often used for the low-voltage supply, more details of its operation are illustrated in Fig. 12-18. In *a*, the circuit is shown for $D1$ alone. The other diode $D2$ is effectively out of the circuit because its anode is reverse-biased on this half-cycle when the ac input voltage is negative. The electron flow I_1 then charges C_1 with the positive side at the cathode of $D1$. This dc voltage has the same polarity on both half-cycles of the ac input.

In *b*, when the ac input voltage is positive, $D2$ conducts without $D1$. Note the series-aiding polarities, as the minus side of C_1 is at the positive side of the ac input. This connection is the same as series-aiding batteries. The effect, then, is to lift the ac input voltage to a higher level on the axis of the dc voltage across C_1. This effect is shown by the waveform in *b*.

The peak of the combined voltage for $D2$ is almost twice the peak of the ac input voltage alone. As a result, $D2$ charges C_2 to double the

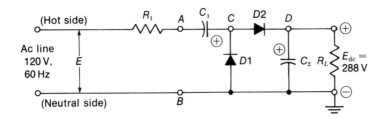

FIGURE 12-17 TYPICAL VOLTAGE DOUBLER. THIS TYPE IS A HALF-WAVE OR CASCADE DOUBLER CIRCUIT.

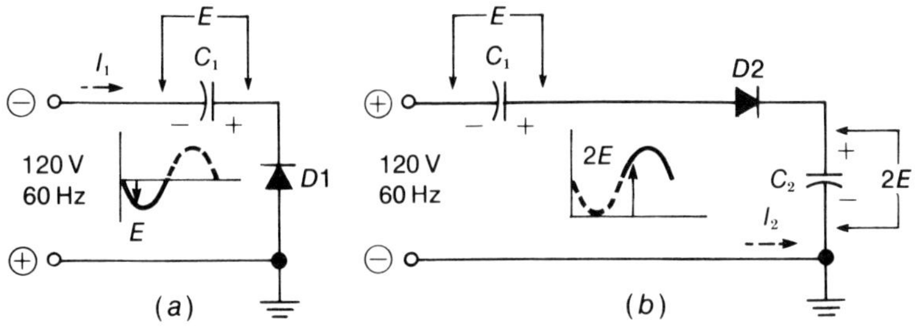

FIGURE 12-18 OPERATION OF HALF-WAVE VOLTAGE DOUBLER. (a) CIRCUIT FOR D1 CHARGING C_1, WITHOUT D2. (b) CIRCUIT FOR D2 CHARGING C_2, WITHOUT D1.

dc output voltage of one half-wave rectifier alone. If we consider the value of E across C_1 in Fig. 12-18a as $1.2 \times 120 = 144$ V, then the doubled output across C_2 in b is $2E = 288$ V.

Full-wave doubler. See Fig. 12-19. On the half-cycle when point A is positive, D1 conducts to charge C_1. On the next half-cycle D2 conducts to charge C_2. The dc output voltage is taken across C_1 and C_2 in series. Therefore, the dc output is doubled. The ripple frequency is 120 Hz, since both half-cycles of the ac input produce dc output. The full-wave voltage doubler circuit is seldom used, however, because it does not have a common connection for ac input and dc output.

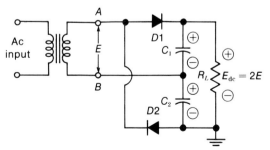

FIGURE 12-19 CIRCUIT OF FULL-WAVE VOLTAGE DOUBLER.

12-9 Voltage Triplers and Quadruplers

For a tripler, the voltage doubler is combined with a single half-wave rectifier. For a quadrupler, two doublers are connected in cascade.

Referring to the half-wave voltage tripler in Fig. 12-20, D1 and D2 form a cascade voltage doubler. Furthermore, the doubled voltage $2E$ across C_2 is applied to D3 in series with C_3 and the ac input voltage. On the half-cycle when point B is positive, the ac voltage across the AB terminals is series-aiding with the dc voltage $2E$. As a result, conduction in D3 charges C_3 to $3E$, which is triple the input voltage.

For the half-wave voltage quadrupler in Fig. 12-21, D1 and D2 form one voltage doubler. Also, D3 and D4 form a voltage doubler. C_3 is the series capacitor for D3, corresponding to C_1 for D1. The output is $4E$ because the input to D4 is the dc voltage $3E$ across C_3 in series with the ac input voltage.

12-10 Heater Circuits

For receivers that use tubes, the heaters can be connected either in series or in parallel. With a transformer to step down the ac line voltage from 120 to 6.3 V, the heaters are in parallel across the filament winding (Fig. 12-22). Then all

242 BASIC TELEVISION

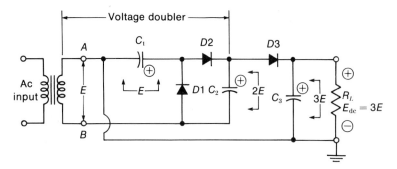

FIGURE 12-20 CIRCUIT OF HALF-WAVE VOLTAGE TRIPLER.

the parallel heaters have the same 6.3-V rating, although the individual branch currents can be different. In the line-connected supply without a power transformer, all the heaters are connected in series to the ac power line (Fig. 12-23). Then the current is the same in all parts of the series string. Therefore, heaters with the same current rating must be used, although the individual heater voltages can be different.

The current is usually either 450 or 600 mA for series heater strings. All the series voltage drops add to equal the power-line voltage of 110 to 120 V. It is important to remember that when one heater is open, there cannot be any current in the entire series string.

Parallel heaters. In Fig. 12-22, one end of the filament winding L_f is tied to chassis ground to save wiring. The opposite end is the high side connecting to all the heaters in parallel. Each heater returns to L_f by an individual ground connection. Note that the heater pins are usually 3 and 4, or 4 and 5 on the amplifier tubes. With parallel connections, when one heater is open, the other heaters are not affected.

Filament chokes. The 1-μH coil L_1 in Fig. 12-22a is a few turns of thick wire to serve as a choke for rf or IF signals. With the rf bypass C_1 of 0.047 μF, L_1 and C_1 provide a decoupling filter to prevent feedback of signals between rf and IF amplifiers through the common heater line. The choke L_1 has high reactance for the signal frequencies but practically none for the 60-Hz heater current. Therefore, the heaters are still effectively in parallel as shown in b. Such rf

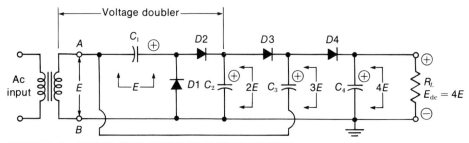

FIGURE 12-21 CIRCUIT OF HALF-WAVE VOLTAGE QUADRUPLER.

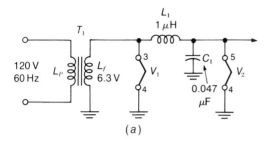
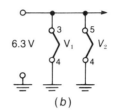

FIGURE 12-22 PARALLEL CONNECTIONS FOR 6.3-V HEATERS. (a) WITH RF DECOUPLING FILTER L_1C_1. (b) EQUIVALENT CIRCUIT FOR 60 Hz.

decoupling filters are also used in series heater circuits. Ferrite beads may be used instead of the rf choke.

Series heaters. In Fig. 12-23, all the heaters including the picture tube are in series across the power-line voltage of 110 to 120 V. The current is 450 mA in all these series heaters. The sum of the heater voltages here is 109 V. With the 6.8 V across R_{114}, the total of all the IR drops then is 115.8 V, approximately equal to the power-line voltage. The high-voltage rectifier is not in the series string, as it receives filament power from the horizontal output transformer.

Color receivers also have series heaters, but the color picture tube often has its own 6.3-V filament transformer. This heater usually has a dc bias voltage added to raise its potential close to the cathode voltage. The purpose is to minimize internal arcing.

Center-tapped heater. Some tubes have the heater in two sections that can be wired either in series for 12.6 V or in parallel for 6.3 V. Examples of such tubes with three heater pins are the 12AU7, 12AT7, and 12BY7.

Controlled warmup time. All the tubes in a series string should reach operating temperature at the same time. Otherwise, the cold heaters have too little resistance, allowing too much voltage across the hot tubes just after the set is turned on. The uneven warmup time can cause repeated troubles of a burned-out heater in the series string. This used to be a common problem with series heaters. However, now tubes for series strings are designed to reach 80 percent of rated heater voltage in 11 s, with specified test conditions. In actual operation, typical warmup time for a series string without a thermistor is about 50 s for a full raster. The warmup is much shorter for parallel heaters.

Some heater strings have a *thermistor* in series to limit the current until the tubes become hot. A thermistor has a negative temperature coefficient, meaning its resistance decreases with heat. Typical values are 100 Ω for the cold resistance, decreasing to 10 Ω for the hot resistance. The thermistor for the heater string is also called a *tube saver*. With a thermistor, the warmup time for the receiver can be as long as 2 min.

Order of series heater connections. All the series heaters have the same current, but a point closer to the grounded end of the string has less potential difference to chassis ground for 60-Hz voltage. Therefore, less hum voltage can be coupled into the signal circuits by heater-cathode leakage in the tube. This factor is why the tubes in the VHF tuner are last in the string, while the horizontal deflection amplifier and damper tubes are first, as in Fig. 12-23.

It is important to note the difference between measuring the 60-Hz ac voltage across

244 BASIC TELEVISION

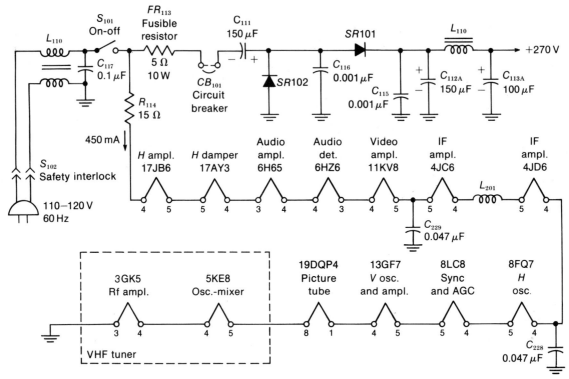

FIGURE 12-23 SERIES HEATER STRING WITH CASCADE VOLTAGE DOUBLER FOR LOW-VOLTAGE SUPPLY. *(RCA MONOCHROME CHASSIS KCS 142)*

the heater pins or measuring from one pin to chassis ground. For instance, in Fig. 12-23 the ac voltage across the heater pins 4 and 5 of the 17JB6 is 17 V. However, from pin 4 to ground, you measure 109 V of the ac line voltage. At pin 4 to ground, the ac voltage is 17 V less, or 92 V.

12-11 Filter Circuits

The function of the filter is to remove the ac ripple. The two basic types for power supplies are the capacitor input filter in Fig. 12-24a and the choke input filter in b. The difference is that C_1 in a can be charged by the rectifier current without the series inductance of the choke L_1. C_1 is the *input filter capacitor* at the rectifier, while C_2 is the *output filter capacitor* across R_L. Because of the peak charging current for C_1, the capacitor input filter produces more dc output voltage than the choke input filter. There is also less ripple voltage, for moderate values of load current. The choke input filter has better voltage regulation. Also, the amount of ac ripple voltage does not increase with the load current. The capacitor input filter is generally used in receivers, however, because of its higher output voltage.

Typical values of C are 50 to 2,000 μF while L is 2 to 10 H. The dc resistance is 250 Ω for a 5-H choke rated at 200 mA. A filter resistor of 100 to 1,000 Ω can be used instead of the choke in some power supplies with relatively small values of load current. Large filter capaci-

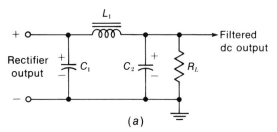
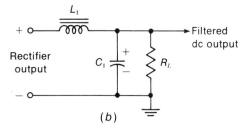

FIGURE 12-24 FILTER CIRCUITS. (a) CAPACITOR INPUT. (b) CHOKE INPUT.

tances of 500 to 2,000 µF are needed in transistorized receivers because of more dc load current and smaller dc voltages, compared with vacuum tubes.

Percent ripple. This is the ratio of the rms value of the ac ripple voltage to the dc output voltage, times 100 for the percentage. As an example, with 1.4 V for the ac ripple and 140-V dc output,

$$\% \text{ ripple} = \frac{1.4}{140} \times 100 = 1\%$$

This value is typical at the input filter capacitor. The output filter capacitor C_2 has even less ripple in the B+ voltage for the load.

Active filter. A transistor can be used as an electronic filter, as shown in Fig. 12-25. Note that Q1 is effectively in series with the power supply and the load, corresponding to a series choke. The electronic filter removes ripple variations in the load current because the transistor has relatively constant current in the collector output circuit. In the base circuit, the ripple is easy to filter by C_1 because of the small value of current. Forward bias for the base is provided by R_3 from the positive collector voltage. The diode D1 clamps the bias at approximately 0.7 V for constant saturation current in the collector circuit. In the output circuit, additional filtering is provided by the π-type filter consisting of C_2, R_2, and C_3.

12-12 Voltage Regulators

Regulation in a power supply means keeping the dc voltage output constant with variations either

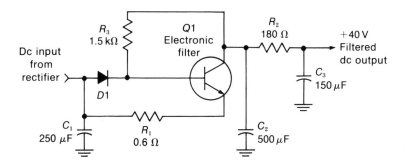

FIGURE 12-25 CIRCUIT FOR ACTIVE FILTER. SEE TEXT FOR EXPLANATION.

in dc load current or in ac input voltage. As one example, a voltage regulator is generally used for the anode voltage of a color picture tube. Constant high voltage is needed to maintain focus and convergence with changes in brightness. In the low-voltage supply of solid-state receivers, regulation is important in order to maintain constant raster size.

In general, voltage regulation is useful because it also improves the filtering of ac ripple and reduces the ac output impedance of the power supply. Low impedance minimizes the coupling between amplifier stages through the common B+ line.

The types of voltage regulator circuits in television receivers range from the relatively simple zener diode to more complicated feedback circuits. The following list illustrates the common methods of regulation.

1. *Zener diode.* Remember that with reverse breakdown, the voltage across the zener diode is constant for a wide range of current values through the diode. Zener regulators are commonly used in voltage ratings of 3 to 180 V.
2. *Voltage-regulating power transformer.* This special transformer in the low-voltage supply regulates the ac line voltage input. Regulation is accomplished by using saturation of the core to limit increases and by resonance in the secondary to limit decreases. An external 3.5-μF oil-filled capacitor tunes the secondary winding to 60 Hz.
3. *Feedback regulators.* In this method, a sample of the output is fed back to a control stage, which can then regulate the circuit for constant output. The current in a tube or transistor can be controlled, as a variable resistance, by varying the bias. A *shunt regulator* is in parallel with the output circuit to bypass more current or less current around the load. A *series regulator* provides a variable voltage drop in series with the output voltage.

Zener-diode regulator. See Fig. 12-26. Note the schematic symbol for the zener diode Z1. Reverse bias is used with positive voltage at the cathode. This zener diode is rated for 12-V breakdown voltage, with a maximum diode current of 150 mA and 10-W power dissipation. Smaller units have a power rating of 400 mW or 1 W. The series regulating resistance R_s has the function of providing a voltage drop that varies with the load current. However, there must be at least 12 V across the diode to provide current through Z1 so that it can operate in its breakdown mode.

The 6BK4 shunt regulator. See Fig. 12-27. This circuit is used for regulating the high voltage supply in most color receivers with vacuum tubes. The shunt current through the 6BK4 regulator V3 and the load current, which is the beam current in the picture tube, together remain constant as a total current. With constant load current in the HV rectifier, its output of 25 kV remains constant.

The regulation of the 6BK4 results from varying its bias. In Fig. 12-27, the grid-cathode voltage is $380 - 404 = -24$ V. Note that the 380

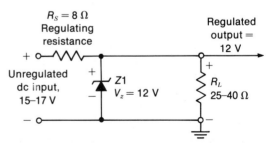

FIGURE 12-26 ZENER-DIODE VOLTAGE REGULATOR CIRCUIT. NOTE 12 V ACROSS Z1 IS REVERSE BIAS.

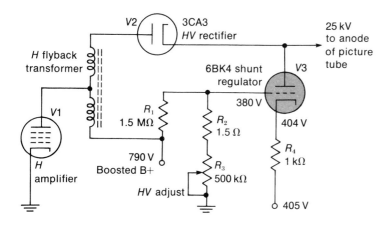

FIGURE 12-27 THE 6BK4 HIGH-VOLTAGE SHUNT REGULATOR CIRCUIT.

V at the grid is obtained from the boosted B+ of 790 V in the horizontal output circuit. This voltage depends on the load for the horizontal amplifier V1.

Suppose that the beam current in the picture tube increases with more brightness. This factor will tend to decrease the high voltage. Then the boosted B+ voltage decreases slightly because of the added load current for V1. As a result, assume that the positive grid voltage for V3 decreases by 5 V. This means the grid-cathode bias is 5 V more negative. The result is less current in the 6BK4 regulator. For the opposite case, less beam current for the load allows more beam current in the shunt regulator circuit. Together, the load current for the anode of the picture tube and the shunt regulator current remain at a constant value for the total current supplied by the HV rectifier.

The high-voltage adjustment R_3 in Fig. 12-27 is set as follows:

1. Turn brightness and contrast controls to minimum for a black screen and maximum high voltage.
2. Connect high-voltage probe (Fig. 12-28) to measure anode voltage.
3. Set the variable resistance R_3 for the specified anode voltage, which is usually 25 kV for a 21-in. color picture tube.

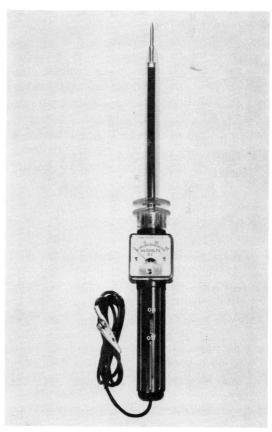

FIGURE 12-28 VOLTMETER FOR MEASURING CRT HIGH VOLTAGE ON ANODE. LENGTH IS 15 IN. MULTIPLIER RESISTANCE IN PROBE IS 600 MΩ. *(POMONA ELECTRONICS)*

With this adjustment, the regulator should be drawing approximately 1-mA cathode current. Then the *IR* drop would be 1 V across the 1-kΩ cathode resistance R_4 in Fig. 12-27.

Feedback circuit with series regulator transistor. See Fig. 12-29. Q1 is the series regulator that provides output to the load. Q2 is the driver to control the bias on Q1. The driver has two input voltages. One is a reference of 11.2 V stabilized by the zener diode in its emitter circuit. The other is the sensing voltage of approximately 11.8 V, from a tap on the $R_1R_2R_3$ voltage divider across the output.

The amount of conduction in the driver Q2 determines the bias on the regulator Q1. Furthermore, the driver Q2 compares the reference of 11.2 V in the emitter to the sample voltage from R_2, which is approximately one-tenth the output, applied to the base. Any difference is amplified to change the bias and conduction for the series regulator Q1. The adjustable resistor R_2 for the driver Q1 is set for exactly 115 V at the output of the series regulator Q1.

12-13 Low-Voltage Supply with Series Regulator Circuit

This circuit for a solid-state color television receiver is shown in Fig. 12-30. The rectifier uses a full-wave bridge, with CR1, CR2, CR3, and CR4. R_2 is a surge-protective resistance of 2.2 Ω. C_2 and C_3 are hash suppressors for the diodes. The B+ supply is line-connected, although the autotransformer T_1 supplies an isolated 6.3 V for the heater of the picture tube. In the ac input, a 5-A fuse and 1.25-A circuit breaker provide protection against overload. L_1 and C_1 are hash filters for the power line. The degaussing coils L_2 and L_3 are fed ac power from the line, through the thermal unit consisting of R_3 and R_4.

The on-off S_1 is a double-pole switch. S_{1A} connects the ac line to the bridge rectifier for B+ voltage. However, S_{1B} only has the function of shorting R_1 in series with the line. When the receiver is off, S_{1B} is open. Then R_1 is in the circuit to decrease the line current for reduced heater power on the picture tube. When the receiver is on, S_{1B} is closed to short R_1. The

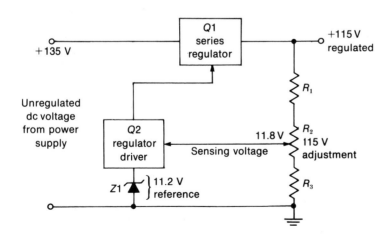

FIGURE 12-29 SERIES REGULATOR FOR LOW-VOLTAGE POWER SUPPLY.

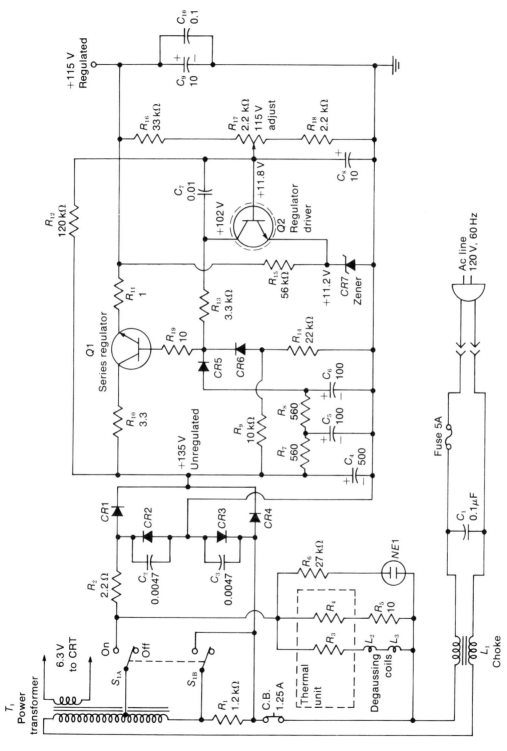

FIGURE 12-30 LOW-VOLTAGE POWER SUPPLY WITH FULL-WAVE BRIDGE RECTIFIER AND SERIES REGULATOR. C IN μF. (SONY CHASSIS KV 1210 U)

result is full heater power for the picture tube and approximately 110-V ac input for the bridge rectifier with S_{14} closed. The neon bulb NE1 indicates that voltage is applied to the B+ circuit.

The bridge rectifier produces 135 V, at the 500-μF input filter capacitor C_4. The two-section filter consisting of R_7 and R_8 with C_4, C_5, and C_6 removes the 120-Hz ripple. In addition, the output of 135 V goes to the regulator circuit consisting of Q1 and Q2. This is the same series regulator illustrated in Fig. 12-29. The dc output from the emitter of Q1 is a well regulated +115 V. R_{17} is adjusted to produce this output voltage. Additional filtering in the output is provided by the 10-μF C_9 and 0.1-μF C_{10} in parallel. The purpose of the small paper capacitor is to provide low reactance for rf signal frequencies, across the self-inductance of a large electrolytic capacitor.

12-14 Troubles in the Low-Voltage Supply

A trouble here is common to all sections of the receiver because it produces multiple effects. Two common indications are: (1) no brightness and no sound because of no B+ voltage, and (2) a small raster because of low B+ voltage. Typical troubles are listed in Table 12-1.

No B+ voltage. The receiver will be dead, without any brightness or sound, although the heaters can light. Remember that the B+ line usually has a protective fuse or circuit breaker that may be open. Also, in many receivers the B+ line is opened when the yoke plug is out of its socket. Troubles that can cause no B+ voltage include an open diode in a half-wave rectifier circuit, an open filter choke or resistor, and a shorted filter capacitor.

Low B+ voltage. As an example, if the B+ voltage is reduced 25 percent, the raster will be about 10 percent smaller. The focus may also be affected. The sound will probably be normal, as the volume is not reduced until the B+ is down about 40 percent. The same effects are produced by a low value of line voltage for the ac input.

A common cause of low B+ voltage is an open input filter capacitor. Furthermore, a diode can be open in a doubler or a bridge rectifier to cause reduced B+ voltage, instead of no output. It should be noted that when one side of a full-wave bridge is open, the ripple frequency is 60 instead of 120 Hz.

Open in series heater line. An open in one heater means no current in the entire series string. The full ac line voltage is across the open heater because its resistance is infinitely high. If you check each heater with an ac voltmeter, the open heater has 120 V. If you use an ohmmeter

TABLE 12-1 TROUBLES IN THE B+ VOLTAGE

NO B+ VOLTAGE	LOW B+ VOLTAGE
No ac line voltage	Low ac line voltage
Open diode in half-wave rectifier	Open diode in doubler or bridge circuit
Open fuse or circuit breaker	Leakage in diode rectifier
Open in series filter R or L	High resistance in series filter R or L
Short in shunt filter capacitor	Open in shunt filter capacitor

with power off, the open heater reads infinite ohms. Start with the tubes for the power stages, such as audio output, vertical output, horizontal output, and damper. These tubes cause the most trouble because they generate the most heat.

Open filter capacitors. What happens, usually, is that the electrolyte dries out. Then the capacitor becomes open or practically open with reduced capacitance. With an open input filter capacitor, the result is reduced B+ voltage and excessive hum. When the output filter capacitor is open, the result is poor regulation of the B+ voltage for the load.

Effects of poor regulation in B+ voltage. The circuits that draw the most load current from the B+ supply are the deflection amplifiers and audio output stage. When the B+ voltage varies with the load current in the audio amplifier, the result can be sound bars in the picture. These sound bars disappear when you turn down the volume. When the B+ voltage varies with the load current in the vertical amplifier, the result can be vertical shading in the picture. Finally,

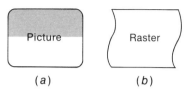

FIGURE 12-31 EFFECTS OF HUM. (a) 60-Hz HUM BARS IN PICTURE. (b) 60-Hz HUM BEND IN RASTER.

when the amount of B+ voltage changes, the size of the raster can increase or decrease, especially with changes in the ac line voltage.

12-15 Hum in the B+ Voltage

Hum is excessive ac ripple in the dc output of the low-voltage power supply. The cause is poor filtering, usually as a result of open filter capacitors. The ripple frequency is 60 Hz in a half-wave rectifier or 120 Hz in a full-wave circuit. In the sound, the ac ripple produces hum. On the screen of the picture tube, the hum can cause horizontal black-and-white *hum bars* or bend at the sides. See Fig. 12-31. Typical hum troubles are summarized in Table 12-2.

TABLE 12-2 HUM TROUBLES

HUM	60 Hz	120 Hz
In video signal	One pair of horizontal black-and-white bars	Two pairs of horizontal black-and-white bars
In horizontal scanning	One cycle of sine-wave bend at sides of raster	Two cycles of sine-wave bend at sides of raster
In horizontal sync	One cycle of sine-wave bend at sides of picture	Two cycles of sine-wave bend at sides of picture
In vertical sync	Poor vertical synchronization in picture	Poor vertical synchronization in picture
In vertical scanning	Nonlinear vertical scanning	Nonlinear vertical scanning

For all these hum effects, remember that the vertical scanning frequency for color television is exactly 59.94 Hz, not 60 Hz. As a result, the 60-Hz hum patterns do not stay still. Hum bars drift slowly up the picture. Also, hum bend at the sides crawls upward on the screen.

Hum in the video signal. Hum voltage can be amplified by the video section supplying signal for the picture tube. The result is black-and-white bars across the screen, as shown in Fig. 12-31a. One pair of bars is shown here for 60-Hz hum. With 120-Hz ripple, the result is two pairs of hum bars.

Hum in the horizontal scanning. The result is sine-wave bend at the sides of the raster (Fig. 12-31b). One cycle of bend is produced by 60 Hz; two cycles show 120 Hz. This bend is in the raster and in the picture.

Hum in the horizontal sync. The result is sine-wave bend in the picture but not in the raster. If you remove the picture, the raster will have straight sides.

Hum in the vertical sync. This trouble makes it difficult to lock in the picture vertically. One specific effect is a picture halfway out of position, as the vertical oscillator is synchronized by the hum voltage instead of vertical sync. Then the black vertical blanking bar is across the middle, while the top and bottom of the picture are reversed.

Hum in the vertical scanning. The result is nonlinearity from top to bottom of the picture. The reason is that the sine-wave ac ripple combined with the sawtooth sweep results in nonlinear vertical scanning.

12-16 Flyback High-Voltage Supply

Practically all television receivers use this circuit because the required high-voltage ac input for the high-voltage rectifier is produced by the fast drop of current in the horizontal output transformer during horizontal retrace. Details of the horizontal output circuit are explained in Chap. 22, but the basic requirements for flyback high voltage are illustrated in Fig. 12-32. T_1 is an autotransformer for the horizontal output. This is called the *flyback transformer.*

Terminals 1 and 2 on T_1 supply output to the horizontal coils in the deflection yoke for horizontal scanning. Terminals 1 and 3 serve as the primary for the horizontal deflection amplifier. This can be a tube or a transistor. Terminal 4 has the stepped-up voltage for the high-voltage rectifier. Figure 12-33 shows a typical flyback transformer.

When the output current drops to zero for horizontal flyback, the fast change in current produces a high value of induced voltage at terminal 4 of T_1 for the high-voltage rectifier. The voltage peak has positive polarity to oppose the decrease in current. With the rectifier plate positive, it conducts to charge the filter capacitance C_1. This is the anode capacitance of the picture tube. The result is 18-kV positive dc voltage at the directly heated cathode of the high-voltage rectifier.

The flyback transformer and rectifier are generally in a separate metal cage. In addition to protection against shock, the enclosure is a dust cover and shield against x-rays. When the receiver has a fuse such as F_1 for the horizontal output circuit, it is generally in or near the cage. An open fuse for the horizontal amplifier can be a simple cause for the trouble of no high voltage and no brightness, while the sound is normal. A practical point to remember is that in some

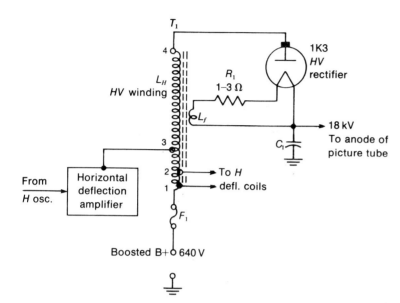

FIGURE 12-32 BASIC CIRCUIT OF FLYBACK HIGH-VOLTAGE SUPPLY.

receivers, the high-voltage output is disconnected when the cage cover is off.

High-voltage rectifiers. Tubes commonly used for anode voltage of 10 to 20 kV for monochrome picture tubes are the 1B3, 1G3, and 1K3. The top cap is the plate on these tubes. All have essentially the same ratings of 30,000-V peak inverse voltage, heater voltage of 1.25 V at 0.2 A, and dc output of 0.5 mA or less for the load current. Because of the small heater power, the filament voltage can be taken from the horizontal output transformer by the winding L_f in Fig. 12-32. This is only one turn of heavy wire wrapped on the core of the transformer. A separate filament transformer would require enough insulation to withstand the high dc voltage at the rectifier filament without arcing to chassis ground. Note that R_1 drops the heater voltage to 1.25 V. This may be a carbon resistor, or resistance wire can be used.

For color picture tubes, a larger tube is needed for the high-voltage rectifier because of more load current. Tubes commonly used are the 3A2 and 3CA3. The rated heater voltage is 3.15 V at 0.22 A, with a maximum load current of 2 mA.

For either monochrome or color picture tubes, a silicon-diode rectifier can be used. These are special high-voltage diodes with the required rating for peak inverse voltage. See Fig. 12-34.

High voltage for focusing. Color picture tubes with high-voltage electrostatic focusing generally require 5 to 7 kV for the focus grid. This voltage is taken from the horizontal output circuit. Either a separate rectifier can be used as in Fig. 12-35, or a tap on the high-voltage supply can provide the focusing voltage, as in Fig. 12-36. In Fig. 12-35, the silicon diode $X1$ is the *focus rectifier*. Tubes also used for this function are the 1V2 and 1X2. These high-voltage rectifiers do not have a top cap for the anode and are smaller than the 1K3.

254 BASIC TELEVISION

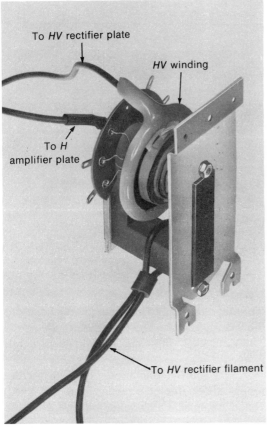

FIGURE 12-33 HORIZONTAL FLYBACK TRANSFORMER.

In Fig. 12-36 the high-voltage rectifier is a solid-state tripler circuit with the silicon diodes, capacitors, and resistors all in one unit. A tap at the output of the first rectifier supplies 7.7 kV to the focus control circuit.

Blooming. This term describes a picture that increases in size, with defocusing in white areas, as the brightness is increased. The blooming results from a sharp decrease in the amount of anode voltage when the beam current is increased to turn up the brightness. It should be noted that the scanning current produces more deflection, and the raster becomes bigger, with less high voltage. A common cause of blooming is a weak high-voltage rectifier, which has excessive internal resistance. Another reason is too little bias on the picture tube. Color receivers generally have a *brightness-limiter* circuit to prevent blooming, by limiting the bias set with the brightness control.

Breathing. This term refers to blooming at a slow, regular rate. The size of both the raster and picture changes with variations in the high voltage. The cause is usually trouble in the high-voltage regulation.

Corona and arcing. A point at high potential can ionize the surrounding air to produce a visible corona effect, which is light blue in color.

(a)

(b)

FIGURE 12-34 SILICON HIGH-VOLTAGE RECTIFIERS FOR COLOR RECEIVERS. (a) 30- TO 45-kV RATING FOR ANODE VOLTAGE OF CRT. LENGTH IS 4 IN. THIN END AT RIGHT IS CATHODE SIDE. (b) 5- TO 10-kV RATING FOR FOCUS VOLTAGE. LENGTH IS 2 IN. (*ERIE TECHNOLOGICAL PRODUCTS, INC.*)

The corona causes loss of power and eventual insulation breakdown with arcing. In addition, precipitation of ionized dust particles is a result of the corona.

To minimize corona, it is important to eliminate sharp edges in the wiring. Also, all soldered joints must be smooth and round. Thick wires have less corona, as the voltage gradient of the conductor surface to the surrounding air is reduced. The high-voltage rectifier socket often has a thick metal ring, called a *corona ring*, for the cathode connection to the dc high-voltage output.

Corona and arcing in the high-voltage supply can produce streaks in the picture. To recognize these effects, arcing can usually be heard as a snapping noise, while corona produces a sizzling sound. Often you can see the

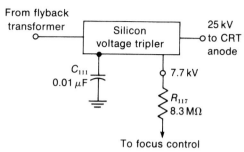

FIGURE 12-36 SILICON VOLTAGE TRIPLER FOR ANODE VOLTAGE, WITH TAP FOR FOCUS VOLTAGE. *(RCA COLOR CHASSIS CTC 58)*

corona or arcing by looking into the high-voltage cage. It can be helpful to note that dc arcing is white, but the corona effect from ac voltage looks blue.

All high-voltage connections must be well separated from the chassis to eliminate arcing. If we take the voltage breakdown rating of air at 20 kV/in., the high voltage of 25 kV can jump a gap of $1\frac{1}{4}$ in.

Brightness troubles. The brightness on the screen of the picture tube depends directly on the amount of high voltage. No high voltage means no brightness. Insufficient high voltage causes low brightness, usually with poor focus. Remember that the horizontal scanning circuits must be operating to produce the flyback high voltage. Also, trouble in the regulator can cause no high voltage or reduced voltage with poor focus.

12-17 Fuses and Circuit Breakers

Typical examples are shown in Fig. 12-37. The function of these safety devices is to open the circuit when a trouble causes excessive current. Sometimes the cause is a temporary surge of line voltage or circuit overload. However, when

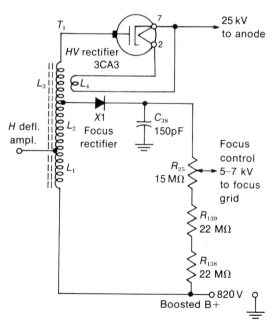

FIGURE 12-35 HIGH-VOLTAGE POWER SUPPLY WITH SEPARATE FOCUS RECTIFIER. *(ZENITH COLOR CHASSIS 23XC38)*

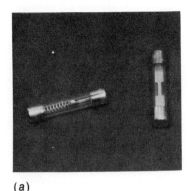  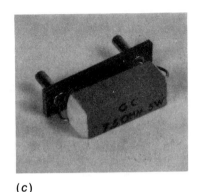

FIGURE 12-37 FUSES AND CIRCUIT BREAKERS. LENGTH ABOUT 1.5 IN. (a) GLASS CARTRIDGE FUSES. (b) CIRCUIT BREAKER WITH RESET BUTTON. (c) FUSIBLE RESISTOR.

a fuse is replaced and blows again, the trouble is probably a short circuit that must be corrected.

Do not use a fuse higher than the specified current rating. The purpose is to have the fuse open before the series components can burn out with too much current. It is the current rating of the fuse that provides the protection.

Fuses with just a straight wire link inside are fast-acting. They will blow in less than 0.3 s with a 400 percent overload. A "slow-blow" fuse has a coiled construction. The purpose is to prevent the fuse from blowing with just a temporary current surge. As an example, such a 1-A fuse will hold a 400 percent overload up to 2 s. Typical fuses are shown in Fig. 12-37a.

A circuit breaker (Fig. 12-37b) has a manual reset button to close the contacts after it has been tripped open by an overload. A bimetallic switch is used that opens with too much heat from excessive current. A 1-A circuit breaker will trip open with an overload current of 1.75 A in about 10 s, as typical values.

The effect of an open fuse or circuit breaker depends on the circuit. A circuit breaker is usually in the line for B+ voltage, without the heaters. Therefore, an open causes no raster and no sound, but the heaters light. An open fuse in the main line to the ac input results in no B+ voltage and no heater power. This fuse is usually rated at 2 to 5 A. A fuse rated 1 to 2 A or less is usually just for the B+ voltage. Remember that an open fuse in the horizontal output circuit causes no high voltage and no brightness, but the sound is normal.

Testing fuses with an ohmmeter. A good fuse reads 0 Ω, showing continuity through the wire link. A blown fuse reads infinite ohms, showing an open circuit because the wire link is broken. The power must be off, or the fuse removed from the circuit, to check with an ohmmeter.

Testing fuses with a voltmeter. It is important to realize that the test is different with power on. Remember that a fuse is just a short piece of wire, with practically no IR drop. This applies whether the fuse is in an ac circuit or dc circuit. When you measure between both ends of a good fuse, the voltage should be zero across the fuse. With an open fuse, however, the voltmeter will read full voltage across the open circuit. The reason is that the high resistance of the voltmeter takes the place of the fuse.

If you measure the voltage from one end of the fuse to chassis ground, and then from the other end to chassis, the two readings should be exactly the same. A good fuse is practically a short circuit, and the voltage to chassis ground cannot be different at the two ends. With an open fuse, however, one end will read the full voltage to chassis ground while the other end reads 0 V. The reason is that one end is connected to the source of power but the other end is disconnected because of the open fuse.

Wire links. Heater circuits often have just a 1 to 2-in. length of thin, bare wire as the fuse. This type of wire fuse is usually soldered between two posts of a terminal strip in the chassis. A 2-in. length of No. 24 gage wire, as an example, can hold a current of 450 mA. Excessive current caused by a short circuit will burn open the wire fuse link. To replace it, just solder back another link, using the same gage wire.

Fusible resistors. These are wire-wound resistors used as a combination fuse and surge-limiting resistor for a silicon diode. A fusible resistor is often used in a line-connected supply without a power transformer, connected in series with the ac input for the diode, but not in the heater line. Typical resistance values are 3 to 10 Ω as a surge-limiting resistor for the rectifier. The current rating is about 1 A as a fuse to protect the components in the B+ circuit. When the fuse opens, there is no B+ voltage, but the heaters light. The fusible resistor usually has plug-in connectors, so that it can easily be replaced, as in Fig. 12-37c.

SUMMARY

The main characteristics of rectifier circuits are summarized in Table 12-3, for 60-Hz single-phase input from the ac power line. The ripple frequency is 60 Hz for a half-wave rectifier or 120 Hz for a full-wave rectifier. The function of the filter is to remove the ac ripple from the dc output voltage. The function of a regulator is to keep the dc output voltage constant with a change in the amount of load current.

Self-Examination   (Answers at back of book)

Match the numbered statements at the left with the letters at the right.

1. Requires four diodes.
2. Maintains constant output voltage.
3. Reduces rf interference.
4. Keeps heaters warm when set is off.
5. Allows current in one direction.
6. Ratio of ripple to dc voltage.

(a) Open heater in series string.
(b) Surge-protective resistor.
(c) Connected to one side of ac power line.
(d) 5 kV.
(e) Flyback high-voltage supply.
(f) Low B+ voltage.

7. Isolates set from ac power line.
8. Hot chassis.
9. Limits diode current.
10. Anode voltage for picture tube.
11. No raster and no sound.
12. Ripple frequency of 120 Hz.
13. Small raster.
14. Secondary winding not isolated.
15. Zero resistance.
16. Open filter capacitors.
17. Shunt regulator tube.
18. Focus voltage for color picture tube.
19. Manual reset button.
20. Fuse in ac power line.

(g) Full-wave rectifier.
(h) Hum.
(i) Instant-on operation.
(j) 6BK4.
(k) Bridge rectifier.
(l) 4 A.
(m) Good fuse.
(n) Zener diode.
(o) Hash filter.
(p) Power transformer with separate secondary.
(q) Circuit breaker.
(r) Percent ripple.
(s) Silicon-diode rectifier.
(t) Autotransformer.

TABLE 12-3 RECTIFIER CIRCUITS

CIRCUIT	DIAGRAM	DIODES	RIPPLE FREQUENCY (Hz)	FILTER	DC OUTPUT*	CAPACITOR VOLTAGE RATINGS†
Half-wave	Fig. 12-8	1	60	Capacitor input	1.2E	$C_1 = 1.4E$
Half-wave doubler	Fig. 12-17	2	60	Capacitor input	2.4E	$C_1 = 1.4E$ $C_2 = 2.8E$
Half-wave tripler	Fig. 12-20	3	60	Capacitor input	3.6E	$C_1 = 1.4E$ $C_2 = 2.8E$ $C_3 = 4.2E$
Half-wave quadrupler	Fig. 12-21	4	60	Capacitor input	4.8E	$C_1 = 1.4E$ $C_2 = 2.8E$ $C_3 = 4.2E$ $C_4 = 5.6E$
Full-wave doubler	Fig. 12-19	2	120	Capacitor input	2.4E	$C_1 = 1.4E$ $C_2 = 1.4E$
Full-wave center-tap	Fig. 12-13	2	120	Capacitor or choke input	1.2E	$C_1 = 1.4E$
Full-wave bridge	Fig. 12-14	4	120	Capacitor or choke input	1.2E	$C_1 = 1.4E$

*Normal dc output voltage with typical load and capacitor input filter. E is rms value of ac input.
†Minimum dc working voltage for peak dc output without load.

Essay Questions

1. Define the following terms: autotransformer; isolation transformer; line-connected, line-isolated, and ac-dc power supply.
2. Give the function of the following components in a power supply: power transformer, silicon diode, 10-H series choke, and 100-μF shunt capacitor.
3. With 60-Hz input, give the ripple frequency for each of the following rectifier circuits: half-wave diode, full-wave center-tap; full-wave bridge; cascade doubler; and full-wave doubler.
4. Give two comparisons between a capacitor input filter and choke input filter.
5. What is the purpose of a voltage regulator?
6. Give one advantage and one disadvantage of series heaters, compared with parallel heaters.
7. Describe briefly how "instant-on" operation can be provided for a receiver with series heaters.
8. Why is there no brightness and no sound with an open heater in a series string?
9. Why is no brightness the result when: (a) high-voltage rectifier does not operate; (b) horizontal oscillator does not operate?
10. Give two effects of hum in the television receiver, as it shows on the screen of the picture tube. How do you distinguish between 60 and 120 Hz?
11. What is meant by "blooming" in the picture?
12. Describe briefly two insulation problems in a high-voltage supply.
13. Give two functions of the fusible resistor in a line-connected power supply.
14. How does the high-voltage rectifier obtain its heater voltage?
15. Referring to Fig. 12-29, how is R_2 adjusted?
16. Referring to Fig. 12-27, how is R_3 adjusted?
17. Give two ways in which a zener diode is different from a conventional diode rectifier.
18. Assume that a solid-state receiver rectifies the horizontal scan voltage for the low-voltage supply, in addition to the high-voltage supply for anode voltage on the picture tube. What will be the effects when the horizontal oscillator does not operate?
19. Draw the schematic diagram of a line-connected low-voltage power supply with a half-wave diode rectifier and series heaters. Show a π-type filter and indicate the dc output voltage.
20. Draw the schematic diagram for a full-wave bridge rectifier with a 1:1 power transformer. Show a π-type filter and indicate the dc output voltage.
21. Draw the schematic diagram for a line-connected voltage doubler, with π-type filter, and give dc output voltage.
22. Compare a good fuse and open fuse for (a) resistance and (b) voltage.
23. Refer to the half-wave rectifier circuit in Fig. 12-11. Give the function for R_1, C_1, CR1, C_2, C_{3A}, L_2, and R_2.
24. Refer to the voltage doubler circuit in Fig. 12-23. Give the function for R_{114}, C_{111}, L_{110}, CB_{101}, $SR101$, and $SR102$.

Problems

1. A series string totals 90 V with 600 mA in all the heaters. (a) What size voltage-dropping resistor is needed with 120-V input? (b) How much power is dissipated in this R?
2. Refer to the series heater string in Fig. 12-23. List the ac voltages to ground at each tube, starting with pin 4 of the 17JB6.
3. Refer to the B + circuit in Fig. 12-23. If the dc load current is 160 mA and the resistance of L_{110} is 140 Ω, how much is the dc voltage at the input filter capacitor C_{112A}?
4. The current through a 90-mH L drops from 400 mA to 0 in 8 μs. Calculate the induced voltage. [Hint: $e = L(\Delta i/\Delta t)$.]
5. Show how to connect the components in Fig. 12-38 in a cascade voltage doubler circuit.

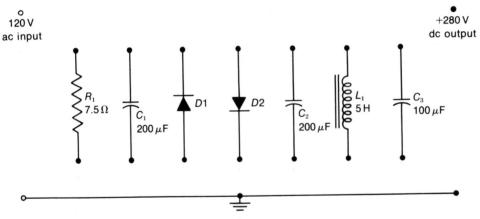

FIGURE 12-38 FOR PROBLEM 5.

13 Video Circuits

As shown in Fig. 13-1, the video section amplifies the output from the video detector. With signal of 2 V, peak to peak, one or two video amplifier stages can provide output of 100 V to the cathode of the picture tube for good contrast. Typical video signal for the picture tube is shown in Fig. 13-2. In color receivers, the video section amplifies the luminance (Y) signal, with essentially the same requirements as in monochrome receivers. The video amplifier is often dc-coupled from the video detector to the picture tube, in order to preserve the dc component for correct brightness, as explained in the next chapter. Here, we analyze the problems of amplifying the ac video signal with frequency components from 30 Hz to 3.2 MHz, approximately. The main features of the video signal and video amplifiers are explained in the following topics:

- 13-1 Requirements of the video amplifier
- 13-2 Polarity of the video signal
- 13-3 Amplifying the video signal
- 13-4 Manual contrast control
- 13-5 Video frequencies
- 13-6 Frequency distortion
- 13-7 Phase distortion
- 13-8 High-frequency response of the video amplifier
- 13-9 Low-frequency response of the video amplifier
- 13-10 Video amplifier circuits
- 13-11 The video detector stage
- 13-12 Luminance video amplifier in color receivers
- 13-13 Functions of the composite video signal
- 13-14 The 4.5-MHz sound trap

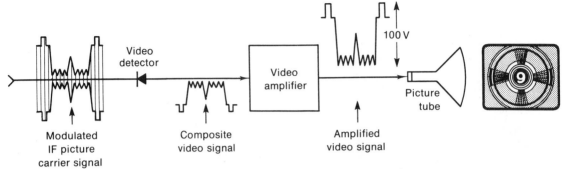

FIGURE 13-1 THE VIDEO SIGNAL COUPLED TO THE CATHODE-GRID CIRCUIT OF THE PICTURE TUBE REPRODUCES THE PICTURE INFORMATION ON THE RASTER.

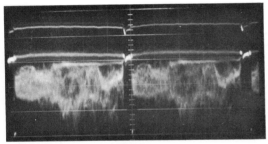

FIGURE 13-2 OSCILLOSCOPE PHOTO OF VIDEO SIGNAL VOLTAGE AT CATHODE OF PICTURE TUBE. AMPLITUDE IS 100 V p-p.

13-1 Requirements of the Video Amplifier

The circuit is basically an *RC*-coupled stage, as shown in Fig. 13-3, operating class A for minimum distortion. In *a*, the 6.8-kΩ R_L is the plate load resistor with a 0.1-μF coupling capacitor. Similarly in *b*, the common-emitter amplifier has 6.8 kΩ for R_L and 0.1 μF for C_c. The *RC* values are the same here because both stages are considered as video output stages driving the picture tube. The contrast control varies the gain and the amount of video signal. In *a*, the cathode resistance is varied, or the emitter resistance is varied in *b*.

The peaking coils L_o and L_c in the output circuit are small inductances of 100 to 250 μH that resonate with the stray capacitances to boost the response for high video frequencies. The transmitted video signal includes frequencies up to 4 MHz, for horizontal detail. However, most receivers limit the video bandwidth to 3.2 MHz. The purpose is to minimize interference with the 3.58-MHz chrominance signal.

Video signal and picture reproduction. The function of the video signal is to vary the amount of beam current in the picture tube. This intensity modulation reproduces the light or shade of the picture elements. Figure 13-4 shows how the input video signal voltage for one line, impressed between grid and cathode of the picture tube, results in reproduction of the picture elements for that line. This operating characteristic, or transfer curve, is for the picture tube, not for the video amplifier.

The beam current and spot illumination vary with the grid voltage of the picture tube. As the control grid (G1) goes more negative, the result is less beam current and reduced light intensity for a darker picture element. When the grid goes as far negative as cutoff, the beam current is zero and there is no spot of light on the

VIDEO CIRCUITS 263

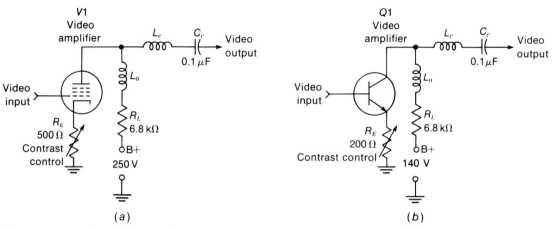

FIGURE 13-3 VIDEO AMPLIFIER CIRCUITS. (a) PENTODE TUBE. (b) CE TRANSISTOR STAGE.

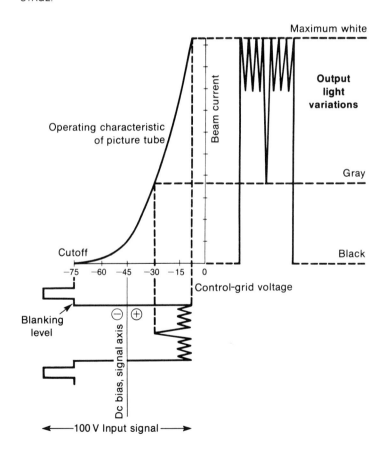

FIGURE 13-4 VIDEO SIGNAL VARYING THE GRID VOLTAGE OF PICTURE TUBE.

screen. This corresponds to black in the picture. As the positive voltage variations of the ac video signal make the grid less negative than the dc bias, the beam current increases to produce a brighter spot on the screen. This corresponds to white in the picture. Maximum white signal drives the grid voltage almost to zero for maximum beam current.

Although the effect of the white parts of the video signal is to drive the control grid in the positive direction, the grid-cathode voltage remains negative because of the negative dc bias for the grid. It should also be noted that the video signal is shown here as negative voltages at the control grid, with respect to cathode. However, the results are the same when the video signal provides positive voltage variations at the cathode, with respect to the control grid.

Video signal amplitude determines contrast. How much the video signal swings away from its average-value axis determines the contrast of the picture. Suppose that the signal is reduced from 100, to 50 V peak-to-peak amplitude. Then the peak white of the weaker signal will not be so light because of less beam current. This means there will be less difference between the maximum white parts of the picture and the black level.

The same idea applies to color in the picture. More color video signal results in more intensity, or color saturation. When you turn up the video control for more luminance signal, the picture has more contrast. When you turn up the color control for more color signal, the picture has stronger colors.

The video signal amplitude must be specified in peak-to-peak voltage. Remember that the common ac relations—that the average is 0.637 and the effective or rms value is 0.707 of the peak—apply only to sine waves. Since the positive and negative half-cycles of the video signal do not even have the same waveform, any notation other than peak-to-peak value is useless.

Black level. The grid-cutoff voltage for the picture tube corresponds to black. Therefore, the part of the video signal that corresponds to black should drive the grid voltage to cutoff. Any grid voltage more negative than cutoff is called "blacker than black." Actually, there is nothing to see in black, but this notation describes the sync voltage amplitude. At the grid of the picture tube, the sync pulses really have no function. However, the sync is used in the synchronizing section of the receiver to time the deflection circuits for horizontal and vertical scanning.

13-2 Polarity of the Video Signal

If the polarity is opposite from that shown in Fig. 13-4 for control-grid voltage, the result will be a negative picture in the same sense as a photographic negative. The dark parts of the actual scene will be white on the screen of the picture tube. Also, the white parts of the picture will be reproduced as black.

The correct polarity of video signal is necessary not only for picture reproduction but also for blanking and synchronization. In order to be definite, the signal is specified in terms of its polarity for sync, as follows:

1. *Negative sync polarity.* See the video signal in Fig. 13-4. The sync voltage is the negative peak of the signal, while the white video is the most positive part.
2. *Positive sync polarity.* See the video signal coupled to the picture tube in Fig. 13-1 for cathode drive. The sync voltage is the positive peak of the signal, while the white video is the most negative part.

For the picture tube, the correct polarity of video signal depends on the following three factors:

1. Whether the picture tube has cathode drive with positive sync polarity or grid drive with negative sync polarity.
2. The number of video amplifier stages, since an RC-coupled stage inverts the polarity of its input signal. A CE transistor amplifier inverts the polarity. However, a CC stage, or emitter-follower, does not.
3. Polarity of video detector output. When the video signal is taken from the cathode of the detector diode, the sync polarity is positive. Output from the anode, however, provides negative sync polarity in the detector output signal.

Many combinations of these factors are possible, but receivers generally use the method in Fig. 13-1. Video signal with negative sync polarity from the detector is inverted by the video amplifier to provide positive sync polarity in the video signal driving the cathode of the picture tube.

13-3 Amplifying the Video Signal

Figure 13-5 illustrates how the stage operates as a class A amplifier. The 3-V dc bias is less than cutoff. Also, the ±2-V ac drive for the grid allows plate current for the full cycle of video signal. The resulting variations in plate current produce amplified video output voltage across the plate load R_L. The signal output is 140 V peak to peak. Note the polarity inversion.

With 140-V output for 4-V input, the voltage gain of this stage is 140/4 = 35. The output can be varied by the contrast control, which is not shown here. Also omitted are the peaking

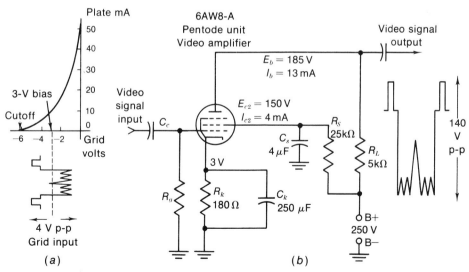

FIGURE 13-5 VIDEO AMPLIFIER OPERATION. (a) GRID INPUT SIGNAL AND TRANSFER-CHARACTERISTIC CURVE OF AMPLIFIER. (b) BASIC CIRCUIT OF RC-COUPLED AMPLIFIER, WITH DC ELECTRODE VOLTAGES AND AC SIGNAL WAVESHAPES. PEAKING COILS OMITTED.

coils, but they do not affect the dc voltages or the peak-to-peak signal amplitudes.

Dc voltages. If you measure with a dc voltmeter from plate to chassis ground, the meter will read 185 V for the average dc plate voltage E_b, as indicated in Fig. 13-5b. This value equals the B+ voltage minus the average $i_b R_L$ drop. Similarly the screen-grid voltage equals the B+ voltage minus the IR drop across R_S. The cathode bias of +3 V makes the control grid 3 V negative with respect to cathode.

Ac voltages. Checking with an oscilloscope at the grid of the amplifier, you can see the composite video signal waveform. The oscilloscope can be calibrated to read the 4-V peak-to-peak amplitude of the ac signal input. With a gain of 35 for the stage, the output video signal equals 140 V peak-to-peak.

Amplitude distortion. If the dc voltages are not correct in the video amplifier, the ac video signal can be distorted. Typical problems are limiting and clipping of the video signal amplitudes or weak signal output. If the sync voltage is compressed, synchronization can be lost because the video amplifier usually provides composite video signal for the sync circuits.

Load line. The amplitude variations of the video signal can be analyzed by a load line for this amplifier, as shown in Fig. 13-6. One end of the load line is at E_{bb} of 250 V, equal to B+. This is the plate voltage when plate current is zero. The other end is at 50 mA, which is the plate current equal to 250 V/5,000 Ω. When the plate voltage is zero, all the B+ voltage is dropped across the plate load R_L.

All instantaneous values of grid voltage e_c, plate current i_b, and plate voltage e_b are on the load line. The quiescent point marked Q is set by the 3-V bias. As grid signal swings e_c 2 V less negative to -1 V and 2 V more negative to -5 V, i_b varies between 32.5 and 1.5 mA. The corresponding peak values of e_b are 90 and 230 V. Therefore, the amplified ac signal voltage at the plate is 140 V peak-to-peak, equal to $230 - 90$ V (see Fig. 13-7).

By means of the values on the load line, we can calculate the dc voltages. The Q point for 3-V bias sets the average plate current I_b at 13 mA. The $I_b R_L$ voltage drop equals $5{,}000 \times 0.013$,

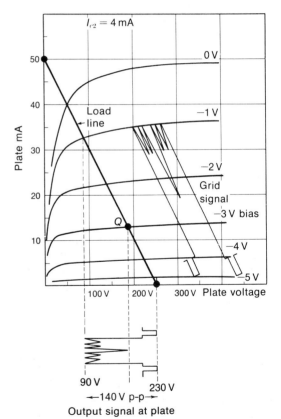

FIGURE 13-6 LOAD-LINE CALCULATIONS FOR VIDEO AMPLIFIER CIRCUIT IN FIG. 13-5. PLATE-CHARACTERISTIC CURVES FOR 6AW8-A PENTODE SECTION.

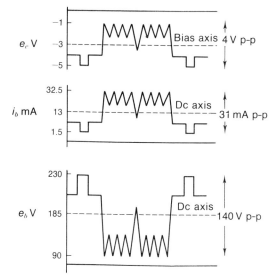

FIGURE 13-7 VIDEO SIGNAL WAVESHAPES FOR LOAD-LINE OPERATION IN FIG. 13-6.

or 65, V. Subtracting this 65-V drop from the E_{bb} supply of 250 V results in 185 V for the average dc plate voltage.

With 13-mA I_b and 4-mA screen-grid current, the total average cathode current equals 17 mA. Therfore, R_k has the value of 3/0.017, or approximately 180 Ω for the 3-V bias.

Similarly, R_s is 25,000 Ω to produce a voltage drop of 100 V with 4-mA screen current. This 100-V IR drop across R_s when subtracted from the supply of 250 V provides 150 V for the screen-grid voltage.

13-4 Manual Contrast Control

As the control is turned clockwise, more ac video signal is provided for the picture tube. Greater variation between black and the whitest parts of the picture results, increasing the contrast.

Any control that varies the amount of ac video signal will operate as a contrast adjustment. Therefore, the contrast varies when the gain is varied in either the picture IF section or the video amplifier. The contrast control is in the video amplifier, however. The reason is that all receivers have automatic gain control to adjust the bias on the picture IF amplifiers automatically according to the signal level. Also, with intercarrier sound a variation of IF bias affects the sound volume.

The most common methods of contrast control in the video amplifier are shown in Fig. 13-8. In a the variable cathode resistor R_1 varies the bias for the video amplifier. The control is unbypassed in order to provide degeneration of the ac signal. Then the amount of degeneration also varies as the bias resistor is varied. The degeneration is important for varying the gain because with linear amplification little change in gain results from changing the bias. Moving the variable arm of R_1 closer to the cathode end reduces the bias and the degeneration. This allows more gain to increase the video signal amplitude and the contrast. The variable emitter resistor R_1 for the transistor amplifier in b controls the contrast in essentially the same way.

The variable-bias method of contrast control has the disadvantage of changing the amplifier operating characteristics, which can introduce amplitude distortion. To overcome this problem, the arrangement in c taps off the desired amount of ac signal output without changing the bias. Now the contrast control R_2 is a potentiometer in the video signal circuits. As the variable arm of R_2 is moved up to terminal 3 for maximum resistance, the video output increases for more contrast. This method is the same as a volume control for audio signal.

The stray capacitance of the potentiometer and its connecting leads can reduce the high-frequency response of the amplifier.

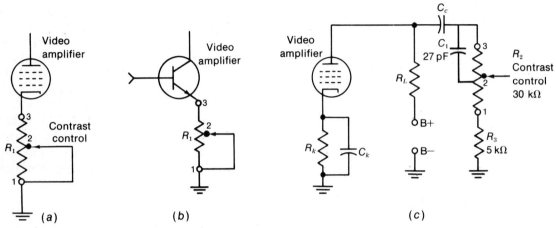

FIGURE 13-8 METHODS OF CONTRAST CONTROL IN VIDEO AMPLIFIER. (a) VARIABLE CATHODE RESISTANCE. (b) VARIABLE EMITTER RESISTANCE. (c) VARIABLE POTENTIOMETER TO TAP OFF AC SIGNAL VOLTAGE.

To minimize shunt capacitance, the control is usually mounted near the video amplifier with the shaft mechanically linked to the front panel of the receiver. In addition, the 27-pF C_1 is used as a compensating capacitor to keep the same frequency response at different settings of the contrast control.

13-5 Video Frequencies

The video stage amplifies signal voltage that may have frequency components from 30 Hz to 4 MHz. High frequencies are produced because the video signal contains, within a line, rapid changes in voltage that occur during very much less time than the active line-scanning time of 53.3 µs. These fast variations in signal voltage can correspond to frequencies infinitely high, but are limited to approximately 4 MHz by the restriction of a 6-MHz transmission channel. Furthermore, receivers generally cut off the video frequency response at approximately 3.2 MHz in order to minimize interference with the chrominance signal at 3.58 MHz.

How to determine the period and frequency. See Fig. 13-9. Note that the visible picture width in a is translated into 53.3 µs of time. This value is the total line time of 63.5 µs minus the horizontal blanking time of 10.2 µs. Similarly, the visible picture height in b is $1/60$ s minus 7 percent vertical blanking.

To convert the size of an element of picture information into frequency, it is first necessary to calculate the time required to scan the element. This time can be considered as the period of a half-cycle of the video signal required to reproduce the information. Then you can

1. Multiply the time by 2 to obtain the period of a full cycle.
2. Take the reciprocal of the period. The answer is the desired frequency. With time in microseconds, the frequency is in megahertz.

Horizontal information. In Fig. 13-9a, the top black bar has a width slightly less than one-tenth

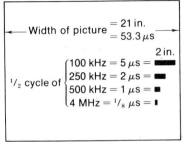
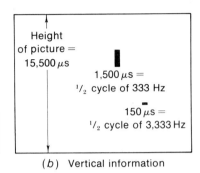

FIGURE 13-9 RELATION OF VIDEO SIGNAL FREQUENCIES TO SIZE OF PICTURE INFORMATION SHOWN AS DARK BARS. THESE ARE DRAWN TO SCALE WITH RESPECT TO FULL WIDTH AND HEIGHT OF PICTURE. (a) IN HORIZONTAL DIRECTION. (b) IN VERTICAL DIRECTION.

of the picture width. Therefore, the bar is scanned horizontally in 5 μs. This time is a little less than one-tenth of 53.3 μs for the entire picture width. As a result, the video signal for this black bar preceded and followed by white information corresponds to one-half cycle of a signal variation with a total period of 10 μs for the full cycle. The corresponding frequency is $1/10 = 0.1$ MHz, or 100 kHz. Note that the width of this bar is 2 in. on a screen 21 in. wide.

The lowest frequency for picture information in the horizontal direction can be considered as 9.4 kHz, corresponding to a line all white or all black, producing video signal with a half-period of 53.3 μs. The highest frequencies are produced by thin bars with the fastest variations in video signal.

Vertical information. The signal frequencies corresponding to picture information scanned in the vertical direction can be calculated in a similar manner (see Fig. 13-9b). Here the height of the black bars is converted to frequency in terms of active vertical scanning time equal to 0.0155 s. This is relatively low frequency information compared with details reproduced within a line. If the video voltage is taken from top to bottom, through all the horizontal lines in a field, this variation will correspond to a half-cycle of a signal with a frequency of approximately 30 Hz.

When the brightness of the picture varies from frame to frame, the resultant signal frequency is lower than 30 Hz. However, this is considered as a change in dc level, corresponding to a change in brightness of the scene.

In summary, then, the ac video signal can be regarded as a complex waveform, not being a sine wave, containing signal voltages that range in frequency from 30 Hz to 4 MHz, approximately. Figure 13-10 illustrates typical video signal waveshapes. Note that the high-frequency variations correspond to white, gray, and black shadings across every line. The lower-frequency variations correspond to vertical shadings of gray, black, and white.

Wideband amplifiers. The video amplifier is a good example of a wideband amplifier, which can be defined as a stage that amplifies both audio frequencies and radio frequencies. To see the difference in frequency response between a wideband amplifier and a tuned amplifier with wide bandwidth, refer to Fig. 13-11. In a, the response curve is for an IF amplifier tuned to 43 MHz. Its bandwidth is 4 MHz, meaning it can amplify the band of frequencies from 41 to 45 MHz with 70.7 percent response or more. However, the video amplifier frequency response in b is entirely different, although it also has a bandwidth of 4 MHz. Here the bandwidth in-

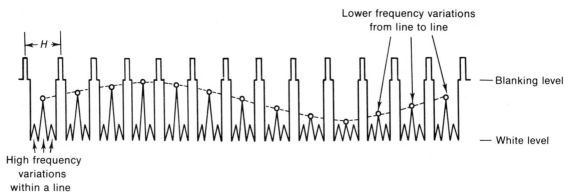

FIGURE 13-10 VIDEO SIGNAL WAVEFORMS. HIGH-FREQUENCY COMPONENTS ARE VARIATIONS WITHIN EACH LINE OF HORIZONTAL SCANNING. LOWER-FREQUENCY VARIATIONS DURING VERTICAL SCANNING ARE FROM LINE TO LINE WITHIN A FIELD.

cludes audio frequencies from 30 Hz and radio frequencies up to 4 MHz. This response is really an example of amplifying a wide range of frequencies because the ratio of the highest to the lowest frequency is very great.

For wideband amplifiers it is convenient to consider the bandwidth in octaves. An *octave* is a range of 2 to 1 in frequency. As examples: from 100 to 200 Hz is 1 octave; 400 Hz is 1 octave above 200 Hz. Note that 100 to 400 Hz is a range

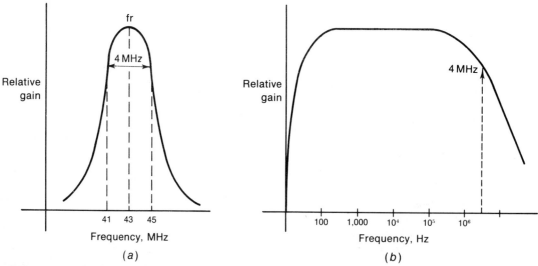

FIGURE 13-11 COMPARISON OF RESPONSE CURVES WITH 4-MHz BANDWIDTH. (a) TUNED RESPONSE CENTERED AT 43 MHz. (b) WIDEBAND RESPONSE FROM LOW AUDIO FREQUENCIES UP TO 4 MHz.

TABLE 13-1 VIDEO AMPLIFIER SPECIFICATIONS

APPLICATION	BANDWIDTH (MHz)	OUTPUT VOLTS (PEAK TO PEAK)
Black-and-white receiver	3.2	80–120
Luminance signal	3.2	80–140
TV camera viewfinder	8	60
TV studio monitor	8	100

of 2 octaves, as each octave means doubling the frequency.

The applications of wideband amplifiers include circuits that must amplify sine waves of audio and radio frequencies, or amplification of complex waveforms having a wide range of sine-wave frequency components. Common examples are amplifiers for the input signal in an oscilloscope, pulse amplifiers, and video amplifier circuits. The bandwidth requirements for typical video amplifier applications are listed in Table 13-1.

13-6 Frequency Distortion

The amplifier has more gain for some frequencies than for others. Excessive frequency distortion cannot be tolerated because it changes the picture information. As shown in the amplifier response curve of Fig. 13-12, the amplifier response should be flat within a tolerance of about ±10 percent. Note that the frequency units are marked off on the horizontal axis in powers of 10, making the spacing logarithmic. This spacing is necessary for a graph of reasonable size that will still show the ends of the response curve.

When the amplifier has a flat response curve, the relative gain of the amplifier is the same for all signal frequencies. Then the amplifier introduces no frequency distortion. As an example, the 100 percent response in Fig. 13-12 may correspond to a voltage gain of 35 at the middle frequency of 10 kHz. Then the flat response means the gain is 35 for all frequencies from 30 Hz to 4 MHz. With an actual video signal containing typical picture information coupled to the amplifier, the different frequency components do not all have the same amplitude. When the amplifier response is flat, however, all frequencies are amplified equally well without any frequency distortion. Then each frequency component in the amplified output signal has the same relative amplitude as in the input signal.

Usually the response of the uncompensated video amplifier is down for the high video frequencies of 1 MHz and above. High-frequency compensation is necessary, therefore, for a flat frequency response up to 3.2 MHz. Video frequencies up to 4.2 MHz are broadcast by the transmitter, but the video amplifier in the receiver generally cuts off at 3.2 MHz, in order to minimize interference with the 3.58-MHz chroma signal.

At the low-frequency end, the video amplifier response can be down at about 100 Hz and below. Then low-frequency correction of

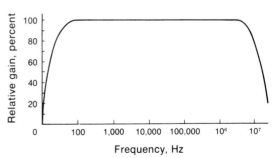

FIGURE 13-12 DESIRED FLAT FREQUENCY RESPONSE OF VIDEO AMPLIFIER.

the video amplifier may also be necessary. The response of the video amplifier over the middle range of frequencies is normally flat and requires no compensation.

Loss of high video frequencies. If these signal frequency variations are lost, the rapid changes between black and white for small adjacent picture elements in the horizontal lines cannot be reproduced, with the resultant loss of horizontal detail. Figure 13-13 shows the effect of loss of the high video frequencies on the reproduced test pattern. Notice the lack of separation between the black-and-white divisions in the top and bottom wedges. The extent to which the divisions in either of these wedges can be resolved indicates the high-frequency response. The divisions all the way in to the center circle indicate video-frequency response up to 4 MHz. The divisions in the widest part of the vertical wedges correspond to a video frequency of about 2 MHz.

FIGURE 13-13 TEST PATTERN ILLUSTRATING LOSS OF HIGH VIDEO FREQUENCIES. NOTE LACK OF SEPARATION FOR THE LINES IN TOP AND BOTTOM WEDGES. ALSO, THEY HAVE WEAKER INTENSITY THAN THE SIDE WEDGES. *(RCA PICT-O-GUIDE)*

In a televised scene, loss of the high video frequency information is evident as reduced detail. The picture does not appear sharp and clear. Small details of picture information, such as individual hairs in a person's eyebrows and details of the eye, are not reproduced. In addition, the edges between light and dark areas, as in the outline of lettering or in the outline of a person's face, are not reproduced sharply but trail off gradually instead.

The effects of reduced high-frequency response causing insufficient detail in the picture are not so noticeable when the camera at the studio presents a closeup view. Then the subject occupies a large area of the picture, and the highest video signal frequencies are not so necessary.

Loss of low video frequencies. The video frequencies from about 100 kHz down to 30 Hz represent the main parts of the picture information, such as background shading, lettering, and any other large areas. This follows from the fact that it takes a longer period of time for the video signal to change over large areas. Frequencies from 100 down to about 10 kHz correspond to black-and-white information in the horizontal direction having a width one-tenth or more of a horizontal line. Frequencies from 10 kHz down to 30 Hz can represent changes of shading in the vertical direction. If a solid white frame is scanned, the signal is a 30-Hz square wave. If this low-frequency square wave is not amplified with its waveshape preserved, the reproduction will show a white screen having a gradual change of intensity from top to bottom.

Figure 13-14 shows a test pattern reproduced with insufficient low-frequency response. The background is dull gray, instead of white, and the lettering is not solid. Actually, the picture as a whole is weak with poor contrast because low video frequencies represent the

13-7 Phase Distortion

The effect of phase distortion is time delay. The relative time delay of some parts of the video signal with respect to the others is important because one element of the picture is being reproduced at a time as the scanning beam traces out the frame. As a result, a great enough time-delay distortion in the video signal can have the effect of displacing the picture information on the picture tube screen. The result is smear in the picture.

Time delay. Phase delay is equivalent to time delay. If one signal voltage is 10° out of phase with another and lagging, as shown in Fig. 13-15, it reaches its maximum and minimum values at a later time. The delay is the amount of time that corresponds to 10° of the cycle. This time varies with the frequency. For a signal voltage having a frequency of 100 Hz, it takes $1/100$ s for one complete cycle of 360°. The amount of time equivalent to 10° in the cycle is $10/360 \times 1/100$ s which is approximately 0.000278 s or 278 μs. In this time the scanning beam can be displaced in a vertical direction by more than four lines.

Phase distortion is very important at low video frequencies, therefore, because even a small phase delay is equivalent to a relatively large time delay. For the extremely high video

FIGURE 13-14 TEST PATTERN ILLUSTRATING LOSS OF LOW VIDEO FREQUENCIES. BACKGROUND IS WEAK, AND LETTERING IS NOT SOLID. SIDE WEDGES HAVE WEAKER INTENSITY THAN TOP AND BOTTOM WEDGES. *(RCA PICT-O-GUIDE)*

main areas of picture information. Notice that the side wedges, which represent low video frequencies, are weaker than the top and bottom wedges, which correspond to the high video signal frequencies. The changes from black to white in the vertical direction between the divisions in the side wedges represent frequencies of 4 to 8 kHz, approximately.

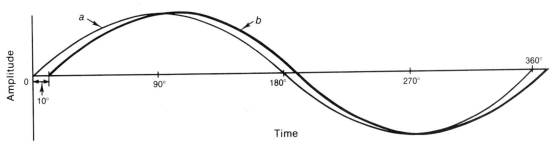

FIGURE 13-15 PHASE DELAY. WAVE *b* LAGS BEHIND WAVE *a* BY THE AMOUNT OF TIME EQUAL TO 10° OF THE CYCLE.

frequencies, the effects of phase distortion are not as evident on the screen because the time delay at these high frequencies is relatively small. Normally, a video amplifier stage that has flat-frequency response up to the highest useful video frequency has negligible time-delay distortion for the high frequencies.

Square-wave response. Consider the waveshape shown at the right in Fig. 13-16a. This wave is actually composed of two sine waves. One is the fundamental, with the same frequency as the combined waveform. The other sine wave is the third harmonic, having a frequency 3 times the fundamental frequency and an amplitude one-third of the fundamental. Addition of the two sine waves, fundamental and third harmonic, produces the resultant nonsinusoidal wave. This wave has the same frequency as the fundamental but tends to become more of a square wave than a sine wave. If enough odd harmonics are added to the fundamental, the result will be a square wave.

The waveshape resulting from combining the fundamental and its harmonics is critically dependent on the phase relation between the waves. Figure 13-16b shows the combination of the same fundamental and third harmonic as in a, the amplitude and frequency of each being preserved, but with a different phase angle between the two waves. Assume that in a stage amplifying such a signal the fundamental and third harmonic are originally in phase, as shown in a. Then because of phase distortion, the fundamental is made to lag by 60° of its cycle, and the third harmonic is made to lag by 120° of its cycle, with the result shown in b. The resultant waveshape is distorted from the original wave because the original phase relations between the fundamental and harmonic are not maintained.

Phase distortion always produces an unsymmetrical distortion of the waveshape, as in b. Notice that the peak amplitudes of the wave in b have different values than the original wave, and they occur at different times than the peaks for

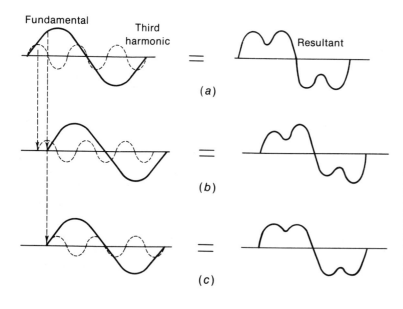

FIGURE 13-16 HOW PHASE DISTORTION CHANGES WAVESHAPE OF NONSINUSOIDAL SIGNAL. (a) FUNDAMENTAL AND THIRD HARMONIC IN PHASE. (b) FUNDAMENTAL DELAYED BY 60° AND THIRD HARMONIC BY 120°. PHASE DELAY NOT PROPORTIONAL TO FREQUENCY. (c) FUNDAMENTAL DELAYED BY 60° AND THIRD HARMONIC BY 180°, PROPORTIONAL TO FREQUENCY.

the undistorted wave. When the signal amplitudes are distorted this way, the corresponding picture information has incorrect light values and is displaced in time, producing smear in the reproduced image (see Fig. 13-17).

Phase shift should be proportional to frequency. It is important to note that the phase distortion is introduced because the amount of phase shift is not proportional to frequency. The distortion is not caused by the phase shift in itself. When the second harmonic is delayed twice as much as the fundamental, the third harmonic 3 times as much, and so on, then the phase shift is proportional to frequency and there is no phase distortion. As an example, in Fig. 13-16c the third harmonic is delayed by 180° which is 3 times 60°. Then the phase shift of the fundamental and third harmonic is proportional to frequency. As a result, there is no phase distortion, and the wave maintains its original shape.

FIGURE 13-17 SEVERE SMEAR CAUSED BY EXCESSIVE LOW-FREQUENCY RESPONSE WITH PHASE DISTORTION. (RCA PICT-O-GUIDE)

To see why the phase shift should be proportional to frequency, the phase delay must be translated into time delay. Consider a 1,000-Hz signal, corresponding to the fundamental frequency in Fig. 13-16, delayed by 60°. This is a delay of $^{60}/_{360}$ of the complete cycle that takes $^{1}/_{1,000}$ s. The amount of time delay can be calculated from the formula

$$T_D = \frac{\theta}{360} \times \frac{1}{f} \qquad (13\text{-}1)$$

In this example,

$$T_D = \frac{60}{360} \times \frac{1}{1,000} = \frac{1}{6,000} \text{ s}$$

When a 3,000-Hz signal corresponding to the third harmonic of the fundamental is delayed by 180°, the amount of time delay is

$$\frac{180}{360} \times \frac{1}{3,000} = \frac{1}{6,000} \text{ s}$$

This is the same amount of time delay as for the fundamental. With phase shift proportional to frequency, therefore, the time delay is uniform.

The time delay is not harmful if all frequency components have the same amount of delay. The only effect of such uniform time delay would be to shift the entire signal to a later time. No distortion results because all components would be in their proper place in the video signal waveshape and in the picture. Therefore, the phase angle should be proportional to frequency, as illustrated by the linear graph in Fig. 13-18a, which provides the uniform time delay in b.

It should be noted that signal inversion of exactly 180° in the video amplifier does not mean phase distortion. This is no time delay, only a polarity reversal.

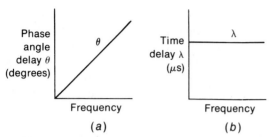

FIGURE 13-18 PHASE RESPONSE. (a) PHASE ANGLES PROPORTIONAL TO FREQUENCY. (b) CORRESPONDING TIME DELAY THAT IS CONSTANT.

In practical terms, low-frequency distortion in the video amplifier can be caused by leakage in the coupling and bypass capacitors. The reduced capacitance results in reactance too high for low-frequency signals.

Square-wave testing. The best check for phase distortion is to test the amplifier with input from a square-wave generator at the desired frequency. This waveform is critically dependent on the phase-angle response of the amplifier. Figure 13-19 shows the input square wave in a and distorted output in b or c. The typical low-frequency droop in b indicates too little capacitance in a coupling circuit. The nonlinear top and bottom on this waveform are actually parts of the capacitor charge and discharge curves, as the RC time constant is not long

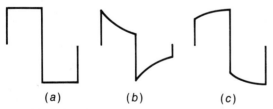

FIGURE 13-19 SQUARE-WAVE TESTING. (a) INPUT FROM SQUARE-WAVE GENERATOR. (b) LOW-FREQUENCY DROOP. (c) LOW-FREQUENCY RISE.

enough. The opposite effect is shown by the rise at the top and bottom of the waveform in c. This distortion can be caused by excessive gain for low frequencies, usually with reduced response for high frequencies.

13-8 High-Frequency Response of the Video Amplifier

As shown in Fig. 13-20, the resistance load R_L makes the response flat over a wide range of middle frequencies. Therefore, an RC-coupled stage is best for amplifying the wide band of video frequencies with minimum frequency and phase distortion. Audio-transformer coupling would not be suitable for the high video frequencies that correspond to radio frequencies. A video frequency of only 2 MHz is higher than any rf signal in the standard radio broadcast band. This entire band from 535 to 1,605 kHz would occupy only a small part of the video spectrum from 30 Hz to 4 MHz. Coupling with rf transformers would not be suitable for the lowest video frequencies that correspond to audio frequencies.

The examples shown in Figs. 13-20 to 13-24 apply to video amplifiers using either tubes or transistors. However, wideband response for high frequencies is less of a problem with transistor amplifiers because of their relatively low input and output resistances.

Total shunt capacitance. The video amplifier gain is down at the high-frequency end because of the shunting effect of capacitances to chassis ground, in parallel with R_L. All these capacitances add in parallel to provide the total value labeled C_t in Fig. 13-20. Typical values for C_t are 15 to 30 pF. This may seem small, but remember that capacitive reactance is inversely proportional to frequency. For the high video frequen-

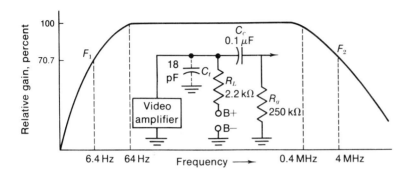

FIGURE 13-20 FREQUENCY RESPONSE WITH SPECIFIC VALUES FOR RC-COUPLED AMPLIFIER.

cies the capacitive reactance of C_t is low enough to reduce the load impedance Z_L. This consists of R_L in parallel with the reactance of C_t. The gain of the amplifier is down in the same proportion as the decrease in Z_L.

Miller effect. This is a dynamic increase of input capacitance in an amplifier. The cause is capacitive feedback of output signal to the input circuit. This is the reason why triodes are seldom used for video amplifiers, as they have a large grid-plate capacitance, compared with pentodes. The Miller effect increases with more gain. As an example, for a triode amplifier with 3 pF for C_{gp} and a gain of 20, the capacitance added to the input circuit is approximately $20 \times 3 = 60$ pF.

Load resistor. Video stages use a low value of R_L, compared with audio amplifiers. Typical values are 2,000 to 8,000 Ω. Although the gain is reduced, lowering R_L extends the high-frequency response.

The effect of C_t does not become important until its reactance is low enough to be comparable with the resistance of R_L. Then the shunt reactance X_{C_t} lowers the impedance Z_L. As examples, suppose R_L is 2,000 Ω and X_{C_t} is 1 MΩ. Then Z_L for the two in parallel is practically 2,000 Ω, the same as R_L. The reason is that a parallel path has little effect when it has very high ohms and very little current. However, if R_L were also 1 MΩ, the combined impedance would then be reduced to 70.7 percent of either one, for equal resistance and reactance in parallel.

High-frequency cutoff F_2. Specifically the gain is down to 70.7 percent of maximum response for the frequency where $X_{C_t} = R_L$. This frequency, called the *cutoff frequency F_2*, can be calculated from this formula:

$$F_2 = \frac{1}{2\pi R_L C_t} \qquad (13\text{-}2)$$

R_L is in ohms and C_t in farads to find F_2 in Hz. For the example in Fig. 13-20 with R_L of 2.2 kΩ and C_t of 18 pF,

$$F_2 = \frac{1}{2\pi R_L C_t} = \frac{1}{2\pi \times 2.2 \times 10^3 \times 18 \times 10^{-12}}$$

$$= \frac{0.16}{39.6 \times 10^{-9}} = \frac{0.16 \times 10^6}{0.0396}$$

$$F_2 = 4 \times 10^6 = 4 \text{ MHz}$$

R_L has the relatively low value of 2,200 Ω here for the F_2 of 4 MHz, where the gain is down 30 percent. Note that the response curve in Fig.

FIGURE 13-21 TYPICAL VIDEO PEAKING COIL. WIDTH IS $\frac{1}{2}$ IN. L IS 180 μH.

13-20 shows the gain is flat to the frequency $0.1 F_2 = 0.4$ MHz $= 400$ kHz.

The lower R_L is and the smaller the shunt capacitance C_t, the better is the high-frequency response. However, practical video amplifiers generally use higher values of R_L with peaking coils to boost the gain for high frequencies. A typical peaking coil is shown in Fig. 13-21.

Shunt peaking. See Fig. 13-22. This circuit is shunt peaking because L_0 in the branch with R_L is in parallel with C_t. L_0 can be connected on either side of R_L. The peaking coil resonates with C_t to boost the gain for high frequencies, where the response of the uncompensated RC-coupled amplifier would drop off. With shunt peaking, the response is practically flat up to F_2 instead of 70.7 percent gain at this frequency. Because it is a small coil with little resistance, the peaking coil does not affect the dc voltages and the response at the middle frequencies. A peaking coil is effective for frequencies above 400 kHz, approximately.

Series peaking. See Fig. 13-23. In this circuit L_c is in series with the two main components of C_t. At one side of L_c is C_{out} for the video amplifier, and on the other side is C_{in} of the next stage. The series peaking circuit has more gain than shunt peaking because there is less shunt capacitance across R_L, while L_c provides a resonant rise of voltage across C_{in}. A series peaking coil usually has a shunt damping resistor, such as the 10-kΩ R_D. Its function is to prevent oscillations or ringing in the coil with abrupt changes in signal. For the same frequency response, series peaking can use R_L 50 percent higher with 50 percent more gain, compared with shunt peaking.

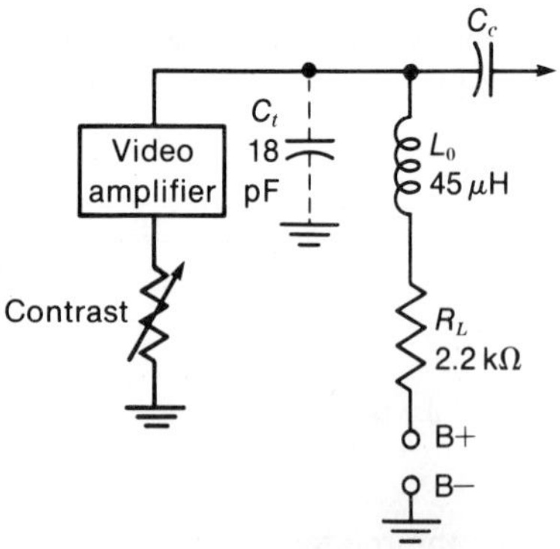

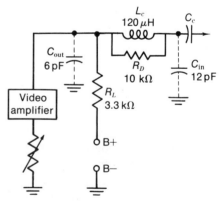

FIGURE 13-22 VIDEO AMPLIFIER WITH SHUNT PEAKING COIL L_0.

FIGURE 13-23 VIDEO AMPLIFIER WITH SERIES PEAKING COIL L_c.

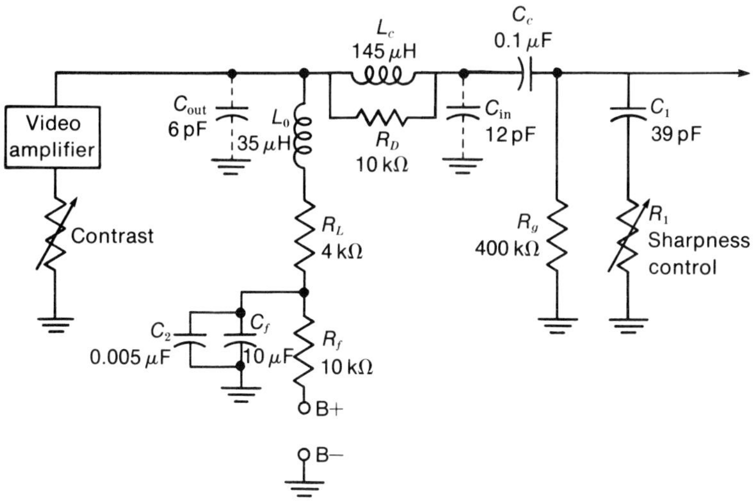

FIGURE 13-24 VIDEO AMPLIFIER WITH SERIES-SHUNT COMBINATION PEAKING FOR HIGH-FREQUENCY COMPENSATION AND R_fC_f DECOUPLING FILTER TO BOOST LOW-FREQUENCY RESPONSE.

Combination peaking. The circuit in Fig. 13-24 combines shunt and series peaking. Of the three methods, combination peaking has the most gain. For the same frequency response R_L can be 80 percent higher with 80 percent more gain, compared with shunt peaking. Values for these three peaking circuits are compared in Table 13-2.

Sharpness control. The circuit in Fig. 13-24 is shown with a variable resistor R_1 that adjusts the high-frequency response of the video amplifier. This is similar to a tone control in an audio amplifier to vary the amount of bypassing at the high-frequency end of the response. R_1 is generally called a *sharpness, fidelity, or peaking control*. Its purpose is to reduce the effect of interfering noise at high video frequencies, if necessary. The interference can be noise pulses, with weak signal, or beat frequencies between the luminance signal and 3.58-MHz chrominance signal. In Fig. 13-24, less resistance for R_1 results in more bypassing through the 39-pF C_1. The result is reduced high-frequency response. With R_1 at maximum, there is no bypassing for maximum sharpness in the picture.

TABLE 13-2 COMPARISON OF HIGH-FREQUENCY PEAKING METHODS

TYPE	R_L	L_0	L_c	RELATIVE GAIN AT F_2
Uncompensated	$1/(2\pi F_2 C_t)$			0.707
Shunt	$1/(2\pi F_2 C_t)$	$0.5 C_t R_L^2$		1.0
Series ($C_{in}/C_{out} = 2$)	$1.5/(2\pi F_2 C_t)$		$0.67 C_t R_L^2$	1.5
Combination ($C_{in}/C_{out} = 2$)	$1.8/(2\pi F_2 C_t)$	$0.12 C_t R_L^2$	$0.52 C_t R_L^2$	1.8

A choke can also be used, instead of the capacitor, in series with the variable resistor as the sharpness control.

13-9 Low-Frequency Response of the Video Amplifier

Remember that for lower frequencies the capacitive reactance increases. This is a problem for coupling and bypass capacitors, which should have very low reactance.

Low-frequency cutoff F_1. With an $R_g C_c$ coupling circuit, specifically, the signal voltage across R_g for the next stage is reduced to 70.7 percent of maximum at the frequency when $X_{C_c} = R_g$. This frequency can be calculated from the formula

$$F_1 = \frac{1}{2\pi R_g C_c} \tag{13-3}$$

R_g is in MΩ and C_c in μF to find F_1 in Hz. For the example in Fig. 13-20 with 0.1 μF for C_c and 0.25 MΩ for R_g,

$$F_1 = \frac{1}{2\pi R_g C_c} = \frac{1}{2\pi \times 0.25 \times 0.1}$$

$$= \frac{0.16}{0.025}$$

$$F_1 = 6.4 \text{ Hz}$$

In this example, then, the low-frequency gain is down 30 percent at 6.4 Hz because of the reactance of C_c. Note that the response curve in Fig. 13-20 shows the gain is flat to $10 \times F_1 = 64$ Hz.

Low-frequency compensation. The low-frequency response can be improved by using the largest possible values for the bypass capacitors and RC coupling circuit. In addition, the decoupling filter $R_f C_f$ in the B+ supply line in Fig. 13-24 can be used to boost the gain and reduce phase distortion for very low frequencies. This low-frequency compensation can be used in any RC-coupled amplifier. Note the 0.005-μF C_2 in parallel with the 10-μF C_f. C_2 is a small paper or ceramic capacitor for the high video frequencies not bypassed because of the small inductance of C_f, which is a tubular electrolytic capacitor.

For the very low frequencies, C_f has enough reactance to provide appreciable impedance Z_f for the $R_f C_f$ combination. Then Z_f in series with R_L boosts the gain for low frequencies. The reason for the increased gain is a higher Z_L. The rise in gain compensates for the reduced signal output caused by the increasing reactance of C_c. Furthermore, the phase shift caused by the shunt C_f is opposite from the phase angle of the series C_c. As a result, C_f corrects phase distortion introduced by the coupling circuit. Typical values for C_f are 2 to 10 μF. R_f should be 10 times larger than X_{C_f}, or more, at the lowest frequency. However, too high a value for R_f drops the supply voltage too much. The time constant of $R_L C_F = R_g C_c$ for low frequency compensation.

13-10 Video Amplifier Circuits

Input signal of 1 to 3 V is supplied by the video detector. This composite video signal must be amplified for about 100-V output, peak to peak. The signal from the video detector usually has negative sync polarity, which is inverted by the video amplifier to positive sync polarity for the cathode of the picture tube. The circuits in Figs. 13-25 and 13-26 illustrate video amplifiers in monochrome receivers. The idea of how the video section serves as the amplifier for luminance signal in color receivers is illustrated in Fig. 13-31.

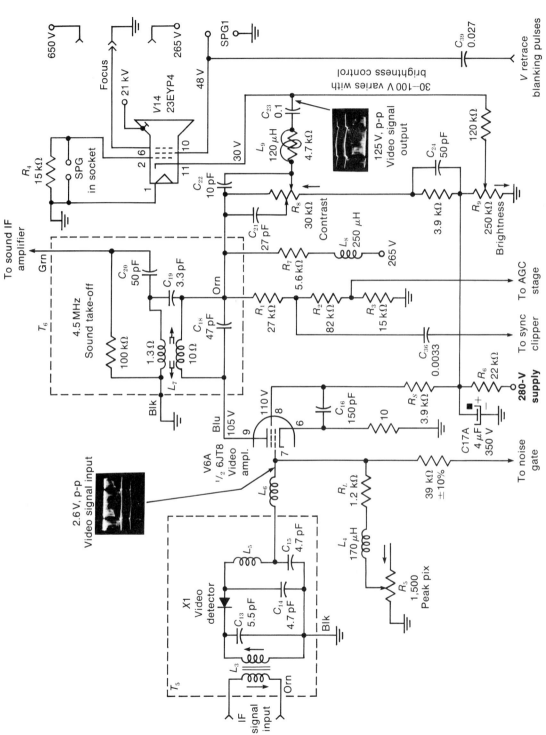

FIGURE 13-25 SINGLE-TUBE VIDEO AMPLIFIER. C VALUES LESS THAN 1 IN μF. "SPG" IS SPARK GAP. *(FROM ZENITH CHASSIS 14M23)*

Single-tube video amplifier. In Fig. 13-25, the video detector diode X1 is dc-coupled to the video amplifier stage. Then the required negative grid bias of 2 V is supplied by the dc component of the detected video signal. In the plate circuit, the 5.6-kΩ R_7 is the load resistor. The shunt peaking coil is L_8, with L_9 used for series peaking. Note that L_9 is wound on a 4.7-kΩ resistor for damping. The amplified video output is coupled to the cathode of the picture tube by the 0.1-μF C_{23}. Since the control grid returns to ground through the 15-kΩ R_4, the video signal is applied to the cathode-grid circuit. With peak-to-peak signal output of 125 V for input of 2.6 V, the voltage gain of the video amplifier stage is $125/2.6 = 48$.

Note that the 4.5-MHz intercarrier sound signal is taken from the plate circuit of the video amplifier. The primary winding of T_6 tunes with C_{18} as a 4.5-MHz trap to keep the sound out of the picture. However, the secondary winding L_7 has the 4.5-MHz sound signal coupled by C_{20} to the grid of the first sound IF amplifier. The dc R of each coil is marked on the diagram.

The contrast control is the 30-kΩ R_8 in the plate circuit of the video amplifier. It varies the amount of video signal coupled by C_{23} to the picture tube. The small capacitors C_{21}, C_{22}, and C_{24} compensate for the capacitance of the contrast control as its resistance is varied. The peaking control is the 1,500-Ω R_5 in series with L_4 in the video detector output circuit. The brightness control is the 250-kΩ R_9. It varies the dc voltage at the cathode of the picture tube, from the 280-V supply.

The video amplifier has outputs for several circuits in addition to the picture tube. High-frequency peaking is not necessary for these connections because only the signal amplitudes are important. The voltage divider consisting of R_1, R_2, and R_3 in the plate circuit supplies video signal for the AGC stage and sync clipper. Dc coupling is used for the AGC circuit, which rectifies the video signal to provide AGC bias proportional to signal strength. The sync clipper separates the sync for the deflection circuits.

An additional output in Fig. 13-25 is video signal from the detector to a noise gate. This circuit eliminates the effect of noise pulses in the sync and AGC circuits. Video signal from the detector is used for the noise gate because this polarity is opposite from the output of the video amplifier.

For the picture tube in Fig. 13-25, note that vertical blanking pulses are coupled to the screen grid. These are negative pulses, taken from the vertical deflection circuit, to cut off beam current during vertical retrace time. Most receivers have this internal vertical retrace suppressor circuit. The vertical blanking pulses can also be coupled to the control grid, or cathode of the picture tube, or into the video amplifier to become part of the video signal. In addition, the receiver may use blanking pulses from the horizontal deflection circuit to eliminate horizontal retraces.

Emitter-follower driving CE amplifier. In Fig. 13-26, signal from the video detector is dc-coupled to the base of Q1. This is an NPN transistor that combines the functions of an emitter-follower and CE amplifier. The emitter load resistor R_2 of 470 Ω results in a high input impedance at the base, to minimize loading of the video detector. Remember that the emitter output signal has the same polarity as the base input signal. The collector load resistor is the R_1 of 1.8 kΩ. There are two output signals, isolated by the separate output circuits. The video signal from the collector goes to the sync clipper. The video signal from the emitter goes to the base of Q2. This is a 5-W power transistor, with a heat sink mounted on the case. The collector supply is 140 V, to provide enough voltage swing for the 80-V peak-to-peak video signal output. In addi-

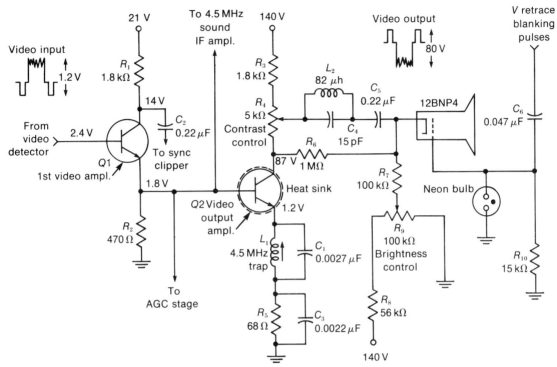

FIGURE 13-26 TRANSISTORIZED VIDEO CIRCUIT WITH EMITTER-FOLLOWER DRIVING CE AMPLIFIER FOR VIDEO OUTPUT STAGE. (FROM RCA HSK T1 CHASSIS)

tion to driving Q2, the emitter output of Q1 supplies video signal for the AGC stage and for the 4.5-MHz intercarrier sound signal.

In the output circuit of Q2, the contrast control R_4 and R_3 in series provide a total collector load resistance of 6.8 kΩ. The video output signal is coupled by the 0.22-μF C_5 to the cathode of the 12-in. picture tube. L_2 is a series peaking coil. The control grid returns to ground for the video signal through the 15-kΩ R_{10}. The neon bulb in the grid circuit provides the protection of a spark gap, as the bulb ionizes and shorts to ground with excessive voltage. For brightness control, the 100-kΩ R_9 varies the dc voltage at the cathode, from the 140-V supply in series with the 56-kΩ R_8.

Note the dc voltages for the two transistors. Q1 has forward bias of $2.4 - 1.8 = 0.6$ V. The 2.4 V at the base is the dc output of the video detector. The 1.8 V at the emitter is the IR drop across R_2 produced by emitter current. This 1.8 V at the emitter of Q1 is the base voltage at Q2 because of the direct coupling. For Q2 its forward bias is $1.8 - 1.2 = 0.6$ V. The value of 0.6 V is typical forward bias for a silicon transistor amplifier operating class A.

13-11 The Video Detector Stage

As shown in Fig. 13-27, the input to this diode detector is the modulated IF picture carrier signal, which is rectified and filtered. Rectification

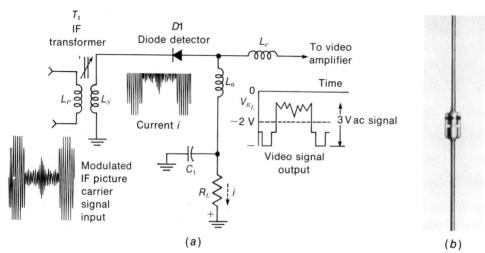

FIGURE 13-27 VIDEO DETECTOR. (a) DIODE CONNECTED FOR NEGATIVE OUTPUT ACROSS R_L IN ANODE CIRCUIT. (b) TYPICAL DETECTOR DIODE. LENGTH IS 1/2 IN. BAND AT CATHODE END. (GE)

allows the amplitude variations to be extracted, while the filtering removes the carrier. As a result, the detected output corresponds to the modulation envelope of the IF signal. This envelope is the original composite video signal with all the information needed for the picture, including sync and blanking. A diode needs about 3-V IF signal for linear detection, without distortion.

The video detector is a *signal diode*, usually silicon or germanium, in a glass package 1/2 in. long. A stripe at one end marks the cathode side. One type is the 1N3064, rated at 60 V for reverse breakdown and 10 mA maximum forward current. The detector is often inside the shield can of the last IF transformer, as indicated for the diode X1 in Fig. 13-25. Shielding is generally used for the detector stage in order to prevent harmonics of the detected signal from radiating back to the rf and IF stages. With an integrated circuit for the IF amplifier, the video detector is usually part of this IC unit.

Diode detection. The video detector action is the same as any diode detector for an AM signal. In Fig. 13-27, the signal input provides ac voltage to drive the diode into conduction as a halfwave rectifier. No dc supply voltage is used for the detector because the signal makes the diode conduct. The signal can be applied either to the anode or to the cathode. At the anode, the positive half-cycles of input signal produce forward current in the diode. At the cathode, the negative half-cycles of IF signal are detected. Negative polarity is shown in Fig. 13-27 because the video detector is often connected this way. Electron flow for i is from the top of L_S, from cathode to anode in $D1$, through R_L, and returning to the grounded side of L_S.

With more ac input, the diode current increases. The tip of sync in the signal produces maximum current. Less input for white signal amplitudes reduces the current. As a result, the current i consists of half-cycles of the IF carrier signal with varying amplitudes. These variations correspond to the envelope of the AM carrier signal.

Since the voltage across R_L is iR_L, the output V_{R_L} has the same signal variations as the current. The capacitance C_1 across R_L filters out variations at the IF rate so that the output voltage includes only the lower-frequency variations of the envelope. This is the composite video signal output voltage of the detector. In this example, the variations between tip of sync and white in the signal provide output of 3 V, peak-to-peak.

The detector load resistance R_L has shunt capacitance C_t, with the peaking coils L_0 and L_c that provide the same response for high video frequencies as required in the video amplifier. Similarly, the relatively low value of 2 to 8 kΩ is used for R_L for good high-frequency response. The video detector is the input circuit for the video amplifier section.

It is necessary to rectify the modulated carrier signal because its amplitude variations have an average value of zero, as a result of the symmetrical envelope. In terms of the desired video signal information, the carrier is just as much positive as negative at any one instant of time. However, after rectification the amplitude variations of the modulation can be obtained as variations in the amount of rectified carrier signal. The detector output voltage across R_L, therefore, is the desired video signal.

Detector polarity. Both sides of the modulation envelope have the same amplitude variations. Therefore, either polarity of the IF signal can be rectified by the detector. Depending on whether R_L is in the anode or cathode circuit, a diode can produce either negative or positive dc output voltage. In Fig. 13-27, R_L is in the anode return to ground. Then the detected signal is a varying dc voltage of negative polarity. The sync produces the most output, resulting in negative sync polarity for the composite video voltage. Also, the average level of the detector output is a negative dc voltage. In this example, the dc output voltage of the detector is −2 V.

If the diode were reversed in Fig. 13-27, the detector output would be positive. Then the IF signal would be applied to the anode, with R_L in the cathode circuit. The average level of the detector output would still be 2 V, but positive. Also, the video signal would have positive sync polarity.

It may be of interest to note that detector polarity does not matter in a sound system. The reason is that the loudspeaker is driven by ac audio signal, without any dc bias. For the picture tube, though, polarity inversion of the video signal produces a negative picture.

Dc level of the detector output. An important feature of the detector output is that it consists of a varying dc voltage. The detector diode is just a low-power rectifier, or *signal diode*, to change the ac input to dc output.

More details of the detector output voltage are shown in Fig. 13-28. In *a*, the dc axis of −2 V is the average level. A dc voltmeter can be used to read this dc level of −2 V.

Peak-to-peak ac signal. The video signal is in the variations of the dc voltage. In Fig. 13-28*a* the peak voltage is −4 V and the minimum is −1 V. Therefore, the peak-to-peak amplitude for the signal voltage variations is 3 V. No polarity is assigned to the ac component.

In Fig. 13-28*b*, the ac signal is shown without the dc level, which is blocked by a coupling capacitor. The dc axis in *a* becomes the average level of zero in *b* for the ac signal. In both cases, the peak-to-peak amplitude is 3 V for the variations in signal voltage. The 3-V signal is typical for driving a video amplifier tube. For transistors, the signal output of the video detector is about 1 V, peak-to-peak.

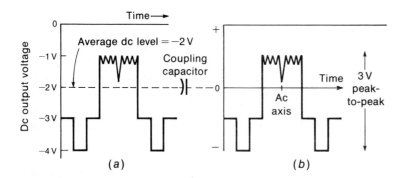

FIGURE 13-28 VIDEO DETECTOR OUTPUT VOLTAGE. (a) DC FORM. (b) AC FORM AFTER AVERAGE DC COMPONENT IS BLOCKED BY COUPLING CAPACITOR.

Video detector test point. Most receivers have a test point (TP) to measure the video detector output from the top of the chassis. As shown in Fig. 13-29, R_1 of about 10 kΩ is used to isolate TP. A dc voltmeter here is a convenient measurement because a dc meter is used to check IF signal. The dc level of the rectified signal is typically 2 or 3 V for vacuum tube amplifiers. With transistors, this dc voltage is about 0.5 V. When the receiver is switched off channel to remove the signal input, the detector output drops to a lower value. The small output voltage without signal is rectified receiver noise.

When the detector dc output voltage has its normal value on channel and drops close to zero off channel, these checks mean the IF amplifier is operating with normal signal. If the output is close to zero, both off channel and on channel, there is no detected IF signal. If the detector output is very high on channel or off channel, this means the IF amplifier is oscillating.

An oscilloscope at the detector test point will show the composite video signal output. A typical value is 3-V peak-to-peak ac signal for a video amplifier tube. For a transistor video amplifier, the detector output is approximately 1 V, peak-to-peak. When the oscilloscope internal sweep frequency is at 30 Hz, you see two fields of video signal with the vertical blanking and sync pulses. When the sweep is at 7,875 Hz, you see two lines of video signal with the horizontal blanking and sync pulses. For a color broadcast, you see the colorplexed composite video signal, with 3.58-MHz color burst on the back porch of horizontal sync and 3.58-MHz chrominance signal in the video. It should be noted, though, that individual cycles at 3.58 MHz are not visible unless you raise ths sweep frequency and expand the sweep to magnify the pattern.

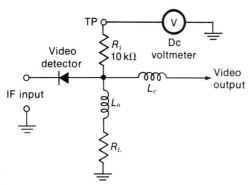

FIGURE 13-29 MEASURING RECTIFIED IF SIGNAL WITH DC VOLTMETER AT VIDEO DETECTOR TEST POINT.

Detecting the 4.5-MHz intercarrier beat. In addition to the fact that the video detector diode rectifies the modulated IF picture carrier to re-

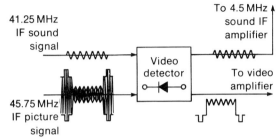

FIGURE 13-30 THE VIDEO DETECTOR CAN PRODUCE THE 4.5-MHz INTERCARRIER SOUND SIGNAL, IN ADDITION TO THE COMPOSITE VIDEO SIGNAL.

cover the composite video signal, this stage serves as a frequency converter to produce the 4.5-MHz second sound IF signal (see Fig. 13-30). In terms of intercarrier sound, the 45.75-MHz picture carrier in the detector stage is equivalent to a local oscillator. The heterodyning action of the 45.75-MHz picture carrier beating with the 41.25-MHz center frequency of the sound signal results in the lower center frequency of 4.5 MHz. This FM signal can then be coupled to the 4.5-MHz sound IF section for amplification and detection to recover the audio modulation.

In monochrome receivers, the video detector is used to supply the 4.5-MHz intercarrier sound signal. In color receivers, though, a separate diode is generally used as the 4.5-MHz sound converter. The reason is to reduce interference of the sound signal in the color picture.

13-12 Luminance Video Amplifier in Color Receivers

This section amplifies the Y signal without the 3.58-MHz chroma signal. Therefore, its function is the same as the video amplifier in a monochrome receiver. The video bandwidth is cut off at 3.2 MHz in both cases to eliminate crosstalk from the 3.58-MHz color signal.

Figure 13-31 shows a block diagram of the video section for Y signal, from a solid-state color receiver. There are five video stages here, all dc-coupled from the detector to the picture

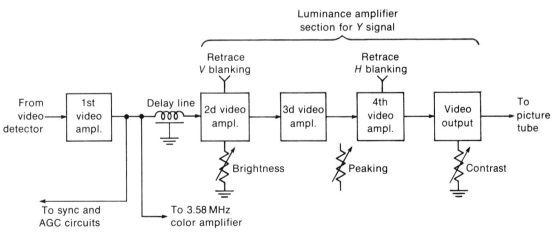

FIGURE 13-31 LUMINANCE SECTION TO AMPLIFY Y VIDEO SIGNAL IN COLOR RECEIVER. THIS IS A TRANSISTORIZED CIRCUIT. *(FROM RCA CTC 40 CHASSIS)*

tube. To provide a high impedance for the detector output, the first video stage is an emitter-follower. This stage is often called a *video preamplifier*. It supplies output to both the 3.58-MHz color amplifier and the Y video amplifier. The second stage is a common base circuit to provide a low-impedance input to the *delay line*. A delay of approximately 1 μs is always used in the Y video amplifier so that the luminance and color signals will arrive at the picture tube at the same time. The circuits for luminance video signal have less delay because they are resistive, compared with the tuned circuits for the 3.58-MHz chroma signal. Note that the color signal is taken off before the delay line. In addition, video signal for the sync and AGC circuits is taken from the first video amplifier.

The third stage in Fig. 13-31 is a common-emitter amplifier for the Y video signal. The fourth stage is an emitter-follower again, feeding the video output stage. This last stage is a power amplifier that supplies 120-V peak-to-peak video signal to drive the three cathodes of the color picture tube. The signal has positive sync polarity here, resulting from positive polarity out of the video detector and two polarity inversions by the common-emitter amplifiers.

The contrast control is in the emitter circuit of the video output circuit to vary the gain for Y video signal. The peaking control in this circuit varies the amount of feedback for high video frequencies from the fourth stage back to the second stage. The brightness control varies the dc bias on the second video amplifier. Because of the dc coupling, any change in dc level in the video circuits is coupled to the cathode of the picture tube. Similarly, instead of coupling the retrace blanking pulses to the picture tube, they are fed into the video amplifier. This way the pulses are amplified while being coupled to the picture tube through the video section.

13-13 Functions of the Composite Video Signal

Figure 13-32 illustrates four paths for the colorplexed signal obtained from the video detector. We can consider that the signal is coupled to these parallel branches for the different functions. Therefore, each circuit can operate independantly of the others. For instance, clipping the synchronizing pulses in the sync separator stage does not interfere with the video amplifiers supplying signal for the picture tube. Similarly, the AGC circuit rectifies the video signal for AGC bias, but the video amplifier provides the complete composite video signal. In color receivers, the luminance section amplifies the Y video signal without the 3.58-MHz chroma signal, while the 3.58-MHz bandpass amplifier

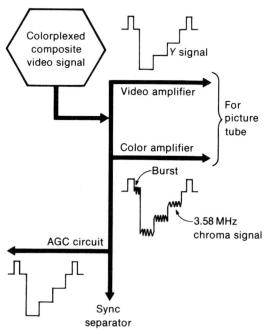

FIGURE 13-32 FUNCTIONS OF THE COMPOSITE VIDEO SIGNAL.

supplies chroma signal for the color demodulators.

In summary, the four functions of the colorplexed composite video signal are as follows:

1. Video signal for the picture tube to put the picture on the raster
2. Supply vertical and horizontal synchronizing pulses for the sync circuits that time the deflection oscillators
3. Supply signal for the AGC circuit that controls the gain of the rf and IF stages to prevent overload distortion on strong signals
4. Supply 3.58-MHz chroma signal for the color amplifier to put color in the picture

A comparison of the video circuits for monochrome and color receivers is shown in Fig. 13-33. Note the following additions for color receivers:

1. Separate diode converter stage for 4.5 MHz intercarrier sound.
2. Video preamplifier for sync and AGC takeoff circuits.
3. Delay line for Y video signal.
4. Y video amplifier to drive all three guns of the color picture tube with luminance signal.
5. Brightness control varies dc bias on Y video amplifier, dc-coupled to the color CRT. This method is used so that one control determines the bias on all three guns of the picture tube.

13-14 The 4.5-MHz Sound Trap

The video amplifier usually has a trap tuned to the intercarrier sound frequency of 4.5 MHz to keep the sound signal out of the picture. In monochrome receivers, the 4.5-MHz trap is also the takeoff circuit for signal to the 4.5-MHz sound IF amplifier. The 4.5-MHz trap may be in the video detector or in the video amplifier. Generally, the trap is in the video output in order to provide more 4.5-MHz signal for the sound

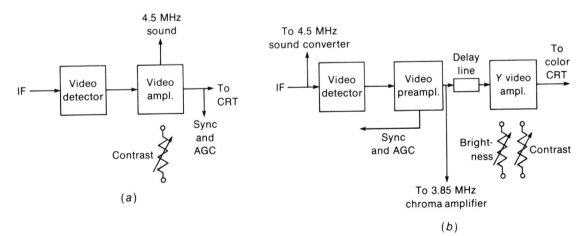

FIGURE 13-33 COMPARISON OF VIDEO CIRCUITS. (a) MONOCHROME RECEIVER. (b) COLOR RECEIVER.

section. This arrangement is shown in the video amplifier circuit in Fig. 13-25.

In color receivers, the luminance amplifier still has a 4.5-MHz trap to reduce 4.5-MHz beat in the picture. However, the 4.5-MHz trap is not used here for the sound takeoff. Color receivers generally have a separate sound converter stage to provide the 4.5-MHz intercarrier sound signal. The reason is to reduce interference between the 4.5-MHz sound signal and 3.58-MHz chroma signal, which together produce a 920-kHz beat.

The 4.5-MHz sound signal in the picture produces about 225 pairs of thin, diagonal lines, with small wiggles like a fine herringbone weave. This effect in Fig. 13-34 is also called a "wormy" picture. The fine lines are produced by the 4.5-MHz carrier while the weave is the result of frequency variations in the FM sound signal. When you observe the pattern closely, the herringbone effect disappears when there is no voice or music, leaving just the straight lines corresponding to the 4.5-MHz carrier without modulation. If necessary, the 4.5-MHz trap can be tuned for minimum interference in the picture. The trap is just an *LC* circuit tuned to 4.5 MHz, as shown in Fig. 13-35.

Two trap circuits are illustrated in Fig. 13-36. In a, L_1 and C_1 form a parallel resonant circuit tuned to 4.5 MHz. The trap is in series with the output. Furthermore, the trap has maximum impedance at parallel resonance. Therefore, the trap frequency is removed because maximum voltage is developed across the trap itself. Note that L_1 in the trap is coupled to L_2 to provide the 4.5-MHz sound signal for the 4.5-MHz sound IF amplifier.

In Fig. 13-36*b*, *L* and *C* form a series resonant circuit tuned to 4.5 MHz. This trap circuit is in shunt with the load. Furthermore, the trap is practically a short circuit at series resonance. As a result, the 4.5-MHz intercarrier sound is removed from the video output signal.

The trap has a coil with a variable slug. Adjust for minimum 4.5-MHz beat in the picture.

FIGURE 13-34 FINE HERRINGBONE WEAVE PRODUCED BY 4.5-MHz INTERCARRIER SOUND IN PICTURE.

FIGURE 13-35 TYPICAL 4.5-MHz TRAP WITH *L* AND *C*. HEIGHT IS $1^1/_2$ IN.

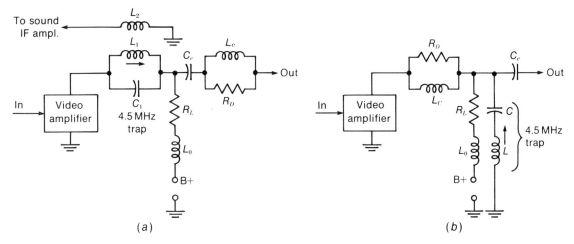

FIGURE 13-36 VIDEO AMPLIFIER CIRCUITS WITH 4.5-MHz TRAP. (a) PARALLEL RESONANT TRAP IN SERIES WITH OUTPUT. TRAP HERE IS ALSO USED FOR SOUND TAKEOFF. (b) SERIES RESONANT TRAP IN SHUNT WITH LOAD.

SUMMARY

1. Composite video signal is coupled to the cathode-grid circuit of the picture tube to vary the beam current and screen illumination. This intensity modulation reproduces the picture information. Maximum beam current produces white; cutoff corresponds to black. About 100-V peak-to-peak video signal is needed for good contrast. Average brightness is set by the dc bias.
2. The video stage is a class A amplifier for minimum amplitude distortion, with RC coupling for wide frequency response. Peaking coils are used to boost the gain for high video frequencies.
3. The two most common methods of manual contrast control are: (a) variable resistor either in the cathode or in the emitter circuit to vary the gain; (b) variable potentiometer to tap off the desired amount of signal voltage. See Fig. 13-8.
4. In the range from 30 Hz to 3.2 MHz, higher video frequencies correspond to smaller details of picture information.
5. In an RC-coupled amplifier, the gain is down for high frequencies because the shunt capacitance C_t has decreasing reactance that bypasses the load R_L. The frequency at which the gain is down to 70 percent response is $F_2 = 1/(2\pi R_L C_t)$.
6. The response is down for low frequencies mainly because the increasing reactance of the coupling capacitor C_c results in less signal across R for the next stage. The frequency at which the gain is down to 70 percent response at the low end is $F_1 = 1/(2\pi R_g C_c)$.

7. Frequency distortion means unequal gain for different frequencies. Insufficient gain for high video frequencies causes loss of horizontal detail in the picture. Low gain for low video frequencies causes weak contrast.
8. Phase distortion is important for low video frequencies because the large time-delay distortion causes smear in large areas of the picture.
9. For good high-frequency response, the video amplifier requires minimum C_t, relatively low R_L of 2 to 8 kΩ, and peaking coils. A shunt peaking coil L_o is in series with R_L on either side, but in shunt with C_t. A series peaking coil L_c is in series with the coupling to the next stage.
10. To improve the low-frequency response, large capacitors are used for coupling and bypassing. In addition, an RC decoupling filter in series with R_L can boost the low-frequency response.
11. In monochrome receivers, only one video amplifier stage is generally used to drive the picture tube. In color receivers, more stages are needed because of the video preamplifier and losses in the delay line.
12. With transistors in the video amplifier section, the first stage is generally an emitter-follower to provide a high input impedance for the video detector.
13. The video detector is a semiconductor diode that rectifies the picture IF signal to provide composite video signal output.
14. The output at the video detector test point can be measured to check IF signal input. A dc voltmeter reads 0.5 to 3 V, which is rectified IF signal. An oscilloscope shows composite video signal of 1 to 5 V, p-p for the video amplifier.
15. The luminance amplifier for the Y signal in a color receiver has the same function as the video amplifier in a monochrome receiver. However, the luminance amplifier has a delay line, after the takeoff point for the 3.58-MHz color signal.
16. A 4.5-MHz trap is generally used in the video amplifier to keep the intercarrier sound out of the picture. In monochrome receivers, the 4.5-MHz trap is also used to take off the sound for the sound IF amplifier.

Self-Examination (Answers at back of book)

Answer True of False.
1. The input signal for the video amplifier is obtained from the video detector.
2. The amplitude of this detector signal is about 100 V peak-to-peak.
3. The output of the video amplifier drives the cathode-grid circuit of the picture tube.
4. The video amplifier gain determines contrast in the picture.
5. Composite video signal of positive sync polarity is needed for cathode drive at the picture tube.

6. If the video amplifier gain drops after 1 MHz, the picture will have weak horizontal detail.
7. A value for R_L of 1 MΩ is too high for good high-frequency response in a video amplifier.
8. A coupling capacitor that is too small causes poor low-frequency response.
9. The dc component of a fluctuating dc voltage is blocked by a coupling capacitor.
10. The luminance amplifier is tuned to 3.58 MHz.
11. Table 13-2 shows that series peaking has more gain than shunt peaking.
12. The composite video signal is generally coupled to the sync separator and AGC circuit, in addition to the picture tube.
13. In Fig. 13-25, the diode X1 is the video detector.
14. In Fig. 13-25, the 250-μH L_s is a shunt peaking coil.
15. In Fig. 13-25, the video signal for cathode drive at the picture tube is 125 V, peak-to-peak.
16. In Fig. 13-26, the emitter of Q1 supplies video signal for Q2.
17. In Fig. 13-26, Q1 and Q2 are dc-coupled.
18. The video detector is a half-wave diode rectifier.
19. White parts of the picture provide maximum IF signal into the video detector.
20. The 4.5-MHz trap to keep the intercarrier sound out of the picture can also be used for the sound takeoff.

Essay Questions

1. What determines contrast of the picture?
2. What determines brightness of the raster?
3. How does the video signal produce maximum white in the picture?
4. How does the video signal produce black in the picture?
5. (a) Draw composite video signal of 100-V peak-to-peak amplitude with the polarity for cathode drive at the picture tube. (b) Do the same for grid drive.
6. Show two circuits for manual contrast control.
7. Draw the circuit of a CE transistor video amplifier with series peaking.
8. Define frequency distortion in an amplifier.
9. Define phase distortion in an amplifier.
10. Why is phase distortion more important for low video frequencies than high video frequencies?
11. What is the effect on the picture of loss of the high video frequencies?
12. For the load line in Fig. 13-6 give the following values: (a) dc grid bias; (b) average dc plate volts; (c) average dc plate mA.

13. If an amplifier has a load resistance of 0 Ω: (a) Will there be any voltage gain? (b) Will there be any polarity inversion of the signal? (c) Will there be any output current?
14. If the total shunt capacitance in the video amplifier were zero, would high-frequency compensation be necessary? Why?
15. Why does a low value for R_L in the video detector and video amplifier improve the response for high frequencies?
16. State the formula for the frequency F_2, where high-frequency gain is down 3 dB. What two factors in this formula determine the high-frequency response?
17. State the formula for the frequency F_1, where low-frequency gain is down 3 dB. What two factors in this formula determine the low-frequency response?
18. What is the function of the luminance amplifier?
19. Why does the luminance amplifier have a delay line?
20. Give four functions of the colorplexed composite video signal.
21. Give two functions of the video detector.
22. (a) Draw a video detector circuit with series peaking and video signal output having negative sync polarity. (b) Do the same for output with positive sync polarity.
23. What two input signals are necessary for the video detector to produce the 4.5-MHz intercarrier sound signal?
24. Explain briefly why any modulated carrier signal must be rectified in order to recover the modulation?
25. Describe how you can check the IF signal output with a dc voltmeter.
26. Referring to the video amplifier circuit in Fig. 13-25, give the function of each of the following: L_3, X1, R_L, and L_6 in the video detector. Also, R_7, R_S, R_6, C_{17A}, L_9, and C_{23} in the video amplifier.
27. Referring to the video amplifier circuit in Fig. 13-26, give the function of each of the following: R_1, R_2, R_8, and C_5.
28. Draw the schematic diagram of a video detector, dc-coupled to a pentode video output amplifier with combination peaking. Include a contrast control, sharpness control, and 4.5-MHz trap. Indicate dc voltages for the pentode.

Problems

1. An RC-coupled amplifier has R_L of 5 kΩ and C_t of 20 pF. (a) Calculate F_2. (b) Calculate the reactance of C_t at this frequency to see that it equals the R_L of 5 kΩ.
2. An $R_g C_c$ coupling circuit has 0.5 MΩ for R_g and 0.1 μF for C_c. (a) Calculate F_1. (b) Calculate the reactance of C_c at this frequency to see that it equals the R_g of 1 MΩ.
3. An audio amplifier tube has R_L of 1 MΩ and C_t of 40 pF. At what frequency will the gain be down to 70.7 percent of the midfrequency response?

4. How much is the time delay in microseconds for the phase angle of 36° at the frequencies of (a) 4 MHz; (b) 40 Hz?
5. A video amplifier has R_L of 3 kΩ and C_t of 20 pF. From Table 13-2 calculate the value of the series peaking coil L_c for an F_2 of 3.2 MHz.
6. Draw the frequency-response curve for an RC-coupled amplifier having the voltage gain values listed below.

FREQUENCY, Hz	VOLTAGE GAIN
0	0
1×10^2	16
1×10^3	32
1×10^4	32
1×10^5	32
1×10^6	16
10×10^6	4

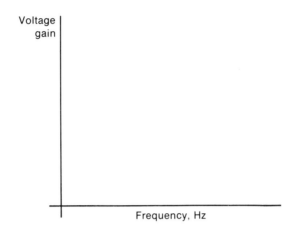

FIGURE 13-37   FOR PROBLEM 6.

14 DC Level of the Video Signal

The dc level in the video signal indicates the difference in brightness between light and dark scenes. Remember that brightness is the average illumination or background level. In the picture, a scene may be in bright sunlight or in a dark room at night. For both cases, the details of picture information could be the same, but the brightness level is different.

At the grid of the picture tube, its dc bias voltage determines the brightness. The manual brightness control can set the dc bias for the correct background with any scene. However, when the background becomes lighter or darker, the dc bias should be different. The dc component of the video signal can shift the dc bias to show the change in brightness. More details are explained in the following topics:

14-1 Changes in brightness
14-2 Definitions of terms for the dc component
14-3 How a coupling capacitor blocks the average dc voltage
14-4 Dc coupling and ac coupling
14-5 Average value of the video signal
14-6 Dc insertion
14-7 Video amplifier dc-coupled to picture tube

14-1 Changes in Brightness

A common example occurs when the program goes to black before a commerical or a change of scene. Similarly, the change can be from a bright scene in daylight to a dark scene at night. If the dc component of the video signal does not shift the bias on the picture tube for less brightness, the reproduced picture will not be dark enough. The opposite effect occurs when the bias is right for a dark scene but the scene changes to a light background. With too much bias for the light scene, whatever should be white becomes gray and the gray tones go black. With color especially, the dc component of the video signal is needed for correct reproduction of the color values.

The signal is transmitted with its dc component. At the receiver, the dc component is in the varying dc voltage output of the video detector. If the video amplifier has dc coupling, from the detector to the picture tube, the dc component will be preserved. However, capacitive coupling in the video amplifier removes the dc component in the signal.

To be more specific about brightness changes, Fig. 14-1 illustrates the effects of negative dc bias voltages at the control grid of the picture tube. The same idea applies to positive dc bias voltages at the cathode. In terms of the control grid, more negative bias decreases the average brightness. Remember that the dc bias voltage is the axis for the ac video signal variations. The dc bias should be the amount that

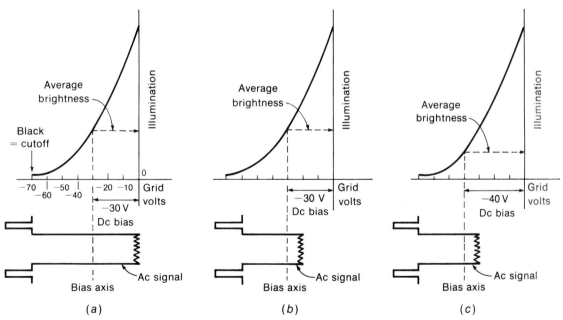

FIGURE 14-1 HOW DC GRID VOLTAGE OF PICTURE TUBE SETS AVERAGE BRIGHTNESS FOR LIGHT AND DARK SCENES. (a) LIGHT SCENE WITH CORRECT GRID BIAS OF −30 V. (b) BIAS OF −30 V IS TOO FAR FROM CUTOFF FOR DARK SCENE. (c) CORRECT BIAS OF −40 V FOR SAME SIGNAL AS IN b.

allows the blanking level in the video signal to swing the grid voltage to cutoff for black.

In Fig. 14-1a, the bias of −30 V is correct for a light scene. The average brightness or background illumination corresponds to a little less than one-half the maximum beam current. Then the blanking voltage in the video signal drives the grid voltage to cutoff so that the blanking level is black. Also, maximum white in the video signal produces the peak beam current.

In Fig. 14-1b, the dc bias is still −30 V, but the ac video signal corresponds to a darker scene. The fact that this signal is darker is evident from the fact that the signal variations are not so far from the black level, compared with a. It should be noted that this is not a question of less ac signal, since the sync amplitude is the same in a and b. Now the grid bias of −30 V is not negative enough for the darker scene. As a result, black shows as gray. On the whole, the reproduced picture is not dark enough.

The darker video signal in Fig. 14-1b needs the more negative grid bias of −40 V shown in c. Now the average brightness level is lower for less illumination with a darker background. Also, the blanking level is at cutoff for black.

It is important to realize that the change in video signal from a to b or c represents a typical line of information from different frames. Any change in background is a change in signal for the frames, not for the lines. Changes of information from line to line are part of the ac video signal.

Figure 14-2 shows a picture with the brightness too high. The diagonal lines upward to the right are the vertical retrace lines. These lines are produced by the horizontal scan, but during the vertical retrace time. They are visible here because the pedestal level is not at cutoff

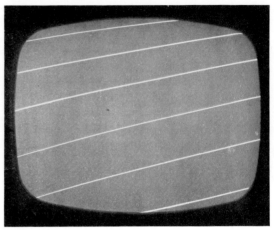

FIGURE 14-2 BRIGHTNESS TOO HIGH- BLACK VALUES NOT DARK ENOUGH. VERTICAL RETRACE LINES ARE SHOWN HERE, BUT THEY ARE NORMALLY NOT VISIBLE BECAUSE OF RETRACE SUPPRESSOR CIRCUIT.

for blanking. However, normally you would not see them because receivers have retrace suppressor circuits to blank the retrace lines at any setting of the brightness control.

14-2 Definitions of Terms for the DC Component

Referring to Fig. 14-3, note the amplitude levels. The blanking level is at 75 percent. Sync amplitudes are between this level and 100 percent for tip of sync. Maximum white is at 12.5 percent, as an average between 10 and 15 percent.

The black level in picture information is at the 68.5 percent level. This value is 7.5 percent from the blanking level, as an average between 5 to 10 percent. The *black setup* of 7.5 percent is used to separate very dark picture information from the blanking level, especially in a colorplexed video signal. In studio work with television cameras, the black setup is also called *ped-*

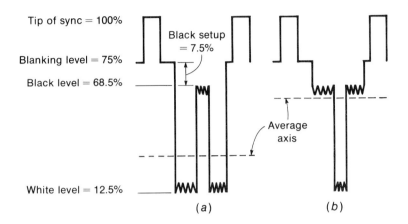

FIGURE 14-3 AVERAGE LEVEL IN THE VIDEO SIGNAL. (a) CLOSE TO WHITE. (b) CLOSE TO BLACK.

estal voltage. The pedestal control to vary the amount of black setup determines the sensitivity to black and dark gray.

The average level in Fig. 14-3 divides the signal amplitudes into equal amounts above and below the axis. In a, the average axis is close to white because the picture information is mostly white, except for the dark bar in the middle. In b, the average axis is close to black because the signal is mostly at the black level. In the ac form of the signal, the average level is the zero axis. As a fluctuating dc signal voltage, the average axis is a dc level.

Dc insertion. In the transmitted signal, the tip of sync is always at 100 percent amplitude. This is necessary to have a reference amplitude for AGC circuits. The fact that the sync pulses have the same amplitude for light or dark picture information means that the transmitter adds a dc component to the average axis. This process is dc insertion. When the receiver adds a dc component after it has been lost by capacitive coupling in the video amplifier, this process is dc reinsertion.

Dc restorer. This is a diode rectifier circuit that has the function of inserting the dc component. The video signal itself can be rectified to produce a dc voltage proportional to the level of brightness.

Clamping. This term means keeping one amplitude at a constant level. For instance, the blanking level can be clamped at cutoff grid voltage for the picture tube. The clamping function is really the same as dc insertion or dc reinsertion, as a dc component must be added to clamp different ac signal waveforms.

Retrace suppressor circuit. Instead of relying on the blanking in the video signal, receivers provide their own blanking pulses for the picture tube to make sure retrace lines are not visible. Vertical flyback pulses from the vertical deflection circuit are used to suppress the vertical retrace lines. Practically all receivers use a vertical retrace suppressor circuit. In addition, horizontal flyback pulses can be used for a horizontal retrace suppressor. Color receivers generally use both horizontal and vertical flyback pulses for blanking.

14-3 How a Coupling Capacitor Blocks the Average DC Voltage

Refer to Fig. 14-4, where C_c is a coupling capacitor with its series R to form an RC coupling circuit. The signal applied to terminals 1 and 2 is a pulsating or fluctuating dc voltage. It varies between 10 and 30 V with an average of 20 V. This average value is the steady dc level. The variations of ± 10 V around this axis form the ac component. C_c will charge to the steady dc level of 20 V because this value is the average charging voltage. As a result, the steady dc component is blocked, since it cannot produce voltage across R. The ac component, however, is developed across R between output terminals 3 and 4. Note that the zero axis of the ac voltage output across R corresponds to the average axis of the fluctuating dc voltage input.

The dc component across C_c. The voltage across C_c is the steady dc component of the input voltage because the variations of the ac component are symmetrical above and below the average level. Furthermore, the series resistance is the same for charge and discharge. As a result, any increase in charging voltage above the average level is counteracted by an equal discharge below the average. In Fig. 14-4, for example, when e_{in} increases from 20 to 30 V, this effect of charging C_c is nullified by the discharge when e_{in} decreases from 20 to 10 V. At all times, however, e_{in} has a positive value that charges C_c in the polarity shown.

The net result is that only the average voltage level is effective in charging C_c, since the variations from the axis neutralize each other. After a period of time that depends on the RC time constant, C_c will charge to the average value of the pulsating dc voltage applied, which is 20 V here.

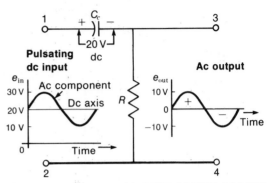

FIGURE 14-4 HOW C_c BLOCKS THE AVERAGE DC LEVEL OF A FLUCTUATING DC VOLTAGE.

The ac component across R. Although C_c is charged to the average dc level, when the pulsating input voltage varies above and below this axis, the charge and discharge current produces IR voltage corresponding to the fluctuations of the input. When e_{in} increases above the average level, C_c takes on charge, producing charging current through R. Even though the charging current may be too small to affect the voltage across C_c appreciably, the IR drop across a large value of resistance can be practically equal to the ac component of the input voltage. In summary, a long RC time constant is needed for good coupling.

Polarity of the ac component. In Fig. 14-4, when e_{in} is above 20 V, the charging current produces electron flow from the low side of R to the top, adding electrons to the negative side of C_c. The voltage at the top of R is then positive. When e_{in} decreases below the average level, C_c loses charge. The discharge current then is in the opposite direction through R. The result is negative polarity for the ac voltage coupled across R.

When the input voltage is at its average level, there is no charge or discharge current. Then the voltage across R is zero. The zero level in the ac voltage across R corresponds to the average level of the pulsating dc voltage applied to the RC circuit.

With positive pulsating dc voltage applied, the values above the average produce the positive half-cycles of the ac voltage across R. The values below the average axis produce the negative half-cycles.

Voltages around the RC coupling circuit. If you measure the pulsating dc input voltage across points 1 and 2 in Fig. 14-4 with a dc voltmeter, it will read the average level of 20 V. A voltmeter that reads only ac values across these same points will read the fluctuating ac component, equal to 7 V rms.

Across points 1 and 3 for C_c, a dc voltmeter reads the steady dc value of 20 V. An ac voltmeter here reads zero.

An ac voltmeter for the output across R between points 3 and 4 will read the ac voltage of 7 V rms. A dc voltmeter here reads zero.

There is practically no phase shift. This rule applies to all RC coupling circuits, since R must be 10 or more times X_C. Then the reactance is negligible compared with the series resistance, and the phase angle of less than 5.7° is practically zero.

Once the dc component is blocked by C_c, it does not matter how many more capacitive couplings are used. The ac component always has an average axis of zero, which cannot be changed by the coupling. Furthermore, direct coupling in a succeeding stage does not put back the dc component that was blocked. Only a dc restorer circuit can do that.

14-4 DC Coupling and AC Coupling

These terms indicate whether the dc component is coupled with the ac signal. In ac coupling, only the ac signal is coupled, while any dc component is blocked. In dc coupling, both the ac signal and its dc component are coupled to the next stage.

Ac coupling. An RC circuit provides ac coupling for the signal, as the steady dc component is blocked by C_c. A transformer with an isolated secondary winding also blocks the dc component, but this method is not used for coupling the video signal. Transformer coupling is generally used in the rf and IF circuits for the modulated picture carrier signal. However, in these stages, the dc component of the video signal is not affected. The dc component is available only after the modulated signal is detected. The advantage of an RC circuit for ac coupling of the video signal is that the dc level of the fluctuating dc voltage in the output of one stage does not affect the dc bias for the input of the next stage.

Dc coupling. This term means a direct connection for the signal, without any series capacitor. The advantage is that the lowest frequencies are coupled, including the steady dc component. With direct coupling, however, all the electrode voltages in the amplifier depend on the dc voltages of the previous stage. Any change in supply voltage or dc bias affects all the stages interconnected by dc coupling. A particular problem is drift in the dc supply voltage, which can alter the bias voltages.

Partial dc coupling. See Fig. 14-5. C_1 is used to couple all the ac video signal. R_1 and R_2 form a voltage divider for the dc component. The dc voltage across R_2 is the dc component for the picture tube. In this example, the voltage across R_2 is $8/10$, or 80 percent, of the total dc voltage. Typical values are 50 to 80 percent of the dc component for the picture tube, with partial dc coupling.

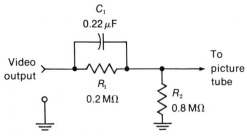

FIGURE 14-5 PARTIAL DC COUPLING. R_2 HAS 80 PERCENT OF THE DC COMPONENT, WHILE C_1 COUPLES ALL THE AC COMPONENT.

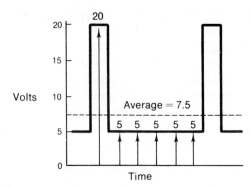

FIGURE 14-6 AVERAGE VALUE OF A FLUCTUATING DC VOLTAGE.

14-5 Average Value of the Video Signal

For any signal, its average value is the arithmetical mean of all the values taken over a complete cycle. With fluctuating dc voltage, the average is some dc level. In Fig. 14-6, as an example, if we add the five values of 5 V each and the 20-V value, the sum is 45 V. The instantaneous dc values are taken from the zero base line. This sum

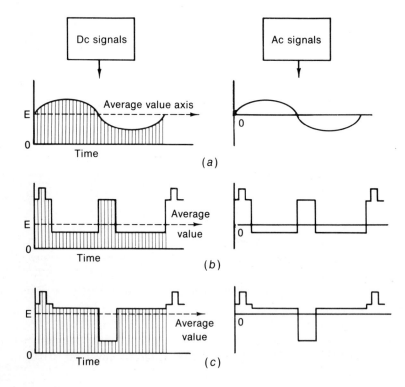

FIGURE 14-7 AVERAGE-VALUE AXIS FOR DIFFERENT WAVEFORMS. DC FORM IS AT LEFT AND AC FORM AT RIGHT. (a) SYMMETRICAL SINE WAVE WITH AVERAGE AXIS IN CENTER. (b) UNSYMMETRICAL VIDEO SIGNAL, MOSTLY WHITE. AVERAGE AXIS IS CLOSE TO WHITE-SIGNAL AMPLITUDES. (c) UNSYMMETRICAL VIDEO SIGNAL, MOSTLY BLACK. AVERAGE AXIS IS CLOSE TO BLACK LEVEL.

divided by 6 for the values included equals $^{45}/_6$, or 7.5 V for the average value.

Notice that the average axis of 7.5 V is close to the 5-V level, not in the middle. The reason is that this signal voltage is 5 V for most of the cycle. The average-value axis divides the signal into two equal areas above and below the axis.

When fluctuating dc voltage is coupled by an *RC* circuit, the average dc axis becomes the zero axis in the ac signal. This idea is illustrated for three types of signals in Fig. 14-7. The sine wave in *a* has an average-value axis exactly in the center because it is a symmetrical waveform. In *b* the video signal corresponds to a black bar down the center of a white frame. Now the average-value axis is closer to white level because the signal is white for most of the cycle. The opposite case in *c* is for a white bar down the center of a black frame. Then the average-value axis is close to black level, because the signal is mostly black.

In all cases the average dc axis becomes the zero ac axis, with equal areas of positive and negative signal above and below the axis. However, notice that for the dark signal in *c* its axis is much closer to the black level than in *b*.

If we put the dark and light video signals on a common axis, the sync pulses are out-of-line, as shown in Fig. 14-8. The common axis for the signals in their ac form is the zero level. Furthermore, the zero axis of the ac signal corresponds to the dc bias at the grid of the picture tube. When the blanking level has different voltage variations from the common axis, one setting of the brightness control cannot be correct for both dark and light scenes. The bias illustrated in Fig. 14-8 is right for the light video signal at the top. Then the blanking voltage drives the grid voltage to cutoff. However, for the dark video signal below, the bias indicated is not negative enough. In this case the blanking level is

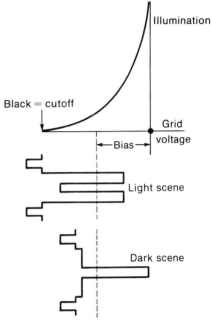

FIGURE 14-8 SYNC PULSES OUT OF LINE AT GRID OF PICTURE TUBE, SHOWN FOR THE TWO LINES OF VIDEO SIGNAL IN FIG. 14-7*b* AND *c*. A SINGLE VALUE OF BIAS CANNOT BE CORRECT FOR BOTH SIGNALS.

not black, and the dark scene is reproduced too light.

14-6 Dc Insertion

The basic principle is illustrated in Fig. 14-9*a*. Note that the output voltage e_R is the sum of the ac generator voltage e_a and the battery voltage E_C because they are in series with each other across *R*. The effect of E_C is to shift the entire ac signal to a new axis of -5 V. Although the variations in the ac signal are still the same, they now vary above and below the negative 5-V axis, instead of the original zero axis.

The output e_R is a fluctuating dc voltage of negative polarity. This is negative dc insertion. If the polarity of the battery is reversed, it will

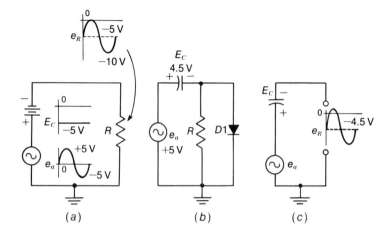

FIGURE 14-9 ILLUSTRATING NEGATIVE DC INSERTION. POSITIVE PEAK OF SIGNAL e_a IS CLAMPED AT ZERO VOLTAGE. (a) BATTERY SUPPLIES DC COMPONENT E_C. (b) DIODE RECTIFIES SIGNAL TO PRODUCE E_C FOR DC COMPONENT. (c) EQUIVALENT CIRCUIT OF E_C IN SERIES WITH e_a.

produce positive dc insertion. Then the variations will be around a new axis of +5 V.

In the waveshape for e_R, notice that the positive peak of the ac signal is kept at the zero level, because of the −5 V added by E_c. This idea of shifting the signal axis to keep one point at a fixed voltage level is clamping. In this case, the positive signal peak is clamped at 0 V. Actually the signal can be clamped at any desired level, depending on the amount of dc voltage inserted.

It is more useful to accomplish the dc insertion for clamping by using a diode to rectify the ac signal, as in Fig. 14-9b. The diode D1 is a rectifier to produce dc voltage proportional to the ac signal. When its anode is driven positive by input signal, the diode conducts to charge C. Between signal peaks when D1 does not conduct, C can discharge through the high resistance of R. With an RC time constant very long compared with the time between positive peaks of the signal, C discharges very slowly, compared with the fast charge of diode current. As a result, C accumulates a negative charge, equal to approximately 90 percent of the positive peak of the ac signal. The resulting dc voltage E_c can

be considered a dc component in series with the signal, as shown in c.

Positive dc restorer. At the grid of the picture tube, the polarity of the dc insertion is positive, to reduce the negative bias for light scenes. For positive dc insertion in Fig. 14-9, the battery would be reversed in a or the diode inverted in b. The results are shown in Fig. 14-10. Here the negative peak of ac signal drives the diode cathode into conduction. The diode current charges C with the cathode side positive. Finally, the waveform here shows the negative peak of the signal clamped at zero in the positive dc restorer circuit.

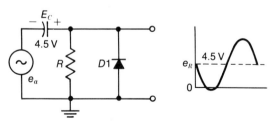

FIGURE 14-10 POSITIVE DC INSERTION WITH DIODE DC RESTORER.

The restored voltage will be clamped for any level of ac signal. If the signal changes, the diode restorer will provide the dc component needed to clamp the signal peak at one level. For television, the composite video signal is rectified to provide the dc component. Once the dc component has been inserted, direct coupling must be used to keep the dc axis of the signal.

Dc restorer time constant. Differences in brightness level correspond to a shift from a light scene to a dark scene, or vice versa. Such variations occur at a rate lower than the frame frequency of 30 Hz. Furthermore, if the brightness of the scene is constant, there is no change at all in the dc level. Then the manual brightness control could set the correct bias without the need for any dc reinsertion. As an example, when a test pattern is transmitted, the dc level of the video signal is constant.

The reinserted dc voltage should vary from frame to frame when the average brightness changes, but not from line to line. Therefore, the time constant of the dc restorer is very long compared with the horizontal-line period, but not with respect to the frame time. A suitable value is 0.03 to 0.1 s. The *RC* time constant includes the coupling capacitor for the diode rectifier and the diode shunt resistor.

14-7 Video Amplifier DC-coupled to Picture Tube

The video output stage is often directly coupled to the picture tube, in order to keep the dc component. See Fig. 14-11. In this case, the dc electrode voltages for the video amplifier affect the bias for the picture tube. In fact, trouble in the video amplifier can cut off the picture tube, resulting in no brightness.

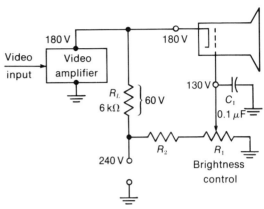

FIGURE 14-11 DC VOLTAGES FOR VIDEO AMPLIFIER DIRECTLY COUPLED TO CATHODE OF PICTURE TUBE.

Normal dc voltages. In Fig. 14-11 the cathode bias for the picture is 50 V. This value is the potential difference between 180 V at the cathode and 130 V at the grid. The cathode voltage is the same 180 V as on the video amplifier because of the direct coupling. This 180 V is the result of the 60-V *IR* drop across R_L from the 240-V supply.

The 130 V at the control grid of the picture tube is set by the brightness control R_1. This is in series with R_2 as a voltage divider across the 240-V supply. The bypass capacitor C_1 is the ac return for video signal coupled from the video amplifier to the picture tube. The cathode-to-grid voltage then is $180 - 130 = 50$ V.

How the picture tube can be cut off by the video amplifier. Suppose that the video amplifier does not conduct. With a tube, the heater can be open. With a transistor, the collector can be open, or there is not enough forward bias. When the video amplifier does not conduct, the result is no *IR* drop across R_L. Then the dc voltage at the output of the video amplifier is 240 V. This value is equal to the supply voltage. The cathode voltage at the picture tube is then 240 instead of 180 V. As a result, the cathode-to-grid voltage is 240 − 130 = 110 V. This bias of 110 V will cut off the picture tube.

When the picture tube is cut off, the result is no brightness. Most often, though, the trouble of no brightness is caused by no high voltage. A useful method of distinguishing between the two troubles is to turn the receiver off, after it has been on for a few minutes. The picture tube usually has some afterglow on the screen just after the receiver is turned off, when there is high voltage on the anode. If the trouble is no brightness, but you can see afterglow, then the picture tube is cut off.

SUMMARY

1. The brightness or average background level of the reproduced picture is determined by the dc bias for the cathode-grid circuit of the picture tube. More negative bias at the grid or more positive bias at the cathode reduces the brightness.
2. The transmitted signal has a dc component to indicate changes in scene brightness. The dc component clamps the blanking at 75 percent carrier level. In the receiver, this dc component is maintained in the rf and IF amplifiers and is present in the video detector output.
3. In a video amplifier with capacitive coupling, the dc component is blocked. With direct coupling, the dc component is preserved. Partial dc coupling is often used for 50 to 80 percent of the dc component.
4. The dc component in the composite video signal is indicated by how far the average axis is from black level.
5. Dc insertion or reinsertion means adding a dc component to an ac signal for the purpose of keeping one peak at a fixed level. This is also clamping.
6. A dc restorer is a diode that rectifies the signal for dc insertion. A positive restorer inserts positive dc voltage; a negative restorer inserts negative dc voltage.
7. The basic purpose of having the dc component of the video signal for the grid-cathode circuit of the picture tube is to clamp the blanking level at cutoff. When black is reproduced correctly, all other light values are correct for different degrees of scene brightness.
8. Receivers have suppressor circuits to blank out retraces, independent of the brightness setting.
9. When the video amplifier is dc-coupled to the picture tube, trouble in the video amplifier can result in cutoff bias for the picture, causing no brightness.

Self-Examination (Answers at back of book)

Answer True of False.

1. The average dc voltage for the cathode-grid circuit of the picture tube determines brightness.
2. When the cathode is 140 V and the grid is at 100 V, the grid bias is −240 V.
3. The correct dc component is always present in the output of the video detector.
4. At the picture tube, the blanking level should be clamped at zero grid-cathode voltage.
5. If you measure the waveshape in Fig. 14-6 with a dc voltmeter, it will read 7.5 V.
6. With an RC coupling circuit, the average dc voltage is across C while the ac component is across R.
7. For the two video signals in Fig. 14-7, the dark signal in c has an average level close to black.
8. In Fig. 14-11, the voltage across the 6-kΩ R_L is 60 V because the current through it equals 10 mA.
9. In Fig. 14-5, if R_1 were increased to 0.8 MΩ to equal R_2, the dc coupling would be 50 percent.
10. When the picture changes from daylight to a night scene, the average axis moves closer to black level.

Essay Questions

1. What is the effect on brightness when the picture tube has (a) too much positive bias at the cathode; (b) too little bias?
2. What is meant by clamping the blanking level at cutoff for the grid-cathode circuit of the picture tube?
3. A video detector is directly coupled to the video output tube, which is RC-coupled to the cathode of the picture tube. Is the dc component of the video signal present at the picture tube? Why?
4. Referring to the diagram in Fig. 14-11, give the function of R_1, R_2, and C_1.
5. In the diagram of Fig. 14-10, give the function of R, C, and D1.
6. Redraw the waveshape in Fig. 14-6 with different amplitudes and show the new average value.
7. Referring to the RC-coupling circuit in Fig. 14-4, how much would a dc voltmeter read across C_c and across R?
8. Define the following: dc insertion, dc restorer, diode clamp, and black setup.

308 BASIC TELEVISION

Problems (Answers to selected problems at back of book)

1. If the waveshape in Fig. 14-6 has a peak value of 30 V and minimum of 6 V, how much will this average value be?
2. Draw an ac sine wave having a peak value of 100 V. Give its amplitudes at 0, 30, 60, 90, 120, 150, and 180°. Calculate the average of these values. How much is the average taken over the complete cycle of 360°?
3. A symmetrical square wave of fluctuating dc voltage varies between 180 and 120 V. (a) How much is the average dc level? (b) If this voltage is applied to an RC coupling circuit, how much will E_C be? (c) How much is the peak-to-peak value of the ac voltage across R?
4. Draw the composite video signal waveshapes for one line all white and another line all black, with 5 percent black setup. Calculate the average value of each signal, using 100 V for peak of sync and 10 V for white.
5. For each of the three waveforms in Fig. 14-12, (a) indicate the average dc level; (b) show the ac form with peak-to-peak amplitude.

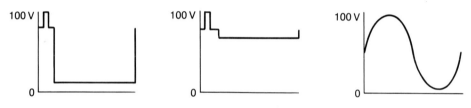

FIGURE 14-12 FOR PROBLEM 5.

15 AGC Circuits

The automatic gain control (AGC) circuit varies the gain of the receiver according to signal strength. Less gain is needed for strong signals. Therefore, the AGC reduces the gain by changing the bias on the IF and rf stages to maintain a relatively constant level for the video detector output. Then the manual contrast control can be in the video amplifier for easy control of the contrast in the picture. The AGC bias is a dc voltage obtained by rectifying the video signal. Practically all receivers use AGC to prevent overload distortion. In Fig. 15-1 the overloaded picture has reversed black-and-white values and is out of sync as a result of too much signal. The details of automatic gain control are explained in the following topics:

15-1 Requirements of the AGC circuit
15-2 Airplane flutter
15-3 AGC bias for tubes
15-4 AGC bias for transistors
15-5 Keying or gating pulses for the AGC rectifier
15-6 AGC circuits
15-7 Twin pentode for AGC and sync separator
15-8 AGC circuit in IC chip
15-9 Transistorized AGC gate and amplifier
15-10 Dc voltages in the AGC circuit
15-11 AGC adjustments
15-12 AGC troubles

310 BASIC TELEVISION

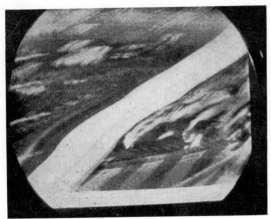

FIGURE 15-1 OVERLOADED PICTURE, OUT OF SYNC WITH BLACK AND WHITE REVERSED. THE DIAGONAL WHITE BAR IS HORIZONTAL BLANKING.

15-1 Requirements of the AGC Circuit

The circuit of Fig. 15-2 illustrates how the AGC bias is used to reduce the receiver gain for strong signals. This idea is the same as automatic volume control (AVC) in radio receivers. With AVC, the purpose is to have relatively constant volume for different signal levels at the antenna. With AGC for picture signal, the purpose is to have relatively constant contrast.

The stages connected to the AGC bias line are usually the rf amplifier with the first and second IF amplifiers. The last IF stage usually does not have AGC because varying the bias with a large signal swing can cause amplitude distortion.

AGC rectifier. In Fig. 15-2, the AGC rectifier stage changes the ac signal input to dc output for the AGC bias. More signal input produces more AGC bias; less signal means less bias. The ac input to the AGC rectifier can be either IF signal or video signal. The load resistor for the AGC rectifier is R_L, which produces dc voltage output proportional to the ac signal input.

The AGC bias voltage has the polarity needed to reduce the gain of the IF and rf stages connected to the AGC line. For a vacuum tube amplifier the AGC bias is negative voltage applied to the control grid. The polarity of the AGC bias for transistors can be either positive or negative, depending on the circuit. In all cases, though, more signal produces more AGC bias to reduce the gain of the receiver.

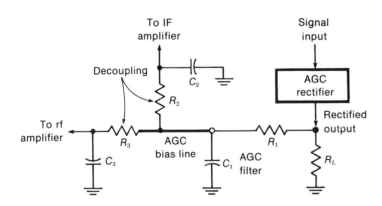

FIGURE 15-2 BASIC REQUIREMENTS OF THE AGC CIRCUIT.

AGC filter time constant. The rectified voltage across the AGC load resistor R_L must be filtered to remove the signal variations, since a steady dc voltage is needed for bias. This is the function of the AGC filter R_1C_1 in Fig. 15-2. The R_1C_1 time constant* is about 0.2 s. A shorter time constant will not filter out the low-frequency variations in the rectified signal. Specifically, the vertical sync voltage can vary the AGC bias, which causes bend in the picture. Too long a time constant will not allow the AGC bias to change fast enough when the receiver is tuned to stations having different signal strengths. In addition, a long time constant will not remove the variations of picture intensity caused by fading of the signal when an airplane is flying nearby.

With tubes, a typical AGC filter has 0.1 μF for C_1 and 2 MΩ for R_1. For transistors, typical values are 20 kΩ for R_1 and 10 μF for C_1. Note that 10 μF requires an electrolytic capacitor for the AGC filter in transistor circuits.

The AGC line. Each stage controlled by AGC has a return path to the AGC line for bias. Then the voltage on the AGC line varies the bias on the controlled stages. The filtered voltage across C_1 is the source of AGC bias to be distributed by the AGC line.

AGC decoupling filters. R_2C_2 and R_3C_3 in Fig. 15-2 isolate each amplifier stage from the common AGC line. The isolation results from the series resistance of R_2 or R_3. This resistance reduces feedback between stages through the AGC filter capacitor as a common impedance. However, each decoupling resistor need its bypass capacitor C_2 or C_3. The low capacitive reactance completes the ac signal path to ground in the amplifier circuit, without shorting the dc bias voltage from the AGC line.

AGC action. The AGC circuit provides relatively constant output voltage from the video detector, with wide variations in antenna signal. As an example, suppose that the rf tuner and IF section have an overall voltage gain of 100,000 for a weak antenna signal of 20 μV, resulting in 2 V of video signal at the detector output. Now assume that a much stronger antenna signal of 20 mV produces enough AGC bias to reduce the overall gain to 100. Note that 20 mV is 20,000 μV, or 0.02 V. The video detector output then is still 2 V, equal to the gain of 100 × 0.02 V. The AGC action reduced the voltage gain from 100,000 to 100. This ratio of 100,000/100 for voltage gain corresponds to 60 dB* of AGC action in reducing the receiver gain.

Types of AGC circuits. The circuit in Fig. 15-2 is called simple AGC. In *delayed AGC*, there is no AGC bias at all unless the antenna signal is above a specific level, generally about 1 mV. The purpose is to prevent any reduction in gain by AGC action when the signal is weak. Usually, the AGC bias for the rf amplifier is delayed. The result is a better signal-to-noise ratio, for weak signal, to minimize snow in the picture.

In *amplified AGC*, the AGC voltage is increased so that a small change in signal level can produce a large change in bias. An amplifier for the AGC bias must have dc coupling for the input and output. Also, the polarity of the AGC bias can be inverted by the AGC amplifier. In *keyed AGC* or *gated AGC*, the AGC rectifier is keyed on by horizontal flyback pulses to conduct only during retrace time. The advantage is that noise pulses in the signal have little effect on the AGC bias voltage.

---

*See Appendix D on *RC* Time Constants.

*See Sec. 26-7 on Decibels.

15-2 Airplane Flutter

This is a rise and fall of picture intensity, making the picture fade in and out, when airplanes are flying nearby. The cause is fading of the picture carrier signal as its amplitude increases and decreases because of the moving airplane.

For the carrier frequencies used in television broadcasting, the airplane is several wavelengths long and can act as an antenna, especially for the UHF channels. Since it is in the field of the radiated wave, the airplane intercepts some signal, current flows as on an antenna, and signal is reradiated. The reradiated signal may aid or oppose the original transmitted signal picked up by the receiving antenna. Furthermore, the phase relations change continuously as the airplane moves. The resultant fading occurs at a rate of about 10 to 25 fluctuations per second. This is the beat frequency between the direct and reflected signals.

In addition, the reradiated signal can produce a ghost in the picture. The sound may also be distorted and pulsate in volume. If the airplane is transmitting its own radio signal, this rf interference can cause diagonal bars in the picture.

The effects of fading in the antenna signal caused by airplane flutter can be practically eliminated by the picture AGC circuit, when the filter time constant is not too long. The required time constant is about 0.2 s or less.

15-3 AGC Bias for Tubes

The grid-plate transconductance (g_m) is reduced as the negative bias increases. As an example, g_m can be reduced from 10,000 μmhos for maximum gain to practically zero for minimum gain. For a pentode IF amplifier, the gain of the stage varies directly with the g_m of the tube. As a result, the gain is reduced as the negative AGC bias makes the control-grid voltage more negative.

Series feed for AGC bias. In Fig. 15-3a, the AGC bias voltage is in series with the secondary L_S of T_1 which supplies the IF signal to the control grid. Effectively, the AGC circuit is in the return path of L_S to chassis ground. Then the AGC line has a dc path to the control grid to vary its bias. Since there is no grid current and L_S has practically zero resistance, the AGC bias is the dc grid voltage. In addition, R_K provides about 1 V of cathode bias, with its bypass C_K.

For the tuned circuit, the stray capacitance C_S tunes with L_S for resonance at the IF signal frequencies. In the decoupling filter, R_1 isolates the low side of the tuned circuit from the AGC line. The bypass C_1 returns L_S to chassis ground. Then the IF signal has a return path through C_1 to cathode, without the series resistance of R_1.

Shunt feed for AGC bias. In Fig. 15-3b, the AGC voltage is connected to the control grid through R_2 to supply grid bias. In this case, R_2 is used to prevent shorting the IF signal of the tuned circuit. Here C_2 couples the IF signal and blocks the AGC voltage from a dc short to ground through L_S. Again, R_K and C_K provide cathode voltage for a minimum bias.

Comparison of series and shunt feed. Note that with the shunt feed in Fig. 15-3b, R_2 in the AGC line is across the tuned circuit. Then R_2 affects the damping and bandwidth for the IF signal. With the series feed in Fig. 15-3a, C_1 bypasses the AGC isolating resistor R_1, but actually C_1 is part of the IF tuned circuit with L_S and C_S. An open in C_1 will result in practically no IF signal because of the high series resistance of R_1.

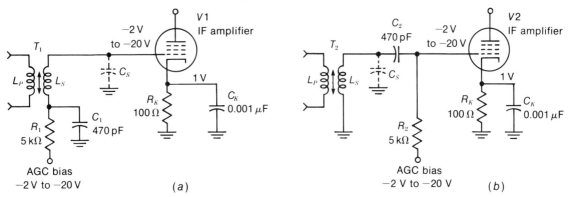

FIGURE 15-3 HOW CONTROL GRID IS RETURNED TO AGC LINE FOR NEGATIVE BIAS.
(a) SERIES FEED. (b) SHUNT FEED.

15-4 AGC Bias for Transistors

For a transistor its beta (β) determines the gain, in a common-emitter (CE) amplifier. The beta is the ratio of a change in collector current to a change in base current, or $\beta = \Delta I_C / \Delta I_B$. As an example, β may be 200 for maximum gain, which is reduced by AGC bias to practically zero for minimum gain. However, there are two opposite methods of using the AGC bias to reduce the beta. These are reverse AGC and forward AGC for transistors.

Reverse AGC. In this method, the transistor gain is decreased by reducing the forward bias at the base, toward cutoff. For instance, more negative AGC voltage at the base of an NPN transistor is reverse AGC, as the positive forward base bias is reduced. Remember that a transistor is cut off with zero forward voltage. As the base-emitter bias approaches cutoff, the changes in I_C are smaller for changes in I_B, reducing the beta.

Actually, reverse AGC corresponds to the idea of increasing the negative grid bias toward cutoff for a tube. However, the polarity for reverse AGC bias can be either negative or positive to approach cutoff for transistors. Reverse AGC voltage is negative for the base of an NPN transistor but positive for a PNP transistor.

Forward AGC. In this method, the transistor gain is decreased by increasing the forward bias at the base, toward saturation. For instance, more positive voltage at the base of an NPN transistor is forward AGC. It should be noted that rf and IF stages controlled by forward AGC generally use special transistors with the required beta characteristic for less gain with more forward bias. This is accomplished by a special shape in the emitter construction.

The polarity of forward AGC at the base can be positive for NPN transistors or negative for PNP transistors. Practically all small-signal transistors are NPN, however, so that we can think of positive voltage at the base as forward AGC in a CE amplifier. Furthermore, forward AGC is generally used more than reverse AGC. The reason is that operation near cutoff with reverse AGC makes the receiver more susceptible to overload and cross-modulation distortion on strong signals.

Forward AGC bias with series feed. See the IF amplifier in Fig. 15-4. When the AGC voltage becomes more positive, it increases the forward bias at the base of the NPN transistor. The R_1R_2 voltage divider provides a fixed positive base bias, from the collector supply V_{CC}. R_3 is the AGC decoupling resistor, with its bypass C_1.

The dc voltage at the junction of R_1, R_2, and R_3 is the base voltage, as the IF transformer has very little resistance. This 2.6 V at the base is the dc voltage to chassis ground. Also, the 2 V at the emitter is to ground. This voltage is the IR drop across R_E produced by the emitter current. Therefore, the actual base-emitter bias is $2.6 - 2 = 0.6$ V as a typical forward voltage for an NPN silicon transistor. When the AGC voltage increases with more signal, the result is more forward bias toward saturation to decrease the gain.

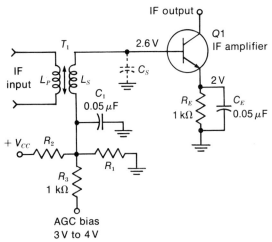

FIGURE 15-4 HOW BASE OF NPN TRANSISTOR IS RETURNED TO AGC LINE. BIAS VOLTAGE HERE IS POSITIVE FOR FORWARD AGC.

Typical changes in AGC bias. When the AGC bias changes, the actual change in base voltage (V_B) is generally less than 0.1 V. However, the change in base current (I_B) can be appreciable. As an example, a change in V_B from 0.6 to 0.65 V can change I_B from 100 to 200 μA. Furthermore, this increase in I_B raises the average dc collector current (I_C) and emitter current (I_E). The change in I_E can be seen by measuring the emitter dc voltage across R_E. Similarly, when the collector circuit has a series decoupling resistor, its dc voltage can be measured to see the change in I_C.

15-5 Keying or Gating Pulses for the AGC Rectifier

Most AGC circuits use flyback pulses from the horizontal output circuit for keying or gating the AGC rectifier into conduction. The video signal is also coupled into the rectifier to provide AGC voltage proportional to signal strength. However, the AGC stage is generally biased to cutoff so that it conducts only for the short time the pulse is applied. Then the signal is rectified only when the blanking and sync pulses are on. As shown in Fig. 15-5, the time of the flyback pulses corresponds to the time of sync and blanking, assuming the picture is in sync. The gating function means that both inputs must be on at the same time to produce output.

The advantage of keyed AGC is that the rectified voltage is free from noise pulses that can occur between the horizontal sync pulses. In addition, there is no error in AGC bias from the vertical sync pulses, because their greater width is not measured by the keyed AGC rectifier. Therefore, the time constant of the AGC filter can be shorter for a faster response to airplane flutter. Finally, the AGC rectifier can be a peak rectifier, producing dc output only for the sync peaks of the ac signal input. This factor helps in

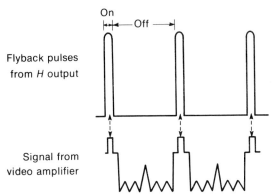

FIGURE 15-5 KEYING PULSES AT HORIZONTAL SYNC RATE FOR AGC CIRCUIT.

preventing the AGC bias from changing with different dc levels of scene brightness in the video signal. The AGC bias should vary only with the strength of rf carrier signal on the antenna.

Keying a triode AGC tube. See Fig. 15-6. Video signal is directly coupled from the video output circuit to the grid of the AGC tube. The dc voltage here is +115 V because of the dc coupling. The grid-cathode bias is 115V − 140 V = −25 V. Without signal, then, the −25 V biases AGC tube beyond cutoff. Furthermore, the plate does not have any positive dc supply voltage.

However, the horizontal flyback pulses drive the plate positive during the sync pulse time. Therefore, the tube conducts plate current when the video signal with 80-V peak-to-peak amplitude makes the grid more positive than cutoff. The horizontal sync pulses at the positive peak of the video signal drive the grid voltage to zero so that the tube can conduct as the plate is pulsed positive at this time.

The plate circuit of the AGC tube in Fig. 15-6 must produce negative voltage for the AGC bias line. This is why there is no B+ voltage for the plate. Instead, the plate is pulsed positive by the flyback pulses with a peak amplitude of 400 V to produce plate current. Furthermore, the plate current produces negative AGC voltage by charging C_2. When plate current flows, it charges C_2 with the plate side negative. The path for charging current includes the cathode-to-plate circuit in the tube, C_2 and the AGC winding in the horizontal output transformer, with electron flow returning to the cathode.

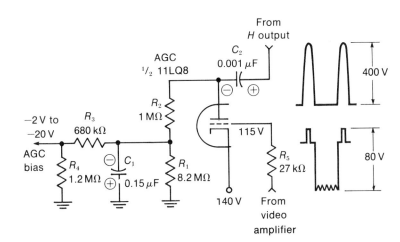

FIGURE 15-6 HORIZONTAL FLYBACK PULSES APPLIED TO PLATE OF TRIODE FOR KEYED AGC.

Between pulses, when the tube does not conduct, C_2 can discharge through R_2 to charge the AGC filter C_1. The time constant of the AGC filter with C_1, R_1, R_3, and R_4 is much longer than the 63.5 μs between pulses. Therefore, the voltage across C_1 is a relatively steady dc voltage for the AGC bias.

The function of R_5 in Fig. 15-6 is decoupling to isolate the AGC tube from the video amplifier supplying the signal input. R_2 in the plate circuit isolates the AGC circuit from the horizontal output circuit supplying the flyback pulses. R_1 and C_1 form the AGC filter. Finally, R_3 and R_4 form a voltage divider, with the voltage across R_4 the filtered AGC voltage for the bias line.

Keying a transistor AGC stage. See Fig. 15-7. The idea of keyed AGC here is the same as for the triode in Fig. 15-6, but smaller voltage amplitudes are used for transistors. The collector is pulsed, instead of the plate. Furthermore, the required polarity of the flyback pulses is negative for the PNP transistor in Fig. 15-7. Note that the base voltage is actually past cutoff. The base-emitter voltage of $1.8 - 1.6 = 0.2$ V is actually a small reverse bias because it is positive voltage for the base of a PNP transistor.

However, the video signal with 3-V peak-to-peak amplitude drives the transistor into conduction when the collector is pulsed on. The video signal input here has negative sync polarity for the base of the PNP transistor. Then the negative sync voltage is in the forward direction to produce collector current. The polarity of the flyback pulses is also negative to be reverse collector voltage for a PNP transistor.

In Fig. 15-7, the 100-V flyback pulse is supplied by L_1, which is a separate winding on the horizontal output transformer. With this AGC winding, either polarity of flyback pulses can easily be obtained. In addition, the separate AGC winding isolates the AGC circuit from the horizontal output circuit. This method of obtaining the flyback pulses can be used with transistors or tubes.

The diode D1 is necessary for isolation in the transistor AGC stage because the AGC bias voltage is the opposite polarity from the required collector voltage. In Fig. 15-7, the AGC voltage is positive, which would put forward voltage on the collector of the PNP transistor. However, D1 isolates the positive AGC voltage. Remember that the arrow on the diode is the anode, while the bar is the cathode. Therefore, D1 can couple the negative flyback pulses from cathode to anode for the collector. However, the positive AGC voltage cannot produce current from cathode to anode in the diode. As a result, D1 prevents C_1 from discharging through the AGC transistor.

It may be interesting to note that an NPN transistor could be used in Fig. 15-7. Then all the polarities would be reversed. The flyback pulses at the collector would be positive. The video signal would have positive sync polarity.

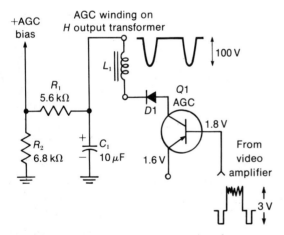

FIGURE 15-7 HORIZONTAL KEYING PULSES AT COLLECTOR OF AGC STAGE. NEGATIVE PULSES ARE USED HERE FOR PNP TRANSISTOR.

The base-emitter voltage for reverse bias would be negative. Finally, the AGC voltage across C_1 in the collector circuit would be negative, with an NPN transistor for the AGC stage.

15-6 AGC Circuits

Typical circuits are shown in Figs. 15-6 to 15-11. Either the modulated picture IF carrier signal or the detected video signal can be rectified by the AGC stage to supply the AGC bias voltage. In most receivers, however, the AGC circuit uses signal from the video amplifier because the higher signal level allows more control by the AGC bias.

With video signal input, it must have dc coupling to the AGC stage in order to preserve the dc component. Then the pedestals of the video signal are in line so that a peak rectifier can measure the sync voltage level to measure the carrier strength correctly.

It should be noted that the picture IF signal has the dc component of the video signal in the modulation. The simplest method of using the IF signal for AGC is to take the AGC bias from the video detector load resistor, as shown in Fig. 15-8. With integrated circuits, the AGC circuit can be part of the IC chip for the IF amplifier.

The stages controlled by AGC bias are generally the rf amplifier in the tuner and one or two IF stages. The last IF stage usually does not have AGC bias. This IF stage has a relatively high signal amplitude, which can easily be distorted by a change in bias. Also, the effect of the AGC is proportional to signal gain from the controlled stage to the AGC rectifier. Therefore, AGC on the earlier stages has more control on the overall receiver gain.

Separate rf and IF bias lines. There are usually separate bias lines for the IF amplifier and to the rf amplifier because the needs are different. With weak signal, the rf bias should allow maximum gain, for a good signal-to-noise ratio and minimum snow in the picture. The snow is produced by internal noise generated in the rf amplifier and mixer stages. The only way to eliminate the snow is to have a large enough signal from the rf tuner, before the IF amplification.

With very strong signal, cross modulation can be produced in the mixer stage, because it is a nonlinear amplifier. Cross modulation results when the modulation of a weak signal is impressed on the carrier of a strong signal. The result of cross modulation is two signals that produce two pictures on the screen. Therefore, with a strong signal the AGC bias must reduce the gain for the rf tuner to minimum, in order to prevent cross modulation in the mixer.

These requirements for rf and IF gain can be summarized as follows:

1. Weak signal—up to about 0.1 mV. Rf gain is maximum for minimum snow. IF gain is the amount needed for constant video detector output.
2. Normal signal—about 0.1 to 100 mV. A typical value is 1 mV. Rf and IF gain are reduced to provide constant output from the video detector.
3. Very strong signal—above 100 mV. Rf gain is minimum to prevent cross modulation, and IF gain is minimum to maintain constant video detector output.

AGC delay. This delay is in voltage, not in time. The purpose is to hold off any AGC bias until the signal is strong enough for a picture without snow. The method is to keep the delay stage cut off until a specific amount of AGC voltage is available. Usually, when an AGC delay stage is used, it is in the rf bias line for the tuner.

Bias clamp. This means clamping, or holding, the AGC bias at a specific voltage level. Generally, a diode is used for the clamp circuit. When the diode conducts, it connects the AGC bias line to a fixed voltage source, preventing any increase in the AGC bias.

Noise inverter. Also called a *noise gate*, this circuit inverts the polarity of noise pulses by reversing the polarity of its video signal. The purpose is to cancel noise pulses in the video signal to the AGC stage. The noise inverter is also used for the sync separator circuit. As an example, the circuit in Fig. 15-9 combines the AGC stage and sync separator, with a noise inverter circuit common to both.

Triode AGC circuits. Referring back to the triode tube in Fig. 15-6 and the transistor in Fig. 15-7, these circuits illustrate a common type of keyed AGC. This stage is generally called the *AGC gate* or *AGC keyer*. The gating means that the stage is keyed on to conduct only during the horizontal flyback pulses.

AGC bias from the video detector output. Refer to Fig. 15-8. Actually, $D1$ is the video detector, but the dc output voltage across the load resistor R_L is proportional to signal strength. Therefore, the voltage across R_L is filtered and used for AGC bias. R_1 and R_2 isolate the AGC filter capacitors C_1 and C_2 from the detector load R_L. The combination of video detector and AGC rectifier is seldom used because the diode cannot be a peak rectifier to measure signal strength correctly. Also, the video detector cannot be keyed for the AGC function. However, this circuit illustrates a method of rectifying the IF signal for the AGC voltage.

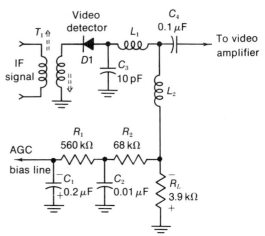

FIGURE 15-8 AGC BIAS FROM VIDEO DETECTOR LOAD RESISTOR R_L.

15-7 Twin Pentode for AGC and Sync Separator

See Fig. 15-9. The special tube for this application has one cathode and one screen grid common to both sharp-cutoff pentodes. Also common is the first control-grid pin 7, which serves as a noise gate. The AGC stage at the left has a separate grid connection at pin 6 for video signal input, with the AGC output at plate pin 3. The pentode at the right has its own grid pin 9 for video input and plate pin 8 for sync output. This sync separator is explained in more detail with Fig. 16-29 in Chap. 16 on Sync Circuits. The twin pentode tubes generally used for this combined sync and AGC circuit are the 4HS8, 6HS8, 3BU8, and 6BU8.

The plate of the AGC section is keyed into conduction by horizontal flyback pulses coupled through C_{416}. When plate current flows, it charges C_{419} through R_{422} and C_{421} through R_{423}. The plate side is negative for negative AGC bias

AGC CIRCUITS 319

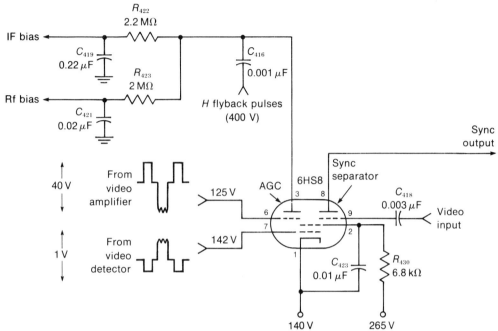

FIGURE 15-9 TWIN PENTODE TUBE FOR AGC AND SYNC SEPARATOR. *(FROM ADMIRAL CHASSIS K16)*

to the rf and IF amplifier tubes. Separate bias lines are used to isolate the rf and IF stages.

The AGC grid at pin 6 has the dc voltage of 125 V because of direct coupling from the video amplifier. The cathode is at 140 V. Therefore the grid-cathode voltage equals 125 $-140 = -15$ V. This bias is more negative than cutoff. With video signal input of 40 V peak-to-peak, though, the positive sync peaks produce plate current when the plate is keyed on by horizontal flyback pulses. These keying pulses have a peak amplitude of 400 V. As a result, the AGC output in the plate circuit is a measure of signal strength.

The noise inverter grid at pin 7 has composite video signal of only 1 V, with negative sync polarity. Note this is the opposite of the sync polarity at the AGC grid pin 6. The dc voltage at the noise inverter grid is set 2 V more positive than the cathode, so that the negative sync cannot cut off plate current. However, noise pulses of high amplitude can drive the grid more negative than cutoff. Then noise pulses in the video signal cannot produce output in either the AGC stage or the sync separator.

15-8 AGC Circuit in IC Chip

See Fig. 15-10. The integrated circuit labeled IC1A here is one-half the chip that includes the picture IF amplifier, video detector, and keyed

320 BASIC TELEVISION

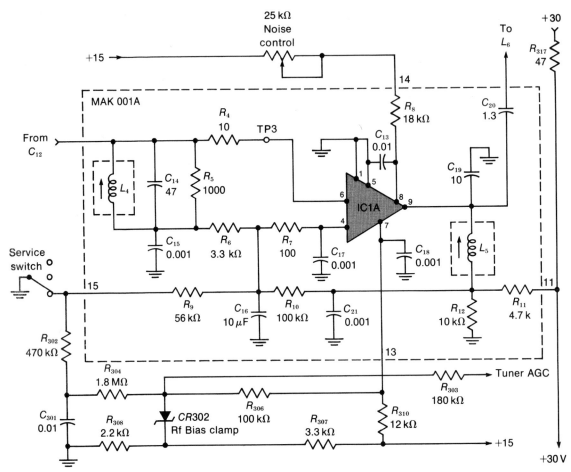

FIGURE 15-10 INTEGRATED CIRCUIT FOR AGC STAGE. *(FROM RCA CHASSIS CTC 49)*

AGC circuit. The IC1B half, not shown, has the last IF stage and video preamplifier, with the input of horizontal flyback pulses needed for the AGC.

We can consider pin 4 as the collector of an AGC transistor in the IC chip. The 100-Ω R_7 is the load resistor in series with the 56-kΩ R_9 to ground, through the service switch at the left. The 10-μF C_{10} is the AGC filter capacitor. Collector supply voltage is obtained through the 100-kΩ R_{10}.

The series feed for the IF AGC is from pin 4, through R_7 and R_6, to the low end of the IF coil L_4 and back into the IC chip through R_4 to terminal 6 for the IF input. The rf AGC line is out at pin 7 of the chip, through R_{306} and R_{303} to the rf tuner. Note that $CR302$ is a zener diode, with the function of clamping the rf bias voltage. The noise control at the top of the diagram adjusts the rf AGC bias for minimum snow with weak signal.

The normal-service switch is shown in the position for a normal picture with ground at the junction of R_9 and R_{302}. When the ground point is removed, the AGC bias cuts off the IF amplifier for a clean raster without snow. This service position is for purity adjustments on the color picture tube.

15-9 Transistorized AGC Gate and Amplifier

Figure 15-11 shows the AGC circuit of a transistorized receiver, with the details of AGC amplification and delay. Starting with the AGC gate Q3 at the left, the video input signal is rectified to produce AGC output voltage across the filter capacitor C_2. Note that Q3 is a PNP transistor. Then the video input signal has negative sync

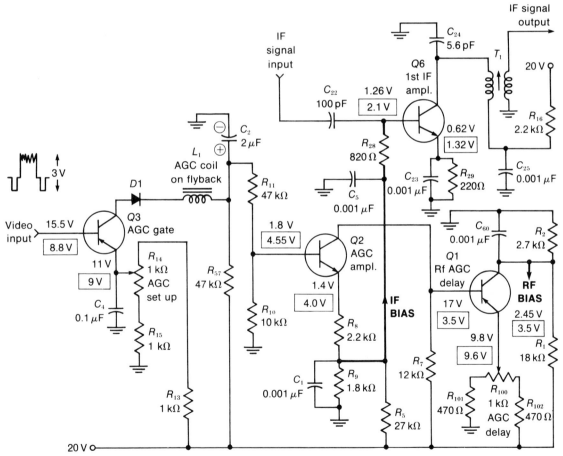

FIGURE 15-11 TRANSISTORIZED AGC GATE AND AMPLIFIER WITH DELAY STAGE FOR RF BIAS. DC VOLTAGES IN BOXES ARE WITH SIGNAL INPUT TO RECEIVER. *(FROM MOTOROLA CHASSIS CTV5-S5)*

polarity, the keying pulses are negative at the collector, and the AGC output voltage is positive.

The AGC bias voltage from Q3 is dc-coupled to the base of the AGC amplifier Q2. This is an NPN transistor. The positive input at the base is inverted to negative voltage at the collector output for the next stage Q1. However, the AGC voltage at the emitter of Q2 has the same positive polarity as the input, since the emitter voltage follows the base voltage. This positive AGC voltage at the junction of R_8 and R_9 in the emitter circuit of Q2 is used for AGC bias on the IF amplifier Q6. Since Q6 is an NPN transistor, the positive bias at its base is forward AGC.

The AGC amplifier Q2 also serves as a buffer stage to separate the IF bias line at the emitter of Q2 from the rf bias line at the collector of Q1. The output at the collector of Q2 is negative AGC voltage to the base of the rf AGC delay stage Q1. Note that Q1 is a PNP transistor. For either PNP or NPN types, though, the polarity at the collector is inverted from the input at the base. Therefore, the negative AGC bias into Q2 is inverted to positive at the collector for the rf bias line. The AGC delay control R_{100} in the emitter circuit is set to keep this stage cut off until the signal can produce enough AGC voltage at the base to make Q1 conduct.

The AGC gate circuit. Now we can consider the function of each component in Fig. 15-11. For the AGC gate Q3, which produces the AGC voltage, R_{57} is the collector load resistor in series with the AGC coil L_1 and D1 as a blocking diode. L_1 supplies negative flyback pulses with a peak amplitude of 40 V from the horizontal output transformer. The 2-μF electrolytic capacitor C_2 is the AGC filter with R_{10} and R_{11}. The positive AGC voltage is blocked by D1 from the collector of Q3, which needs negative collector voltage because it is a PNP transistor. Most important, the diode D1 prevents the AGC filter capacitor C_2 from discharging through the AGC transistor Q3. In the emitter circuit of Q3, the voltage divider of R_{13}, R_{14}, and R_{15} provides the emitter bias. R_{14} sets this bias for the desired amount of AGC voltage output. C_4 is an rf bypass capacitor.

The AGC amplifier circuit. For the AGC amplifier Q2, the collector load resistor is R_7. In the emitter circuit, the voltage across R_9 supplies the AGC bias voltage for the IF bias line. Although R_9 is part of a voltage divider with R_5 from the 20-V supply, the voltage across R_9 varies with the emitter current through R_9 in series with R_8. More conduction in Q2 produces more AGC voltage across R_9, with positive polarity in the emitter circuit.

The AGC delay stage. The negative AGC voltage at the collector of Q2 is dc-coupled to the rf AGC delay stage Q1. In the collector circuit of Q1, the voltage divider of R_1 and R_2 provides collector voltage from the 20-V supply. C_{60} is an rf bypass capacitor. The positive AGC voltage at the collector feeds the rf bias line to the tuner. In the emitter circuit of Q1, the point of conduction is set by the AGC delay control R_{100}. This adjusts the emitter voltage provided by the voltage divider of R_{102}, R_{100}, and R_{101} from the 20-V supply line at the bottom of the diagram.

15-10 DC Voltages in the AGC Circuit

The AGC bias itself is a dc voltage, and the entire AGC circuit is a network of dc values from input to output.

Dc level of the video signal. The video signal to be rectified must have its dc component. Otherwise, the AGC bias would vary with scene brightness, instead of indicating the strength of carrier signal at the antenna. For this reason, the video signal has dc coupling, without any series capacitors, from the video amplifier to the AGC stage.

Dc voltage on the AGC bias line. The dc bias on the AGC line cannot have any series capacitors in the feed line to the rf and IF controlled stages. When shunt feed is used for the tuned amplifier, this stage has a series capacitor to couple the signal and block the AGC bias from shorting through the rf or IF coil. With series feed for the AGC bias, the bypass capacitor in the decoupling filter completes the tuned circuit for the rf or IF stage, without shorting the dc voltage on the bias line.

Dc voltages in the AGC amplifier. When an AGC amplifier is used, this stage must be a dc amplifier, without any series coupling capacitors. The AGC amplifier can provide a reversal of polarity for the AGC bias, in addition to the amplification. Also, the AGC amplifier often serves as a buffer between AGC takeoff points in the output and input for rf and IF bias voltages.

Dc coupling in the controlled stages. In many circuits the AGC bias is on an rf or IF amplifier that has direct coupling to the next stage. Then the dc connection puts the AGC on both amplifiers. In fact, the AGC bias is amplified and usually inverted for the next amplifier.

Typical AGC voltages for transistors. We can use Fig. 15-11 to see the dc voltages in a transistorized AGC circuit. For Q3 the base is at 8.8 V with signal. Note that the values in the boxes are with signal. The 8.8 V at the base of Q3 is from the dc coupling to the video amplifier. The emitter of Q3 is at 9 V, from the 20-V supply. The actual base-emitter bias V_{BE} then is 8.8 V − 9.0 V, which equals −0.2 V. This value is normal forward bias on the PNP germanium transistor for Q3.

At the base of Q2, the AGC bias voltage is 4.55 V. The emitter is at 4.0 V. Then V_{BE} is 4.55 − 4.0 = 0.55 V. This is normal forward bias for the silicon NPN transistor for Q2.

Notice that the emitter voltage of Q2 changes from 1.4 V without signal to 4.0 V with signal. This change of 4.0 V − 1.4 V is an increase of 2.6 V in positive AGC bias.

At the rf AGC delay stage Q1, note that the collector-emitter voltage V_{CE} is actually negative for this PNP transistor, although the collector-to-ground voltage is positive. For the voltage values with signal, V_{CE} is 3.5 − 9.6 V = −6.1 V. The positive 3.5 V at the collector provides forward AGC voltage for the rf amplifier on the tuner.

15-11 AGC Adjustments

When the receiver has controls for adjusting the amount of AGC action, they have a big effect on the picture and sound because the AGC bias sets the overall rf and IF gain. These controls are provided to adjust the amount of AGC for strong signals or weak signals in different areas.

With tubes, the AGC level control usually can be varied to cut off the picture and sound at one end or produce an overloaded picture (Fig. 15-1) at the opposite extreme. This control can be set by adjusting for an overloaded picture, on the strongest station, and then backing off a little for a normal picture with good contrast.

With transistors, the AGC level control may not be able to produce an overloaded picture. The corresponding idea is to adjust for maximum snow in the picture and then to back off a little. To set the AGC level exactly, the service notes for the receiver give a specific value of AGC bias needed for the signal level of 1 V peak-to-peak out of the video detector.

When the receiver has two AGC adjustments, usually one is for AGC level and the other for rf delay. The AGC level is adjusted with a strong signal, just below overload. However, the rf delay is adjusted with a weak signal, for the best signal-to-noise ratio with minimum snow in the picture.

15-12 AGC Troubles

The AGC circuit is a common cause of troubles in the picture, while the raster is normal. Too much AGC bias reduces the receiver gain, causing a weak picture. This weak picture is without snow. A weak antenna signal can also produce a weak picture but with the snow caused by a poor signal-to-noise ratio in the rf tuner.

When the AGC bias is great enough to cut off an rf or IF amplifier, the result is no picture, with a blank raster. The sound is also cut off because of the intercarrier sound circuit. Troubles in the AGC amplifier can cause too much AGC bias.

For the opposite case, when there is not enough AGC bias, the receiver has too much gain. Then the rf mixer, an IF stage, or the video amplifier can have too much signal for the bias. The result is overload distortion. This term just means that the amplifier has amplitude distortion caused by too much signal input, by too little bias, or by both.

Overloaded picture. See Fig. 15-1. This picture has reversed black-and-white values and is out of sync. The white diagonal bar is horizontal blanking, which should be black and at the side. Usually, the picture will roll vertically also. The picture is out of sync horizontally and vertically because the sync pulses are either compressed or lost completely with the severe amplitude distortion.

The reversed black-and-white values are caused by reversal of the polarity of modulation for the video signal as a result of rectification in an overloaded amplifier, which is usually the last IF stage. The buzz in the sound is caused by the vertical blanking pulses of the video signal modulating the sound signal in an overloaded IF stage common to both picture and sound. The AGC troubles of too much or too little bias are summarized in Table 15-1.

It should be noted that too little AGC bias can also cause high contrast and bend in the picture, just before the point of a reversed picture out of sync. Furthermore, if the receiver tends to overload on normal signal levels but has a normal picture with a very weak signal, this can be a clue to the trouble of not enough AGC bias.

Localizing picture troubles to the AGC circuit. A test of the AGC action can be made by shorting the AGC bias to ground, temporarily, either on the AGC bias line or the AGC filter capacitor. The result is zero AGC bias. The result of zero AGC bias with tubes is an overloaded picture. Similarly, with reverse AGC on

TABLE 15-1 AGC TROUBLES

TROUBLE	EFFECT
Too little AGC bias	Overload distortion, with reversed picture out of sync and buzz in the sound.
Too much AGC bias	No sound and no picture, with normal raster.

transistors the result is a stronger picture or overload. With forward AGC on transistors, though, the result of zero AGC bias is less picture or no picture. In all cases, though, shorting the AGC bias should have a big effect on the picture. If there is no effect, the trouble is either a short in the AGC line or a defect in the rf and IF amplifiers.

Battery bias box. A more general method of localizing troubles to the AGC circuit requires the use of a bias box, as shown in Fig. 15-12. This provides an adjustable bias voltage to take the place of the AGC bias. Connect one output lead directly to the AGC line and the opposite lead to chassis ground. The low resistance of the battery effectively shorts the high-resistance AGC circuit, so that it need not be disconnected. If the picture and sound are normal when the battery supplies the bias, but not with the AGC bias, the trouble must be in the AGC circuit. The bias box is also useful for supplying the fixed bias that is necessary in rf and IF alignment.

The polarity of the bias box is shown with negative output in Fig. 15-12, for negative grid bias on tubes. With transistors, though, the required polarity may be either negative or positive. For positive bias voltage, connect the positive output lead to the AGC line and ground the negative lead.

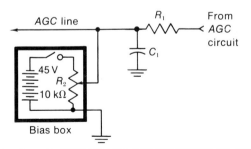

FIGURE 15-12 BATTERY BIAS BOX CONNECTED TO AGC LINE FOR MANUAL CONTROL OF BIAS.

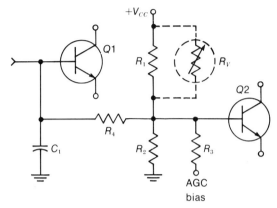

FIGURE 15-13 VARYING THE TRANSISTOR IF BIAS MANUALLY WITH R_V, TO CHECK FOR AGC TROUBLE.

Manual control of bias line for transistors. Figure 15-13 shows Q1 and Q2 dc-coupled at the base to share the AGC bias. The function of the AGC voltage here is to vary the positive base bias. Positive forward bias at the base is provided by the R_1R_2 divider from the supply voltage V_{CC}. If you connect a variable resistor R_V of about 25 kΩ across R_1, then R_V can be used to control the bias manually. This method does not require a bias box. If a normal picture can be obtained with R_V but not with the AGC alone, this shows trouble in the AGC bias.

AGC filter capacitor. This capacitor in transistorized AGC circuits is an electrolytic one. A short causes the trouble of zero AGC bias. If the capacitor is leaky, with insufficient capacitance, the result will be too little AGC bias.

AGC blocking diode. An example of an AGC blocking diode is D1 in Fig. 15-7 for transistor Q3. If the diode is open, there will be no AGC voltage on the AGC bias line. If the diode shorts, there will be forward voltage on the AGC stage, making it conduct too much. The result can be either a short on the AGC line or the AGC transistor becoming defective.

Troubles in the AGC decoupling filters. With series feed for the AGC line to the rf or IF amplifier, the bypass capacitor in the decoupling filter is part of the tuned circuit for signal. An example is C_1 in Fig. 15-4. If C_1 is open, the tuned circuit with L_S for IF signal will be practically open for ac signal, because of the resistance of R_1. The result is very weak signal. If C_1 is shorted, the AGC bias on Q1 will be zero.

SUMMARY

1. The AGC circuit rectifies the signal to produce dc bias proportional to signal strength so that the AGC bias can control receiver gain. More signal produces more AGC bias to reduce the rf and IF gain for constant output at the video detector.
2. The AGC filter removes the signal frequencies to make the AGC voltage a dc bias. A typical filter time constant is 0.2 s. An electrolytic capacitor of 5 to 10 μF is generally used for the AGC filter in transistor circuits.
3. The filtered AGC bias is distributed by the AGC line, through an RC decoupling filter to each controlled stage.
4. In delayed AGC, no bias is produced until the signal is strong enough to overcome the delay voltage on the AGC stage. The result is a better signal-to-noise ratio for less snow with weak signals. The rf amplifier usually has delayed AGC.
5. In keyed or gated AGC, the AGC stage is pulsed into conduction by horizontal flyback pulses. This stage is generally called the *AGC keyer* or *AGC gate*. The polarity of the flyback pulses is positive on tubes and NPN transistors, or negative for PNP transistors, at the collector.
6. A transistor for the AGC stage needs a diode to block the AGC voltage from its collector and to prevent the AGC filter capacitor from discharging through the AGC transistor.
7. For forward AGC on transistors, the forward voltage at the base is increased by the AGC bias. On an NPN transistor, forward AGC voltage at the base has positive polarity.
8. For reverse AGC on transistors, the forward voltage at the base is reduced toward cutoff by the AGC bias to reduce the amplifier gain. On an NPN transistor, reverse AGC voltage at the base has negative polarity.
9. For AGC on tubes, the bias makes the control grid more negative to reduce the amplifier gain.
10. Too little AGC bias can cause an overloaded picture and 60-Hz buzz in the sound. The overloaded picture (Fig. 15-1) has reversed black-and-white values and is out of sync.

11. Too much AGC bias results in no picture and sound, caused by cutoff in one or more of the rf and IF amplifiers.
12. The AGC level control sets the amount of bias for the rf and IF stages controlled by AGC bias. When the AGC control can produce an overloaded picture, set the adjustment just below the point of overload on the strongest signal. If there is no overload, set the AGC for minimum snow on the weakest signal.

Self-Examination   (Answers at back of book)

Answer True or False.

1. The AGC bias increases with more picture carrier signal at the antenna.
2. The AGC filter removes video signal variations from the dc bias voltage.
3. A keyed AGC stage conducts only during the time of the H flyback pulses.
4. With a PNP transistor for the keyed AGC stage, the H flyback pulses have negative polarity at the collector.
5. With a PNP transistor for the keyed AGC stage, the AGC voltage output at the collector has positive polarity.
6. For forward AGC, the bias is negative at the base of an NPN amplifier.
7. For reverse AGC, the bias is negative at the base of an NPN amplifier.
8. For AGC bias on tubes, the AGC bias is negative at the control grid.
9. Too much AGC bias can cause an overloaded picture with buzz in the sound.
10. The problem of airplane flutter is helped by a shorter time constant for the AGC filter.
11. A voltage gain of 10,000 equals approximately 1,000 dB.
12. With positive AGC voltage into the base of an AGC amplifier, the AGC bias output at the collector has negative polarity.

Essay Questions

1. What is the main advantage of AGC for the picture signal?
2. In an AGC circuit, give the function of (a) AGC rectifier; (b) AGC filter; (c) AGC line; (d) AGC decoupling filters.
3. Why is the video signal coupled to the AGC stage without any series blocking capacitors?
4. What signal stages are usually controlled by the AGC bias?
5. What is the purpose of delayed AGC?
6. When the carrier signal level increases, what is the effect on the amount of negative AGC bias on tubes? Give the effect on receiver gain.

7. Give the polarity required for the following examples of AGC bias at the base of a transistor IF amplifier: (a) forward AGC with an NPN transistor; (b) reverse AGC with an NPN transistor; (c) forward AGC with a PNP transistor; (d) reverse AGC with a PNP transistor.
8. Show how to connect a battery bias box to the AGC line, for the case of forward AGC on NPN amplifier stages.
9. What is the advantage of keyed AGC?
10. What two input voltages are necessary for a keyed AGC stage to conduct?
11. Give the polarity of flyback pulses needed for (a) plate of AGC tube; (b) collector of PNP AGC stage; (c) collector of NPN AGC stage.
12. What are two characteristics of an overloaded picture?
13. With an overloaded picture, why is there usually buzz in the sound?
14. Describe briefly why zero AGC bias can cause an overloaded picture.
15. Describe briefly why too much AGC bias can cause no picture and no sound.
16. Describe how to adjust the AGC level control, for the case where it can produce an overloaded picture.
17. What is meant by snow in the picture? What is the main cause?
18. Refer to the dc voltages in the AGC stage in Fig. 15-9. Give the values for: (a) pin 1, pin 7, and pin 6 to ground; (b) grid-cathode bias at pin 7, with polarity; (c) grid-cathode bias at pin 6, with polarity.
19. Give the function of each component for the AGC stage in Fig. 15-7.
20. In Fig. 15-7, give the dc values for (a) V_E to ground; (b) V_B to ground; (c) V_{BE} with polarity.

Problems

1. Calculate the time constant in seconds for the following RC combinations: (a) 2 MΩ and 0.1 µF; (b) 20 kΩ and 10 µF.
2. What value of R is needed with C of 5 µF for a time constant of 0.1 s?
3. A signal voltage of 1,000 µV at the antenna is amplified to 2-V output from the video detector. How much is the overall voltage gain?
4. How much is the dB gain for Problem 3? (Decibels are explained in Sec. 26-7.)

16 Sync Circuits

The synchronizing pulses, generally called *sync*, are part of the composite video as the top 25 percent of the signal amplitude. Included are the horizontal, vertical, and equalizing pulses. These are clipped from the video signal by the sync separator. Then the sync is coupled to the horizontal and vertical deflection oscillators to time the scanning frequencies. As a result, the picture information reproduced by the camera signal variations is in the correct position on the raster. The horizontal sync pulses hold the line structure of the picture together by locking in the frequency of the horizontal oscillator. The vertical sync pulses hold the picture frames locked in vertically by triggering the vertical oscillator. The equalizing pulses help the vertical synchronization be the same in even and odd fields for good interlacing.

The operation of the sync in timing the deflection oscillators is the same for monochrome or color. The only difference is that, in color, the horizontal line frequency for sync and scanning is 15,734 Hz instead of 15,750 Hz, while the vertical field frequency is 59.94 Hz.

The time for the transmitted signal to travel to the receiver has no effect on synchronization. The reason is that the sync pulses must be present at the same time as the camera signal variations in the composite video signal. Furthermore, the sync is for the picture, not the raster, as the deflection circuits can produce vertical and horizontal scanning with or without sync. In short, the sync times the scanning in the raster with respect to the picture information in the video signal. The effects of no sync are shown in Figs. 16-1 and 16-2. Details of the sync circuits are explained in the following topics:

16-1 Vertical synchronization of the picture
16-2 Horizontal synchronization of the picture
16-3 Separating the sync from the video signal
16-4 Integrating the vertical sync
16-5 Noise in the sync
16-6 Horizontal AFC
16-7 Sync separator circuits
16-8 Phasing between horizontal blanking and flyback
16-9 Sync and blanking bars on the screen
16-10 Sync troubles

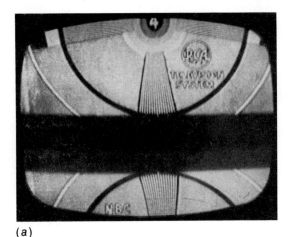
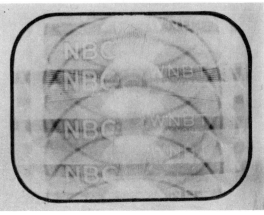

FIGURE 16-1 PICTURE ROLLING WITHOUT VERTICAL SYNC. (a) PICTURE SLOWLY SLIPS FRAMES VERTICALLY. (b) PICTURE ROLLS FAST UP OR DOWN.

16-1 Vertical Synchronization of the Picture

When every vertical scanning field is produced at the correct time, the frames are superimposed on each other. The result is a steady picture, locked in frame. If the vertical scanning is not locked in at the 60-Hz sync frequency, successive frames cannot overlap. Instead, the frames are displaced either above or below the first frame. Without vertical synchronization, therefore, the picture on the screen appears to roll up or down.

The faster the picture rolls, the farther the vertical scanning frequency is from 60 Hz. If the vertical scanning frequency is just slightly off the 60-Hz synchronizing rate, the picture will be recognizable, as shown in Fig. 16-1a, but it slips out of frame slowly. In b the picture is rolling fast. The wide black bar across the picture in Fig. 16-1a is produced by vertical blanking, which now occurs during vertical trace time because the scanning is out of sync.

When the picture rolls up or down, the vertical hold control is varied to lock in the picture. The hold control adjusts the frequency of the oscillator close enough to 60 Hz to enable the vertical sync voltage to trigger the oscillator at the sync frequency. The picture locks in when it is rolling upward because the oscillator frequency must be slightly less than the sync frequency for triggering.

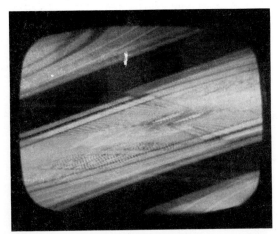

FIGURE 16-2 PICTURE IN DIAGONAL BARS WITHOUT HORIZONTAL SYNC.

16-2 Horizontal Synchronization of the Picture

When every scanning line is produced at the correct time, the line structure of the reproduced picture holds together to provide a complete image that stays still horizontally. If the horizontal scanning is just slightly off, the 15,750-Hz sync frequency, the line structure is complete, but the picture slips horizontally, as the picture information on the lines is displaced horizontally in successive frames. The faster the picture slides horizontally, the farther the scanning frequency is from 15,750 Hz. When the horizontal scanning frequency departs from the 15,750-Hz synchronizing frequency by 60 Hz or more, the picture tears apart into diagonal segments, as shown in Fig. 16-2. The black diagonal bars represent parts of horizontal blanking, which is normally at the sides of the picture. Each diagonal bar represents a 60-Hz difference from 15,750 Hz. In Fig. 16-2, the two bars sloping down to the left show the horizontal scanning frequency is 120 Hz below 15,750 Hz.

When the picture is in diagonal bars, the horizontal hold control is adjusted to make a complete picture. Then the horizontal AFC circuit keeps the picture locked in horizontally.

16-3 Separating the Sync from the Video Signal

Figure 16-3a shows oscilloscope waveforms of composite video signal input to the separator stage and its clipped sync output in b. Note that

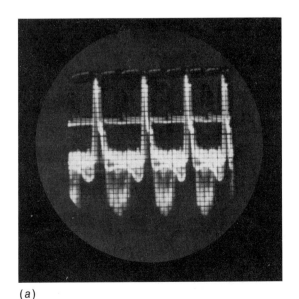

(a)

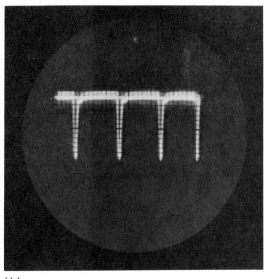

(b)

FIGURE 16-3 INPUT AND OUTPUT OF SYNC SEPARATOR. ONLY HORIZONTAL SYNC CAN BE SEEN WITH OSCILLOSCOPE SWEEP FREQUENCY AT 15,750/4 Hz. (a) COMPOSITE VIDEO SIGNAL INPUT WITH POSITIVE SYNC POLARITY. AMPLITUDE IS 40 V p-p. (b) SEPARATED SYNC WITH NEGATIVE POLARITY. AMPLITUDE IS 60 V p-p.

the clipped sync is actually amplified, from an amplitude of about 10 V in the input to 60 V in the sync output. Horizontal sync is shown here with the oscilloscope internal sweep frequency at $^{15,750}/_4$ Hz. When the scope is set at $^{60}/_4$ Hz, however, you see four vertical sync pulses.

The sync separator can be a triode or pentode tube, a transistor, or a diode. All do the clipping the same way, basically. They have dc bias beyond cutoff. However the ac signal drive is enough to produce output only for the sync tips of the video signal input. This sync clipping is illustrated in Fig. 16-4 by the input-output characteristic of a separator stage. The horizontal, vertical, and equalizing pulses are all clipped from the peak amplitudes of the video signal. The sync can be clipped at any point above the blanking level, as the leading and trailing edges of the pulses actually supply the timing information.

If there is a question of what happens to the camera signal in the video input, the answer is "nothing." The input to the sync separator has the composite video with sync, blanking, and camera signal, but the output is only sync. Remember that the camera signal with the information for the picture tube is in a separate path with the video amplifier.

Signal bias for the sync separator. Signal bias means that the dc bias voltage is produced by the ac signal input. The requirements are to rectify the input signal and have an *RC* time constant long enough to maintain the bias between peaks of the ac input. A perfect example shown in Fig. 16-5a is gridleak bias. Without signal, there is no bias. However, the positive half-cycle of the ac input drives the grid positive to produce grid current. Then C_1 is charged with the grid side negative. Between peaks of the input signal, C_1 can discharge slightly through

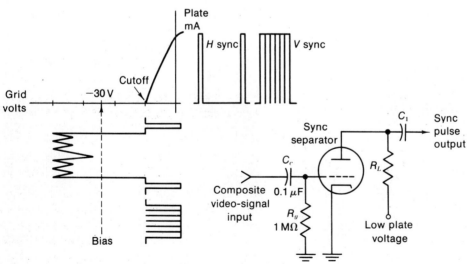

FIGURE 16-4 OPERATION OF SYNC SEPARATOR WITH SIGNAL BIAS. ONLY THE SYNC TIPS PRODUCE OUTPUT. EQUALIZING PULSES OMITTED. *V* AND *H* PULSES NOT SHOWN TO SCALE.

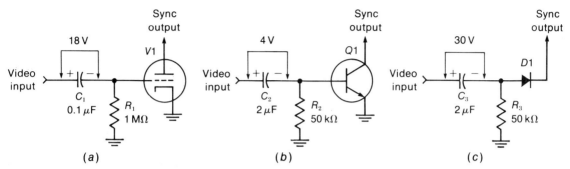

FIGURE 16-5 SIGNAL BIAS FOR SYNC SEPARATOR STAGES. DC BIAS VOLTAGE ACROSS C_1, C_2, AND C_3 IS PRODUCED BY THE INPUT SIGNAL. (a) GRID-LEAK BIAS ON TRIODE TUBE V1. (b) REVERSE VOLTAGE AT BASE OF NPN TRANSISTOR Q1. (c) REVERSE VOLTAGE AT ANODE OF DIODE D1.

R_1. However, the R_1C_1 time constant is made long enough to keep C_1 charged to about 90 percent of the peak positive input. The grid-cathode circuit is serving as a diode rectifier to produce the negative dc bias voltage across C_1, proportional to the peak signal input.

Not only is the dc bias across C_1 negative for the control grid, but for a sync separator the amount of bias is more negative than cutoff. However, the peak positive drive is always enough to produce output plate current for the positive peaks of the ac input. Grid-leak bias automatically adjusts itself to this level. The reason is that the bias increases or decreases with the amount of signal just enough to keep the positive peaks at zero grid voltage and maximum plate current. This action is called *clamping*, which means keeping the tips lined up at one level.

In Fig. 16-5b, the R_2C_2 coupling provides signal bias for the base of an NPN transistor. The polarities correspond exactly to grid-leak bias with a tube. Actually, grid-leak bias can be considered reverse voltage for the control grid. The negative voltage on C_2 is reverse voltage for the P-base. However, positive sync voltage in the signal input drives the transistor into conduction to produce the sync output in the collector current. The voltage values in b are less than for the tube in a because the transistor requires only 0.5 V or more for forward voltage at the base input to produce collector current output. The RC time constant in b is the same as a but with larger C and smaller R because of the lower input resistance of a transistor.

The clipping action with the diode in c is similar to the other two circuits. The diode corresponds to the base-emitter junction of the transistor. More bias with more signal input is shown in c, as there is no amplification of the clipped sync with a diode.

Note that for all three circuits in Fig. 16-5 the dc signal bias voltage always has the polarity away from the direction of conduction. Then the video signal is applied in the polarity that allows only the sync to produce output current.

Amplitude separation. This term means clipping all the sync pulses from the composite video signal. One or two stages may be used. With two stages, both the top and bottom of the sync can be clipped in opposite polarities.

334 BASIC TELEVISION

Waveform separation. This means separating the vertical sync from the horizontal sync. The total separated sync voltage includes horizontal, vertical, and equalizing pulses, as shown in Fig. 16-6. All have the same amplitude, but they differ in their repetition rate, or frequency, and in pulse width. We can think of the width as the time the pulse is on. The horizontal sync with a narrow pulse width of 5 μs, repeated at 15,750 Hz, represents a high-frequency signal compared with vertical sync. The vertical pulses are repeated at the slow rate of 60 Hz. The width is 27.3 μs between the narrow serrations with a total width of 190.5 μs or three horizontal lines.

Therefore, the vertical and horizontal sync can be separated from each other by *RC* filters. The vertical sync requires a low-pass filter, with series *R* and shunt *C*, to bypass the horizontal sync but pass the vertical synchronizing voltage. This is the function of the R_1C_1 integrator* in Fig. 16-7. The R_1C_1 time constant of 100 μs is too long for the horizontal sync to produce any appreciable voltage across C_1. In a separate path, the R_2C_2 coupling circuit in Fig. 16-7 is a high-pass filter for the horizontal sync into the horizontal AFC circuit. The R_2C_2 time constant

*See Appendix D for explanation of *RC* integrator and differentiator circuits.

of 0.82 μs is too short for the vertical sync to produce any appreciable voltage across R_2.

Sequence of sync separation. As shown in Fig. 16-7, first the sync in the composite video signal is separated by a sync clipper. The total separated sync is then coupled to two parallel paths for waveform separation. The R_1C_1 integrator receives all the sync pulses, but C_1 builds up charge only during the relatively wide vertical sync pulse. The equalizing pulses help in making the integrated voltage across C_1 the same for even and odd fields. As a result, the vertical deflection oscillator is triggered at 60 Hz to hold the picture locked in vertically.

All receivers use automatic frequency control for the horizontal deflection oscillator. The horizontal AFC circuit is better than triggered sync for holding the horizontal oscillator at the correct frequency in the presence of noise pulses.

16-4 Integrating the Vertical Sync

With the relatively large value of 0.01 μF for C_1 in Fig. 16-8, it bypasses the horizontal sync pulses but not the vertical sync. More exactly, this R_1C_1 low-pass filter integrates, or adds up, the effect

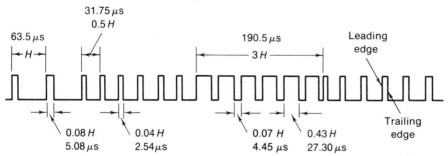

FIGURE 16-6 WAVEFORM OF THE SYNCHRONIZING PULSES. *H* IS HORIZONTAL LINE TIME OF 63.5 μs.

SYNC CIRCUITS 335

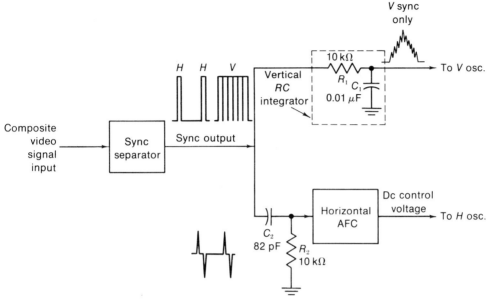

FIGURE 16-7 SEQUENCE FOR SYNC SEPARATION. CLIPPER STAGE FOR AMPLITUDE SEPARATION IS FOLLOWED BY WAVEFORM SEPARATION FOR VERTICAL AND HORIZONTAL SYNC.

of the serrated vertical sync pulse. The requirements for integrating the vertical sync are (1) take the voltage output across C_1, not R_1, and (2) have the R_1C_1 time constant long compared with the H pulse width but not the V pulse width. As a result, the total sync voltage is coupled to the R_1C_1 integrator, but the output voltage across C_1 is vertical sync alone. This voltage triggers the vertical deflection oscillator at the vertical sync frequency.

The time constant of the R_1C_1 integrator in Fig. 16-8 is 100 μs. The serrated vertical pulse consists of six individual pulses, each of approximately 27 μs. In this time, C_1 charges to 27 percent of the applied voltage. This value is obtained from the universal charge curve in Appendix D. During the serration, applied voltage is removed and the capacitor discharges. This is only for 4.4 μs, however. Therefore, C_1 loses little of its charge before the next 27-μs pulse provides sync voltage to recharge the capacitor. As a result, the integrated voltage across C_1 builds up toward maximum amplitude at the end

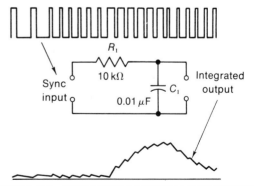

FIGURE 16-8 INTEGRATION OF VERTICAL SYNC VOLTAGE ACROSS C_1. THE R_1C_1 TIME CONSTANT HERE IS 100 μs.

of the vertical pulse, followed by a decline practically to zero for the equalizing pulses and horizontal pulses that follow. The result is a pulse of the triangular waveshape shown for the complete vertical synchronizing pulse.

You can see the 60-Hz integrated vertical sync pulses with an oscilloscope at the grid of the vertical oscillator (Fig. 16-9). However, the oscillator must be off, as its grid voltage is much larger than the sync voltage. Or, with the oscillator on but slightly out of sync, you can see the vertical sync pulse moving across the grid-voltage waveform.

The amplitude in Fig. 16-9 is 10 V, but the integrated sync usually triggers the vertical oscillator before the peak. How much sync voltage is necessary depends on the oscillator frequency. The closer the oscillator is to 60 Hz, the less sync voltage needed for triggering. Generally, the vertical hold control sets the oscillator at 50 to 59 Hz, to be locked in sync at 60 Hz at the time of the third or fourth division of the vertical sync. The main requirement for timing is that the triggering occur at exactly the same time of the vertical sync pulse for every field.

Effect of the equalizing pulses. Their function is improving the accuracy of vertical synchro-

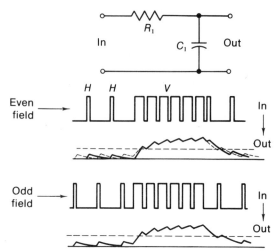

FIGURE 16-10 INTEGRATED OUTPUT ACROSS C_1 FOR EVEN AND ODD FIELDS WITHOUT EQUALIZING PULSES. DOTTED WAVEFORM IN UPPER FIGURE IS THE LOWER WAVEFORM SUPERIMPOSED. THE DASHED LINES ACROSS BOTH WAVEFORMS SHOW THE LEVEL TO WHICH C_1 VOLTAGE MUST RISE TO TRIGGER VERTICAL OSCILLATOR.

nization in even and odd fields for good interlace (see Fig. 16-10). Note that with the equalizing pulses, the voltage across C_1 can adjust itself to practically equal values for even and odd fields, although there is a half-line difference in time before the vertical pulse begins. Then the integrated output for vertical sync is the same for even and odd fields.

Cascaded integrator sections. A very long time constant for the integrating circuit removes the horizontal sync pulses but reduces the vertical sync amplitude across the integrating capacitor. The rising edge on the integrated vertical pulse is not sharp enough. With a time constant that is not long enough, the horizontal sync pulses cannot be filtered out, and the serrations in the vertical pulse produce notches in the integrated output. The notches should be filtered out because they give the integrated ver-

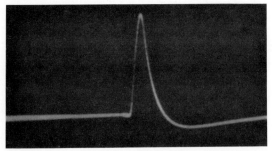

FIGURE 16-9 OSCILLOSCOPE PHOTO OF 60-Hz INTEGRATED VERTICAL SYNC. AMPLITUDE IS 10 V.

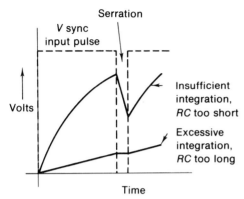

FIGURE 16-11 INTEGRATED VERTICAL SYNC VOLTAGE FOR TIME CONSTANT TOO LONG AND TOO SHORT.

tical synchronizing signal the same amplitude value at different times. Figure 16-11 illustrates too much integration and not enough integration.

Figure 16-12 shows a two-section integrating circuit with each RC section having a time constant of 50 μs. The operation of the circuit can be considered as though the R_1C_1 section provided integrated voltage across C_1 that is applied to the next integrating section R_2C_2. The overall time constant for both sections is long enough to filter out the horizontal sync, while the shorter time constant of each section allows the integrated voltage to rise more sharply because each integration is performed with a time constant of 50 μs. If one section is open,

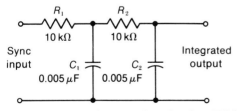

FIGURE 16-12 TWO-SECTION INTEGRATOR FOR VERTICAL SYNC.

there will be less filtering but actually more vertical sync voltage because of a shorter time constant.

16-5 Noise in the Sync

Noise pulses are produced by ignition interference from automobiles, arcing brushes in motors, and by atmospheric static. The noise is either radiated to the receiver or coupled through the power line. Especially with weak signal, the noise can act as false synchronizing pulses. In the vertical circuit, the noise pulses can trigger the vertical oscillator too soon. See Fig. 16-13. Then the noise makes the picture roll temporarily, until the sync locks in the oscillator again. For horizontal synchronization, the noise can be mistaken for horizontal sync pulses. Then the picture tears into diagonal segments during the loss of horizontal sync.

Furthermore, when the noise pulses have much higher amplitude than the sync voltage,

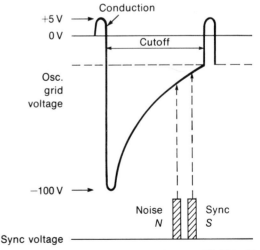

FIGURE 16-13 HOW OSCILLATOR CAN BE TRIGGERED INTO CONDUCTION BY A NOISE PULSE INSTEAD OF THE SYNC PULSE.

the signal bias on the sync separator is too high. This effect of increased bias by the noise voltage is called *noise setup*. Then the sync separator has too much bias for the actual signal level. The result is weak sync or no sync at all. During the noise, then, the picture does not hold still until the synchronization is established again.

In order to reduce the effects of noise, the sync circuits generally include one or more of the following:

1. Sync clipper stage for the sync alone, after the sync is separated from the video signal. The additional clipping means that both the top and bottom of the sync pulses can be clipped by cutoff, because of the opposite polarities for the sync voltage in successive stages. Then the noise pulses cannot have higher amplitude than the sync pulses.
2. Double time constant for the signal bias in the sync separator. See Fig. 16-14. A long time constant is needed for the vertical sync, but a shorter time constant reduces noise setup.
3. Noise-inverter stage, noise-canceller, noise switch, or noise gate. Such a stage can remove high-amplitude noise pulses by cutting off the sync separator, just during the time of a noise pulse. See Fig. 16-15.

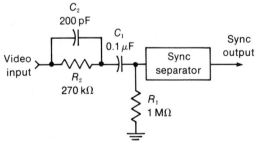

FIGURE 16-14 DOUBLE TIME CONSTANT FOR SIGNAL BIAS ON SYNC SEPARATOR. R_2C_2 HAS SHORT TIME CONSTANT TO REDUCE EFFECT OF NOISE.

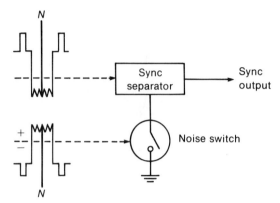

FIGURE 16-15 NOISE SWITCH TO CUT OFF SYNC SEPARATOR, TEMPORARILY, DURING THE TIME OF NOISE PULSE *N*.

4. Sine-wave stabilizing circuit. The horizontal deflection oscillator usually has an *LC* circuit to produce sine-wave voltage that is added to the oscillator voltage. As shown in Fig. 16-16, the result is a sharper slope in the oscillator grid voltage, just before conduction time when the oscillator is easy to trigger. Then any noise pulse must have a high amplitude, or it cannot trigger the oscillator into conduction. The sine-wave coil, often called the *stabilizing coil*, also stabilizes the oscillator against changes in supply voltage and transistor characteristics.

It should be noted that reducing the noise in the sync does not prevent noise pulses in the video signal from producing horizontal streaks in the picture. The purpose of the noise-reduction circuits in the sync section is to allow the picture to hold still in the presence of noise.

How noise triggers an oscillator. Figure 16-13 illustrates the effect of a noise pulse in the sync voltage. Consider this example for a vertical oscillator using a vacuum tube for a blocking oscillator or a multivibrator. The oscillator is held cut off by its own grid voltage for a relatively long

SYNC CIRCUITS 339

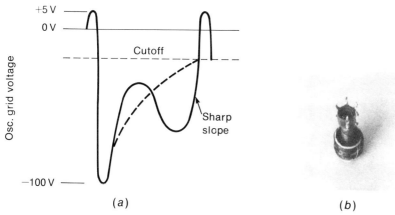

FIGURE 16-16 (a) SINE WAVE ADDED TO GRID-VOLTAGE WAVEFORM FOR STABILIZING OSCILLATOR AGAINST TRIGGERING BY NOISE PULSES. DASHED LINE SHOWS GRID VOLTAGE WITHOUT SINE WAVE. (b) PHOTO OF STABILIZING COIL. HEIGHT IS $1\frac{1}{2}$ IN.

time and conducts for a short time, at the repetition rate for vertical scanning. Toward the end of the cutoff time, a small positive voltage at the grid can trigger the oscillator into conduction. Normally, the sync pulse S triggers the oscillator for every cycle to lock it in at 60 Hz. However, if the noise pulse N occurs just before the sync pulse, the noise can trigger the oscillator. Then the sync pulse has no effect. As a result, the vertical scanning is off 60 Hz, and the picture rolls, until the vertical sync triggers the oscillator again.

It should be noted that, without the noise, the grid voltage in Fig. 16-13 would be normal, remaining beyond cutoff until the sync triggers the oscillator. This means that canceling the noise pulses is the best way to eliminate the effects of noise in the sync.

Sync separator time constant. The time constant of the grid-leak bias circuit in the input to the separator must be long enough to maintain the bias from line to line and through the time of the vertical synchronizing pulse, in order to maintain a constant clipping level. Typical values for grid-leak bias with a tube are 0.1 μF for C_1 and 1 MΩ for R_1, providing a time constant of 0.1 s (Fig. 16-14). These values allow the bias to vary from frame to frame for different brightness values, keeping the tip of sync clamped at zero grid voltage. However, a time constant of 0.1 s is too long for the bias to follow amplitude variations produced by noise pulses.

In order to reduce the effect of high-frequency noise pulses on the grid-leak bias for the sync separator, an RC circuit with a short time constant can be added to the grid circuit of the sync separator, as shown in Fig. 16-14. The grid coupling circuit R_2C_2 provides normal grid-leak bias, with a time constant of 0.1 s, for the sync signal. The small 200-pF C_2 and the 270-kΩ R_2 provide a short time-constant combination. Then the grid-leak bias can change fast enough to reduce the effect of noise pulses in the input to the sync separator.

The smaller C_2 can charge faster than C_1 when noise pulses produce grid current. Then most of the noise pulse voltage is across C_2. However, C_2 discharges fast through R_2 between noise pulses. As a result, the signal bias produced by R_1C_1 remains at the level determined by the sync voltage.

Noise switch. In Fig. 16-15, the noise switch illustrates a tube or transistor that is part of the return circuit to chassis ground for the sync separator. With the switch closed, the separator conducts to produce sync output. When the switch is open, though, the sync separator is cut off. Closed or open for the switch corresponds to its being conducting or cut off.

The switch operation is controlled by video signal with opposite sync polarity from the signal input to the sync separator. Usually, the noise switch has input from the video detector, while the video amplifier feeds the sync separator. Then a noise pulse with the polarity that would increase conduction in the separator has the reverse polarity for the switch to cut it off. As a result, the switch can eliminate noise in the sync separator output by cutting off the output, temporarily, during the time of noise pulses. It is important to note, though, that the separator operates normally for the sync voltage to provide separated sync in the output.

Stabilizing coil. See Fig. 16-16. This coil is usually in the plate circuit of the horizontal oscillator, to feed back sine-wave voltage to the grid. The sharp slope as the grid voltage returns to cutoff means that the oscillator cannot be triggered easily. Horizontal AFC is used to hold the oscillator on frequency, while the stabilizing coil prevents false triggering by interfering pulses. In some circuits, the variable slug in the stabilizing coil serves as the horizontal hold control.

16-6 Horizontal AFC

A deflection oscillator triggered by sync pulses for each cycle is capable of exact synchronization, if there is no noise interference. However, noise pulses can be mistaken for sync and trigger the oscillator at the wrong time. In order to make the synchronization more immune to noise, an AFC circuit is used for the horizontal deflection oscillator in all television receivers. This is generally called *sync lock, stabilized sync,* or *horizontal AFC.*

The AFC circuit is generally the only section used for horizontal sync alone. Therefore, when the picture tears into diagonal bars too easily, the trouble is likely to be in the horizontal AFC, or the horizontal oscillator. AFC is generally not used for the vertical deflection oscillator. The main reason is that the vertical *RC* integrator is a low-pass filter that removes high-frequency noise pulses.

AFC requirements. The typical arrangement of an AFC circuit for the horizontal deflection oscillator is illustrated in Fig. 16-17. The operation can be considered in the following steps:

1. Horizontal sync voltage and a fraction of the horizontal deflection voltage are coupled into the sync discriminator. A discriminator is a dual-diode circuit that can detect a difference in frequency. The deflection voltage is needed as a sample of the oscillator frequency. It can be taken either from the oscillator or horizontal output circuit.
2. The sync discriminator circuit produces dc output voltage proportional to the difference in frequency between its two input voltages.
3. The dc control voltage indicates whether the oscillator is on or off the sync frequency. The greater the difference between the oscillator and sync frequencies, the larger is the dc control voltage.
4. The dc control voltage is filtered by an *RC* integrator. With a shunt bypass capacitor, noise pulses cannot affect the dc control voltage appreciably.

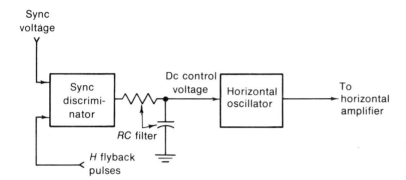

FIGURE 16-17 BLOCK DIAGRAM OF AFC CIRCUIT FOR HORIZONTAL OSCILLATOR.

5. The filtered dc control voltage changes the oscillator frequency by the amount necessary to make the scanning frequency the same as the sync frequency.

It is important to note that the AFC circuit needs two ac voltage inputs, from sync and deflection, to produce the correct dc voltage output. If one ac voltage input is missing, the dc voltage output cannot measure the frequency correctly.

The horizontal deflection oscillator generally uses either the multivibrator or blocking oscillator circuit. For these pulse oscillators, the frequency depends on the dc operating voltages, especially the bias. The oscillator frequency is changed by 50–100 Hz for 1 V of change in dc voltage at the control grid of a triode tube. Positive grid voltage increases the frequency for a blocking oscillator. At the synchronizing grid of a cathode-coupled multivibrator, negative voltage increases the frequency.

Sync discriminator. A dual-diode unit is used, as shown in Fig. 16-18. There are only three leads, as one is common to the two diodes. In the schematic symbols, the arrow is the anode side with the bar for the cathode. In a, the anode of one diode is common to the other cathode, leaving an anode and cathode for the other two leads. This type is for a balanced discriminator circuit, where push-pull sync voltages of opposite polarities are required for input to the two diodes. The type in b is for a single-ended discriminator with negative sync voltage for the common cathodes. It should be noted that when the receiver has a multipurpose tube that in-

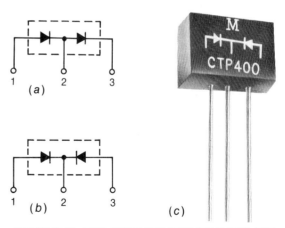

FIGURE 16-18 DUAL-DIODE UNIT COMMONLY USED FOR SYNC DISCRIMINATOR. (a) ONE ANODE AND CATHODE COMMON AT TERMINAL 2 FOR PUSH-PULL SYNC INPUT TO TERMINALS 1 AND 3. (b) COMMON CATHODES FOR SINGLE-ENDED SYNC INPUT AT TERMINAL 2. (c) PHOTO OF UNIT FOR SINGLE-ENDED CIRCUIT.

342 BASIC TELEVISION

cludes twin diodes, they are often used for the sync discriminator instead of a semiconductor unit.

Push-pull sync discriminator. In Fig. 16-19, opposite polarities of sync voltage are coupled to the two diodes. R_2C_2 couples positive sync to the $D2$ anode, while R_1C_1 couples negative sync to the $D1$ cathode. Remember that negative voltage at the cathode makes diode current flow, just like positive voltage at the anode. As a result, the sync input voltage drives both diodes into conduction.

Deflection voltage is needed as a sample of the oscillator frequency. Therefore, flyback pulses from a winding on the horizontal output transformer are applied to the R_3C_3 network. The polarity is chosen to make the slope of the flyback on the sawtooth increase in the positive direction. Each positive pulse charges C_3 during the fast flyback. Then C_3 discharges during the relatively long time between pulses. The result is the sawtooth voltage shown across C_3, which is applied to the $D1$ anode and the $D2$ cathode.

The sawtooth input voltage is applied in the same polarity to both diodes. Note that when this voltage makes the $D1$ anode positive, it aids the diode current. However, this positive voltage at the $D2$ cathode opposes diode current, like negative anode voltage.

The ac input is rectified by the diodes to produce the required dc control voltage in the output. Each diode is a peak rectifier for the sync pulses. This means the amount of diode conduction depends on the peak value of the ac input voltage. The amount of conduction depends on how the sync input voltage is phased with respect to flyback on the sawtooth voltage. Note that C_4 in series with R_4 is in a common path for both diodes. R_5 provides a dc return to chassis ground.

Conduction in $D1$ makes the voltage across C_4 more positive. When $D2$ conducts, however, the current is in the opposite direction,

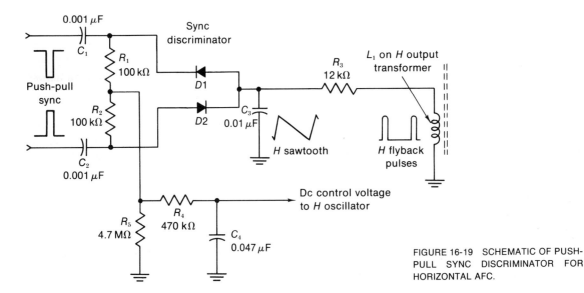

FIGURE 16-19 SCHEMATIC OF PUSH-PULL SYNC DISCRIMINATOR FOR HORIZONTAL AFC.

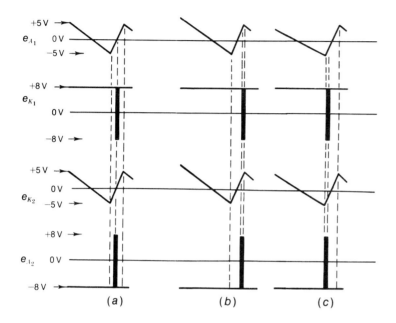

FIGURE 16-20 WAVESHAPES FOR PUSH-PULL SYNC DISCRIMINATOR IN FIG. 16-19. ANODE VOLTAGE IS e_A; CATHODE VOLTAGE IS e_K. (a) NORMAL CONTROL WITH SYNC AND SAWTOOTH SAME FREQUENCY. SYNC PULSE IN MIDDLE OF FLYBACK. (b) OSCILLATOR TOO FAST WITH SHORTER SAWTOOTH CYCLE. SYNC PULSE AT END OF FLYBACK. (c) OSCILLATOR TOO SLOW WITH LONGER SAWTOOTH CYCLE. SYNC PULSE AT START OF FLYBACK.

and the voltage across C_4 becomes more negative. When both diodes conduct the same amount of current, the net voltage across C_4 is zero. This condition of zero dc control voltage occurs when the sync and scanning frequencies are the same. When the oscillator is below or above the sync frequency, C_4 provides negative or positive dc control voltage to correct the oscillator frequency.

The waveforms in Fig. 16-20 show how the sync and sawtooth voltages combine to control conduction in the discriminator diodes. In a each sync pulse occurs in the middle of flyback time because the oscillator is on frequency. At this time, the sawtooth input voltage is at the same value of zero for both diodes. The zero level is the average-value axis of the ac sawtooth wave. With equal sync voltages, therefore, the peak input voltage has the same amplitude for either diode. Then they conduct equal currents to provide a net voltage of zero across C_4.

The oscillator frequency is too high in Fig. 16-20b. As the sawtooth cycle takes less time, the sync pulse occurs later, at the end of the flyback voltage. Then D1 conducts more than D2, and the result is positive dc control voltage across C_4.

In c the oscillator frequency is too low. As the sawtooth cycle takes more time, the sync pulse now occurs earlier, at the start of flyback voltage. Now D2 conducts more than D1, and the result is negative dc control voltage across C_4 for the horizontal oscillator.

As a result, the sync discriminator continuously measures the frequency difference between the sync and sawtooth waves to produce the dc correction voltage that locks in the horizontal oscillator at the synchronizing frequency.

Single-ended sync discriminator. The AFC circuit in Fig. 16-21 does not require push-pull sync input. Instead, negative sync voltage is coupled by C_1 to the common cathode of the two diodes D1 and D2. Effectively, the diodes are in

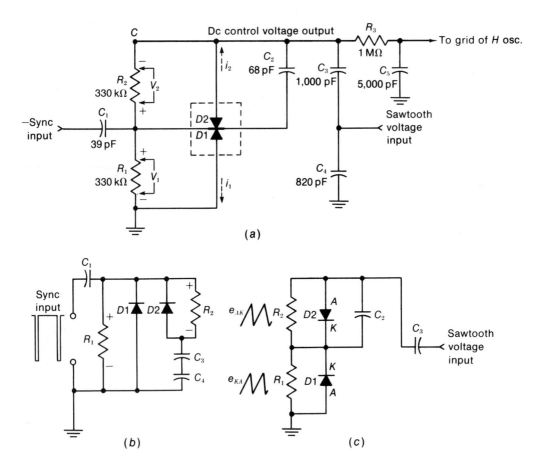

FIGURE 16-21 SCHEMATIC OF SINGLE-ENDED DISCRIMINATOR FOR HORIZONTAL AFC. (a) ACTUAL CIRCUIT WITH NEGATIVE SYNC INPUT THROUGH C_1 AT THE LEFT TO COMMON CATHODES. THE DC CONTROL VOLTAGE IS TAKEN OUT FROM POINT C AT THE TOP THROUGH R_3. (b) EQUIVALENT CIRCUIT SHOWING DIODES IN PARALLEL FOR SYNC VOLTAGE INPUT. (c) DIODES IN SERIES FOR SAWTOOTH VOLTAGE INPUT.

parallel for the sync input, as shown in b. The R_1C_1 coupling circuit provides negative sync voltage across R_1 for D1. Similarly, R_2C_1 couples sync voltage for D2. The top end of R_2 is returned to ground through the series combination of C_3 and C_4.

Sawtooth voltage is also coupled to the diodes, to compare the oscillator frequency with sync. As shown in c, the sawtooth voltage coupled by C_3 is applied across the two diodes in series. Therefore, each diode has one-half the sawtooth input voltage.

Notice that the sawtooth voltage is applied anode to cathode for D2 but cathode to anode for D1. Let us consider the sawtooth as cathode-to-anode voltage for D2 the same way sync is applied to both diodes. Then the sawtooth voltage for D2 can be shown inverted, for the discriminator waveshapes in Fig. 16-22. In effect, the two diodes have equal and opposite sawtooth input voltages. Notice that this circuit has push-pull sawtooth input, compared with push-pull sync input for the discriminator in Fig. 16-19.

The negative sync voltage input at the cathode, combined with sawtooth voltage input, makes the diodes conduct. Both i_1 and i_2 are shown as electron flow in Fig. 16-21, opposite from the direction of the arrow in each diode. When $D1$ conducts, the result is dc voltage V_1 across R_1 with the polarity shown in a. Similarly, conduction in $D2$ results in the dc voltage V_2 shown across R_2. For electron flow, the direction is the opposite of hole current, but the polarities of V_1 and V_2 are still the same.

The dc control voltage, taken from point C with respect to ground, consists of V_1 and V_2 in series opposition. When the two voltages are equal, the net output at C is zero. When V_2 is larger than V_1, negative output voltage results, but when V_1 is larger, the control voltage is positive. The dc control voltage is filtered by R_3C_5 for the horizontal oscillator.

Summarizing the functions of the components in Fig. 16-21a, R_1 and R_2 are diode load resistors. C_1 is the coupling capacitor for negative sync input to both diodes in parallel. The anode of $D2$ is returned to chassis ground through C_3 and C_4 in series. Also, C_4 is needed for a ground return without shorting B+ of the tube supplying sawtooth voltage. C_3 couples sawtooth voltage input to the two diodes in series, without shorting the dc control voltage. C_2 equalizes the amount of sawtooth voltage across the diodes. In order to help balance the input to the diodes, C_2 is larger than C_1 but smaller than the series combination of C_3 and C_4. Finally, R_1 can be smaller than R_2, but proportioned to compensate for unbalance in the diode circuits.

Because of unbalance in the single-ended discriminator circuit, the dc control voltage is 1 to 2 V when the horizontal oscillator is on frequency at 15,750 Hz. This voltage can vary between +6 and −6 V, approximately, to correct the oscillator frequency.

Filtering the dc control voltage. The RC time constant of the integrating filter determines how fast the dc control voltage can change its ampli-

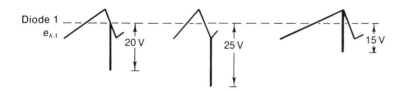
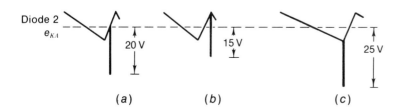

FIGURE 16-22 WAVESHAPES FOR SINGLE-ENDED SYNC DISCRIMINATOR IN FIG. 16-21. VOLTAGES e_{KA} ARE SHOWN WITH POLARITY AT CATHODE WITH RESPECT TO ANODE. (a) NORMAL CONTROL. OSCILLATOR AT 15,750 Hz. (b) OSCILLATOR TOO FAST. (c) OSCILLATOR TOO SLOW.

tude to correct the oscillator frequency. A time constant much longer than 63.5 μs is needed for the shunt capacitor to bypass the horizontal sync and sawtooth components in the control circuit, while filtering out noise pulses. However, the oscillator should be able to pull into sync within a fraction of a second when sync is temporarily lost in changing channels. Also, if the time constant is too long, the dc control voltage may be affected by vertical sync, causing bend at the top of the picture. A typical value for the AFC filter time constant is approximately 0.005 s. This time includes about 80 horizontal scanning lines.

Hunting in the AFC circuit. The shunt capacitor in the filter introduces time delay in the effect of dc control voltage on oscillator frequency. The reason is that the output filter capacitor takes time to charge to the applied dc control voltage from the AFC circuit, or discharge down to a lower voltage. As a result, the filtered voltage is still correcting the oscillator after it has been pulled in to the correct frequency. Each succeeding step of overcorrection is less until, finally, the oscillator operates at 15,750 Hz. This action of the control circuit in varying the oscillator frequency within a smaller and smaller range around the correct frequency is called *hunting*. Excessive hunting in the AFC circuit can cause scalloped edges in the picture, generally called *piecrust* or *gear-tooth* effect (Fig. 16-23).

In order to prevent hunting, a double-section filter is commonly used for the dc control voltage, as shown in Fig. 16-24. The R_1C_1 time constant of 0.005 s is long enough to filter out noise and horizontal sync or deflection voltages. The relatively low resistance of R_2 serves as a damping resistor across C_1, making the output voltage more resistive and less capacitive to reduce time delay in the control voltage. C_2

FIGURE 16-23 THE GEAR-TOOTH OR PIECRUST EFFECT IN PICTURE CAUSED BY HUNTING IN HORIZONTAL AFC CIRCUIT.

blocks the dc control voltage from shorting to chassis ground.

Dc control stage. In some AFC circuits, the dc correction voltage is amplified and then connected to the horizontal oscillator (Fig. 16-25). Several uses for such a dc amplifier include: (1) amplify the dc correction voltage from the sync discriminator for better control of the horizontal oscillator; (2) dc control tube for a blocking oscillator; (3) reactance tube for a sine-wave oscilator, called *synchrolock* AFC.

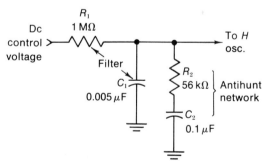

FIGURE 16-24 AFC FILTER WITH ANTIHUNT NETWORK R_2C_2 ACROSS FILTER CAPACITOR C_1.

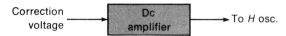

FIGURE 16-25 DC AMPLIFIER FOR AFC CORRECTION VOLTAGE.

16-7 Sync Separator Circuits

The *RC*-coupled amplifier is generally used with a resistance plate load of about 50 kΩ for tubes or 1.5 kΩ for transistors. Peaking coils may be needed to boost the high-frequency response, for sharp edges on the sync pulses. Signal bias for the input circuit clamps the tip of sync at the desired level for clipping the pulses. The required polarity of the video input signal has positive sync for a tube or an NPN transistor, with negative sync polarity for the input to a PNP transistor. A diode may be used as a clipper, for a noise switch, or to block the wrong polarity of collector voltage on a transistor stage.

Triode separator with single-ended sync discriminator. In Fig. 16-26, composite video with positive sync polarity is coupled from the output of the video amplifier to the input of the triode sync separator. The 47-kΩ R_3 is the plate-load resistor that develops clipped sync of negative polarity. The sync output is coupled by C_6 to the integrator for the vertical deflection oscillator.

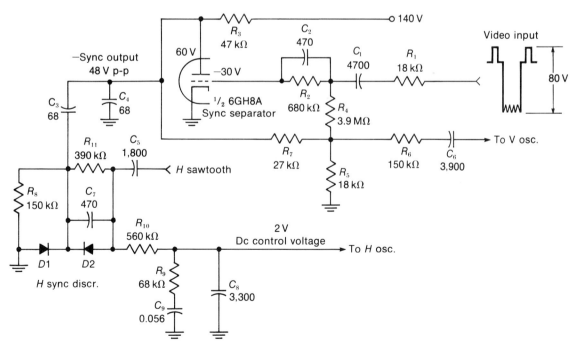

FIGURE 16-26 TRIODE SYNC SEPARATOR CIRCUIT WITH SINGLE-ENDED SYNC DISCRIMINATOR. *R* VALUES IN OHMS. *C* VALUES MORE THAN 1 IN pF OR LESS THAN 1 IN μF.

In a separate path, the sync output is also coupled by C_3 to the sync discriminator with diodes D1 and D2. The filtered dc control voltage in the output is directly connected to the horizontal deflection oscillator to control its frequency.

In the input to the sync separator, C_1 is the coupling capacitor, with R_4 and R_5 the grid resistance. R_1 is a decoupling resistance to isolate the sync input circuit from the video amplifier. The R_2C_2 combination in the grid circuit provides a short time constant to prevent noise setup in the bias voltage. Note the dc grid voltage of -30 V. This value is the net voltage resulting from the negative grid-leak bias and some positive voltage from the R_7R_5 divider.

In the single-ended sync discriminator, horizontal sync voltage of negative polarity is coupled by C_3 to the common cathodes of D1 and D2. The bypass C_4 filters out noise. R_8 is the load resistor for D1, with R_{11} the load resistor for D2. Sawtooth voltage as a sample of the horizontal oscillator frequency is coupled by C_5 into the discriminator. C_7 helps balance the diode voltages. The dc output is taken from the anode of D2 with R_{10} and C_8 as the filter to remove noise and sync from the dc control voltage for the horizontal oscillator. C_9 with R_9 form the antihunt network. The normal value of dc control voltage is 2 V from the unbalanced discriminator, with the oscillator on frequency at 15,750 Hz.

Phase splitter for push-pull sync discriminator. In Fig. 16-27, the clipped sync output from Q1 is coupled to the phase splitter Q2. Note that Q1 is a PNP transistor, which requires negative sync for the video input to produce positive clipped sync in the collector output. The phase splitter Q2 is an NPN transistor, however. Output is taken from both the emitter and collector. The collector, shown at the bottom, has a supply voltage of 30 V, with the 510 Ω of R_8 as the collector load resistance. In the emitter circuit, the 820 Ω of R_4 is the load.

R_8 and R_4 provide push-pull sync voltage output to the balanced sync discriminator using diodes D1 and D2. With positive sync into the base of Q2, the collector signal is negative sync coupled by C_2 to the D2 cathode, while the emitter signal is positive sync coupled by C_1 to the D1

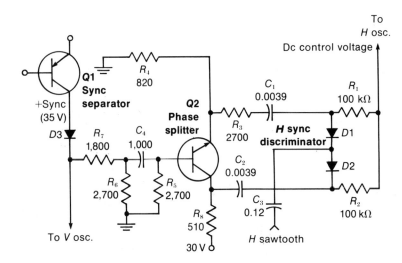

FIGURE 16-27 TRANSISTOR SYNC PHASE SPLITTER FOR PUSH-PULL DISCRIMINATOR. R VALUES IN OHMS. C VALUES MORE THAN 1 IN pF OR LESS THAN 1 IN μF.

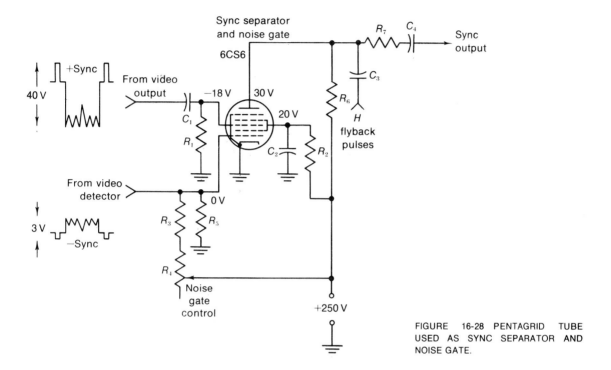

FIGURE 16-28 PENTAGRID TUBE USED AS SYNC SEPARATOR AND NOISE GATE.

anode. R_3 helps balance the input to the two diodes.

In the coupling between Q1 and Q2, sync for the vertical oscillator is taken from the junction of R_7 and diode D3. R_7 isolates the vertical oscillator from Q2. The diode D3 blocks any positive voltage spikes from the vertical oscillator into Q1, which requires negative collector voltage because it is a PNP transistor. C_4 couples horizontal sync to the phase splitter.

In the sync discriminator, R_1 and R_2 are the balanced load resistors for D1 and D2. Sawtooth input as a sample of the oscillator frequency is coupled by C_3 to the common anode-cathode junction of the diodes. The balanced output from the junction of R_1 and R_2 is the dc control voltage for the horizontal oscillator. The normal value of dc control voltage is zero from the balanced discriminator, with the oscillator on frequency at 15,750 Hz. Although not shown here, an RC filter and antihunt network are needed for the dc control voltage.

Pentagrid sync separator and noise gate. The circuit in Fig. 16-28 uses a tube having two control grids with sharp cutoff. The first grid next to the cathode is one control grid and the other is grid 3. Grids 2 and 4 are connected internally to serve as a screen grid, and grid 5 is the suppressor. Cutoff for each control grid is approximately −2 V. It is important to note that either control grid can cut off plate current. We can consider this a gated circuit, therefore. Both control-grid gates must be on to produce output in the plate circuit.

Video signal is applied to each control grid. Grid 3 has 40 V of composite video signal to allow clipping the sync voltage. Grid 1 has 3 V

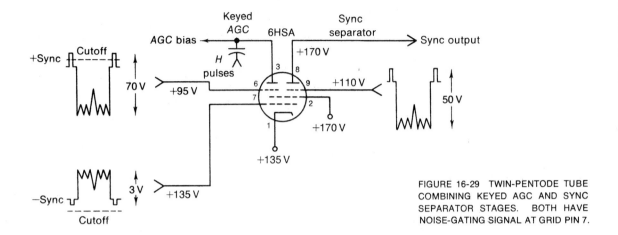

FIGURE 16-29 TWIN-PENTODE TUBE COMBINING KEYED AGC AND SYNC SEPARATOR STAGES. BOTH HAVE NOISE-GATING SIGNAL AT GRID PIN 7.

of video signal of opposite polarity, to serve as a noise gate. Negative noise pulses at this grid can cut off plate current, reducing noise voltages in the sync output.

For grid 3, composite video signal with positive sync polarity from the plate of the video amplifier is coupled by C_1 and R_1. The negative grid-leak bias allows only the peak positive sync voltage to produce plate current. The result is separated sync output across the plate load resistor R_6. The sync amplitude in the output is about 25 V peak-to-peak.

Grid 1 has composite video signal of low amplitude, with negative sync polarity, from the video detector. The bias on grid 1 is set by R_4 to allow sync voltage for normal signal to remain less negative than cutoff. This adjustment allows plate current to flow for producing separated sync output from the signal at grid 3. However, high-amplitude noise pulses can drive the grid 1 voltage more negative than cutoff. For the noise, then, plate current is cut off. As a result, noise pulses in the composite video signal input are not present in the separated sync output. The noise-gate control R_4 can be set approximately by adjusting to the point where the picture is out of sync for strong signal, and then back off just enough for normal sync.

Sync separator and gated AGC. The noise-gating function can also be helpful in reducing the effects of noise in the automatic gain control circuit of the receiver. Therefore, special tubes are manufactured to combine the keyed AGC stage and sync separator stage with a noise-gate circuit common to both. These tubes are twin sharp-cutoff pentodes, as shown in Fig. 16-29. The cathode is common to both sections; grid 1 serves as a noise gate for both the sync separator and AGC stages; grid 2 is a common screen grid. Grid 3 in each section is an individual control grid, and there are two plates for separated sync output from one pentode unit and AGC voltage from the other. The sync separator circuit is the same as in Fig. 16-28, and the keyed AGC circuit is the same as explained previously for Fig. 15-9.

Note the dc voltages in Fig. 16-29. Pin 6 is at +95 V because of dc coupling to the video amplifier plate. The cathode is at +135 V to make the grid bias at pin 6 equal to $95 - 135 = -40$ V, for the AGC grid. Pin 7 is also at +135 V making the grid-cathode bias zero for the noise-gate video signal voltage, at the common control grid. However, this grid voltage can be varied slightly by the noise-gate adjustment. The plate of the AGC section is keyed into conduction by horizontal flyback pulses. In the separator, its plate is at $+170$ V to provide $170 - 135 = 35$ V from plate to cathode. The screen grid has the same voltage. Pin 9 at $+110$ V has grid-cathode bias equal to $110 - 135 = -25$ V for the sync separator grid.

16-8 Phasing between Horizontal Blanking and Flyback

While the deflection circuits are scanning the raster, the composite video signal is varying the intensity of the electron scanning beam. The blanking on the picture tube screen has the timing of the sync and blanking pulses in the transmitted signal, but the flyback is determined by the deflection circuits in the receiver. In a triggered system, the synchronized flyback starts during blanking time automatically because each sync pulse begins the retrace. With AFC, however, the horizontal oscillator produces scanning independently of individual synchronizing pulses. In fact, horizontal retrace can start with the front porch of blanking, allowing more time for flyback within blanking.

In Fig. 16-30 normal phasing of the flyback within blanking time is shown in a. It should be noted that a small change of phasing can put more or less blanking at the left and right edges of the raster. This is why the horizontal centering of the picture usually shifts with respect to the raster when the hold control is varied.

The other examples in Fig. 16-30 illustrate incorrect phasing between blanking and horizontal flyback. This occurs because the AFC circuit can lock in the oscillator at 15,750 Hz but with retraces not completely within blanking. Then some picture information may be reproduced during flyback time. In b, horizontal flyback starts just before blanking. Then some of the picture information that should be at the extreme right side of the trace in the picture is reproduced during the flyback to the left. With picture information reproduced during both retrace and trace at the right, this side can appear folded over or under itself, usually brighter than normal.

When the horizontal flyback starts too late after blanking, as in c, the retrace cannot be completed before picture information starts for the trace at the left side of the next line. Then

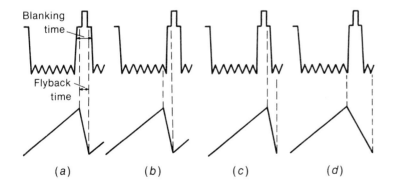

FIGURE 16-30 PHASING OF HORIZONTAL BLANKING AND FLYBACK. (a) NORMAL FLYBACK WITHIN BLANKING TIME. (b) FLYBACK STARTS TOO SOON. (c) FLYBACK STARTS LATE. (d) FLYBACK TIME IS TOO LONG.

352 BASIC TELEVISION

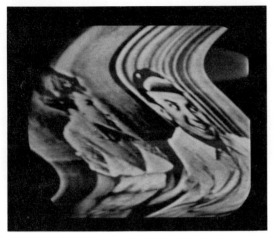

FIGURE 16-31 BEND IN PICTURE, BUT NOT IN RASTER, CAUSED BY 60-Hz HUM IN THE HORIZONTAL SYNC. NOTE WHITE SPEAR AT TOP RIGHT, PRODUCED BY WHITE PICTURE INFORMATION IN FLYBACK TIME.

the left side of the picture appears folded. If the horizontal retrace time is too long, as in d, the flyback can start before blanking but still continue after blanking. Then both the left and right sides of the picture may be folded.

Whenever the flyback occurs during picture information time, there may be a light haze in the background, pointing outward from either side of the picture, like a big spear. This is white picture information spread out by fast flyback, instead of being reproduced correctly during trace time. The picture usually has this effect just before horizontal hold is lost. As an example, note the white haze in Fig. 16-31 extending out from the right side of the picture near the top. The bend in the picture is caused by 60-Hz hum in the horizontal sync.

16-9 Sync and Blanking Bars on the Screen

The entire composite video signal with sync and blanking pulses is coupled to the picture tube. Therefore, amplitude variations of the sync and blanking pulses can be seen as relative intensities on the picture tube screen (Fig. 16-32). This can be seen by varying the vertical hold control to allow the picture to roll slowly, out of sync, so that the vertical blanking bar is in the middle of

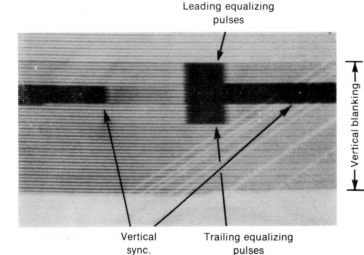

FIGURE 16-32 HAMMERHEAD PATTERN ON PICTURE TUBE SCREEN SHOWING VERTICAL SYNC PULSES WITH ITS SERRATIONS AND THE EQUALIZING PULSES, WITHIN THE VERTICAL BLANKING TIME. (RCA)

the picture instead of at the top and bottom. Brightness is turned up higher than normal to make the blanking level gray instead of black. The sync amplitude, which is 25 percent above the blanking level, then becomes black.

The black bar within the vertical blanking in Fig. 16-32 is often described as the "hammerhead" pattern. It shows the equalizing and vertical sync pulses occurring during vertical blanking time. The black hammerhead is produced by the equalizing pulses at the head and the vertical pulses for the handle. There are really two hammerheads, but one head is lost in flyback time.

The appearance of the horizontal sync within the horizontal blanking bar on the picture tube screen is shown in Fig. 16-33. This can be seen by adjusting the phasing of the horizontal deflection oscillator to put the horizontal blanking bar in the picture, with the brightness higher than normal. Normally the sync and blanking bars are at the edges of the picture behind the mask of the screen and are not visible.

The intersection of the horizontal and vertical black bars in Fig. 16-33 is called a *pulse-cross display*. The bars can be examined to check the sync voltage. It should be blacker than blanking and the darkest parts of the picture. Normal bars on the screen show normal sync voltage in the composite video signal input to the sync separator. If the sync is not blacker than blanking, this indicates compression of the sync pulses in the video or IF section.

16-10 Sync Troubles

Vertical rolling and a picture in diagonal segments show there is no synchronization of the vertical and horizontal deflection oscillators. When both occur at the same time, the trouble is in the sync separator circuits, as both the vertical and horizontal sync are missing. You can tell if the picture is rolling vertically, even with diagonal bars, by watching the horizontal black bar across the screen produced by vertical blanking.

When the whole picture rolls up or down, the trouble is just vertical synchronization, as the horizontal AFC is holding the picture together. When the picture is in diagonal bars, without vertical rolling, the trouble is just the horizontal synchronization.

It is important to remember that the sync is part of the signal and, therefore, that the receiver must have enough signal to provide a good picture for good synchronization. When the signal is weak, the sync is weak also. Then noise voltages can easily interrupt the synchronization. In addition, hum in the receiver can easily affect weak sync, especially for triggering the vertical oscillator.

No vertical hold. Vertical rolling of the picture, as shown in Fig. 16-1, means no vertical synchronization. Generally, the circuits for vertical sync alone are just the *RC* integrator and the vertical oscillator. There is seldom any trouble with the *RC* integrator, however. Vertical rolling,

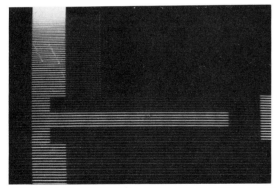

FIGURE 16-33 HORIZONTAL SYNC AND BLANKING BARS ON PICTURE TUBE SCREEN. THE INTERSECTION IS A PULSE-CROSS DISPLAY. (*RCA*)

then, is usually a trouble in the vertical oscillator. With tubes, the first step is to try a new tube for this stage. With transistors, the isolation diode for the vertical oscillator can cause troubles. For tubes and transistors, poor low-frequency response in the video amplifier caused by a defective bypass capacitor can cause weak vertical sync.

In order to hold an oscillator synchronized, not only is sync voltage necessary, but the oscillator frequency must also be close to the sync frequency. To check whether a trouble of rolling picture is caused by no sync or incorrect oscillator frequency, vary the vertical hold control. The important thing to see is whether the picture can be made to roll both up and down. If it does, this shows the oscillator can be set to the correct frequency. The trouble, then, is in the vertical sync. If not, the trouble is incorrect oscillator frequency. For deflection oscillator circuits, the *RC* values in the coupling circuits are the main factors in determining the frequency.

No horizontal hold. The picture tears apart in diagonal segments, as shown in Fig. 16-2. Again the trouble can be in either the oscillator or the sync. In this case, though, there is the horizontal AFC stage for horizontal sync alone. Vary the horizontal oscillator frequency to see if a whole picture can be produced. If not, the trouble is incorrect oscillator frequency. When a whole picture can be produced but it does not stay still horizontally, the trouble is usually in the horizontal AFC. A common trouble is a defective diode, which changes the dc control voltage.

Poor interlace. Inaccurate timing of the vertical scanning in even and odd fields causes poor interlacing of the scanning lines. The result is partial pairing or complete pairing, which reduces the detail in the picture. Stray pickup of pulses generated by the receiver's horizontal deflection circuits and coupled into the vertical sync and deflection circuits can cause interlace troubles.

Hum in the sync. Excessive hum in the horizontal sync produces bend in the picture as shown in Fig. 16-31. The hum in the sync bends the picture, but the edge of the raster is straight. This shows that the hum is in the horizontal sync but not in the raster circuits.

With excessive hum in the vertical sync, the picture tends to show the vertical blanking bar across the middle of the picture, as in Fig. 16-1a, drifting slowly upward on the screen. Especially with weak signal, the oscillator can easily be triggered by 60-Hz hum voltage instead of vertical sync.

SUMMARY

1. The synchronizing pulses time the receiver scanning to make the reproduced picture hold still on the raster. Without vertical hold the picture rolls up or down. Without horizontal hold, the picture is torn apart with diagonal black bars.
2. The sync separator clips the sync voltage from the composite video signal. Then the integrator filters out all but the vertical sync voltage, which triggers the vertical deflection oscillator at 60 Hz. Also, the horizontal sync is coupled to the AFC circuit to control the frequency of the horizontal deflection oscillator.
3. The *RC* integrator for vertical sync has a time constant very long for the horizon-

tal pulse width but not for the vertical pulses. This is the vertical sync section of the receiver.
4. Noise in the sync can cause loss of synchronization, especially with weak signal.
5. A noise gate or switch cuts off the sync separator for noise pulses.
6. A common circuit for the sync separator uses a triode tube or transistor with signal bias beyond cutoff (Fig. 16-26).
7. Another common circuit is the gated sync separator. Either a pentagrid tube can be used, as in Fig. 16-28, or a twin pentode combining the sync separator and AGC stages, as in Fig. 16-29.
8. When the picture rolls vertically but it can be stopped temporarily with the hold control, this means no vertical synchronization. Check the integrator and vertical oscillator.
9. If varying the horizontal oscillator frequency control can produce a picture but it drifts into diagonal bars, this means no horizontal hold. Check the horizontal AFC circuit.
10. When the picture has normal contrast but it rolls vertically and tears horizontally on all channels, the trouble is no sync. Check the sync separator and amplifier stages.
11. Hum in the horizontal sync can cause sine-wave bend from top to bottom of the picture, as in Fig. 16-31. With hum in the vertical sync, the blanking bar drifts slowly from bottom to top of the picture. See Fig. 16-1a.
12. Automatic frequency control is used to lock in the horizontal deflection oscillator at 15,750 Hz. The AFC circuit compares the oscillator frequency with the sync frequency to produce dc output voltage proportional to their difference. This is the horizontal sync section of the receiver.
13. With a dual-diode sync discriminator for the horizontal AFC circuit, each diode has sync input and sawtooth voltage that indicates oscillator frequency. The dc output from two diodes measures any difference between the sync and scanning frequencies to hold the oscillator locked in at 15,750 Hz.
14. The time constant for the AFC filter is about 0.005 s to filter noise but allow fast lock-in when changing channels.
15. Overcorrection of the oscillator frequency can cause hunting in the AFC circuit. In the picture, the result is *piecrust* or *geartooth* effect (Fig. 16-23). A series *RC* combination is usually connected across the AFC filter capacitor as an antihunt network.
16. With horizontal AFC, picture information may be reproduced during horizontal retrace time. This effect can cause a white haze in the background corresponding to white picture information spread out by the fast flyback (Fig. 16-31).

Self-Examination (Answers at back of book)

Answer True or False.
1. Without vertical sync, the bar produced by vertical blanking rolls up or down the screen.

2. The diagonal black bars resulting from no horizontal hold are produced by horizontal blanking.
3. The integrated vertical sync voltage triggers the vertical deflection oscillator.
4. The horizontal deflection oscillator frequency is controlled by an AFC circuit.
5. A grid-leak bias sync separator requires composite video signal with positive sync polarity at the grid.
6. Signal bias is reverse dc voltage proportional to the input voltage.
7. When a noise pulse triggers the vertical oscillator, the picture can roll out of sync.
8. Video signal voltage at the picture tube grid is about 3 V peak-to-peak.
9. The hammerhead pattern in Fig. 16-32 is a reproduction of the serrated vertical sync and equalizing pulses.
10. If the sync separator stage does not function, the picture will roll vertically and have diagonal bars at the same time.
11. Negative signal bias at the base of an NPN transistor is reverse voltage.
12. Weak picture signal can cause poor vertical and horizontal synchronization.
13. In Fig. 16-26, the diodes $D1$ and $D2$ form an unbalanced sync discriminator with negative sync voltage input.
14. In Fig. 16-27, the output at the collector of phase splitter $Q2$ is negative sync for the cathode of diode $D2$.
15. The dc output of the sync discriminator stage is directly coupled to the horizontal deflection oscillator.
16. Hunting in the sync discriminator stage causes a rolling picture.
17. In Fig. 16-27, if diode $D1$ in the sync discriminator is defective, the picture will have diagonal bars.
18. Figure 16-31 shows 60-Hz hum in the horizontal sync but not in the raster.
19. The horizontal AFC is the only part of the receiver circuits operating just for horizontal sync.
20. The output from the *RC* integrator for vertical sync is a dc control voltage directly coupled to the vertical deflection oscillator.

Essay Questions

1. Give the specific function for the horizontal sync pulses, vertical pulses, and the equalizing pulses.
2. Describe briefly the sequence of separating the sync from the composite video signal to time the frequency of the vertical and horizontal deflection oscillators.
3. Give the function of the *RC* integrator for vertical sync.
4. Give the function of the horizontal AFC circuit.
5. Draw the diagram of a sync separator stage with grid-leak bias using a triode tube. Show input and output waveshapes.
6. (*a*) Do the same as in Question 5 for an NPN transistor. (*b*) Do the same with a PNP transistor.

7. Give two reasons for vertical rolling in the picture.
8. Give two reasons for diagonal bars in the picture.
9. In Fig. 16-26, give the function for C_1, R_3, C_3, and C_6.
10. In Fig. 16-27 give the function for R_4, R_8, C_1, and C_2.
11. For the gated sync separator in Fig. 16-28, what input signals are applied to the two grids?
12. In Fig. 16-28, if the screen bypass C_2 is shorted, what will be the effect in the picture?
13. What is the hammerhead pattern on the screen of the picture tube?
14. What is the pulse-cross display on the screen of the picture tube?
15. Describe briefly the difference between triggered sync and AFC.
16. Name the three input voltages to the push-pull horizontal sync discriminator in Fig. 16-19.
17. Name the two input voltages to the single-ended horizontal sync discriminator in Fig. 16-21.
18. In the balanced sync discriminator in Fig. 16-19, give the function for C_1, R_1, C_2, R_2, R_4, and C_4.
19. In the unbalanced sync discriminator in Fig. 16-21: (a) Give the function for C_1, R_1, and R_2. (b) Why will the diodes $D1$ and $D2$ not conduct if the sync input has positive polarity?
20. (a) What is meant by hunting in the AFC circuit? (b) What is the effect in the picture? (c) What components form the antihunt network in Fig. 16-26? (d) Show where to add a filter and antihunt network for the sync discriminator output voltage in Fig. 16-27.
21. If C_2 in Fig. 16-12 opens, why will the result be more vertical sync voltage?
22. Give the effect on the raster and picture for the following troubles in Fig. 16-26: (a) R_3 open; (b) C_3 open; (c) C_6 open; (d) C_4 shorted.
23. Give the effect on the raster and picture for the following troubles in Fig. 16-27: (a) open collector in $Q2$; (b) shorted junction in $Q1$; (c) open diode $D2$.
24. Give two effects in the picture when 60-Hz hum is in the video signal for the picture tube and in the horizontal sync.

Problems

1. Calculate the time constant for each RC section of the vertical integrator in Fig. 16-12, with R of 10 kΩ and C of 0.005 μF.
2. Calculate the time constant of the RC coupling circuits in Fig. 16-5a, b, and c.
3. A grid-leak bias circuit has 0.05-μF C and 2-MΩ R. (a) Calculate the time constant. (b) After C has been charged, how long will it take to discharge down to one-half of its voltage? (c) To 37 percent? (d) To zero? (Hint: See universal RC curve in Appendix D.)
4. A picture is out of horizontal hold with six diagonal bars. How far off is the oscillator frequency from 15,750 Hz?

17 Color Circuits

The color section in the television receiver amplifies the 3.58-MHz chroma (*C*) signal, detects the modulation, and provides red, green, and blue video signals for the three guns of the tricolor picture tube. The color circuits are usually on a plug-in board, as shown in Fig. 17-1. In addition, the luminance (*Y*) signal is amplified to form a monochrome picture on the raster. Then the color information can be superimposed on the black-and-white picture. The theory of color television and block diagrams of color receivers are described in Chaps. 8 and 9, but here we analyze the basic circuits in the color section. Automatic circuits that are generally used to obtain the best color picture are described in the next chapter. The fundamental requirements for color can be seen from this list of circuits:

- 17-1 Signals for the color picture
- 17-2 The video preamplifier stage
- 17-3 The *Y*-channel video amplifier
- 17-4 Chroma bandpass amplifiers
- 17-5 The burst amplifier stage
- 17-6 Color oscillator and synchronization
- 17-7 AFPC alignment
- 17-8 Chroma demodulators
- 17-9 Matrixing the *Y* video and color video signals
- 17-10 Video amplifiers for color
- 17-11 The blanker stage
- 17-12 Troubles in the color circuits

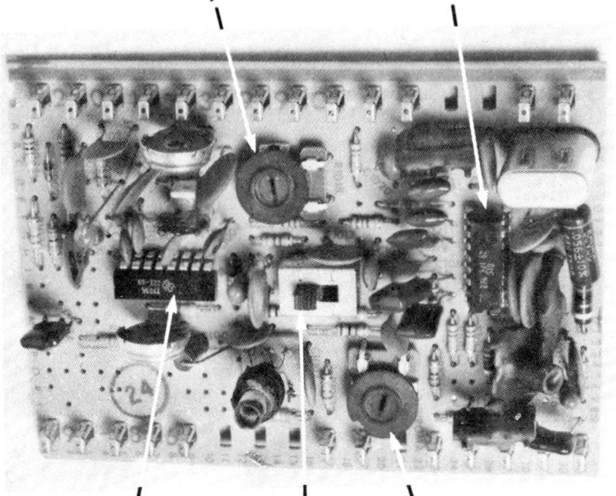

COLOR CIRCUITS 359

FIGURE 17-1 COLOR CIRCUITS ON PLUG-IN BOARD. WIDTH IS 5 IN. *(ZENITH CORPORATION)*

Labels: APC adjust; Color oscillator IC1002; Chroma IC1002; Color setup; ACC adjust.

17-1 Signals for the Color Picture

The colorplexed composite video signal out of the video detector stage includes the Y signal for luminance multiplexed with the 3.58-MHz C signal for color information. The modulated C signal is amplified in the chroma bandpass amplifier. This stage is tuned to 3.58 MHz with a total bandwidth of 1 MHz. The amplitude of C signal determines color saturation. Its phase angle determines the hue or tint, with respect to the reference of burst phase.

The colorplexed composite video signal also includes the burst of 8 to 11 cycles of 3.58 MHz on the back porch of each horizontal blanking pulse. This burst is the reference phase for color synchronization, to determine hue or tint. The phase of burst is yellow-green, exactly 180° opposite from $B - Y$ phase. In the color circuits, the burst is separated and amplified by the burst amplifier stage. Then the burst is used for automatic frequency and phase control (AFPC) on the 3.58-MHz color oscillator.

The color oscillator regenerates the suppressed subcarrier to supply 3.58-MHz cw for the synchronous demodulators. They also have C signal input. Then each demodulator detects the phase of C signal that is synchronous with the phase of its oscillator cw input.

The detected color signals generally have the phase of $R - Y$ and $B - Y$ video, which are then combined to form $G - Y$ video. These color-difference signals are matrixed with the Y signal to form the red, green, and blue video signals for the picture tube.

Frequencies for chroma and color video. It is important to note the difference between modulated chroma signal at 3.58 MHz and the detected color video signals with a bandwidth of 0 to 0.5 MHz. In brief, 3.58 MHz is the frequency for color before demodulation of the chroma signal. Actually, 3.58 MHz is the color IF of the receiver, and a chroma bandpass stage is a color IF amplifier. The bandwidth of these tuned stages in the receiver is ±0.5 MHz.

After demodulation of the 3.58-MHz C signal, the color video signals have the baseband frequencies of 0 to 0.5 MHz. This bandwidth of 0.5 MHz for the detected color video applies to $R - Y$, $B - Y$, $G - Y$, R, G, and B signals.

All colors are in the 3.58-MHz chroma signal. The purpose of broadcasting the 3.58-MHz C signal is to have all the color information in one modulated signal. Amplifier stages for chroma signal affect all colors, therefore. These include red, green, and blue with their complementary colors and color mixtures. It should also be noted that the 3.58-MHz oscillator must be operating for all the colors because the C signal cannot be demodulated without reinsertion of the subcarrier.

Specific colors are in the video signals. As examples, $B - Y$ is mainly blue, with yellow for the opposite hue; $R - Y$ includes red and cyan; $G - Y$ includes green and magenta.

Comparison of Y and C signals. The Y signal for black-and-white information is transmitted with a bandwidth of 0 to 4.2 MHz. However, the Y video amplifier in the receiver cuts off at 3.2 MHz in order to reduce interference with the color signal. The luminance signal is a wideband signal, therefore, without one specific frequency. The C signal, though, consists of modulation side bands around the 3.58-MHz suppressed subcarrier. This specific frequency is needed for chroma signal. Circuits for the C signal, therefore, are tuned to 3.58 MHz, with a bandwidth of ±0.5 MHz.

Comparison of chroma and burst signals. These both have the frequency of 3.58 MHz. However, the C signal is present during trace time with the modulation information for color in the picture. The burst is on during horizontal blanking time only. There is no picture information in the burst signal, as its only function is to lock the color oscillator into the correct phase.

The phase and amplitude of the 3.58-MHz chroma signal are varying continuously with the color information in the picture. The phase indicates hue or tint. The amplitude is the color saturation. However, the phase and amplitude of the 3.58-MHz burst remain constant, for the station tuned in. The burst is the reference for synchronizing the phase of the color oscillator.

Keying pulses. The color circuits make use of flyback pulses from the horizontal output circuit for the purpose of turning a stage either on or off just during horizontal retrace time. For an amplifier that is normally conducting, a negative pulse at the control grid can cut the tube off for horizontal flyback. Or, for a stage that is normally cut off, a positive flyback pulse at the plate can make the tube conduct only during horizontal blanking time.

As an example, the burst separator can be keyed on by horizontal flyback pulses in order to amplify only the burst. This signal is present during retrace time. For the opposite case, the chroma bandpass amplifier is usually keyed off during horizontal flyback time.

CRT blanking. This technique for eliminating vertical and horizontal retraces can be considered a form of keying the picture tube off during blanking time. In color receivers, retrace blanking is generally used for both vertical and horizontal flyback. Usually, the blanking pulses are applied in the Y video amplifier. Then the blanking is automatically part of the video signals driving the three guns of the picture tube. As an example, positive pulses at each cathode can cut off the beam current for the red, green, and blue guns during retrace time.

Burst amplitude. In addition to using the phase of burst as a reference for hue, the color circuits use the amplitude of burst as a measure of how much color signal is present. First, if there is no burst signal, the program must be in monochrome. Then the color killer (CK) stage operates to disable the color circuits. Also, the relative amplitude of the burst signal indicates the strength of the color signal. This feature is used to control the gain of the bandpass amplifier for automatic chroma control (ACC).

17-2 The Video Preamplifier Stage

This circuit is the first video amplifier which is part of the Y-channel amplifiers. However, the preamplifier is shown separately in Fig. 17-2 to illustrate its special functions:

1. To amplify the colorplexed composite video, including Y signal and C signal
2. To supply Y signal to the delay line for the Y video amplifiers
3. To supply Y signal as composite video to the AGC and sync separator circuits
4. To supply C signal to the 3.58-MHz chroma amplifier

With transistors, this stage is generally an emitter-follower, for impedance matching. The high input impedance is a good load for the video detector, isolating it from the video amplifier. The low output impedance can feed the delay line for Y signal.

In Fig. 17-2, the output of the video detector is composite video, with the Y and C signals. The takeoff for the 4.5-MHz sound is normally ahead of this point, with a separate sound converter from the picture IF amplifier. Therefore, the video detector can use traps for the 41.25-MHz sound IF carrier. In addition, the video amplifier uses 4.5-MHz traps to keep the sound signal from beating with the color at 3.58 MHz. Trapping the sound signal is critical because the 920-kHz beat produces a herringbone interference pattern in the picture.

Transistor video preamplifier. In Fig. 17-3, the signal from the video detector is fed through a 4.5-MHz sound trap to the base of $Q2$. This trap is a bridged-T, consisting of C_{29}, T_5, and R_{20}. The peaking coil L_{11} is not part of the trap. $Q2$ is a common-collector or emitter-follower stage because the collector is grounded and the load is in the emitter circuit.

The dc voltage of 3.5 V at the base of $Q2$ is developed mostly by output current of the integrated circuit IC1-B, which contains the picture IF amplifier, not shown here. The 4.1 V at the emitter is produced by I_E through R_{324} from the 30-V supply. As a result, the net forward bias is $3.5 - 4.1 = -0.6$ V, at the base of the PNP transistor.

Output signals from $Q2$ are taken from the top of R_{324} for sync, AGC, and the delay line. The third output is from the chroma takeoff circuit to the input of the chroma amplifier, which is on an IC unit. The chroma peaking coil L_7, together with C_{32} and C_{33}, forms a low-Q circuit resonant at 4.1 MHz. This circuit compensates for the tilt

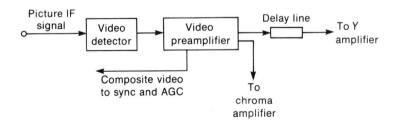

FIGURE 17-2 FUNCTIONS OF THE VIDEO PREAMPLIFIER STAGE.

362 BASIC TELEVISION

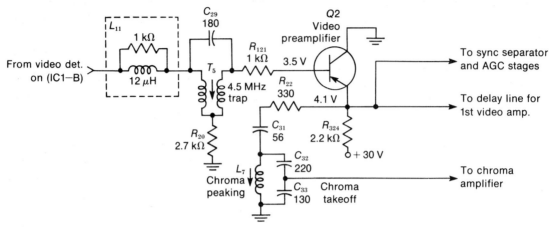

FIGURE 17-3 TRANSISTOR VIDEO PREAMPLIFIER CIRCUIT. R IN OHMS AND C IN PICOFARADS. (FROM RCA CHASSIS CTC-59)

of the chroma side bands in the response curve of the picture IF amplifier.

Delay line. The narrower bandwidth of 1 MHz for the chroma circuits delays the signal more than the 3.2-MHz bandwidth of the Y amplifiers. Without a time delay for the Y signal, it will arrive at the CRT ahead of the color image and the two will not be exactly superimposed. All that is required to prevent this problem is approxi-

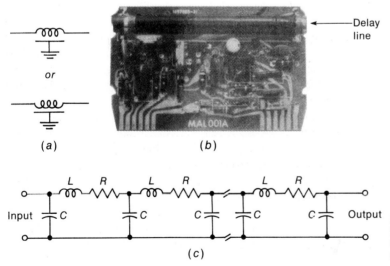

FIGURE 17-4 DELAY LINE FOR Y SIGNAL. (a) SYMBOL; (b) MOUNTING ON PLUG-IN MODULE; (c) EQUIVALENT CIRCUIT.

mately 0.8 μs delay in the Y signal. This is accomplished by installing an LC delay line, usually in the output of the video preamplifier.

The delay line is usually a coil about 8 in. long (Fig. 17-4). It consists of an insulating tube covered with a foil strip over which is a tightly wound coil. The entire unit is covered with an insulating coating. Terminals are brought out from the coil and the foil strip, which is grounded.

A delay line is similar to a transmission line in that it must be terminated with a resistance equal to its characteristic impedance (Z_0). A typical value of Z_0 for the delay line is 680 Ω.

17-3 The Y-Channel Video Amplifier

The composite video must be amplified enough to provide about 120 V p-p of Y signal to drive the picture tube. An example of a hybrid Y channel is shown in Fig. 17-5. It includes a PNP transistor for the Q304 preamplifier or first video amplifier and a vacuum tube V301 for the second amplifier, which is the video output stage. The delay line for Y signal is between the first and second amplifiers. The Y amplifier has the contrast control to vary the amount of luminance signal.

It is important to note that the video signal in the plate circuit of the output tube is dc-coupled to the cathodes of the color picture tube. The direct coupling means that dc voltages on the output tube affect the cathode-grid bias on the CRT. This is why the brightness control for biasing the picture tube can be in the grid circuit of the video output stage. An advantage of this method for color picture tubes is that one brightness control in the video amplifier controls the bias for all three guns.

In addition, vertical blanking pulses are applied in the grid circuit of the video output stage. This way the pulses are amplified before being coupled to the CRT cathodes, to blank out the vertical retrace lines. The polarity of the blanking pulses is negative at the grid of the output tube. This is inverted to positive-going pulses in the output for the picture tube. Positive polarity is needed at the cathode to cut off the beam current for blanking.

For the video preamplifier in Fig. 17-5, composite video from the detector is coupled to the base of Q304. This signal has positive sync polarity. The amplified output is available at the emitter and collector. The 330-Ω R_{324} is the emitter load resistor. There is no phase inversion here. This path acts as an emitter-follower to supply signal for the AGC circuit and for the chroma bandpass amplifier.

In the collector circuit, the 3.3-kΩ R_{325} is the load resistor. This amplified signal has inverted polarity, with negative sync phase. From the collector, the Y signal goes through 3.58-MHz traps, which keep the chroma signal out of the Y channel, and the delay line DL101. The diode D302, coupling capacitor C_{333}, and peaking coil L_{314} with damping resistor R_{334} complete the signal path to the grid of the video output tube. Also coupled to this grid are the vertical blanking pulses, which are isolated by D302. The negative pulses cannot flow from anode to cathode. However, the diode is turned on by positive voltage at the anode to conduct the video signal from cathode to anode.

In the output circuit of V301, the 5.6-kΩ R_{338} is the plate load. T_{302} provides high-frequency peaking. The drive controls R_{341} and R_{342} are adjusted to balance the Y video signals for the blue and green guns to the full output for the red gun. All video signal is removed by the service switch S_{102} to provide a blank raster for purity adjustments.

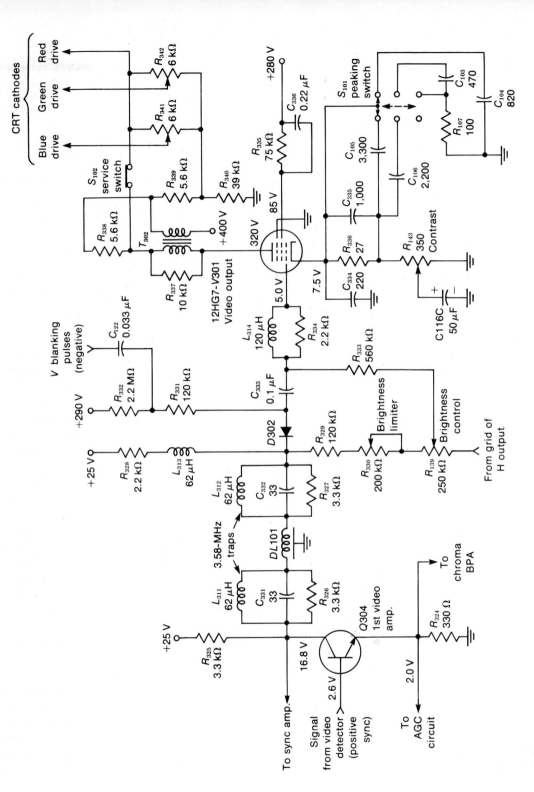

FIGURE 17-5 PENTODE VIDEO OUTPUT AMPLIFIER FOR Y SIGNAL, WITH TRANSISTOR VIDEO PREAMPLIFIER. R IN OHMS AND C IN PICOFARADS. DC VOLTAGES AT MAXIMUM BRIGHTNESS. (FROM ADMIRAL CHASSIS 2K20)

In the cathode circuit of V301, the 350-Ω R_{143} varies the cathode bias to adjust gain for the Y signal and contrast in the picture. The peaking or sharpness switch S_{101} adjusts the bypassing in the cathode circuit. This controls the high-frequency response by varying the amount of degeneration. The top portion of S_{101} provides maximum sharpness, with minimum degeneration of the high video frequencies for maximum peaking.

The brightness control R_{139} is in the grid circuit of the video output tube, instead of the CRT circuit, because of the dc coupling. Positive voltage on R_{139} comes from the 25-V source through the brightness limiter adjustment. Negative voltage is obtained from the grid-leak bias on the horizontal output tube. To decrease the brightness, the grid bias on V301 is made more negative. Then its plate current decreases and the plate voltage increases. This increase in positive voltage at the cathodes increases the bias of all three guns to decrease the beam current in the CRT for less brightness. The brightness limiter R_{330} is a setup adjustment on the chassis which prevents blooming at the highest level of brightness for R_{139}.

17-4 Chroma Bandpass Amplifiers

This section uses one or two stages tuned to 3.58 MHz to amplify the modulated chroma signal. The input is colorplexed video signal from the video preamplifier, as shown in Fig. 17-6.

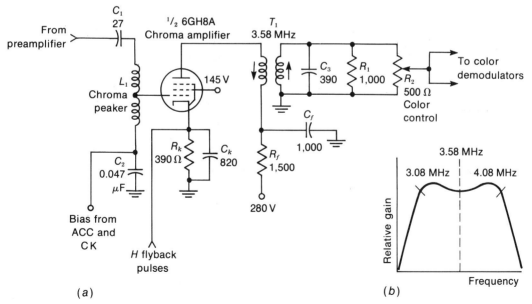

FIGURE 17-6 CHROMA BANDPASS AMPLIFIER USING PENTODE TUBE. R IN OHMS AND C IN PICOFARADS. (a) CIRCUIT; (b) OVERALL CHROMA RESPONSE FROM MIXER GRID TO CHROMA OUTPUT.

The amplified output is 3.58-MHz chroma signal to the color demodulators. The chroma section is considered a bandpass amplifier because its total bandwidth of 1 MHz is a relatively large percentage of the resonant frequency of 3.58 MHz. Specifically, the bandwidth ratio is 1/3.58 = 0.28 or 28 percent. This stage is generally called the bandpass amplifier (BPA), the color IF amplifier, the chroma amplifier, or simply the color amplifier.

Requirements of the chroma amplifier. Whether the amplifier has one or two stages, or uses tubes, transistors, or an integrated circuit (IC), the following features are necessary:

1. The amplifier must be tuned to 3.58 MHz with the required bandwidth. In Fig. 17-6a, the peaker coil L_1 and transformer T_1 are tuned to provide the desired chroma response. T_1 is resonant at 3.58 MHz, but L_1 peaks at higher frequencies to compensate for the sloping response for the chrominance signal in the picture IF amplifier. The overall response of the picture IF amplifier and chroma amplifier is shown by the curve in Fig. 17-6b.
2. The chroma signal level is varied to control the color saturation in the picture. In Fig. 17-6a, the potentiometer R_2 is the color control that varies the amount of chroma signal to the demodulators.
3. Automatic chroma or color control (ACC) is used to determine the bias on the amplifier, in accordance with signal strength. As a result, the color level in the picture remains constant when stations are changed. In Fig. 17-6a, the grid bias for the pentode tube is controlled by the ACC circuit. The burst amplitude determines the ACC bias.
4. When no color signal is received, the chroma amplifier is cut off by a bias voltage applied from the color killer (CK) circuit. In Fig. 17-6a, the chroma amplifier is cut off by negative bias of about −30 V at the control grid. The chroma circuits must be killed because high-frequency components of the monochrome signal can be demodulated to produce color at vertical edges in a black-and-white picture.
5. The chroma amplifier is keyed off during horizontal retrace time, in order to remove the burst from the chrominance signal to the demodulators. In Fig. 17-6a, positive flyback pulses are applied to the cathode to cut off plate current. The burst is removed here because different levels of burst for different stations can change the dc clamping level for the detected color video. This affects the color temperature adjustments for the picture tube.

Pentode chroma amplifier. In Fig. 17-6a, the single pentode tube is the complete chroma amplifier. Cathode bias is provided by R_k and C_k, in addition to the grid bias from the ACC circuit. Typical bias is −2 V, operating as a class A amplifier for the 3.58-MHz chroma signal input from the video preamplifier. The amplified output is 6 V p-p, which can be varied by the color control. The tuning adjustments are L_1 in the grid input and T_1 in the plate output circuit. R_f and C_f form a decoupling filter for the 280-V supply to the plate. C_3 tunes T_1 to 3.58 MHz. R_1 is a damping resistor to provide the required bandwidth. A typical 3.58-MHz bandpass transformer is shown in Fig. 17-7.

IC chroma amplifier. The integrated circuit in Fig. 17-8 includes two chroma amplifier stages and the color killer, as shown in the block diagram in Fig. 17-9. Referring to Fig. 17-8, colorplexed composite video signal is applied to

COLOR CIRCUITS 367

FIGURE 17-7 A 3.58-MHz BANDPASS TRANSFORMER. HEIGHT IS $1\frac{1}{2}$ IN.

pin 2 of IC301, through the 4.5-MHz traps. Amplified chroma signal at 3.58 MHz is available at pin 6. The gain is controlled by ACC bias developed in another IC unit, not shown.

Before the chroma signal from the first amplifier out at pin 6 is coupled to the second amplifier in at pin 7, a peak chroma control (PCC) stage is used. Its function is to prevent excessive color saturation by clipping the signal peaks if the chroma amplitude exceeds a specified level. The output from the second chroma

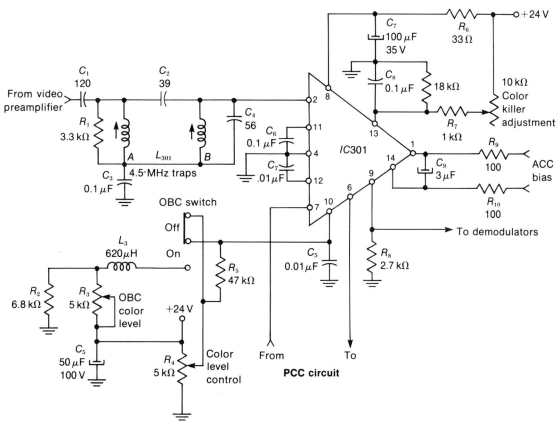

FIGURE 17-8 CHROMA AMPLIFIER USING IC UNIT. R IN OHMS AND C IN PICOFARADS. (FROM SEARS MODEL 4360)

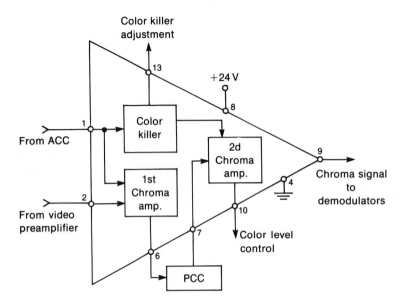

FIGURE 17-9 BLOCK DIAGRAM OF THE *IC*301 UNIT FOR THE CHROMA AMPLIFIER IN FIG. 17-8.

amplifier is out at pin 9 to the chroma demodulators. A 24-V source is applied to pin 8 to supply dc power for all stages, with respect to ground at pin 4.

Note the one-button control (OBC) switch at pin 10 of *IC*301. With this switch in the off position, the manual color level control R_4 adjusts the gain for the chroma level. When OBC is turned on, however, the 47-kΩ resistance of R_5 is added in series. Then the manual control has little effect and the color level is determined by the preset OBC adjustment R_3. The color killer connection at pin 13 provides a dc voltage to cut off the second chroma amplifier on a monochrome program.

Chroma alignment. The response curves in Fig. 17-10 apply. The picture IF response out of the video detector is shown in Fig. 17-10a. The color subcarrier at 42.17 MHz is at 50 percent response on the slope opposite the side for the picture IF carrier. This method is used because it would be too difficult to have the 42.17 MHz at full gain and keep the sound signal at 41.25 MHz from interfering with the color signal. Figure 17-10b shows the compensating response of the chroma amplifier itself. This curve peaks a little above 4.1 MHz. The opposite slopes of Fig. 17-10a and b result in the overall response in c for the 3.58-MHz chroma signal. This curve is at the input to the chroma demodulators, including the gain of the picture IF amplifier and chroma amplifier.

If you were to align only the chroma bandpass amplifier, with its peaker circuit, you would look for the sloping response in Fig. 17-10b. This method is rarely used, however, because it is difficult to get the response exactly right. Usually, the chroma bandpass is part of the overall picture IF alignment.

17-5 The Burst Amplifier Stage

The reference phase for regenerating the subcarrier in the color oscillator is transmitted as a burst of 8 to 11 cycles of 3.579545-MHz signal on the back porch of each horizontal blanking pulse. (See Fig. 17-11.) Since the burst occurs during horizontal blanking time, it is easily separated from the chroma signal. The burst amplifier is simply a stage that is normally off but is keyed on during horizontal retrace. Two methods of keying on the burst separator are:

1. Flyback pulses from the horizontal output circuit.

2. Horizontal sync pulses from the sync separator. The sync pulses must be delayed approximately 4 μs so that they occur during the back porch of horizontal blanking, when the burst is present.

The method of delayed sync has the advantage of making the burst separation independent of the flyback pulses. Their timing can vary with the horizontal hold control and with the way the AFC circuit locks in the horizontal oscillator.

This stage is generally called the burst amplifier, burst separator, or burst gate. The signal that has the burst can be taken from either the video preamplifier or a chroma stage

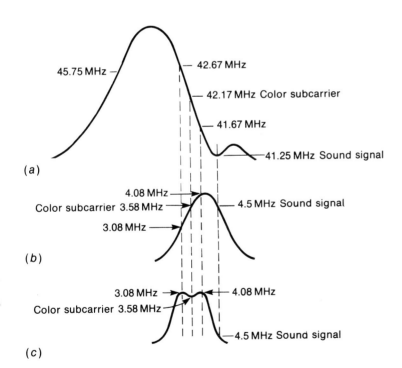

FIGURE 17-10 ALIGNMENT CURVES FOR CHROMA SIGNAL. (a) SLOPING RESPONSE IN PICTURE IF AMPLIFIER; (b) OPPOSITE SLOPE IN CHROMA AMPLIFIER; (c) COMBINED RESULT WITH 4.5-MHz TRAP.

370 BASIC TELEVISION

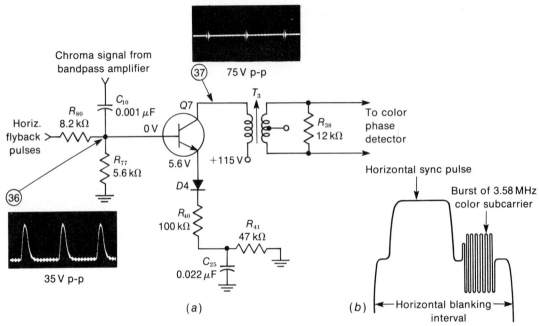

FIGURE 17-11 (a) BURST AMPLIFIER CIRCUIT TO SEPARATE 3.58-MHz BURST SIGNAL SHOWN IN (b).

before blanking. A typical 3.58-MHz transformer for the burst amplifier is shown in Fig. 17-12.

For the transistor $Q7$ in Fig. 17-11, note that the dc voltages normally hold the stage cut off. There is no forward bias at the base. The 5.6 V at the N emitter of the NPN transistor is a reverse bias. However, the 35-V positive flyback pulses applied to the P base are able to key the transistor on during the time the burst signal is present. Therefore, only the burst is amplified to appear in the collector output with an amplitude of 75 V. The burst transformer T_3 couples the burst signal to the color phase detector used for AFPC on the color oscillator. When there is no AFPC circuit, the amplified burst can ring the oscillator crystal directly to lock it in the correct phase.

The amplifier in Fig. 17-11 is basically a common-emitter stage, with input signal to the base and the output signal transformer-coupled from the collector. In the base circuit, R_{80} isolates the burst amplifier from the horizontal output circuit. In the emitter circuit, the split resistance of R_{40} and R_{41} has several functions.

FIGURE 17-12 A 3.58-MHz TRANSFORMER FOR BURST AMPLIFIER. HEIGHT IS $1\frac{1}{2}$ IN.

First, the resistors help to limit the base current to a safe value. Note that D4 has the polarity to block reverse current, but that it conducts for forward electron flow into the emitter. In addition, R_{40} and R_{41} provide dc bias and stabilization in the emitter circuit. Finally, the 100-kΩ R_{40} is not bypassed in order to provide ac feedback for degeneration of the input signal. The purpose is to prevent the high-amplitude flyback pulses in the input from appearing in the output circuit.

17-6 Color Oscillator and Synchronization

The color oscillator regenerates the 3.58-MHz color subcarrier needed in the receiver for the demodulators. This function is necessary because the subcarrier is suppressed in transmission of the color signal.

The input to the color oscillator is the synchronizing voltage needed to hold the hues steady. Although the oscillator uses a 3.579545-MHz crystal for frequency stability, it must be locked into the phase of the burst signal. Either the burst voltage is injected directly into the oscillator circuit or an AFPC circuit compares the oscillator and burst phase to produce a dc control voltage for the oscillator. The manual hue or tint control is a phase adjustment in the oscillator or its control circuit.

The oscillator output is a 3.58-MHz *continuous wave*, or *cw*, signal. This means it has no modulation.

The oscillator synchronizing circuits have the additional functions of turning the color killer off for a color program and providing ACC bias for the chroma bandpass amplifier. The color killer depends on the presence or absence of burst signal. The ACC bias depends on the burst amplitude. The amount of burst can be determined by a control voltage from either the burst phase detector or the bias on the oscillator. These features can be seen in the block diagrams of Fig. 17-13, illustrating different types of circuits for the color oscillator and its synchronization.

For the ringing circuit in Fig. 17-13a, the burst signal shocks the crystal into oscillation at exactly the same frequency and phase as the color sync. Because the crystal has very high Q, its oscillations continue through the time from one pulse of burst to the next. The amplitude of oscillations depends on the amplitude of burst signal. Actually, the stage labeled as the oscillator is an amplifier for the crystal oscillations. The next amplifier stage is also a limiter to provide constant output of oscillator cw signal for the demodulators.

The circuit in Fig. 17-13b uses a free-running crystal oscillator, but the phase and frequency are controlled by the reactance tube. Its reactance is varied by a dc control voltage from the phase detector. Two diodes are used in a discriminator circuit that compares the color burst signal to a sample of the color oscillator output. Essentially the same circuit is shown in Fig. 17-13c, but a semiconductor capacitive diode, varactor, or varicap, is used instead of the reactance tube.

Figure 17-13d shows a combination of ringing with a free-running oscillator. The burst pulses excite the crystal, and the amplified output synchronizes a Hartley or Colpitts type of oscillator.

The last example, Fig. 17-13e, uses a single IC unit, which is the subcarrier regenerator. It contains all the semiconductor components necessary to generate and synchronize the color oscillator. The connections required include the dc voltage source, labeled V, chroma signal input with the burst, horizontal keying pulses, and an external 3.579545-MHz crystal. The output includes the necessary oscillator cw signal for the color demodulators, plus a dc volt-

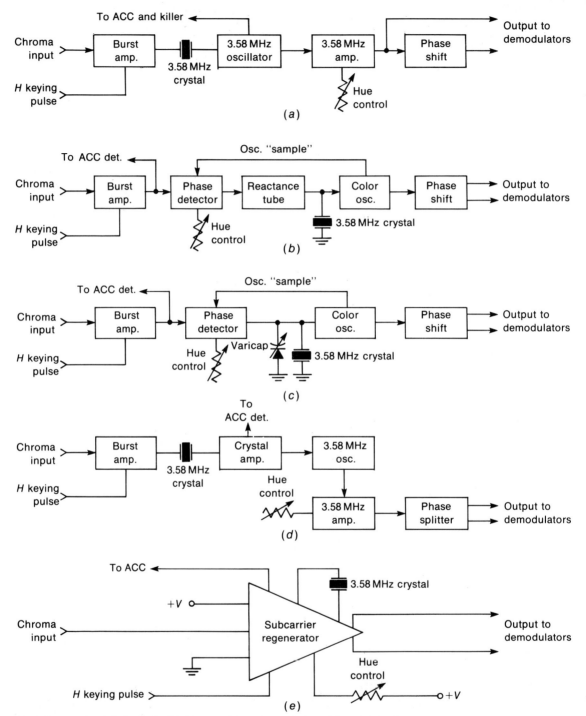

FIGURE 17-13 TYPES OF CIRCUITS FOR 3.58-MHz COLOR OSCILLATOR AND ITS SYNCHRONIZATION. (a) RINGING OSCILLATOR LOCKED IN DIRECTLY BY BURST SIGNALS; (b) FREE-RUNNING OSCILLATOR CONTROLLED BY REACTANCE TUBE; (c) OSCILLATOR CONTROLLED BY VARICAP INSTEAD OF REACTANCE TUBE; (d) COMBINATION OF RINGING AND FREE-RUNNING OSCILLATOR; (e) IC UNIT.

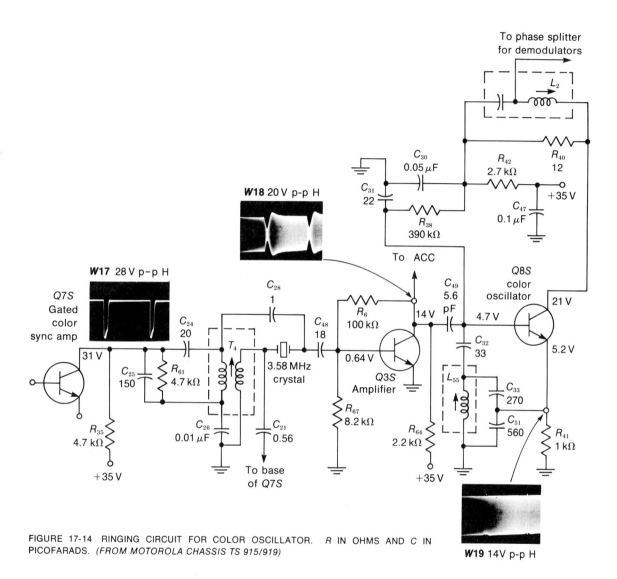

FIGURE 17-14 RINGING CIRCUIT FOR COLOR OSCILLATOR. R IN OHMS AND C IN PICOFARADS. (FROM MOTOROLA CHASSIS TS 915/919)

age proportional to the burst signal that can be used for ACC bias and the color killer.

Ringing circuit for color oscillator. This method can be used with tubes or transistors, but a transistorized circuit is shown in Fig. 17-14. The separated burst signal from $Q7S$ shock-excites the 3.58-MHz crystal. This output is amplified by $Q3S$ to drive $Q8S$ which is a Colpitts oscillator circuit.

The burst transformer for the first stage is T_4. Note that C_{28} is used to provide some regenerative feedback to help maintain crystal oscillations between burst pulses. The crystal output amplifier $Q3S$ serves as a buffer stage to isolate the crystal from loading by the oscillator stage.

Except for the crystal input signal, the color oscillator $Q8S$ is a typical modified Colpitts circuit. Feedback is from the emitter into

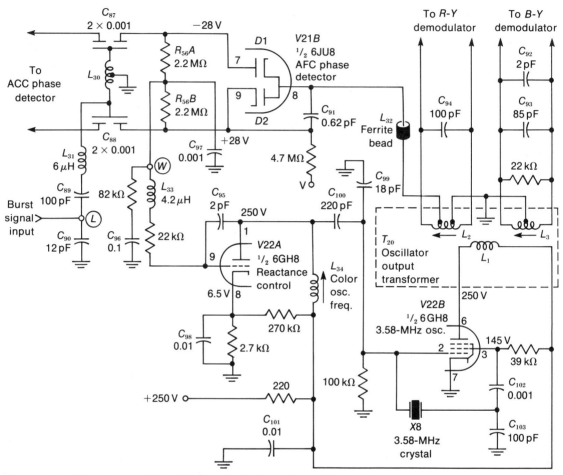

FIGURE 17-15 COLOR OSCILLATOR CONTROLLED BY REACTANCE TUBE. R IN OHMS AND C IN MICROFARADS. *(FROM ZENITH CHASSIS 25LC30)*

the oscillator tank circuit at the junction of C_{33} and C_{51}, across the oscillator coil L_5. Waveform W19 at the emitter shows how the amplitude of the oscillator signal is now constant compared with W18. The output is taken from the collector through L_2 to a phase splitter stage, which supplies oscillator cw to the color demodulators.

The most important change in the waveform cannot be identified in the photos. That is the phase of the oscillator, which is synchronized exactly to the crystal input signal. This synchronization occurs by brute force as the amplitude of the crystal input signal is larger than the free-running amplitude of the oscillator.

Color oscillator with reactance tube. See Fig. 17-15. The dual diode $V21B$ is a phase detector for color AFC. It detects the phase difference between color burst and a sample of the oscillator output. The output is a dc control voltage for bias on the reactance tube* V22A. This stage holds the color oscillator frequency locked in with a specific phase compared with the burst signal.

The color phase detector operates in practically the same way as a sync discriminator for horizontal AFC. However, the color AFC circuit compares the burst signal with a sample of the oscillator output 90° out of phase with burst. When the color oscillator signal is correct, the output of the phase detector is zero. Then both diodes conduct the same amount.

If the phase of the oscillator drifts, its signal will not be exactly in quadrature with the burst signal. Then one diode conducts more than the other and the detector output will go either positive or negative. The polarity indicates whether the oscillator output changed above or below 90°.

The dc control voltage is coupled to the control grid of the reactance tube. The change in bias controls the reactance to correct the oscillator. Phasor diagrams of the signals into the color AFC phase detector are shown in Fig. 17-16.

The color oscillator V22B is a Pierce circuit, with the screen grid as the oscillator anode. This makes the stage an electron-coupled oscillator, isolating the oscillator circuit from the output load in the plate circuit. The windings on the oscillator output transformer supply cw input to the $R - Y$ and $B - Y$ demodulators. The 22-kΩ resistance across L_3 allows this phase to be opposite burst phase for the $B - Y$ signal. The 100-pF capacitance across L_2 makes this circuit resonant to provide quadrature phase for the $R - Y$ signal.

The hue control is not shown in Fig. 17-15 because it is in the output of the burst amplifier.

---

*Operation of the reactance tube is the same as described for an FM signal in Chap. 27.

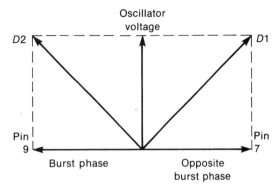

FIGURE 17-16 PHASOR DIAGRAM FOR THE OSCILLATOR AND BURST VOLTAGES APPLIED TO THE AFC PHASE DETECTOR IN FIG. 17-15.

Variation in hue is accomplished by slightly shifting the phase of the burst signal out of the amplifier and into the phase detector. This phase change causes a shift in dc control voltage. The change in bias on the reactance tube shifts the phase of the oscillator output. Finally, the phase shift is in the $R - Y$ and $B - Y$ cw signals to the demodulators, resulting in different hues in the detector output.

17-7 AFPC Alignment

Either misalignment or trouble in the automatic frequency and phase control circuit shows as a loss of color sync in the picture. The result is color bars, as shown in color plate VI. Other symptoms are wrong hues in the picture or a narrow range with the tint control. If the color locks in well for a strong signal but not for weaker stations, this may be caused by poor AFPC action. Normally, if there is color to see in the picture, it should lock in.

You can estimate the frequency error in the 3.58-MHz oscillator by counting the number of color bars. Each pair of horizontal bars corresponds to 60 Hz off the correct frequency. In color plate VI, the five pairs of color bars show that the color oscillator is $5 \times 60 = 300$ Hz off the frequency of 3.58 MHz. As the oscillator frequency comes closer to 3.58 MHz, there are fewer bars. For errors of less than 60 Hz, you can see diagonal colors. When the oscillator frequency is at 3.58 MHz but not locked in sync, the colors change from a solid picture to drifting bars. This condition is called *zero beat* for adjusting the color oscillator, meaning that it is on frequency although not locked in.

In addition to the fact that the colors must be locked in, the tint control should vary the hues from purple at one end to green at the opposite end, with correct flesh tones in the center. The range of the tint control is determined by the AFPC adjustments.

The following general procedures for AFPC adjustments apply to most receivers:

1. Use a color bar generator to see all the hues in the picture. With sync, the vertical color bars are locked in. However, without color sync you see diagonal colors across the vertical bars.
2. Set the manual hue or tint control to the center of its range.
3. Disable the color killer. This turns on the chroma bandpass amplifier. Have the color control at maximum.
4. Turn automatic color circuits off.
5. Tune the 3.58-MHz burst circuits for maximum output.
6. Tune the 3.58-MHz color oscillator for maximum output at the correct frequency. Without any burst input, the oscillator is on frequency at the point of zero beat, when the color bars drift very slowly.

After the AFPC adjustments, make sure all controls are back to normal and the color is locked in. If necessary, the color oscillator can be readjusted slightly to give the tint control the desired range. Set the color killer to remove color snow on an unused channel.

17-8 Chroma Demodulators

The chroma signal from the bandpass amplifier consists of the side bands produced by modulation of the 3.58-MHz color subcarrier. However, the color subcarrier itself is suppressed in transmission. This 3.58-MHz cw signal is reinserted in the demodulators by the output of the color oscillator. Then the chroma side bands can beat with the 3.58-MHz subcarrier so that the color video signal can be detected as the difference frequencies between the side bands and the subcarrier. An important factor in this method is that the detected output depends on the phase of the reinstated subcarrier. This is why the color demodulator is called a *synchronous detector*.

The basic arrangement for the color demodulator circuits is illustrated by the block diagram in Fig. 17-17. Tubes or transistors can be used, including diodes and integrated circuits. With two demodulators, two color-difference signals are detected. These are usually $R-Y$ and $B-Y$ video, which are combined to form $G-Y$. Then the color-difference signals are matrixed to form R, G, and B video signals. The end result required is red, green, and blue video for the picture tube. Some circuits use R, G, and B demodulators to detect the red, green, and blue video directly. The hue of the detected output signal from the color demodulator depends on the phase of its oscillator cw input. Any hue has its own phase angle.

Note the multiplying factors shown for the amplifiers in Fig. 17-17. These are correction values to compensate for the relative amplitudes used in the modulation at the transmitter. The relative gain for the $R-Y$ video amplifier is 1.14 because the signal is reduced by the factor of $1/_{1.14}$ to prevent overmodulation. Similarly, the relative gain for the $B-Y$ amplifier is 2.03. In brief, the $B-Y$ detector output is boosted by almost double the gain of the $R-Y$ signal. However, the

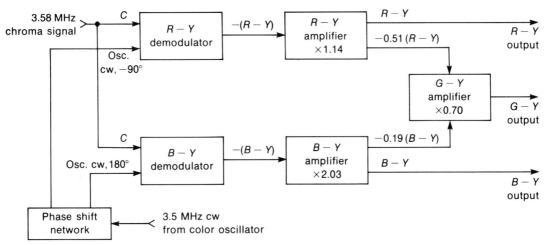

FIGURE 17-17 BLOCK DIAGRAM OF $R-Y$, $B-Y$ DEMODULATION SYSTEM WITH COLOR-DIFFERENCE AMPLIFIERS.

$G-Y$ gain is reduced by the factor of 0.70 or $1/1.14$ because this signal is increased by the factor 1.4 in modulation. This amplitude correction is applied before the $R-Y$ and $B-Y$ signals are combined to form $G-Y$. For the corrected signals, $G-Y$ is equal to 51 percent of $-(R-Y)$ and 19 percent of $-(B-Y)$.

Phase angles for the demodulators. The combination of $R-Y$ and $B-Y$ demodulators is most common because these two axes are in quadrature. $B-Y$ phase is 180° from burst phase. The hue for $R-Y$ is at 90°, exactly between the phase of burst and $B-Y$. The quadrature phase is preferred because a synchronous detector produces maximum signal output for the phase of its cw input but minimum for any signal 90° out of phase. Signal that is 180° out of phase also produces maximum output, but of opposite polarity.

For R, B, and G demodulators, three detectors are needed. These have oscillator cw input at $-77°$, $-193°$, and $-299°$, as shown in Fig. 17-18. Note that these angles are with respect to burst as the reference at 0°. The negative sign indicates the clockwise direction.

Another combination uses X and Z demodulators. The X axis is at $-282°$, close to the

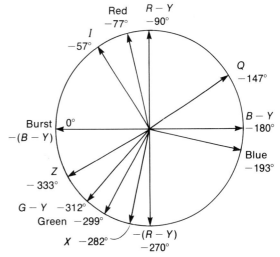

FIGURE 17-18 HUE PHASE ANGLES FOR COLOR DEMODULATORS. VALUES ARE SHOWN NEGATIVE IN THE CLOCKWISE DIRECTION FROM BURST AS THE REFERENCE.

phase of $-(R-Y)$. The Z axis is at 333°, close to $-(B-Y)$ or burst phase. Actually, the X and Z demodulators just form a balanced circuit, with a $G-Y$ adder to provide $R-Y$, $B-Y$, and $G-Y$ video signals in the output. This result is determined by the relative amplitudes for the detector output voltages.

The I and Q modulation axes for the transmitted chroma signal are practically never used at the receiver because of the extra bandwidth of the I signal. The receiver circuits are much simpler when the detected color video signals all have the same bandwidth of 0.5 MHz.

Basic demodulator requirements. Figure 17-19 shows the circuit of a $B-Y$ demodulator using a pentode tube. The same circuit can be used for $R-Y$, X, or Z demodulators, the only difference being the phase of the oscillator cw input. Also, the circuit can use a triode tube, transistor, or diode. It is only required to have two inputs, the 3.58-MHz chroma signal from the bandpass amplifier and the 3.58-MHz cw subcarrier from the color oscillator.

For the pentode in Fig. 17-19, the oscillator cw is applied to grid 1. This signal at the control grid can be considered as keying on or off the conduction of plate current (I_b) for the chroma signal applied to grid 2. Note that the $B-Y$ phase of the oscillator cw is close to the phase of blue in the chroma signal but opposite from yellow. Maximum I_b flows for blue, then, as both signals have the same phase. Minimum I_b flows for yellow. Therefore, the demodulator has its maximum peak-to-peak value for hues on the axis of blue and yellow. Chroma signal of the quadrature phase has little effect on the output.

In the demodulator output circuit, the variations in I_b produce video signal voltage across R_L. The polarity of the output is inverted

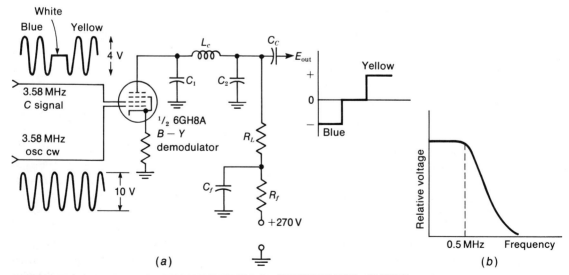

FIGURE 17-19 (a) CIRCUIT WITH PENTODE TUBE FOR $B-Y$ DEMODULATOR; (b) VIDEO-FREQUENCY RESPONSE FOR DETECTED OUTPUT.

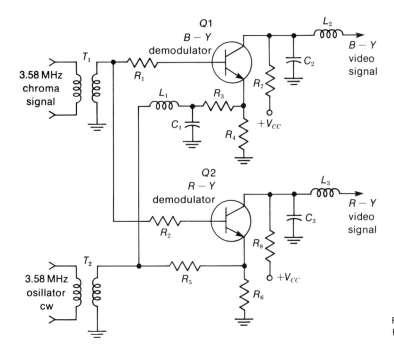

FIGURE 17-20 TRANSISTOR CIRCUITS FOR $B-Y$ AND $R-Y$ DEMODULATORS.

from the input as the plate voltage decreases because of a larger voltage drop across R_L with more I_b. The video peaking coil L_c with the shunt capacitance of C_1 and C_2 provides the video-frequency response up to 0.5 MHz shown in Fig. 17-19b. In brief, the color information is in the 3.58-MHz chroma signal before the demodulators, but after detection the color video signals have the bandwidth of 0 to 0.5 MHz.

Transistors for the demodulator circuits. See Fig. 17-20, with NPN transistors Q1 and Q2 used as $B-Y$ and $R-Y$ demodulators. For each stage, chroma signal is coupled to the base by T_1. Also, oscillator cw signal is applied to both emitters by T_2. Note that there is no forward bias voltage at either base. Therefore, the ac input signals drive the transistor into conduction. Effectively, the base-emitter serves as a diode rectifier for synchronous demodulation, and the detected resultant is amplified in the collector output circuit. For Q1, R_7 is the load resistance and L_2 the peaking coil for the $B-Y$ color video signal. Similarly, R_8 is the collector load and L_3 the peaking coil for $R-Y$ signal from Q2. Then the $R-Y$ and $B-Y$ signals can be combined in a $G-Y$ adder, not shown here.

IC demodulator module. See Fig. 17-21. The integrated circuit here is one of two small modules with IC units that contain all the chroma stages. The other IC, not shown, has the bandpass amplifier, color killer, burst amplifier, color oscillator, and AFPC circuits. In Fig. 17-21, this IC unit has $R-Y$ and $B-Y$ demodulators, the $G-Y$ adder, and three emitter-follower stages to isolate the output signals. Each goes to a separate module for matrixing and video drive to the red, blue, and green guns of the picture tube.

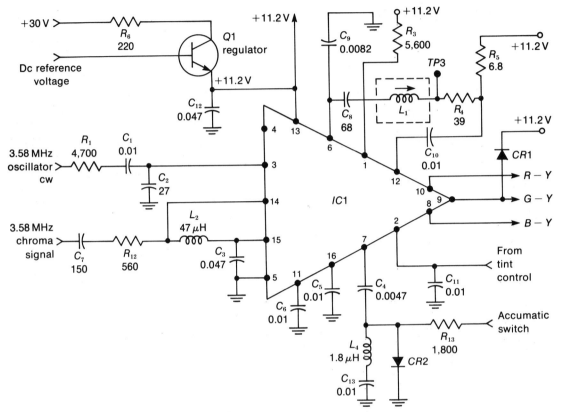

FIGURE 17-21 IC DEMODULATOR. R IN OHMS; C MORE THAN 1 IN PICOFARADS (pF), LESS THAN 1 IN MICROFARADS (μF). (FROM RCA CHASSIS CTC 59)

The 3.58-MHz chroma signal for demodulation is applied to pin 14 of $IC1$ in Fig. 17-21. This is applied directly to the $R-Y$ and $B-Y$ demodulators. The oscillator cw signal input at pin 3 is amplified and limited for constant amplitude before being applied to the demodulators with the required quadrature phases. The only external adjustment is the tuning coil L_1, which is set for maximum 3.58-MHz oscillator cw at test point $TP3$. This circuit provides the 90° phase shift for the $R-Y$ demodulator.

Control of the tint or hue is accomplished at pins 2 and 7. A dc voltage from the tint control on the front panel is applied to pin 2. This voltage controls the phase of the oscillator cw signal amplified within the IC unit. At pin 7, the voltage depends on the Accumatic switch, which is the RCA name for one-button color. Normally, the signal at pin 7 is grounded through C_4 and $CR2$. The diode is biased into conduction by 11.2 V at its anode with the Accumatic switch off. When the switch is on, this positive voltage is removed and the diode is turned off. Now pin 7 goes to ground through C_4, L_4, and C_{13}. This network establishes a fixed phase angle for the oscillator cw signal for proper flesh tones, even if the manual tint control is set wrong.

The regulator stage Q1 in Fig. 17-21 supplies 11.2 V regulated dc collector voltage for the stages in IC1 and other modules. At the base of the regulator the reference is 11.2 V from a zener diode. C_{12} is a bypass capacitor for the 11.2-V supply line from the emitter of Q1. Similarly C_3, C_5, and C_6 are bypass capacitors for the internal stages of the IC unit.

Diode demodulators. In Fig. 17-22 two dual-diode units D1 and D2 are used for $B-Y$ and $R-Y$ demodulators. In each unit, the diodes are series-aiding, with the anode connected internally to the cathode. Externally, the three terminals include a cathode at 1, the opposite anode at 2, and the common center connection at 3. The oscillator cw signal is applied push-pull to terminals 1 and 2 by the input transformer T_1. This is connected to both the $R-Y$ and $B-Y$ demodulators without any 90° phase shift. Instead, the chroma signal from input transformer T_2 is shifted 90° for the $R-Y$ demodulator D2. The quadrature phase is provided by L_4 with C_{45} and C_{16}. The chroma signal is applied to the center terminal 3 of the demodulators.

In this circuit, the chroma signal has the quadrature phase for the $R-Y$ demodulator, instead of using 90° for the oscillator cw signal. However, the results are the same. The advantage is that the push-pull oscillator cw signal can be balanced for zero output when the chroma signal input is zero.

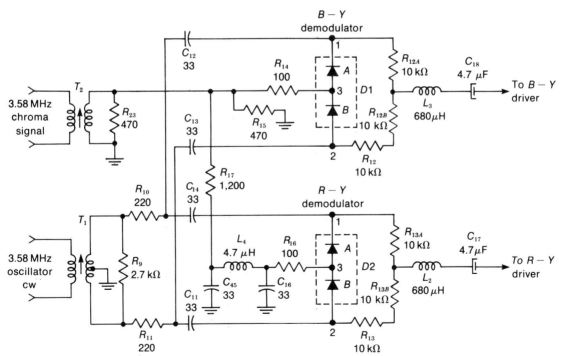

FIGURE 17-22 DIODE DEMODULATOR CIRCUITS. R IN OHMS AND C IN PICOFARADS. (FROM MAGNAVOX CHASSIS T979)

Each diode pair detects its phase of chroma signal to produce color video output. This signal is taken from the center connection of either R_{12} or R_{13} as a center-tapped diode load resistor. The voltage at this point corresponds to the voltage at terminal 3 between the diodes A and B in the demodulator. When the oscillator cw signal is in phase with the chroma signal, one diode conducts more than the other. For opposite phases, the opposite diode conducts more. The result is maximum output signal for $B - Y$ phase from $D1$ and for $R - Y$ from $D2$. In either demodulator, there is no crosstalk between quadrature signals. Then both diodes conduct the same, which results in zero output because of the balance.

Summary of demodulator action. For any type of synchronous demodulator, two input signals are required: a 3.58-MHz modulated chroma signal and a 3.58-MHz oscillator cw signal. The result is synchronous detection in the following steps:

1. When the oscillator cw and chroma input have the same phase, the output amplitude increases.
2. When the oscillator cw and chroma input have opposite phases, the output amplitude decreases below the medium level.
3. When the oscillator cw and chroma input are 90° out of phase, the output amplitude corresponds to zero oscillator cw signal. The reason is that for 90° phase, one signal is at zero when the other is maximum.
4. With varying amplitudes of chroma signal, the amount of detector output varies in proportion to the amplitude.
5. With varying phases of chroma signal, the detector output varies in proportion to the phase change.
6. As a result, the demodulator detects only the phase of chroma signal corresponding to the phase of the oscillator cw input.

It is important to note that the color demodulator is where the color video signals are extracted from the 3.58-MHz chroma signal. For the input there is just one chroma signal. Its frequency is 3.58 MHz with a bandwidth of ±0.5 MHz. In the detected output, there are separate video signals for red, green, or blue with a bandwidth of 0 to 0.5 MHz.

17-9 Matrixing the Y Video and Color Video Signals

Matrixing means combining or adding signals. An example is adding $R - Y$ and $B - Y$ to form $G - Y$. The most important application in the receiver, though, is matrixing the color-difference signals with Y video to form the red, green, and blue video signals needed for the three guns of the picture tube. This matrixing for R, G, and B can be done either in the picture tube or in the color video circuits. Examples are shown in Figs. 17-23 to 17-25.

Matrixing in the picture tube. This method in Fig. 17-23 is commonly used with vacuum tube amplifiers. The control grid of each gun in the picture tube has $R - Y$, $B - Y$, or $G - Y$ video signal. All three cathodes have the same $-Y$ video signal. The polarity is negative for Y signal at the cathode because the effect of cathode voltage on beam current is opposite that of grid voltage. Effectively, then, each grid has R, G, or B video signal to control beam current for the color information, as the Y component in the color signal at the grid is canceled by the Y signal at the cathode. Algebraically, the matrixing for the color video signals can be added as follows:

$(R - Y) - (-Y) = R - Y + Y =$ red video
$(G - Y) - (-Y) = G - Y + Y =$ green video
$(B - Y) - (-Y) = B - Y + Y =$ blue video

The same algebraic combinations can be used when the Y signal and color video signals are all applied to the cathodes for the picture tube, with the grids grounded. This method is often used with transistor amplifiers as protection against arcing at the control grid in the picture tube. The color-difference signals at the cathode then have negative polarity, like the $-Y$ signal. For red, as an example, $-(R - Y) + (-Y) = -R$ at the cathode.

Matrixing in color video amplifiers. Figure 17-24 shows an example of matrixing Y signal in an $R - Y$ amplifier. The NPN transistor Q1 is a color video amplifier for $-(R - Y)$ signal at the base, but Y signal is also injected at the emitter. In terms of the input color signal, it is amplified and inverted to produce $R - Y$ signal at the collector. However, the Y signal at the emitter is amplified without phase inversion. The resultant output signal from the collector is equivalent to $R - Y + Y = R$.

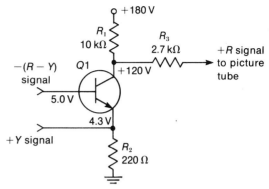

FIGURE 17-24 CIRCUIT FOR MATRIXING $-(R - Y)$ AND Y SIGNALS IN A COLOR VIDEO AMPLIFIER.

For the circuit in Fig. 17-24, the 10-kΩ R_1 is the collector load resistance. The 220-Ω R_2 is the emitter resistance for Y input. For Y signal, the stage is effectively a common-base amplifier, without phase inversion. For the $-(R - Y)$ signal at the base, the stage is a common-emitter amplifier. Both the Y and $R - Y$ signals are developed in the collector output circuit. The 2.7-kΩ R_3 is an isolating resistance for the coupling circuit to the control grid of each gun in the picture tube.

Three stages such as Q1 would be used for the three color video signals. The output of each is positive R, G, or B video for the control grid in the picture tube. When the color video signals are applied to the cathodes in the picture tube, they have negative polarity, with $-R$, $-G$, and $-B$ for cathode drive.

Matrixing in the demodulators. In Fig. 17-25, the Y signal is applied to the center tap of T_1 for the chroma input. As a result, the Y signal is in series with the chroma signal for D1 and D2. The two diodes form one demodulator circuit. Three of these demodulators are used to detect $R - Y$, $B - Y$, and $G - Y$. However, the matrixing of Y signal results in R, G, and B video output.

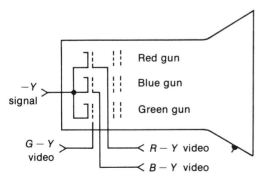

FIGURE 17-23 MATRIXING IN THE PICTURE TUBE.

384 BASIC TELEVISION

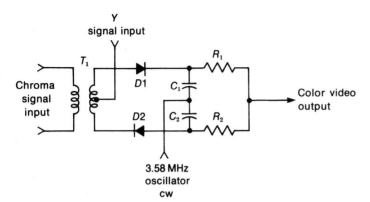

FIGURE 17-25 CIRCUIT FOR MATRIXING Y SIGNAL IN A COLOR DEMODULATOR. $D1$ AND $D2$ FORM ONE DEMODULATOR CIRCUIT FOR DETECTING $R-Y$, $B-Y$, OR $G-Y$.

The Y signal will be passed through to the output whether or not a chroma signal is being received. If no signal is applied to the primary of T_1, the Y signal will be the only signal applied to the anode of $D1$ and the cathode of $D2$ in the same phase. However, when chroma signal is present at T_1, the Y signal will represent a changing voltage level at the secondary center tap. In this way, the chroma signal will be superimposed on the Y signal. Both signals will act on the two diodes. As a result, the Y signal will become part of the output.

The 3.58-MHz oscillator cw signal is connected to both diodes through C_1 and C_2. As the oscillator signal alternately turns the diodes on and off, the average current will depend on the polarity of the Y signal. When the Y signal is positive, $D1$ will conduct more than $D2$ and the output will be positive. When the Y signal is negative, $D2$ will conduct more than $D1$ and the output will be negative.

The 3.58-MHz chroma signal is smaller in amplitude than the Y signal. As a result, the demodulated color video will ride on the Y signal, much as an ac signal rides on a dc level when both are in the same circuit. This is desirable because the Y signal represents the brightness component, while the chroma is the color.

The result, then, is that the correct brightness level is restored to the color video signal.

17-10 Video Amplifiers for Color

These circuits include color-difference stages such as the $R-Y$, $B-Y$, and $G-Y$ amplifiers and the output stages for red, green, and blue video. When the Y video signal is matrixed in the picture tube, it is only necessary that the color video amplifiers have a bandwidth of 0.5 MHz for the color information. When Y signal is combined with the color signal, however, the video amplifier must have the bandwidth of 3.2 MHz to include the details of luminance. An advantage of matrixing the Y and color video signals in the color video circuits is the fact that the red, green, and blue video drive controls then proportion both the luminance and color video signals for the picture tube.

Color-difference amplifier stages. In Fig. 17-26, triple-triode 6MJ8 is used for all three color video amplifiers. The $G-Y$ signal is obtained by combining signals from the $R-Y$ and $B-Y$ stages. One source is the common-cathode resistor R_{773}, which is in parallel with the

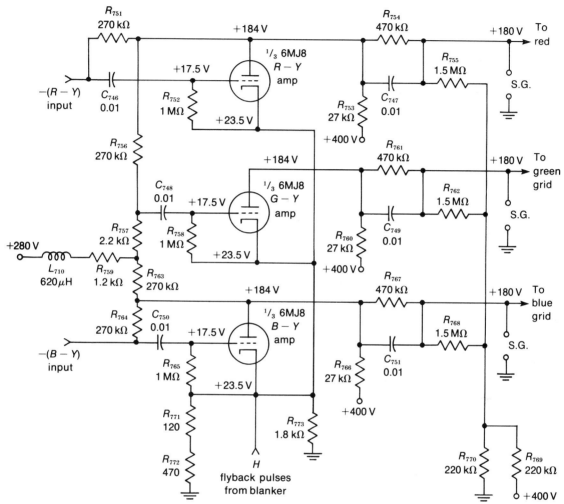

FIGURE 17-26 CIRCUIT FOR COLOR-DIFFERENCE AMPLIFIERS. R IN OHMS AND C IN MICROFARADS. S.G. IS SPARK GAP FOR GRID OF PICTURE TUBE. (FROM ADMIRAL CHASSIS 2K20)

series combination of R_{771} and R_{772}. This group is shown in the cathode circuit of the $B - Y$ amplifier at the bottom of the diagram, but the combined resistance has the plate signal current of the three amplifiers. The other source for $G - Y$ signal is from the plate circuits of the $R - Y$ and $B - Y$ amplifiers through the resistor network of R_{756}, R_{757}, and R_{763} at the top left in the diagram. The overall result is the proper amount of combined signal to form $G - Y$. Each triode is a video amplifier, with response up to 0.5 MHz for the $R - Y$, $B - Y$, or $G - Y$ video signal.

The input signals are ac-coupled to the amplifiers by C_{746}, C_{748}, and C_{750}. In each stage, the coupling capacitor blocks any dc component of the signal level. However, the dc level is restored, effectively, by grid-leak bias. Then the amplified plate signal for each amplifier is directly coupled to its respective grid in the picture tube. The dc coupling means that the dc component reinserted in the grid circuit is applied from the plate circuit as part of the video signal drive for the picture tube. Grid-leak bias results from horizontal flyback pulses applied to the cathodes from a transistor blanker stage. More details of the blanking will be shown in Fig. 17-29.

Output stage for blue video signal. The circuit in Fig. 17-27 amplifies the blue signal, with two more identical stages used for red and green video output. The Y signal has already been matrixed with the color video. As a result, the input here is one signal that has everything needed to drive one gun in the color picture tube, including the luminance and blue information. The amplitude of the video output signal is about 120 V p-p. Furthermore, the dc component has been reinserted in the signal by a previous stage, so that the video output stage is dc-coupled to the picture tube.

The input signal is applied to the base of Q8 as a CE amplifier. Base bias is provided by the dc level of the signal taken from an IC demodulator unit not shown here. The emitter voltage is made variable with R_{54} to control the gain. This is the blue drive control. Because of the dc coupling of the collector voltage, the drive control also varies the bias on the picture tube.

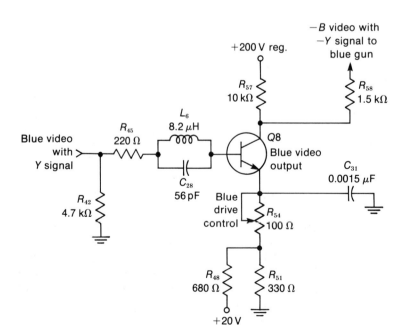

FIGURE 17-27 COLOR AMPLIFIER CIRCUIT FOR BLUE VIDEO DRIVE. (FROM MOTOROLA CHASSIS TS938)

COLOR CIRCUITS 387

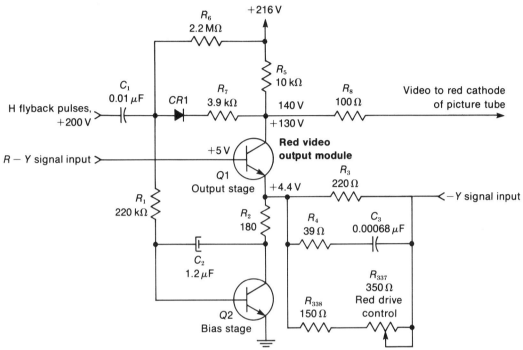

FIGURE 17-28 COLOR VIDEO AMPLIFIER THAT INCLUDES MATRIXING AND BLANKING. (FROM RCA CHASSIS CTC 59)

Red video output module. In Fig. 17-28, Q1 is the video output transistor supplying all the signal needed for the red gun, including the red video, matrixed Y video, and blanking pulses. Q2 is a bias transistor to control the current in the output stage. Both Q1 and Q2 are on a small module board, with three separate modules (for red, green, and blue video output) to drive the picture tube.

17-11 The Blanker Stage

Figure 17-29 shows the same color-difference amplifiers as Fig. 17-26 but with the blanker Q710 added. This stage provides horizontal flyback pulses for the cathode of each amplifier. Consider the $R - Y$ amplifier at the top of the diagram. Since its output is coupled to the control grid of the red gun in the picture tube, blanking pulses in this signal eliminate horizontal retraces in the picture. Furthermore, the blanking pulses in the input circuit are used to clamp the bias for the color video signal. This dc level is maintained at the picture tube because of the dc coupling.

The purpose of the dc clamping level is to keep the bias on each gun constant despite different levels of color video signal and a possible drift in the dc supply voltages. The same method of blanking and clamping by means of grid-leak bias is used in the three color-difference amplifiers.

388 BASIC TELEVISION

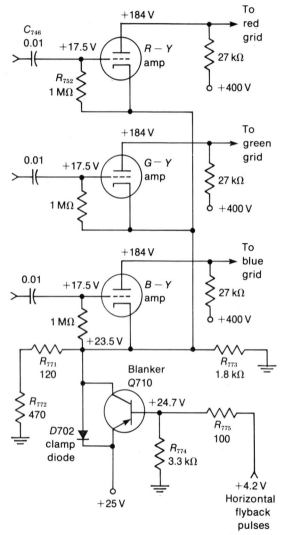

FIGURE 17-29 DETAILS OF BLANKER STAGE, FROM CIRCUIT IN FIG. 17-26. R IN OHMS AND C IN MICROFARADS.

Grid-leak bias for the $R-Y$ amplifier, as an example, is produced by the flyback pulses at the cathode from the collector of the blanker stage $Q710$. Into the base are pulses of positive polarity. Out from the collector are amplified and inverted negative flyback pulses. The negative pulse at the cathode of the $R-Y$ amplifier corresponds to positive grid voltage, to produce grid current. Then the 0.01-μF coupling capacitor C_{746} is charged. The RC time constant with the 1-MΩ R_{752} is long enough to maintain the bias between pulses.

In addition, the flyback pulses at the cathode of the color-difference amplifier are amplified but not inverted in the plate circuit. Then the negative pulses to the control grid of the picture tube cut off beam current during horizontal retrace time. The guns are cut off only for the period of the flyback pulses.

The blanker $Q710$ is a PNP transistor. V_{CE} is -1.5 V, equal to the difference between 25 V on the emitter and 23.5 V at the collector. The transistor is normally conducting full saturation current, but the positive flyback pulses at the base cut the stage off. Then the collector voltage rises to provide positive flyback pulses at the cathode of each color-difference amplifier. The clamp diode $D702$ prevents the cathode voltage from exceeding 25 V.

17-12 Troubles in the Color Circuits

To save time in localizing troubles, it is important to keep in mind the fact that the ac color signals from all the color circuits have only one function in the overall picture. The color information is important, but it is just superimposed on a monochrome picture formed on a white raster. The gray scale or color temperature of the raster depends on dc voltages for the picture tube. The monochrome picture is produced by the ac luminance signal.

The color information is a result of correct operation of the 3.58-MHz chroma amplifier, 3.58-MHz burst amplifier, 3.58-MHz color oscillator, color AFPC circuit, color demodulators, and color video amplifiers. Normal oscilloscope waveshapes for these ac color signals can be seen by referring back to Fig. 9-4. These waveshapes are obtained by tracing the signal from a color-bar generator.

Is the raster normal? Its gray scale depends on the dc grid bias and screen voltage for each gun in the picture tube. Because of dc coupling from the color video amplifiers, a change in dc voltage here can change the bias for the raster.

In addition, a short circuit from grid to cathode or cathode to heater in the picture tube affects the bias. However, normal gray scale without color in the raster and brightness that can be varied by the brightness control indicate that the picture tube is good and the dc bias voltages are normal.

Is the black-and-white picture normal? This shows normal Y video signal with balanced drive for the red, green, and blue guns. The contrast control should vary the amount of contrast in the picture. A normal monochrome picture on a white raster indicates a good picture tube. See if the screen controls vary the color of the raster.

Is the color picture tube good? The conventional trouble of weak emission from one gun in an old tube shows as a weak, silvery picture for that color. However, even a new picture tube can have internal short circuits that affect the bias and the brightness. This often is indicated by flashing in the picture, with visible retrace lines. The possibility of a cathode-to-heater short can be checked with an isolation-booster transformer for the heater. However, when the monochrome picture is normal on a good raster, any color trouble is not likely to be in the picture tube because its most difficult function is to make a good black-and-white picture.

Dc-voltage troubles. In the picture tube, color video amplifier, or Y video amplifier, any trouble that changes the dc bias for the picture tube affects the gray scale of the raster. Furthermore, a bias trouble can be confused with no high voltage because of no brightness. The screen may be very dark, but any illumination at all shows that high voltage is present.

A dc-voltage trouble that causes excessive color in only one hue is illustrated by the circuit in Fig. 17-30 for a red video amplifier Q1. The collector voltage of Q1 determines the control-grid voltage for the red gun. The cathode voltage is set by the brightness control at 200 V. Normally, the grid voltage is 150 V. This results

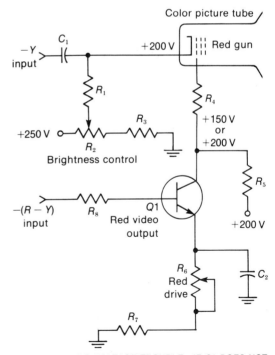

FIGURE 17-30 DC VOLTAGE TROUBLE. IF Q1 DOES NOT CONDUCT, THE GRID OF THE RED GUN WILL BE TOO POSITIVE.

from the 50 V drop of I_C across R_5 from the 200-V collector supply. The net grid-cathode voltage then is $150 - 200 = -50$ V for the control grid.

However, if Q1 is not conducting, its collector voltage rises to 200 V. Then the grid bias is $200 - 200 = 0$ V. With zero bias, the red gun conducts maximum current. Since the blue and green guns have only normal bias, the entire screen will be red. Probably the raster will bloom also because of excessive beam current. This trouble of too much red will be in the raster, monochrome picture, and color picture.

Troubles in the ac color signal. These effects are just in the color; the monochrome picture is normal. The different possibilities are no color at all, weak color, no color sync, wrong hues, and trouble in just one color. These are considered separately in the next sections. Remember that the 3.58-MHz signal is for all the colors. Only the individual color video signals after the demodulators have separate colors. However, these stages may all be in one common tube or IC unit. Also note that the color video amplifiers in Fig. 17-26 have one common-cathode resistor, which is R_{773}.

No color. Check for output in the 3.58-MHz chroma amplifier and oscillator circuits. Make sure the color killer or ACC circuit is not cutting off the bandpass amplifier. Note that if there is no burst signal, the color killer will cut off the chroma signal, either in the bandpass amplifier or in the demodulators.

Color snow. See color plate V. With the color level control at maximum, the color killer off, and the receiver tuned to an unused channel, the screen should show color snow. This is random noise at frequencies near 3.58 MHz that is amplified in the color circuits. The presence of color snow but no color picture shows that the color circuits are operating, but there is no 3.58-MHz chroma signal.

No color sync. See color plate VI. Drifting colors or color bars indicate that the color oscillator is not being synchronized by the color burst signal. The cause is always the AFPC section. If there is no color sync burst, however, the result will be no color at all, as the color killer cuts off the chroma signal.

One color missing. This symptom is usually caused by a problem in a single color video channel. Keep in mind the hues for the color-difference signals. $B - Y$ is essentially blue and yellow; $R - Y$ is red and cyan; $G - Y$ is green and magenta; X signal is close to $-(R - Y)$ with cyan and red; and Z signal is close to $-(B - Y)$ with yellow and blue.

SUMMARY

1. The color circuits in the receiver use the 3.58-MHz modulated chroma signal to provide red, green, and blue video signals for the picture tube.
2. The input to the video preamplifier stage is the colorplexed composite video signal from the video detector. The output supplies Y signal to the delay line into the video amplifier, composite video signal for the deflection sync and AGC stages, plus 3.58-MHz chroma signal for the bandpass amplifier.

3. The Y video amplifier supplies the luminance signal of 80 to 140 V p-p needed to drive the picture tube. The delay line in the input to the Y amplifier delays the luminance signal approximately 0.8 μs with respect to the color signal.
4. The 3.58-MHz bandpass amplifier, or color amplifier, supplies chroma signal to the color demodulators. The color amplifier generally has the color level control to vary the saturation in the picture by varying the amount of color signal.
5. The burst amplifier separates the 3.58-MHz color sync burst on the back porch of the horizontal blanking pulses. This is accomplished by keying the stage to be on for horizontal retrace time only.
6. The color oscillator is controlled by a 3.579545-MHz crystal. Its function is to regenerate the 3.58-MHz subcarrier suppressed in transmission. The output of the color oscillator, with the phases required to detect the desired hues in the 3.58-MHz modulated chroma signal, goes to the synchronous demodulators.
7. The automatic frequency and phase control (AFPC) circuit locks in the color oscillator to hold the hues steady in the reproduced picture.
8. A color demodulator is a synchronous detector that requires two input signals. One is the 3.58-MHz modulated chroma signal. The other is 3.58-MHz oscillator cw of a particular phase to detect specific hues in the chroma signal. The detected output is color video signal.
9. The Y video signal and color video signals may be matrixed in the picture tube to produce red, green, and blue video, or the matrixing can be in the color video circuits.
10. Color troubles can be localized in terms of the raster and monochrome picture. A good raster shows normal dc voltages. A good monochrome picture shows normal Y signal and correct dc voltages for the picture tube. Usually, this indicates that the picture tube is not defective. No color at all results from no ac chroma signal at 3.58 MHz or no oscillator cw output. Troubles with one color are in the ac signal from a color video amplifier. Color sync troubles are generally caused by the AFPC circuit.

Self-Examination (Answers at back of book)

1. Which of the following is *not* tuned to 3.58 MHz: (a) burst amplifier; (b) video preamplifier; (c) chroma amplifier; (d) color demodulator input.
2. When $B - Y$ and Y signal are combined, the result is: (a) blue video; (b) 3.58-MHz chroma; (c) red video; (d) green video.
3. The phase angle between $B - Y$ and $R - Y$ is: (a) 180°; (b) 57°; (c) 0°; (d) 90°.
4. Which of the following applies for a monochrome program: (a) chroma amplifier on; (b) Y video amplifier off; (c) color killer on; (d) picture tube off.

5. The manual color control is generally in which circuit: (a) red video output; (b) Y video output; (c) chroma bandpass amplifier; (d) R − Y demodulator.
6. The contrast control is generally in which circuit: (a) burst amplifier; (b) Y video amplifier; (c) blue video output; (d) color oscillator.
7. A manual control in the AFPC circuit is for (a) tint; (b) saturation; (c) contrast; (d) brightness.
8. If the color oscillator does not operate, the result will be: (a) no picture; (b) no color; (c) incorrect hues; (d) no color sync.
9. The hue of the color sync burst phase is: (a) red; (b) blue; (c) magenta; (d) yellow-green.
10. The balance for Y video signals to the three guns in the picture tube is set by the: (a) drive controls; (b) screen controls; (c) contrast control; (d) color control.

Essay Questions

1. Give three requirements for a picture reproduced in color.
2. List the input and output signals for the following stages: video preamplifier, Y video amplifier, chroma bandpass amplifier, burst amplifier, R − Y demodulator, color oscillator, and G − Y adder.
3. Give the circuits and functions for the manual contrast, color, and tint controls.
4. Give two differences between the sets of controls for R, G, B drive and R, G, B screen-grid adjustments.
5. What is the difference in color synchronization between a ringing oscillator circuit and an oscillator with a reactance tube?
6. Approximately how much video signal voltage is required to drive a 21-in. color picture tube?
7. Give the polarity of Y video signal and color video signal required at the cathode of the picture tube.
8. List the video signals with their polarity for matrixing in the picture tube.
9. Give one advantage and one disadvantage of matrixing in a video stage, before the picture tube.
10. Give one stage in the color circuits that must be conducting during horizontal flyback time and one that is cut off.
11. Draw a phasor diagram, showing two sets of color axes that can be used in the demodulator circuits.
12. Draw frequency response curves for the chroma bandpass amplifier, without compensation for the picture IF amplifier and with compensation.
13. What is the function of the color killer circuit?
14. Make a list of the color circuits that are tuned to 3.58 MHz.

COLOR CIRCUITS 393

15. In Fig. 17-3, give the functions for T_5 and R_{324}. How much is the collector-emitter voltage?
16. In Fig. 17-11, what are the functions of T_3, R_{38}, and C_{10}?
17. In Fig. 17-15, list the three stages for AFPC and give the function of each.
18. In Fig. 17-20, list the signals into the two demodulators Q1 and Q2.
19. In Fig. 17-28, give the signals into and out of the red video output stage Q1.
20. How can the trouble of no color sync burst cause no color in the picture?
21. Give two troubles that can cause no color with a normal monochrome picture.
22. How will the reproduced picture look with no Y video signal?
23. At all three cathodes of the picture tube the bias voltage is 184 V, with 150 V at the control grids. The cathode voltage then shifts to 210 V. How much is the bias now?
24. How does the picture look without color sync?
25. What is the effect of no high voltage for the anode of the color picture tube?
26. Show arrows for the block diagram in Fig. 17-31 below, to indicate the sequence for Y and $B - Y$ signals to the picture tube.

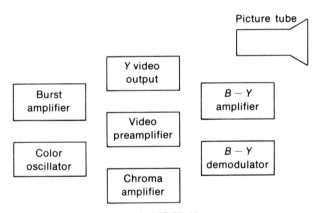

FIGURE 17-31 FOR ESSAY QUESTION 26.

18 Automatic Color Circuits

These circuits make it much easier to obtain a good color picture. Starting with the rf tuner, the automatic fine tuning (AFT) sets the local oscillator frequency to tune in the color signal. In the chroma circuits, the automatic chroma control (ACC) provides automatic bias for constant output from the color amplifier. Also, the automatic tint control (ATC) shifts the hues toward red for more pleasing flesh tones. Furthermore, the color killer circuit automatically cuts off the 3.58-MHz chroma section when a monochrome program is received. Finally, the one-button tuning automatically connects preset controls for good contrast, brightness, and color in the picture. The automatic color circuits generally use small transistors and integrated circuit (IC) units, as there is little power involved. An alphabetical listing of the automatic features is given in the summary at the end of this chapter; the circuits explained here are:

18-1   Color killer and automatic chroma control (ACC)
18-2   Color killer circuit
18-3   Color killer adjustment
18-4   ACC circuit
18-5   Peak chroma control (PCC)
18-6   Automatic tint control (ATC)
18-7   Automatic brightness limiter (ABL)
18-8   One-button tuning

18-1 Color Killer and Automatic Chroma Control (ACC)

For a monochrome picture, the color receiver uses only the amplifiers for Y luminance signal. The 3.58-MHz chroma bandpass amplifier and chroma demodulators are not needed because there is no color signal to amplify. When the 3.58-MHz color circuits are on without any signal applied, they generate receiver noise. This appears in the picture as color snow or "confetti." In a monochrome picture, crosstalk between the color snow and high-frequency components of the luminance signal produces color sparkle at the edges of black-and-white objects in the scene.

The function of the color killer is to cut off the 3.58-MHz chroma circuits when there is no color signal. Either the bandpass amplifier or the demodulators can be turned off by the color killer. A threshold or level control is provided to adjust the point where the killer circuit automatically cuts off the chroma circuits for a black-and-white program.

The color killer circuit recognizes a monochrome program by the absence of the color burst signal. Any time there is no separated burst for color synchronization, the color killer will cut off the color amplifier.

Automatic chroma control (ACC) is to the chroma bandpass amplifier as the AGC bias circuit is to the picture IF amplifier. The ACC bias varies the gain for the 3.58-MHz bandpass amplifier inversely as the strength of the received color signal. The signal strength is determined from the relative amplitude of burst. More burst means more color signal, which needs less gain in the bandpass amplifier. As a result, the ACC circuit maintains a constant color level in the picture when changing stations.

The color killer and ACC circuits usually function together, as illustrated in Fig. 18-1. The reason is that they both depend on the 3.58-MHz color sync burst transmitted on the back porch of the horizontal blanking pulses. Burst amplitude determines the ACC bias. No burst at all means the color killer operates to cut off the color amplifier.

In Fig. 18-1, the amount of 3.58-MHz oscillator cw input to the killer detector depends on the burst input to the oscillator. The detector rectifies this ac input to produce a dc bias voltage. For the killer, the bias is amplified to cut off the second bandpass amplifier. For the ACC circuit, the bias is amplified to control the gain of the first bandpass amplifier.

18-2 Color Killer Circuit

In Fig. 18-2, the signal from the 3.58-MHz oscillator is coupled to the base of the killer detec-

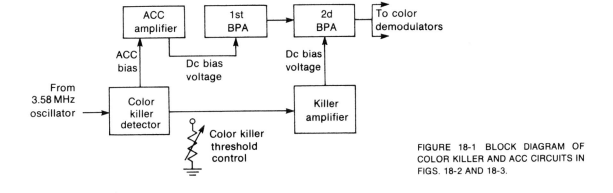

FIGURE 18-1 BLOCK DIAGRAM OF COLOR KILLER AND ACC CIRCUITS IN FIGS. 18-2 AND 18-3.

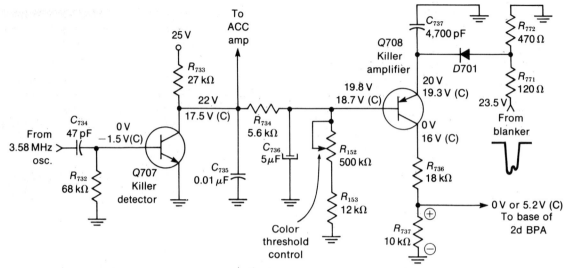

FIGURE 18-2 COLOR KILLER CIRCUIT. VOLTAGES MARKED (C) ARE WITH COLOR SIGNAL. ACC CIRCUIT IN FIG. 18-3. *(FROM ADMIRAL CHASSIS 2K20)*

tor Q707 by C_{734}. The oscillator operates all the time, but its output is not large enough to turn on Q707 until a color signal is received. Burst input causes the amplitude of the color oscillator to increase. Now the positive half-cycles of the oscillator output will turn on Q707. The signal is rectified by the base-emitter diode of the transistor, producing a signal bias of -1.5 V. Then only the positive tips of the signal produce I_C in Q707. The resulting average I_C causes the collector voltage to drop from 22 V to 17.5 V.

A voltage divider from the collector of Q707, through R_{734}, R_{152}, and R_{153}, determines the base voltage of Q708. When the collector voltage of Q707 decreases, the base voltage of Q708 also goes less positive and Q708 turns on. Note that Q708 is a PNP transistor drawn with the emitter at the top. The N base needs negative forward bias.

The emitter voltage of Q708 is at a relatively constant value from the blanker circuit. When Q708 conducts, it produces a voltage drop of 5.2 V across R_{737} in the collector circuit. The IR drop has positive polarity with a PNP transistor. This voltage is applied to the base terminal on the second bandpass amplifier to allow conduction. In short, the killer detector, amplifier, and bandpass amplifier are all on for a color program with burst signal into the color oscillator.

When there is no burst signal, the killer detector Q707 is off. Then the amplifier Q708 is also off because it does not have enough forward bias at the base. With Q708 not conducting, its collector is at 0 V. The absence of 5.2 V for the base of the bandpass amplifier cuts it off.

18-3 Color Killer Adjustment

In Fig. 18-2, the base voltage on the killer amplifier Q708 is set by the variable resistance R_{152}. This is the color threshold or *color killer control*.

It is set here at 19.8 V for no color. With the base at 19.8 V and the emitter at 20 V, the net bias equals 0.2 V. This is not enough forward voltage for the silicon transistor, which needs approximately 0.6 V base-emitter bias for conduction. As a result, Q708 is off, its collector voltage is zero, and the base of the second bandpass amplifier is cut off.

In general, the color threshold or killer control is adjusted as follows:

1. Select an unused channel.
2. Set the color level to maximum.
3. Adjust the control until color snow appears on the screen. Back off the control slightly until the snow just disappears.
4. Check all channels to be sure color is on for all color programs. If not, readjust the control slightly.

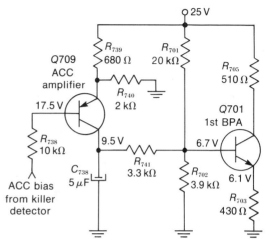

FIGURE 18-3 ACC CIRCUIT FOR THE RECEIVER IN FIG. 18-2.

The color killer control should be set at the threshold point, even though it can cut off color for a wider range of settings. This adjustment can have an important effect on how the color is tuned in by the fine tuning control without the 920-kHz beat. If the color cuts out too soon, the correct tuning is more difficult.

18-4 ACC Circuit

In Fig. 18-3, the amplitude of the chroma signal is controlled by using automatic bias from the ACC amplifier to vary the gain of the first bandpass amplifier. Q709 is a dc amplifier for the ACC bias. Its input is dc voltage from the collector of the killer detector Q707, shown in Fig. 18-2. Remember that this collector voltage varies with the amount of oscillator signal, which in turn depends on burst amplitude. More signal causes more conduction in Q707, reducing the collector voltage. Then the input to the base of Q709, through R_{738}, goes less positive.

Note that Q709 is a PNP transistor, with the emitter shown at the top. The emitter voltage is determined by R_{739} and R_{740} from the 25-V supply. When Q709 has base voltage of 17.5 V with color signal, the base-emitter voltage is $17.5 - 18.2 = -0.7$ V. This forward bias then enables the ACC amplifier Q709 to conduct.

With Q709 conducting, its collector current produces an IR drop across R_{702}. This increase in positive base voltage for the NPN transistor allows Q701 to conduct. The base voltage on the bandpass amplifier then varies in accordance with the strength of the color signal to control the gain.

18-5 Peak Chroma Control (PCC)

This circuit can be compared with delayed AGC, in that the control bias is not developed until the signal reaches a specific amplitude. Another name is *color overload control*. When the chroma amplitude exceeds a maximum level,

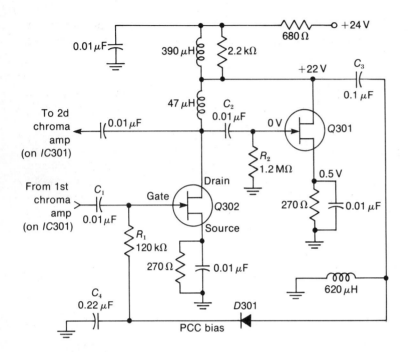

FIGURE 18-4 PEAK CHROMA CONTROL (PCC) CIRCUIT. Q301 AND Q302 ARE JUNCTION FIELD-EFFECT TRANSISTORS. (FROM SEARS MODEL 4360)

the PCC circuit clips the peaks of the signal. As long as the chroma is below this point, though, the PCC circuit functions only as a chroma amplifier.

In Fig. 18-4, Q301 and Q302 are junction field-effect transistors. The channel is N-type, which requires negative voltage for reverse bias at the gate. This operation is similar to negative grid bias without grid current for a triode tube. Q302 is the PCC amplifier. Q301 supplies chroma signal for diode D301. Its rectified signal is the bias feedback to Q302.

The chroma signal, already amplified by the first color amplifier, is coupled through C_1 to the gate of Q302. Operating class A, this stage amplifies the signal and couples it back to the second color amplifier in the IC301 unit, not shown here. However, the output signal from the drain of Q302 is also coupled by C_2 to the gate of Q301. Then this signal is coupled through C_3 to the rectifier D301. The diode rectifies the signal to produce negative voltage that is the operating bias for the gate of Q302. C_4 with R_1 filters the rectified signal to provide a dc bias voltage.

The bias on Q302 results from rectified chroma signal. Therefore, the bias changes in value with the signal amplitude. Still, this does not affect Q302 as an amplifier until the chroma amplitude reaches a value high enough to cause excessive color saturation. At this point, the bias on Q302 is negative enough to make it clip the signal peaks. The clipping continues until the chroma amplitude drops. In this way, the peak chroma signal is limited to a specific maximum value.

18-6 Automatic Tint Control (ATC)

In many cases, the hue has to be adjusted for proper flesh tones in the picture. Most people prefer more red. Also, the hue can change slightly for different programs because of a shift in burst phase. The ATC circuit compensates for these problems by automatically emphasizing the red hues. Two methods used are: (1) reduce the gain for blue in the color video amplifier; (2) shift the axis of the blue demodulator closer to the phase angle of red.

Referring to Fig. 18-5, Q6 is used as a gate and Q7 is a solid-state switch. If Q6 is to conduct, it must have the proper ACC voltage input and the Instamatic switch must be in the automatic position. Then the base of Q6 is at 7.4 V, with the emitter at 8 V. This bias of −0.6 V on the PNP transistor allows collector current. As a result, the base of Q7 is at 0.7 V, which drives this NPN transistor into saturation.

The low resistance of the conducting switch Q7 now provides the effect of ground on C_{21} in the base circuit and R_{32} in the collector circuit. Grounding C_{21} changes the demodulation angle for blue, shifting the output to more red. In addition, grounding one end of R_{32} puts it in parallel with the emitter resistance of the red video stage, not shown here. This change in emitter bias shifts the dc voltage for the red gun of the picture tube. The background bias and gray-scale tracking are then shifted toward red.

18-7 Automatic Brightness Limiter (ABL)

A number of different methods are used to control brightness. Some maintain a constant brightness despite changes in ac line voltage. Others hold the brightness constant despite large changes in the average CRT anode current, as would occur when the scene changes from day to night. The circuit of Fig. 18-6 does both.

The base voltage for the video amplifier Q4 is the dc voltage at the variable arm of the brightness control R_{4202}. This voltage controls the emitter current of Q4 and the average voltage drop across R_{330}. Therefore the dc level of the video output signal is controlled.

Since the signal is dc-coupled to the three cathodes of the picture tube, the dc level determines the average beam current and therefore the brightness. The end result is that when the base of Q4 goes more positive, its emitter volt-

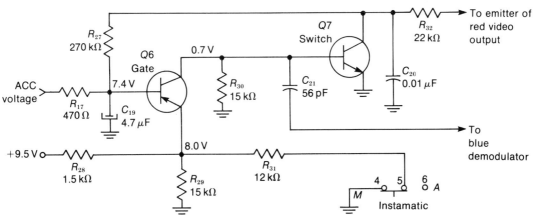

FIGURE 18-5 AUTOMATIC TINT CONTROL, WITH Q7 USED TO CONTROL THE PHASE ANGLE OF THE BLUE DEMODULATOR. *(FROM MOTOROLA CHASSIS TS938)*

400 BASIC TELEVISION

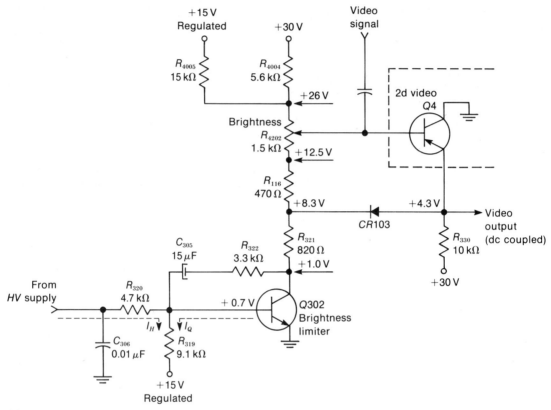

FIGURE 18-6 AUTOMATIC BRIGHTNESS LIMITER CIRCUIT. *(FROM RCA CHASSIS CTC-59)*

age and the cathode voltage on the CRT also go more positive to decrease the brightness. Note that Q4 is a PNP transistor.

Any change in ac line voltage will affect the voltage at the top end of the brightness control. This voltage is approximately +26 V, derived from the voltage divider with R_{4004} and R_{4005}. One end of the divider is connected to +15 V, regulated. The other end is connected to the +30-V output of the low-voltage power supply. If the ac line voltage changes, the +30-V line will also change, but the regulated 15 V will not. An increase in line voltage would tend to increase brightness. However, the effect is also more positive voltage at the top of the brightness control. This makes the base of Q4 more positive to decrease the brightness. When the circuit is operating properly, the decreased brightness from Q4 will exactly offset the increase due to line voltage and no change will actually occur.

Assume 26 V at the top of the brightness control. Now the base voltage of Q4 is determined by the current through the control R_{4202}. This current is the I_C of Q302. More current through the brightness control makes the Q4 base less positive and results in more brightness; less current through the brightness control means less brightness. Therefore, the collector current of Q302 is directly proportional to CRT brightness.

Normally, Q302 is saturated, resulting in maximum brightness for that position of R_{4202}. This transistor is a brightness limiter. Its purpose is not to increase brightness, but rather to decrease it when necessary, to prevent blooming in the picture.

The collector current of Q302 is determined mainly by its base voltage because the emitter is grounded. Collector voltage has little effect on I_C. The base voltage is 0.7 V because of the 14.3-V drop across R_{319}. Its current is the sum of two others: the base current of Q302 and the CRT beam current from the high-voltage power supply through R_{320}. Normally, the sum of these two currents is the required amount. However, if the CRT beam current changes, the base current on Q302 will change and the brightness is affected.

The integrator circuit of C_{305} and R_{322}, connected between base and collector of Q302, prevents the circuit from responding to rapid changes in brightness level. If the circuit responds too fast, it could reduce the white highlights in the scene. By responding more slowly, it changes only when the high beam current persists long enough to represent a true change in average level.

The diode CR103 is normally off. It conducts only if the emitter of Q4 goes more positive than the cathode of CR103. As a result, the diode limits the positive emitter voltage on Q4.

18-8 One-Button Tuning

This feature makes operation of the color receiver very simple. You push the button in and a multipurpose switch does all or most of the following:

1. Sets contrast to a preset level for Y video.
2. Sets brightness to a preset level for dc bias voltage.
3. Sets color level to a preset gain for chroma signal.
4. Sets hue or tint to a preset phase.
5. Turns on automatic tint control (ATC) circuit.
6. Turns on automatic fine tuning (AFT) circuit for local oscillator on rf tuner.

Four of these functions are illustrated by the circuits in Fig. 18-7. In addition, a 6.3-V indicator light is turned on by the switch. When the preset adjustments have been made correctly, the one-button tuning automatically results in a good color picture, regardless of the settings of the manual controls.

Preset contrast. Figure 18-7a shows that SW1 changes the output connection for video signal from potentiometer control R_1 to R_2. The switch is shown in the manual (M) position for R_1, the manual contrast control. Video signal is applied across R_1, with one end grounded by 1 and 2 on the switch, through R_3. The amount of video signal set by the wiper contact on R_1 is coupled out through 7 and 8 on the switch.

When SW1 is turned to the automatic (A) position, it moves up in the diagram, joining 2 to 3 and 8 to 9. Then the preset contrast control R_2 is grounded through R_3, while R_1 is disconnected. Also, the wiper arm of R_2 is connected to the output through 8 and 9 on the switch.

The preset control R_2 has previously been adjusted for strong contrast without overload distortion. Therefore, the switch automatically provides the proper contrast in the picture.

Preset brightness. For the circuit in Fig. 18-7b, the bias on the picture tube is determined by a dc voltage applied to the color demodulator IC unit. These circuits are dc-coupled to the CRT.

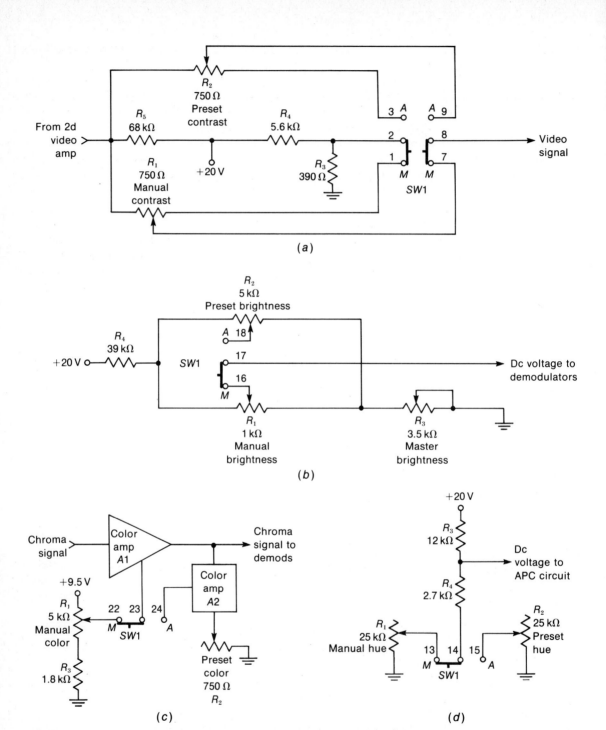

FIGURE 18-7 CIRCUITS FOR ONE-BUTTON TUNING. SW1 IS THE INSTAMATIC SWITCH THAT CHANGES FROM MANUAL TO PRESET CONTROLS. (a) CONTRAST; (b) BRIGHTNESS; (c) COLOR LEVEL OR INTENSITY; (d) HUE OR TINT. (MOTOROLA INC.)

Note that terminals 16 and 17 on SW1 select a dc voltage from the manual brightness control R_1. When the switch is moved up, the preset control R_2 is connected through 17 and 18. The master control R_3 is used to set the maximum brightness just below the point of blooming, for either the manual or the preset adjustment.

Preset color level. The color intensity or saturation depends on the amplitude of the 3.58-MHz chroma signal. In Fig. 18-7c the gain of the color amplifier A1 is controlled by dc bias from the manual control R_1. In the automatic position, though, SW1 disconnects R_1. Then the gain of A1 is controlled by A2. This stage, in turn, has bias from the preset color control R_2.

Preset hue or tint. For the circuit in Fig. 18-7d, the 3.58-MHz color oscillator and its automatic phase control (APC) circuit are in an IC color processor unit, not shown here. Only an external dc voltage for the IC unit is needed to change the oscillator phase, which determines hue or tint. The 20-V source feeds a voltage divider with R_3, R_4, and either of the hue controls. For manual control, R_1 is connected to the voltage divider through 13 and 14 on SW1. For the preset hue or tint, R_2 is connected to the voltage divider through 14 and 15 on the switch.

In addition to connecting the preset controls, the one-button tuning turns on the ATC and AFT circuits. As a result, the ATC circuit shifts the tint for more red and less blue. The AFT circuit provides automatic fine tuning, to tune in the color exactly without 920-kHz beat in the picture.

SUMMARY

The following alphabetical list of abbreviations includes the main types of automatic color circuits. Also included are symbols often used for the circuits in color receivers.

ABL is automatic brightness limiter. It controls bias on the picture tube.
ACC is automatic chroma control. It determines bias on the 3.58-MHz color amplifier.
ADG is automatic degaussing. It demagnetizes the picture tube when the receiver is
    turned on or off (see Chap. 11).
AFC is automatic frequency control on an oscillator.
AFPC is automatic frequency and phase control.
AFT is automatic fine tuning. It is AFC on the local oscillator in rf tuner (see Chap. 24).
APC is automatic phase control on the color oscillator for hue or tint.
ATC is automatic tint control. It shifts hues toward red.
BPA is bandpass amplifier for the 3.58-MHz chroma signal.
C is 3.58-MHz chrominance, chroma, or color signal.
CK is color killer. It cuts off BPA with monochrome signal.
CO is color oscillator. It regenerates the 3.58-MHz subcarrier for color demodulators.
FET is field-effect transistor.
IC is integrated circuit.

Instamatic is a trademark for one-button tuning.

OBT is one-button tuning (see Fig. 18-7).

PCC is peak chroma control. It clips the peaks of the chroma signal to prevent overload distortion.

Videomatic is a trademark for automatic control of brightness, contrast, and color according to viewing light in the room.

Self-Examination (Answers at back of book)

Answer True or False.
1. The color killer can cut off the bandpass amplifier.
2. The ACC circuit controls the gain of the chroma amplifier.
3. The ABL circuit prevents blooming in the picture.
4. Changing the phase of the color oscillator changes hue or tint.
5. Changing the gain of the chroma amplifier changes color level or saturation.
6. A PNP transistor needs positive base voltage for forward bias.
7. In an FET, the gate corresponds to the base of a junction transistor.
8. The ATC circuit shifts the tint toward more blue.
9. In one-button tuning, the contrast, brightness, and color intensity are set to preset levels.
10. The color sync burst is needed for operation of the color killer and ACC circuits.

Essay Questions

1. List five abbreviations for automatic color circuits and give their functions.
2. Give the function of the color killer circuit.
3. Explain how to set the color killer threshold control.
4. Which stages determine color intensity or saturation?
5. Which stages determine hue or tint?
6. Referring to Figs. 18-2 and 18-3, give the specific functions for Q707, Q708, Q709, and Q701.
7. What is meant by blooming in the picture and what is the cause?
8. List at least five functions of one-button tuning.
9. Referring to Fig. 18-7, give the specific connections of SW1 in a, b, c, and d for one-button tuning.

# 19 Troubles in the Raster and Picture

Troubles in the raster, such as insufficient height or width, show in the picture also, as the picture can be considered as an overlay on the raster. The scanning circuits produce the raster. Then the video signal reproduces the picture information. Trouble in the raster circuits, therefore, will be evident even without a picture on the screen. In general, problems with size or shape are raster troubles. The raster circuits include the vertical and horizontal deflection circuits, with the common deflection yoke for scanning, as shown in Fig. 19-1. The sync circuits hold the picture still on the raster. Problems of picture quality, including interference, are troubles in the signal. More details are described in the following topics:

    19-1 Raster circuits
    19-2 Troubles in height and width
    19-3 No brightness
    19-4 Picture troubles
    19-5 Bars in the picture
    19-6 Sound in the picture
    19-7 Ghosts in the picture
    19-8 Interference
    19-9 Hum troubles

19-1 Raster Circuits

The block diagram in Fig. 19-1 illustrates the stages for vertical and horizontal scanning. For vertical deflection, only a vertical oscillator operating at 60 Hz and the amplifier are necessary. The output stage is similar to a class A amplifier to supply enough sawtooth scanning current in the yoke. With tubes, the vertical oscillator and amplifier are usually combined in just one stage.

The horizontal circuits are more complicated because they include the damper stage and flyback high-voltage rectifier, in addition to the oscillator operating at 15,750 Hz which drives the output stage. The horizontal amplifier is always a separate stage because of its power requirements. It is a class C pulse amplifier, operating with the damper for more efficient scanning. In effect, the horizontal output stage serves as a switch that is pulsed into saturation output current. With transistors, a buffer amplifier is generally used to isolate the oscillator from the output stage, for both the vertical and horizontal deflection circuits.

Vertical scanning circuits. The vertical oscillator is triggered by vertical sync pulses to lock in at 60 Hz. This stage is either a multivibrator or a blocking oscillator. It produces sawtooth voltage output to drive the vertical amplifier that supplies the sawtooth current in the yoke. Note that flyback pulses are coupled to the picture tube to blank out vertical retrace lines.

The controls in the vertical deflection cir-

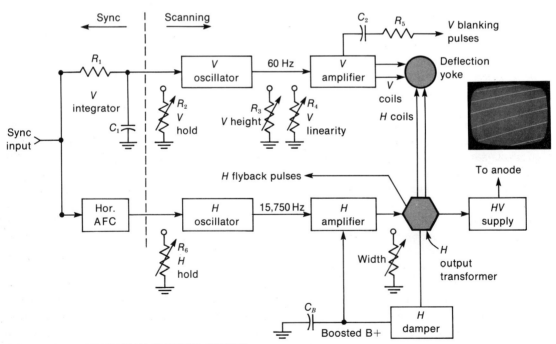

FIGURE 19-1 BLOCK DIAGRAM OF RASTER CIRCUITS.

cuits include the height and linearity adjustments. These are on the back of the receiver chassis. Normally, they can be adjusted so that the raster fills the screen top to bottom, with equally spaced scanning lines. In addition, the vertical hold control adjusts the frequency just below 60 Hz to allow each vertical sync pulse to trigger the oscillator and hold it at the vertical scanning frequency.

Horizontal scanning circuit. The horizontal oscillator also is generally a multivibrator or blocking oscillator. However, the frequency is held at 15,750 Hz by the AFC circuit, instead of triggered sync. The oscillator drives the horizontal amplifier with its output transformer, often called the *flyback transformer*. The functions of the horizontal output circuit include supplying the following:

1. Horizontal scanning current in the horizontal deflection coils of the yoke.
2. Input to the high-voltage rectifier for anode voltage on the picture tube.
3. Input to the horizontal damper. This stage damps oscillations or ripples in the horizontal scanning current that occur just after flyback. In addition, the damper operates in conjunction with the horizontal amplifier to provide full cycles of horizontal scanning current. The damper scans the left side of the raster, about one-third to one-half the way across, and the amplifier finishes the traces to the right edge.

In addition, the damper serves as a rectifier to add dc voltage to the B+ supply. For example, with 140 V for B+, the damper can add 250 V, for a total of 390 V. This is called *boosted B+ voltage*. It is used for plate voltage on the horizontal and vertical amplifier tubes, for screen-grid voltage on the picture tube, and for other tubes requiring a high B+ voltage. The boost voltage is developed across capacitor C_B in Fig. 19-1. The presence of the boosted B+ voltage shows that the damper is operating normally.

The horizontal output transformer generally has voltage taps or extra windings to supply flyback pulses for some stages in the signal circuits. Their function is to key on the stage just during horizontal retrace time. Two examples are keying pulses for the AGC stage and keying pulses for the burst separator in color receivers.

The horizontal hold control can be in either the oscillator or the AFC circuit. If there is a width control, it often varies the screen-grid voltage on the horizontal output tube. There are few, if any, controls to adjust width and linearity, because the horizontal output circuit is designed for maximum efficiency in scanning and high voltage.

19-2 Troubles in Height and Width

When the raster does not have enough height, it looks like Fig. 19-2. The black bars at the top and bottom are the unused screen areas not scanned. This means that the sawtooth current in the vertical deflection coils does not have enough peak-to-peak amplitude to scan the full height of the raster. Assuming that the vertical height and linearity controls cannot be adjusted to fill the screen, insufficient height usually means trouble in the vertical output stage. If this is a tube, try a new one. The vertical oscillator is necessary for deflection, but it usually does not cause weak output because its only function is to drive the amplifier.

In many cases, the full height can be obtained only with poor linearity. The extreme case of vertical nonlinearity is a white bar at the bottom or top, caused by foldover of the scan-

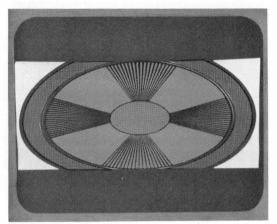

FIGURE 19-2 INSUFFICIENT HEIGHT IN RASTER AND PICTURE.

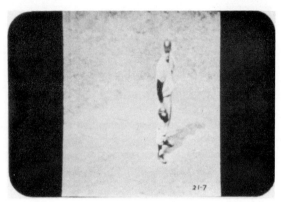

FIGURE 19-4 INSUFFICIENT WIDTH IN RASTER AND PICTURE.

ning lines. The vertical nonlinearity is really an example of weak output, as too much drive is used on the vertical amplifier.

If there is no vertical output at all, the result is just a horizontal line across the center (Fig. 19-3). No vertical deflection means no output, either from the oscillator or from the amplifier.

Width troubles in the horizontal scanning are usually associated with the output stage and damper. When the whole raster is narrow, without any vertical white bars at the left side, this indicates weak horizontal output (Fig. 19-4). However, when the width is drastically reduced at the left side, with a wide, white bar as in Fig. 19-5, this indicates trouble in the damping circuit.

One very definite trouble is the trapezoidal raster shown in Fig. 19-6. This is caused by a

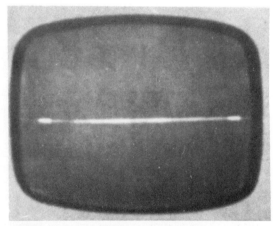

FIGURE 19-3 NO HEIGHT.

FIGURE 19-5 REDUCED WIDTH AND WIDE WHITE BAR AT LEFT, CAUSED BY OPEN B+ BOOST CAPACITOR.

defect in the deflection yoke. The raster can be keystoned either vertically or horizontally. It is important to note that the deflection yoke has the balance in the coils that provides symmetry in the raster. Any trouble that makes the raster unsymmetrical is likely to be the result of a defective yoke.

19-3 No Brightness

This means no light from the screen on the picture tube, not to be confused with no picture on a good raster. Assuming that the low-voltage supply is operating and the sound is normal, no brightness is usually the result of no high voltage for the anode of the picture tube. For flyback high voltage the following circuits must be normal:

1. Horizontal oscillator driving the amplifier
2. Horizontal output stage driving the high-voltage rectifier, through the high-voltage winding on the flyback transformer
3. Damper diode operating to produce boosted B+ voltage for the horizontal output stage

To check for high voltage, work backward from the high-voltage rectifier. Check this stage first for dc output to the anode of the picture tube. If the horizontal output is not driving the rectifier with high-voltage ac input, the trouble can be in the amplifier, transformer, or damper. You can measure the boosted B+ voltage to see if the damper is operating normally. For the last possibility, the horizontal oscillator may not be operating to drive the amplifier.

It is still possible to have no brightness with normal high voltage on the anode. Assuming that the heater of the picture tube is lit, the trouble can be in the electrode voltages. The screen-grid voltage must be normal for there to

FIGURE 19-6 TRAPEZOIDAL RASTER, CAUSED BY DEFECTIVE YOKE.

be beam current. With color picture tubes, there is a separate focus voltage supply. If this voltage is zero or too low, there will not be any beam current. Finally, the grid-cathode bias can cut off the beam current. With cutoff voltage, however, you can usually see some afterglow on the screen of the picture tube just after the receiver is turned off. Check for afterglow in a dark room. If you see the afterglow, the picture tube has high voltage on the anode.

In color receivers, the screen can be very dark without Y video signal when there is dc coupling. Then a dc trouble in the Y video output stage changes the grid-cathode bias on the picture tube.

19-4 Picture Troubles

We consider the picture separate from the raster. The scanning circuits can produce a perfect raster but trouble in the signal circuits still can cause no picture, no color, snow in the picture, or a distorted picture.

No picture. It is useful to check whether the sound is normal. No picture with no sound means there is no signal through the rf tuner and the common IF amplifiers up to the sound take-off point. The video detector and video amplifier may also be common to the picture and sound in monochrome receivers. Remember that the AGC circuit can cut off picture and sound by biasing off the rf and IF amplifiers.

Even though there is no picture, turn up the contrast control to see if this produces white speckles of snow on the raster. Actually, the snow is signal from noise generated in the mixer stage. The presence of snow shows that the signal from the mixer is being amplified in the IF and video amplifier. Then the trouble is in the antenna circuit, transmission line, and rf amplifier ahead of the mixer. If there is no snow, the trouble is in the stages after the mixer, including the IF and video stages. It should be noted that when the AGC circuit cuts off the picture, the raster is clean, without snow. When the antenna is not supplying signal, the raster has snow. Also, there is a loud rushing noise in the sound with the volume at maximum.

If there is no picture but normal sound, the trouble is in the stages after the sound take-off circuit. This section usually includes the video detector and video amplifier. In color receivers, if the Y video channel is open, the chroma section can still produce color on the screen of the picture tube.

Snow in the picture. See Fig. 19-7. Usually, the snow is a result of not having enough antenna signal. There may be trouble in the rf amplifier section, or the transmission line from the antenna may be open. A helpful check here is the fact that with an open line the picture is usually better on some channels when only one side of the transmission line is connected.

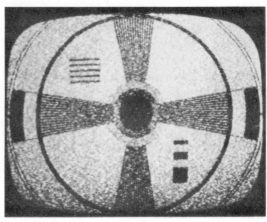

FIGURE 19-7 SNOWY PICTURE, CAUSED BY WEAK ANTENNA SIGNAL.

Weak picture with low brightness. This is often caused by low emission in an old picture tube. It can be recognized by a silvery bloom in white highlights of the picture, especially when the brightness is increased. This applies to both color and monochrome picture tubes.

No color in the picture. Assuming a normal monochrome picture on a white raster, this trouble means there is no color signal from the chroma section. Likely sources of trouble are the 3.58-MHz chroma amplifier, color oscillator, and burst amplifier stages. More details are given in Chap. 17.

Picture does not hold still. Vertical rolling is a trouble in the vertical sync or the vertical oscillator, operating at the wrong frequency. Trouble in the horizontal hold, with diagonal bars in the picture, is often caused by defective diodes in the horizontal AFC circuit. When the picture does not hold both vertically and horizontally, the trouble is no output from the sync separator.

TROUBLES IN THE RASTER AND PICTURE 411

Overloaded picture. As shown in Fig. 15-1 and explained in Chap. 15, overload distortion is usually an AGC trouble. The overload is caused by incorrect bias on the rf and IF amplifiers.

Bend in the picture. The bend is actually a case of weak horizontal synchronization just before the point where the picture tears into diagonal bars. The trouble in the horizontal sync can result from hum, overload distortion, interference, or a defect in the horizontal AFC circuit.

Smear and streaking. See Fig. 19-8. The streaking is especially evident trailing to the right after the edges of numbers or letters in the picture. The trail may be black after white, or the same as the original image. Any motion up or down of the smear and streaks makes this effect very obvious. The cause is phase distortion with time delay for low video frequencies up to about 200 kHz.

The phase distortion is often caused by a leaky or open capacitor used for coupling or to

FIGURE 19-9 MULTIPLE OUTLINES IN PICTURE, CAUSED BY RINGING IN IF AMPLIFIER.

bypass the video signal. Remember that a bypass is only a coupling capacitor to the chassis ground return for signal. In either case, the capacitor is in the signal path. A common example is the ac bypassing for dc voltage in the brightness control circuit. When these capacitors are leaky, they have too little capacitance and too much ac impedance, which changes the response for low video frequencies.

Ringing in the picture. See the multiple outlines in Fig. 19-9. The ringing means there are self-excited oscillations caused by an abrupt change in signal voltage. When the outlines change with the fine tuning control, this means the trouble is in the picture IF amplifier. In the video amplifier, the cause of ringing is excessive high-frequency response. A possible source is an open damping resistor across a series peaking coil.

19-5 Bars in the Picture

The way horizontal and vertical bars are produced on the screen of the picture tube can be illustrated experimentally by using an audio sig-

FIGURE 19-8 LOW-FREQUENCY SMEAR, WITH STREAKS AFTER LETTERING, CAUSED BY LEAKY COUPLING CAPACITOR IN VIDEO AMPLIFIER.

nal generator to supply video signal to the receiver. The variable-frequency generator has a range of 20 Hz to 200 kHz. The transmitted signal is not used. Instead, the output of the signal generator is coupled into the video amplifier, where the signal is amplified for the picture tube. As the generator signal varies the grid-cathode voltage of the picture tube, pairs of dark and light bars are formed on the raster (Fig. 19-10).

Horizontal bars are produced when the frequency of the signal at the picture tube grid is less than 15,750 Hz; above 15,750 Hz the bars are vertical or diagonal. Since the sync for the deflection oscillators is usually taken from the video amplifier in the receiver, the generator signal also supplies synchronization. The bars can be locked when the generator frequency is an exact multiple of the scanning frequency. Just vary the signal generator frequency to obtain the desired number of bars and adjust the receiver hold control to make the bars stay still.

Horizontal bars. Suppose that a 60-Hz sine-wave signal is varying the control-grid voltage in synchronism with the vertical scanning motion at the field frequency of 60 Hz. During the positive half-cycle the signal makes the grid more positive, increasing the beam current and screen illumination. The negative half-cycle reduces the beam current and screen illumination. Since it takes $1/_{120}$ s for a half-cycle of the 60-Hz signal, the scanning beam moves halfway down the screen during this time. Therefore, if the positive half-cycle of the sine-wave signal coincides with the first half of the vertical scan, the top half of the picture will be brighter than the bottom half. The pattern on the screen then is a pair of horizontal bars, one bright and the other dark.

When the signal generator output frequency is increased to multiples of 60 Hz, ad-

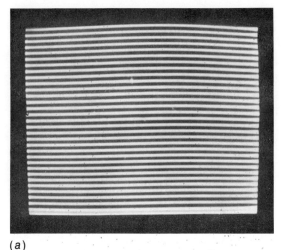

(a)

(b)

FIGURE 19-10 BAR PATTERNS PRODUCED BY SIGNAL GENERATOR. (a) HORIZONTAL BARS WHEN FREQUENCY IS BELOW 15,750 Hz; (b) VERTICAL BARS WHEN FREQUENCY IS ABOVE 15,750 Hz.

ditional pairs of narrower horizontal bars will be formed on the screen, as shown in Fig. 19-10a. The number of pairs of bars is equal to the signal generator frequency divided by the vertical scanning frequency. However, subtract any bars that may be produced during vertical retrace time if the signal frequency is high enough to produce more than about 20 pairs of bars. As an example, a frequency of 240 Hz results in 240/60 = 4 pairs of horizontal bars.

Vertical scanning linearity is indicated by the spacing between the parallel horizontal bars. If the vertical scanning motion is linear, the bars will be equally spaced. Otherwise, the bars will be spread out or crowded together.

Vertical bars. When the frequency of the modulating signal becomes equal to the horizontal line-scanning frequency, vertical bars are formed instead of horizontal bars. Consider the case of a 15,750-Hz sine-wave signal varying the grid voltage in phase with the horizontal scanning. During one horizontal line the screen is made brighter for approximately one-half the picture width, as the positive half-cycle of the grid voltage increases the beam current; the negative half-cycle makes the screen darker. The same effect occurs for succeeding horizontal lines, and the result is a pair of vertical bars on the screen, one bar white and the other dark. If the frequency of the signal generator is increased to multiples of 15,750 Hz, additional pairs of narrower vertical bars will be produced on the screen, as shown in Fig. 19-10b. Their spacing indicates linearity of the horizontal scanning motion.

The number of pairs of vertical bars is equal to the frequency of the applied signal divided by the horizontal scanning frequency. As an example, a frequency of 157.5 kHz results in 10 pairs of vertical bars. However, not all the bars may be visible. Some are formed during the horizontal retrace time if the signal frequency is high enough to produce more than about 10 pairs of bars. These bars formed during the flyback are wider because of the fast retrace. They appear as variations of shading in the background, as can be seen in Fig. 19-10b. It is possible to determine the retrace time by counting the bars visible during the trace time and comparing this with the total number that should be produced.

Diagonal bars are produced when the signal frequency is higher than the horizontal line-scanning frequency but is not an exact multiple. In this case, the light and dark parts of each line are regularly displaced in successive order at different positions with respect to the start of the trace, instead of lining up one under the other. The diagonal bars usually do not stay still because the signal frequency is not an exact multiple and does not synchronize the deflection oscillator.

19-6 Sound in the Picture

There are three types of sound interference: (1) horizontal bars from the audio signal; (2) a fine-line pattern from the 4.5-MHz intercarrier sound signal; (3) a 920-kHz diagonal bar pattern as interference between the 4.5-MHz intercarrier sound and the 3.58-MHz color subcarrier. The 920-kHz beat is the most common form in which the sound signal interferes with the picture.

920-kHz beat. This effect can be seen in color plate XIII. The 920 kHz is the difference frequency of 4.5 − 3.58 = 0.92 MHz. With an interfering signal at 920 kHz, the number of bars is 920,000/15,734, which equals approximately 60.

Audio sound bars. Slope detection can convert the modulation of the FM sound signal to audio voltage in the video detector output even though the diode is an AM detector. Then the audio combines with the video at the grid-cathode circuit of the picture tube. The result is horizontal

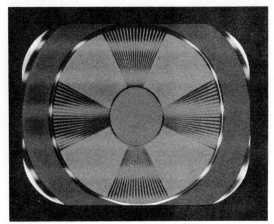

FIGURE 19-11 SOUND BARS IN THE PICTURE. THE WIDTH AND INTENSITY VARY WITH THE AUDIO MODULATION.

bars, as in Fig. 19-11. The sound bars can be recognized, as they vary with the audio modulation and disappear when there is no voice or music.

4.5-MHz beat. This effect is shown in Fig. 19-12. The interference pattern has about 225 pairs of thin black and white lines with small

FIGURE 19-12 FINE HERRINGBONE OR "WORMY" EFFECT OF 4.5-MHz BEAT IN PICTURE.

"wiggles" like a herringbone weave. This effect is called a "wormy" picture. The number of pairs of lines equals 4.5 MHz/15,750 Hz, which is approximately 285. The fine lines are caused by the 4.5-MHz carrier, while the herringbone weave is a result of frequency variations in the FM sound signal. When there is no voice or music, the herringbone effect disappears but the thin lines remain because they are produced by the sound carrier.

19-7 Ghosts in the Picture

A duplicate of the reproduced picture, to the side, as in Fig. 19-13, is called a *ghost*. The multiple pictures are caused by multiple signals. The main picture is the direct signal received by the antenna from the transmitter. In addition, the antenna can receive signals produced by reflections of the transmitted signal by buildings, bridges, or other large metal structures.

The reflected signal takes longer to reach the receiver than the direct signal. As a result, a multipath signal can produce a ghost a little to the right of the main image. Reducing these ghosts caused by multipath signals is mainly a question of using a directional antenna, which receives signals best through a small angle from the front and very little from the back.

Built-in ghosts. In most cases, the cause of ghosts is an antenna problem. However, multiple images can be produced by either the receiver or a very long transmission line. These built-in ghosts can be recognized by the fact that usually there are three, four, or more images with uniform spacing.

Excessive response for the high video frequencies in the IF amplifier or video amplifier can cause built-in ghosts. These are the same on all channels. If the excessive response is in

FIGURE 19-13 GHOST CAUSED BY REFLECTED SIGNAL AT ANTENNA.

the picture IF amplifier, varying the fine tuning will affect the ghosts.

Ghosts caused by reflections of the antenna signal in a long transmission line change when hand capacitance is added by holding the line. These ghosts are not the same on all channels.

Leading and trailing ghosts. A ghost resulting from multipath reception by the antenna is a trailing ghost, to the right of the main image. However, a leading ghost can be produced to the left of the main image. One cause of leading ghosts is direct pickup of signal, especially in strong signal areas. In this case, the direct pickup can provide a picture before the antenna signal delivered by a long transmission line. With direct pickup, the ghost will vary when people walk near the receiver.

Eliminating a leading ghost, to the left of the image, is a problem of reducing stray pickup. Try shielding the transmission line and rf tuner, if necessary, and supplying more signal from the antenna. The antenna system and the rf amplifier can be checked for the possibility of a trouble that causes weak antenna signal. With an antenna distribution system, it must supply much more antenna signal than the direct pickup, at each receiver outlet.

19-8 Interference

Frequencies that are not even in the television broadcast band can produce interference in the picture because of heterodyning effects that cause interfering beat frequencies. As one example, the rf interference can beat with the local oscillator in the rf tuner to produce difference frequencies that are in the IF passband of the receiver. Then the interfering signal beats with the picture carrier in the video detector to produce interference in the video signal. This causes interference in the picture. In addition, adjacent television broadcast channels can cause interference. Also, the interference can have frequencies in the IF band of 40 to 46 MHz.

All these are examples of interfering signals external to the receiver. The receiver can also generate its own interference as harmonics of the detected video or sound signals and the horizontal scanning current, which are radiated back to the signal circuits. Typical examples of interference patterns are shown in Fig. 19-14. A list of radio services that can cause interference is included in Appendix B.

Diagonal lines. The uniform diagonal lines in Fig. 19-14a are caused by an unmodulated carrier wave (cw). They are of uniform thickness because there is no modulation. Usually, the bars shift slowly from one diagonal position, through the vertical, and to the opposite diagonal as the interfering carrier wave drifts in frequency. This type of interference can be caused by local oscillator radiation from a

FIGURE 19-14 INTERFERENCE EFFECTS IN THE PICTURE. *(a)* UNIFORM DIAGONAL BARS CAUSED BY CW INTERFERENCE; *(b)* HERRINGBONE WEAVE CAUSED BY FREQUENCY MODULATION; *(c)* DIATHERMY INTERFERENCE.

nearby receiver or from any unmodulated carrier wave. The number of bars and their thickness depends on the beat frequency produced by the cw interference. If the beat frequency resulting from the rf interference is less than 15,750 Hz, it will produce uniform horizontal bars, in a venetian-blind effect.

Herringbone weave. An interfering FM carrier wave produces interfering beats that vary in frequency. With a center frequency high enough to produce a fine-line interference pattern, the frequency modulation adds a herringbone weave to the vertical or diagonal bars. When the beat frequencies are too low for a diagonal-line pattern, instead of horizontal bars the FM interference produces a watery effect through the entire picture. Then the picture looks as if it were covered with a silk gauze, as in Fig. 19-14*b*.

Diathermy interference. Diathermy machines and other medical or industrial equipment usually produce the rf interference pattern shown in Fig. 19-14*c*. There may be two dark bands across the screen instead of the one shown, and the bars will be darker if the interference is stronger. This rf interference pattern is produced because diathermy equipment is effectively a transmitter. The waving effect is caused by frequency variations. The single band shows that there is a strong 60-Hz component in the interference; with 120-Hz modulation, there would be two bands.

Strong rf interference. In addition to bar patterns in the picture, the light values are altered by the rf interference when it has enough amplitude. A very strong interfering signal, therefore, can produce a negative picture or black out the picture completely. The negative picture is a result of detection in an overloaded rf or IF amplifier, with the video signal modulating the picture carrier in reverse polarity.

How interference enters the receiver. An interfering rf signal can enter the receiver through the antenna and transmission line, through direct pickup by the chassis, or from the power line. Rf interference at the antenna input of the receiver must go through the rf tuner and, therefore, usually appears on specific channels. Direct pickup of IF interference by the chassis results in the same interference on all channels. If the rf interference is from the power line, reversing the plug and adding hand capacitance by holding the line should affect the interference. This can be reduced by an rf filter in the power line.

FM broadcast interference. An FM interference pattern only on channel 2 is usually caused by an FM broadcast station in the 88- to 108-MHz band. You may also hear the audio modulation. An FM trap in the antenna input circuit can be tuned to reject the interfering frequency and eliminate the interference.

Sound IF harmonics. Harmonics of the associated sound IF signal in the receiver can cause an FM interference pattern. Sound IF harmonic interference in the picture can be identified by the fact that it varies with the audio modulation of the associated sound signal and will disappear if a sound IF tube is removed.

Picture IF harmonics. Harmonics of the picture IF carrier signal are usually generated in the video detector. They can be coupled back to the rf tuner and cause interference in the picture. This may appear as diagonal lines, depending on the beat frequency. There is usually a grainy effect due to amplitude modulation of the video signal by the high-frequency components. The frequency of the beat interference will vary widely with a slight change in the oscillator fine tuning control.

Cochannel interference. Stations broadcasting on the same channel are separated by 150 miles or more, but it is still possible for cochannel stations to interfere with each other, especially in fringe locations between cities. If the interfering signal is strong enough, its picture will be superimposed on the desired picture. In addition, there is usually a bar pattern resulting from the beat between the two picture carriers because of their offset frequencies. The beat is an audio frequency that produces a horizontal bar pattern, generally called *venetian-blind effect* (Fig. 19-15a). This interference usually ap-

(a)

(b)

FIGURE 19-15 CHANNEL INTERFERENCE. (a) VENETIAN-BLIND EFFECT CAUSED BY COCHANNEL; (b) WINDSHIELD-WIPER EFFECT CAUSED BY ADJACENT CHANNEL.

pears on the low-band VHF channels 2 to 6. The remedy is a more directional antenna, especially for the front-to-back ratio, as interfering co-channel stations are often in opposite directions. Or, an antenna rotator can be used.

Adjacent channel interference. When the picture signal of the upper adjacent channel is strong enough, the side bands corresponding to low video frequencies can beat with the desired picture carrier. Then picture information of the interfering station is superimposed on the desired picture. Most noticeable is the vertical bar produced by horizontal blanking, as it drifts from side to side because of the slight difference in horizontal phasing between the two signals. This is generally called *windshield-wiper effect* (Fig. 19-15b). The remedy is a more directional antenna, but the IF selectivity can also be improved by wavetraps tuned to reject the adjacent channel intermediate frequencies because they are outside the required IF passband. In the lower adjacent channel, its sound signal can interfere with the picture. Figure 19-16 illustrates how channels 2 and 4 are adjacent to channel 3, as the station tuned in.

Television receiver interference in radio receivers. Harmonics of the 15,750-Hz horizontal deflection current in the television receiver can cause beat-frequency whistles in nearby radio receivers. The whistles appear every 15 kHz with maximum points every 70 to 100 kHz, approximately, on the dial of the radio receiver. This interference can be reduced by inserting an rf filter in series with the power cord of the television receiver. These harmonics can also radiate into the signal circuits and cause interference in the picture.

External noise. Interfering pulse voltages can be produced by ignition systems in automobiles and by arcing in electric motors. This interference shows up as short, horizontal black streaks in the picture (Fig. 19-17). The picture may roll vertically and tear apart horizontally if the noise is strong enough to interfere with synchronization. It is important to notice that when the noise streaks are in a cluster forming a pair of horizontal hum bars, the arcing is from a device operating on the 60-Hz power line. Ignition-noise streaks appear at random through the picture.

Arcing in the high-voltage power supply of the receiver can also produce noise streaks. Usually, you can see the arcing in the high-voltage cage.

External noise interference can enter the receiver as pickup by the antenna or an unshielded transmission line, as direct pickup by

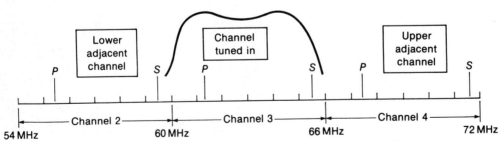

FIGURE 19-16 EXAMPLE OF CHANNELS 2 AND 4 AS ADJACENT CHANNELS WHEN RECEIVER IS TUNED TO CHANNEL 3.

TROUBLES IN THE RASTER AND PICTURE   419

FIGURE 19-17 HORIZONTAL BANDS OF NOISE INTERFERENCE IN THE PICTURE, WITH VERTICAL ROLLING.

also. Hum at 60 Hz produces one pair of bars, as in Fig. 19-18; two pairs of bars result from 120-Hz hum.

When the hum frequency is 120 Hz, this is caused by excessive ripple in the B+ voltage from a full-wave rectifier with insufficient filtering. The electrolytic capacitors can become dry with age and lose their ability to remove ac ripple from the rectified dc output. When the hum frequency is 60 Hz, this can be caused by heater cathode leakage in a tube. Also, 60 Hz is the ripple frequency in the B+ output from a half-wave rectifier. Insufficient filtering here allows excessive 60-Hz hum voltage.

The hum bars would stay still on the screen if the 60-Hz vertical scanning frequency were locked to the power-line frequency. However, with 59.94 Hz for vertical scanning, the hum bars for 60 Hz and 120 Hz drift slowly upward on the screen.

Modulation hum. Hum can be present in high-frequency signal circuits, even though they are not able to amplify the hum frequency, if the

the chassis, or through the ac power line. Minimizing the interference effect in the picture involves supplying more antenna signal or reducing the noise pickup, or doing both, to provide a suitable signal-to-noise ratio. This may be accomplished by using a high-gain directive antenna, with vertical stacking. Moving the antenna out of the noise field by either increasing the antenna height or just finding an antenna placement farther from the noise source is often helpful. It may be necessary to use shielded transmission line to prevent noise pickup by the line.

To reduce pickup from the power line, a low-pass filter can be used. This unit consists of 1-mH rf chokes in series in each side of the line cord and 0.01-μF bypass capacitors across the line. Connect the filter at the point closest to the interference source in order to minimize radiation from the power line.

19-9 *Hum Troubles*

In addition to hum in the sound, excessive hum voltage can cause one or two pairs of horizontal hum bars. Usually, there is bend in the picture

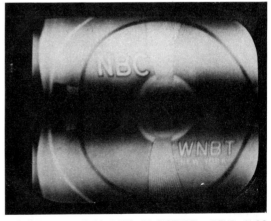

FIGURE 19-18 HUM BARS IN PICTURE. BEND ALSO PRESENT BECAUSE OF HUM IN THE HORIZONTAL SYNC.

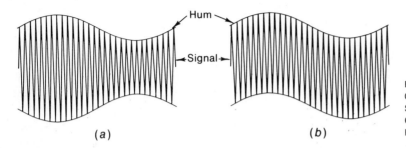

FIGURE 19-19 WAVEFORMS OF HUM COMBINING WITH HIGH-FREQUENCY SIGNAL. (a) MODULATION HUM WITH OPPOSITE ENVELOPES; (b) ADDITIVE HUM WITH PARALLEL ENVELOPES.

hum voltage modulates the high-frequency signal. In a picture IF amplifier tuned to 45.75 MHz, for example, leakage between heater and cathode can allow the filament voltage to modulate the grid signal voltage. The result is IF output amplitude-modulated by the 60-Hz hum. Succeeding stages amplify the modulated IF signal, and when it is rectified the detected output includes the 60-Hz hum in the video signal. Modulation hum is often called *tunable hum*.

Additive hum. In a circuit that has a plate load impedance for 60-Hz or 120-Hz ac voltages, the hum can be present in addition to the desired signal. Modulation is not necessary, since the circuit can amplify the low-frequency hum voltage itself. The audio amplifiers, video amplifiers, sync amplifiers, and vertical deflection circuits can amplify the hum voltage.

The difference in the waveforms for modulation hum and additive hum is shown in Fig. 19-19. Because of the modulation in Fig. 19-19a, the hum is evident only with a picture produced by the signal. In Fig. 19-19b, though, the hum is riding on the signal. Therefore, additive hum can also be seen in the raster even without a picture.

Hum bend in the raster. See Fig. 19-20. The edges of the raster have sine-wave bend because of hum. One cycle of bend is shown here for 60-Hz hum. With 120-Hz hum there are two cycles of bend at the edges of the raster. In Fig. 19-20a, the width of the horizontal scanning lines is varying at the hum frequency. This can be caused by hum in the horizontal scanning current. In Fig. 19-20b, the lines are displaced horizontally. Note that the two sides of the raster are parallel. This effect can be caused by hum in the horizontal AFC or in the horizontal oscillator.

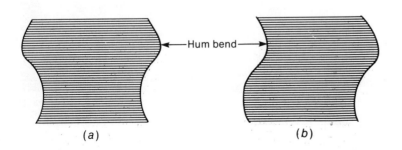

FIGURE 19-20 BEND IN RASTER, CAUSED BY HUM. (a) WIDTH MODULATED AT 60 Hz; (b) LINES DISPLACED HORIZONTALLY AT 60 Hz.

SUMMARY

1. Troubles in the height, width, or shape of the raster are in the vertical and horizontal scanning circuits, including the deflection yoke.
2. The damper diode in the horizontal output circuit: (a) damps oscillations; (b) provides the left side of horizontal scanning; and (c) rectifies the deflection voltage to produce boosted B+ voltage.
3. The boosted B+ is generally used for plate voltage on the horizontal amplifier tube. Therefore, this stage cannot operate without the damper.
4. A trapezoidal raster is caused by a defective deflection yoke.
5. The usual cause of no brightness with normal sound is no high voltage. The trouble can be in the high-voltage rectifier or the horizontal scanning circuits.
6. Ghosts in the picture are usually caused by multipath signals received by the antenna. There is no way to eliminate these ghosts except by eliminating multipath reception at the antenna.
7. Random noise generated in the mixer stage is amplified by the receiver to produce snow in the picture. A snowy picture indicates there is not enough rf signal.
8. Hum in the video signal produces hum bars, often with bend in the picture.
9. The three ways of having sound interference in the picture are (a) horizontal bars from the audio, (b) 4.5-MHz beat, and (c) 920-kHz beat.
10. Troubles in the coupling and bypass capacitors in the video amplifier cause smear and streaking in the picture. The capacitor can be leaky or open.
11. Cochannel interference can produce a venetian blind effect, similar to the effect of audio signal in the picture. The remedy is a more directional antenna.
12. The upper adjacent channel can produce an interfering image superimposed on the main picture, with a windshield-wiper effect caused by the horizontal blanking bar drifting across the screen. The remedy is a more directional antenna.

Self-Examination (Answers at back of book)

Match the troubles at the left with a possible cause or source at the right.

1. No height
2. No brightness
3. Poor vertical linearity
4. Trapezoidal raster
5. Insufficient width
6. Sound in picture
7. Two hum bars and bend
8. Smear with streaking

a. Weak horizontal amplifier
b. 4.5-MHz trap
c. Filter capacitor in full-wave power supply
d. Weak high-voltage rectifier
e. Shorted turns in yoke coil
f. Video bypass capacitor
g. Cochannel interference
h. Audio output stage

9. Windshield-wiper effect
10. Snowy picture
11. No picture and no sound
12. Ghost
13. Blooming
14. Raster too blue
15. No color
16. No sound, but normal picture

i. Too much AGC bias
j. Vertical oscillator not operating
k. Weak vertical amplifier
l. Screen controls for picture tube
m. Antenna
n. 3.58-MHz bandpass amplifier
o. Horizontal damper not operating
p. Rf amplifier

Essay Questions

1. Give the function and circuit for each of the following controls: height, vertical linearity, width, vertical hold, focus, and horizontal hold.
2. Give three functions of the horizontal output transformer.
3. What is meant by boosted B+ voltage?
4. Give two functions of the damper diode.
5. Give the functions of the vertical integrator circuit and horizontal AFC circuit.
6. Give an example of adjacent channel interference, other than the example of channels 2, 3, and 4 in Fig. 19-16.
7. Give an example of cochannel interference. How can this be reduced?
8. Give three ways in which the sound signal can interfere in the picture.
9. What is the difference between additive hum and tunable hum?
10. The picture has two pairs of hum bars, with bend in the picture but not in the raster. What is the most likely cause of this trouble?
11. The picture has a slight vertical bend, without any hum bars. What is a possible cause of this trouble?
12. The picture is reversed in black and white and torn apart in diagonal segments, with buzz in the sound. What is this trouble?
13. List the circuits that would be for sound alone in a color receiver.
14. Give three possible troubles that would cause no brightness, with normal sound.
15. Give two possible troubles that would cause no brightness and no sound.
16. Why does a weak rf amplifier cause a picture that has snow?
17. What is the main cause of ghosts in the picture?
18. Why is the video detector circuit usually shielded?
19. What is a possible cause of smear and streaking in the picture?
20. Describe the difference between ghosts from multipath signals and multiple outlines in the picture caused by ringing.
21. What is venetian-blind effect?
22. What is windshield-wiper effect?

Problems (Answers to selected problems at back of book)

1. Refer to the example of adjacent channels in Fig. 19-16. (a) Calculate the carrier frequencies of P and S for channels 2, 3, and 4. (b) With the rf local oscillator at 107 MHz to tune in channel 3, calculate the IF values of P and S for the three channels.
2. Repeat Problem 1 for the example of channels 7 and 9 adjacent to channel 8 as the station tuned in.
3. Approximately how many pairs of horizontal bars can be produced by an interfering audio frequency of (a) 60 Hz, (b) 420 Hz?
4. Approximately how many pairs of vertical bars can be produced by an interfering frequency of (a) 920 kHz, (b) 4.5 MHz?
5. If a video signal is delayed 2 μs, how many inches will it be displaced horizontally to the right on a screen 20 in. wide?
6. Refer to Table 19-1 below and describe briefly the trouble for each figure listed and give a possible cause.

TABLE 19-1 TROUBLES IN THE RASTER AND PICTURE

FIGURE	TROUBLE	CAUSE
19-2		
19-3		
19-4		
19-6		
19-7		
19-8		
19-12		
19-13		
19-17		

20 Deflection Oscillators

The scanning of the electron beam in the picture tube is made possible by the deflection oscillator stages in the receiver. As illustrated in Fig. 20-1, the deflection voltage generated by the vertical oscillator at 60 Hz* is coupled to the vertical deflection amplifier. This output stage supplies the amount of current needed for the vertical scanning coils in the deflection yoke mounted on the neck of the picture tube. Similarly, the horizontal oscillator produces the deflection voltage at 15,750 Hz needed for the horizontal output stage. The two types of oscillator circuits generally used are the blocking oscillator (Fig. 20-7) and the multivibrator (Fig. 20-18). These are free-running oscillators, which operate with or without sync input. However, the sync locks the oscillator into the correct frequency. The advantage of using the vertical and horizontal oscillators, instead of driving the deflection circuits directly with sync voltage, is that the raster is still produced between stations when the receiver has no signal. More details are described in the following topics:

20-1 The sawtooth deflection waveform
20-2 Producing sawtooth voltage
20-3 Blocking oscillator and discharge tube
20-4 Analysis of blocking oscillator circuit
20-5 Blocking oscillator sawtooth generator
20-6 Transistorized blocking oscillator circuit
20-7 Frequency and size controls
20-8 Synchronizing the blocking oscillator

*Broadcast stations generally use the scanning frequencies of exactly 59.94 Hz and 15,734 Hz for color and monochrome programs. However, the nominal values of 60 Hz for V and 15,750 Hz for H are used here. In either case, the deflection oscillators will be locked in at the sync frequency.

20- 9   Types of multivibrator circuits
20-10   Plate-coupled multivibrator
20-11   Cathode-coupled multivibrator
20-12   Multivibrator sawtooth generator
20-13   Synchronizing the multivibrator
20-14   Frequency dividers
20-15   Transistors in multivibrator circuits
20-16   Sawtooth, trapezoidal, and rectangular voltages and currents
20-17   Incorrect oscillator frequency

20-1 The Sawtooth Deflection Waveform

Linear scanning requires the sawtooth waveform shown in Fig. 20-2 because it has: (1) a linear rise in amplitude which will deflect the electron beam at uniform speed for a linear trace without squeezing or spreading the picture information and (2) a sharp drop in amplitude for the fast retrace or flyback. These requirements apply to horizontal and vertical scanning. Note the following factors about the sawtooth waveform in Fig. 20-2:

1. One sawtooth cycle includes both trace and retrace. The frequency is 60 Hz for vertical scanning and 15,750 Hz for horizontal scanning.
2. The waveform is an ac voltage (E) or current (I), with an average-value axis of zero. The dotted lines in the figure show equal amplitudes above and below the zero axis. The electron beam can be centered by dc voltages or by a steady magnetic field. Then the ac sawtooth waveform of E or I deflects the beam to both sides of center. Zero amplitude on the sawtooth waveform is the time when the beam is at the center, without deflection.
3. The peak-to-peak amplitude of the sawtooth wave determines the amount of deflection away from center. The electron beam is at the extreme left and right edges of the raster when the horizontal-deflecting sawtooth wave has its peak positive and negative amplitudes. Similarly, the beam is at the top and bottom for the peak amplitudes of the sawtooth waveform for vertical deflection.

The sawtooth waveform can have opposite polarities, as shown in Fig. 20-3. In Fig. 20-3a, the waveform is a positive-going sawtooth, as the trace amplitudes increase in the positive direction. In Fig. 20-3b, the amplitudes increase in the negative direction for the trace. In both cases, the trace includes the linear rise from the start at point 1 to the end of trace at the peak at point 2. This point also is the start of retrace, finishing at point 3 for a complete sawtooth cycle.

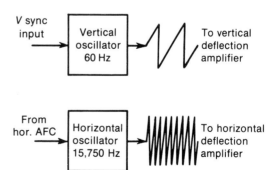

FIGURE 20-1 THE HORIZONTAL AND VERTICAL DEFLECTION OSCILLATORS GENERATE SAWTOOTH VOLTAGE FOR THE SCANNING CIRCUITS.

426 BASIC TELEVISION

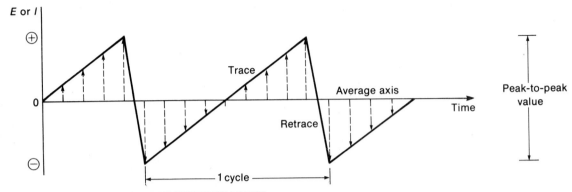

FIGURE 20-2 THE SAWTOOTH DEFLECTION WAVEFORM.

For the specific case of horizontal scanning, the start of the trace at point 1 is when the beam is at the left edge of the raster. At the time of point 2 on the sawtooth, the beam is at the right edge. Then the retrace between points 2 and 3 makes the fast flyback to the left edge. Point 3 marks the start of another line for the next cycle of horizontal scanning.

Similarly, for vertical scanning point 1 is when the beam is at the top of the raster. At the time of point 2, the beam is at the bottom. Then the retrace between points 2 and 3 makes the vertical flyback to the top. Point 3 marks the start of another field for the next cycle of vertical scanning.

20-2 Producing Sawtooth Voltage

Usually, sawtooth voltage is obtained as the voltage output across a capacitor that is alternately charged slowly and discharged fast (see Fig. 20-4). There are two requirements:

1. Charge the capacitor through a high resistance in series with C_s for a long time constant. The charging produces a linear rise of voltage across C_s for the trace part of the sawtooth wave.
2. Discharge the same capacitor through a much smaller resistance for a relatively fast time constant. The sharp drop in capacitor voltage as it discharges is the retrace part of the sawtooth wave.

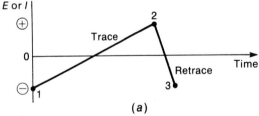

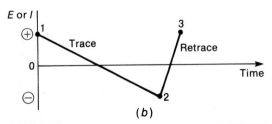

FIGURE 20-3 INVERTED POLARITIES OF THE SAWTOOTH WAVEFORM. (a) POSITIVE-GOING TRACE; (b) NEGATIVE-GOING TRACE.

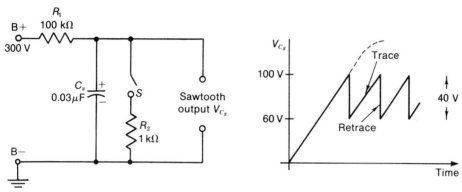

FIGURE 20-4 FUNDAMENTAL METHOD OF PRODUCING SAWTOOTH VOLTAGE. C_s CHARGES SLOWLY THROUGH HIGH RESISTANCE OF R_1 FOR LINEAR RISE. THEN S IS CLOSED TO DISCHARGE C_s FAST THROUGH R_2 FOR RETRACE.

This sawtooth waveform is also called a *ramp voltage, sweep voltage,* or *time base.*

In Fig. 20-4, when dc voltage is applied the capacitor charges toward the B+ voltage applied. The result is an exponential charge curve, at a rate determined by the RC time constant.[1] The path for charging current is from B− to accumulate negative charge at the grounded side of C_s. Electrons are then repelled from the opposite plate to produce current through R_1, returning to the B+ terminal of the voltage source. As charge builds up in the capacitor, the voltage across C_s increases. The rate of increase in voltage depends on the RC time constant. This equals 0.003 s for charging C in the example here.

Let us assume that, when the voltage across C_s builds up to 100 V, we close the switch. Then C_s discharges through the low-resistance path in parallel. The path for discharge current is from the negative side of C_s through R_2 and the closed switch back to the positive side of C_s. Practically no current flows through the R_1 path because of its high resistance compared with R_2. Since electrons lost from the negative side are added to the positive side, the charge in the capacitor is neutralized. Then the voltage across C_s decreases toward zero. The R_2C_s time constant of 0.00003 s for discharge is only one-hundredth of the time constant for charge. Therefore, the decrease in capacitor voltage, with a short time constant on discharge, is much faster than the voltage rise on charge.

If we open the discharge switch when the voltage across C_s has discharged down to 60 V, the capacitor will start charging again. From now on, the capacitor starts charging from a 60-V level. However, we can still discharge the capacitor by closing the switch every time V_c reaches 100 V. Also, we open the switch to allow recharging when V_c is down to 60 V. The result is the sawtooth waveform shown for the output voltage across C_s. One cycle of the sawtooth wave includes a linear rise for trace and the sharp drop in amplitude for the retrace. The frequency of the sawtooth wave here depends on the rate of switching. Note that the peak-to-peak amplitude of this sawtooth voltage waveform is 40 V, equal to the difference between the maximum at 100 V and the minimum at 60 V.

---

[1] See Appendix D for details of RC time constant.

Actually, the switch S in Fig. 20-4 represents a gas tube, vacuum tube, or transistor that can be switched on or off at almost any speed. When the tube or transistor is not conducting, this condition is the off position. Maximum conduction is the on position. Connected in parallel with C_s, the tube or transistor conducting maximum current provides a low-resistance discharge path to allow a fast decay of capacitor voltage for the retrace. In this application, we can call the tube a *discharge tube*. C_s is often called the *sawtooth capacitor* or *sweep capacitor*.

The voltage rise on the sawtooth wave is kept linear by using only a small part of the exponential RC charge curve. As shown in Appendix D, the first 40 percent of the curve is linear within 1 percent. To limit the voltage rise to this linear part of the curve, the charging time should be no more than one-half the RC time constant. A higher B+ voltage also improves linearity, as a specific amount of V_c is a smaller percentage of the applied voltage. Then a smaller part of the RC charge curve is used, the part where it is most linear.

20-3 Blocking Oscillator and Discharge Tube

Figure 20-5 shows how a vacuum tube can be used as a discharge tube in parallel with the sawtooth capacitor C_s. While its grid voltage is more negative than cutoff, the discharge tube cannot conduct plate current. Then it is an open circuit. While the discharge tube is cut off, therefore, the sawtooth capacitor C_s in the plate circuit charges toward the B+ voltage, through the series resistance R, to produce the linear rise on the sawtooth voltage wave. When the grid voltage drives the discharge tube into conduction, its plate-to-cathode circuit becomes a low resistance equal to several hundred ohms. Then C_s discharges quickly from cathode to plate through the discharge tube, producing the rapid fall in voltage for the flyback on the sawtooth voltage wave. Therefore, by applying narrow positive pulses to the grid of the discharge tube and keeping it cut off between pulses, a sawtooth wave of voltage is produced in the output.

Figure 20-6 illustrates how the linear rise of voltage in the output corresponds to the time when the grid is more negative than cutoff, while the flyback time coincides with the positive grid pulse. The result is a sawtooth wave of voltage output from the plate of the discharge tube with the same frequency as the grid-voltage pulses from the blocking oscillator.

Blocking oscillator circuit. Essentially Fig. 20-7 is a transformer-coupled oscillator with grid-leak bias. The transformer provides grid feedback voltage with the polarity required to reinforce grid signal and start the oscillations. When the oscillator feedback drives the grid positive, grid current flows to develop grid-leak bias. This regenerative circuit could oscillate with continuous sine-wave output at the natural resonant frequency of the transformer, depending on its inductance and stray capacitance. However, several factors enable the oscillator to cut itself off with high negative grid-leak bias. A large amount of feedback is used. Also, the $R_g C_c$ time constant is made long enough to allow the grid-leak bias to keep the tube cut off for a relatively long time. Finally, the transformer has high internal resistance for low Q, so that after the first cycle the sine-wave oscillations do not have enough amplitude to overcome the negative bias.

The tube remains cut off until C_c can discharge through R_g to the point where the grid-leak bias voltage is less than cutoff. Then plate current can flow again to provide feedback

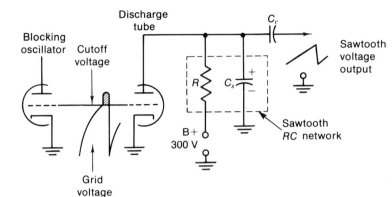

FIGURE 20-5 BLOCKING OSCILLATOR GRID VOLTAGE DRIVING DISCHARGE TUBE TO PRODUCE SAWTOOTH VOLTAGE ACROSS C_s. SHADED AREA IN GRID VOLTAGE INDICATES TIME OF PLATE-CURRENT CONDUCTION.

signal for the grid, and the cycle of operation repeats itself at the blocking rate. Therefore, the circuit operates as an intermittent or blocking oscillator. The tube conducts a large pulse of plate current for a short time and is cut off for a long time between pulses.

Pulse repetition frequency. As shown in Fig. 20-8, one sine-wave cycle of high amplitude is produced at the blocking rate. The succeeding sine waves, in dotted lines, do not have enough amplitude to overcome the negative grid-leak bias blocking the oscillator. The number of times per second that the oscillator produces the pulse and then blocks itself is the pulse repetition rate, or frequency. This blocking frequency is much lower than the frequency of the sine waves, which is the ringing frequency or resonant frequency of the transformer (Fig. 20-9).

For a deflection oscillator, the pulse repetition rate is what we consider to be the oscilla-

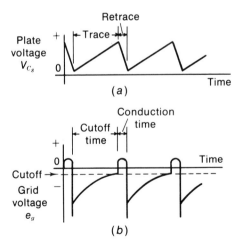

FIGURE 20-6 (a) HOW SAWTOOTH VOLTAGE ACROSS C_s CORRESPONDS TO (b) GRID VOLTAGE.

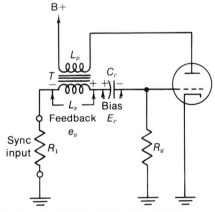

FIGURE 20-7 BLOCKING OSCILLATOR CIRCUIT.

430 BASIC TELEVISION

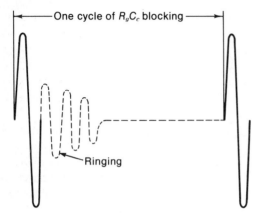

FIGURE 20-8 RINGING OF BLOCKING OSCILLATOR TRANSFORMER AT RESONANT FREQUENCY OF ITS INDUCTANCE AND STRAY CAPACITANCE.

tor frequency. This is the rate at which the tube oscillates between conduction and cutoff. The blocking frequency is determined mainly by the $R_g C_c$ time constant. In a horizontal deflection oscillator, its $R_g C_c$ time constant allows the blocking oscillator to operate at a pulse repetition frequency of 15,750 Hz; a vertical blocking oscillator operates at 60 Hz. The type of circuit that oscillates between conduction and cutoff is often called a *relaxation oscillator,* as the stage is relaxing while it is not conducting.

20-4 Analysis of Blocking Oscillator Circuit

The waveform of instantaneous grid voltage e_c can be considered in two parts. One is the ac feedback voltage e_g for the grid, induced across the transformer secondary by a change in primary current. The other is the dc bias voltage E_c, developed when the feedback signal drives the grid positive to produce grid current. E_c can increase fast when grid current charges the coupling capacitor to produce the grid-leak

FIGURE 20-9 VERTICAL BLOCKING OSCILLATOR TRANSFORMER. HEIGHT IS 2 IN. *(STANCOR)*

bias. The bias voltage must decrease slowly, however, as C_c discharges through the high resistance of R_g when the feedback signal voltage decreases.

At any instant, the net grid voltage e_c is equal to the algebraic sum of the bias E_c and the signal drive e_g. The grid signal voltage caused by feedback can drop to zero instantaneously when the feedback ceases, but the bias cannot. This is why the oscillator can cut itself off with the grid-leak bias produced by its own feedback.

It is worth reviewing the fundamentals of induced voltage with a transformer to see how the feedback voltage can increase, decrease, or reverse polarity. First, the primary current must change to induce voltage in the secondary. The amount of induced voltage in the secondary winding increases with a sharper rate of change in primary current. A slower rate of current change means less induced voltage. When the current stops changing, there is zero induced voltage for the steady primary current.

Furthermore, we must consider the polarity of induced voltage. In the blocking oscillator transformer, plate current in the primary can flow in only the one direction, from the plate to the B+ voltage source. However, the current can either increase or decrease. Whatever polarity the induced voltage has for an increase of current, the polarity reverses for a decrease in current. The polarity reversal results from the fact that increasing current has an expanding magnetic field cutting across the winding, but the field collapses into the winding with decreasing current. The polarity of induced voltage reverses across both the primary and the secondary when the current change reverses from increasing to decreasing values, or vice versa.

The cycle of operations can be followed from the time power is applied to produce plate current. Note the plate and grid waveforms in Fig. 20-10. The numbers in the e_c waveform correspond to the following steps in the circuit analysis:

1. Increasing plate current. Plate current flows immediately because the grid-leak bias is zero at the start. Now the plate current through L_p is increasing from zero. The windings are poled to make the resultant induced voltage positive at the grid side of L_s. Therefore, increasing plate current induces feedback voltage that drives the grid positive. The positive feedback increases the plate current still more. As a result of the regeneration, the plate current increases very quickly to its maximum value.

During this time when the feedback is driving the grid positive, grid current flows to develop negative bias a little less than the peak positive drive. As an example, for +35 V at the peak, the grid-leak bias may be about −30 V. Then the net grid voltage is +5 V. The positive grid voltage is maintained while e_g is positive and more than E_c. This part of the cycle occurs while the plate current is increasing to produce positive grid feedback voltage.

2. Maximum plate current. The rise in plate current is limited by saturation. Then the current increases at a slower rate. The result is less induced feedback voltage. However, the negative bias remains. With less positive grid drive, now the plate current starts to decrease. At this turning point, the grid feedback voltage changes in polarity from positive to negative. Plate current is still flowing, but it is decreasing instead of increasing.

3. Negative grid feedback. As the grid becomes less positive, the negative grid feedback decreases the plate current still further. The amplification of the tube then makes the plate current drop sharply toward zero. This fast drop in current produces a very large negative voltage at the grid.

4. Maximum negative grid feedback. The fastest drop in plate current produces the maximum negative voltage at the grid, which can be

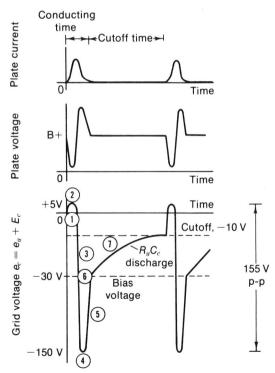

FIGURE 20-10 PLATE AND GRID WAVEFORMS IN BLOCKING OSCILLATOR CIRCUIT. SEE TEXT FOR EXPLANATION OF NUMBERED STEPS IN GRID VOLTAGE.

−150 V, or more. This large peak voltage in the negative direction is the result of two factors. First, there is no grid current now. Without the load of grid current in the secondary, the primary current can drop faster than it rises. Also, the negative grid feedback is series-aiding with the negative grid-leak bias so that now e_g is added to E_c.

5. Decreasing negative grid feedback. Because inductance opposes a change in current, the plate current in L_p decreases at a slower rate as it drops toward zero. The current decay in an inductance corresponds to a capacitor discharge curve, with the sharpest slope at the start. Therefore, the amount of negative feedback voltage decreases as the drop in plate current approaches zero. Plate current can flow even though the grid voltage is very negative, as the self-induced voltage across L_p makes the plate voltage rise to a positive value much higher than B+.

6. Plate current drops to zero. Now there is no feedback voltage at all. However, the grid-leak bias produced while the grid was positive still remains after the feedback stops. This bias voltage keeps the tube cut off.

7. C_c discharges through R_g. As C_c loses charge, its negative bias voltage decreases. If we take the example of −30 V maximum grid-leak bias, and a grid-cutoff voltage of −10 V, the tube is cut off for the time it takes C_c to discharge 20 V, from −30 to −10 V. Therefore, the cutoff time depends on the R_gC_c time constant.

When the grid-leak bias becomes less than the grid-cutoff voltage, plate current can flow again to produce the next pulse and the cycle is repeated. Therefore, the blocking oscillator continuously generates sharp positive grid pulses followed by relatively long periods of cutoff. The blocking rate is faster for a higher frequency, with a shorter time constant for R_gC_c. The blocking oscillator frequency also depends on the tube characteristics, especially its grid-cutoff voltage, the plate voltage, and the transformer. Frequency increases with less plate voltage because the grid cutoff voltage is reduced. However, the oscillator frequency is usually varied by adjusting either the R_gC_c time constant or the bias voltage in the grid circuit.

20-5 Blocking Oscillator Sawtooth Generator

Instead of having a separate discharge tube, the blocking oscillator itself can have the sawtooth capacitor C_s in the output circuit, as shown in Fig. 20-11. This circuit can be operated as a sawtooth generator for either vertical deflection at 60 Hz or horizontal deflection at 15,750 Hz. Either a tube or a transistor can be used. The stage is a *free-running oscillator,* meaning it operates with or without sync input.

Single-triode circuit. In Fig. 20-11, C_s is in the oscillator plate circuit. While the oscillator is cut off by its blocking action, C_s charges through the series resistance of R_3 and R_4, toward the B+ voltage. When plate current flows during the oscillator pulse, C_s discharges through the tube. The discharge path for C_s is from cathode to plate and through the plate winding of the transformer T_1. Note that the discharge current of C_s is in the same direction as normal plate current during oscillator conduction. The frequency of the sawtooth voltage output is the frequency of the blocking oscillator.

It should be noted that changes in plate voltage vary the oscillator frequency. This effect occurs when R_4 in the plate circuit is varied to

DEFLECTION OSCILLATORS 433

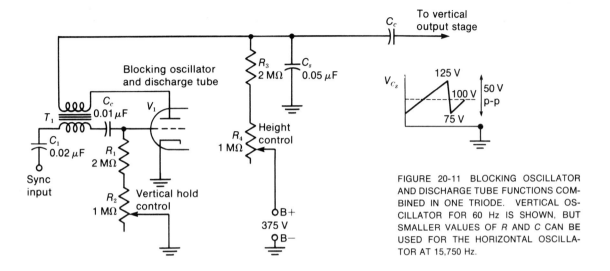

FIGURE 20-11 BLOCKING OSCILLATOR AND DISCHARGE TUBE FUNCTIONS COMBINED IN ONE TRIODE. VERTICAL OSCILLATOR FOR 60 Hz IS SHOWN, BUT SMALLER VALUES OF R AND C CAN BE USED FOR THE HORIZONTAL OSCILLATOR AT 15,750 Hz.

adjust the height of the raster. The resulting change in E_b has the side effect of changing the oscillator frequency, which usually makes the picture roll. However, the frequency can be brought back to normal by readjusting the hold control R_2.

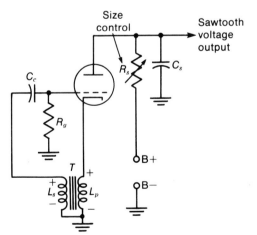

FIGURE 20-12 BLOCKING OSCILLATOR CIRCUIT WITH CATHODE FEEDBACK. POLARITIES SHOWN FOR INCREASING CATHODE CURRENT.

Cathode feedback. In Fig. 20-12, the blocking oscillator transformer supplies feedback from cathode to grid. The plate current returns through L_p, from the ground side to cathode. Note that the voltage polarities shown for increasing cathode current induce positive grid feedback. The sawtooth capacitor C_s is in the plate circuit without the transformer. While the tube is cut off, C_s charges through R_s. Then C_s discharges through L_p and the tube when it conducts at the blocking oscillator frequency.

20-6 Transistorized Blocking Oscillator Circuit

As shown in Fig. 20-13, this collector-coupled stage is similar to the plate-coupled oscillator in Fig. 20-7. The blocking oscillator transformer can also be emitter-coupled to the base, similar to the cathode coupling in Fig. 20-12.

In transistor circuits, capacitance values are much larger because the resistances are smaller. The 100-μF C_1 is the sawtooth capaci-

434 BASIC TELEVISION

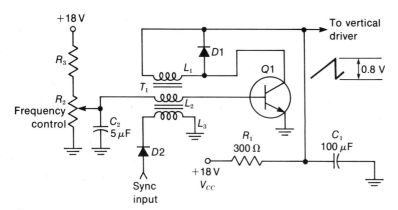

FIGURE 20-13 TRANSISTORIZED BLOCKING OSCILLATOR AS SAWTOOTH VOLTAGE GENERATOR FOR VERTICAL DEFLECTION AT 60 Hz. THIS CIRCUIT CAN ALSO BE USED WITH SMALLER R AND C VALUES FOR HORIZONTAL OSCILLATOR AT 15,750 Hz.

tor with the 330-Ω R_1 in the collector circuit. Also, the collector supply is much less than typical plate voltage for tubes. However, the required amount of sawtooth voltage for the driver stage is less than 1 V. The driver isolates the oscillator from the very low input resistance of the power output stage.

In Fig. 20-13, the oscillator frequency control R_2 varies the amount of positive forward voltage at the base of the NPN transistor. In order for a transistor oscillator to be free-running it must have forward bias, provided by the $R_2 R_3$ voltage divider. Unlike a tube, a transistor is cut off without any bias.

The protective diodes $D1$ and $D2$ in Fig. 20-13 are needed with transistors. Remember that semiconductor diodes conduct only with positive anode voltage at the arrow terminal, or negative at the cathode. The diode is an open circuit for the opposite polarity. $D1$ is in parallel with the collector winding to short-circuit transient negative voltage pulses that would put forward voltage on the collector. In the base circuit of the oscillator, $D2$ prevents positive voltage pulses from being coupled back into the sync circuits. The sync is inductively coupled to the base of $Q1$ by means of a separate winding L_3 on the blocking oscillator transformer T_1. The winding L_1 has the collector current, while L_2 provides feedback to the base.

20-7 Frequency and Size Controls

The frequency and amplitude of the oscillator output can be adjusted by variable resistances. Referring to Fig. 20-11, R_2 varies the time constant of the grid-leak bias circuit to adjust the oscillator frequency. R_4 in the plate circuit determines the time constant for charging C_s to adjust the peak-to-peak amplitude of the sawtooth voltage output.

Size control. Decreasing the resistance of R_4 in Fig. 20-11 makes the time constant shorter for charging C_s. Then the voltage across C_s increases at a faster rate. However, the amount of time allowed for charging is equal to the cutoff period, as determined by the oscillator frequency. For any one frequency, therefore, C_s can charge to a higher voltage because of the faster charging rate with a shorter time constant. As a result, decreasing the resistance of the size control increases the amplitude of sawtooth voltage output to increase the size of the scanning raster. Increasing the resistance of the size control decreases the scanning amplitude. Usually, a fixed resistance is in series with the control to limit the range of variations for easier adjustment.

This method of size control is generally used in the vertical deflection generator to adjust the height of the raster. Then it is called the *height control*.

Effect of C_s time constant on amplitude and linearity. Figure 20-14 shows the effect on sawtooth voltage amplitude when the height control in Fig. 20-11 is varied. For zero resistance in R_4, the time constant of R_3C_s equals 0.1 s. This is approximately six times longer than the vertical scanning period of 0.016 s. Therefore, C_s can take on enough charge to raise its voltage by one-sixth the net charging voltage, since the charging time equals one-sixth of the time constant.

Let us assume 75 V across C_s at the start of charge, after the first few cycles of charge and discharge. With B+ voltage of 375 V, the net charging voltage equals 375 − 75, or 300 V. Therefore, C_s charges an additional 50 V, or one-sixth of 300 V. This voltage equals the peak-to-peak amplitude of sawtooth output, between 75 V and 125 V.

At the opposite extreme, with the height control at maximum resistance, the time constant for charge equals 0.15 s. This is approximately ten times longer than 0.016 s. Therefore, C_s takes on additional charge to raise its voltage by one-tenth of 300 V. Then the peak-to-peak sawtooth voltage equals 30 V, between 75 V and 105 V. Note the better linearity but smaller amplitude with the longer time constant.

Frequency control. To control the oscillator frequency, the $R_g C_c$ time constant in the oscillator grid circuit is varied. This determines how fast the negative bias voltage can discharge down to cutoff. A shorter time constant with smaller values for R_g and C_c allows a faster discharge for a higher frequency. Making the $R_g C_c$ time constant longer lowers the oscillator frequency. The range of frequency control for the vertical oscillator is usually about 40 to 90 Hz.

The oscillator frequency control is adjusted to the point where the sync voltage can lock in the oscillator at the sync frequency to make the picture hold still. For this reason the frequency adjustment is generally called the *hold control*.

In the vertical oscillator, the vertical hold control generally is the oscillator frequency adjustment, as in Fig. 20-11. However, since the horizontal oscillator usually has automatic frequency control, the horizontal hold adjustment may be in the AFC circuit.

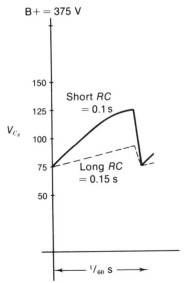

FIGURE 20-14 EFFECT OF VARYING THE TIME CONSTANT FOR C_s IN FIG. 20-11, WITH HEIGHT CONTROL AT MINIMUM AND MAXIMUM R.

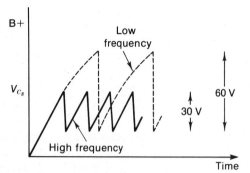

FIGURE 20-15 EFFECT OF FREQUENCY ON AMPLITUDE OF SAWTOOTH VOLTAGE OUTPUT.

Effect of oscillator frequency on sawtooth amplitude. It should be noted that the frequency is determined by the oscillator grid circuit, while the sawtooth capacitor in the plate circuit of the discharge tube determines amplitude and linearity of the sawtooth voltage output. However, changing the frequency will affect sawtooth amplitude and linearity. This idea is illustrated in Fig. 20-15. With any given time constant for C_s on charge, if the frequency is higher, less time is available for charging. Then C_s is charged to a lower voltage at the time of discharge. For the opposite case, a lower frequency allows C_s to charge to a higher voltage before discharge occurs, producing more sawtooth voltage output.

20-8 Synchronizing the Blocking Oscillator

A circuit like this is called a *soft oscillator* because its frequency is easily changed by variations in the electrode voltages. Its advantage, however, is that the oscillator can easily be synchronized to lock in at the sync frequency. The frequency can be synchronized either by sync pulses that trigger the oscillator into conduction at the sync frequency or by dc voltage to control the grid bias. The vertical deflection oscillator is usually locked in with triggered sync at 60 Hz. However, the horizontal oscillator frequency is usually synchronized with dc control voltage produced by the horizontal AFC circuit.

Triggered sync. The blocking oscillator can be synchronized by small positive pulses injected in the grid circuit to trigger the oscillator into conduction at the frequency of the sync pulses. The sync voltage is applied in series with the grid winding of the transformer. Then the positive sync voltage cancels part of the grid bias voltage produced by the oscillator. For the vertical oscillator, the sync pulse from the integrator has a peak amplitude of 1 to 10 V.

In Fig. 20-16, the positive sync pulses arrive at the times marked S, when the declining grid voltage is close to cutoff. Then a small sync voltage will be enough to drive the grid voltage momentarily above the cutoff voltage. As soon as plate current starts to flow, the oscillator goes through a complete cycle. With another positive sync pulse applied at a similar point of the following cycle, the oscillator again begins a new cycle at the time of the pulse. As a result, the

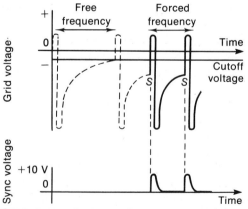

FIGURE 20-16 SYNCHRONIZING THE BLOCKING OSCILLATOR WITH POSITIVE TRIGGER PULSES IN THE GRID CIRCUIT.

pulses force the oscillator to operate at the sync frequency. The frequency of the oscillator without sync is called the *free-running frequency;* the synchronized oscillator frequency is the *forced frequency.*

The free-running frequency must be lower than the sync frequency. Then the sync pulses will drive the grid voltage in the positive direction when the oscillator is ready for triggering. This is the time when the grid bias has declined practically down to cutoff by itself, and needs only a slight additional positive voltage to start the flow of plate current and the beginning of a cycle. A positive sync pulse that occurs in the middle of the oscillator cycle must have a much higher value to drive the grid voltage to cutoff. The peak negative swing of the grid-voltage wave may be more than 150 V, but toward the end of the cycle a few volts of positive sync voltage can be enough to trigger the oscillator into conduction.

Sync voltage of negative polarity at the grid cannot trigger the blocking oscillator. Also, the oscillator cannot be triggered if the free frequency is slightly higher than the synchronizing frequency. Then the sync pulses will occur after the oscillator has started to conduct by itself, and they will have no effect.

Operating the oscillator at the same frequency as the synchronizing pulses does not provide good triggering because the oscillator frequency can drift above the sync frequency, resulting in no synchronization. For best synchronization, the free-running oscillator frequency is adjusted slightly lower than the forced frequency, so that the time between sync pulses is shorter than the time between pulses of the free-running oscillator. Then each synchronizing pulse occurs just before an oscillator pulse and forces the tube into conduction, thereby triggering every cycle to hold the oscillator locked in at the sync frequency.

Effect of noise. The triggering action can be made less sensitive to noise pulses by returning the grid to a positive voltage, instead of chassis ground. The added positive voltage has two effects on the grid discharge. First, the negative bias declines to cutoff with a sharper slope because a smaller part of the $R_g C_c$ discharge curve is used. With a sharper slope of declining grid voltage, an interfering noise pulse just before the sync pulse will need much more amplitude to trigger the oscillator. The second effect is that the added positive voltage makes the negative bias decline to cutoff in less time, raising the oscillator frequency. However, the frequency control can be adjusted to bring the oscillator to the frequency desired.

Dc control voltage. In Fig. 20-17 positive dc control voltage is directly coupled to the blocking oscillator grid. R_1 varies the amount of dc voltage used to control the oscillator frequency. In the waveshape below, the +10 V added decreases the negative grid voltage. As a result, less time is necessary for the grid leak bias to decline to cutoff to start the next cycle. The blocking oscillator frequency is raised, therefore, by inserting positive dc control voltage to make the grid voltage less negative.

Synchronizing transistor oscillators. The same ideas apply for triggered sync and dc control voltage. Positive sync voltage is needed at the base of an NPN transistor to trigger a blocking oscillator. Or, negative sync triggers a PNP transistor. Similarly, more positive dc control voltage at the base of an NPN transistor increases the forward bias to raise the frequency.

Methods of coupling the sync. In Fig. 20-11, the 0.02-μF C_1 couples the vertical sync pulses from the sync separator to the grid of the triode tube. In Fig. 20-13, a separate winding on the blocking oscillator transformer couples the sync

438 BASIC TELEVISION

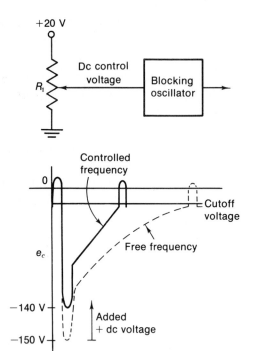

FIGURE 20-17 ADDING POSITIVE DC CONTROL VOLTAGE AT THE GRID TO RAISE THE FREQUENCY OF THE BLOCKING OSCILLATOR.

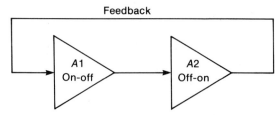

FIGURE 20-18 SWITCHING ACTION BETWEEN THE TWO AMPLIFIER STAGES IN A MULTIVIBRATOR.

pulses to the transistor base. Either capacitive or inductive coupling can be used for the sync, with either tubes or transistors. However, transistor circuits generally have a separate sync winding on the blocking oscillator transformer in order to provide isolation between the oscillator and sync circuits.

20-9 Types of Multivibrator Circuits

The multivibrator (MV) is a pulse generator like the blocking oscillator, but no transformer is needed. As shown in Fig. 20-18, two amplifier stages A1 and A2 are used for the MV. They can be tubes or transistors. Feedback is used to make the output of one stage drive the input of the other. Since each amplifier inverts the polarity of its signal, the feedback is positive in the same polarity as the original input. Therefore, oscillations can be produced. The oscillations are in the on-and-off conditions for each stage. When one stage conducts, it cuts off the other. For instance, A1 conducting forces A2 into cutoff. Then, as soon as A2 starts to conduct again it cuts off A1. The rate at which the stages are cut off is the oscillator frequency. One cycle includes cutoff for both stages. In the output circuit, the MV is basically a square-wave or rectangular-wave generator.

Multivibrator circuits can be classified according to the method of feedback. Then we have the following types:

Plate-coupled MV. The plate of A1 drives the control grid of A2, and the plate of A2 drives the grid of A1, through the feedback line.

Cathode-coupled MV. The plate of A1 drives the control grid of A2, but A2 is coupled back to A1 only through a common-cathode resistor for both stages.

Collector-coupled MV. This transistor circuit is similar to the plate-coupled MV. The collector corresponds to the plate and the base to the control grid.

Emitter-coupled MV. This transistor circuit is similar to the cathode-coupled MV. The emitter corresponds to the cathode, with an emitter resistance that is common to both stages.

The many applications of multivibrators include sawtooth generators, square-wave generators, trigger switches, and frequency dividers.

Multivibrators are also classified according to stability. A stable stage will remain in the off condition until it is triggered into conduction by a pulse from an external circuit. Three classes are:

Astable MV. Neither stage is stable, as one can cut off the other, at the MV repetition rate. This type is simply a free-running multivibrator.

Monostable MV. This circuit has one of the stages stable.

Bistable MV. This circuit has both stages in the stable condition.

The bistable MV and monostable MV are *trigger circuits.* This means they need an input pulse to upset the stable stage. Two important types are the *Schmitt trigger* circuit and the *Eccles-Jordan trigger* circuit. The Schmitt trigger circuit is similar to a cathode-coupled or emitter-coupled MV, but the trigger is not free-running. Also, the Eccles-Jordan trigger is similar to a plate-coupled or collector-coupled MV, but the trigger circuit needs to be driven by input pulses. These triggers are also called *flip-flop* circuits, indicating that they can change abruptly from cutoff to conduction. After one stage flips into conduction because of the input pulse, it flops back into cutoff because of drive from the opposite stage. One application, of trigger circuits is producing square-wave output from sine-wave input pulses.

20-10 Plate-coupled Multivibrator

The basic circuit in Fig. 20-19 illustrates how A1 cuts off A2 and then A2 cuts off A1 at the oscillating frequency of the MV. The essential facts in the operation are:

1. Conditions are reversed when the stage that was cut off starts to conduct.
2. With capacitive coupling to each grid circuit, the period of cutoff depends mainly on the $R_g C_c$ time constant.

The reason why conduction in one stage can cut off the other is the fact that when the plate current i_b increases from zero, the $i_b R_L$ drop reduces the plate voltage e_b. This drop in positive e_b is a negative-going voltage pulse to the next grid.

At the start of operation in Fig. 20-19, plate voltage is applied, plate current begins in both tubes, and the MV immediately begins oscillating. The amount of plate current cannot be exactly the same in A1 and A2 even when both use the same supply voltage and equal values of R_L. No matter how small this difference in plate current may be, it is immediately amplified, and the result is that one tube conducts while the other is cut off. Assume that A1 conducts slightly more than A2 when plate voltage is applied. This drives the grid A2 slightly more negative. The negative signal is amplified and inverted to provide feedback that drives the grid of A1 more positive, which in turn allows the first stage to drive the grid of A2 still more negative. The amplification of the unbalance in the stage takes place almost instantaneously to drive the grid of A2 to cutoff immediately. The tube remains cut off for a period of time that depends upon the $R_2 C_2$ grid time constant, as C_2 discharges.

The grid voltage of A2 then declines

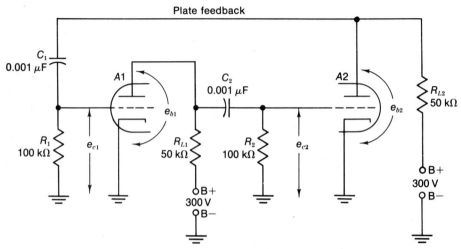

FIGURE 20-19 BASIC PLATE-COUPLED, FREE-RUNNING, SYMMETRICAL MULTIVIBRATOR CIRCUIT.

toward zero. As soon as the grid voltage is reduced to less than cutoff, plate current begins to flow. Then conduction in A2 cuts off A1. Now the cutoff time depends on the R_1C_1 time constant. As a result, the slight initial unbalance sets up a regenerative switching action, with first one stage conducting and then the other. Each stage conducts for a period equal to the time the other is cut off.

Analysis of multivibrator waveforms. Beginning at time A in Fig. 20-20, A2 has just been cut off because of conduction in A1. Assume that 4 mA of i_b flows through the 50-kΩ R_L of the first tube, producing a 200-V drop in plate voltage. Since the plate-to-cathode voltage applied across the R_2C_2 coupling circuit is reduced abruptly from 300 V to 100 V, C_2 must discharge. Instantaneously, the entire 200-V drop in voltage is developed across R_2, with the grid side negative, cutting off A2.

As C_2 discharges, the voltage across R_2 declines toward zero with the typical RC discharge curve shown. When the grid voltage is down to cutoff, A2 begins to conduct plate current. The resultant decrease of e_b in A2 drives the grid of the first tube more negative, reducing its plate current and increasing the plate voltage. This drives the grid of A2 more positive, further reducing the plate voltage and allowing A2 to drive A1 still more negative. This amplification of the unbalance in the stage reverses the action of the two tubes almost instantaneously, with A1 now cut off and A2 conducting, as shown at time B.

The plate-to-cathode voltage of the first tube, now cut off, rises immediately to B+, driving the grid of A2 positive. C_2 charges rapidly through the low resistance of the grid-to-cathode circuit of A2, and the grid voltage is reduced to zero very soon, as C_2 becomes completely charged. The grid voltage for A2 remains at zero and the zero-bias plate current flows as long as A1 remains cut off.

Meanwhile, the coupling capacitor C_1 for the first tube is discharging through its grid

DEFLECTION OSCILLATORS 441

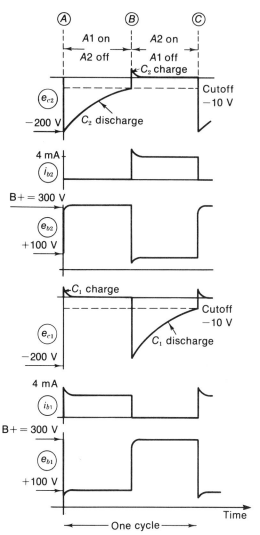

FIGURE 20-20 PLATE AND GRID WAVEFORMS FOR THE PLATE-COUPLED MV IN FIG. 20-19.

resistor R_1, and the negative grid voltage of $A1$ declines toward zero. When cutoff voltage is reached at time C in the illustration, conduction begins again in the first stage, cutting off $A2$ to repeat the cycle.

The waveforms for both stages are exactly the same but of opposite polarity, since one is conducting while the other is cut off. The period of conduction for each stage is equal to the cutoff time of the other. It is the change from cutoff to conduction that initiates the switching operation.

The output voltage from either plate is a symmetrical square wave, as the plate voltage rises sharply to B+, remains at that value for a period equal to the cutoff time, and then drops sharply to some low value resulting from plate-current flow. The slight departure from square corners is caused by charging of the coupling capacitors. The output is symmetrical because both tubes are cut off the same amount of time.

Multivibrator frequency. The time from A to C in Fig. 20-20 is one complete cycle, including a complete flip-flop of operating conditions. The frequency may be from 1 to 100,000 Hz, depending primarily on the RC time constant of the grid coupling circuits. The period of one cycle is exactly equal to the sum of the cutoff periods for both stages. From the grid waveshapes in Fig. 20-20, the cutoff time can be calculated as 0.0003 s, which is a half-cycle. For both tubes in this symmetrical MV, the total cutoff period is $2 \times 0.0003 = 0.0006$ s. The frequency then is 1/0.0006 s, or 1,667 Hz, approximately.

20-11 Cathode-coupled Multivibrator

As shown in Fig. 20-21, the coupling for feedback from $A2$ to $A1$ is produced by the common-cathode resistor R_k. As a result, this MV circuit is unsymmetrical. $A2$ must be cut off for a longer time than $A1$ because $A2$ has the R_gC_c coupling circuit, while $A1$ does not.

How $A2$ is cut off. Referring to Fig. 20-21, when plate voltage is applied, both tubes start to conduct. The flow of plate current in $A1$ reduces its plate voltage, driving the grid of $A2$ negative.

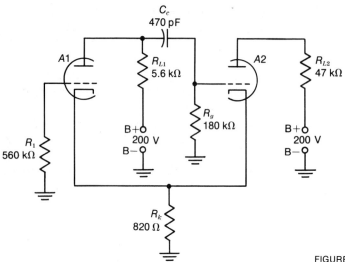

FIGURE 20-21 CATHODE-COUPLED MULTIVIBRATOR.

The plate current in A2 is reduced because of the negative grid signal. This decreases the cathode voltage across R_k. Then A1 conducts more plate current, driving the grid of A2 more negative. The unbalance is amplified to drive A2 to cutoff almost instantaneously.

A2 is held cut off during the time C_c discharges through R_g, R_k, and the plate-to-cathode resistance of A1, which is now conducting. The negative grid voltage across R_g declines exponentially with the normal capacitor discharge curve until the grid voltage for A2 has been reduced to the grid cutoff voltage. Then A2 starts to conduct.

How A1 is cut off. With the plate current of A2 now flowing through the common cathode resistor, the cathode bias for A1 is increased, driving its grid negative. Plate current in A1 is reduced because of the additional cathode bias, allowing its plate voltage to rise toward the B+ of 200 V. The increase in plate voltage drives the grid of A2 more positive as C_c charges from the B supply through the grid-to-cathode circuit of A2 and R_{L_1}. With A2 driven more positive, its plate current increases, and more bias is developed across R_k as the cathode voltage follows the applied grid voltage. The action is cumulative and results in A1 being cut off almost instantaneously by the plate current of A2. Now A1 is cut off. This cutoff time depends on how long it takes C_c to charge. Until C_c charges, the grid of A2 is positive enough to produce the amount of plate current needed to raise the cathode voltage enough to cut off A1.

It should be noted that the grid of A1 is at ground potential, since there is no grid-coupling circuit from the other tube. Therefore, A1 remains cut off as long as the cathode voltage exceeds its cutoff voltage, because this is the only input voltage.

When the cathode voltage drops below the cutoff value, because of decreasing plate current in A2, then A1 can start to conduct again. Then conduction in A1 cuts off A2 again to repeat the cycle. The circuit operates as a free-running multivibrator, therefore, as each tube alternately conducts to cut off plate current in the other tube. The waveshapes are shown in Fig. 20-22.

Note the following points:

1. The e_{c_1} voltage is zero with respect to ground because there is no input to this grid. Therefore, no waveshape is shown for the e_{c1} grid voltage.
2. Instead, the e_{gk} voltage is shown for A1. This is the same as e_k but with inverted polarity.
3. When C_c charges by grid current, in this example e_{c_2} is reduced from +12 V to +4 V. At this e_{c_2}, the plate current of A2 is reduced enough to drop e_k to 6 V. This is not enough to keep A1 cut off. Then A1 conducts to produce negative grid drive that cuts off A2.

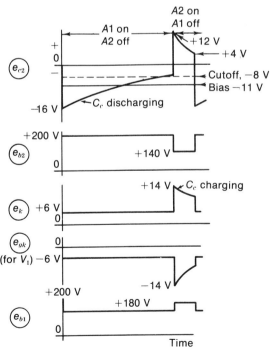

FIGURE 20-22 WAVEFORMS FOR THE CATHODE-COUPLED MV IN FIG. 20-21.

Summary of operation. A2 is cut off by negative grid voltage when the plate voltage of A1 drops with conduction. Therefore, the cutoff period for A2 depends on the grid time constant for C_c to discharge through R_g. However, A1 is cut off by the cathode voltage across R_k produced by maximum plate current in A2. The reason why A2 is not cut off when the increased voltage across R_k cuts off A1 is that this is time when the grid of A2 is driven positive from the plate of A1. In fact, it is the positive grid drive at A2 that makes the cathode voltage rise.

The cathode voltage e_k remains high enough to cut off A1 until C_c is charged by grid current to the point when the positive grid drive on tube 2 drops close to zero. Therefore, the cutoff period for tube 1 depends on the time constant for charge of C_c. This charging path is through the low resistance of approximately 2 kΩ internal grid-cathode resistance of A2 when grid current flows, through R_k and R_{L_1} in the A1 plate circuit while it is cut off. As a result, C_c charges fast for a short period of cutoff in A1 compared with the slow discharge of C_c through R_2 for a long period of cutoff in A2. The cathode-coupled multivibrator, therefore, automatically produces unsymmetrical output, as A2 must be cut off for a much longer time than A1.

20-12 Multivibrator Sawtooth Generator

An unsymmetrical multivibrator can be used as a sawtooth generator by connecting a sawtooth capacitor in the plate circuit of the tube that is cut off a long time and conducting a short time. Then this stage functions as a discharge tube. In the cathode-coupled multivibrator this tube is the one with the RC coupling circuit for grid drive from the plate of the opposite tube.

Figure 20-23 shows a typical circuit for a horizontal sawtooth oscillator. C_2 is the saw-

444 BASIC TELEVISION

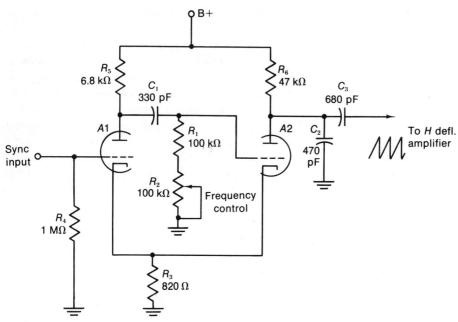

FIGURE 20-23 HORIZONTAL SAWTOOTH GENERATOR USING CATHODE-COUPLED MV.

tooth capacitor in the plate circuit of $A2$. While negative drive from $A1$ holds $A2$ cut off for a relatively long time C_2 charges toward B+ through R_6. The charging action produces a linear rise for the trace part of the sawtooth voltage across C_2. When the grid of $A2$ is driven positive, because of cutoff in $A1$, then C_2 can discharge fast for the retrace part of the sawtooth wave. The path of discharge current is through R_3 and the low internal plate-cathode resistance in $A2$, which is now conducting. The sawtooth voltage output across C_2 is coupled by C_3 to the grid circuit of the horizontal output stage. Figure 20-24 illustrates how the sawtooth voltage output corresponds to cutoff and conduction of the $A2$ stage.

Note that the variable resistance R_2 in the grid circuit of $A2$ serves as the frequency control. Decreasing R_2 reduces the cutoff time for $A2$ to raise the free frequency of the oscillator.

Increasing R_2 lowers the oscillator frequency. The grid resistance for $A1$ does not control the oscillator frequency because its grid voltage is determined by the cathode voltage across R_3. However, this grid is the best connection for injecting sync because of isolation from the oscillator voltages.

20-13 Synchronizing the Multivibrator

Either positive or negative sync polarity can be used with multivibrators. A positive pulse applied to the grid of a cutoff tube can cause switching action if the pulse is large enough to raise the grid voltage above cutoff. This idea corresponds to triggering the grid of a blocking oscillator with positive pulses. However, negative trigger pulses can be used, particularly in the cathode-coupled multivibrator. The re-

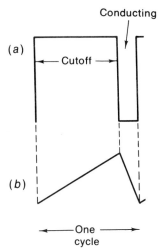

FIGURE 20-24 HOW SAWTOOTH VOLTAGE IS PRODUCED BY THE CATHODE-COUPLED MV IN FIG. 20-23. (a) RECTANGULAR PLATE VOLTAGE OF A2 WITHOUT SAWTOOTH CAPACITOR C_2; (b) CORRESPONDING SAWTOOTH VOLTAGE WITH C_2.

quirement is that negative pulses applied to the conducting tube be amplified and inverted to produce positive pulses large enough at the grid of the cutoff tube to make it conduct.

Synchronization with negative trigger pulses is illustrated in Fig. 20-25. At time A, the negative pulse at the grid of A1 reduces its plate current, but the amplified pulse is not positive enough at the grid of A2 to make it conduct. Therefore, the pulse has no effect on switching. The negative pulses B and C also have no effect because they are applied to A1 while it is cut off. However, at time D, the negative sync pulse is inverted and amplified enough to drive A2 from cutoff into conduction. Then operating conditions are switched as A2 conducts and cuts off A1. Once A2 is triggered by the sync voltage, the oscillator operates at the sync frequency as each sync pulse triggers each cycle.

Just as in blocking oscillator synchronization, the free frequency of the multivibrator must be slightly lower than the sync frequency. Then the pulses occur just before the natural switching action would take place by itself, when the oscillator is ready for triggering.

Negative dc control voltage. The multivibrator frequency can also be varied by controlling its dc grid voltage. This method applies when the oscillator is controlled by an AFC circuit for horizontal synchronization. In the cathode-coupled multivibrator, negative dc control voltage at the

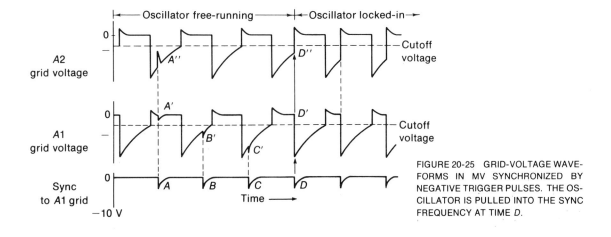

FIGURE 20-25 GRID-VOLTAGE WAVEFORMS IN MV SYNCHRONIZED BY NEGATIVE TRIGGER PULSES. THE OSCILLATOR IS PULLED INTO THE SYNC FREQUENCY AT TIME D.

A1 grid raises the oscillator frequency. Its added negative grid voltage reduces plate current when A1 conducts. The result is a smaller drop in plate voltage, and less negative drive at the A2 grid. Then less time is needed for C_c to discharge down to cutoff for conduction in A2. With a shorter cutoff time for A2, the multivibrator frequency is increased.

20-14 Frequency Dividers

Either the multivibrator or the blocking oscillator can be triggered at a submultiple of the sync frequency to act as a frequency divider. In Fig. 20-26, as an example, the input sync frequency of 600 Hz is divided by 3 to provide 200 Hz for the frequency of the MV output. The circuit is a divider, as the oscillator output frequency is an exact submultiple of the sync frequency input.

The sync voltage is coupled to the grid that needs positive pulses for triggering. For minimum sync voltage input, the free frequency of the oscillator is set slightly below the desired submultiple of the sync frequency. Then the oscillator is forced to lock in at the exact submultiple frequency because of the trigger pulses. Note that the third sync pulse labeled C in Fig. 20-26 has enough amplitude to trigger the oscillator from cutoff into conduction. Pulses A and B do not affect the oscillator because they do not have enough amplitude to raise the grid voltage above cutoff. The reason is that the free frequency slightly under 200 Hz has been set much lower than the sync frequency of 600 Hz. However, the third pulse C does trigger the oscillator into conduction. Then every third pulse triggers one oscillator cycle to produce output at one-third the sync frequency.

The voltages in Fig. 20-26 indicate how the frequency-division factor can be changed with more sync input. If the sync is increased to

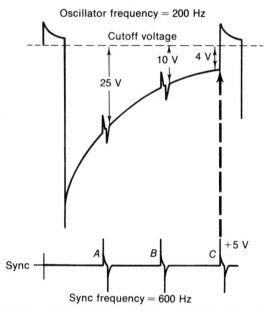

FIGURE 20-26 TRIGGER PULSES FORCING MV TO OPERATE AS FREQUENCY DIVIDER AT ONE-THIRD THE SYNC FREQUENCY. EVERY THIRD PULSE WILL TRIGGER THE MV.

more than 10 V, pulse B will have enough amplitude to trigger the oscillator. Then it divides by 2 with output at 300 Hz, locked in by every second sync pulse. Or, sync voltage above 25 V could lock in the oscillator at the sync input frequency, without any frequency division. This relatively large sync voltage to trigger the oscillator for this example is necessary because the oscillator free frequency is not close to the sync frequency.

20-15 Transistors in Multivibrator Circuits

Two common-emitter stages are used for the amplifiers. Collector coupling for feedback corresponds to a plate-coupled MV. A common resistor for both emitter circuits corresponds to cathode coupling. Remember, though, that a

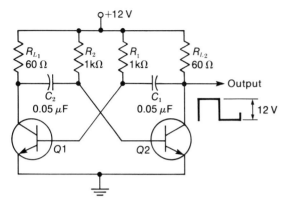

FIGURE 20-27 FREE-RUNNING, COLLECTOR-COUPLED, SYMMETRICAL MULTIVIBRATOR.

transistor is cut off with zero bias. About 0.6-V forward voltage is needed for conduction in a silicon transistor. A constant value of saturation collector current flows with about 0.8-V forward bias, assuming a load resistance that drops the collector voltage below the base voltage.

The collector-coupled MV in Fig. 20-27 is a free-running, symmetrical MV with square-wave output from either stage. C_2 and R_2 provide coupling to the base of Q2. Feedback to the base of Q1 is provided by C_1 and R_1. Note that R_1 and R_2 return to the +12-V supply in order to provide forward bias at the base for the NPN transistors. However, when the collector voltage of one stage drops, the negative drive at the next base produces cutoff. The multivibrator oscillates, with Q1 and Q2 alternately off and on as conduction in one stage cuts off the other.

20-16 Sawtooth, Trapezoidal, and Rectangular Voltages and Currents

In magnetic scanning, it is the current in the deflection coils that must have a sawtooth waveshape. The reason is that the magnetic field of the current deflects the electron beam. With sawtooth current through the coils, the voltage across the inductance is not the same waveshape because inductance opposes any change in current.

We can make a comparison with the more familiar example of inductance L in sine-wave circuits. The self-induced voltage across L produced by changes in current is always 90° out of phase with the current through L. This concept of phase angle, however, only applies to sine waves. The corresponding idea in non-sinusoidal circuits is that inductance changes the waveshape of the voltage, compared with the current. Similarly, capacitance C can produce a 90° phase angle with sine waves, but the waveshape is changed in nonsinusoidal circuits. For resistance R, however, there is no change in phase angle or waveshape between the current and voltage waveforms. More details of waveshaping with RL and RC circuits are described in Appendix D, on RC Time Constant. However, the main facts that apply to deflection circuits are illustrated in Figs. 20-28 to 20-31.

Sawtooth i_R and v_R for resistance. Figure 20-28 shows that with sawtooth current through R, the voltage across R also has the sawtooth waveform. Resistance does not concentrate the magnetic field like an inductance, or store electric charge like a capacitance. Therefore, R does not have any reaction to a change in voltage or current.

Sawtooth i_L and rectangular v_L for inductance. See Fig. 20-29. The amount of induced voltage across L depends on the rate of change of current. A faster change in i_L produces more self-induced voltage v_L. Furthermore, for a constant rate of change in i_L, the amount of v_L is constant. As a result, v_L in Fig. 20-29 is at a relatively low level during the slow rise in i_L during trace time. Then the fast drop in i_L produces a

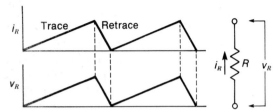

FIGURE 20-28 SAWTOOTH CURRENT i_R AND VOLTAGE v_R FOR RESISTANCE R.

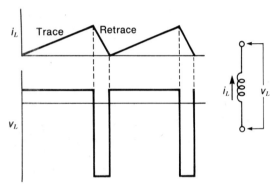

FIGURE 20-29 SAWTOOTH CURRENT i_L AND RECTANGULAR VOLTAGE v_L FOR A PURE INDUCTANCE L.

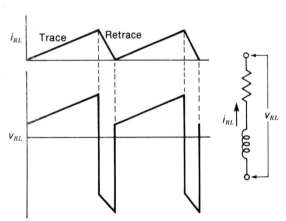

FIGURE 20-30 TRAPEZOIDAL VOLTAGE v_{RL} WITH SAWTOOTH CURRENT FOR AN RL CIRCUIT.

sharp voltage peak, or spike, for the fast retrace or flyback. The polarity of the flyback pulse must be opposite to the trace voltage because i_L is decreasing instead of increasing. Therefore, sawtooth current in L produces rectangular voltage. Or, to produce sawtooth current in an inductance, it needs rectangular voltage applied.

Trapezoidal voltage for an RL circuit. This waveform combines the sawtooth voltage needed for R with the rectangular voltage needed for L, as shown in Fig. 20-30. With sawtooth current through a circuit having inductance and resistance, therefore, the voltage waveform is a trapezoid. Or to produce sawtooth current in a circuit with R and L, trapezoidal voltage must be applied.

Voltage waveshapes for horizontal deflection. The horizontal scanning coils are mainly inductive. Because of the large self-induced voltage produced by the fast horizontal flyback, the effect of L is much greater than the effect of R. Therefore, the voltage waveform is rectangular across the horizontal yoke coils and output transformer. The polarities of i and v may be as shown in Fig. 20-29 or inverted with positive flyback pulses.

Voltage waveshapes for vertical deflection. The self-induced voltage is smaller at the lower scanning frequency. Therefore, both R and L must be considered. With sawtooth current, the result is trapezoidal voltage across the RL circuit. The polarity may be as in Fig. 20-30, but can be inverted. Figure 20-31a illustrates a negative-going sawtooth waveform, and Fig. 20-31b, rectangular v_L with a positive spike for flyback. The resultant trapezoid in Fig. 20-31c has the inverted polarity usually produced in the vertical output circuit.

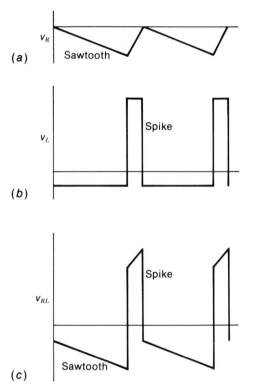

FIGURE 20-31 INVERTED POLARITIES OF DEFLECTION VOLTAGE WAVEFORMS. (a) SAWTOOTH; (b) RECTANGULAR; (c) TRAPEZOIDAL.

Trapezoidal voltage generator. This voltage waveform is often needed from the deflection oscillator to drive the output stage. A circuit for generating trapezoidal voltage is shown in Fig. 20-32. It is the usual circuit for producing sawtooth voltage, but with the peaking resistor R_p added in series with C_s. The peaking resistor produces voltage spikes that are combined with the sawtooth voltage across C_s for trapezoidal voltage output.

The value of 8 kΩ for R_p is small compared with 2 MΩ for R_1. Therefore, the voltage across R_p is very small while C_s is charging from B+ through R_1 when the discharge tube is cut off. On discharge, though, R_p is relatively large compared with the low resistance of the conducting discharge tube. Then C_s discharges fast, with a high value of discharge current, to develop a large negative pulse of voltage across R_p. This narrow pulse or spike is a negative voltage because it is produced by discharge current.

The voltages across C_s and R_p are in series with each other across the output circuit. Therefore, the trapezoidal voltage at point T to ground is the sum of the positive-going sawtooth voltage across C_s and the negative voltage spikes across R_p.

S-shape correction. In many cases the desired scanning current is not a perfectly linear sawtooth. Actually, this would stretch the raster at the edges for a wide-angle picture tube with a

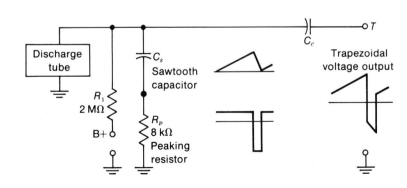

FIGURE 20-32 SAWTOOTH CAPACITOR WITH PEAKING RESISTOR TO PRODUCE TRAPEZOIDAL VOLTAGE.

450 BASIC TELEVISION

flat faceplate. The reason is that the edges of the screen are further away from the point of deflection, compared with the middle. To correct the problem, an *S-shaping capacitor* is often used for the sawtooth current. This may be in series with the yoke coils as a coupling capacitor, or in a shunt path. For both cases, the S capacitor resonates with the yoke inductance to add a sine-wave component. The result is to round off the peak amplitudes at the top and bottom of the sawtooth wave, corresponding to the edges of the raster. The S-shaping can be applied to vertical scanning and horizontal scanning.

20-17 Incorrect Oscillator Frequency

If the vertical hold control cannot stop the picture from rolling, even for an instant, this indicates the vertical oscillator is operating at the wrong frequency since it cannot be adjusted to 60 Hz. With the oscillator frequency lower than 60 Hz, the picture rolls upward; above 60 Hz the picture rolls downward.

Vertical roll. The reason for the rolling picture is illustrated in Fig. 20-33 for the case of vertical oscillator frequency too high. Notice the relative timing of vertical blanking in the composite video signal and vertical retrace in the sawtooth vertical deflection current. When both are at the same frequency, every vertical retrace occurs within vertical blanking time. Then vertical blanking is not visible at the top and bottom edges of the frame. However, with the vertical frequency too high, the sawtooth cycles advance in time with respect to the 60-Hz blanking pulses. Then vertical blanking occurs during trace time, instead of during retrace. Furthermore, each sawtooth advances into trace time for succeeding blanking pulses. As a result, the black bar produced across the screen by the ver-

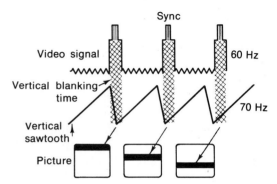

FIGURE 20-33 WHY THE PICTURE WITH BLANKING BAR ROLLS UP OR DOWN WHEN VERTICAL SCANNING FREQUENCY IS NOT AT 60 Hz.

tical blanking pulse appears lower and lower down the screen for successive cycles.

Remember that the information for the top of the picture as it is transmitted always comes immediately after vertical blanking. When the vertical oscillator is locked in sync, each frame is reproduced over the previous frame and then the picture holds still. However, when the picture information and blanking in each frame are reproduced lower on the screen than the previous frame, the picture appears to roll down. The same idea applies to rolling up.

The farther the vertical scanning frequency is from 60 Hz, the faster the picture rolls. The upward rolling is usually slower than the downward rolling because the oscillator frequency changes more gradually at the high-resistance end of the hold control for low frequencies. If the vertical oscillator frequency can be made as low as 30 Hz or 20 Hz, which are submultiples of 60 Hz, two or three duplicate pictures will be seen one above the other. If the frequency is raised to 120 Hz, the bottom of the picture will be superimposed on the top, usually with reduced height in the raster.

The main factor in the frequency of a blocking oscillator or multivibrator is the $R_g C_c$

time constant in the grid circuit. When either R_g or C_c has a value too small, the oscillator frequency is too fast. When the $R_g C_c$ time constant is too long, the oscillator frequency is too slow.

Diagonal bars. When the horizontal oscillator is locked into the sync frequency of 15,750 Hz, the line structure holds together to show a complete picture and horizontal blanking is invisible at the left and right edges. If the oscillator is off the correct frequency, the picture will tear into diagonal segments. The diagonal black bars are produced by horizontal blanking pulses. The picture is in segments because the horizontal AFC circuit prevents individual horizontal lines from tearing apart, as the frequency cannot change from line to line. When the number of diagonal bars is continually changing, this shows the AFC circuit is not controlling the oscillator. When the bars are steady, the AFC circuit is holding the oscillator but at the wrong frequency.

In either case, the horizontal blanking pulses produce diagonal black bars when the horizontal oscillator is off the correct frequency. If the frequency differs from 15,750 Hz by 60 Hz, there will be one diagonal bar. Every 60-Hz difference between the oscillator frequency and 15,750 Hz results in another diagonal bar. As the bars increase in number, they become thinner with less slope. The bars slope down to the left when the oscillator frequency is below 15,750 Hz, or up to the left above 15,750 Hz.

The reason for the diagonal bars is illustrated in Fig. 20-34, for the case of horizontal oscillator frequency too high. Notice how successive sawtooth cycles advance in time with respect to the blanking pulses transmitted at 15,750 Hz. The blanking pulse then goes into trace time. Each blanking pulse is 10 μs wide,

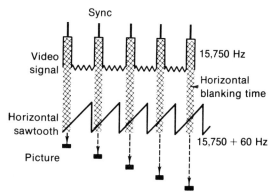

FIGURE 20-34 WHY THE PICTURE TEARS DIAGONALLY WHEN THE HORIZONTAL SCANNING FREQUENCY IS NOT AT 15,750 Hz.

reproducing black for about one-sixth of every line. Remember that the left edge of the picture is always immediately after horizontal blanking. Furthermore, the blanking goes more into trace time for successive sawtooth cycles. For each line, then, the black is more to the right.

Since vertical scanning is occurring at the same time, the black area moves down as it progresses to the right. Only five scanning lines are illustrated here, but if all the lines were shown the result would be one diagonal black bar from the top left corner to the bottom right corner. Then, the same action is repeated over the previous diagonal bar. The same idea applies to the case of oscillator frequency too low, but the black would start at the top right corner and progress diagonally down to the left.

Between the black diagonal bars, the picture information is reproduced in the wrong position to such an extent that the picture usually cannot be recognized. Near the bottom of each bar, the picture information is actually reversed in its left-right position. Also, the horizontal flyback during visible time stretches black or white information all the way across the screen.

SUMMARY

1. Sawtooth voltage is produced by charging a capacitor C_s slowly through a high resistance for the slow linear rise and discharging the capacitor fast through a low resistance for the retrace. One complete cycle includes trace and retrace.
2. A peaking resistor in series with C_s on discharge produces a voltage spike that combines with the sawtooth to provide trapezoidal voltage waveshape.
3. The blocking oscillator is cut off for a relatively long time and conducting for a short time. Therefore, it can control the charge and discharge of a sawtooth capacitor. In the plate circuit, C_s charges while the oscillator is cut off and discharges through the oscillator while it is conducting. The sawtooth voltage output is at the frequency of the blocking oscillator.
4. The blocking oscillator can be synchronized by positive trigger pulses at the grid. The free-running frequency is adjusted slightly below the sync frequency to allow every pulse to force the tube into conduction.
5. The multivibrator uses two stages that are alternately conducting and cut off. The period of conduction for each depends on the cutoff time of the other. One cycle includes cutoff for both stages.
6. In the plate-coupled multivibrator in Fig. 20-19, the plate circuit of each tube is RC-coupled to drive the grid of the opposite tube. Conduction in one tube drops its plate voltage to drive the next grid negative beyond cutoff. The collector-coupled MV with transistors in Fig. 20-27 corresponds to the plate-coupled MV.
7. In the cathode-coupled multivibrator in Fig. 20-21, A1 cuts off A2 through the R_gC_c coupling circuit, but A2 cuts off A1 through R_k. The output is unsymmetrical, as A2 is cut off a relatively long time and conducting a short time. Therefore, a sawtooth capacitor can be connected in the plate circuit of A2, serving as a discharge tube, to provide sawtooth voltage output at the multivibrator frequency. An emitter-coupled MV with transistors corresponds to the cathode-coupled MV.
8. The cathode-coupled multivibrator can be synchronized by negative pulses at the A1 grid, which are amplified and inverted to become positive pulses at the grid of the A2 stage.
9. The free frequency of a blocking oscillator or cathode-coupled multivibrator is usually adjusted by making the grid resistor variable, to set the R_gC_c time constant. Larger values for R_g or C_c lower the oscillator frequency; decreasing R_gC_c raises the frequency.
10. The sawtooth voltage output can be adjusted by varying the resistance in series with C_s on charge. This size control increases the amplitude when the resistance is decreased; more resistance decreases the amplitude. Lower amplitudes allow better linearity, as a smaller part of the RC charge curve is used.
11. R has sawtooth current and voltage waveshapes; L has sawtooth current and rectangular voltage; an RL circuit has sawtooth current and trapezoidal voltage.
12. When the vertical oscillator is not at 60 Hz, the picture, with the black horizontal bar produced by vertical blanking, rolls up or down the screen. When the hori-

zontal oscillator is not at 15,750 Hz, the picture tears into diagonal segments, with black diagonal bars produced by the horizontal blanking pulses.

Self-Examination (Answers at back of book)

Part A Answer True or False.
1. One cycle of the sawtooth waveform includes trace and retrace.
2. In a sawtooth wave of negative polarity, the trace is faster than the retrace.
3. The sawtooth waveform can be used for either current or voltage.
4. Sawtooth current through an inductance produces rectangular voltage with a big pulse during retrace.
5. The sawtooth capacitor charges through the discharge tube while it is cut off.
6. The vertical oscillator cannot operate without sync input.
7. The horizontal oscillator must be operating for the receiver to have flyback high voltage.
8. If the vertical oscillator does not operate, there will be just a bright line across the center of the screen.
9. Increasing the charging resistance for C_s reduces the sawtooth voltage amplitude.
10. The free frequency of the oscillator must be set lower than the sync frequency.
11. When plate voltage drops, C_c and R_g couple negative voltage drive to the next grid.
12. The cathode-coupled multivibrator can be used for a horizontal sawtooth generator.
13. In the cathode-coupled multivibrator, C_s is in the plate circuit of the tube that has $R_g C_c$ coupling into its grid.
14. In a multivibrator, the sync voltage must drive the cutoff tube into conduction to lock in the oscillator at the sync frequency.
15. In a cathode-coupled multivibrator, the free frequency depends on the $R_g C_c$ time constant.
16. In a cathode-coupled multivibrator, the voltage across R_k cuts off both tubes at the same time.
17. When the picture rolls up, the vertical oscillator frequency is more than 60 Hz.
18. Two diagonal bars sloping down to the left indicate the horizontal oscillator frequency is 120 Hz below 15,750 Hz.
19. When the picture can be made to roll up and down with the vertical hold control but does not lock in, this indicates the trouble is no vertical sync.
20. If the oscillator frequency is increased, the sawtooth voltage across C_s will decrease.

Part B *Fill in the required value.*

1. A 0.05-μF sawtooth capacitor charging through 3 MΩ has a charge time constant of _____ s.
2. The same capacitor discharging through 8 kΩ and 1 kΩ in series has a discharge time constant of _____ s.
3. A 0.01-μF capacitor charged to -60 V will discharge through a 3-MΩ resistor down to -6 V in _____ s. (See Appendix D.)
4. If the vertical blocking oscillator transformer has a resonant frequency of 2,000 Hz, the time for one-half cycle of ringing is _____ μs.
5. In the plate-coupled MV of Fig. 20-19, if each tube is cut off for 500 μs, the oscillator frequency is _____ Hz.
6. In the cathode-coupled MV of Fig. 20-23, if A1 is cut off 7.5 μs and A2 is cut off 58 μs, the sawtooth voltage output frequency is _____ Hz.
7. In Fig. 20-23, when the cathode current through R_3 is 20 mA, the cathode voltage equals _____ V.
8. In the collector-coupled MV of Fig. 20-27, if the supply voltage is increased to 28 V, the peak-to-peak output will be approximately _____ V.
9. In Fig. 20-11, if the sawtooth voltage across C_s varies between 140 V and 110 V, the peak-to-peak output equals _____ V.
10. In Fig. 20-26, if the free frequency is set at 295 Hz, the oscillator frequency will lock in at _____ Hz.

Essay Questions

1. What is the function of the vertical deflection oscillator? The horizontal deflection oscillator?
2. Why is the sawtooth waveshape required for linear deflection? In magnetic deflection, why is the sawtooth waveshape required for current in the coils?
3. Give two factors that affect linearity of the voltage rise across C_s.
4. Give three factors that affect cutoff time of the blocking oscillator. Give one factor affecting conduction time.
5. Referring to the numbered steps in the e_c waveshape in Fig. 20-10, which numbers occur during conduction and which during cut off
6. Draw the schematic diagram of a blocking oscillator–discharge tube trapezoidal

voltage generator for vertical deflection, using a single triode. Indicate typical values of all components. Label the height and hold controls.
7. Why is the free-running frequency of a blocking oscillator lower than the sync frequency? Why is the polarity positive for sync voltage at the grid? Why does the frequency increase when positive dc grid voltage is added?
8. Referring to the cathode-coupled multivibrator in Fig. 20-21, what determines cutoff time and conduction time for A2? Cutoff and conduction time for A1?
9. Referring to the cathode-coupled multivibrator in Fig. 20-23, draw the voltage waveshapes at each electrode of A1 and A2 in a ladder diagram, with one under the other to show the different voltages at similar times.
10. In a cathode-coupled multivibrator, why is the polarity negative for sync voltage at the grid of A1? Why is the free frequency set lower than the sync frequency? Why does negative dc voltage added to the A1 grid increase the frequency?
11. Describe briefly the function of the height control and vertical hold control in a vertical deflection oscillator.
12. In a circuit with the sawtooth capacitor in the plate of the blocking oscillator, why does the picture usually roll when the height control is adjusted?
13. Explain briefly why the height of the raster can decrease when the vertical oscillator frequency is much higher than 60 Hz.
14. Show the waveform of sawtooth current through a pure inductance and the corresponding voltage.
15. Give the function of the peaking resistor in a trapezoidal voltage generator.
16. If the vertical oscillator locks in at 30 Hz, what will be the effect on the picture? What will be the effect at 120 Hz?
17. Name four types of multivibrators based on the method of coupling, using tubes and transistors.
18. Name three types of multivibrators based on stability.
19. What is the difference between a free-running multivibrator and a trigger circuit?
20. Name two types of MV trigger circuits.
21. Name five components or voltages that affect the frequency of a multivibrator or blocking oscillator.
22. Referring to the blocking oscillator in Fig. 20-11, give the functions of all components.
23. Referring to the plate-coupled MV in Fig. 20-19, give the functions of all components.
24. Referring to the collector-coupled MV in Fig. 20-27, give the functions of all components.
25. If R_3 in Fig. 20-23 opens, what will be seen on the screen in a receiver using flyback high voltage?
26. If the primary winding opens in the blocking oscillator transformer in Fig. 20-11, what will be seen on the screen?

Problems (Answers to selected problems at back of book)

1. With 100 V applied, C initially uncharged, and a time constant of 2 s, calculate V_c for the following charging periods: (a) 0.2 s; (b) 1.0 s; (c) 1.14 s; (d) 2 s; (e) 4 s; (f) 10 s. (*Hint:* See Appendix D.)
2. For the examples in Prob. 1, draw a graph showing V_c versus time in seconds.
3. (a) A capacitor charged to 10 V is connected to a 110-V source. How much is V_c after one time constant? (b) C is then discharged for one time constant. How much is V_c?
4. A capacitor has a time constant of 0.045 s for charge. A separate discharge path has a time constant of 0.0002 s. The capacitor is charged from a 100-V source for 0.009 s and then discharged for 0.001 s. Draw the resulting sawtooth voltage across C, to scale, in voltage versus time. What is the frequency of the sawtooth voltage?
5. A 1-μF capacitor charged to 100 V is discharged through a 1-MΩ resistor. (a) How much is the peak discharge current at the start of discharge? (b) How much is the resistor voltage then? (c) How much is the discharge current after 0.707 s? (d) How much is the resistor voltage now?
6. R_g is 1 MΩ, with 1-μF C_c charged to an average plate voltage of 150 V. Then the plate voltage drops to 50 V and remains at that value for 2 s. Draw a graph with values of current and voltage across R_g during the discharge.
7. Calculate the voltage induced across a 1-H inductance when the current: (a) increases from 0 to 20 mA in 100 μs; (b) decreases from 20 mA to 0 in 100 μs; (c) decreases from 20 mA to 0 in 10 μs. [*Hint:* $e = L\,(di/dt)$]
8. A sawtooth waveform of current through an 8-mH inductance increases from 0 to 300 mA during 50 μs, then drops to zero in 5 μs. Draw the waveforms of current and corresponding induced voltage, with values.

21 Vertical Deflection Circuits

Both the vertical and horizontal deflection circuits produce the scanning raster on the screen of the picture tube, as illustrated in Fig. 21-1. For horizontal scanning, the horizontal oscillator drives the horizontal output stage to produce the scanning lines at 15,750 Hz. At the same time, the vertical oscillator generates 60-Hz deflection voltage to drive the vertical output stage. As a result, the vertical scanning fills the screen from top to bottom with horizontal lines to form the raster. The height depends on the amount of vertical output; the width depends on the horizontal output. The requirements of the horizontal output circuit, with flyback high voltage and the damper stage, are analyzed in the next chapter. Here we consider just the circuits involved in vertical scanning. These topics include:

- 21-1 Triode vertical output stage
- 21-2 Transistorized vertical output stage
- 21-3 Vertical output transformers
- 21-4 Vertical linearity
- 21-5 Internal vertical blanking
- 21-6 Multivibrator for combined vertical oscillator and amplifier
- 21-7 Transistorized blocking oscillator and vertical amplifier
- 21-8 Miller feedback integrator circuit
- 21-9 Transistor pair in vertical output circuit
- 21-10 Vertical deflection troubles

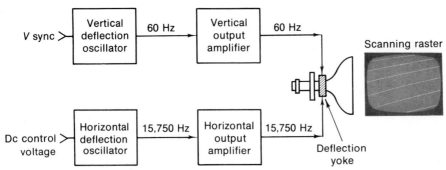

FIGURE 21-1 THE VERTICAL AND HORIZONTAL DEFLECTION CIRCUITS PRODUCE THE SCANNING RASTER ON THE SCREEN OF THE PICTURE TUBE.

21-1 Triode Vertical Output Stage

This is a power amplifier to supply the required amount of sawtooth current. The frequency of the amplifier output is 60 Hz, the same as the drive from the vertical deflection oscillator. With tubes, a triode amplifier is often used because it provides better linearity in the sawtooth current output than a pentode. However, a beam-power pentode with negative feedback can be used to reduce nonlinear distortion. With transistors, a PNP or NPN power transistor in the common-emitter circuit corresponds to a triode output amplifier. However, a transistorized circuit needs a driver stage to serve as a buffer and isolate the low impedance of the output stage from the sawtooth generator.

As shown in Fig. 21-2, the main components of the triode vertical amplifier are the power output tube V2 and the stepdown output transformer T_0. Like an audio output transformer, T_0 matches the relatively high impedance with low plate current in the primary to the low secondary impedance and high current needed for the scanning coils in the deflection yoke. Actually, the vertical output stage is an audio power amplifier for the 60-Hz deflecting signal. Its operation in providing 60-Hz sawtooth current in the vertical deflection coils can be summarized as follows:

1. Trapezoidal voltage from the vertical oscillator provides a linear rise in grid voltage to produce a linear rise in plate current during trace time. When the negative spike on the trapezoidal voltage input drives the grid voltage more negative than cutoff for retrace, the plate current drops to zero. The result is sawtooth plate current.
2. The sawtooth plate current of the amplifier flows through the primary winding of T_0.
3. The plate-current variations in the primary induce voltage in the T_0 secondary that produces sawtooth secondary current.
4. Since the T_0 secondary winding is across the vertical deflection coils, they also have sawtooth current.

Referring to Fig. 21-2, C_c and R_g couple the 60-Hz trapezoidal voltage from the plate of the vertical oscillator to the grid of the amplifier. Cathode bias is provided by R_{k_1} and R_{k_2}, both bypassed by C_k. The variable resistance R_{k_2} ad-

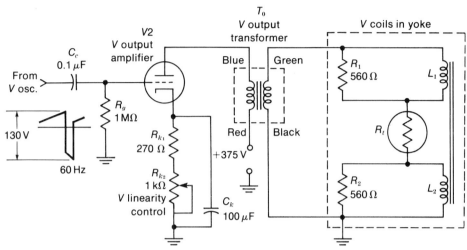

FIGURE 21-2 TRIODE VERTICAL DEFLECTION AMPLIFIER CIRCUIT.

justs the amplifier bias to control the linearity of the sawtooth rise in plate current. Minimum bias is provided by R_{k_1} when the linearity control is at zero resistance.

The cathode bias voltage V_k equals $I_k \times R_k$. If we assume R_{k_2} is set for 730 Ω, the total cathode resistance equals 1,000 Ω. With an average plate current of 30 mA, this is also the cathode current for a triode stage without grid current. Therefore, V_k is 1,000 × 0.03, or 30 V.

Normally there is no grid-leak bias on the vertical output stage. The only bias is the cathode voltage, for minimum amplitude distortion in the sawtooth output current. Grid current would limit the peak plate current, causing poor linearity with a white bar at the bottom of the raster.

Typical voltage and current waveshapes for a triode vertical output stage are shown in Fig. 21-3. Note that the negative spike in the trapezoidal grid voltage drives e_c far negative to cut off plate current, even though the plate voltage rises to a high positive value. This rise in e_b is the self-induced primary voltage in T_0 when i_b drops to zero for the retrace. The function of the pulse in the trapezoidal waveform, therefore, is to cut off the output stage during retrace time.

The trapezoidal waveshape of the output e_b is an amplified and inverted form of the input e_c. The sawtooth plate current i_b is a fluctuating dc waveform, but the sawtooth current in the secondary is an ac waveform, as only the variations of current are coupled by the transformer.

The shunt damping resistors R_1 and R_2 across the vertical scanning coils are in the yoke housing. Their function is to prevent ringing at the start of vertical flyback.

The thermistor R_t is also in the yoke. A typical value is 8 Ω, cold. As the yoke heats up with scanning current, the wire resistance increases but the thermistor resistance decreases to maintain constant height in the raster.

Typical values for a vertical deflection coil are an L of 3.5 to 50 mH, R of 4 to 60 Ω, and sawtooth scanning current of 400 mA to 3 A, peak-to-peak. The two coils are usually in series for a vacuum tube amplifier but in parallel for a transistor vertical output circuit.

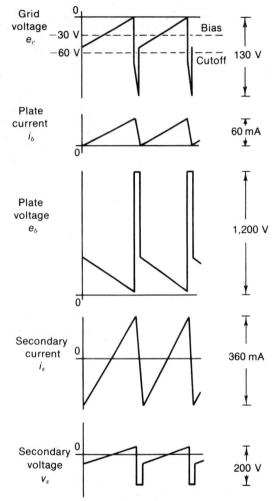

FIGURE 21-3 WAVEFORMS FOR TRIODE VERTICAL OUTPUT CIRCUIT IN FIG. 21-2. FREQUENCY IS 60 Hz. VOLTAGE OUTPUT SHOWN FOR TURNS RATIO OF 6:1 IN T_0. THE POLARITY OF V_s MAY BE INVERTED.

21-2 Transistorized Vertical Output Stage

The circuit in Fig. 21-4 has the same function as the triode amplifier in Fig. 21-2, supplying sawtooth current at 60 Hz to the vertical scanning coils in the deflection yoke. With transistors, though, the driver Q2 is a buffer stage to isolate the power amplifier Q3 from the deflection oscillator. Q2 is an emitter-follower dc-coupled to the common-emitter (CE) amplifier Q3. Both are NPN transistors, requiring positive voltage at the base for forward bias. R_2 is the emitter load resistance for Q2, while the inductance L_0 is a choke in the collector output of Q3.

The trapezoidal voltage to the base of Q2 is 0.8 V, peak-to-peak. Actually the sawtooth component is only about ±0.1 V, to drive a transistor between saturation and cutoff. Note the polarity, with a positive-going sawtooth for trace and the negative spikes that produce cutoff for retrace. The polarity out of the emitter of Q2 is the same as into the base, because the stage is an emitter-follower. Therefore, the base of Q3 has the same polarity of drive voltage as the driver. In the collector output of Q3 the trapezoidal voltage is amplified and inverted by the CE output stage.

Because of the low impedance of the transistor power amplifier Q3, the collector output voltage can be choke-coupled, with L_0 as the collector load. C_0 couples the vertical scanning signal to the yoke coils and blocks the dc collector voltage. A typical value is 500 μF, which requires an electrolytic capacitor. The blocking function is necessary, as any direct current through the deflection coils would move the electron beam off center. The vertical scanning coils L_1 and L_2 are shown connected in parallel for a transistor output circuit.

The *varistor* SC1 across the inductance L_0 protects the output transistor Q3 against voltage transients in the collector circuit. A varistor provides a short circuit above its breakdown voltage, in either polarity. During retrace, the high positive spike on the trapezoidal voltage output may exceed the collector voltage break-

VERTICAL DEFLECTION CIRCUITS 461

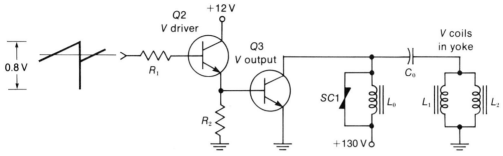

FIGURE 21-4 TRANSISTOR VERTICAL DEFLECTION AMPLIFIER WITH EMITTER-FOLLOWER Q2 DRIVING CE OUTPUT STAGE Q3.

down rating. For the opposite polarity, any transient negative voltage would put forward bias on the collector, causing excessive current.

21-3 Vertical Output Transformers

A typical unit is shown in Fig. 21-5. The transformer may have an isolated secondary winding as shown for T_0 in Fig. 21-2, or an autotransformer can be used. With separate windings, any dc voltage in the primary is blocked from the secondary, and so it is not necessary to use a coupling capacitor. With an autotransformer, however, a capacitor is used to prevent the dc level of the scanning current from the changing the centering of the electron beam.

Common values for the primary L_p are 12 to 40 H, with a turns ratio ranging from 40:1 to 6:1. Typical dc resistances are 500 Ω for L_p and 6 Ω for L_s. The leads may be color-coded, red for B+ voltage and blue for the plate or collector in the primary, while the secondary wires are green, brown, and black.

Autotransformer for vertical output. In Fig. 21-6, L_p includes all the turns but L_s is tapped down on the coil. The result is voltage stepdown and current stepup from primary to secondary. L_p is between points 1 and 3, while L_s is between 2 and 3. The varying current in the primary generates self-induced voltage in all the turns. Since this is an iron-core transformer with unity coupling, the voltage and current ratios are proportional to the number of turns from 1 to 3 for L_p and from 2 to 3 for L_s, the same as with a separate secondary winding. However, the B+ voltage for the primary of the autotransformer is not

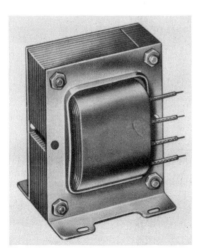

FIGURE 21-5 VERTICAL OUTPUT TRANSFORMER. BASE IS 2 IN. SQUARE. *(STANCOR).*

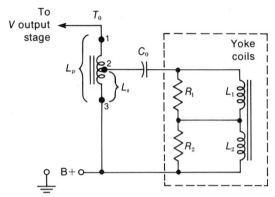

FIGURE 21-6 AUTOTRANSFORMER FOR VERTICAL OUTPUT CIRCUIT.

isolated from the secondary. The deflection coils in the yoke then have B+ voltage. C_0 blocks the path for direct current, though, to prevent a shift in vertical centering.

Vertical output connections for color picture tubes. Scanning is not related to picture information. However, color picture tubes need dynamic correction for pincushion distortion, because permanent magnets cannot be used.

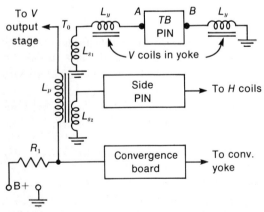

FIGURE 21-7 VERTICAL OUTPUT TRANSFORMER WITH CONNECTIONS OF PINCUSHION (PIN) CORRECTION CIRCUITS FOR COLOR PICTURE TUBE. TB IS TOP AND BOTTOM.

The pincushion (PIN) circuits are usually on a separate board, but they are connected to the vertical deflection coils in the yoke to modify the scanning with a waveform that compensates for the pincushion distortion. In addition, although the convergence circuits are on a separate board, they need the vertical signal from the output circuit for dynamic convergence correction at the top and bottom of the picture.

In Fig. 21-7, note that the pincushion correction circuits for the top and bottom (TB) of the raster are directly in series with the two vertical deflection coils in the yoke. This connection means that troubles in the TB pincushion circuits can affect the vertical scanning. To localize this problem, the two connections at points A and B can be shorted temporarily to check the vertical scanning without the TB pincushion circuits.

The vertical output transformer T_0 also has a separate winding L_{s_2} to provide vertical scanning signal to the pincushion correction circuits[1] for the left and right sides of the raster. However, these connections are separate from the vertical scanning coils. In addition, the sawtooth voltage across R_1 in the output circuit is used for vertical dynamic convergence[1] at the top and bottom of the picture.

21-4 Vertical Linearity

Nonlinear vertical scanning is more noticeable than horizontal nonlinearity because we can easily see the effects. When a person's head is too flat, this means the scanning lines at the top are compressed or crowded together. Stretching the scanning lines at the top makes the head too long. At the bottom, stretching or compressing the lines can make a person's legs too long or too short. With a test pattern, the top and bot-

---

[1] Pincushion correction and dynamic convergence circuits are described in Chap. 11.

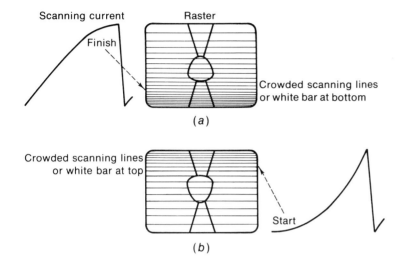

FIGURE 21-8 VERTICAL NONLINEARITY. (a) FLAT FINISH ON SAWTOOTH CAUSES CROWDING AT BOTTOM OF RASTER; (b) FLAT START ON SAWTOOTH CAUSES CROWDING AT TOP OF RASTER.

tom wedges can be too short or too long, as in Fig. 21-8. Also, circles are distorted.

The three main factors affecting linearity of the vertical scanning are:

1. Flat finish toward the end of the sawtooth waveform, as in Fig. 21-8a. This can result from using too much of the RC charge curve, from a short time constant, or from a leaky sawtooth capacitor.
2. The same type of nonlinear distortion can be caused by saturation in the iron core of the vertical output transformer or by a weak amplifier.
3. The input-output operating characteristic of the vertical output amplifier usually has the opposite distortion shown in Fig. 21-8b, with a flat start on the sawtooth waveform.

Remember that the finish of the sawtooth at the top of the waveform corresponds to the bottom of the raster, just before retrace. The start of the sawtooth waveform is the top of the raster. By adjusting the operating characteristic of the vertical amplifier, the two opposite curvatures in Fig. 21-8 can be made to balance each other for linear scanning. For this reason, the vertical linearity control is in the vertical output circuit to adjust the operating characteristic. In Fig. 21-2, the vertical linearity control varies the cathode bias on the vertical amplifier. Any method of varying the bias affects the linearity of the vertical amplifier. Another common method of adjusting linearity is to adjust the amount of feedback in the vertical circuit.

Since the linearity control changes the gain of the vertical amplifier, it also changes the height. Similarly, the height control in the oscillator circuit changes linearity. Both controls should be adjusted together to fill the screen top to bottom with uniform spacing of the scanning lines. It can be useful to note that the linearity control generally will stretch the top more than the height control, which stretches the bottom. A common problem is that the height control cannot pull the bottom of the raster down enough. Then good linearity can only be obtained with too little height. Or, the correct height can be produced only with stretching at the top. This trouble is really insufficient height, with weak vertical output, rather than poor linearity. With tubes, try a new vertical amplifier. For actual linearity troubles, check for the correct dc operating voltages on the vertical output stage and for a leaky coupling capacitor.

Vertical linearity can be checked by using a bar generator. With about 10 to 20 horizontal bars across the screen, adjust for uniform spacing. Another method is to check the width of the black horizontal blanking bar across the screen as it drifts up or down. Turn the vertical hold control slightly off frequency to make the picture roll slowly. Incidentally, the picture will usually roll when the vertical linearity and height controls are varied, but this is no problem as long as the picture can be locked in with the vertical hold control.

21-5 Internal Vertical Blanking

The voltage pulse produced during flyback in the vertical output circuit is coupled to the picture tube for the purpose of providing additional blanking during vertical retrace time. The circuit is also called a *retrace suppressor*. This is in addition to the blanking voltage at the grid-cathode circuit of the picture tube, which is part of the composite video signal. The advantage of the added blanking voltage is that the retrace lines produced during vertical flyback do not appear on the screen for any setting of the brightness control.

Actually, internal vertical blanking must be used in receivers that do not have the dc component of the video signal at the picture tube. Otherwise, the vertical retrace lines would show when the background in the picture changed for a darker scene. Just about all receivers use a vertical retrace suppressor circuit. Color receivers usually have a horizontal retrace suppressor circuit also.

The vertical flyback pulse for blanking can be either applied to the picture tube or injected into the video amplifier. At the picture tube, the required amplitude is 50 to 150 V to cut off beam current. The polarity for blanking is positive at the cathode, as shown in Fig. 21-9a. The inverted trapezoid in Fig. 21-9b has negative voltage spikes for vertical blanking at the control grid or the screen grid. In both coupling circuits, C blocks any dc voltage.

When the flyback pulses for blanking are injected into the video amplifier, the required amplitude is only a few volts. This method is often used in transistor receivers. In color receivers, a separate blanker stage is generally used to amplify both vertical and horizontal flyback pulses, which are coupled into the video amplifier for blanking.

The methods of obtaining the vertical flyback pulses for internal retrace blanking include the following:

1. From plate or collector of vertical output stage (Fig. 21-9a). The polarity of the flyback pulse here is positive at the plate of a tube or at the collector of an NPN transistor.
2. From secondary of output transformer (Fig. 21-9b).
3. From a separate winding on the output transformer for vertical blanking.
4. From the spiking resistor in series with the sawtooth capacitor in the vertical oscillator circuit. This method provides a sharp peak alone, without the sawtooth, but the amplitude is low compared with the output circuit.

The *RC* network for coupling the vertical retrace pulses has two functions: (1) block dc voltage; and (2) flatten the sawtooth tilt. Otherwise, the sawtooth voltage can modulate the beam intensity to vary the shading in the picture from top to bottom.

VERTICAL DEFLECTION CIRCUITS

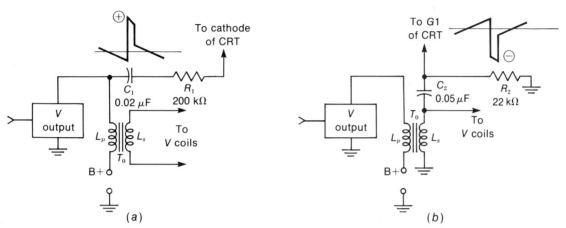

FIGURE 21-9 CIRCUITS FOR INTERNAL VERTICAL BLANKING. (a) POSITIVE RETRACE PULSES FROM PLATE OF OUTPUT TUBE; (b) NEGATIVE RETRACE PULSES FROM TRANSFORMER SECONDARY.

It should be noted that the retrace suppressor circuits provide a connection that links the deflection circuits with the picture tube and its video signal circuits. A trouble in the retrace suppressor, or the blanker stage, can affect the picture.

21-6 Multivibrator for Combined Vertical Oscillator and Amplifier

As illustrated in Fig. 21-10, a free-running, plate-coupled, unsymmetrical MV can be used for the complete vertical deflection circuit. Only one dual-section tube is necessary in the MV, and the feedback from the output circuit eliminates the need for a blocking oscillator transformer. Either a dual-triode or a triode-pentode is used. Common dual-triodes in this circuit are the 8CG7, 10DR7, 13EM7, and 15FM7, for series heaters. Triode-pentodes often used are the 21LR8 and 25J28. The pentode section is the output stage. The pentode output is preferred with lower values of B+ supply voltage.

In Fig. 21-10, the plate of V1 drives the grid of V2 through an RC coupling circuit, and the plate of V2 is coupled to V1 through the R_1C_1 feedback circuit. Conduction in V1 cuts off V2. Then conduction in V2 cuts off V1. The MV is unsymmetrical, as V2 conducts for a longer time to supply sawtooth output current for the trace. Then V1 conducts for a short time to cut off V2 for the retrace. The sawtooth capacitor is in the plate circuit of V1, serving as the discharge tube. A pentode for V2 in the vertical output stage usually has negative feedback from plate to its own grid to improve linearity.

The schematic diagram of this MV circuit[1] for vertical scanning is shown in Fig. 21-11 with the 15FM7 dual-triode. The V9A section at the left serves as the vertical deflection oscillator, while the V9B section is the vertical amplifier. The autotransformer T_9 supplies sawtooth output current to the vertical deflection coils. The

---

[1]The circuit in Fig. 21-10 can also be considered as V1 for a blocking oscillator with feedback from the amplifier V2.

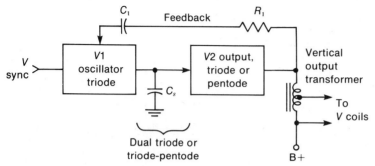

FIGURE 21-10 DUAL-SECTION TUBE AS PLATE-COUPLED MV FOR VERTICAL DEFLECTION.

variable cathode resistance R_{22} varies the bias on V9B to control the linearity of the sawtooth output.

In the plate of V9A from R_{18}, the 0.1-µF C_{41} is the sawtooth capacitor, charging toward +695 V while V9A is cut off. The height control R_{20} varies the RC time constant on charge to adjust the sawtooth amplitude. This voltage drive is coupled to the grid of V9B by C_{42} and R_{23}.

In the output circuit, the plate of V9B provides feedback through C_{44}, A1, and C_{43} to the grid of V9A. The printed circuit A1 is a compact unit with series resistances and shunt capacitances to shape the feedback voltage. In the grid circuit of V9A, the vertical hold control R_{21} varies the RC time constant, which determines the cutoff time, to adjust the frequency.

Vertical sync voltage from the A2 integrator is applied to the plate of V9A or the grid of V9B. The sync polarity is negative to reduce the output plate current of V9B, which increases its plate voltage. This positive feedback to the grid of V9A triggers V9A into conduction to start the vertical retrace.

The waveshapes in Fig. 21-11b show trapezoidal voltge drive at the grid of the vertical amplifier, with a peak-to-peak amplitude of 160 V. In the plate circuit, this trapezoid is amplified and inverted with 680-V amplitude. The positive spike in the plate circuit results from the high value of self-induced voltage produced across the output transformer T_9 when the plate current is cut off. Since the plate voltage of V9B is fed back to V9A, the spike is also in the trapezoidal plate voltage of V9A. This negative spike in the plate circuit of V9A is a flyback pulse that is coupled by C_{40} to the screen grid of the picture tube for vertical retrace blanking.

21-7 Transistorized Blocking Oscillator and Vertical Amplifier

See Fig. 21-12. Q1 is a blocking oscillator, synchronized at 60 Hz by the vertical sync input. Q2 is a silicon controlled rectifier (SCR) that serves as a switch to discharge the sawtooth capacitance, consisting of C_7 and C_8 in series. The collector output voltage of Q1 drives the gate electrode of Q2. The sawtooth voltage output is directly coupled to the base of Q3. This stage is an emitter-follower with high input impedance that isolates the oscillator and switch from the low impedance of the vertical output stage Q301. This is a CE (common-emitter) amplifier.

In the Q1 circuit, T_1 is the blocking oscillator transformer which supplies feedback from collector to base. The negative dc voltage of

VERTICAL DEFLECTION CIRCUITS 467

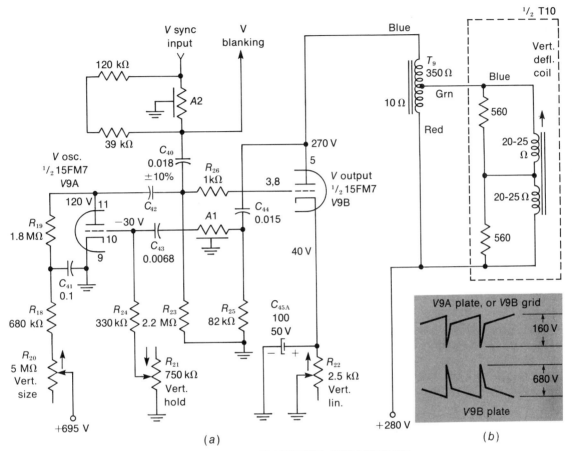

FIGURE 21-11 DUAL-TRIODE AS PLATE-COUPLED MV FOR VERTICAL OSCILLATOR AND OUTPUT STAGES. R IN Ω; C IN μF. A1 AND A2 ARE PRINTED-CIRCUIT RC INTEGRATORS. ARROW ON CONTROLS INDICATES CLOCKWISE ROTATION. (a) CIRCUIT; (b) TRAPEZOIDAL VOLTAGE WAVESHAPES. (FROM ZENITH CHASSIS 14M29)

−5.3 V at the base results from the positive feedback voltage from the transformer, producing base current and reverse bias, similar to the grid-leak bias with a tube. C_5 and the series resistance in the base circuit provide the RC time constant for discharge of the reverse bias, which determines the free frequency. The vertical hold control R_{115} varies the RC time constant and the amount of positive base voltage from the +12-V supply to control the frequency of the oscillator. Positive sync voltage from the sync separator and vertical integrator is coupled to the base by C_4 to trigger the oscillator at 60 Hz.

In the vertical switch Q2, the gate voltage determines when the SCR conducts between anode and cathode. The collector output of Q1 is a series of positive pulses at 60 Hz to gate Q2 on. Without the gate pulse, the SCR is off. Then C_7 and C_8 can charge. The charging voltage is ob-

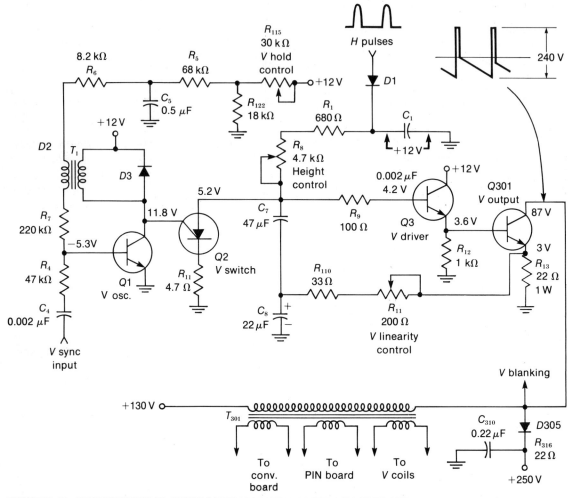

FIGURE 21-12 TRANSISTORIZED BLOCKING OSCILLATOR WITH EMITTER-FOLLOWER AND CE VERTICAL AMPLIFIER. *(FROM MAGNAVOX CHASSIS T979)*

tained from the 12 V across C_1 as a dc source. This equivalent voltage source is produced by horizontal flyback pulses coupled by $D1$ to charge C_1. These pulses are positive, charging C_1 to a positive 12 V.

The time constant for charging the sawtooth capacitors C_7 and C_8 is varied by the height control R_8, to adjust the sawtooth amplitude. The resulting sawtooth voltage is directly connected to the base of the vertical driver $Q3$.

The reason for dividing the sawtooth capacitance of C_7 and C_8 is to provide feedback at the junction, from the emitter of the vertical amplifier, to improve the linearity of the sawtooth rise. The linearity control R_{11} varies the amount of feedback.

The reason for using horizontal flyback pulses to provide the dc voltage across C_1 for charging the vertical sawtooth capacitors is to make the vertical output dependent on the horizontal scanning amplitude. If excessive load current reduces the horizontal scanning width, then the vertical height will also decrease.

In the vertical driver Q3, the emitter-load resistance is R_{12}. This is an emitter-follower, without any ac load impedance in the collector circuit. Note the base-emitter bias of 4.2 − 3.6 = 0.6 V positive at the base of a silicon NPN transistor serving as a linear amplifier. There is no polarity inversion in an emitter-follower. Here, the waveform is a positive-going sawtooth voltage from the switch Q2 into the base of Q3 and out from the emitter to drive the base of the output amplifier Q301. This voltage is amplified and inverted to produce the trapezoidal voltage output shown at the collector with a peak-to-peak amplitude of 240 V.

In the vertical amplifier Q301, the sawtooth current output is transformer-coupled by T_{301} to the vertical scanning coils in the deflection yoke. In addition, separate windings are used for vertical deflection signals to pincushion circuits on the deflection board and to the convergence correction circuits on a separate panel. Also, the output from the collector of Q301 goes to a blanker stage on the video amplifier board for vertical retrace blanking.

The unbypassed resistance R_{13} has 3-V emitter bias for Q301 and sawtooth voltage that is fed back to C_8, through R_{110} and the linearity control R_{11}. The net bias from base to emitter is 3.6 − 3.0 = 0.6 V for this silicon transistor, positive at the base for NPN.

The vertical output stage Q301 is a 25-W power transistor in a metal TO-3 case with a heat sink. Supply voltage for the collector is +130 V. However, the self-induced voltage in the primary of T_{301} produces a flyback pulse of 240 V.

The diode D305 in the line to +250 V protects the collector against higher transient voltages, which can be more than the breakdown voltage rating. D305 is biased off by +250 V at the cathode. Any voltage more than 250 V at the diode anode and Q301 collector, however, makes the diode conduct. Then the collector is clamped at 250 V.

21-8 Miller Feedback Integrator Circuit

One problem with a vertical sawtooth voltage generator in transistor circuits is the need for a large, electrolytic sawtooth capacitor because of the low shunt resistance. One solution is to use the feedback circuit with the Miller integrating capacitor C_M illustrated in Fig. 21-13. Q2 represents an amplifier with feedback from output to input, usually including several stages. C_M is a 0.47-μF paper capacitor in the feedback path. Q1 is directly coupled to Q2 in order to serve as a switch that controls the collector voltage of Q2. The sync input to Q1 reduces its I_C to increase I_C in Q2. When the collector voltage of Q2 decreases with more I_C, the feedback capacitor C_M discharges. The result is a negative-going sawtooth for the output of Q2, as shown at the collector. This sawtooth voltage has very low amplitude of the order of millivolts, but is extremely linear.

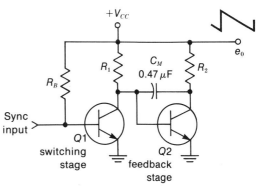

FIGURE 21-13 BASIC CIRCUIT OF MILLER FEEDBACK INTEGRATOR FOR LINEAR SAWTOOTH OUTPUT PRODUCED BY C_M.

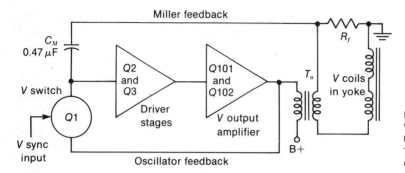

FIGURE 21-14 BLOCK DIAGRAM OF VERTICAL DEFLECTION CIRCUIT USING MILLER FEEDBACK INTEGRATOR WITH C_M. (FROM RCA CHASSIS CTC49)

The name "Miller integrator" indicates two important effects. In general, for any amplifier, the Miller effect means that the input capacitance is multiplied by the gain of the amplifier. As a result, a relatively small capacitance in the input circuit can be equivalent to a much larger capacitance, multiplied by a factor as high as 100.

An integrating circuit has the characteristic that the rate of change in the output is proportional to the amount of input. In Fig. 21-13, the voltage change across C_M as it discharges is very linear because the amount of discharge current is kept constant by the varying resistance of Q2 in the discharge path. C_M charges through Q1 but discharges through Q2.

A practical vertical deflection circuit using the Miller integrator principle is illustrated by the block diagram in Fig. 21-14. Q101 and Q102 are a pair of power transistors in the vertical deflection output circuit. Q2 and Q3 are amplifiers to drive the power output stage. All the amplifier stages are in the feedback loop, from the secondary of the output transformer through C_M and back to the input of Q2. In addition, the feedback to the switch Q1 makes the circuit free-running. The adjustments for this circuit include a height control and vertical hold for frequency control, but there is no vertical linearity control.

21-9 Transistor Pair in Vertical Output Circuit

Two transistors are often used in push-pull to provide the power output required for vertical deflection. Furthermore, transistor circuits can be arranged without the need for a center-tapped push-pull output transformer. One method uses PNP and NPN transistors in a *complementary-symmetry* output stage (Fig. 21-15). Another method uses a pair of NPN power transistors for the output stages, but the driver stages are opposite transistor types (a circuit called *quasi-complementary symmetry*).

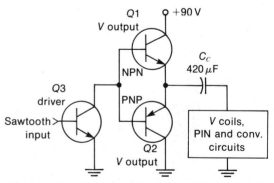

FIGURE 21-15 BASIC CIRCUIT FOR A TRANSISTOR PAIR AS PUSH-PULL VERTICAL OUTPUT STAGE. NPN AND PNP TRANSISTORS IN COMPLEMENTARY SYMMETRY PROVIDE PUSH-PULL OUTPUT.

These push-pull circuits are similar to the audio output stages in high-fidelity amplifiers.

In Fig. 21-15 the collector of the NPN transistor Q1 uses + 90 V for V_{CC}. Q1 and Q2 are in series across the 90-V supply. Therefore, the dc voltage at the emitter of Q1 serves as the emitter voltage for Q2. The supply voltage is inverted for the Q2 collector, since positive voltage at the emitter corresponds to negative collector voltage for this NPN transistor. The output deflection signal to the vertical deflection coils is taken from the junction of the emitters for Q1 and Q2. The 470-μF C_C blocks any dc voltage from the vertical scanning coils, which would shift the centering.

The push-pull action of Q1 and Q2 results because they have opposite drives. Assume a positive-going sawtooth rise into the driver Q3, resulting in negative drive during trace time at the base for both Q1 and Q2. Since Q1 is PNP with an N base, this driving signal increases the collector current. At the base of Q2, however, the negative drive reduces the forward voltage at the base for less collector current in the NPN transistor.

A push-pull stage can be operated class B or AB for greater efficiency, compared with class A. However, good regulation in the power supply is necessary. Each transistor supplies one-half the ac cycle of sawtooth current in the yoke. One transistor is effective in scanning the top half of the raster, and the other half is produced by the opposite transistor. Figure 21-16 shows a vertical output module board with push-pull transistors.

21-10 Vertical Deflection Troubles

These include stretching and compression at the top or bottom, not enough height, or just a horizontal line across the screen. You can see the troubles in the raster and in the picture.

Horizontal line only. Just a bright line or bar across the middle of the screen, as in Fig. 21-17, means there is no output from the vertical deflection circuits. Often, you can see dark and light picture information moving across the line, which shows video signal is present. This effect of no height can be caused by trouble in any stage of the vertical deflection circuits, including the oscillator, driver, and output stages.

Insufficient height. This trouble is generally caused by a weak amplifier in the vertical output circuit. As shown in Fig. 21-18, the screen has black bars at the top and bottom. These areas are not scanned and, therefore, not illuminated. The reduced height can also be caused by reduced dc supply voltage for the vertical amplifier.

White bar across bottom. This effect, shown in Fig. 21-19, is actually an extreme case of vertical nonlinearity. The end of the sawtooth trace is too flat, compressing lines at the bottom to form the white bar. With a picture, the information at the bottom is folded over. This trouble is generally caused by a weak output tube or incorrect bias. A common problem is a leaky coupling capacitor, which shifts the bias. As another cause, a low value of dc supply voltage can reduce the peak sawtooth current needed to pull the scanning down to the bottom of the screen.

Vertical jitter. This trouble is really in vertical synchronization, and caused by noise. We assume the AGC is normal and there is no overload distortion of the vertical sync. The jitter results because the picture starts to roll slightly but is then locked in by the vertical sync.

The problem of horizontal flyback pulses coupled into the vertical deflection oscillator can also produce poor interlacing. To correct this problem of crosstalk between the horizontal

FIGURE 21-16 VERTICAL OUTPUT MODULE BOARD WITH PUSH-PULL TRANSISTORS. WIDTH IS 6 IN. *(GE)*

FIGURE 21-17 NO VERTICAL DEFLECTION.

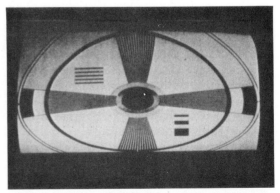

FIGURE 21-18 INSUFFICIENT HEIGHT IN RASTER CAUSED BY WEAK VERTICAL OUTPUT.

FIGURE 21-19 WHITE BAR AT BOTTOM, WITH FOLDOVER IN PICTURE, CAUSED BY COMPRESSION AT END OF SAWTOOTH WAVE FOR VERTICAL SCANNING.

FIGURE 21-20 TRAPEZOIDAL RASTER, WITH KEYSTONING CAUSED BY SHORT ACROSS ONE VERTICAL COIL IN YOKE.

and vertical circuits, check the lead dress, make sure the cover is on the high-voltage cage in the horizontal output circuit, and check the RC decoupling filters in the vertical oscillator.

Opens and shorts in the vertical scanning coils. When the two coils are in series, an open in one means no deflection and no height. With parallel coils, one section can affect the top or bottom part of the raster. Shorted turns in either coil, however, produce a trapezoidal raster (Fig. 21-20). Any form of trapezoidal distortion is practically always caused by a defective yoke, or by a trouble in the pincushion correction circuits in color receivers.

Oscillator frequency troubles. If the hold control cannot make the picture roll both up and down, the free frequency cannot be set to 60 Hz. The trouble can be a defective oscillator stage but is often in the RC time constant of dc bias voltage for the oscillator. When you see two or three complete pictures, the oscillator frequency is at 30 Hz or 20 Hz. For 120 Hz and above, you can see the bottom part of the picture folded over the top.

SUMMARY

1. The vertical deflection circuits consist of an oscillator synchronized at 60 Hz driving the power output stage coupled to the vertical scanning coils in the deflection yoke.
2. The height or vertical size control varies the amount of drive to the output amplifier to control the height of the raster.
3. The vertical linearity control either varies the bias in the vertical output stage or adjusts the feedback.

4. Both the linearity and height controls are adjusted fo fill the screen top to bottom with equally spaced scanning lines.
5. For internal retrace blanking, the vertical flyback pulses are coupled either to the picture tube or to the video amplifier to cut off beam current during vertical retrace time.
6. With tubes, the plate-coupled multivibrator shown in Fig. 21-11 is generally used for the vertical deflection circuit. A dual-triode or triode-pentode serves as the oscillator and power amplifier.
7. With transistors for vertical deflection, a typical circuit includes a blocking oscillator and emitter-follower driving the power output stage (Fig. 21-12).
8. In transistorized vertical deflection circuits, the Miller feedback integrator in Fig. 21-14 is often used. This circuit allows a smaller sawtooth capacitor, with good linearity. Usually, there is no linearity control.
9. The vertical output stage may use PNP and NPN transistors in complementary symmetry (Fig. 21-15). This circuit is a push-pull amplifier, similar to the power output stage in audio circuits.
10. With tubes in the output stage, the vertical scanning coils are in series. With transistors, the vertical coils are usually in parallel.
11. In color receivers, the vertical output feeds the pincushion (PIN) correction circuits and convergence correction circuits. The PIN circuits are in series with the vertical scanning coils.
12. Troubles in the vertical deflection circuits affect the height of the raster. Two common troubles: (a) just a horizontal line as in Fig. 21-17, caused by no vertical output; (b) insufficient height as in Fig. 21-18, caused by weak vertical output.

Self-Examination (Answers at back of book)

1. The voltage waveshape in the output of the vertical amplifier is a (a) trapezoid; (b) sawtooth; (c) rectangle; (d) square.
2. An autotransformer in the vertical output circuit: (a) steps up the voltage for the scanning coils; (b) isolates the scanning coils for B + voltage in the primary; (c) isolates the oscillator and output stages; (d) does not isolate the secondary from dc voltage in the primary.
3. Vertical flyback pulses at the plate of the vertical output tube are for retrace blanking at the picture tube: (a) cathode; (b) control grid; (c) screen grid; (d) anode.
4. The top of the picture is stretched with too much height. To correct this: (a) vary the vertical hold control; (b) reduce height with the vertical linearity control; (c) increase height with the size control; (d) replace the vertical oscillator tube.
5. The vertical deflection circuit in Fig. 21-11 is a: (a) plate-coupled multivibrator; (b)

cathode-coupled multivibrator; (c) driven blocking oscillator; (d) Schmitt trigger.
6. In Fig. 21-11, during vertical trace time: (a) the oscillator section V9A conducts; (b) the amplifier section V9B is cut off; (c) V9B conducts; (d) the sawtooth capacitor discharges.
7. In Fig. 21-12, the forward bias between base and emitter of the vertical output stage Q301 equals: (a) 3.0 V; (b) 3.6 V; (c) 0.6 V; (d) 130 V.
8. Peak-to-peak sawtooth scanning current in the vertical coils can be: (a) 0.7 mA; (b) 0.7 A; (c) 50 A; (d) 150 μA.
9. A push-pull amplifier in the vertical output circuit: (a) usually operates class C; (b) can use PNP and NPN transistors; (c) cannot be used; (d) generally uses two beam-power pentodes.
10. In the transistorized vertical deflection circuit in Fig. 21-12, if R_{13} in the emitter of Q301 opens, the result will be: (a) poor linearity; (b) vertical rolling; (c) pincushion distortion; (d) just a horizontal line across the center of the screen.

Essay Questions

1. Describe briefly how the scanning raster is produced. Which stages determine height? Which stages determine width?
2. Why can the deflection circuits produce the raster with or without sync?
3. How is the vertical oscillator synchronized at 60 Hz with (a) positive sync; (b) negative sync?
4. Why does the plate current of the vertical output tube have the sawtooth waveshape?
5. Give two requirements of the vertical output transformer.
6. List three controls for vertical deflection, giving the function of each, and describe briefly how to adjust them.
7. Draw the schematic diagram of a triode output stage, with autotransformer coupling to the vertical deflection coils. Include internal vertical blanking for the picture tube. Give typical values of components. Show voltage waveshapes with typical amplitudes at grid, at plate, and across yoke coils.
8. Give two component troubles that can cause a white bar at the bottom of the raster, with insufficient height.
9. Give two component troubles that can make the oscillator frequency too high.
10. Give two component troubles that can cause no vertical scanning, with just a white line across the center of the screen.
11. What is the function of a thermistor in the deflection yoke?
12. What is the function of a varistor at the collector of the vertical output stage?
13. In the vacuum tube circuit in Fig. 21-11, give the function for each of the following components: C_{40}, C_{41}, C_{42}, R_{23}, T_9, A2, R_{20}, R_{21}, and R_{22}.

14. In the transistor circuit in Fig. 21-12, give the function for each of the following components: C_4, T_1, $D3$, $Q2$, C_1, R_{12}, R_{13}, R_{115}, R_8, and R_{11}.
15. Give one advantage of the Miller feedback circuit for vertical deflection.
16. Refer to the transistor pair for push-pull output in Fig. 21-15. Why would the collector-emitter voltage of Q2 be -45 V when the emitter voltage of Q1 is $+45$ V to chassis ground?

Problems (Answers to selected problems at back of book)

1. How much cathode bias voltage is produced by an R_k of 2,000 Ω with an average cathode current of 20 mA?
2. Referring to the transistor circuit in Fig. 21-12, how much is the emitter current for: (a) Q3; (b) Q301.
3. Vertical scanning coils have an inductance of 40 mH with R of 50 Ω. Calculate the inductive reactance for a sine-wave current at 60 Hz.
4. Horizontal scanning coils have an inductance of 10 mH with R of 12 Ω. Calculate the inductive reactance for a sine-wave current at 15,750 Hz.

22 Horizontal Deflection Circuits

The horizontal output stage is a power amplifier which supplies the scanning current needed in the horizontal deflection coils to fill the width of the raster. Since the horizontal output circuit is primarily inductive, it can also generate flyback pulses of 18 kV or more. This voltage is used for the high-voltage rectifier that produces anode voltage for the picture tube. Furthermore, the high value of self-induced voltage makes the output circuit oscillate. To control this sine-wave ringing, a diode damper is used in the horizontal output circuit. The operation of the horizontal amplifier must be considered along with the damper, as both stages are needed to produce horizontal scanning. It is important to remember that without horizontal scanning there is no high voltage and no brightness. More details of the horizontal output circuit are described in the following topics:

22-1 Functions of the horizontal output circuit
22-2 The horizontal amplifier
22-3 Damping in the horizontal output circuit
22-4 Horizontal scanning and damping
22-5 Boosted B+ voltage
22-6 Flyback high voltage
22-7 Horizontal deflection controls
22-8 Deflection yokes
22-9 Horizontal output transformers
22-10 Analysis of horizontal output circuit
22-11 Complete horizontal deflection circuit
22-12 Protective bias on horizontal output tube
22-13 Transistorized horizontal deflection
22-14 SCR horizontal output circuit
22-15 High-voltage limit control
22-16 Troubles in the horizontal deflection circuits

478 BASIC TELEVISION

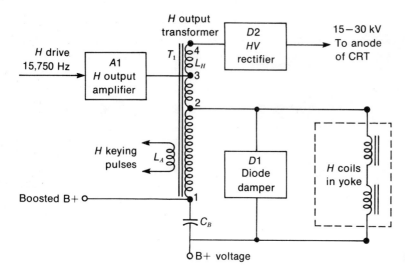

FIGURE 22-1 BLOCK DIAGRAM OF HORIZONTAL OUTPUT CIRCUIT.

22-1 Functions of the Horizontal Output Circuit

As shown in Fig. 22-1, the essential stages are the amplifier, damper diode, and high-voltage rectifier, operating with the horizontal output transformer (H.O.T.). This is also called the *flyback transformer*. These stages can use either tubes or transistors and semiconductor diodes. Also, the output stage can use semiconductor switching devices, such as the SCR or gate-controlled switch.

Typical beam-power pentodes for the horizontal amplifier are the 6DQ5, 17BQ6, and 6GW6 octal tubes with a top cap for the plate connection. The 6JR6, 22JR6, 6LQ6, and 24LQ6 are examples of tubes with a novar base and no top cap. Typical diode damper tubes are the 6AX4, 6CL3, 12CL3, 17CT3, and 25CM3. In addition, pentode-diode types such as the 33GY7 and 38HE7 combine the horizontal amplifier and damper in one envelope.

Horizontal scanning. The output stage is a power amplifier which supplies 1 to 5 A p-p scanning current in the yoke coils. In Fig. 22-1, the winding between terminals 3 and 1 is the primary on the autotransformer T_1. The winding between 2 and 1 is the secondary for voltage stepdown and current stepup to the horizontal scanning coils.

Horizontal damping. The damper D1 has the primary purpose of damping sine-wave oscillations, or ringing. Immediately after the fast flyback, the diode conducts to serve as a low shunt resistance that stops the ringing of the inductance. Damping is needed because the oscillations produce white vertical bars at the left side of the raster. It should be noted, though, that the damper does not conduct during retrace. At this time, damping is not desired, in order to allow a fast flyback and maximum amount of high voltage.

Using the damper for scanning. For greater efficiency, the damped output current is used to produce about one-third of the horizontal trace at the left side of the raster. During this time, immediately after flyback, the damper conducts but the amplifier is cut off. The average amplifier current is reduced appreciably, therefore, improving the efficiency. Since the damped current produces part of each horizontal trace, both the damper and amplifier stages are needed for horizontal deflection. This system is often called *reaction scanning*, and the damper is an *efficiency diode*.

Boosted B+ voltage. When the damper conducts, its current charges C_B in Fig. 22-1. This capacitor is in series with $D1$ and the supply voltage. As a result, the voltage across C_B becomes higher than B+ by the amount of deflection voltage rectified by the damper. As an example, 280 V for B+ can be boosted to 550 V across C_B. Or, 140 V for B+ can be boosted to 300 V. This boosted B+ is the plate supply voltage for the horizontal amplifier tube, which is the reason why the damper must be operating for there to be horizontal output. In fact, all the tubes in the vertical and horizontal deflection circuits generally use boosted B+ for higher plate voltage.

Scan rectification. In solid-state receivers, part of the horizontal output is rectified to provide dc operating voltages. These include screen-grid voltage for the picture tube and collector voltage for the transistors. In this case, it should be noted that the signal circuits depend on the horizontal output for dc supply voltage.

Flyback high voltage. The high-voltage pulse produced across the primary of the horizontal output transformer is stepped up, rectified, and filtered to provide anode voltage of 15 to 30 kV for the picture tube. In Fig. 22-1, the turns of L_H between terminals 3 and 4 are the high-voltage winding for $D2$. It should be noted that the high-voltage rectifier is the only stage conducting in the horizontal output circuit during retrace time.

Although not shown here, the high-voltage supply can also include a circuit to provide about 5 kV to the focus grid in a color picture tube. There may be a separate focus rectifier or the focus voltage can be tapped down on the anode supply.

Horizontal keying pulses. The horizontal output transformer usually has one or more auxiliary windings, indicated by L_A in Fig. 22-1, to supply flyback pulses at the horizontal rate. These pulses are used for the AGC circuit, retrace blanking, horizontal AFC stage, and in the color circuits.

22-2 The Horizontal Amplifier

The block diagram in Fig. 22-1 corresponds to the schematic in Fig. 22-2, using tubes in the horizontal output circuit. $V1$ is the amplifier, transformer-coupled at terminal 2 to the deflection coils in the yoke. $V2$ is the diode damper. It is tapped higher at terminal 3 in order to provide more boosted B+ voltage. $V3$ is the high-voltage rectifier, with its filament winding L_5 on T_1. The auxiliary winding L_A supplies horizontal flyback pulses for the AGC stage. The high-voltage rectifier and flyback transformer are usually in a metal cage on the receiver chassis (Fig. 22-3).

Circuit components. In Fig. 22-2, C_1 and R_1 couple the sawtooth voltage from the horizontal oscillator to the output tube. The grid-driving voltage is 120 V p-p. Grid-leak bias of −40 V is developed because grid current flows when the positive peak of input voltage drives the grid positive. In fact, measuring this bias with a dc voltmeter is a good way to see if the output tube has grid drive from the oscillator. Normal dc bias means normal ac drive. With less drive, the bias is reduced. Zero bias means no drive. The output tube should not be operated without grid

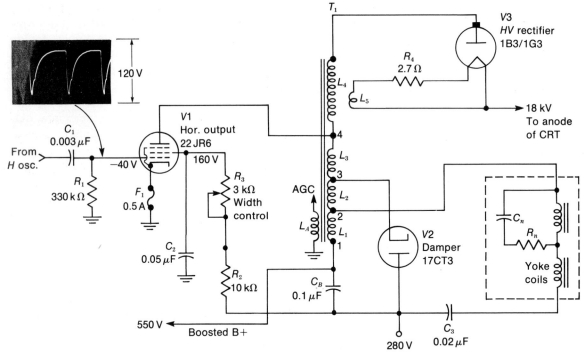

FIGURE 22-2 SCHEMATIC DIAGRAM OF HORIZONTAL OUTPUT CIRCUIT USING TUBES FOR AMPLIFIER, DAMPER, AND HV RECTIFIER.

drive, as there is no safety bias in the cathode circuit. The 0.5-A fuse in the cathode protects the tube and transformer against excessive current.

The plate current of V1 can flow from terminal 4 on T_1 through L_3, L_2, and L_1 to C_B, which provides boosted B+ as the source of plate supply voltage. Current in the output tube is a load on C_B, making it discharge. However, C_B is recharged when the damper conducts. Note that C_3 blocks direct current from the deflection coils, which would shift the horizontal centering.

The width control R_3 varies the screen-grid voltage on the output tube. Increasing the screen voltage increases the width. In many circuits, the screen-grid voltage is adjusted by taps to change the series resistance, instead of by a

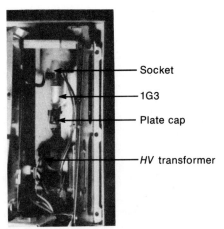

FIGURE 22-3 HV RECTIFIER AT TOP AND FLYBACK TRANSFORMER AT BOTTOM IN CAGE ON RECEIVER CHASSIS. THE 1G3 IS MOUNTED UPSIDE DOWN.

variable control. C_2 is the screen-grid bypass capacitor.

It should be noted that the suppressor grid usually has a small positive bias voltage to prevent parasitic oscillations in the horizontal output tube. In addition, the control grid may have 50- to 100-Ω series resistance as a parasitic suppressor. The parasitic oscillations reduce the gain of the output tube and may produce thin, black vertical lines at either side of the raster.

Harmonic tuning. The horizontal output transformer often has a capacitor of about 150 pF across part of the winding to resonate at an odd harmonic of 15,750 Hz. Either the third or the fifth harmonic is generally used. The harmonic component is phased to reduce the peak flyback voltage on the amplifier.

Circuit operation. As the linear rise of sawtooth input voltage drives the grid in the positive direction, the amount of plate current increases. The result is a sawtooth rise of current in T_1 and the yoke coils. At the peak of the sawtooth input, the grid voltage drops sharply to a value more negative than cutoff. This negative grid drive cuts off the amplifier. As a result, V1 stops supplying current to the output circuit. The output tube remains cut off while the grid-driving voltage completes its swing in the negative direction for the flyback. Also, the tube remains cut off until the linear rise in the positive direction makes the instantaneous grid voltage less negative than the cutoff voltage. Then the sawtooth rise in grid voltage produces a sawtooth rise of current again in the output circuit to produce the next cycle of operation.

Figure 22-4 shows the grid-plate transfer characteristic curve of the horizontal output tube. The grid voltage is a sawtooth wave. However, the plate current flows in pulses corresponding to only part of the sawtooth voltage input. When the grid voltage due to the bias and

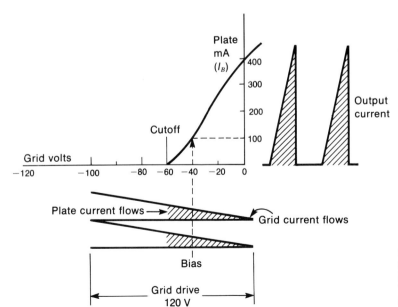

FIGURE 22-4 OPERATION OF HORIZONTAL OUTPUT TUBE. SAWTOOTH PLATE CURRENT FLOWS FOR ONLY THE PART OF THE GRID VOLTAGE INDICATED BY THE SHADED AREA.

driving voltage is more negative than cutoff, no plate current flows. For approximately two-thirds of the cycle, however, the sawtooth input voltage drives the grid voltage less negative than cutoff to produce a linear rise in plate current. This time is indicated by the shaded area of grid voltage. As an example, when the ac input voltage drives the grid voltage 30 V more positive than the bias of -40 V, the instantaneous grid voltage is -10 V, allowing plate current to flow. The amount of grid-leak bias is approximately 90 percent of the peak positive drive.

Horizontal output tube. This is a beam-power amplifier which supplies the required amount of scanning current. As an example, the 6DQ5 maximum ratings are 24-W plate dissipation, 315-mA average cathode current with 1,100-mA peak value at the end of trace, and 6.5-kV peak positive pulse at the plate during flyback. Maximum dc plate supply voltage is 990 V, including boosted B+.

22-3 Damping in the Horizontal Output Circuit

The inductance of the output transformer and scanning coils, with their distributed and stray capacitances, provides a tuned output circuit that can oscillate at its resonant frequency. Oscillations occur because the flyback on each sawtooth wave of current produces a rapid change in current that generates a high value of induced voltage across the inductance. The output circuit then oscillates at its natural resonant frequency by shock excitation. This effect is called *ringing*.

The oscillations, at about 70 kHz in the horizontal output circuit, would continue past flyback time and produce ripples on the scanning-current waveform, as shown in Fig. 22-5a. Since the scanning current produces deflection, the electron beam also oscillates back and forth in accordance with the oscillatory ripples. As a result, the oscillations can produce one or more white bars at the left side of the raster, as shown in Fig. 22-5b. The bars are at the left because the oscillations occur immediately after flyback time. The bars are white because the electron beam scans these areas several times during every horizontal line traced, as the oscillations make the horizontal scanning current repeat equal amplitudes at different times. Ringing occurs in the horizontal output circuit, primarily, rather than in vertical scanning, because the fast horizontal flyback produces high values of induced voltages.

Shock-excited oscillations. The action of shock-exciting a tuned circuit into oscillations is illustrated in Fig. 22-6. In *a* the switch S has been closed and current flows through L and R, while the distributed capacitance C charges to the applied voltage. When the switch is opened in *b*, the voltage E is disconnected from the tuned circuit. Then the change in applied voltage can make the tuned circuit start oscillating.

The tuned circuit oscillates without the battery because the energy stored in the electromagnetic field of L and the electrostatic field of C while the battery was connected provides the required current and voltage. The current in the tuned circuit and the voltage across it oscillate with decreasing amplitudes until the stored energy is dissipated in the resistance. R is the internal resistance of the coil. The switch S corresponds to the action of the deflection amplifier. It conducts during trace time to supply current to the output circuit but then is cut off at flyback time by the grid-driving voltage.

The current in the coil decays toward zero when the switch is opened. However, the sharp change as the current starts to decline produces a large self-induced voltage across L. Now the

HORIZONTAL DEFLECTION CIRCUITS 483

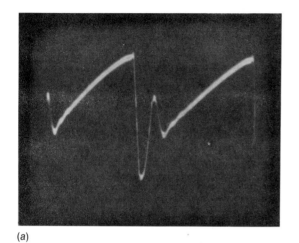 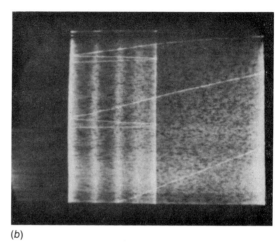

(a) (b)

FIGURE 22-5 EFFECT OF INSUFFICIENT DAMPING IN HORIZONTAL OUTPUT CIRCUIT. *(a)* OSCILLATIONS IN SCANNING CURRENT. *(b)* WHITE BARS AT LEFT SIDE OF RASTER CAUSED BY OSCILLATIONS.

coil is a generator producing voltage that keeps the current flowing in the same direction through the inductance as when the switch was closed. This is illustrated in *b* of Fig. 22-6. At the same time the capacitor discharges through *L* and *R*, reducing the voltage across *C*. Its discharge current is in the same direction as the current in the coil.

A little later the capacitor is completely discharged and the voltage across *C* is zero. The current i_L produced by the coil still is in the same direction, however, charging *C* to produce a voltage of opposite polarity from the original battery voltage. As the current in the coil decays to zero, the voltage across *C* charges to its maximum value.

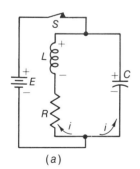 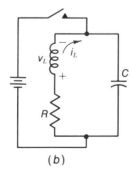 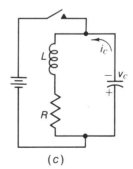

(a) (b) (c)

FIGURE 22-6 SHOCK-EXCITING A TUNED CIRCUIT INTO OSCILLATION. *(a)* SWITCH IS CLOSED FOR BATTERY *E* TO SUPPLY CURRENT AND VOLTAGE FOR *L* AND *C*. *(b)* SWITCH IS OPENED AND SELF-INDUCED v_L ACTS AS GENERATOR. *(c)* REVERSED POLARITIES ONE-HALF CYCLE LATER WITH v_C AS GENERATOR.

Then the capacitor voltage becomes the generator as it discharges to produce the current i_c through L and R, as shown in Fig. 22-6c. This current through the coil is in the opposite direction from the original current from the battery. When the capacitor voltage is down to zero, the coil again acts as the generator to maintain the current. C then charges with its original polarity. As a result, the inductance and capacitance interchange energy, making the tuned circuit oscillate.

The waveforms of the oscillations are shown in Fig. 22-7. Notice that the current and voltage are 90° out of phase with each other, as the current is maximum when the voltage is zero. During the first half-cycle of oscillations:

1. The current reverses from maximum in one direction to maximum in the opposite direction.
2. The voltage reaches its maximum negative value and returns to zero.

The LC circuit oscillates at its resonant frequency, with reduced amplitude for successive cycles until the stored energy is dissipated in the resistance. The frequency is approximately $1/(2\pi\sqrt{LC})$, when R is small. How long the oscillations continue depends upon the Q of the circuit. The higher the series R, the lower the Q and the sooner the oscillations decay to zero. A low value of resistance in parallel with the circuit has the same effect as a high series resistance in reducing the Q and damping the oscillations.

Damping methods. The undesired oscillations can be eliminated from the linear rise on the sawtooth wave by connecting a damping resistance R_D in parallel with the scanning coils as shown in Fig. 22-8. Because of its low value, which is about 1,000 Ω or less, R_D lowers the Q of the output circuit, reducing the amount of self-induced voltage. With just a simple resistance for damping, however, some of the scanning current is shunted through the parallel resistor. By the insertion of a capacitor C_1 in series with R_D, as in Fig. 22-8b, the scanning current can be prevented from flowing through the damping resistor while the high-frequency oscillations are damped out.

Damping with a shunt resistor is usually employed in the vertical output circuit. This method is not suitable for the horizontal output circuit, however, because there should be no damping during flyback time.

For horizontal damping, a diode damper is used, as illustrated in Fig. 22-8c. The diode can be an open circuit when it is not conducting, to allow a fast horizontal flyback. Then it conducts to act as a low damping resistance during trace time. The first half-cycle of oscillations makes the diode plate voltage negative. This prevents conduction in the damper. Meanwhile, the current in the horizontal deflection coils drops rapidly to produce the fast horizontal flyback. After the first half-cycle of oscillations,

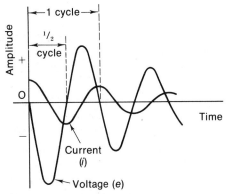

FIGURE 22-7 CURRENT AND VOLTAGE WAVEFORMS IN OSCILLATING LC CIRCUIT. NOTE THAT i AND e ARE 90° OUT OF PHASE.

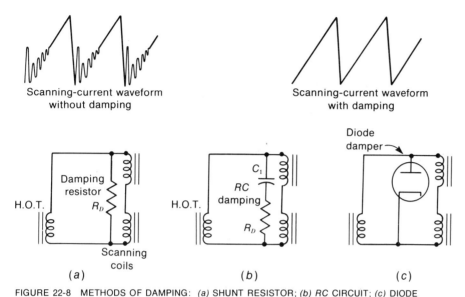

FIGURE 22-8 METHODS OF DAMPING: (a) SHUNT RESISTOR; (b) RC CIRCUIT; (c) DIODE DAMPER.

the damper plate voltage is positive, causing conduction. Then the diode is a relatively low resistance across the horizontal deflection coils that can damp out the oscillations.

22-4 Horizontal Scanning and Damping

When the horizontal output tube conducts to supply current for the horizontal scanning coils, the linear rise in current deflects the electron beam to the right side of the raster. During this time, energy is stored in the inductance and capacitance of the output circuit. Then, when the deflection amplifier tube is cut off, the output circuit starts oscillating.

The first half-cycle of oscillations is allowed to continue undamped for a fast flyback from the right side of the raster to the left side. After the retrace, though, the damper conducts to make the oscillations decay to zero. The damped current in the horizontal scanning coils deflects the electron beam from the left side of the raster toward the center. Just before the damped current decays to zero, the deflection amplifier starts conducting again to finish the trace to the right edge of the raster.

The operation of the diode damper in the horizontal output circuit is illustrated in Fig. 22-9. The first half-cycle of oscillations makes the damper plate negative, as shown in Fig. 22-9a. Then the diode cannot conduct. As a result, the undamped current in the deflection coils drops sharply from its maximum value to zero and reverses to its maximum in the opposite direction very quickly for a fast flyback. With a frequency of 70 kHz for the oscillations in the horizontal output circuit, the period of a half-cycle is $1/0.14$ μs, resulting in a flyback time of approximately 7 μs. This half-cycle is from maximum positive i to maximum negative i for the waveforms in Fig. 22-10.

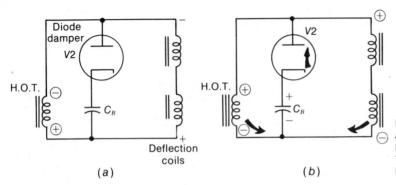

FIGURE 22-9 DAMPER DIODE OFF AND ON. (a) PLATE NEGATIVE TO PREVENT CONDUCTION DURING RETRACE. (b) PLATE POSITIVE FOR CONDUCTION AT START OF TRACE.

The voltage across the oscillating circuit is 90° out of phase with the current, as shown in Fig. 22-10. Therefore, the polarity of damper plate voltage reverses to become positive just as the current starts to decrease from its maximum negative value. This is the time for the start of the trace. Now the damper conducts because of positive plate voltage.

Conduction in the diode is illustrated in Fig. 22-9b. Note that C_B is charged by the damper current at this time. As the damped current declines to zero, the electron beam is deflected from the left side of the raster toward the center. Finally, the voltage across C_B biases the diode damper out of conduction. With zero current in the scanning coils, the electron beam would be undeflected at the center. Before the damped current declines to zero, however, the output tube starts conducting. Its linear rise of current then completes the trace to the right side of the raster.

Figure 22-11 illustrates how the current supplied by the deflection amplifier and the damped current in the output circuit combine. The result is sawtooth current in the deflection coils for the full horizontal trace from left to right across the scanning raster. Note that both components of the scanning current for the trace deflect the electron beam toward the right. The decrease in damped current of negative polarity varies in the same direction as the positive rise of current produced by the output tube.

In summary, the horizontal scanning occurs in three steps:

1. The output stage supplies current for the horizontal scanning coils to deflect the electron beam to the right edge of the raster.
2. The undamped oscillation in the horizontal output circuit, while the output stage and damper are not conducting, produces the

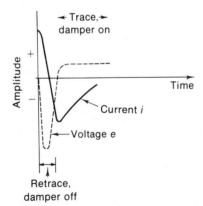

FIGURE 22-10 WAVEFORMS OF VOLTAGE e AND CURRENT i FOR HORIZONTAL DEFLECTION COILS WHILE OUTPUT TUBE IS CUT OFF.

rapid reversal of scanning current required for the fast flyback from right to left.

3. Immediately after the flyback, the damper conducts while the output stage is cut off, to provide the reaction scanning current for the trace at the left side.

The damper tube must withstand the high negative voltage pulse at the plate, have good cathode-to-heater insulation because of the B+ boost voltage at the cathode, and be able to conduct the peak values of scanning current. Typical maximum ratings for the 6CL3 half-wave rectifier as a diode damper are 5.5-kV peak inverse plate voltage, 1,300-mA peak plate current, −900-V average cathode-heater voltage, and 8.5-W plate dissipation.

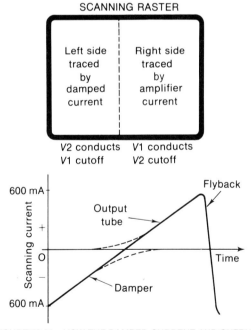

FIGURE 22-11 HOW THE DAMPED CURRENT AND OUTPUT CURRENT COMBINE TO PRODUCE HORIZONTAL SCANNING IN THE RASTER.

22-5 Boosted B+ Voltage

The damper diode is a half-wave rectifier for the ac deflection voltage produced during trace time in the horizontal output circuit. Its dc output voltage then is added to the B+ supply voltage. The sum of the two voltages is boosted B+.

Damper in secondary. In Fig. 22-12a, 270 V is applied to the low end of L_S, while the other side is connected to the damper plate. Therefore, when the damper conducts, it charges C_B in the cathode circuit to the B+ voltage. The path of charging current is through L_S and from cathode to plate in the damper, charging C_B to 270 V with the cathode side positive.

In addition, however, the ac deflection voltage e_{LS}, which produces sawtooth current in the scanning coils, makes the diode plate 350 V more positive during horizontal trace time. As a result, C_B charges to a voltage about 330 V higher than 270 V, or to 600 V. The path of charging current is the same as when the B supply charges C_B. In effect, the ac deflection voltage is in series with the B supply voltage and the damper tube to produce the boosted B+ voltage across C_B.

The damper circuit in Fig. 22-12a is for an output transformer with an isolated secondary, which can have the polarity of deflection shown with negative flyback pulses. In an autotransformer, however, any part of the winding must have deflection voltage with positive flyback pulses. This is the same polarity as the primary winding. Then the damper circuit with an inverted diode, shown in Fig. 22-12b, is used.

Damper in primary. In Fig. 22-12b, the diode is inverted. Then the ac input voltage is applied to the cathode. This connection is necessary because the deflection voltage is negative in the

488 BASIC TELEVISION

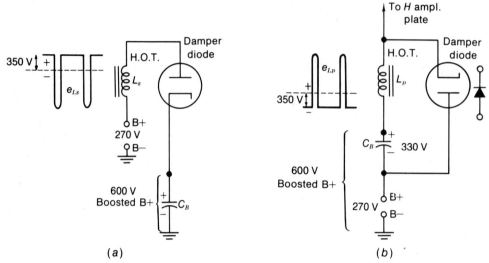

FIGURE 22-12 PRODUCING THE BOOSTED B+ VOLTAGE. EITHER A TUBE OR A SILICON DIODE CAN BE USED. (a) DAMPER IN SECONDARY TO RECTIFY POSITIVE POLARITY OF THE AC DEFLECTION VOLTAGE. (b) INVERTED DIODE IN PRIMARY TO RECTIFY NEGATIVE POLARITY OF THE AC DEFLECTION VOLTAGE.

primary during trace time. The negative polarity of ac deflection voltage at the cathode produces plate current in the diode to charge C_B. Note that the rectifier plate current still charges C_B with its cathode side positive. C_B is in series with the B supply voltage. The potential difference between the positive side of C_B and chassis ground, therefore, is 600 V. Note that in both a and b of Fig. 22-12, the damper plate or anode must have a dc connection to the B+ supply.

The damper in the primary is the circuit generally used. Tubes now have the voltage rating to prevent arcing between cathode and heater, while semiconductor diodes do not have a heater.

Uses for boosted B+. This voltage is commonly used for the horizontal output tube, horizontal oscillator, vertical oscillator, and vertical amplifier. The higher plate supply voltage allows more power output with improved linearity. In addition, the screen-grid supply of the picture tube is generally boosted B+ voltage. It is important to note that the B+ boost circuit must be operating in order to produce normal plate voltage for all these stages.

A question often asked is, "How can the horizontal amplifier operate when the receiver is first turned on, before boosted B+ voltage is available across C_B?" In effect, this is a "bootstrap circuit," where the output stage increases its own plate supply voltage. Notice in Fig. 22-12 that for the circuits in both a and b the amplifier plate circuit can return to B+ through the cathode-to-plate path in the conducting damper tube. Therefore, the horizontal output tube starts conducting with just B+ voltage for the plate supply. After a few cycles of operation, boosted B+ voltage is available across C_B.

22-6 Flyback High Voltage

The voltage pulse produced in the horizontal output circuit during the first half-cycle of oscillations for flyback provides the dc high voltage required for the CRT anode. As illustrated in Fig. 22-13, operation of the flyback high-voltage supply can be summarized as follows:

1. The flyback pulse has positive polarity at the plate of the horizontal amplifier in the primary of the output transformer, as shown by waveshape 1. Both the damper and amplifier tubes are cut off for the retrace. The first half-cycle of undamped oscillations produces the high-voltage flyback pulse. A typical value is 6,000 V at the amplifier plate.
2. The high-voltage winding L_s steps up the flyback pulse for the plate of the high-voltage rectifier V2, as shown by waveshape 2. This amplitude is 18 kV because of the 3:1 voltage stepup.
3. With the stepped-up ac deflection voltage at its plate, V2 conducts to produce positive dc output voltage at the cathode. This voltage is filtered to provide the steady dc output shown by waveshape 3. The dc output voltage is approximately equal to the peak positive ac input of 18 kV.

When the high-voltage rectifier conducts, current can flow from cathode to plate in V2, through L_s and L_p in the output transformer, discharging C_B slightly and charging the filter C_1. This is usually the anode capacitance of the picture tube. Since the ripple frequency is 15,750 Hz for the half-wave rectifier, 2,000 pF is enough capacitance to provide the required fil-

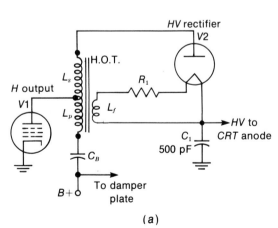

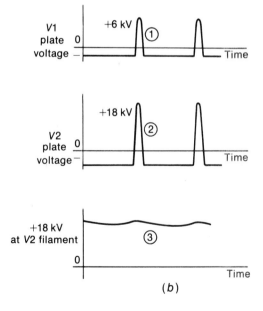

FIGURE 22-13 PRODUCING THE FLYBACK HIGH VOLTAGE. EITHER A TUBE OR A SILICON DIODE CAN BE USED FOR THE RECTIFIER. (a) TYPICAL CIRCUIT; (b) VOLTAGE WAVESHAPES WITH NUMBERS FOR THE STEPS EXPLAINED IN TEXT.

tering. R_2 is a voltage-dropping resistor for the heater of V3. Its heater power is supplied by L_f, which is a single-turn loop on the transformer core.

The amount of high voltage is maximum when the brightness is at a minimum. The reason is that there is less load current on the high-voltage rectifier. It is important to note that with a flyback high-voltage supply, the dc output for anode voltage cannot be produced without the ac deflection voltage input to the plate of the high-voltage rectifier. Therefore, the horizontal deflection circuits must be operating to have brightness on the screen of the picture tube.

22-7 Horizontal Deflection Controls

The only adjustments usually are for width and linearity. However, even these may be omitted in some receivers. The reason is that incorrect adjustments can cause too much average dc plate current in the horizontal output stage.

Width control. Varying the screen-grid voltage of the output tube, as in Fig. 22-2, is a common method. Instead of the variable resistance, though, variable taps on a voltage divider can be used. Another method uses a width coil in parallel with part of the secondary winding of the output transformer. A typical coil is shown in Fig. 22-14.

Efficiency coil. L_1 in Fig. 22-15 forms a parallel resonant circuit with C_B and C_1, tuned to 15,750 Hz approximately. Adjusting L_1 shifts the phase of the 15,750-Hz ripple in B+ boost voltage for the horizontal amplifier. The parallel resonance allows minimum average dc plate current in the output tube. This is the condition for maximum efficiency. The efficiency coil L_1 is generally adjusted for minimum cathode current on the hori-

FIGURE 22-14 COIL FOR WIDTH OR LINEARITY CONTROL. L IS 2 TO 18 mH. R IS 14 Ω.

zontal output tube. Some receivers provide connections for inserting a dc milliammeter in the cathode circuit.

In addition, L_1 affects the voltage at the transition period when the damper finishes conduction and the amplifier starts. Therefore, L_1 adjusts linearity at the center of the raster. The main requirement, however, is minimum cathode current in the output tube for maximum efficiency.

Horizontal linearity. Three examples of nonlinear horizontal scanning are illustrated in Fig.

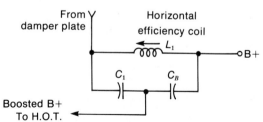

FIGURE 22-15 CIRCUIT FOR HORIZONTAL EFFICIENCY COIL WITH L OF 1 TO 8 mH.

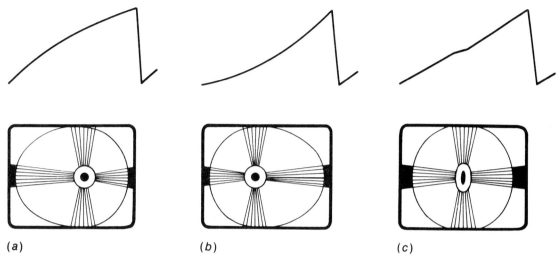

FIGURE 22-16 NONLINEAR HORIZONTAL SCANNING. *(a)* STRETCHING AT LEFT AND CROWDING AT RIGHT; *(b)* REVERSE EFFECT; *(c)* CROWDING AT CENTER.

22-16. To relate the current waveshapes to the scanning, remember that:

1. Damped current produces the left side of the trace.
2. The beam is at the center with zero current, when the damped current approaches zero and the output stage starts to conduct.
3. Plate current of the output stage completes the trace at the right side.

How fast the damped current decays to zero depends on the damper tube and B+ boost capacitor. The time when the output tube starts to conduct depends on its grid-leak bias and plate voltage. Conduction is delayed until the driving voltage is positive enough to make the instantaneous grid voltage less than cutoff. How far the beam is deflected to the right edge depends on the peak plate current in the output stage.

In Fig. 22-16*a*, the picture information is stretched at the left and crowded at the right. With people in the picture, they look too broad at the left or too thin at the right. This type of nonlinearity can be caused by too much horizontal drive and too little inductance in the width coil. A weak horizontal output tube produces the same nonlinearity, but with reduced width. The reverse happens for the nonlinear scanning in Fig. 22-16*b*, where the left side is crowded and the right side stretched. This effect can be caused by a weak damper tube or too much load current for C_B. With either type of nonlinear horizontal scanning, a person at the center of the screen appears to have one shoulder broader than the other.

Figure 22-16*c* illustrates crowding at the center. This can be caused by damper current decaying to zero before the output tube starts to conduct.

22-8 Deflection Yokes

Since the current in the scanning coils deflects the electron beam, the yoke is rated in terms of

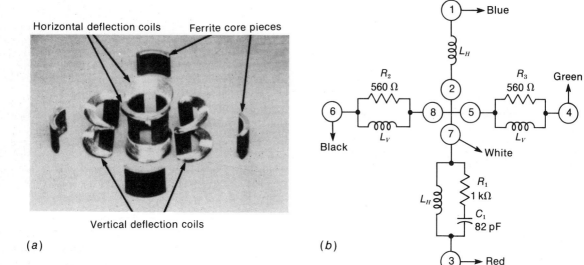

FIGURE 22-17 A 90° YOKE FOR MONOCHROME PICTUE TUBE. DIAMETER, 3 IN. (a) CUTAWAY VIEW; (b) WIRING DIAGRAM.

deflection angle. Typical values are 70°, 90 to 92°, and 110 to 114°.

Monochrome yoke. See Fig. 22-17. Typical values for a 90° yoke may be L_H of 12 to 30 mH with R of 25 Ω in the horizontal scanning coils. For the vertical coils, L_V is about 42 mH with R of 40 Ω. It should be noted that a 110° yoke would be about 1 in. shorter, with a smaller opening for the narrow neck on the picture tube. The construction in Fig. 22-17a shows why this type is called a *saddle yoke*.

In the wiring diagram in Fig. 22-17b, R_2 and R_3 are damping resistors in parallel with the vertical coils. R_1 and C_1 form an antiringing network across one horizontal coil. Without the proper values for balancing capacitances, the horizontal ringing produces thin, white vertical lines at the left side of the raster. All these components are inside the yoke housing.

Color yoke. See Fig. 22-18. This is a 90° yoke for color picture tubes. The characteristics are L_H of 12 mH with R of 9 Ω for the horizontal coils, and L_V of 24 mH with R of 13 Ω for the vertical coils. The yoke diameter is 6½ in., length 3¾ in., and weight 3¾ lb. Although it is larger than the yoke in Fig. 22-17, it also uses the saddle construction for the deflection coils.

Toroidal yoke. A toroid is an inductance wound in a closed loop to concentrate the magnetic field. It can be constructed for either a 90° or a 110° deflection angle. The 90° yoke in Fig. 22-19 is made to mount directly on the neck of the color picture tube. This yoke has an extra winding for dynamic convergence. The horizontal coils have L_H of 1 mH series-connected, but they can be in parallel for L_H of 0.5 mH. The vertical coils have L_V of 2 mH in series. The dc resistance of the coils is only 1 to 2 Ω. The relatively low impedance, compared with saddle yokes, makes the toroidal yokes suitable for deflection circuits using transistors or thyristors.

HORIZONTAL DEFLECTION CIRCUITS 493

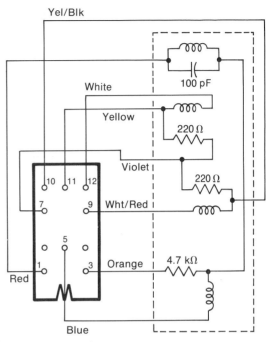

FIGURE 22-18 A 90° YOKE FOR COLOR PICTURE TUBE. (a) PHOTO; DIAMETER IS 6½ IN.; (b) WIRING DIAGRAM. (TRIAD-UTRAD).

FIGURE 22-19 TOROIDAL YOKE, FOR 90° DEFLECTION ANGLE. DIAMETER IS 4 IN. (GTE SYLVANIA INC.)

22-9 Horizontal Output Transformers

Typical units for the H.O.T. or flyback transformer are shown in Figs. 22-20 and 22-21.

Autotransformer windings. In Fig. 22-20a, the total winding between 1 and 5 is the primary L_P for plate current of the amplifier. This includes L_2, L_1, and L_S. The tapped winding between 3 and 1 is the secondary L_S to step down the voltage for the deflection coils. Note that the turns ratio is between all the turns in L_P and the number of turns tapped for L_S.

Since there is no isolated secondary winding for polarity inversion, all the taps on an autotransformer have voltage of the same polarity as the plate voltage on the amplifier. In Fig. 22-20a, for instance, the voltage at terminal

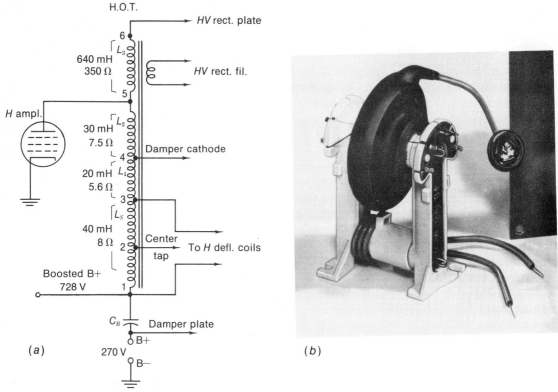

FIGURE 22-20 FLYBACK TRANSFORMER FOR MONOCHROME RECEIVER. (a) SCHEMATIC OF AUTOTRANSFORMER; (b) PHOTO. (TRIAD-UTRAD)

3, 4, or 5 has a rectangular waveshape with positive flyback pulses. The drop in plate current for retrace induces voltage of positive polarity across the inductance.

The positive flyback pulses require that the damper diode be inverted. These connections are shown in Fig. 22-12b for flyback pulses to drive the damper cathode positive. This corresponds to negative anode voltage to prevent the damper from conducting during flyback.

With an autotransformer, the damper can be tapped up on the winding for a higher boosted B+ voltage. As shown in Fig. 22-20a, the tap at 4 for the damper has more deflection voltage than the tap at 3 for the scanning coils. As a result, the damper charges C_B to a higher voltage. Some receivers have an extra diode to rectify more deflection voltage that is added in series with C_B. This output is called *boosted boost voltage*. It can be as high as 1,100 V.

The winding L_3 at the top of Fig. 22-20a steps up the primary voltage to supply ac input to the high-voltage rectifier. For a tube, its filament power is also taken from the output transformer. Positive flyback pulses of about 6 kV at the amplifier plate can be stepped up to 18 kV. The amount of stepup is limited by the fact that increased L and stray C reduce the resonant frequency of the output circuit. L_3 is a separate winding with many turns of fine wire, which is the reason for its relatively high R of 350 to 790 Ω. This winding is in series with L_P on the autotransformer. For color receivers, the high-voltage supply generally includes a focus rectifier also, as shown in Fig. 22-21.

Auxiliary windings. Although they are not shown in Fig. 22-20, the horizontal output transformer usually has connections that provide flyback pulses for auxiliary functions in the receiver. These connections may be taps on the autotransformer or isolated windings. Typical functions include keying pulses for the AGC stage, sample of the scanning frequency for horizontal AFC, and horizontal retrace blanking for the video amplifier or picture tube. Color receivers also need the horizontal flyback pulses to key on the burst separator. In addition, the horizontal output is used for pincushion correction and dynamic convergence circuits.

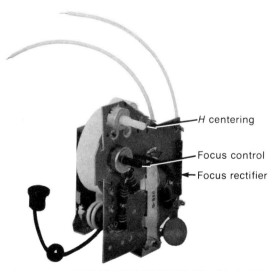

FIGURE 22-21 FLYBACK TRANSFORMER FOR COLOR RECEIVER. NOTE NUMBERED TERMINALS FOR TAPS. *(TRIAD-UTRAD)*

22-10 Analysis of Horizontal Output Circuit

The inductance values in Fig. 22-20a can be used for some approximate calculations to illustrate more details of how the horizontal output circuit functions. Unity coupling between windings is assumed because of the powdered-iron or ferrite core. Also, the coil resistances can be neglected, as the induced voltages are much greater than the IR drops.

In analyzing the inductance values, we can start with the secondary winding L_S for the deflection coils, as they require a specified amount of current for full deflection. The secondary inductance here is made 40 mH to be twice the total inductance of 20-mH deflection coils connected across L_S. Higher values of L_S allow more inductance in the primary for a given turns ratio. However, too much inductance can make the flyback time too long. The value of 20 mH, then, is suitable for L_S. In the primary, the inductance L_P for the autotransformer includes all the turns of L_2, L_1, and L_S. The total primary inductance L_P then is 90 mH, equal to 30 mH for L_2 plus 20 mH for L_1 plus 40 mH for L_s.

The turns ratio for voltage stepup to L_P from L_S is equal to the square root of their inductance ratio, or $\sqrt{90/40}$. The square root is based on the fact that inductance of a coil increases as the square of its turns. This ratio of $\sqrt{90/40}$ equals $\sqrt{2.25}$, which is 1.5 for the turns ratio of L_P to L_S. With one and one-half the number of turns for L_P, its voltage is 50 percent more than the voltage across L_S.

In order to calculate the induced voltages we can assume a flyback of 8 μs. This time corresponds to one-half cycle of undamped oscillations at 62.5 kHz. Trace time equals 8 μs subtracted from the line period of 63.5 μs. Then the remainder of time for trace is 55.5 μs.

Let the peak-to-peak yoke current be 1,200 mA in the 20-mH horizontal deflection coils. This value is derived from 400-mA peak plate current in the output tube stepped up to 600 mA in the yoke by the turns ratio. The opposite peak of 600 mA is a result of oscillation in the output circuit after the amplifier cuts off.

During trace, the rate of change of current in the yoke is 1,200 mA during 55.5 μs. However, for flyback the rate of change is 1,200 mA in 8 μs. These values are necessary to calculate the induced voltages as equal to $L\,di/dt$.

The voltages induced by the rise of current during trace are negative with respect to chassis ground at all terminals on the autotransformer. During flyback, the fast drop of current induces positive voltages of much higher amplitude. These voltages are shown in Fig. 22-22.

Induced voltages during retrace. The flyback voltage induced across the yoke coils by a 1,200-mA change in 8 μs is

$$+e_y = L\frac{di}{dt}$$
$$= 20 \times 10^{-3} \times \frac{1,200 \times 10^{-3}}{8 \times 10^{-6}}$$
$$+e_y = 3,000 \text{ V}$$

This voltage is also across L_S. Since the turns ratio of L_P to L_S is 1.5:1, the voltage across L_P is stepped up by 1.5 to be 4,500 V at the plate of the output tube.

The high-voltage winding L_3 develops $4 \times 3,000 = 12,000$ V during flyback. This voltage is four times the secondary voltage because L_3 is sixteen times greater than L_S. The 12,000 V across L_3 is in series with the 4,500-V flyback pulse across L_P to provide 16,500-V peak positive ac input to the plate of the high-voltage rectifier.

Induced voltages during trace. These are much smaller because of the slower rate of change of current. Also, the increasing current

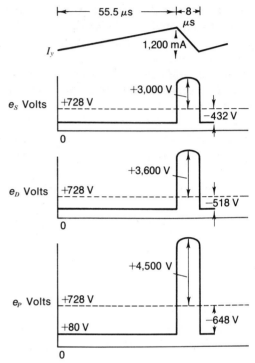

FIGURE 22-22 INDUCED VOLTAGES PRODUCED BY YOKE CURRENT, CALCULATED FOR THE TRANSFORMER IN FIG. 22-20a.

induces negative voltage. The amount of voltage induced across the yoke coils by a 1,200-mA change in 55.5 μs is

$$-e_y = L\frac{di}{dt}$$
$$= 20 \times 10^{-3} \times \frac{1,200 \times 10^{-3}}{55.5 \times 10^{-6}}$$
$$-e_y = 432 \text{ V approx}$$

This voltage is also across L_S during trace time. Across L_P the voltage is stepped up by 1.5, for -648 V. The net plate voltage is still positive, however, because of the boosted B+ supply of 728 V at terminal 1.

Notice that the damper cathode at terminal 4 is tapped higher than L_S. This connection results in more boosted B+ voltage, as more deflection voltage is applied to the damper for rectification. The combined turns of L_1 and L_S provide a voltage stepup of approximately 1.2 with respect to L_S. Therefore, the damper voltage e_D is 20 percent more than e_y. Then e_D equals 518 V. This much voltage is available to drive the damper cathode negative for diode current to charge C_B.

Note that the 518 V across C_B is for the boost only. In addition, C_B is in series with the B+ of 270 V. Therefore, the boosted voltage at terminal 1 with respect to chassis ground would be 270 + 518 = 788 V. With a typical load of 10 mA average current on the boost supply, however, its dc voltage is a little less than the peak ac input. In this example, the boost voltage alone is 458 V. This value added to the B+ of 270 V provides the 728-V average axis shown in Fig. 22-20. Then 728 V is the amount of boosted B+ voltage available from terminal 1 at the low end of the autotransformer.

22-11 Complete Horizontal Deflection Circuit

As shown in Fig. 22-23 for a monochrome receiver with tubes, the stages included are the horizontal oscillator with its AFC, the output amplifier, the damper, and the high-voltage rectifier. A transistorized circuit is shown in Fig. 22-28. A horizontal output circuit using silicon-controlled rectifiers is shown in Fig. 22-29. Integrated circuit (IC) units are not used for deflection because of the high power requirements.

In Fig. 22-23, the horizontal oscillator V1 is a cathode-coupled multivibrator driving the deflection amplifier V2 with autotransformer coupling to the horizontal deflection coils in the yoke. The damper V3 is an inverted diode rectifier producing boosted B+ voltage of 660 V. The high-voltage rectifier V4 supplies 18 kV for the picture tube, with its anode capacitance as the high-voltage filter. Note that the octal yoke plug has a jumper between pins 1 and 8, which closes the B− return to chassis ground for operation of the power supply. Removing the yoke plug makes the entire receiver dead, therefore, because the B-supply circuit is open.

The oscillator is locked in sync by dc control voltage from a sync discriminator connected to grid pin 2. In the plate circuit, R_{454} is the load resistor in series with the sine-wave stabilizing circuit consisting of L_{401} and C_{452}. The voltage peak across the common-cathode resistor R_{457} cuts off this triode section. Plate voltage output is coupled by C_{455} to the opposite triode section, which serves as a discharge tube. Its grid resistance is varied by the hold control to adjust the free frequency of the oscillator. Plate pin 6 connects to B+ through R_{455}. The sawtooth capacitor is C_{457}. Note the peaking resistor R_{459} to provide trapezoidal grid voltage for driving the output tube. The peaking resistor can be on either side of the sawtooth capacitor, as they are in series. C_{456} is the coupling capacitor, with grid resistor R_{460}. The series grid resistor R_{464} limits the peak grid current charging the coupling capacitor for grid-leak bias, in order to reduce radiation of horizontal pulses.

The grid-leak bias of −40 V on the output tube is produced by drive from the horizontal oscillator. There is no cathode bias. Screen-grid voltage is obtained from the B+ line, with R_{465} the screen-dropping resistor and C_{460} its bypass. The boosted B+ of 660 V is used for the plate of the output tube, the vertical deflection circuit, through the isolating resistor R_{467}, and for the screen grid of the picture tube.

In the plate circuit of the output tube, the top winding of the autotransformer T_{403} supplies

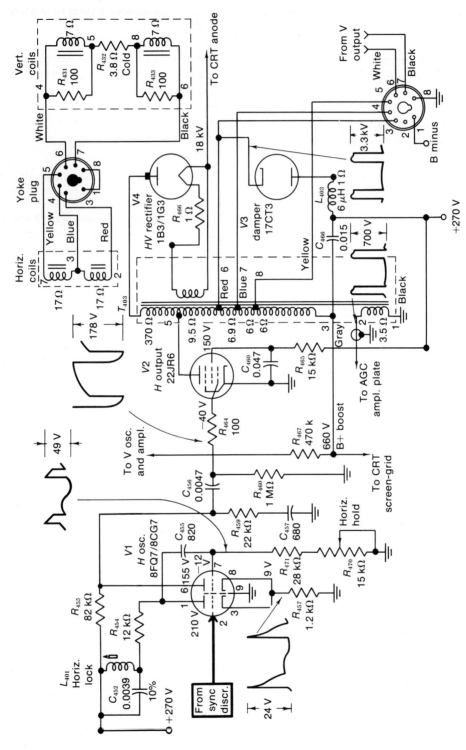

FIGURE 22-23 TYPICAL HORIZONTAL DEFLECTION CIRCUIT USING TUBES, FOR MONOCHROME RECEIVER. C VALUES MORE THAN 1 IN pF. NOTE YOKE PLUG AT TOP RIGHT AND ITS SOCKET CONNECTIONS AT BOTTOM.

positive flyback pulses for the plate of the high-voltage rectifier. The primary winding for the amplifier plate consists of all turns between terminals 5 and 3. Note that terminal 3 at the low end is the source of boosted B+ voltage of 660 V. C_{466} is the boost capacitor. It is in series with terminal 3 and B+ connected to the damper plate. The damper cathode is tapped at terminal 6. This negative voltage during trace makes the damper diode conduct to charge C_{466}. The 6-μH choke L_{403} prevents radiation of horizontal pulses by the peak damper current.

Terminals 6, 7, and 8 on the autotransformer connect to the horizontal coils in the yoke, through pins 3, 4, and 5 on the yoke plug. Pin 4 is a center-tap connection. In this circuit, the ac deflection voltage for boost is higher than the yoke voltage by the amount of voltage across terminals 8 and 3 on the transformer. Finally, the extra winding at the bottom of T_{403} supplies positive flyback pulses for the plate of the keyed AGC stage. A separate winding is necessary so that one end can be grounded.

All waveshapes in the horizontal circuits are obtained with the oscilloscope internal sweep at 15,750/2 Hz. Although not shown here, the positive flyback pulses at the plate of the output tube can be observed by clipping the oscilloscope input cable to the insulation of the wire to the plate.

22-12 Protective Bias on Horizontal Output Tube

When the horizontal oscillator stops for any reason, there is no drive at the grid of the output tube. Its grid bias then drops to zero. Assuming no cathode bias, the tube conducts maximum plate current. The plate usually turns cherry red. If this continues, the output tube will be burned out. To prevent this problem, the circuit in Fig. 22-24 provides protective bias on the horizontal output tube when the drive fails.

Normally, the drive will produce about -50 to -70 V grid-leak bias. This voltage makes A at the D2 anode more negative than B at its cathode. Then D2 is reverse-biased and cannot conduct. D2 locks out the protective bias, therefore, as long as the output tube has normal drive voltage from the oscillator.

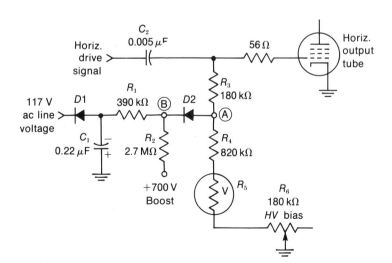

FIGURE 22-24 PROTECTIVE BIAS CIRCUIT FOR HORIZONTAL OUTPUT TUBE.

In addition, the ac line voltage is rectified by $D1$ to charge C_1 to approximately -160 V. This rectified voltage is negative at the $D1$ anode. The -160 V is divided by R_1 and R_2, which return to $+700$ V. As a result, point B at the $D2$ cathode is about -45 V. Then $D2$ is reverse-biased because its cathode voltage from $D1$ is less negative than the negative anode voltage from the grid-leak bias.

When the horizontal drive signal is lost, however, there is no grid-leak bias to make the $D2$ anode negative. $D2$ then can conduct. In effect, the conducting diode connects the negative voltage at B to the grid of the output tube. As a result, the plate current is limited by the protective bias.

The grid will receive more than the normal -45 V at point B because the 700 V from the boost supply drops. When horizontal drive is lost, boost voltage is no longer developed by the damper. Without the positive boost voltage, point B goes even more negative and the output tube's plate current is actually cut off.

In troubleshooting the horizontal circuits, it may be necessary to disable the protective bias circuit. The easiest way to do this is to short point B to ground temporarily with a clip lead. However, you then have to be careful not to let the output tube be damaged by the absence of horizontal drive. If the plate turns cherry red, shut off the power.

22-13 Transistorized Horizontal Deflection

When a transistor is used for the horizontal amplifier, its function is equivalent to that of an on-off switch. When the stage conducts, its very low resistance makes the horizontal output circuit essentially a pure inductance. For sawtooth current the only requirement is a steady applied voltage that is periodically cut off.

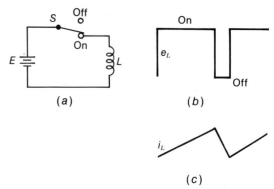

FIGURE 22-25 PRODUCING SAWTOOTH CURRENT IN INDUCTANCE L. (a) BATTERY E WITH SWITCH S; (b) EQUIVALENT RECTANGULAR VOLTAGE; (c) SAWTOOTH CURRENT.

Sawtooth current in L. In Fig. 22-25a the battery supplies the steady voltage E through the closed switch S to the inductance L. Then S is opened to disconnect E. The equivalent result in Fig. 22-25b is the rectangular voltage e_L applied to the inductance. While E is applied, the current i_L in the inductance rises as shown in Fig. 22-25c. The current cannot rise instantly with E because inductance produces a self-induced voltage that opposes the change in i_L. For the retrace part of the sawtooth, i_L drops when e_L is turned off. The trace part of the sawtooth is produced while the output stage is conducting, or on. The retrace part occurs while the output stage is cut off.

Power transistors. The required power rating for the horizontal output stage is 25 to 50 W. Its resistance when conducting maximum current is 1 Ω or less.

The horizontal output transistor is generally cut off for a longer period than retrace time, for better efficiency. The damper diode produces part of the trace at the left. Actually, the cutoff time of the output stage is critical. The reason is that a small increase in conduction

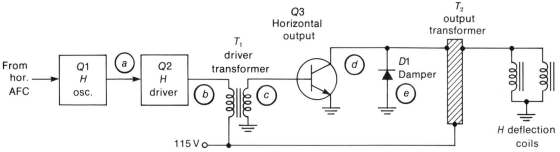

FIGURE 22-26 BLOCK DIAGRAM OF HORIZONTAL DEFLECTION CIRCUIT WITH TRANSISTORS. LETTERS IN BALLOONS ARE FOR WAVESHAPES IN FIG. 22-27.

angle can raise the average I_C to a level exceeding the power rating for collector dissipation.

Typical horizontal circuit with transistors. The block diagram in Fig. 22-26 illustrates the oscillator, driver, and output stages. The driver Q2 has several functions. It is a buffer or isolation stage to prevent the output circuit from changing the oscillator frequency. With the interstage transformer T_1, the voltage drive is stepped down to match the low impedance of the power amplifier Q3. In addition, the Q2 circuit can be used to shape the driving voltage for the output stage. It is often necessary to stretch the cutoff time of Q3 for maximum efficiency. In the output circuit, note that the deflection coils are in parallel, for a lower impedance than with series connections.

Transistor waveshapes for horizontal deflection. The waveforms in Fig. 22-27 match the numbered balloons in Fig. 22-26. The oscillator output voltage is shown in a. Note that the oscillator pulse is for flyback time, which eventually cuts off the output stage. This oscillator pulse is amplified and inverted by the driver. The stretching of the flyback pulse is accomplished by RC networks in the driver circuit and by the interstage transformer T_1.

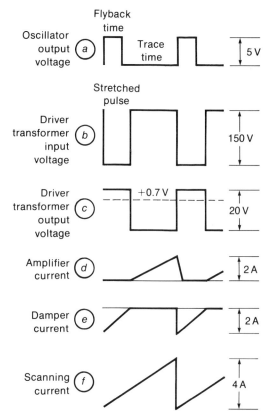

FIGURE 22-27 WAVESHAPES FOR HORIZONTAL DEFLECTION CIRCUIT IN FIG. 22-26.

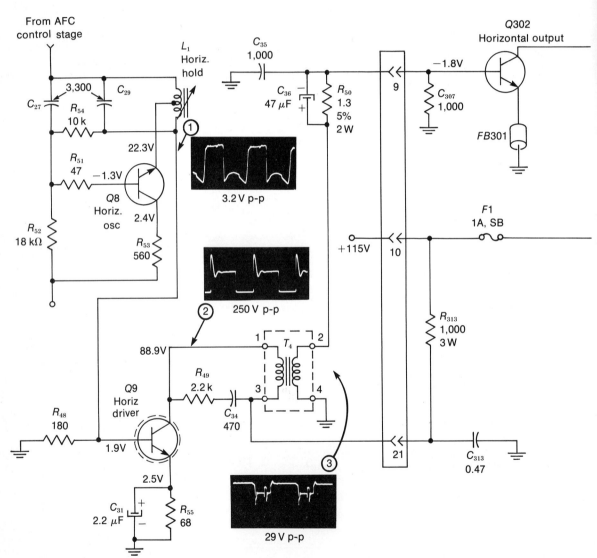

FIGURE 22-28 TRANSISTORIZED HORIZONTAL DEFLECTION CIRCUIT FOR COLOR RECEIVER. R IN OHMS, C VALUES MORE THAN 1 IN pF. *(FROM MAGNAVOX CHASSIS T979)*

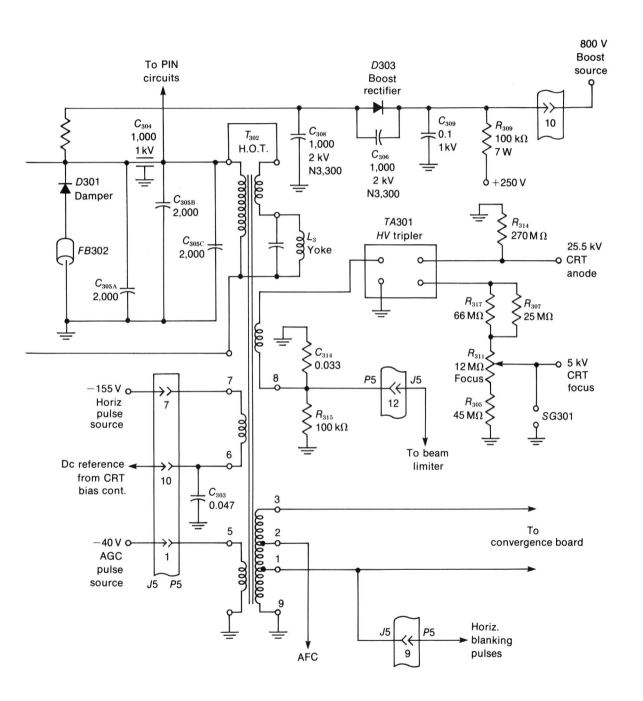

The primary voltage in b is stepped down so that the voltage waveform in c can drive the base of the output transistor. The amplifier current in d shows that Q3 is cut off for retrace plus one-quarter of the trace at the left side of the raster. The damper D1 fills in the missing component for trace, as shown in e. The combined result in f is sawtooth current in the yoke for the complete trace from left to right.

Typical circuit. Refer to Fig. 22-28. The horizontal oscillator Q8 supplies pulses that are amplified by the driver Q9, transformer-coupled to the amplifier Q302. These stages correspond to Q1, Q2, and Q3 in Fig. 22-26. Note the oscilloscope waveshapes shown at 1 for oscillator output, at 2 for driver output, and at 3 for drive to the base of the output stage.

Q8 is a sine-wave oscillator using the Hartley circuit with a tapped coil in the emitter leg. The oscillator output is directly coupled to the base of driver Q9. In its collector circuit, the driver transformer T_4 couples the pulses to the base of the output transistor Q302. Finally, Q302 is coupled to the yoke coils by the horizontal output transformer T_{302}.

The diode D301 is connected across Q302 in the polarity that allows damper current for the left side of trace when the output transistor Q302 is still cut off. Ferrite beads FB301 and FB302 are equivalent to small rf chokes to suppress parasitic oscillations.

The circuit in Fig. 22-28 is from a color receiver. Therefore, another function of the horizontal output is to provide 5-kV focus voltage, in addition to the 25.5-kV anode supply. Also, correction voltages are coupled to the convergence board and the pincushion circuits. Finally, the horizontal flyback pulses are used for blanking, horizontal AFC, and the AGC circuits.

22-14 SCR Horizontal Output Circuit

A silicon-controlled rectifier (SCR) has anode, cathode, and a gate electrode. Positive voltage at the gate turns on the anode current, with positive anode voltage. The SCR then is a closed switch. However, the gate cannot turn off the anode current. Only when the anode voltage drops too low does the SCR open. Then gate voltage can turn on the SCR again.

In Fig. 22-29 two SCR units are used for the horizontal output circuit. SCR_T conducts for the trace and SCR_C for the retrace. Switching on the retrace is called the *commutating* operation.

Each SCR has a damper diode with inverse parallel connections. The purpose is to allow current in both directions from cathode to anode in either the SCR or its diode, depending on which anode is positive. When the diode conducts, its low resistance drops the SCR anode voltage to the cutoff level.

During horizontal trace time, SCR_T is on and SCR_C is off. Then the pulses from the horizontal oscillator at 15,750 Hz turn on the retrace SCR. The conduction in SCR_C then cuts off the trace SCR.

For troubleshooting, note that the retrace section has a dc connection to the 160-V supply through T_1 and the circuit breaker. It will trip open with a short in SCR_C or D_C. The trace section has no dc connection to the B+ voltage. A short in SCR_T or D_T, therefore, will not open the circuit breaker. In both cases, though, the result will be no horizontal output and no brightness.

The test points TP1 and TP2 at each SCR anode can be used to check the resistance of the SCR with its reverse diode. Normal resistance is 1,000 Ω or higher. The positive ohmmeter lead is at the test point. A low resistance means either the diode or the SCR is shorted.

Since the SCR needs an inverse diode for this circuit, units are made with both on a common chip. The result is an *integrated thyristor and rectifier (ITR)*.

22-15 High-Voltage Limit Control

One of the special safety features that must be incorporated in color TV receivers is a means of preventing danger from excessive x-radiation. Manufacturers are required to incorporate some circuit that will make the picture unviewable if the high voltage exceeds specifications.

Several different methods have been employed to meet this requirement. When the high voltage exceeds a predetermined level, some sets will blank the video, while others will cause a loss of sync. In the example of Fig. 22-30, the SCR limit switch disables the horizontal sweep, causing a complete loss of high voltage and no brightness. Not only does this make the picture unviewable, it removes the cause of the potential radiation.

In Fig. 22-30, a positive pulse from the horizontal output transformer is rectified by CR210 to charge C_{246}. During the time the pulse is applied, CR210 is biased on and C_{246} charges rapidly through the low resistance of the diode. Between pulses, C_{246} discharges slowly through the high resistance of R_{331}, equal to 1.8 MΩ. With normal high-voltage levels, the average charge on C_{246} is not high enough to ionize the neon bulb PL1.

If the high voltage exceeds a preset value, however, the increased flyback pulse will charge C_{246} to a higher voltage. Then the neon bulb is ionized on. This causes a pulse of current to flow through R_{329} and R_{330}. The voltage drop across R_{329} makes the gate of Q209 positive. Since the SCR already has a positive voltage applied to its anode from the horizontal module, the positive gate voltage will cause the SCR to conduct. Then it short-circuits the horizontal sweep circuit, removing the high voltage completely. The SCR serves as a limit switch, therefore, normally open but closed to disable the horizontal scanning if there is too much high voltage.

Sometimes an arc in the picture tube or elsewhere in the chassis will trigger the SCR. In that case, it is only necessary to turn the set off for 10 s to discharge C_{246} and reset the limit switch.

22-16 Troubles in the Horizontal Deflection Circuits

These circuits cause troubles in the width of the raster. In addition, there is no high voltage and no brightness without horizontal flyback pulses for the high-voltage supply. The output stage, flyback transformer, and damper are common sources of trouble because of the high currents and voltages.

Insufficient width. Too much black space at the sides of the screen means that the raster is narrow because of insufficient horizontal output (Fig. 22-31). With tubes, this trouble is often a result of a weak horizontal amplifier or damper. The oscillator is needed for horizontal drive, but this stage has little effect on width.

It should be noted that the horizontal output circuit is the biggest load on the B+ supply. The first symptom of low supply voltage, therefore, is usually reduced width on the screen.

White vertical bars. Bright vertical bars at the left side of the raster are produced by oscilla-

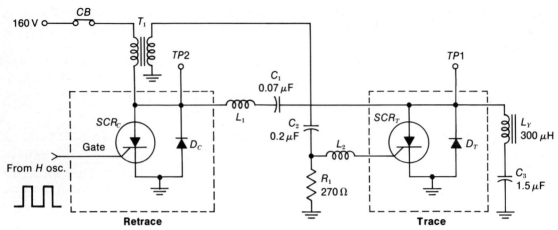

FIGURE 22-29 HORIZONTAL OUTPUT CIRCUIT USING SILICON-CONTROLLED RECTIFIERS (SCRs), SIMPLIFIED TO SHOW SEPARATE TRACE AND RETRACE SECTIONS. D_T and D_C ARE DAMPER DIODES. *(RCA)*

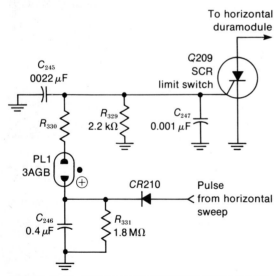

FIGURE 22-30 SCR AS LIMIT SWITCH TO DISABLE HORIZONTAL SCANNING WITH EXCESSIVE HIGH VOLTAGE. *(FROM ZENITH CHASSIS 25DC57)*

tions, or ringing, in the horizontal scanning current. The oscillations are present at the start of trace, immediately after flyback. The bars are white because the same area is scanned more than once. The example in Fig. 22-32 illustrates too much drive. These bars are thin because the damper stage is operating normally. Since this ringing is not severe, the bars may not be noticeable with a picture on the raster. Too much damping to eliminate ringing completely can reduce the amount of flyback high voltage.

The effect of no damping because of an open boost capacitor is shown in Fig. 22-33. In this case, the width of the raster is reduced. Also, the white bar is wide. With a picture, the left side is folded over.

Figure 22-34 shows excessive yoke ringing. This results from trouble in the *RC* damping network for the horizontal scanning coils. The cause may be an open *R* or *C* or incorrect *RC* values.

FIGURE 22-31 INSUFFICIENT WIDTH. *(RCA)*

Black vertical lines. These may be at the left or right side. The effect in Fig. 22-35, called "snivets," is caused by radiation of harmonics of the horizontal scanning current to the signal circuits. Rf chokes of 1 to 10 μH or ferrite beads are used in the horizontal amplifier and damper circuits to prevent the snivets.

Thin, straight, black vertical lines at the left can be caused by *Barkhausen oscillations* in the horizontal output tube radiated to the signal circuits. This stage usually has a 47-Ω parasitic-suppressor resistor in series with the control grid. Also, the suppressor grid is generally at +30 V for the dual functions of preventing Barkhausen oscillations and reducing snivets.

No high voltage. This is the most common cause of no brightness. Either the high-voltage rectifier is defective or it has no input from the flyback transformer. Remember that the damper must be operating with the amplifier stage, and the oscillator must supply drive.

To localize this trouble, first check to see if the rectifier has high-voltage ac input to its anode. If it does and there is no dc output, the trouble is in the high-voltage rectifier. If there is no flyback high voltage into the rectifier, check

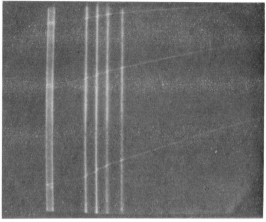

(a)

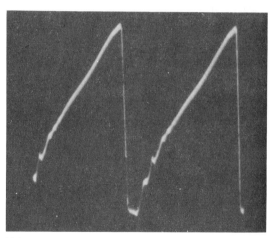

(b)

FIGURE 22-32 EXCESSIVE HORIZONTAL DRIVE. *(a)* THIN WHITE LINES AT LEFT. *(b)* RIPPLES IN SCANNING CURRENT.

the horizontal deflection stages, from the output transformer back to the oscillator.

On the dc output side, the high voltage rectifier tube may be gassy or shorted. Also, disconnect the rectifier lead to the CRT anode to see if the high voltage comes on.

The output stage usually has a fuse, which may be open. A common trouble is a short in the damper, which blows the fuse. See if the output transformer shows burned wires or melted wax. The resistance of the windings can be checked with an ohmmeter. However, a few shorted turns will not show on the ohmmeter. This trouble can still cause no horizontal output because the short loads down the output circuit. Then the Q is too low for a resonance effect.

To decide whether a short is in the deflection yoke or the flyback transformer, disconnect the yoke. If the boosted B+ voltage goes up, the yoke is bad. If not, the short is in the output transformer.

It should be noted that color receivers often have a focus rectifier and a high-voltage regulator. Trouble in these stages also can result in no brightness on the screen of the picture tube.

FIGURE 22-33 REDUCED WIDTH WITH FOLDOVER AT LEFT CAUSED BY OPEN B+ BOOST CAPACITOR.

FIGURE 22-34 EXCESSIVE YOKE RINGING.

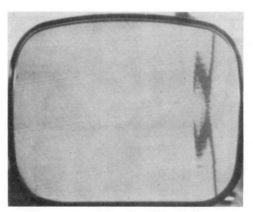

FIGURE 22-35 BLACK "SNIVETS" AT RIGHT.

SUMMARY

1. With tubes the horizontal deflection circuits include the oscillator driving a beam-power amplifier with its diode damper. In addition, the flyback high-voltage supply is in the horizontal output circuit.
2. The oscillator output voltage at 15,750 Hz provides the amplifier grid voltage needed for sawtooth plate current in the output.
3. Autotransformer coupling is generally used from the output stage to the horizontal deflection coils. The amplifier conducts for two-thirds of the cycle to scan the right side of the raster.
4. At the end of trace, the output stage is cut off. Then the output circuit oscillates. The damper is cut off by the first half-cycle of oscillations, allowing a fast flyback.
5. The high-voltage rectifier conducts during flyback to produce anode voltage for the CRT.
6. Immediately after flyback, the damper conducts. As the damped current decays to zero, the left side of the raster is scanned while the amplifier remains cut off. Then the output stage starts conducting again to finish the trace.
7. While the damper is conducting, its current charges the $B+$ boost capacitor. The boosted $B+$ voltage is the plate supply for the horizontal amplifier tube.
8. The voltage waveshape at all points in the horizontal output circuit is rectangular with sharp flyback pulses. For the usual case of an autotransformer, all taps have positive polarity of flyback pulses. Therefore, the diode damper is inverted, with positive pulses at the cathode for cutoff during flyback but negative deflection voltage for conduction during trace.
9. The frequency of the horizontal scanning current is set by the oscillator and its AFC adjustments. The amplitude adjustments to fill the width of the raster are in the horizontal output. These may include drive, width, and linearity controls.
10. The horizontal amplifier tube operates with grid-leak bias approximately equal to the peak positive drive from the oscillator. A typical bias is -30 to -50 V.
11. A transistor or SCR in the horizontal output stage serves as a switch to supply rectangular voltage for sawtooth current in the inductance of the scanning coils. See Fig. 22-25.
12. In transistorized circuits, a driver stage is used as a buffer between the horizontal oscillator and amplifier. See Fig. 22-26.
13. Insufficient width without distortion indicates a weak horizontal amplifier or low $B+$ voltage. Reduced width with foldover at the left is caused by trouble in the damper circuit.
14. The most common cause of no brightness is no high voltage. Either the rectifier circuit is defective or there is no horizontal output. Check the horizontal oscillator, amplifier, fuse, damper, flyback transformer and the deflection yoke for no ac input. Check the high voltage rectifier circuit for no dc output.

Self-Examination (Answers at back of book)

1. Which stage is not necessary for producing horizontal output: (a) horizontal oscillator; (b) damper; (c) horizontal amplifier; (d) horizontal AFC.
2. The frequency of sawtooth current in the horizontal amplifier is : (a) 60 Hz; (b) 10,500 Hz; (c) 15,750 Hz; (d) 70 kHz.
3. When the horizontal amplifier is conducting peak plate current, the electron scanning beam is at the: (a) left edge of the raster; (b) right edge of the raster; (c) center of trace; (d) center of flyback.
4. Which of the following is conducting during flyback: (a) high-voltage rectifier; (b) output stage; (c) diode damper tube; (d) silicon diode damper.
5. With a resonant frequency of 70 kHz for the horizontal output circuit, the flyback time equals: (a) 5 μs; (b) 7 μs; (c) 20 μs; (d) 50 μs.
6. With autotransformer coupling, the voltage waveshape at the cathode of the inverted diode damper is: (a) sawtooth; (b) trapezoidal; (c) rectangular with positive flyback pulses; (d) rectangular with negative flyback pulses.
7. With 120 V p-p sawtooth drive at the grid of the horizontal amplifier, its grid-leak bias equals: (a) zero; (b) 10 V; (c) 54 V; (d) 120 V.
8. The voltage induced across a 240-μH inductance by a 5-A current dropping to zero in 8 μs equals: (a) 80 V; (b) 150 V; (c) 2,000 V; (d) 15,000 V.
9. The B+ boost capacitor charges when the: (a) high-voltage rectifier conducts; (b) beam retraces; (c) damper is cut off; (d) damper is conducting.
10. The damper tube has an open heater. The result is: (a) a wide white bar at the left; (b) thin white bars at the left; (c) foldover at the right; (d) no brightness.
11. For the transistor circuit in Fig. 22-26: (a) Q3 is on for retrace; (b) the scanning coils are in series; (c) Q2 is transformer-coupled to the output stage; (d) the damper D1 is on for retrace.
12. For the SCR circuit in Fig. 22-29: (a) the horizontal oscillator pulses drive the scanning coils; (b) the gate of each SCR is used to turn it off; (c) SCR_T is used to produce retrace. (d) SCR_C has B+ of 160 V at the anode.

Essay Questions

1. What three stages in the receiver determine the width of the raster? What stages determine the height? (Low-voltage power supply is not counted.)
2. How is the horizontal oscillator synchronized at 15,750 Hz?
3. Name two types of horizontal oscillator circuits.
4. What is meant by ringing? What two factors determine the ringing frequency?
5. Give six functions of the horizontal output circuit.

6. In horizontal deflection, describe briefly how the scanning current is produced for the right side of trace, the retrace, and the left side of trace.
7. Why is the damper cut off for flyback?
8. What two voltages add to produce the boosted B+ voltage?
9. What three stages must be operating to produce ac input voltage for the high-voltage rectifier?
10. Draw the schematic diagram of a horizontal amplifier circuit with autotransformer output. Include the damper and high-voltage rectifier circuits and show boosted B+.
11. Name three circuits that may use boosted B+ as a voltage source.
12. Give two methods of checking with a dc voltmeter to see if the horizontal oscillator is operating.
13. Draw a sine wave of current and of voltage on one graph, showing the two waveforms 90° out of phase. State the main characteristic of the 90° phase difference.
14. Draw one cycle of sawtooth current in the horizontal deflection coils. Mark the trace and retrace periods in microseconds. Indicate which tubes in the output circuit are conducting for three parts of the cycle.
15. Give the function of the circuit in Fig. 22-24.
16. Give the function of the circuit in Fig. 22-30.
17. What is meant by scan rectification for dc supply voltage?
18. Give three differences between horizontal deflection circuits using tubes and those using transistors.
19. Name the three electrodes in an SCR and give the function of each.
20. Describe briefly one method of width control.
21. The fuse $F1$ in Fig. 22-2 is open. What is the effect on the raster?
22. In Fig. 22-2, give the functions of R_1, C_1, C_2, C_B, C_3, and C_n.
23. In Fig. 22-28, give the functions of Q8, Q9, Q302, D301, T_4, T_{302}, F1, the TA301 tripler unit, and the R_{311} potentiometer.
24. From waveshape (c) in Fig. 22-27, for drive voltage on the output transistor, this stage is cut off for what percent of a full cycle?
25. For the table below, indicate which stages are on or off during retrace and any part of trace.

	TRACE	RETRACE
Horizontal amplifier		
High-voltage rectifier		
Damper		

Problems (Answers to selected problems at back of book)

1. (a) Calculate the RC time constant of a 0.005-μF coupling capacitor with 1-MΩ grid resistance. (b) How much longer than the horizontal scanning period is this time constant? (c) Can grid-leak bias be developed when grid current flows?
2. Approximately how much grid-leak bias is produced for the following peak-to-peak values of grid drive: 40 V, 80 V, and 160 V?
3. Referring to Fig. 22-4: (a) What is the average I_b set by the grid bias of -40 V? (b) What is the $E_b I_b$ plate power dissipation with a 200-V average plate voltage?
4. A power transistor has a 3-A average collector current I_C. The average collector voltage V_C is 10 V. How much is the $V_C I_C$ power dissipation at the collector?
5. A horizontal output transistor is cut off for retrace and 25 percent of trace. For how many microseconds is the transistor on?
6. For the horizontal output transformer in Fig. 22-36 below, give the functions for the taps at terminals 1, 2, 3, and 4, with the auxiliary windings L_1 and L_2.

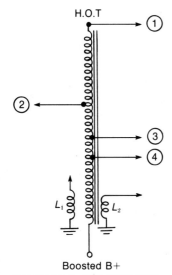

FIGURE 22-36 FOR PROBLEM 6.

23 The Picture IF Section

As shown in Fig. 23-1, this section amplifies the IF signal output from the mixer stage on the rf tuner. A short length of coaxial cable feeds the tuner output to the first IF amplifier. For any channel in the VHF or UHF bands, the rf picture carrier signal is converted by the tuner to 45.75 MHz for the picture IF amplifier. This section is also called the *video IF amplifier*, since the composite video signal is the envelope of the modulated picture IF signal.

Practically all the gain and selectivity of the receiver is provided by the IF section. With tubes, 2 or 3 IF stages are used. Typical miniature glass pentodes are the 4DK6 and 6JC6A. With transistors, 3 or 4 IF stages are needed. With integrated circuits, one IC chip contains all the IF amplifier stages. When solid-state amplifiers are used, the entire picture IF section is generally on one small board or module (Fig. 23-2) with the video detector. Details of the picture IF amplifier circuits are explained in the following topics:

23-1 Functions of the picture IF section
23-2 Tuned amplifiers
23-3 Single-tuned circuits
23-4 Double-tuned circuits
23-5 Neutralization of transistor IF amplifiers
23-6 The picture IF response curve
23-7 IF wave traps to reject interfering frequencies
23-8 Picture IF amplifier circuits
23-9 IC unit for the picture IF section
23-10 Picture IF alignment
23-11 Troubles in the picture IF amplifier

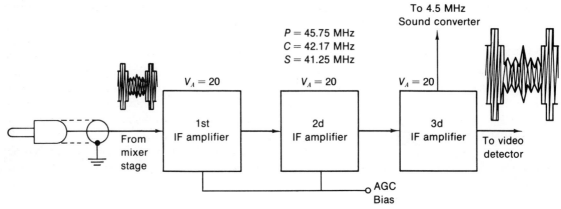

FIGURE 23-1 PICTURE IF AMPLIFIER SECTION WITH THREE TUNED STAGES.

23-1 Functions of the Picture IF Section

The main function is to amplify the modulated IF signal enough to drive the detector. With a 0.5-mV picture IF signal from the mixer, the IF amplifier can provide an overall voltage gain of 8,000, for 4 V into the video detector diode. Then the video signal output of the detector is amplified for 100-V p-p video signal at the picture tube for good contrast.

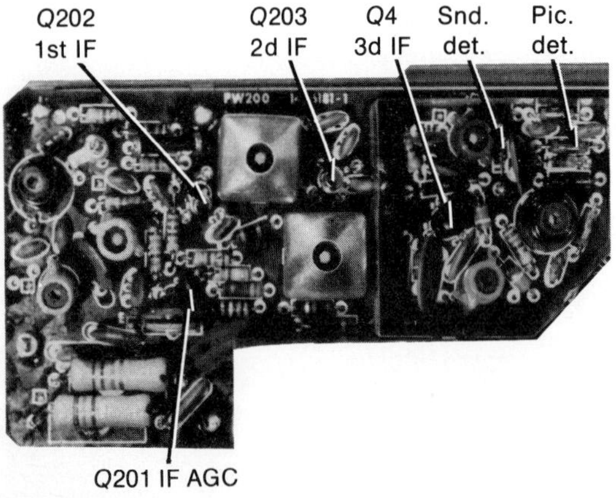

The intermediate frequencies. The standard value specified by the EIA for the IF picture carrier is 45.75 MHz in all television receivers. The associated IF sound carrier is automatically 4.5 MHz below, at 41.25 MHz. These values were adopted in the year 1950, raising the intermediate frequencies from the old value of 25.75 MHz for the IF picture carrier. In general, the intermediate frequencies should be as high as possible but below the lowest rf channel. The IF band at approximately 41 to 47 MHz is just below channel 2 at 54 to 60 MHz.

The IF sound signal. For intercarrier sound, the IF sound signal at 41.25 MHz is amplified with the IF picture signal at 45.75 MHz in the common IF stages. Both the sound and picture IF signals can be amplified just by having enough bandwidth. The relative gain for the IF sound at 41.25 MHz is only 5 to 10 percent in the IF picture amplifier. Color receivers have a separate diode converter for the 4.5-MHz intercarrier sound, as shown in Fig. 23-1. In mono-

FIGURE 23-2 IF BOARD WITH TRANSISTOR AMPLIFIERS. (RCA)

chrome receivers, the 4.5-MHz sound is obtained from the video detector diode.

The IF color signal. With the IF picture carrier at 45.75 MHz, the IF color subcarrier is automatically 3.58 MHz below, at the frequency of $45.75 - 3.58 = 42.17$ MHz. For color receivers, the IF color signal at 42.17 MHz is amplified as part of the IF picture signal. Remember that the 3.58-MHz color subcarrier signal is combined with the luminance signal as part of the composite video signal that modulates the picture carrier. Therefore, the IF picture signal at 45.75 MHz includes the color signal. The only requirement in the IF section is that it have enough bandwidth for the required gain at 42.17 MHz.

In summary, then, the important intermediate frequencies to remember are:

45.75 MHz for picture signal
42.17 MHz for color signal
41.25 MHz for sound signal

23-2 Tuned Amplifiers

Each IF stage is a tuned amplifier. This means the gain is determined by a resonant circuit tuned to the intermediate frequencies. The tuned circuit provides an ac load impedance for the amplifier, only for the IF signal. The dc resistance of an IF coil is practically zero.

The way that a resonant LC circuit is used as the ac load impedance for the output of a tuned amplifier is illustrated in Fig. 23-3 for tubes and Fig. 23-4 for transistors. In both cases the IF coil L_1 tunes with the stray capacitance C_t for resonance. In this example, the resonant frequency f_r is 43 MHz.

It should be noted that f_r usually is not at 45.75 MHz for any of the IF tuned circuits. The reason is that the IF response is only 50 percent for the picture carrier, in order to compensate for the vestigial sideband transmission.

The tuned circuit is an application of parallel resonance, because the source of ac signal is outside the LC circuit. For parallel resonance, the impedance Z_L is maximum at the resonant frequency. Then the resonant circuit provides the load impedance needed for gain in the amplifier stage. A typical value of Z_L is 6,000 Ω. This value is an ac load impedance that cannot be measured with an ohmmeter.

In Fig. 23-3, L_1 tunes with C_t to form the resonant plate load for the pentode IF amplifier. Pentodes are used because their high plate resistance shunting the tuned circuit allows a high Q for parallel resonance. R_f is a plate decoupling resistor which isolates the ac signal of the tuned circuit from the dc supply, in order to prevent feedback between stages. C_f is the bypass for R_f. Note that C_f is actually part of the tuned circuit, as it returns L_1 to C_t for the IF signal. If C_f opens, then the resistance of R_f becomes part of the tuned circuit, resulting in low Q and practically no IF gain.

In Fig. 23-4, the resonant circuit of L_1 with C_t forms the collector load for IF output signal from the transistor amplifier. C_1 couples the IF

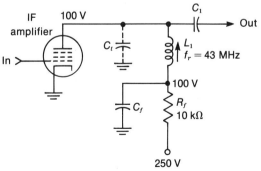

FIGURE 23-3 PARALLEL RESONANT CIRCUIT AS TUNED LOAD IMPEDANCE FOR PLATE OF PENTODE TUBE.

516 BASIC TELEVISION

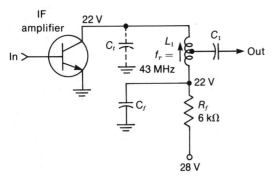

FIGURE 23-4 PARALLEL RESONANT CIRCUIT AS TUNED LOAD IMPEDANCE FOR COLLECTOR OF TRANSISTOR. NOTE TAP ON L_1 FOR BASE OF NEXT STAGE.

signal to the base of the next stage. The CE (common-emitter) circuit is used. The idea of the parallel resonant circuit as the collector load impedance is the same as the plate load impedance for a tube. Note the lower values of dc supply voltage for the transistor, compared with tubes.

The plate or collector side of L_1 is called the *high side* of the tuned circuit. This end has the IF signal voltage. The opposite end is the *low side* because it is grounded for ac signal by the bypass capacitor C_f.

Impedance matching. In Fig. 23-4 the output connection for C_1 is tapped down on the IF coil L_1. The purpose is to provide a stepdown of impedance from the collector to the base of the next transistor. A typical collector output resistance in the CE circuit is 50 kΩ, while the base input resistance may be 1 kΩ.

In Fig. 23-4, L_1 is connected as a stepdown autotransformer. All the turns serve as the primary for the collector load. However, only those turns between the tap and the low end of the coil are used as the secondary for output to the base. Another method is to use a stepdown transformer with separate primary and secondary windings. Also, a tapped voltage divider of R or C can be connected across the IF coil. In all cases, the purpose is to step down the impedance in the collector output circuit to the base input circuit of the next stage.

Dc voltages. In a tuned amplifier, the dc plate voltage or collector voltage has practically the same value as the dc supply voltage at the low side of the tuned circuit. The reason is that the IF coil has no appreciable *IR* drop for dc voltage because of its low resistance.

In Fig. 23-3, the plate supply voltage at the low end of R_f is 250 V. This value is dropped to 100 V for the tuned circuit because of the *IR* drop of 150 V across R_f. The 150-V drop across the 10 kΩ of R_f means the dc plate current I_b is 150 V/10 kΩ = 15 mA. The dc plate voltage of 100 V at the high side of L_1 is the same as the 100 V at the low end because L_1 has a dc resistance of less than 1 Ω.

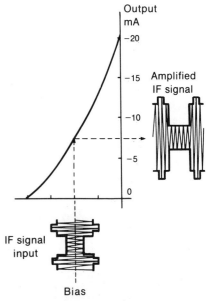

FIGURE 23-5 IF AMPLIFIER OPERATING CLASS A FOR MINIMUM AMPLITUDE DISTORTION.

In Fig. 23-4, the collector supply voltage at the low end of R_f is 28 V. This value is dropped to 22 V because of the *IR* drop of 6 V across R_f. The 6-V drop across the 1,000 Ω of R_f means the dc collector current I_C is 6 V/6,000 Ω = 0.001 A or 1 mA. The collector voltage of 22 V at the high side of L_1 is the same as the 22 V at the low end because of the low resistance of L_1.

Class A amplification. Each IF stage operates class A for minimum distortion. Class A operation means that output current flows for 360°, or the full cycle of the signal.

A class A stage is biased to be on with output current flowing with or without signal input. As shown in Fig. 23-5, the dc bias sets the operating point at the middle of the input-output operating characteristic. The modulated IF picture signal is the ac input. Then the variations of the input signal produce corresponding changes in the output. The envelope of the IF signal is actually the video signal, which must be amplified without amplitude distortion. It is especially important to avoid compressing the sync amplitudes. In addition, linear amplification prevents cross-modulation between the desired signal and an interfering signal.

23-3 Single-tuned Circuits

A single-tuned stage has one *LC* circuit as the load impedance in the output. See Fig. 23-3. The output signal from the high side of the resonant circuit is coupled by C_1 to the next stage. The function of C_1 as a coupling capacitor is to block the dc voltage at the plate or collector, but have low reactance for ac voltage at the IF signal frequencies. Values for C_1 are 100 to 200 pF.

The resonant load impedance. The main features of tuned amplifiers can be illustrated by the resonant circuit in Fig. 23-6a. For a single-tuned stage, this circuit is the load impedance for ac signal in the plate or collector circuit of the amplifier. It is an example of parallel reso-

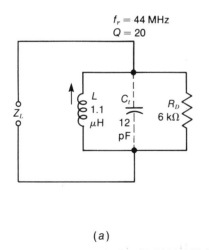
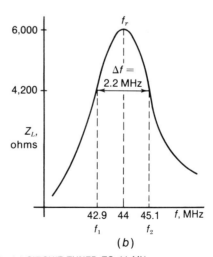

(a)   (b)

FIGURE 23-6 RESPONSE FOR PARALLEL RESONANCE. (a) CIRCUIT TUNED TO 44 MHz, WITH A Q OF 20; (b) RESONANCE CURVE WITH MAXIMUM IMPEDANCE AT 44 MHz AND TOTAL BANDWIDTH OF 2.2 MHz.

nance because the tube or transistor is the signal source outside the tuned circuit. The resonant frequency $f_r = 1/2\pi\sqrt{LC_t}$. C_t is generally about 12 pF. Then the required L is 1.1 μH for the f_r of 44 MHz. At f_r, the impedance Z_L is maximum for parallel resonance. A typical value for Z_L is 6,000 Ω at the resonant frequency.

The tuned circuit provides a resonant rise in Z_L, as shown by the response curve in Fig. 23-6b. Furthermore, the gain is directly proportional to the load impedance. For a pentode tube the voltage gain equals $g_m \times Z_L$. With a g_m of 8,000 μmho and a Z_L of 6,000 Ω, the gain is $0.008 \times 6,000 = 48$.

Q, damping, and bandwidth. Q is a figure of merit, indicating how sharp the resonance is. An IF coil generally has a Q of about 80, either by itself or in a tuned circuit with C. However, a parallel resistance such as R_D in Fig. 23-6a lowers the Q. The smaller the shunt R_D is, the lower the Q becomes. In addition, the bandwidth of the tuned circuit increases with lower Q. This effect of reducing the Q and increasing the bandwidth is called *damping*. In Fig. 23-6a, the 6-kΩ value of R_D reduces the Q of the tuned circuit to 20. This value is calculated as $Q = X_L/R_D$, where X_L is 300 Ω and R_D is in parallel rather than in series with L. The total bandwidth Δf in Fig. 23-6b is shown as 2.2 MHz. This bandwidth is calculated as f_r/Q, which is 44 MHz/20 = 2.2 MHz. R_D is a ½-W carbon resistor in parallel with the tuned circuit.

Overall response. The overall gain of cascaded stages is the product of the values for each stage. When one stage has a gain of 24, the next is 10, and the final-stage gain is 20, the

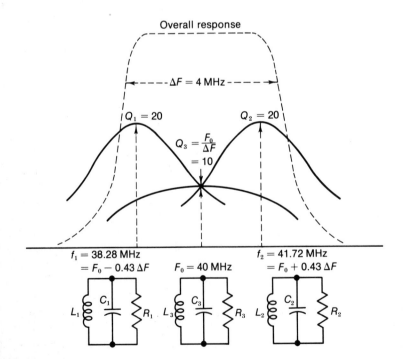

FIGURE 23-7 STAGGER TUNING OF THREE SINGLE-TUNED CIRCUITS FOR OVERALL BANDWIDTH OF 4 MHz, CENTERED AT 40 MHz. THESE FREQUENCIES ARE NOT THE IF PASSBAND BUT JUST A CONVENIENT EXAMPLE OF A STAGGER-TUNED AMPLIFIER.

overall gain is 24 × 10 × 20 = 4,800. Also, the overall response curve for tuned amplifiers is the product of the individual response curves for each of the cascaded stages.

Stagger-tuned stages. When cascaded amplifiers are tuned to the same frequency (synchronous tuning), the overall bandwidth shrinks drastically. The reason is that the overall gain equals the product of individual gain values, not their sum. Then the peak values become more peaked, while the low-gain frequencies become more attenuated. The sharp peak with narrow bandwidth is undesirable for the wide passband needed in the picture IF amplifier. However, the required overall response can be obtained with single-tuned stages by staggering their resonant frequencies. This procedure is illustrated in Fig. 23-7 for three single-tuned stages, staggered around 40 MHz, with an overall bandwidth of 4 MHz. For this triplet, one stage is tuned to the center at 40 MHz, while the other two are above and below by ±1.72 MHz. Staggered tuning can be used for combinations of two, three, four, or five stages.

23-4 Double-tuned Circuits

The advantage of these circuits is more bandwidth, with sharper skirts on the response curve, than with single-tuned circuits. Figure 23-8 shows a typical double-tuned transformer from the output of one IF amplifier to the input of the next stage. No coupling capacitor is necessary, as the transformer couples the IF signal by induction between L_P and L_S, while isolating the dc voltages between the two windings.

The primary and secondary inductances L_P and L_S tune the shunt capacitances C_{out} and C_{in} to the desired frequency in the IF range. C_{out}

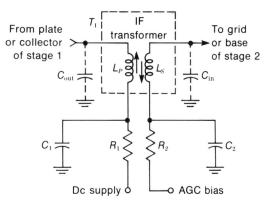

FIGURE 23-8 DOUBLE-TUNED TRANSFORMER FOR IF AMPLIFIER.

is the output capacitance of stage 1, while C_{in} is the input capacitance of stage 2. Usually, L_P and L_S each have a slug for tuning the primary and secondary, so that two adjustments are needed to align the tuned transformer. The primary and secondary are both resonant at the same frequency. Note that $R_1 C_1$ is a decoupling filter for the dc supply voltage and $R_2 C_2$ is a decoupling filter for the AGC bias line.

Double-tuned coupling. More details of resonance in the double-tuned coupling arrangement are shown in Fig. 23-9. Both primary and secondary are tuned to 44 MHz here for comparison with the single-tuned circuit of Fig. 23-6. Coupling in the double-tuned transformer in Fig. 23-9a results from the mutual inductance L_M that links L_P and L_S, providing a mutual coupling impedance common to both circuits. Similar results can be obtained with other types of mutual coupling impedance, indicated by Z_M in Fig. 23-9b. Physically, Z_M can be a coupling coil or capacitor, or a combination of both L and C to function also as a wave trap to attenuate one frequency. Some double-tuned transformers have only one slug, which adjusts the coupling.

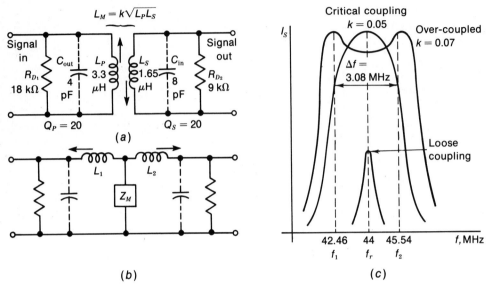

FIGURE 23-9 DOUBLE-TUNED COUPLING. (a) TRANSFORMER WITH L_P AND L_S TUNED TO THE SAME f_r; (b) GENERAL FORM OF A DOUBLE-TUNED CIRCUIT WITH MUTUAL COUPLING IMPEDANCE Z_M; (c) EFFECT OF COUPLING COEFFICIENT k ON THE RESPONSE OF A DOUBLE-TUNED CIRCUIT.

Coefficient of coupling. The effect of coupling between L_P and L_S in a double-tuned transformer is illustrated in Fig. 23-9c. The coupling coefficient k indicates the fraction of total flux from one coil linking the other coil. For loose coupling, the primary and secondary are far apart, with little flux linkage between L_P and L_S. The secondary response then has a single peak at f_r like a single-tuned circuit, with low amplitude. However, as the coupling is increased by having L_P and L_S closer together, the primary produces more secondary output. Also, the secondary current has a greater effect on the primary. With close coupling, then, two peaks result in the secondary output to broaden the frequency response. For a middle value called *critical coupling* (k_C), the secondary has its maximum amplitude and greatest bandwidth without double peaks.

Impedance coupling or bandpass coupling. Examples of using a common impedance Z_M for mutual coupling between two tuned circuits are shown in Fig. 23-10. This method is also called *bandpass coupling* because it usually has the bandwidth of an overcoupled double-tuned transformer. In Fig. 23-10a, the coil L_M is common to both tuned circuits with L_1C_1 and L_2C_2. As a result, signal voltage across L_M from one tuned circuit is coupled into the other. In Fig. 23-10b, capacitive coupling is used, with C_M common to both tuned circuits. In both circuits, R_D is a damping resistor to provide the required bandwidth.

Link coupling. See Fig. 23-11. This method is generally used to couple the IF signal from the mixer stage on the rf tuner to the first IF stage on the main chassis. The link is a 75-Ω coaxial

THE PICTURE IF SECTION

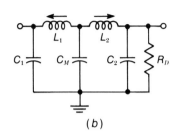

FIGURE 23-10 IMPEDANCE COUPLING WITH DOUBLE-TUNED CIRCUIT. (a) INDUCTIVE COUPLING WITH L_M; (b) CAPACITIVE COUPLING WITH C_M.

cable with shielded plugs at both ends. Its purpose is to couple the signal at a low impedance level. As a result, the effect of stray capacitance is minimized. With the outside shield grounded for the low-impedance line, there is no pickup of interfering signals and no radiation of the mixer output. Radiation of beat frequencies and their harmonics from the mixer can cause interference in the receiver.

Referring to Fig. 23-11, the mixer output transformer T_1 is on the rf tuner. The secondary winding L_2 has just a few turns for a stepdown of the voltage and impedance with reference to the primary. In addition, L_2 is connected to a jack on the tuner that takes the plug of the coaxial cable. At the other end, the plug connects the IF signal to the IF transformer T_2 on the receiver chassis. L_3 here is the primary to step up the signal voltage for the secondary L_4. This way the IF signal is coupled from the mixer output to the IF input by means of the IF cable, which can be 1 to 2 ft or more in length, without radiation or stray pickup. This link coupling generally has the response of an overcoupled stage.

Loading resistor. When either the primary or the secondary is damped with a parallel resistor of about 100 to 300 Ω, the double-tuned circuit then has a single-peak response like a single-tuned circuit. The reason is that the low Q in one side of the double-tuned circuit requires a much higher value of coupling between L_P and L_S for critical coupling. The actual coupling, which is part of the transformer construction, then is equivalent to loose coupling with one resonant peak. For the purpose of alignment,

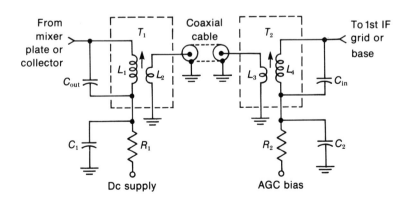

FIGURE 23-11 LINK COUPLING FROM MIXER OUTPUT ON THE RF TUNER TO THE FIRST IF STAGE ON THE MAIN CHASSIS.

therefore, it is possible to peak each side of the transformer at f_r when a small damping resistance is temporarily shunted across the opposite side. This is also called a *swamping resistor,* or *loading resistor.*

Bifilar IF coils. This type of coil has two windings, like a transformer, but it is not double-tuned. As shown in Fig. 23-12, the bifilar coil is wound with twin conductor wires, each insulated from the other. One winding is for the plate or collector, tuned to the resonant frequency as a single-tuned circuit. The other winding is for the grid or base of the next stage. Since the IF signal is inductively coupled to the next stage, no coupling capacitor is needed. Furthermore, the response is the same as that of a single-tuned circuit. In short, the bifilar coil provides a method of having a single-tuned amplifier without using a coupling capacitor C_C.

23-5 Neutralization of Transistor IF Amplifiers

Any tuned amplifier can oscillate when enough signal is fed back from the output circuit to the input circuit in the same phase as the input signal. With vacuum tubes, triodes have appreciable feedback through the internal grid-plate capacitance C_{gp}. Pentodes seldom have feedback problems. However, PNP and NPN transistors are equivalent to triodes with relatively high capacitance between collector and base. Typical values of this internal capacitance are 5 to 20 pF. The technique of canceling the internal feedback is called *neutralization.* The method consists of feeding back signal voltage through an external path to cancel the internal feedback.

Four types of neutralizing circuits are shown in Fig. 23-13. Each is a method of obtaining signal voltage that is the opposite polarity from the collector signal. The neutralizing voltage is coupled back through the feedback

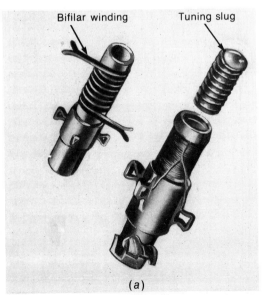

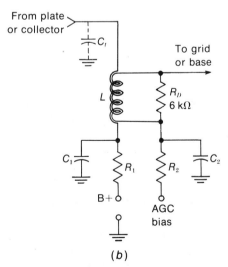

FIGURE 23-12 BIFILAR IF COIL. *(a)* PHOTO; HEIGHT ABOUT $1\frac{1}{2}$ IN.; *(b)* SCHEMATIC SYMBOL.

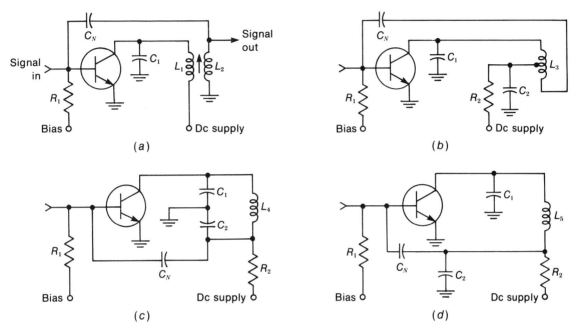

FIGURE 23-13 METHODS OF NEUTRALIZING AN IF AMPLIFIER STAGE. C_N FEEDS BACK NEUTRALIZING VOLTAGE OF OPPOSITE POLARITY FROM OUTPUT SIGNAL. (a) FEEDBACK VOLTAGE FROM SECONDARY OF IF TRANSFORMER; (b) TAPPED PRIMARY COIL; (c) CAPACITIVE VOLTAGE DIVIDER WITH C_1 AND C_2 ACROSS L_4; (d) ANOTHER FORM OF CAPACITIVE VOLTAGE DIVIDER WITH C_1 AND C_2.

capacitor C_N. A typical value for C_N is 2.7 pF. It is usually a fixed value for IF amplifiers but may be variable for rf amplifiers. It should be noted that C_N couples the feedback and blocks the dc collector voltage from the base. However, the feedback can be in a resistive path, without C_N, as long as the dc voltages on the base are arranged to provide the correct bias for the base-emitter circuit.

In Fig. 23-13a the feedback voltage is taken from the secondary L_2 of the IF transformer in the collector output circuit and coupled back by the neutralizing capacitor C_N to the base. The secondary voltage is phased to provide signal of opposite polarity from the collector signal in the primary. Therefore, the neutralizing signal fed back by C_N cancels the internal feedback. In Fig. 23-13b, two opposite polarities of output signal are provided by the tap on the coil L_3. The top of L_3 is the collector side. The bottom end has signal of opposite polarity with respect to the ground at the tap. Therefore, the neutralizing voltage fed back by C_N cancels the internal feedback.

The circuits in Fig. 23-13c and d use a capacitive voltage divider for feedback. In Fig. 23-13c, the divider consists of C_1 and C_2 across the coil L_4. Then C_N at the low end of the coil feeds back the neutralizing signal voltage across C_2, while the IF voltage across C_1 is the output signal. Furthermore, the signal voltage at the ungrounded side of C_2 is the opposite polarity from the signal across C_1. Note that the isolating resistor R_2 is needed in the collector circuit to prevent shorting the signal through the low impedence of the dc supply. Finally, the circuit in Fig. 23-13d corresponds to the capacitive voltage divider in Fig. 23-13c. The reason is that C_1 and C_2 are joined by the ground connection, forming a capacitive voltage divider across L_5. In this circuit, C_2 has the same position as a bypass capacitor. However, its capacitance is small enough to develop neutralizing voltage that is coupled back through C_N to the base.

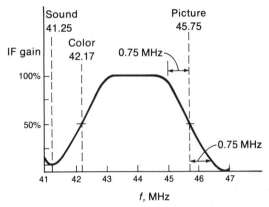

FIGURE 23-14 IDEAL PICTURE IF RESPONSE CURVE.

23-6 The Picture IF Response Curve

The graph in Fig. 23-14 compares how much overall voltage amplification the IF section has for different frequencies in the passband. Note the following features of this ideal IF response curve:

1. The picture IF carrier frequency of 45.75 MHz at the side of the response curve has 50 percent of maximum IF gain. This reduced IF response for the picture carrier, and side frequencies close to it, is opposite to the effect of vestigial-sideband transmission.
2. The sound IF signal at 41.25 MHz is amplified with the picture signal for intercarrier sound. However, the relative gain is only about 10 percent for the sound signal in the picture IF amplifier section.
3. The frequency of the color subcarrier signal in the picture IF amplifier is 42.17 MHz, separated by 3.58 MHz from the picture carrier at 45.75 MHz. Note that the color frequency is at the opposite side of the IF response curve.
4. The IF bandwidth for the luminance signal is approximately 3 MHz, from 45.75 to 42.75 MHz. This IF bandwidth then corresponds to 3-MHz response for video frequencies in

the picture reproduction. The bandwidth is limited to 3 MHz in the receiver, instead of 4 MHz, to minimize interference from the color signal.

How 42.17 MHz corresponds to 3.58 MHz for the color signal. In the video detector, the color signal at 42.17 MHz beats with the IF picture carrier at 45.75 MHz to produce the difference frequency of $45.75 - 42.17 = 3.58$ MHz. The response for 42.17 MHz is generally at 50 percent. In monochrome receivers, though, the response for 42.17 MHz is close to zero to eliminate interference from the color signal.

How 41.25 MHz corresponds to 4.5 MHz for the sound signal. Either in the video detector or in a separate diode detector as the sound converter, the sound signal at 41.25 MHz beats with the IF picture carrier at 45.75 MHz to produce the difference frequency of $45.75 - 41.25 = 4.5$ MHz. For intercarrier sound, the response at 41.25 MHz is only about 10 percent or less in the common IF amplifier. The required attenuation of the sound signal is obtained by one or more wave traps tuned to 41.25 MHz.

Inversion of IF side frequencies. As an example, the rf sound carrier frequency is transmitted 4.5 MHz higher than the rf picture carrier frequency, but in the receiver IF section the IF sound carrier at 41.25 MHz is 4.5 lower than the picture carrier at 45.75 MHz. This frequency inversion results only because the local oscillator in the tuner beats above the rf signal frequencies, which is the usual case. Then higher rf frequencies are closer to the oscillator frequency, resulting in lower values for the IF difference frequencies. Additional examples are listed in Table 23-1.

The same frequency separation is maintained between the picture carrier and its side frequencies in the rf and IF signals. However, those frequencies above the picture carrier in the rf signal are below it in the IF signal. Also, the lower rf side band becomes an upper IF side band.

How the IF side frequencies correspond to video frequencies. The highest video modulating frequencies produce side frequencies farthest from the picture carrier in the transmitted signal. Although the IF side bands are inverted, the same frequency difference compared with the carrier is maintained. Therefore, the IF side frequencies farthest from the picture carrier correspond to the highest video frequencies in the output of the video detector. In the IF response curve, these frequencies have lower numerical values, near the sound IF carrier. However, the IF gain for these frequencies determines the response for the high video frequencies that reproduce fine detail in the picture. At the opposite part of the curve, side frequencies close to the picture carrier represent the low video frequencies. The continuity between rf, IF, and video signal frequencies is illustrated by the resolution chart in Fig. 23-15.

In terms of frequency components in the detected signal, we can consider that each IF side frequency beats with the picture carrier in the video detector to produce the difference frequency. As an example, 42.75 MHz beats with 45.75 MHz in the IF signal to produce the difference frequency of 3 MHz in the video signal out of the detector. Therefore, 3-MHz bandwidth, from 41.75 to 44.75 MHz in the IF curve, corresponds to 3-MHz response for the detected video signal, from 0 Hz or direct current up to 3 MHz for very small picture details.

Compensation for vestigial-sideband transmission. The lower video modulating frequencies

TABLE 23-1 INVERSION OF IF SIGNAL FREQUENCIES

CHANNEL 4, 66–72 MHz	TRANSMITTED RF SIGNAL FREQUENCY, MHz	LOCAL OSCILLATOR FREQUENCY, MHz	INTERMEDIATE FREQUENCY, MHz
Upper edge of channel	72	113	41
Sound carrier	71.25	113	41.25
Color subcarrier	70.83	113	42.17
Side frequency for 3-MHz video modulation	70.25	113	42.75
Picture carrier	67.25	113	45.75
Lower edge of channel	66	113	47

526 BASIC TELEVISION

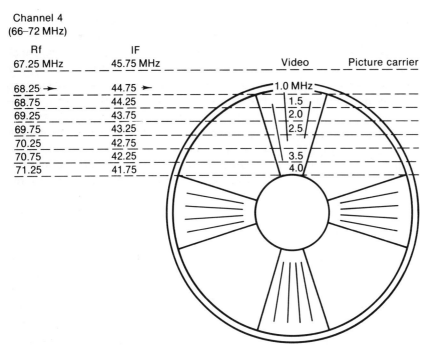

FIGURE 23-15 RESOLUTION CHART TO ILLUSTRATE THE CONTINUITY OF RF, IF, AND VIDEO SIGNAL FREQUENCIES FOR CHANNEL 4.

up to 0.75 MHz are transmitted as double sideband signals, while the higher video frequencies are transmitted as single-sideband signals, as explained in Sec. 6-2. Therefore, video frequencies above 0.75 MHz have only one-half the effective modulation of the lower-frequency double-sideband signals. If the receiver IF response were the same for all signal frequencies, the demodulated output from the video detector would be twice as great for video signals below 0.75 MHz as for the higher video frequencies.

In order to equalize the effect of vestigial-sideband transmission, the overall IF response is aligned to give the picture carrier approximately 50 percent of maximum response, as shown in Fig. 23-14. As a result, the two side bands of a double-sideband signal are given an average response of 50 percent, compared with 100 percent response for the single-sideband frequencies. The relative output from the video detector, therefore, will be the same for all video modulating signals having the same amplitude, whether they are transmitted with single or double sidebands.

When the picture carrier is too high on the IF response curve, with more than 50 percent relative gain, the low video frequencies, up to 0.75 MHz, are emphasized, while higher video frequencies are attenuated. The extra gain for low video frequencies increases the contrast in the reproduced picture, but usually with excessive smear. Attenuation of the high video frequencies reduces the high-frequency detail that should make the picture sharp and clear.

When the IF gain is too low for the picture carrier frequency, the reduced response attenuates low video frequencies and the picture carrier itself. The result is insufficient video signal and weak contrast in the picture. With zero response for the IF picture carrier there would be no picture at all.

23-7 IF Wave Traps to Reject Interfering Frequencies

A wave trap is a resonant circuit tuned to attenuate a specific frequency. Because of the wide passband in the picture IF amplifier, wave traps are needed to reduce interference and to provide the IF response for intercarrier sound. In Fig. 23-16, the trap frequencies are shown on the zero base line of the IF response. This curve is shown downward, as it would be from a negative-polarity detector. Note that the wave traps determine the shape of the overall response at both ends of the curve, for the picture carrier at one side and the color subcarrier at the other side. The trap frequencies in the picture IF amplifier are as follows:

1. 41.25 MHz is the associated sound signal for the channel to which the receiver is tuned. We want to hear the sound, but audio interference in the picture causes horizontal bars that vary with the voice or music.
2. 47.25 MHz is the sound signal of the lower adjacent channel. For instance, when the receiver is tuned to channel 4, the sound signal of channel 3 may interfere with the picture. The lower adjacent sound at 47.25 MHz is always 6 MHz from the associated sound at 41.25 MHz. Since 47.25 MHz is close to the picture carrier at 45.75 MHz, tuning the adjacent sound trap affects the picture IF response for luminance signal.
3. 39.75 MHz is the picture signal of the upper adjacent channel. For instance, when the receiver is tuned to channel 3, the picture carrier of channel 4 may cause interference. The upper adjacent picture at 39.75 MHz is always 6 MHz from the IF picture carrier at 45.75 MHz. Since 39.75 MHz is close to the color IF subcarrier at 42.17 MHz, tuning the adjacent picture trap affects the picture IF response for color signal.

Sound signal in the picture. The audio in the video produces wide horizontal bars in the picture that move in step with the modulation and disappear when there is no voice or music. The 4.5-MHz beat produces a very fine pattern of about 225 pairs of diagonal lines. The 920-kHz beat with the color signal produces approximately 60 pairs of bars. These effects of sound in the picture are illustrated in Chap. 19, on Troubles in the Raster and Picture. This chapter also describes in more detail the effects of interference from adjacent channels.

The effects of sound interference in the picture can usually be produced by varying the fine-tuning control on the rf tuner. The reason is that the rf local oscillator frequency is varied to change the intermediate frequencies and their response in the IF amplifier. When the tuner has automatic fine tuning (AFT), it must be disabled

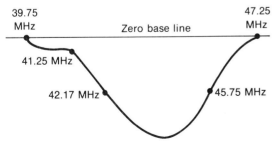

FIGURE 23-16 TYPICAL PICTURE IF RESPONSE CURVE WITH WAVE TRAP FREQUENCIES.

528 BASIC TELEVISION

to vary the oscillator frequency manually. The interference patterns are useful in determining the tuning for the best picture. The oscillator frequency is set to the point that just removes the sound interference. With the correct IF alignment, this adjustment provides good contrast with maximum bandwidth for sharp details in the picture.

Wave-trap circuits. The main types are illustrated in Fig. 23-17. In Fig. 23-17a, the parallel resonant trap is connected in series with the input to the next stage. Parallel resonance provides a very high resistance at the rejection frequency. The series connection for the trap makes it a voltage divider with the following input circuit. Most of the undesired signal voltage is developed across the high resistance of the resonant trap circuit, therefore, with little voltage coupled to the next grid circuit at the rejection frequency. The trap in Fig. 23-17b is a series resonant circuit connected in shunt with the next grid circuit. Series resonance provides a very low resistance at the rejection frequency. At this frequency, the grid input circuit is effectively shorted by the series resonant trap. The result is very little voltage in the grid circuit at the trap frequency.

The absorption trap in Fig. 23-17c is inductively coupled to the plate load inductance. At resonance, maximum current flows within the trap circuit. As a result, maximum current is coupled from the primary. This absorption of power from the primary causes a sharp decrease in the Q of the tuned plate load impedance, only for the trap frequency. The result is reduced gain in the IF amplifier stage for the undesired signal.

The circuit in Fig. 23-17d is a bridged-T trap. The feature of this trap circuit is that it is tuned to the rejection frequency by having the value of shunt resistance R that cancels the in-

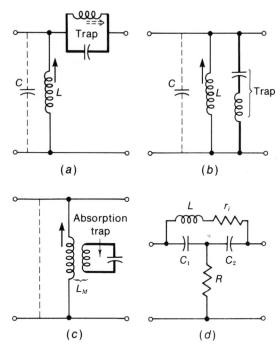

FIGURE 23-17 FOUR TYPES OF WAVE TRAPS TO REJECT AN INTERFERING FREQUENCY: (a) PARALLEL RESONANT CIRCUIT IN SERIES WITH OUTPUT; (b) SERIES RESONANT CIRCUIT IN SHUNT WITH OUTPUT; (c) ABSORPTION TRAP INDUCTIVELY COUPLED TO IF COIL; (d) BRIDGED-T CIRCUIT.

ternal resistance (r_i) of the coil. Then L, C_1, and C_2 form a high-Q parallel resonant circuit in series with the output.

In all cases, the trap is tuned to the frequency to be rejected. As examples, the associated sound traps are aligned at 41.25 MHz to attenuate this frequency in the picture IF response, while the lower adjacent channel sound traps are tuned to reject 47.25 MHz. The coil in the wave-trap circuit has a variable slug to tune the trap for minimum output. Each of the traps can be aligned by coupling in IF signal at the rejection frequency and tuning for minimum dc voltage output across the video detector load resistor. For this part of the IF alignment, the desired result is a dip in the output because the traps have the function of reducing the gain for the rejection frequencies.

23-8 Picture IF Amplifier Circuits

With tubes, there are usually only two stages. Three or four IF stages are needed with transistors. With an IC unit, the entire IF amplifier section is in one chip. It should be noted that either tubes or transistors can be used for monochrome and color receivers. Some IF circuits combine tubes and transistors.

Two-tube IF amplifier for monochrome receiver. Starting at the left in Fig. 23-18, the IF cable from the VHF tuner supplies IF signal to the first IF stage. This link couples IF signal from the mixer plate circuit on the tuner to the IF transformer T_{205}. Note the wave traps at 39.75 MHz for the upper adjacent picture carrier and 47.25 MHz for the lower adjacent sound. The AGC bias line feeds the first IF stage and the rf amplifier in the tuner. The first IF amplifier is coupled to the second IF stage by T_{207}. Only one adjustment is provided in this IF transformer to vary its coupling. The plate circuit of the second IF stage has the double-tuned transformer T_{208} to couple the IF signal into the video detector.

The overall rf and IF response curve shows a bandwidth of about 3 MHz between the 70 percent response points. The test point TP-1 at the plate of the first IF amplifier is used to obtain the response of the link circuit alone, including the mixer plate transformer, not shown here. This circuit provides the required bandwidth, with the slope of the side skirts adjusted by the wave traps. TP-2 is used to feed in signal for aligning the double-tuned transformer T_{208} alone. The overall response shown is at the test point in the video detector output. This curve shows the typical IF response for a monochrome receiver.

The 4JC6A in the second IF stage is a *sharp cutoff pentode*. Its grid cutoff voltage is −3 V for practically zero plate current. The 4KT6 in the first IF stage has a *semiremote cutoff* characteristic, with a grid cutoff of −22 V. This feature is desirable for minimum amplitude distortion as the grid voltage is varied over a wide range by the AGC bias. However, sharp cutoff tubes have higher g_m, for more gain.

Transistorized IF amplifier for color receiver. See Fig. 23-19. The overall IF response for this color receiver is the same as Fig. 23-14, with 45.75 MHz for the IF picture carrier and 42.17 MHz for the IF color subcarrier at 50 percent response at opposite sides of the curve.

The IF cable from the rf tuner supplies IF signal to the base of Q101, which is the first IF amplifier. This stage has a combination of emitter bias with R_{103} and AGC bias at the base through R_{102}. C_{112} is the emitter bypass capacitor for R_{103}. The IF tuning coils are L_{101A} and L_{101B}. There are two traps in this IF input circuit: 47.25 MHz for the lower adjacent sound and 39.75 MHz for the upper adjacent picture. The IF output of Q101 is transformer-coupled by L_{103} to the second IF stage Q102. The feedback capacitor C_{109} from the secondary of L_{103} to the base of Q101 is for neutralization. Collector voltage for Q101 is from the 24-V supply shown at the top of the diagram, through R_{104} and the primary of L_{103}.

The second IF stage Q102 has the input tuned by L_{103} and the output tuned by L_{104}. R_{106} is a damping resistor for L_{103}, and R_{111} damps L_{104}, for more bandwidth. The bias is a combination of emitter bias from R_{110} and fixed bias from R_{108} in the voltage divider, with R_{197} to the 24-V supply. C_{116} is the bypass capacitor for the emitter

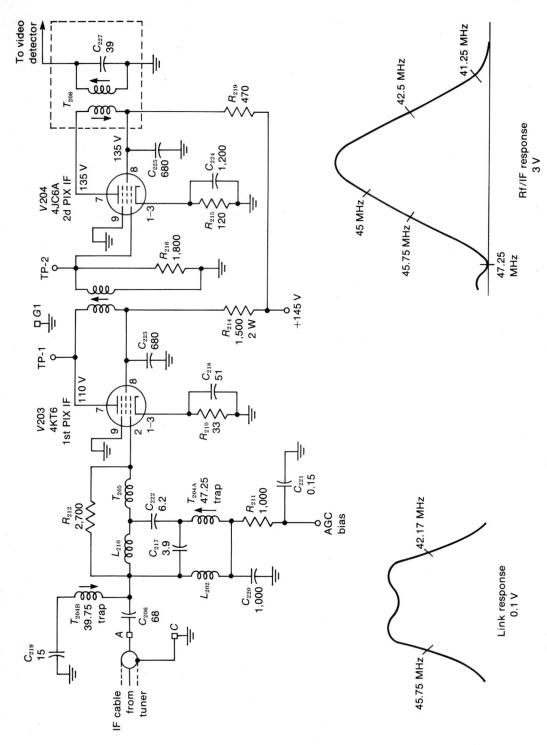

FIGURE 23-18 PICTURE IF AMPLIFIER WITH TWO PENTODE TUBES. R VALUES IN Ω; C MORE THAN 1 pF. *(RCA KCS 179 CHASSIS)*

resistor R_{110}. The collector resistor R_{109} provides shunt feed for the dc collector voltage from the 24-V supply.

The IF output from the collector of Q102 goes to the base of the third IF amplifier Q104 and to a separate sound converter stage Q103, through C_{118}. This sound detector beats the 41.25-MHz IF sound signal with the 45.75-MHz picture carrier to convert to the 4.5-MHz intercarrier sound signal. A separate 4.5-MHz sound converter is generally used in color receivers, instead of using the video detector, in order to reduce beat-frequency interference between the sound and color signals. Since the sound takeoff is in the second IF amplifier here, this stage has the 41.25-MHz trap L_{106} in the output to attenuate the associated sound in the video signal.

The collector of Q104 feeds the IF signal to the video detector through the IF coils L_{109} and L_{110}. This stage also couples IF signal to the AFC module for automatic fine tuning (AFT) on the local oscillator in the rf tuner. The AFC detects any difference between the IF picture carrier frequency and 45.75 MHz and provides a dc voltage to correct the oscillator frequency. This circuit is explained in Chap. 24, The RF Tuner.

23-9 IC Unit for the Picture IF Section

Refer to Fig. 23-20. Although the IC unit is shown in two sections, this is one 20-pin dual-in-line package (DIP) with all the following functions:

1. Three-stage IF amplifier. IF signal from the tuner is in at pin 6 for the first and second IF amplifiers. The amplified IF signal is out at pin 9 for the interstage coupling circuit and back into pin 13 of IC1B for the third IF amplifier.
2. Video detector. The video signal output at pin 19 goes to the video amplifier.
3. AGC. The IF AGC is internal, but the AGC output is also supplied at pin 4 of IC1A for the rf tuner.
4. Sound converter with a 4.5-MHz IF amplifier providing output at pin 2 on IC1B for the audio detector in a separate module for the sound signal.
5. 3.58-MHz chroma signal is out at pin 3 of IC1B for the chroma bandpass amplifier in a separate module for chrominance signal.
6. IF picture carrier signal is out at pin 14 of IC1B for the AFT or AFC module that corrects the frequency of the local oscillator in the rf tuner.
7. The regulator stage Q1 connected between terminals 15 and 18 provides steady dc supply voltage for the amplifier stages in the IC1B section.

Although the IC unit provides miniaturized transistor amplifier circuits, the pin connections must go to external circuits for the dc supply voltages and for the IF tuned circuits. In the IF input circuit for pin 6, the IF coils L_1, L_2, and L_3 are adjusted for the proper bandwidth between the IF picture carrier at 45.75 MHz and the IF color subcarrier at 42.17 MHz. The variable capacitor C_{12} is a trimmer to place these frequencies at 50 percent response. The wave traps T_1 and T_2 are adjusted to attenuate the upper adjacent picture at 39.75 MHz and the lower adjacent sound at 47.25 MHz. In the interstage coupling circuit between pin 9 on IC1A and pin 13 on IC1B, the IF transformer L_6 and IF coil L_5 are adjusted for a symmetrical response curve. This coupling circuit also has the wave trap T_4 to reduce the associated sound signal at 41.25 MHz to 5 to 10 percent response for intercarrier sound.

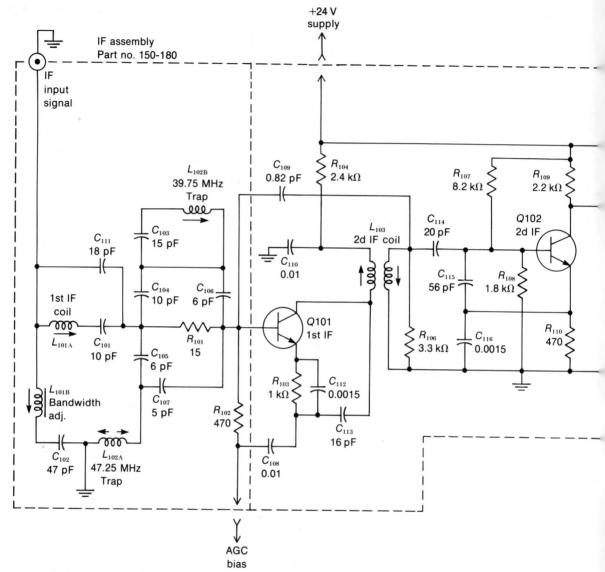

FIGURE 23-19 TRANSISTORIZED PICTURE IF AMPLIFIER CIRCUIT. R VALUES IN Ω, C LESS THAN 1 IN μF. (ZENITH COLOR TV CHASSIS 23 DC14)

23-10 Picture IF Alignment

Correct alignment of the picture IF stages is necessary for a good picture, since the composite video signal is the envelope of picture carrier signal amplified in the IF section. In the modulated signal, those sideband frequencies farthest from the 45.75-MHz picture IF carrier correspond to the highest video frequencies. If these frequencies are missing or attenuated, the picture will not have horizontal detail. In color

THE PICTURE IF SECTION 533

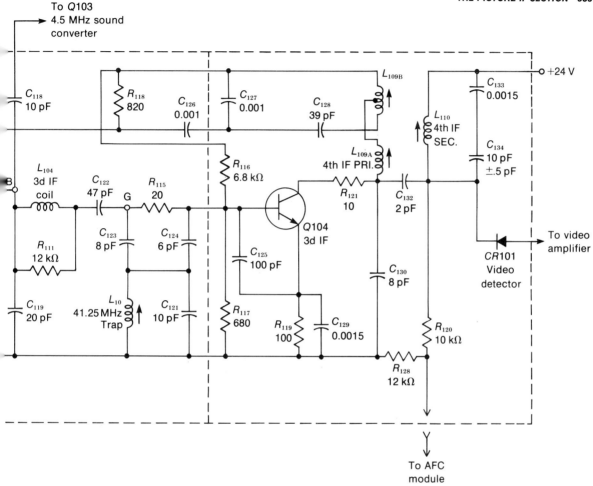

receivers, when the IF color subcarrier at 42.17 MHz is missing or attenuated, there will be no color or weak color.

At the opposite end of the IF response curve, those frequencies close to the IF carrier at 45.75 MHz correspond to the low video frequencies that reproduce large areas of picture information. When the IF response for the picture carrier is less than 50 percent, this attenuation causes a weak picture, with little contrast. Too much gain at 45.75 MHz allows strong contrast, but usually with reduced detail and smear in the picture. In addition, bars in the picture and

920-kHz beat can result from misadjustment of the sound traps.

Actually, the overall response of the IF section is more important than the bandwidth of the video amplifier in determining the bandwidth of the composite video signal and the amount of color signal. The overall IF response curve includes all the IF tuned circuits, from the mixer output in the rf tuner to the video detector.

Visual Response Curve. Figure 23-21 shows the connections for the sweep generator and oscilloscope needed to obtain a visual response

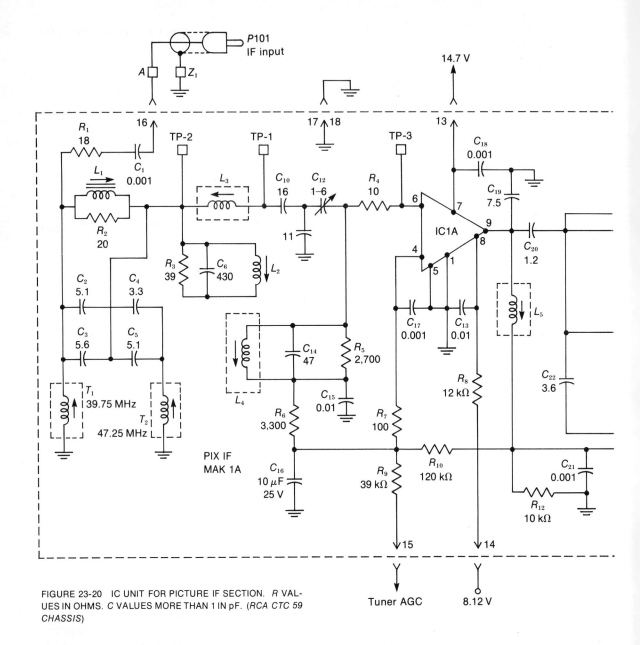

FIGURE 23-20 IC UNIT FOR PICTURE IF SECTION. R VALUES IN OHMS. C VALUES MORE THAN 1 IN pF. (RCA CTC 59 CHASSIS)

curve for the overall gain in the picture IF section. The sweep generator is necessary to supply IF signal input at all frequencies in the passband of the amplifier. The oscilloscope shows the amount of video detector output voltage at the different IF signal frequencies. This voltage indicates IF gain, since the more IF input signal to the detector, the greater its dc output voltage. With the generator supplying signal frequencies that are continuously sweeping through the IF range at the rate of 60 Hz, the detector output is a dc voltage that varies in amplitude at the rate of 60 Hz.

The visual alignment method is generally

necessary to adjust double-tuned stages. Furthermore, we can see the bandwidth and relative gain at the important frequencies of 45.75 MHz for the picture carrier, 42.17 MHz for the color subcarrier, and 41.25 MHz for the associated sound. A typical sweep generator is shown in Fig. 23-22.

The average oscilloscope has high-frequency response that is good enough for alignment. With the vertical input of the oscilloscope connected across the dc load resistance of the dectector, the input to the oscilloscope vertical amplifier is not the IF signal but only dc voltage fluctuating at the 60-Hz modulation rate of

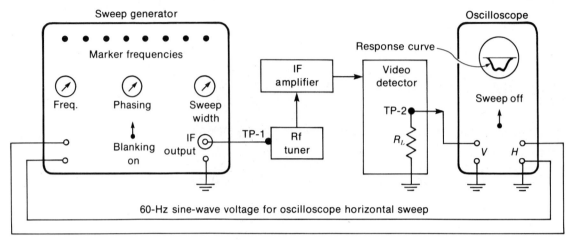

FIGURE 23-21 SWEEP GENERATOR AND OSCILLOSCOPE CONNECTIONS FOR OVERALL IF RESPONSE CURVE. TP-1 IS TEST POINT 1 AT MIXER GRID ON RF TUNER. TP-2 IS DETECTOR OUTPUT. DC BIAS VOLTAGE AND MARKER OSCILLATOR ARE NOT SHOWN SEPARATELY, AS THEY ARE USUALLY SUPPLIED BY SWEEP GENERATOR.

the sweep generator. The amount of dc voltage depends on the amplitude of the signal input to the detector, which is proportional to IF gain. Therefore, detector output indicates the IF amplifier response. Actually, the low-frequency response of the oscilloscope is important because the visual response curve shows a graph of detector dc output voltages, varying at 60 Hz.

Disable the horizontal scanning. This precaution prevents radiation from the horizontal deflection circuits from producing spikes on the response curve, making the curve difficult to mark. However, be careful not to allow the horizontal amplifier to have excessive current without its drive from the horizontal oscillator. You can use a dummy load to keep the B+ voltage from going too high.

Dc bias adjustment. The AGC bias must be disabled so that the alignment can be done with fixed bias. This can be obtained from a bias box. However, many sweep generators also supply the required dc bias. Normal bias and gain can be checked by observing that the oscilloscope screen has ''grass,'' which is receiver noise, without any generator signal.

IF alignment procedure. It is important to realize that just seeing a response curve is not always good enough. You must constantly check to make sure that the curve you see really represents the actual receiver operation with typical signal. Otherwise, you can have a pretty response curve on the oscilloscope screen but a poor picture on the screen of the picture tube. When connecting the test equipment, make the ground connections of the probes as close as possible to the signal ground connections. You can check for proper ground connections by grasping the cables and touching the chassis to be sure your body capacitance does not affect the oscilloscope waveform. All the equipment should warm up for at least 15 min. The following steps illustrate the visual alignment procedure:

FIGURE 23-22 TYPICAL SWEEP GENERATOR FOR RF, IF, AND VIDEO RESPONSE CURVES. (*B AND K DIVISION OF DYNASCAN CORP.*)

1. Connect the oscilloscope vertical input terminals to the video detector output. A test point for the video signal is usually at the top of the chassis. This test point can also be in a video amplifier dc-coupled to the video detector. Use a 10-kΩ isolating resistor.
2. Connect the IF output of the sweep generator to the mixer grid. A test point for this connection is usually at the top of the rf tuner. The sweep cable is usually terminated in 75 Ω. If it is not included, a 2,200-pF blocking capacitor is used.
3. The oscilloscope internal sawtooth sweep is not used. Instead, horizontal deflection for the oscilloscope is a sine wave at 60 Hz. There are two possibilities. If the oscilloscope has a phasing control, switch the horizontal selector switch to 60-Hz line and use this phasing control. Then no external connections to the scope horizontal input terminals are necessary. Usually, though, the oscilloscope horizontal amplifier is used for the 60-Hz voltage from the sweep generator connected to the oscilloscope, as in Fig. 23-21. Then the phasing control on the generator is used.
4. Set the frequency of the sweep generator to obtain an IF response curve on the oscilloscope screen. Use about 10-MHz sweep width to see the entire IF response. The exact frequency of the IF signal does not matter, as the frequencies must be indicated by an accurate marker generator.
5. The IF response curve is downward on the oscilloscope screen, in the negative direction, for negative dc voltage output from the video detector. For positive voltage, the curve is upward. Some oscilloscopes have a polarity-reversing switch to reverse the curve. However, make sure the curve has the correct polarity because overload distortion can turn the curve upside down.
6. Use little generator output to avoid overload distortion. Turn the oscilloscope gain up to make the curve fill the screen. Make sure that varying the generator output voltage

varies the height of the response curve. A basic test for no overload is to reduce the generator signal and see if the curve becomes smaller without changing its shape.

7. Two response curves will be produced by the sweep generator, without blanking. One curve is produced by trace on one half-cycle of the 60-Hz sweep and another by retrace on the next half-cycle. First, you must make the two curves overlap by adjusting the phasing control. Then turn on the blanking to provide one curve with a zero base line. If you turn on the blanking before the curves are superimposed, the trace is a combination of parts of two curves.
8. When you mark the frequencies in the curve, they may increase from left to right or from right to left. Some generators have a reversing switch to make the frequencies increase to the right. However, either method can be used as long as the marker frequencies are correct.

It may be of interest to note that the word "sweep" is used for both the signal generator and the horizontal deflection for the oscilloscope, but the two functions are entirely different. Sweep just means to move or pass over quickly. The signal generator sweeps through a range of different frequencies. In the oscilloscope, the horizontal deflection voltage makes the electron beam move across the screen of the CRT. The internal sweep circuit generates this deflection voltage inside the oscilloscope.

Marker frequencies. The visual response curve means little until frequencies are marked. The marker is a crystal-controlled oscillator that is mixed with the sweep signal to indicate a specific frequency accurately. A separate marker generator can be used, but many sweep generators include the markers.

The IF tuning adjustments. Adjusting the tuned coupling circuits and the wave traps changes the shape of the response curve to fit the markers correctly. It is important to realize that the wave-trap adjustments not only attenuate the undesired frequencies but determine the shape of the IF response at opposite ends of the curve, for the picture carrier and the color subcarrier.

The typical overall IF response curve for a monochrome receiver is shown in Fig. 23-18. Here the main markers are at 50 percent response for 45.75 MHz and 5 to 10 percent response for the sound at 41.25 MHz. The response is low at 42.17 MHz, to eliminate interference from the color signal in monochrome receivers. For color receivers, the typical overall IF response curve is shown in Fig. 23-23a. Here the main markers are at 50 percent response for the picture IF carrier at 45.75 MHz and 50 percent for the color subcarrier at 42.17 MHz, with 5 to 10 percent response for the sound at 41.25 MHz. Some examples of misalignment are shown in Fig. 23-23b, c, and d.

When tuned transformers are being adjusted, the slugs may have two resonance points. They differ, though, in their effect on coupling and bandwidth. Usually, the adjustment farther away from the center of the transformer is correct.

The IF alignment is usually done in three parts:

1. Link alignment. Input signal is at the mixer grid with the oscilloscope at the first IF amplifier. A detector probe is necessary for the oscilloscope here because the IF amplifier does not rectify the signal.
2. Overall IF alignment. Input signal is at the mixer grid with the oscilloscope at the video detector output. No detector probe is needed here.

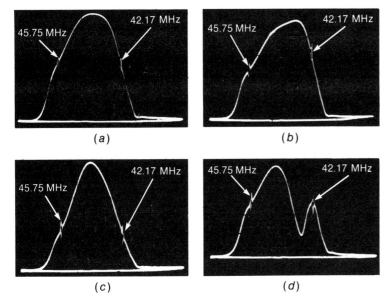

FIGURE 23-23 EFFECTS OF MISALIGNMENT ON OVERALL IF RESPONSE CURVE. (a) CORRECT RESPONSE. NOTE THAT LOWER FREQUENCIES ARE TO THE RIGHT HERE. (b) COLOR SUBCARRIER AT 42.17 MHz TOO HIGH ON CURVE. (c) BANDPASS TOO NARROW WITH RESPONSE TOO LOW FOR PICTURE CARRIER AND COLOR SUBCARRIER. (d) MISTUNED SOUND TRAP AT 41.25 MHz. (RCA)

3. Overall rf and IF alignment. Input signal is at the antenna input with the oscilloscope at the video detector output. This is the final test of how the IF alignment combines with the rf and mixer stages for all the tuned circuits. The input is at the rf channel frequencies, but the markers are at the IF signal frequencies.

Adjusting single-tuned stages. This can be accomplished with just a dc voltmeter and an rf signal generator, such as the marker oscillator. A sweep generator is not needed for test signal at just one frequency. The oscilloscope is not necessary because the dc voltmeter serves as the output indicator. For an IF amplifier circuit, adjust for maximum; for a wave trap, adjust for minimum.

Causes of misalignment. When the rf tuner is replaced, it may be necessary to realign the link-coupling circuit between the mixer output and the IF input. However, it is probably safe to say that misalignment is not a common cause of trouble in receivers. Once the tuned stages have been aligned, they should seldom need readjusting. When a tube or transistor is changed, the different capacitances may affect the tuning, but the IF circuits are usually designed to take this into account. Some types of trouble change the alignment, but in this case realignment is not the solution. An example is an open bypass capacitor for one of the tuned circuits. In general, when there is trouble with the picture, the cause is most likely to be a defective component, rather than misalignment.

23-11 Troubles in the Picture IF Amplifier

Trouble here affects the picture, while the raster is normal. No picture IF signal results in no picture, since there is no video signal for the picture tube. Also, there is no sound when a common IF stage is not operating. If there is no picture because of a defective IF stage, the raster is clean, without any snow. Insufficient gain in the IF section can also cause a weak picture. Misalignment may result in poor picture quality, including color troubles.

The amount of picture IF signal can be checked by measuring the rectified output from the video detector. This dc voltage should be 1 to 3 V. To see the detected picture IF signal, connect an oscilloscope to the video detector output. This amplitude of composite video signal is about 3 V, peak-to-peak.

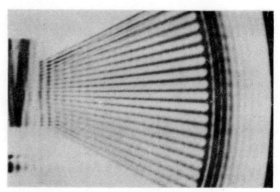

FIGURE 23-24 MULTIPLE OUTLINES CAUSED BY RINGING. (*RCA*)

Tunable smear. This is smear in large areas of the picture that moves when the fine tuning control is varied. As the control varies the local oscillator frequency, the IF picture carrier frequency changes, resulting in a different response for the picture IF signal. Tunable smear may indicate that 45.75 MHz is too high on the IF response curve. High gain for the IF picture carrier and the sideband frequencies close to the carrier allows strong contrast in the picture, but some high-frequency detail is missing. Then the picture is not sharp and large areas may be smeared.

Narrow bandwidth. In this case, the IF response does not have enough bandwidth to provide normal gain for the sideband frequencies farthest from the picture carrier. These side frequencies correspond to high video frequencies. Therefore, the effect is the same as having insufficient gain for high video frequencies in the video amplifier. The result is poor horizontal resolution, with little detail. This picture cannot be sharp and clear.

Grainy picture. If the IF picture carrier is too low on the response curve, the picture will have weak contrast for large areas, with the high-frequency details too strong. This effect is equivalent to having too much IF bandwidth. The picture looks grainy, with the speckles of noise voltages too obvious. In addition, this response makes it easy to have sound interference in the picture.

Color subcarrier too low. Refer to Fig. 23-23c. Here the color at 42.17 MHz is less than the required 50 percent response. As a result, the color in the picture can be weak.

Oscillations in the IF amplifier. Regeneration can make the IF amplifier oscillate, especially when there is an excessive peak in the IF response at one frequency. The effect in the picture is usually multiple outlines, with black streaks caused by radiation of IF signal back to the rf tuner. See Fig. 23-24.

To check whether the IF amplifier is oscillating, measure the rectified IF output across the video detector load resistor. Normally, this dc output is 1 to 3 V with a station tuned in.

When the fine tuning control is varied, the dc voltage varies as the picture carrier moves up or down the side of the IF response curve. Also, the detector output drops practically to zero when the receiver is switched to an unused channel. However, if the detector output voltage is very high and stays about the same regardless of the rf tuning, it indicates that the IF amplifier is oscillating. The trouble can be caused by misalignment or by a defective component that alters the alignment.

No sound with no picture. Usually, all the picture IF stages are common to the sound. As a result, no IF signal means no picture and no sound. In some cases, there may be weak sound, as sound signal can be capacitively coupled through an open stage.

SUMMARY

1. The picture IF section amplifies the IF signal output from the mixer to supply enough IF voltage for the video detector. With 0.5 mV into the IF section, an IF gain of 8,000 provides 4-V output.
2. A coaxial cable connects the IF signal from the mixer on the rf tuner to the first IF amplifier. Link coupling is used for the cable, as shown in Fig. 23-11.
3. With tubes, two pentodes are generally used. With semiconductors, three or four transistors or one IC unit are needed. Usually the whole IF section is a separate module.
4. The standard frequencies are 45.75 MHz for the IF picture carrier, 41.25 MHz for the first IF sound carrier, and 42.17 MHz for the first IF color subcarrier.
5. The IF picture carrier has 50 percent response to compensate for vestigial-sideband transmission. On the opposite end of the IF response curve, the IF color subcarrier also has 50 percent response. The associated sound signal is also amplified for intercarrier sound, with a response of 5 to 10 percent.
6. The IF bandwidth is about 2.5 MHz in monochrome receivers and 3.5 MHz in color receivers, to accommodate the side bands of the modulated picture carrier signal. Side frequencies close to the carrier correspond to the low video frequencies that determine contrast in the picture. Side frequencies farther from the picture carrier correspond to the high video frequencies that determine horizontal detail. The color subcarrier at 42.17 MHz can be considered a side frequency separated from the picture carrier by 3.58 MHz.
7. Wave traps are used in the picture IF amplifier to attenuate the associated sound at 41.25 MHz, the lower adjacent sound at 47.25 MHz, and the upper adjacent picture at 39.75 MHz. These trap frequencies are marked on the IF response curve in Fig. 23-16.
8. Each IF stage is a tuned amplifier. It has gain only for frequencies that provide a resonant rise in impedance. The IF amplifier operates class A for minimum distortion of the video envelope in the AM signal.

9. The dc voltage drop across an IF coil is practically zero. This means the dc voltage at the plate or collector is essentially the same as the supply voltage at the low end of the coil.
10. With transistor amplifiers, the IF coil is often tapped down to match the relatively high collector output resistance to the lower base input resistance of the next stage.
11. Transistor IF amplifiers are often neutralized to prevent the amplifier from oscillating. A neutralizing capacitor C_N feeds back signal from output to input, in opposite polarity from the amplified output. See Fig. 23-13.
12. In a double-tuned transformer, both L_P and L_S are tuned to the same frequency. The advantages are more bandwidth with sharper skirt selectivity than is provided by a single-tuned stage.
13. With staggered single-tuned stages, each is resonant at a different frequency to broaden the overall response.
14. The overall picture IF response can be observed with an oscilloscope connected to the video detector output, while a sweep generator at the mixer grid provides the required frequencies in the IF passband. See Fig. 23-21. Fixed bias must be used instead of AGC bias. Also, the rf local oscillator and horizontal scanning should be disabled to prevent interference in the response curve. Typical curves are shown in Fig. 23-23.
15. No output in a common IF amplifier means no picture and no sound, although the raster is normal. The raster is clean, without any snow. You can check for IF output by measuring the dc voltage across the video detector load resistor; it should be 1 to 3 V, with the rf tuner on a channel.

Self-Examination   (Answers at back of book)

Answer True or False.
1. The IF output of the mixer includes picture carrier signal at 45.75 MHz, sound carrier signal at 41.25 MHz, and color signal at 42.17 MHz.
2. Link coupling is generally used for the IF coaxial cable from the rf tuner to the first IF amplifier on the main chassis.
3. The amount of IF signal into the video detector is approximately 100 V.
4. If the IF picture carrier at 45.75 MHz is too high on the IF response curve, the picture will not have enough contrast.
5. If there is no IF gain at 42.17 MHz, the picture cannot have any color.
6. The trap frequency for the lower adjacent channel sound carrier is 47.25 MHz.
7. The intercarrier sound frequency is 4.5 MHz out of the video detector or out of a separate sound converter stage.
8. The color subcarrier frequency out of the video detector is 3.58 MHz.
9. A tuned circuit has maximum ac impedance at parallel resonance.

10. A single-tuned circuit with f_r of 44 MHz and Q of 20 has a bandwidth of 2.2 MHz.
11. A picture IF stage operates class A for minimum distortion of the envelope of the AM picture signal.
12. A smaller value of shunt damping resistance across a tuned circuit increases the bandwidth.
13. The plate voltage on a pentode IF stage is about one-half the dc voltage at the low side of the IF coil.
14. When an IF coil in the collector circuit is tapped down, this is done for impedance matching to the base of the next transistor.
15. In a double-tuned transformer, both the primary and secondary are tuned to the same frequency.
16. A wave trap is always tuned for minimum output at the trap frequency.
17. Fixed bias must be used instead of AGC bias when an IF stage is aligned.
18. Although the sound is an FM signal, it can produce audio interference in the video signal.
19. When the oscilloscope is connected to the video detector output for the IF response curve, a scope demodulator probe must be used.
20. With a visual response curve, the marker frequencies indicate where the picture carrier is.
21. The overall IF response curve from the video detector in Fig. 23-16 shows negative dc voltage out of the detector.
22. When three IF amplifiers have a voltage gain of 20 for each stage, the overall gain equals 8,000.
23. In a stagger-tuned amplifier, each stage is resonant at a slightly different frequency.
24. When an IF stage common to both the picture and sound signals is cut off, the result is no picture and no sound.
25. When an IF stage after the sound takeoff point is cut off, the result is no picture but normal sound.
26. The base-emitter bias on an NPN silicon transistor is about 0.6 V positive.
27. For forward AGC at the base of an NPN transistor, the bias becomes more positive with more signal.
28. The collector voltage is positive for an NPN transistor.
29. In the common-emitter (CE) circuit, the output resistance in the collector circuit is much higher than the input resistance in the base circuit.
30. In Fig. 23-19, the only IF stage that has AGC bias is $Q101$.

Essay Questions

1. Draw a typical overall IF response curve for a color receiver. Mark the following frequencies: picture carrier, associated sound carrier, and color subcarrier.

2. Give one difference between the IF response for color and monochrome receivers.
3. Repeat Table 23-1, but with the frequencies listed for channel 2 at 54 to 60 MHz.
4. Draw the schematic diagram of a single-tuned IF amplifier, using a pentode tube with a plate decoupling filter.
5. Draw the schematic diagram of a single-tuned IF amplifier, using an NPN transistor with a decoupling filter in the collector circuit.
6. Show a neutralized transistor amplifier.
7. Why is neutralization used with transistor amplifiers but not with pentodes?
8. Why is an IF coil tapped down for the next stage in transistor amplifiers? Why is this not necessary with vacuum tube amplifiers?
9. Draw the schematic diagram of a double-tuned IF stage using an NPN transistor.
10. The receiver is tuned to UHF channel 83 at 884 to 890 MHz. List the following carrier frequencies in the IF amplifier: picture carrier, color subcarrier, associated sound, lower adjacent sound.
11. What is the advantage of stagger tuning compared with synchronous tuning?
12. Give one advantage of a bifilar coil for a single-tuned circuit.
13. What is the advantage of a double-tuned amplifier compared with a single-tuned stage?
14. List three trap frequencies in the picture IF section.
15. What causes audio sound bars in the picture? Give two differences between sound bars and hum bars.
16. Name the two signals into the video detector which produce the 4.5-MHz intercarrier sound and give their frequencies. Do the same for a separate sound converter stage.
17. Why is link coupling used between the mixer output and IF input?
18. Describe briefly how to obtain the overall visual IF response curve. (a) What is the purpose of using fixed bias? (b) What is the function of the marker frequencies? (c) Why is the sweep generator at the mixer grid, instead of at the first IF stage?
19. List five controls or output terminals on a sweep generator with their functions.
20. Why is a demodulator probe needed for sweep alignment of a double-tuned coupling circuit between the mixer output and first IF stage?
21. In the IF amplifier in Fig. 23-18, which pentode stages are common to picture and sound?
22. In the transistor IF amplifier in Fig. 23-19, which IF stage is for picture alone?
23. In Fig. 23-18, give two functions for C_{223} at pin 8 of the first IF amplifier tube.
24. In Fig. 23-18, give the functions for T_{204A}, T_{204B}, and T_{207}.
25. Name the IF stages that have AGC bias in Figs. 23-18 and 23-19.
26. In Fig. 23-20, give the functions for T_1, T_2, T_4, and L_6.
27. In Fig. 23-20, at which pin of the IC chip can you connect an oscilloscope to see the composite video signal?
28. Give five possible effects in the picture which result from incorrect IF response.

Problems (Answers to selected problems at back of book)

1. For the IF response curve in Fig. 23-14, assume the overall voltage gain at 44 MHz is 8,000. How much is the gain at (a) 45.75 MHz; (b) 42.17 MHz; (c) 41.25 MHz?
2. The overall gain for the curve in Prob. 1 is reduced one-half by AGC bias. How much is the gain at 45.75 MHz, 44 MHz, 42.17 MHz, and 41.25 MHz?
3. In Fig. 23-19, when the collector current I_C equals 5 mA for Q102, how much is its collector voltage V_C?
4. Give the local oscillator frequency with rf and IF carrier frequencies for picture and sound of channels 3 and 4 when the receiver is tuned to: (a) channel 4; (b) channel 3.
5. Calculate the inductance L for resonance at 43 MHz with capacitance C of 10 pF.
6. Give the function for each of the frequencies listed below.

FREQUENCY (MHz)	FUNCTION
39.75	
41.25	
42.17	
43.75	
45.75	
47.25	
211.25	
257.00	

24 The RF Tuner

The rf amplifier, mixer, and local oscillator stages form the rf tuning section, generally called the *front end* or *tuner*. This section selects the picture and sound signals of the desired station by converting the rf signal frequencies of the selected channel to the intermediate frequencies of the receiver. Only those frequencies in the IF passband can be amplified enough for you to see a picture and hear the sound.

There are separate VHF and UHF tuners, as shown in Fig. 24-1, each on a separate subchassis. For either tuner, it is the local oscillator that tunes in the desired station. The reason is that the oscillator frequency determines which rf signal frequencies become IF signal frequencies to be amplified by the receiver. The fine tuning control varies the oscillator frequency slightly for exact tuning for the best picture and sound. Color receivers usually have an automatic fine tuning (AFT) circuit to control the oscillator frequency. More details of the tuner circuits are described in the following topics:

24-1 Tuner operation
24-2 Functions of the rf amplifier
24-3 Rf amplifier circuits
24-4 The mixer stage
24-5 The local oscillator
24-6 Automatic fine tuning (AFT)
24-7 Types of rf tuners
24-8 VHF tuner circuits
24-9 UHF tuner circuits
24-10 Cable television channels
24-11 Varactors for electronic tuning
24-12 Rf alignment
24-13 Remote tuning
24-14 Tuner troubles

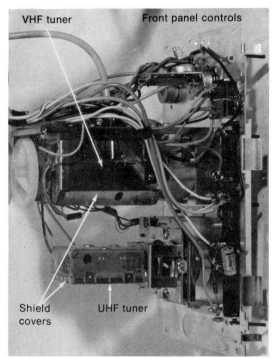

FIGURE 24-1 MOUNTING ASSEMBLY WITH VHF AND UHF TUNERS. LENGTH OF VHF TUNER IS 4 IN. *(MAGNAVOX CO.)*

24-1 Tuner Operation

The VHF tuner is for channels 2 to 13, with frequencies from 54 to 216 MHz in the VHF band of 30 to 300 MHz. Channel 1 is not assigned to television broadcasting anymore, but this position on the station selector switch is generally used to turn on the UHF tuner. The UHF channels are 14 to 83, with frequencies of 470 to 890 MHz in the UHF band of 300 to 3,000 MHz. For UHF operation, the VHF tuner serves as an IF amplifier for the output of the UHF tuner. See the block diagram illustrating tuner operation in Fig. 24-2.

Referring to the rf amplifier on the VHF tuner in Fig. 24-2, its input is the rf signal from the antenna. The input circuit is tuned to the 6-MHz band of the selected channel to amplify its rf picture and sound carrier signals. The amplified output, still at the rf channel frequencies, is coupled into the mixer stage. Included are the FM sound carrier and the AM picture carrier with its multiplexed color subcarrier for a color program. In the mixer stage, the rf signals beat with the oscillator output for conversion to the intermediate frequencies. The one oscillator frequency beats with both picture and sound rf carrier signals to produce the IF signals.

The local oscillator frequency beats above the rf signal frequencies. For the example in Fig. 24-2, tuning in channel 4, the oscillator is at $67.25 + 45.75 = 113$ MHz.

When the station selector switch is varied, the resonant circuits for the rf amplifier and mixer input are changed to the 6-MHz band for each channel. At the same time, the oscillator frequency is changed to the value that can beat the rf picture and sound carrier frequencies to the IF values of 45.75 MHz and 41.25 MHz. The circuits are on a common shaft for ganged tuning.

Tuner connections. As illustrated in Fig. 24-3, the VHF tuner needs connecting leads for B+ voltage, AGC bias, and heater voltage for tubes. The IF output is connected to the IF amplifier by a short length of coaxial cable with a plug at the tuner end. Output from the UHF tuner also has a coaxial cable that plugs into the VHF tuner. Since there is a terminal strip for the electrode voltages and coaxial plug connections for the signal leads, it usually is relatively easy to disconnect the tuner for replacement.

UHF tuner. This tuner is usually solid state, so that it only needs one B+ lead from the VHF tuner. In Fig. 24-2 only a UHF mixer and oscillator are included, without an rf amplifier. The

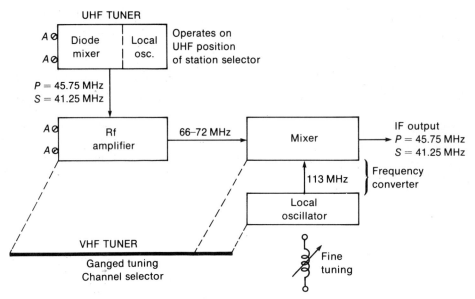

FIGURE 24-2 BLOCK DIAGRAM OF VHF TUNER, SET FOR CHANNEL 4 AT 66 TO 72 MHz. NOTE THAT THE IF OUTPUT OF THE UHF TUNER IS FED TO THE VHF TUNER FOR UHF OPERATION ONLY.

UHF antenna signal is coupled into the diode mixer. The output of the UHF oscillator is also coupled into this stage. The oscillator beats with the UHF channel frequencies to produce IF output at 45.75 MHz for the picture carrier and 41.25 MHz for the sound carrier. For channel 21 at 512 to 518 MHz, as an example, P is 513.25 MHz and S is 517.75 MHz. The UHF oscillator then is at 559 MHz.

Then the IF picture and sound signals from the UHF mixer are coupled to the VHF tuner. These rf and mixer stages operate as IF amplifiers when the VHF station selector is set to the UHF position. The VHF oscillator is disabled and the rf tuned circuits are changed to IF tuned circuits.

Turret tuner. See Fig. 24-4. This VHF tuner has coil strips for each channel mounted with spring clips on a drum or turret. Rotating the drum connects any strip to a set of fixed contacts inside the tuner subchassis. The strip has coils for the rf amplifier, mixer, and oscillator tuned circuits for one channel. On the UHF position, B+ voltage is removed from the VHF oscillator, while the rf amplifier and mixer input circuits are tuned to the IF band.

Rotary-switch tuner. See Fig. 24-5. This VHF tuner has rotary wafer switches for each stage. The oscillator section is at the front. Each wafer has contacts for 12 VHF channels plus the channel 1 position for UHF operation. All the wafers are ganged on one common shaft.

Manual fine tuning control. The usual method of manual fine tuning is to push in the control to engage a plastic wheel that turns a slug in the oscillator coil. Adjust to the point where you see sound bars and then back off the tuning slightly for the best picture. With color, tune for maximum color without 920-kHz beat in the picture.

THE RF TUNER 549

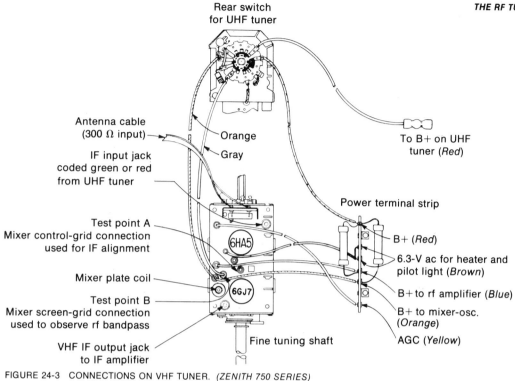

FIGURE 24-3 CONNECTIONS ON VHF TUNER. *(ZENITH 750 SERIES)*

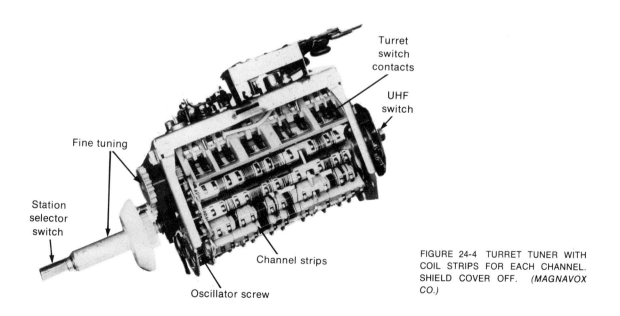

FIGURE 24-4 TURRET TUNER WITH COIL STRIPS FOR EACH CHANNEL. SHIELD COVER OFF. *(MAGNAVOX CO.)*

550 BASIC TELEVISION

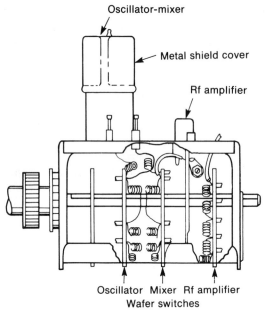

FIGURE 24-5 ROTARY-SWITCH TUNER WITH WAFER SWITCHES FOR EACH STAGE.

Mechanical detent. The VHF tuner has a wheel with notches for each position. Each notch, or detent, holds the switch steady for good contact by means of a detent spring. The tuner should be tight to turn as it clicks into position on each channel. Good connections at the switch contacts are essential for the very high signal frequencies.

Mechanical troubles. The two most common problems with VHF tuners are a weak detent spring and dirty contacts. All the switch contacts are silver plated for very low resistance. However, silver tarnishes as it becomes oxidized by the air. After a year or two, the tuner contacts often must be cleaned, preferably with a commercial tuner spray. Remove the tuner cover, which snaps off, to spray all the switch contacts. Then rotate the switch in both directions a few times.

Indications of poor contact at the switches are: (1) moving the station selector slightly improves the picture, or the switch must be "jiggled"; (2) rotating the switch in one direction tunes in the station better than the opposite rotation. When the switch feels tight, the intermittent connections are probably caused by dirty contacts. If the switch is loose, the detent spring should be replaced.

24-2 Functions of the RF Amplifier

The main purpose of the amplifier is to provide enough rf signal into the mixer for a clean picture without snow. Random noise generated in the receiver circuits is amplified to produce the white speckles called *snow* in the picture (Fig. 24-6). The mixer circuit generates the most noise because of its heterodyning function. Furthermore, the mixer noise is amplified in the IF section. For this reason, the noise generated in the mixer stage is called *IF noise* or *receiver noise*.

If there is sufficient rf signal into the

FIGURE 24-6 SNOW IN PICTURE, CAUSED BY WEAK RF SIGNAL.

mixer, the signal-to-noise ratio will be high enough to produce a picture without visible snow. The required signal-to-noise ratio is approximately 30:1. A typical value of receiver noise at the input is 15 µV. With a low-noise rf amplifier, a signal level of 450 µV or more at the receiver input can provide a picture without snow.

Any time you see snow in the picture, it shows that the mixer noise is being amplified in the IF and video amplifiers. The trouble is weak rf signal, either in the rf amplifier or in the antenna input.

Oscillator radiation. The rf amplifier provides the only isolation between the local oscillator stage and the antenna input connections. Any oscillator output coupled to the antenna can be radiated to produce interference in nearby receivers. Therefore, the rf amplifier also serves as a buffer stage to minimize oscillator radiation.

It should be noted, though, that the oscillator output can also be radiated by the receiver chassis. This is the reason why a separate chassis-ground return is generally used for the tuner circuits. The FCC specifies the maximum radiation at a distance of 100 ft or more from the receiver. These limits for field strength are 50 to 150 µV/m in the VHF band and 500 µV/m in the UHF band.

Oscillator radiation produces diagonal lines in other receivers, just like any cw interference pattern. However, the use of 45.75 MHz for the picture IF carrier in the receiver results in oscillator frequencies that are not in the band of any VHF channel.

Image frequencies. Practically all the receiver selectivity to reject the frequencies of adjacent channels is in the IF section. However, superheterodyne receivers have the problem of rejecting interfering rf signals that can beat with the local oscillator to produce frequencies within the IF passband. These are image frequencies because they are as much above the oscillator frequency as the desired signal frequency is below it. As an example, in Fig. 24-2 the oscillator frequency is at 113 MHz for the picture carrier of $113 - 45.75 = 67.25$ MHz for channel 4. The image frequency then is $113 + 45.75 = 158.75$ MHz.

Rf selectivity. Another function of the rf amplifier is to prevent reception of spurious signals, which have no relation to the desired signal frequency. Especially important is rejecting rf signals in the IF passband. Any interference in the range of 40 to 47 MHz that enters the IF amplifier will be amplified along with the desired signal.

Figure 24-7 illustrates the wave traps and filter generally used in the rf amplifier input to improve rf selectivity and prevent interference. These circuits include:

1. IF wave trap to reject 40 to 47 MHz.
2. High-pass filter cutting off at 50 MHz. All frequencies below 54 MHz for channel 2 should be rejected.
3. FM trap to reject the commercial broadcast band of 88 to 108 MHz. These frequencies can cause interference in channel 2.

The rf tuned circuits are often called the *preselector*, indicating that they select only the desired signal.

Input impedance. A definite value must be provided at the antenna terminals on the receiver, which is the input circuit for the rf amplifier. The purpose is to match the characteristic impedance of the transmission line from the antenna. Without this match, the length of the line would be critical. The standard impedance is

552 BASIC TELEVISION

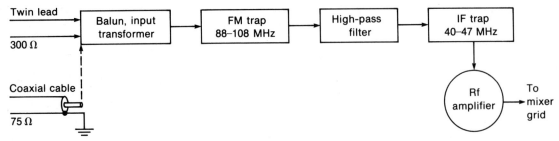

FIGURE 24-7 BLOCK DIAGRAM OF INPUT CIRCUIT FOR RF AMPLIFIER STAGE.

300 Ω for television receivers and FM receivers, to match 300-Ω twin lead. Most receivers now also have a grounded jack with 75-Ω impedance for coaxial cable. The impedance of 300 Ω balanced, or 75 Ω single-ended to ground, is provided by a balancing transformer (Fig. 24-8).

24-3 RF Amplifier Circuits

The transistor stage in Fig. 24-9, the triode tube in Fig. 24-11, the cascode twin-triode in Fig. 24-12, and the dual-gate field-effect transistor (FET) in Fig. 24-13 are typical. Common triode tubes for the rf amplifier are the 6DS4 nuvistor, 6HA5, 2FH5, and 3GK5. Triodes generate less tube noise than pentodes because the internal

FIGURE 24-8 BALANCING TRANSFORMER OR BALUN FOR RF INPUT CIRCUIT.

noise is increased by partition of the space current with more grids.

Transistor rf amplifier. The rf stage in Fig. 24-9 uses an NPN transistor with a grounded shield. The input circuit to the base is tuned to the desired channel by the parallel resonant circuit with L_a, C_1, and C_2. In the collector output circuit, the double-tuned transformer T_r has L_2 and L_3 tuned by C_7 and C_8. The collector supply of 10 V is fed through the RC decoupling filter of R_3 with bypass C_{10}. Circles shown at the connections for L_a, L_2, and L_3 indicate that these coils are switched for each channel. The amplified rf output goes to the input of the mixer stage.

The transistor operates as a class A amplifier, usually with AGC bias. Note that in Fig. 24-9, the AGC bias is fed to the base of Q1 through the RC decoupling filter of R_2 with bypass C_9.

The emitter voltage of 0.5 V on Q1 is produced by I_E through R_1. Also, the AGC bias at the base is 1.1 V. Then the net bias for V_{BE} is $1.1 - 0.5 = 0.6$ V.

Feedthrough bypass capacitors. The symbol for C_9 and C_{10} in Fig. 24-9 indicates a feedthrough construction, generally used for bypassing on the B+ and AGC lines into the rf tuner. As shown in Fig. 24-10, one side of the capacitor is the metal base, which is mounted on

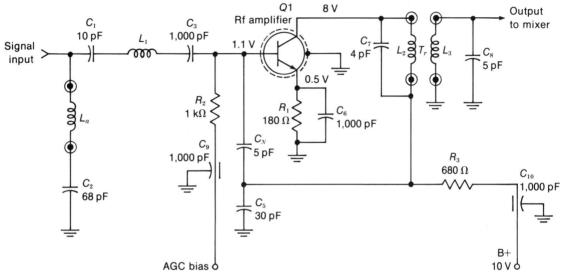

FIGURE 24-9 TRANSISTOR RF AMPLIFIER CIRCUIT.

the chassis. The two end terminals are both connected internally to the opposite capacitor plate. This construction features very low inductance. In addition, it prevents radiation of signal from the leads that must go through the metal case of the tuner. Furthermore, the end terminal outside the chassis can serve as a convenient looker point for connecting test equipment.

Rf neutralization. A triode tube or transistor for the rf amplifier has internal feedback from output to input, which can make the stage oscillate. To prevent oscillations, the rf amplifier often has a neutralizing circuit. Neutralization requires feeding back part of the output signal with opposite polarity from the internal feedback. In Fig. 24-9, C_N is the neutralizing capacitor and feeds signal back from the collector circuit to the base.

Neutralized triode rf amplifier. In Fig. 24-11, the neutralizing capacitor C_N feeds back part of the plate signal to the control grid. Note that the plate coil L_3 is tapped to provide signal at the low end for neutralizing voltage. In the input circuit of the rf amplifier, L_1 and L_2 are tuned by stray capacitances to the frequencies of the desired

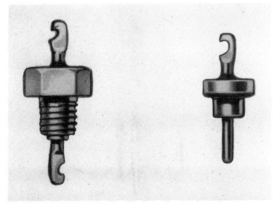

FIGURE 24-10 FEEDTHROUGH BYPASS CAPACITORS. LENGTH ABOUT 3/4 IN. WITH LEADS.

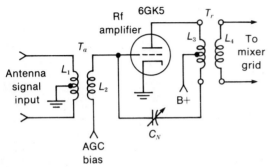

FIGURE 24-11 TRIODE RF AMPLIFIER CIRCUIT WITH NEUTRALIZATION.

channel. Similarly, L_3 and L_4 in the plate circuit provide the amplified rf signal into the mixer stage.

The conventional circuit in Fig. 24-11 is called a *grounded-cathode amplifier*. Input signal is applied to the control grid, with respect to the common cathode, while the amplified output signal is from the plate. It is also possible to use a *grounded-grid amplifier*. Then the input signal is applied to the cathode instead of the control grid, which is grounded for signal. The grounded-grid rf amplifier generally does not require a neutralizing circuit because there is no signal voltage at the control grid.

Cascode rf amplifier circuit. See Fig. 24-12. A twin-triode tube combines a grounded-cathode stage *V1* driving *V2* as a grounded-grid amplifier. The *V1* stage has AGC bias, which is dc-coupled to *V2*. This circuit has the gain of a pentode but the low noise of a triode. The *V1* stage is not neutralized because it has a very low plate load impedance. Most of the gain is in the grounded-grid stage *V2*, which needs no neutralization.

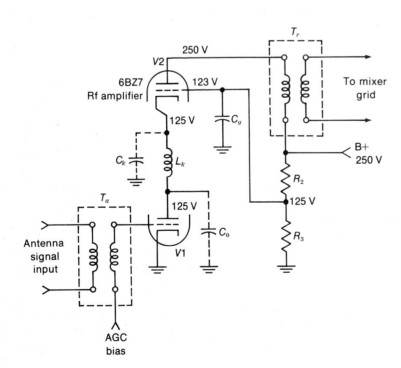

FIGURE 24-12 CASCODE RF AMPLIFIER CIRCUIT USING TWIN-TRIODE WITH DC COUPLING.

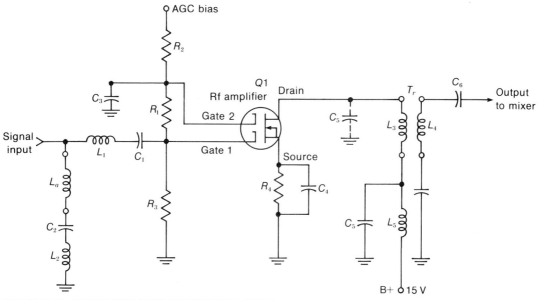

FIGURE 24-13 RF AMPLIFIER USING FET WITH DUAL GATES.

Note that V1 and V2 are in series for dc supply voltage. The plate current of V1 is the cathode current of V2. As a result, the plate-cathode voltage for each tube is one-half the B+ supply voltage. Note that 123 V is provided at the grid of V2 by the R_2R_3 voltage divider. Then the V2 grid-cathode bias is 123 − 125 = −2 V. C_g is a bypass to return the grid of V2 to ground for rf signal.

The signal input is applied by T_a to the grid of V1. Its plate load impedance is the resonant circuit formed by L_k, C_k, and C_o. They form a π-type filter coupling circuit. The resonant rise in voltage across C_k provides signal input for the cathode of the grounded-grid stage V2. In the output circuit, the amplified rf signal in the plate circuit of V2 is coupled by T_r to the mixer grid. The rf tuning for each station is changed only for the transformers T_a and T_r, as the network between V1 and V2 provides a resonant response broad enough to include all 12 VHF channels.

FET rf amplifier. The circuit in Fig. 24-13 uses a field-effect transistor (FET) with dual gates for its N channel. An FET features very high input impedance, compared with junction transistors, and has low noise. The drain electrode corresponds to the plate or collector, the gate is like the control grid or base, and the source has the function of the cathode or emitter. The main advantage of an FET or amplifier is that it can operate without overload distortion for a wide range of signal input voltages.

This dual-gate FET circuit is similar to a cascode amplifier. Gate 1 has input signal as a grounded-source stage, while gate 2 is for a grounded-gate stage. C_3 is an rf bypass. The amplified output signal from the drain is coupled by T_r to the mixer stage. This FET rf amplifier does not need neutralization because of the grounded-gate stage. The AGC bias voltage applied to gate 2 and gate 1 may vary from -5 V for strong signals to $+6$ V for weak signals.

Rf response curve. The graph of relative gain in Fig. 24-14 shows that the rf amplifier has 6-MHz bandwidth for each selected channel. The signals amplified are the sound carrier S and the picture carrier P with its multiplexed 3.58-MHz color subcarrier signal C. This overall rf response is from the antenna input terminals to the mixer grid. Note that the curve is shown for channel 10 at 192 to 198 MHz because this channel is generally used for rf alignment. Each channel should have this response within approximately ± 10 percent for the relative amplitudes of P and S.

24-4 The Mixer Stage

The function of this circuit is to convert the rf signal frequencies to the IF passband of the receiver. The two inputs required are (1) rf signal and (2) oscillator injection voltage. The mixer output is produced by heterodyning or beating the oscillator voltage with the rf signal. As a result, the difference frequencies provide the IF signal.

Note that the one oscillator signal beats with both the picture and sound carriers. Therefore, the separation between the two carrier frequencies in the IF output is the same 4.5 MHz as in the rf signals. Also, the color subcarrier signal is 3.58 MHz from the picture carrier frequency in either the rf signal or the IF signal.

The mixer circuit generally uses a pentode tube as in Fig. 24-15 or a transistor as in Fig. 24-16 for the VHF tuner. In the UHF tuner, the mixer is usually a crystal diode, which has less noise and requires less oscillator voltage. There is no AGC bias on the mixer stage.

Rectification in the converter circuit. The mixer stage combined with the local oscillator can be considered as a *frequency converter.* Rectification in the mixer is necessary, as the nonlinear response produces the beat frequencies. Therefore, the mixer is the first detector in the superheterodyne receiver, rectifying the combination of rf signal and oscillator voltage to provide IF signal output. In a tube, the grid-cathode circuit serves as a diode rectifier; in a transistor, the base-emitter junction is a diode rectifier.

The oscillator must provide enough injection voltage to cause nonlinear amplification. For a pentode mixer tube, the oscillator injection voltage is about 3 V peak value, which is very large compared with millivolts of rf signal.

The ratio of IF signal output to rf signal input is the *conversion gain* of the mixer stage. A typical gain is 6. With 2-mV rf signal input, as an example, the IF signal output is $6 \times 2 = 12$ mV.

Pentode mixer circuit. The pentode in Fig. 24-15 is combined in a single tube with a triode oscillator, not shown here. The two input sig-

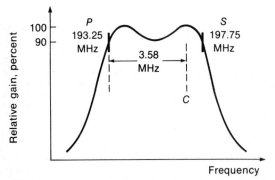

FIGURE 24-14 TYPICAL RF RESPONSE CURVE, FROM ANTENNA TO MIXER. BANDWIDTH OF 6 MHz SHOWN FOR CHANNEL 10 AT 192 TO 198 MHz.

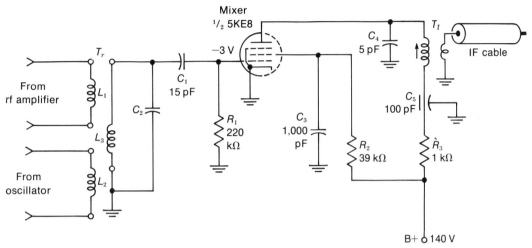

FIGURE 24-15 PENTODE MIXER CIRCUIT.

nals for the mixer are oscillator output coupled by L_2 and rf signal coupled by L_1. Grid-leak bias of -3 V is produced by the oscillator injection voltage. The mixer rf input circuit is tuned to the selected channel by L_3 with C_2.

Because of nonlinear operation in the mixer stage, the output plate current contains the IF signal. The plate circuit of the mixer is tuned to the IF band by the first IF transformer T_I, with the capacitance of C_4. The secondary of T_I supplies IF signal to the tuner output jack for the IF cable to the main chassis.

In the plate circuit, R_3 and C_5 form a decoupling filter in the B+ line. In the screen-grid circuit, R_2 is a dropping resistor to provide the desired screen voltage. C_3 is the bypass for R_2.

Transistor mixer circuit. In Fig. 24-16, rf signal is coupled to Q1 by the rf input transformer T_r. The primary L_1 and secondary L_2 are tuned by C_1 and C_2 for the selected channel. C_4 is a coupling capacitor and blocks the dc bias at the base. Oscillator voltage is injected into the base circuit by capacitive coupling. C_3 and C_5 form a capacitive voltage divider, with the oscillator voltage across C_5 for the mixer.

In the emitter circuit, R_3 and C_6 produce emitter bias. This is reverse voltage for the emitter-base junction of the NPN transistor. However, the R_1R_2 voltage divider provides positive forward bias for the base. The input junction at the base serves as a diode rectifier to heterodyne the rf and oscillator signals.

In the collector output circuit, the IF transformer T_I with C_7 is tuned to the IF signal frequencies. The IF output is connected to the tuner jack for the IF cable. R_4 in the B+ line is a decoupling resistor to isolate the collector signal from the common B+ supply. C_8 is the bypass for R_4.

24-5 The Local Oscillator

The function of the local oscillator is to generate unmodulated sine-wave voltage, or cw output, to beat with the rf signal in the mixer. For each station, the oscillator operates at only the one frequency needed to convert the rf picture and

558 BASIC TELEVISION

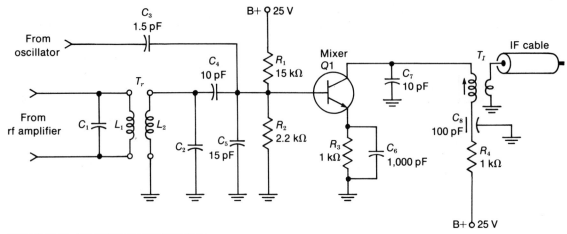

FIGURE 24-16 TRANSISTOR MIXER CIRCUIT.

sound signals to the intermediate frequencies of the receiver.

Table 24-1 lists the oscillator frequencies needed for the 12 VHF channels. These values are calculated as 45.75 MHz above the rf picture carrier in each channel, or 41.25 MHz above the rf sound carrier. The same method can be used to calculate the oscillator frequency for each of the UHF channels listed in Appendix A. It is the oscillator frequency that actually tunes in the channel so that it can be converted to the IF band for the receiver.

Triode oscillator circuit. Most common is the *ultraudion oscillator,* shown in Fig. 24-17a, with a triode tube. This circuit is a modified Colpitts

TABLE 24-1 LOCAL OSCILLATOR FREQUENCIES*

CHANNEL NUMBER	FREQUENCY BAND, MHz	PICTURE CARRIER, MHz	SOUND CARRIER, MHz	OSCILLATOR, MHz
2	54–60	55.25	59.75	101
3	60–66	61.25	65.75	107
4	66–72	67.25	71.75	113
5	76–82	77.25	81.75	123
6	82–88	83.25	87.75	129
7	174–180	175.25	179.75	221
8	180–186	181.25	185.75	227
9	186–192	187.25	191.75	233
10	192–198	193.25	197.75	239
11	198–204	199.25	203.75	245
12	204–210	205.25	209.75	251
13	210–216	211.25	215.75	257

*For VHF channels, with picture IF carrier of 45.75 MHz in receiver.

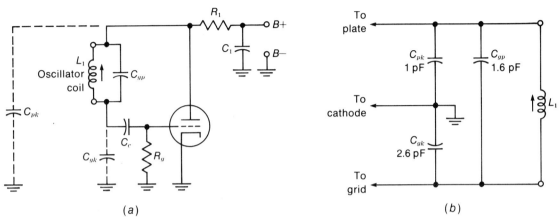

FIGURE 24-17 (a) ULTRAUDION OSCILLATOR AS A MODIFIED COLPITTS CIRCUIT. (b) HOW TUBE CAPACITANCES FORM VOLTAGE DIVIDER FOR FEEDBACK.

oscillator using the tube capacitances as the voltage divider for feedback. The oscillator coil L_1 is connected from plate to grid across C_{gp}. Also, C_{pk} and C_{gk} form a capacitive voltage divider across L_1, as shown in Fig. 24-17b. The oscillator voltage across C_{gk} is the feedback voltage to the grid.

The ultraudion circuit is preferred for the oscillator in television receivers because the stray capacitances are used as part of the tuned circuit. Furthermore, the oscillator coil is not tapped. This feature simplifies switching the coils for every channel.

The oscillator resonant frequency. The output of the oscillator is at the resonant frequency (f_r) of its tuned circuit. Using $f_r = 1/(2\pi\sqrt{LC})$, we can calculate the resonant frequency in Fig. 24-17. Assume L_1 is 0.85 µH. The total capacitance across L_1 is C_{gp} in parallel with the series combination of C_{gp} and C_{pk}. For 2.6 pF in series with 1 pF, the combination is 0.7 pF. This value added to the parallel 1.6 pF makes the total C equal to 2.3 pF.

With 2.3 pF for C across L of 0.85 H, the resonant frequency f_r is 113 MHz. This is the oscillator frequency for tuning in channel 4 at 66 to 72 MHz. The calculations for the resonant frequency are as follows:

$$f_r = \frac{1}{2\pi\sqrt{LC}} = \frac{0.16}{\sqrt{0.85 \times 10^{-6} \times 2.3 \times 10^{-12}}}$$

$$= \frac{0.16 \times 10^9}{\sqrt{0.85 \times 2.3}} = \frac{160}{1.41} \times 10^6$$

$$f_r = 113 \times 10^6 = 113 \text{ MHz}$$

Transistor oscillator circuit. In Fig. 24-18, the NPN transistor $Q4$ is used as an oscillator. The oscillator coil is across a pair of contacts on the rotary switch S_1, shown on channel 13. The arrow on each coil indicates a variable slug for fine tuning.

In the collector circuit of $Q4$, the 24-V B+ is supplied through R_1. The emitter resistor R_2 is not bypassed to provide ac feedback. Also, C_{14} provides feedback from collector to base. Bias for the base is supplied by R_5, through the rf choke L_{10}, from the collector circuit.

This oscillator has automatic fine tuning (AFT) by means of the varactor or capacitive

560 BASIC TELEVISION

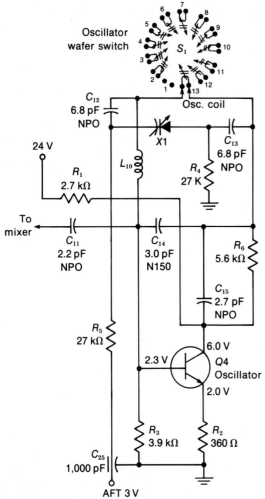

FIGURE 24-18 TRANSISTOR OSCILLATOR CIRCUIT, WITH OSCILLATOR COILS ON WAFER SWITCH. WAFERS FOR RF AMPLIFIER AND MIXER NOT SHOWN HERE. *(FROM ZENITH VHF TUNER 175-2200-40).*

diode $X1$. The dc reference voltage of 3 V is supplied from the AFT circuit on a separate board. When the AFT correction voltage changes, the capacitance of $X1$ varies to change the oscillator frequency.

Capacitors marked NP0 have a zero temperature coefficient. The mark N150 means a negative temperature coefficient with the capacitance decreasing by 150 parts per million per degrees Celsius (°C).

Switching the oscillator coils. In Fig. 24-18, the wafer switch S_1 is rotated to connect one oscillator coil for each channel. The switch is on channel 13, but when rotated one step to the right it connects the coil for channel 12. One step to the left is channel 1, or the UHF position. No coil is used here, as the VHF oscillator is not used for UHF operation. S_1 has the oscillator coils for all 12 VHF channels. However, the coils for the rf amplifier and mixer stages are on separate wafer switches, ganged on a common shaft with S_1. The oscillator is always on the front section for convenience in adjusting the inductance to set the frequency.

Connections for the oscillator coils on the strips of a turret tuner are shown in Fig. 24-19. The coils here are for all the tuner circuits but only one channel. There are 12 strips for all VHF channels plus one strip for the UHF position. Note that on the UHF strip, B+ voltage is supplied to the rf circuits, which operate for the IF band, but there is no VHF oscillator coil.

Oscillator fine tuning. The oscillator coil is usually constructed with an aluminum screw as the core. The fine tuning control turns the screw in or out. Inductance decreases with an aluminum core because of eddy currents in the metal. The screw is turned by a gear wheel that is engaged when the fine tuning control is pushed in against its holding spring. The frequency variation is enough to make it possible to tune in a station on the next channel number.

The gear is disengaged when you release the fine tuning control. As a result, the adjustment for each channel stays the same when

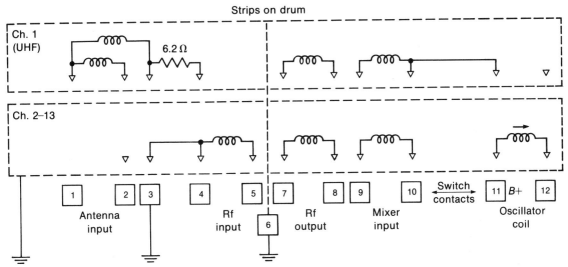

FIGURE 24-19 OSCILLATOR AND RF COILS ON STRIPS FOR TURRET TUNER. ONE STRIP IS USED FOR EACH CHANNEL. *(FROM ZENITH VHF TUNER 175-1814).*

you switch stations. This system is called *preset* or *memory fine tuning*.

Set the fine tuning control for the best picture with normal sound. Adjust until you see sound bars in the picture, then back off the control a little. With color, adjust for 920-kHz beat in the picture, then back off for maximum color without the beat.

24-6 Automatic Fine Tuning (AFT)

The AFT circuit is actually automatic frequency control (AFC) on the local oscillator in the rf tuner. For color receivers, AFT is very useful for tuning in the color without 920-kHz beat. The tuning is generally needed when changing channels, especially with remote control.

The key to AFT operation is having the local oscillator convert the picture carrier to exactly 45.75 MHz in the IF section. This frequency is measured by a discriminator in the AFT circuit, with input from the IF amplifier. The discriminator output is a dc correction voltage that indicates how much off 45.75 MHz the IF carrier frequency is. Since this frequency depends on the oscillator input to the mixer, the AFT voltage indicates the error in oscillator frequency. The AFT circuit can correct a shift in frequency of as little as 50 kHz.

Response curve of AFT discriminator. Two diodes in a balanced detector circuit produce the S response curve shown in Fig. 24-20. The slope is very steep to measure small changes in frequency. The output is zero at balance. For frequency deviations above or below 45.75 MHz, the net dc correction voltage is either positive or negative, depending on whether the frequency is too high or too low.

Correcting the oscillator. This frequency is generally controlled by a varactor for AFT. The varactor is a silicon diode with reverse bias.

562 BASIC TELEVISION

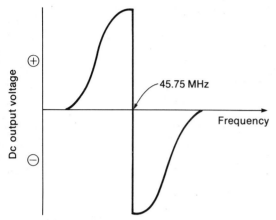

FIGURE 24-20 RESPONSE CURVE OF AFT DISCRIMINATOR. AMPLITUDE IS 8 V p-p.

Then varying the diode voltage changes the junction capacitance. Typical values may be a C of 10 pF with 10 V reverse bias.

AFT circuit. See Fig. 24-21a. Signal from the IF amplifier on the receiver chassis is coupled by C_1 and L_1 to the base of Q1. This stage is an IF amplifier for the AFT discriminator using diodes X1 and X2. The collector voltage and base bias for Q1 are provided by the 24-V supply line.

Amplified IF output is coupled from the collector by the discriminator transformer T_1. The two diodes with T_1 form a Foster-Seeley or phase-shift discriminator circuit.* T_1 proportions the signal voltage for X1 and X2 in accordance with the frequency of the IF signal input. Then the rectified output is the error voltage that indicates by how much the IF signal differs from 45.75 MHz.

A typical value for the AFT error voltage is 3 V. This can be measured with a dc voltmeter at the AFT test point. Actually, the dc output of a balanced discriminator is 0 V on frequency. In Fig. 24-21, though, the dc level of 3 V is inserted from the 24-V supply line, through R_6, L_2, and R_{10}.

It should be noted that in some AFT circuits, the IF signal is first rectified by the discriminator. Then the AFT correction voltage is amplified in a dc amplifier. In any case, the dc voltage output of the AFT circuit is connected to the VHF and UHF tuners to correct the oscillator frequency. The AFT circuit is usually on a separate module board or small subchassis.

AFT defeat switch. This is used to turn off the AFT circuit when necessary for manual fine tuning. In Fig. 24-21a, S_1 grounds the bias line for the base of Q1 to cut off the amplifier. Without IF signal input to the discriminator, there is no rectified output.

AFT action. To check the AFT circuit, turn it off with the defeat switch. Then use the fine tuning control to get a picture with 920-kHz beat. When the AFT is turned on, the beat should disappear.

AFT alignment. The response curve should look like Fig. 24-20. The main effect on balance is provided by the secondary of the discriminator transformer. The response curve is obtained with an oscilloscope at the AFT test point and IF sweep signal input. The discriminator adjustments are shown in Fig. 24-21b.

It may be possible to readjust the AFT circuit by watching the picture. Turn off the AFT and adjust the manual fine tuning control. Then turn on the AFT. If the oscillator is detuned, adjust the secondary of the AFT discriminator transformer for the best picture. When the AFT is turned on and off, there should be no effect on the correct tuning.

*See Chap. 27 for a detailed description of discriminator circuits.

THE RF TUNER 563

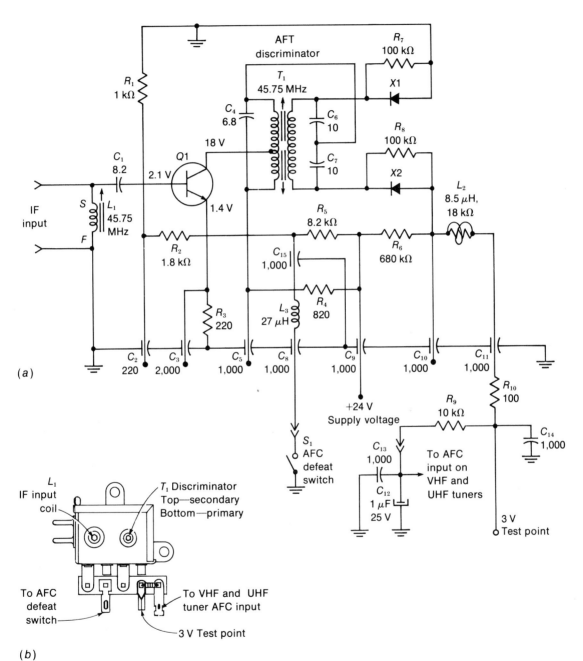

FIGURE 24-21 (a) AFC OR AFT CIRCUIT. R IN OHMS AND C IN PICOFARADS. (b) CONNECTIONS AND ALIGNMENT ADJUSTMENTS. (FROM ZENITH AFC SUBCHASSIS 150-6)

24-7 Types of RF Tuners

In terms of television channels, the receiver has separate VHF and UHF tuners. Older types of VHF tuners have two tubes, but now transistors are used. The UHF tuner is generally solid state, with just a crystal-diode mixer and transistor for the oscillator stage.

The VHF tuner is usually either the turret type shown in Fig. 24-4 or the rotary-switch type shown in Fig. 24-5. The UHF tuner often uses variable capacitors for continuous tuning through channels 14 to 83. Later types have a detent with 70 click stops, one for each UHF channel (Fig. 24-22). This type uses mechanical stops for the variable capacitor. Instead of varying continuously, the capacitor just moves in little jumps.

All these tuners use mechanical tuning, switching coils or varying a capacitor to tune in each channel. For electronic tuning, varactors are used. Changing the dc voltage on the varactor produces the channel tuning. When the UHF tuner uses varactors, it generally has an rf amplifier stage with the mixer and oscillator.

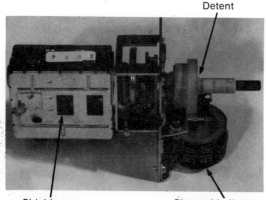

FIGURE 24-22 UHF TUNER WITH DETENT FOR 70 CHANNELS. *(MAGNAVOX CO.)*

24-8 VHF Tuner Circuits

The complete schematic of a VHF tuner is shown in Fig. 24-23. $Q1$ is the rf amplifier stage, $Q2$ the mixer, and $Q3$ the local oscillator. Rotary wafer switches are used for the station selector.

In the path for antenna signal input, the first components are *capristors* CPR-1 and CPR-2. A capristor combines a coupling capacitor with a shunt resistor and possibly a spark gap. The capacitor isolates the antenna terminals from the chassis, for 60-Hz ac line voltage, as a safety feature in transformerless receivers. The shunt resistor discharges any static charge accumulated on the capacitor. A spark gap can also be used across the capacitor as protection against excessive charge on the antenna.

The rf input circuit includes the balun transformer L_1 connected to FM and IF traps before coupling the rf signal through C_5 and L_{10} to the wafer switch S-4F. Its disk conductor connects this line through C_7 to S-3F. From here the signal is connected to the base of the rf amplifier $Q1$. The base also has forward AGC bias, applied through R_1.

The amplified rf output is coupled from the collector of $Q1$ to the base of the mixer $Q2$, through C_{13}. The coils on wafer switch S-2F tune the mixer input circuit for each channel. R_3 is a damping resistance to provide the required 6-MHz bandwidth on channels 2 to 6. In addition, the coils supply a dc path for collector voltage to $Q1$, through R_4 and R_{91}. The bypass capacitors in this line are C_{25} and C_{92}. Note that C_{25} is a feedthrough capacitor, with the double-ended terminal indicated by two small circles. Neutralization of the rf amplifier is provided by the feedback path with L_{13} and C_{10}.

In the oscillator circuit, the tuning coils are on wafer switch S-1F, connected between collector and base of $Q3$. The variable inductance L_{22} is the fine tuning control. L_{27} is used to tune channels 2 to 6 within range of the fine tuning control, while L_{28} has the same function for channels 7 to 13. These are called *oscillator tracking* adjustments.

The oscillator output is injected into the base of the mixer by C_{18}. In the mixer collector circuit, L_{21} is the IF output transformer. The tap between C_{27} and C_{28} in the secondary of L_{21} provides 75-Ω impedance for the cable connected to the IF output jack.

How the switches select the channel. The five wafer switches from right to left in Fig. 24-23 have the following functions:

1. S-1F is the oscillator section. It connects the required inductance for Q3.
2. S-2F tunes the mixer input circuit to the selected channel.
3. S-3F operates only on channel 1 for UHF. It tunes the rf circuits to the IF band for the output of the UHF tuner. This jack is at the top of L_{11}.
4. S-4F tunes the antenna circuit to the selected channel for input to rf amplifier Q1.
5. S-5 connects B+ voltage to the UHF tuner only on the channel 1 position of the selector switch.

The wafers have coils for each channel. For the high-band channels, these inductances are just stamped metal loops.

Shorting switches. This means the switch short-circuits all or part of the inductance on a wafer. As an example, S-4F is set on channel 13. Then the metal disk at the center shorts all the coils on the switch. L_{10} tunes the circuit to channel 13, with the stray capacitance in the circuit. When the switch is moved one position counterclockwise for channel 12, one loop is added to the inductance of L_{10} to lower the resonant frequency. The same idea applies for S-2F, with L_{20} tuning the mixer input circuit to channel 13. Finally, in the oscillator section, L_{28} tunes the circuit for channel 13, and S-1F adds inductance for each of the lower channels. This type of tuner is an incremental-inductance tuner.

Switching to UHF operation. When the station selector is set to the channel 1 position:

S-1F   disconnects B+ from the VHF oscillator Q3 so that it does not operate while the UHF oscillator is on.

S-4F   disconnects the antenna input circuit from the base of the rf amplifier Q1.

S-3F   then connects the channel 1 input jack to the base of Q1. In this path, L_{11}, C_{11}, and L_{12} are added for resonance in the IF band. Also, L_{14} on S-2F is added to the inductance of this section.

S-5   connects the B+ voltage to the oscillator on the UHF tuner.

In summary, then, for UHF operation on channel 1, the VHF oscillator is disabled, the UHF oscillator is turned on, and the mixer input is tuned to the IF band for the IF output of the UHF tuner.

24-9 UHF Tuner Circuits

In Fig. 24-24, the UHF tuner is divided into three metal compartments, or cavities, as indicated by the heavy lines in the diagram. Each cavity corresponds to a resonant transmission line. The equivalent inductances L_2, L_4, and L_8 are tuned by C_2, C_4, and C_{14}. These capacitors are on a common shaft for continuous tuning.

The antenna signal input is coupled by L_1 to the preselector section, which is tuned by C_2

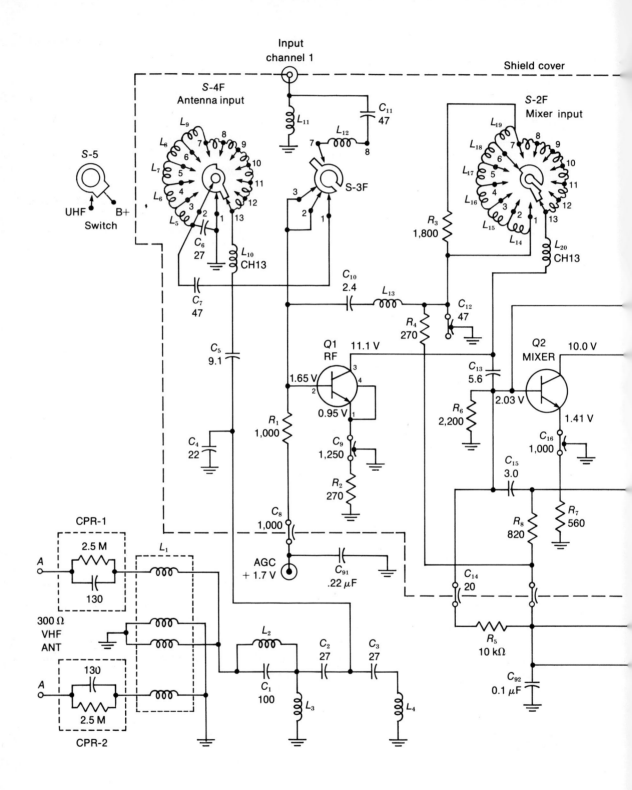

THE RF TUNER 567

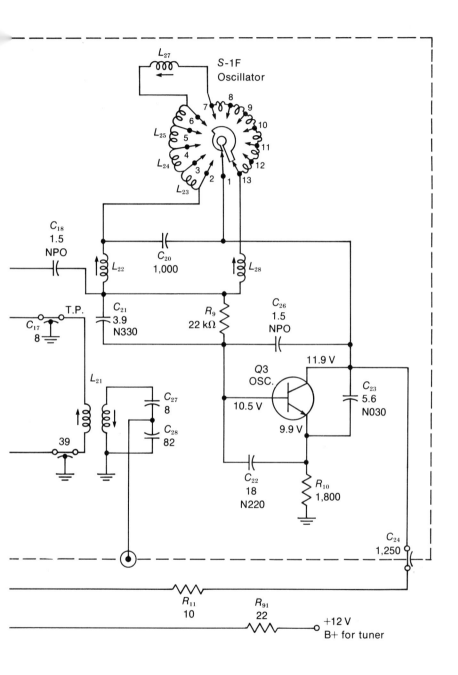

1. EMITTER
2. BASE
3. COLLECTOR
4. CASE

Bottom view of transistors

FIGURE 24-23 TRANSISTORIZED CIRCUITS FOR VHF TUNER. R IN OHMS AND C IN PICOFARADS. (*RCA KRK 157A*)

568 BASIC TELEVISION

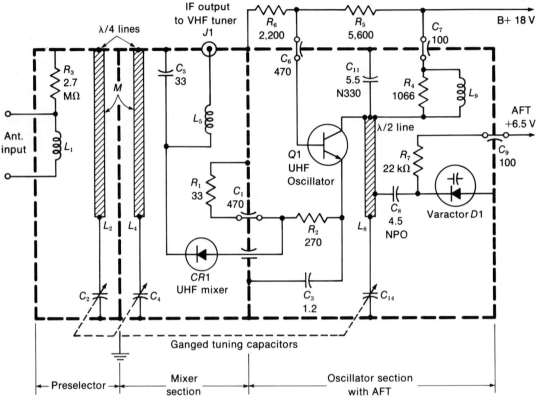

FIGURE 24-24 UHF TUNER WITH GANGED TUNING CAPACITORS. R IN OHMS AND C IN pF. (RCA KRK 185)

to the desired UHF channel. Coupling to the mixer section is provided by a window, marked M for mutual coupling. The mixer input circuit is tuned by C_4 to the channel frequencies. CR1 is the crystal-diode mixer. Also coupled into the mixer through R_2 is output from oscillator Q1, tuned by C_{14}. The output of the diode mixer is connected to the jack J1 through L_5. This IF signal output goes to the VHF tuner, which operates as an IF amplifier on the UHF position.

AFT on UHF tuner. The varactor D1 in Fig. 24-24 is used to control the oscillator frequency.

Note that D1 is connected to the oscillator line through C_8. Since the reverse voltage varies the varactor capacitance, it holds the oscillator at the correct frequency. The nominal value of AFT control voltage, on frequency, is 6.5 V in this circuit. The positive voltage is reverse bias at the cathode of D1.

UHF varactor tuner. In Fig. 24-25, varactors are used to tune Q1, Q2, and Q3, instead of using variable capacitors. Each of the varactors is provided with a dc voltage that tunes the stage to the desired frequency. The dc tuning voltage

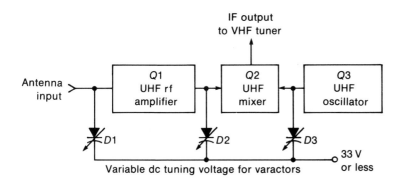

FIGURE 24-25 BLOCK DIAGRAM OF UHF TUNER WITH VARACTOR TUNING. (FROM ADMIRAL M25 CHASSIS)

is determined by the station selector. Different amounts of dc voltage set the tuner for different channels.

Note that the varactor tuner uses an rf amplifier stage for the UHF channels. The extra rf amplification is necessary because varactor tuned circuits have lower Q, compared with a variable air capacitor for the resonant circuit.

24-10 Cable Television Channels

Receivers for cable TV require more than 12 channels in the VHF spectrum. Distribution of VHF channels is preferred because cable losses are much greater in the UHF band. In order to provide additional VHF channels for a wide-band, single-cable system, the frequency allocations shown in Fig. 24-26 are used. In addition to the 12 standard FCC channels for broadcasting, there are 14 cable channels. A cable channel has the 6-MHz bandwidth of an FCC channel but is used only for CATV. It should be noted that EIA recommendations are to add six more cable channels for a total of 20.

Mid-band cable channels. These include nine channels from 120 to 174 MHz. The channels have the letters A to I, or are numbered 84 to 91 starting with channel B. The band of 120 to 174 Mhz is not used for television broadcasting in order to prevent interference with other radio services. However, the signals for cable TV are not broadcast by radiation.

Super-band cable channels. These include five

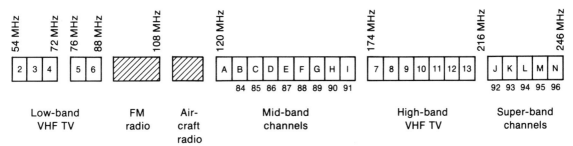

FIGURE 24-26 FREQUENCY ALLOCATIONS OF 24 CHANNELS IN THE VHF SPECTRUM FOR CABLE TELEVISION.

channels from 216 to 246 MHz, just above FCC channel 13. The super-band cable channels have the letters J to N or are numbered 92 to 96.

CATV converter. This unit is essentially an rf tuner for the mid-band and super-band cable channels. The circuit is usually a double superheterodyne, with two local oscillators. One frequency conversion is done to raise the rf signal frequencies. This method reduces the ratio of highest to lowest frequencies for easier tuning over both bands. However, the second frequency conversion produces the standard IF signals at 45.75 MHz and 41.25 MHz for the receiver. When the cable converter is used with VHF and UHF tuners, they all feed IF signal to an IF amplifier on the converter, which has the IF cable to the main chassis. Another method is to convert all cable channels to channel 12 for the VHF tuner, or to any channel not used in the area.

24-11 Varactors for Electronic Tuning

The varactor is a silicon diode that serves as a voltage-variable capacitance. An increase in reverse bias makes the capacitance decrease. The varactor is also called a *varicap* or *capacitive diode*. Its schematic symbol shows a semiconductor diode with a variable capacitor.

In addition to its function in correcting the oscillator frequency in an AFT circuit, the varactor can be used to tune in the desired channel. This method is considered electronic tuning because mechanical rotation of a switch or turret is not necessary. The varactors for each stage are tuned by applying the required dc tuning voltages.

Varactor-tuned circuit. Figure 24-27 illustrates how the varactor $D1$ is used in a resonant circuit. L_1 provides the required inductance, while $D1$ is

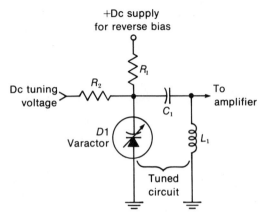

FIGURE 24-27 A BASIC CIRCUIT FOR VARACTOR TUNING.

the voltage-variable capacitance. The coupling capacitor C_1 has little reactance at the resonant frequency but is needed to prevent the dc supply voltage from shorting to ground through L_1. The dc voltage supplied through R_1 provides the reverse bias for $D1$. This voltage is positive at the N cathode of the diode.

Without any other voltage, the tuned circuit is resonant at a particular frequency. However, the dc tuning voltage supplied through R_2 varies the reverse voltage on $D1$. Then its capacitance changes to vary the resonant frequency. As a result, the varactor makes it possible to tune the resonant circuit by varying the amount of reverse bias for $D1$. Typical values are 20 pF with 4-V reverse bias. The C decreases with more reverse voltage.

Varactor tuning for channel selection. See Fig. 24-28. The four circuits tuned by the varactors are the local oscillator, mixer input, and rf amplifier input and output. Each circuit with a varactor diode and inductance L_1 corresponds to the varactor-tuned circuit in Fig. 24-27. As a result, the four varactors provide tuning for all the resonant circuits on the tuner. The rf circuits

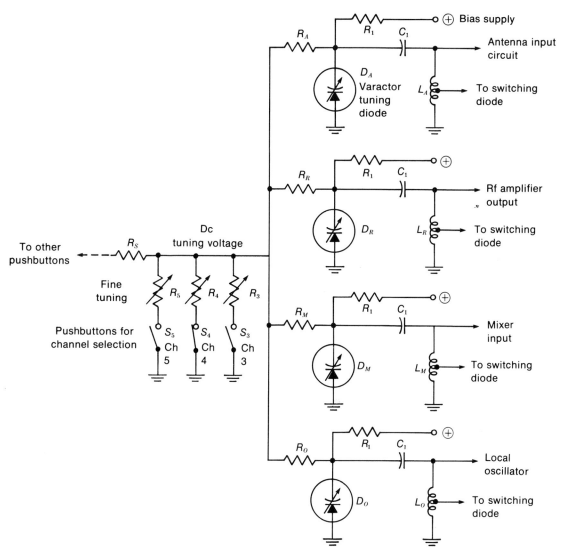

FIGURE 24-28 CIRCUIT WITH FOUR VARACTORS TO CHANGE CHANNELS BY TUNING RF AMPLIFIER, MIXER, AND OSCILLATOR ON TUNER.

are tuned to the channel frequencies, while the oscillator is above the rf signal.

The question of which channel is tuned in depends on the value of the dc control voltage connected to the varactors. One common line is used for the tuning voltage, with a dc voltage divider for each channel.

Pushbuttons are generally used to switch in the channel-tuning voltage. As an example, in Fig. 24-28 the pushbutton is in for channel 4.

Then R_4 forms a voltage divider with R_S to supply the amount of dc voltage needed to tune the varactors for this channel. The fine tuning control R_4 can be varied to make slight changes in the dc tuning voltage. Note that in this case the fine tuning affects all the stages on the tuner, not just the oscillator.

Only three channels are illustrated in Fig. 24-28, but the same idea applies to all channels. The desired channel is selected by its pushbutton, which supplies the required dc tuning voltage. Furthermore, although channel selection is shown here for a VHF tuner, similar circuits are used for the UHF tuner.

Switching diodes. The VHF channels 2 to 13 include frequencies between 54 and 216 MHz. This range has a ratio of 4:1, which is too much of a variation for capacitive tuning. Therefore, part of the inductance is shorted to reduce L for the high-band VHF channels. A switching diode has the function of shorting a tap on the coil when the selector switch is set for a channel from 7 to 13.

In Fig. 24-29, $D1$ is the switching diode connected through C_1 to a tap on the tuned-circuit inductance L_1. When $D1$ is not conducting, it is an open circuit without any effect on the coil. However, when positive voltage is applied to the $D1$ anode, it conducts. Then its low resistance is equivalent to a short circuit across the bottom section of L_1. The blocking capacitor C_1 is needed to prevent the diode voltage from shorting to ground through the coil.

Whether the switching diode conducts or not is decided by the band-switching voltage supplied through S_1. When this switch is open, no voltage is applied to $D1$ and it is an open circuit. All the turns in L_1 are used, then, for the low-band VHF channels 2 to 6. However, when S_1 is closed, a positive voltage is applied to the $D1$ anode to turn on the switching diode. Then only some of the turns on L_1 are used for the high-band VHF channels 7 to 13.

Only one switching diode is shown in Fig. 24-29. However, each of the tuned circuits for all the stages has a switching diode connected to a tap on the coil. The positive band-switching voltage is applied to all the switching diodes. Note that the bandswitch S_1 is ganged with the

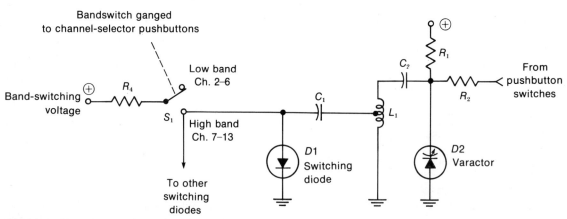

FIGURE 24-29 FUNCTION OF THE SWITCHING DIODE IN CHANGING BANDS FOR VARACTOR TUNING.

channel selector to apply switching voltage for the pushbuttons 7 to 13. In addition to the switching voltage for channels 7 to 13, remember that the pushbuttons apply tuning voltage to the varactors for selecting each of the channels.

24-12 RF Alignment

The exact procedure for aligning the rf tuner is given in the manufacturer's service notes, but the main requirements are as follows:

1. The rf tuned circuits for antenna input, rf amplifier output, and mixer input must be tuned to the 6-MHz band of the selected channel. See the typical rf response curve in Fig. 24-14.
2. The local oscillator must be set to the correct frequency for tuning in each channel.

When aligning the rf tuned circuits, the tuner should have its normal AGC bias. When the oscillator frequency is adjusted, the AFT circuit is off, but the average dc correction voltage must be present at the AFT terminal. The shield cover must be on because its capacitance affects all the tuner adjustments.

Realignment of the rf signal circuits is seldom necessary. However, the oscillator will need frequency adjustments if this tube or transistor is replaced.

Oscillator adjustments. In some receivers, the oscillator adjustment screws are separate from the fine tuning control, which is a variable capacitance. In this case, set the oscillator on each channel for the best picture with normal sound. The fine tuning control should be at midposition. The best picture is obtained by adjusting for sound bars and then backing off a little.

In turret tuners with separate strips for each channel, the order of oscillator adjustments does not matter. One channel can be adjusted without affecting the others. In wafer-switch tuners, though, the oscillator inductance is usually a series of coils. The frequency is reduced for lower channels by adding more turns for the oscillator coil. Then channel 13 must be adjusted first when aligning the oscillator coils. The adjustment for any one channel affects all the lower channels. However, some switch tuners have separate oscillator coils for each channel.

In cases where no broadcast station is available to supply a picture on the channel, or if the tuner is not connected to the receiver to check its IF output, the oscillator frequency can be measured directly. The required oscillator frequencies for the VHF channels are given in Table 24-1, with all the channels listed in Appendix A. One method is to heterodyne the oscillator output against a crystal-calibrated oscillator for zero beat. The marker-generator in Fig. 24-30 has its own audio section so that you can hear the low-frequency beat down to a null at zero beat. This signal generator also can supply marker frequencies for the rf response curve.

Rf alignment. Using a sweep generator and oscilloscope to obtain the rf response curve is similar to the method for IF alignment. However, the sweep generator is connected to the antenna input terminals. The generator must be set to the same channel as the receiver, generally channel 10. It is usually not necessary to align the rf circuits on every channel. A typical sweep generator for IF and rf alignment is shown in Fig. 23-22. The oscilloscope is connected to the mixer output, usually at a test point on the top of the tuner.

The tuned circuits aligned for the rf response curve are the antenna input circuit to the rf amplifier, its output circuit, and the input to

574 BASIC TELEVISION

FIGURE 24-30 CRYSTAL-CALIBRATED SIGNAL GENERATOR. *(RCA)*

the mixer. In the mixer output, its IF circuit is part of the IF alignment, not the rf response.

Specific instructions for rf alignment are given in the manufacturer's service notes. Actually, repair or alignment is generally done by service companies that specialize in rf tuners. This service is also provided for alignment of solid-state IF modules. The tuner or IF module usually can be removed easily from the main chassis.

UHF alignment. A variable air capacitor used in UHF tuners usually has segments on the rotor blade, as shown in Fig. 24-31. These can be adjusted for the rf signal circuits and the oscillator. It should be noted, though, that some UHF tuners are factory-sealed because they are not to be aligned in the field.

24-13 Remote Tuning

Many receivers have provision for remote operation. The basic functions include changing channels, varying the volume, and turning the set on and off. The ultrasonic system in Fig. 24-32 is generally used. A small hand-held remote transmitter, or "clicker," radiates sound waves at frequencies around 40 kHz. These frequencies are too high for the human ear to hear, but the waves of compression and rarefaction through the air operate the microphone at the receiver.

The microphone is a transducer that converts the sound waves to electric signals. The remote-control amplifier is an additional chassis in the receiver cabinet for operation from the remote transmitter. Amplification and rectification of the remote signals are necessary to obtain enough dc control voltage for each function.

Sound waves are used for the remote transmitter, in preference to radio transmission, in order to minimize interference problems. It should be noted that sound waves do not pass easily through the walls of a room. The remote transmitter can operate the control unit from a distance of up to 20 to 30 ft.

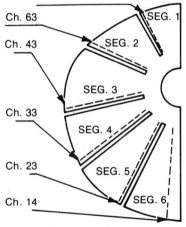

FIGURE 24-31 SEGMENTS ON ROTOR BLADE OF EACH CAPACITOR IN UHF TUNER. *(RCA KRK 152)*

THE RF TUNER 575

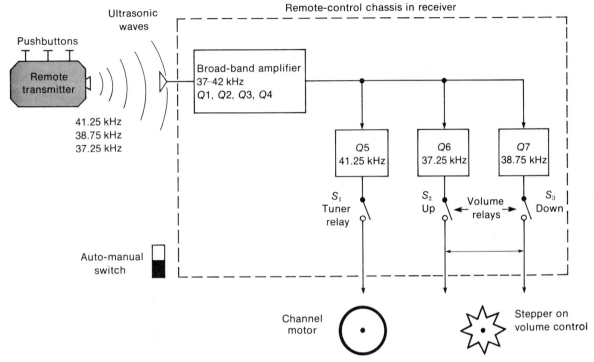

FIGURE 24-32 REMOTE CONTROL FOR THREE RECEIVER FUNCTIONS.

Control frequencies and functions. There are no standard values, but the frequencies in Fig. 24-32 are typical. The main functions are:

41.25 kHz	Turn channel selector
37.25 kHz	Volume up
38.75 kHz	Volume down and turn set off

In this example, the channel selector is rotated in only one direction. To turn it both ways, an additional control frequency can be used. More functions that may be used, with typical frequencies, are:

43.25 kHz	Chroma gain up
40.25 kHz	Chroma gain down
44.75 kHz	Tint toward purple
35.75 kHz	Tint toward green

However, channel selection and volume control are most common for remote operation because they do not have automatic control circuits in the receiver. Remote contrast and brightness controls are usually not included because the AGC circuit automatically controls the receiver gain.

For remote operation, the channel selector is turned one position by a motor on the shaft of the tuner. A relay in the remote-control unit turns on the tuner motor. The volume control can have a small step motor that turns the shaft. For chroma gain, a dc control voltage can supply AGC bias to the chroma bandpass amplifier to vary the color level. For tint control, the phase angle of demodulation can be shifted toward either purple or green.

Channel programming. Unused channels can be skipped by pin settings on an index wheel that turns with the tuner shaft. The channel motor stops when its ground return is opened. If the index pin is turned so that it does not open the ground connection, the motor keeps moving to the next channel until it is stopped by an open ground return.

Transmitter unit. This may be either an electronic oscillator or a mechanical source of vibrating sound waves. Figure 24-33 shows a mechanical unit, often called a "bonger." A separate metal rod, similar to a tuning fork, is used for each frequency. When the control button is pushed, a spring-loaded hammer strikes the rod and it radiates sound through an open grille at the top. No battery is needed for this transmitter, as its operation is entirely mechanical.

The electronic unit contains transistorized oscillator circuits. Operating frequencies are also around 40 kHz, the same as a mechanical unit. When each control button is pushed, it connects a capacitor to change the oscillator frequency. The oscillator drives a ceramic transducer serving as a loudspeaker to produce ultrasonic output. A small battery is the power supply for the transistor. This type transmits the remote signal as long as the control button is held down.

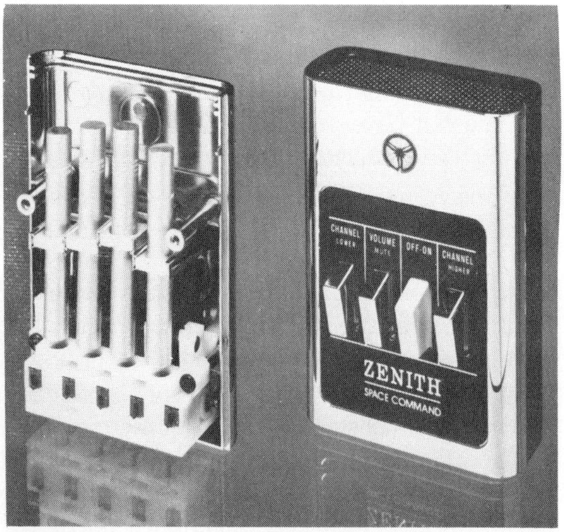

Control chassis. This unit is a small separate subchassis. Referring to Fig. 24-32, the output of the microphone is amplified in a high-gain, broad-band amplifier with four transistors. The amplified output is then fed to a parallel bank of driver stages tuned to the control frequency for each function.

Normally, the output stage is cut off. With enough signal input, though, it is driven into conduction to operate the relay in its output circuit. The auto-manual switch applies power to the remote-control chassis at all times, so that the receiver can be turned on by remote operation.

Memory circuit. In some receivers, dc voltage control may be used instead of a motor for remote operation. An example is a varactor tuner that does not have a rotating switch. For dc voltage control, some method of remembering the desired condition is needed. Otherwise, removal of the control signal would allow the circuit to return to its original condition. Therefore, a memory circuit is used to hold the effect of the control voltage in remote operation.

The module in Fig. 24-34 can hold the control voltage by storing charge in the memory capacitor C_1. A positive or negative voltage in the input makes $D1$ or $D2$ conduct. The input is dc control voltage from the remote-control chassis. When either diode is on, the neon bulb

FIGURE 24-33 MECHANICAL TYPE OF TRANSMITTER UNIT FOR REMOTE CONTROL. (*ZENITH RADIO CORP.*)

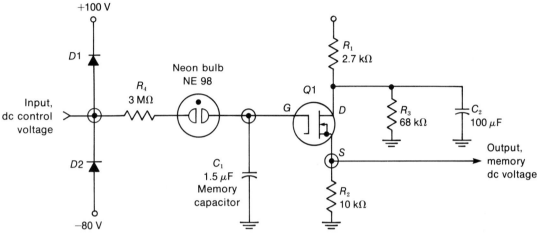

FIGURE 24-34 MEMORY CIRCUIT USED FOR REMOTE SYSTEMS WITH ELECTRONIC CONTROL. (MOTOROLA INC.)

is ionized. The bulb is then a low resistance, allowing C_1 to be charged.

Positive input at the anode makes D1 conduct. Then the +100-V supply charges C_1 positive. Negative input at the cathode of D2 allows it to conduct to charge C_1 negative from the supply of −80 V.

The neon bulb is ionized from the supply voltage of +100 V or −80 V when either D1 or D2 conducts. After the control-voltage input is removed, the diodes cannot conduct and the neon bulb deionizes to become an open circuit. However, the memory capacitor C_1 has become charged to store the effect of the control voltage.

The voltage across C_1 at the gate of Q1 determines how much the FET conducts. For output, R_2 in the source circuit produces dc control voltage that depends on the amount of transistor current. The R_2 voltage is proportional to the amount of input control voltage, therefore, but the output remains the same as long as C_1 holds its charge.

24-14 Tuner Troubles

The three most common problems here are:

1. Mechanical troubles, including dirty switch contacts or a weak detent spring
2. Snow in the picture, although the contrast may be good
3. No picture and no sound, but a normal raster

One way to determine if the tuner is defective is to substitute a good one. Commercial "tuner subber" units such as the one in Fig. 24-35 are available. This is a battery-operated transistor tuner. To substitute tuners, you just switch the IF cable from the tuner in the receiver.

Dirty switch contacts. Typical symptoms are: (1) the channel switch must be rocked to get the picture, (2) there is only picture, no sound, or vice versa, on some channels, or (3) one or two channels cannot be received at all. The dirty

578 BASIC TELEVISION

FIGURE 24-35 PORTABLE TUNER SUBSTITUTE FOR TELEVISION RECEIVERS. *(CASTLE TV TUNER SERVICE INC.)*

contacts are caused by tarnishing of the silver plating.

The contacts can be cleaned and lubricated by commercial sprays especially made for tuners. Just spray all the contacts and rotate the switch in both directions. This cleaning job usually lasts a year or two. In extreme cases of dirty contacts, a turret can be taken out of its metal case by removing a spring clip. Then the fixed switch contacts on the housing can be cleaned thoroughly. The eraser end of a pencil can be used to remove the dark coating down to the shiny silver base.

Snow in the picture. The snow on the screen is the visible effect of shot-effect noise generated in the mixer stage. Therefore, the snow can be used as an indicator to locate troubles in either the rf circuits before the mixer or the IF section after the mixer.

Suppose the trouble is no picture and no sound, or weak picture. Turn up contrast and volume controls to maximum. Note whether there is an increase in snow and the hissing sound of receiver noise. This effect indicates that the mixer stage is operating and all the following stages are amplifying the noise from the mixer. Either the signal from the antenna is too weak or the rf amplifier does not have enough gain.

SUMMARY

1. VHF tuners for channels 2 to 13 are generally the turret or switch type. This is a separate subchassis with rf amplifier, mixer, and oscillator stages. Ganged tuning is provided by the station selector. The fine tuning control varies the local oscillator frequency.
2. UHF tuners use variable-capacitor tuning for channels 14 to 83. Later models have a 70-position detent. The IF output of the UHF mixer is amplified by the VHF tuner on UHF operation.
3. For electronic tuning, capcitive-diode varactors are used for selecting the desired channel. The capacitance is varied by a dc tuning voltage.
4. The varactor is also used for automatic fine tuning (AFT) on the local oscillator. This circuit holds the oscillator at the correct frequency for the best picture without interference from the sound signal.

5. The rf amplifier must have low noise for a high signal-to-noise ratio in the amplified rf signal to the mixer.
6. The input impedance of the tuner is generally 300 Ω at the antenna terminals for twin lead or 75 Ω at a jack for coaxial cable.
7. The mixer stage requires rf signal input and oscillator injection voltage to produce the IF output.
8. The local oscillator generally uses a modified Colpitts circuit called the *ultraudion*. This circuit features one *LC* tuned circuit with a voltage divider formed by capacitances in the tube or transistor. The local oscillator tunes in the desired station, as the oscillator frequency determines which channel can be converted to IF signal.
9. For oscillator alignment, adjust each coil for best picture, without sound bars and with no 920-kHz beat in color receivers.
10. The rf amplifier can be aligned by the visual response-curve method, usually on channel 10, for a symmetrical curve with 6-MHz bandwidth.
11. Cable television uses special CATV channels, illustrated in Fig. 24-26.
12. Wireless remote control of the receiver is generally done with the ultrasonic system shown in Fig. 24-32. The operating frequencies are around 40 kHz.
13. Two common mechanical troubles in tuners are a weak detent spring and dirty switch contacts. In both cases, the station-selector setting is too critical when tuning in a channel.
14. Snow in the picture is the effect of receiver noise. Most of the noise is produced in the mixer stage. The rf amplifier is the most important stage in providing a high signal-to-noise ratio for the signal input to the mixer for minimum snow.

Self-Examination (Answers at back of book)

1. The fine tuning control varies the frequency of the: (a) rf amplifier; (b) mixer input circuit; (c) antenna input circuit; (d) local oscillator.
2. With an IF picture carrier frequency of 45.75 MHz, to tune in channel 13 the local oscillator frequency is: (a) 211.25 MHz; (b) 215.75 MHz; (c) 257 MHz; (d) 302.75 MHz.
3. An oscillator circuit tuned to 160 MHz has an L of 1 μH. If this is increased to 4 μH, the new resonant frequency will be: (a) 80 MHz; (b) 160 MHz; (c) 320 MHz; (d) 400 MHz.
4. The oscillator frequency is set for: (a) maximum sound; (b) minimum sound; (c) sound in the picture; (d) best picture.
5. A varactor is used as a voltage-controlled: (a) capacitor; (b) inductance; (c) resistance; (d) amplifier.
6. The AFT discriminator is tuned to: (a) 41.25 MHz; (b) 45.75 MHz; (c) 113 MHz; (d) 3.58 MHz.

7. A typical frequency in a remote control system is: (a) 40.25 kHz; (b) 45.75 MHz; (c) 10.7 MHz; (d) 54 MHz.
8. Which of the following replacements may require oscillator realignment: (a) rf amplifier; (b) first IF stage; (c) antenna; (d) oscillator stage.
9. Which of the following can cause excessive snow in the picture: (a) excessive antenna signal; (b) weak rf amplifier; (c) weak video amplifier; (d) excessive oscillator injection voltage.
10. Which of the following stages usually has AGC bias: (a) local oscillator; (b) mixer; (c) rf amplifier; (d) AFT discriminator.
11. In electronic tuning with varactors, a switching diode has the function of: (a) switching channels; (b) shorting a tap on L in the tuned circuit; (c) shorting the AFT voltage; (d) switching to UHF operation.
12. For the VHF tuner during UHF operation, which of the following is not true: (a) the VHF oscillator is off; (b) the mixer is tuned to the IF band; (c) B+ voltage is supplied to the UHF tuner; (d) the mixer stage is cut off.

Essay Questions

1. Give the function of each of the three stages in a VHF tuner.
2. What is the function of the station selector? The fine tuning control?
3. Give the input and output signals for the VHF mixer stage. Which is the next stage for the mixer output?
4. Give the input and output signals for the UHF mixer stage. Which is the next stage for the mixer output?
5. Draw the desired rf response curve for channel 10, marking the sound and picture carrier frequencies with the color subcarrier.
6. Draw a block diagram of equipment and connections for obtaining a visual response curve for the rf section.
7. Compare two characteristics for a triode tube, an NPN transistor, and an FET in an rf amplifier circuit.
8. Give three requirements of the rf amplifier stage.
9. Why is neutralization used with triode rf amplifiers?
10. Define the following: balun, feedthrough capacitor, capristor, and detent spring.
11. Referring to the block diagram of a combination VHF-UHF tuner in Fig. 24-2, give the operating frequencies for tuning in channel 7. Do the same for channel 21.
12. Describe the effect on picture, sound, and color when the oscillator frequency is varied to cause: (a) sound carrier too high on IF response curve; (b) picture carrier too high on IF response curve.
13. Why is AFT more important in color receivers than in monochrome?
14. Give two features of a varactor diode. Give two uses for varactor diodes in rf tuners.

15. Give the function of the AFT-defeat switch.
16. Describe briefly how to check whether the AFT circuit is operating normally.
17. What is the reason for using switching diodes in circuits for electronic tuning?
18. Give two differences between ultrasonic waves at 40 kHz and rf signal at the same frequency.
19. Give the function of the following in a remote-control system: transmitter unit, microphone, control chassis; auto-manual switch, and channel-control relay.
20. Name two types of transmitter units for remote operation.
21. What is meant by programming, in remote operation of the channel selector?
22. When is a memory circuit needed in a remote-control system?
23. What is meant by the mid-band and super-band channels for cable television?
24. For a receiver with memory fine tuning, how would you align the oscillator after the tube is changed?
25. Why is minimum noise more important in the rf amplifier than in the last IF stage?
26. For each of the following, state what you think is wrong and why. (a) No picture and no sound for channel 5 only on a turret tuner; (b) channel 11 is received on the channel 10 position; (c) no picture and no sound on all channels, but a clean raster without snow; (d) no picture and no sound on all channels; the picture is very snowy when the contrast is turned up; (e) good contrast but snow in the picture on all channels; (f) good contrast but snow in the picture only on some channels; (g) the picture is weak with no color on some channels, but becomes perfect when the channel-selector switch is moved slightly.
27. Referring to Fig. 24-23, give the effect of the following troubles: (a) R_{10} for the emitter of Q3 is open; (b) L_{20} for the collector of Q1 is open; (c) R_5 for the base of Q2 is open; (d) R_{91} in the B+ line is open.
28. Referring to Fig. 24-28, give the function of R_1, C_1, L_O, and D_O for the oscillator stage.
29. Describe briefly two mechanical troubles in tuners.
30. Why is the problem of snow in the picture associated with the antenna signal input and the rf amplifier?

Problems (Answers at back of book)

1. Calculate the lowest and highest frequencies for a VHF oscillator to tune in channels 2 to 13. Do the same for UHF channels 14 to 83.
2. What oscillator frequency is needed to tune in channel 3? Calculate the L needed with 5-pF C for resonance at this frequency.
3. Calculate the wavelength of an ultrasonic signal at 40 kHz if the velocity of sound waves is 1,130 ft/s.
4. Calculate the capacitive reactance of the 130-pF capristor used in Fig. 24-23 for the frequencies of 60 Hz and 60 MHz.

25 Antennas and Transmission Lines

An antenna or aerial can be simply a length of wire or any conductor which has current induced in it by the transmitted electromagnetic wave. To have enough signal, though, the antenna should either be very long or have a resonant length that magnifies the effect of particular frequencies. The resonance effect for the length is related to the frequency and wavelength of the signal current in the conductor.

For the VHF and UHF television channels, one-half wavelength is a practical size, one which ranges from approximately 8.5 ft for the lowest frequency to 0.5 ft for the highest frequency. Therefore, the basic antenna for television is the half-wave resonant dipole illustrated in Fig. 25-1. The dipole intercepts the radiated electromagnetic wave to provide induced signal current in the antenna conductors. The transmission line conducts the signal current to the antenna input terminals of the receiver. It should be noted that the antenna signal includes both the picture and sound carrier signals, which are received by the same antenna. Also, a combination antenna with multiple dipole elements can receive all the television channels.

The receiver needs a good antenna signal to reproduce a good picture. Although most receivers can produce sufficient contrast with weak antenna signal, the antenna and transmission line determine the important features of no snow and no ghosts. The required amount of antenna signal is about 100 to 2,000 μV. Cable distribution systems are designed for 1,000 μV, or 1 mV, antenna signal at the receiver.

It should be noted that the antenna requirements are essentially the same for color and monochrome television. After all, the color signal is only a side band of the picture carrier signal in the standard 6-MHz channel. The only difference is that poor antenna signal is more obvious in a color picture. In fact, there may be no color if the antenna response does not have the bandwidth to include the color frequencies. More details of antennas and transmission lines are described in the following topics:

- 25-1 Resonant length of an antenna
- 25-2 Definition of antenna terms
- 25-3 How multipath antenna signals produce ghosts
- 25-4 Straight dipole
- 25-5 Folded dipole
- 25-6 Broad-band dipoles
- 25-7 Long-wire antennas
- 25-8 Parasitic arrays
- 25-9 Multiband antennas
- 25-10 Stacked arrays
- 25-11 Transmission lines
- 25-12 Characteristic impedance
- 25-13 Transmission-line sections as resonant circuits
- 25-14 Impedance matching
- 25-15 Antenna installation
- 25-16 Multiple installations
- 25-17 Troubles in the antenna system

25-1 Resonant Length of an Antenna

The wavelength and frequency of the radiated electromagnetic wave are inversely proportional. The higher the frequency, the shorter the corresponding wavelength. In addition, the wavelength depends on the velocity of propagation, which is the speed of light for radio waves in free space. Specifically,

$$\lambda = \frac{\text{velocity}}{\text{frequency}} = \frac{3 \times 10^{10} \text{ cm/s}}{f}$$
$$= \frac{186,000 \text{ mi/s}}{f} \quad (25\text{-}1)$$

where lambda (λ) is wavelength and f is frequency in Hz.

When the physical length of the antenna, in feet or meters, is cut to equal either one-half or one-quarter of the wavelength corresponding to the signal frequency, the antenna is resonant at that frequency. The result is a resonant rise in antenna current. Multiples of these lengths are considered harmonic resonances. The reason why an antenna can be considered resonant is that the length determines how long it takes for current to flow to the open end.

The two basic types of resonant antennas are the grounded quarter-wave Marconi antenna used at lower frequencies and the half-wave Hertz antenna in Fig. 25-1. The half-wave antenna usually consists of two quarter-wave elements insulated from each other, which add to provide a half wavelength. This is called a *dipole*. The dipole antenna operates independently of ground and therefore may be installed far above the earth or other absorbing bodies. Because of this, and because the physical length of a half wave is a practical size at high frequencies, the dipole is the basic antenna in television.

584 BASIC TELEVISION

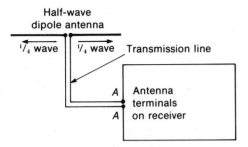

FIGURE 25-1 ANTENNA AND TRANSMISSION LINE CONNECTED TO RECEIVER.

More elaborate antenna arrays are usually just combinations of dipoles.

On the basis of Eq. (25-1), the formula for the length of a half wave in feet is derived as $L = 492/f$, where f is in MHz and L gives the half wavelength in feet, in terms of the electromagnetic field traveling with the speed of light. However, the resonant length of a half-wave conductor is slightly less than a half wave in free space because the antenna has capacitance that alters the current distribution at the ends of the antenna. This end effect requires foreshortening of the conductor length, to provide the resonant current distribution that corresponds to the length of a half wave in free space. Wider antenna conductors and higher frequencies require more foreshortening, but it can be taken as approximately 6 percent for the antennas used in television. Therefore, the length of the half-wave dipole is computed from the formula

$$L(\text{ft}) = \frac{462}{f(\text{MHz})} \qquad (25\text{-}2)$$

L gives the length of the half-wave dipole directly in feet. This is the actual physical length of the half-wave antenna corresponding to an electrical half wave. As shown in Fig. 25-1, one-half this value is used for the length of each of the two quarter-wave poles. The small insulation distance between the two poles can be considered negligible. For a dipole tuned to 60 MHz, as an example, the length of the half wave is 7.7 ft, and each section is made 3.85 ft long.

25-2 Definition of Antenna Terms

Wave polarization. The moving electromagnetic field, which is the radio signal, consists of two components: a magnetic field associated with the current in the transmitting antenna and the electric field associated with its potential. The two fields are perpendicular to each other in space, and both are perpendicular to the direction of travel of the wave. When the electromagnetic wave passes the receiving antenna, it induces antenna current with the same variations as the transmitted radio signal.

The polarization is arbitrarily defined as the direction of the electric field. This is determined by the physical position of the antenna in space. A horizontal dipole is horizontally polarized. Then the magnetic lines of force are in the vertical plane around the conductors and the electric field is horizontal between the conductors. An antenna vertical with respect to earth is vertically polarized. Horizontal polarization is specified by the FCC for transmission in the television and FM broadcast bands. Therefore, the receiving antenna is mounted horizontally for maximum pickup of a horizontally polarized wave. Horizontal polarization is chosen because experimental results show more signal strength and less reflection for frequencies in the VHF and UHF spectrum. Also, the horizontal directivity of the receiver dipole helps reduce ghosts.

Microvolts per meter. This unit is a measure of field strength. A meter is slightly less than 40 in.

As an example, when a resonant half-wave dipole 1 m long provides 300 μV signal to the transmission line, the field strength is 300 μV/m. The antenna height is standardized, often at 30 ft, and the antenna polarization is the same as the wave polarization.

Note that the same field strength produces more antenna signal in a longer antenna, which is resonant for lower frequencies. As an example, assume a field strength of 800 μV/m with a half-wave dipole 4.62 ft long for resonance at 100 MHz. The antenna signal then is 1,120 μV, because the length is 1.4 times more than a meter. A half-wave dipole 2.31 ft long for resonance at 200 MHz in the same field provides one-half the antenna signal, equal to 560 μV, since this antenna has one-half the length.

The field strength of the rf carrier signal at the receiver antenna depends on radiated power and propagation characteristics for the carrier frequency. Even though several stations may transmit from one location, the field strength at the receiver is generally not the same for different channels. In addition, the characteristics of the receiver antenna vary for different frequencies.

Polar directivity patterns. As shown in Fig. 25-2, signal strength is plotted in polar coordinates to show magnitude and direction. The angle gives the direction for which the signal strength is plotted, while the length of the radial arm defines the magnitude. The polar diagram shows the horizontal directivity pattern of the antenna, as determined by the current distribution of the antenna conductor. For a transmitting antenna, the pattern shows in which direction the antenna radiates the most signal; for a receiving antenna the pattern shows the direction from which most signal is intercepted by the antenna conductor. Rotating the antenna for the direction of best signal pickup is called *orientation.*

A half-wave dipole at its fundamental resonant frequency has the *figure-eight* directivity pattern in Fig. 25-2. The antenna receives best from the front and back, broadside to the antenna conductor, with little signal received from directions off the ends. As examples, in Fig. 25-2 the antenna provides 1,000 μV for signal in the broadside direction, about 500 μV signal at 60°, and a null of practically zero signal at 0° and 180°. This pattern applies only for half-wave resonance. For frequencies off resonance, the directional pattern changes because of a different current distribution on the antenna.

Antenna impedance. This impedance Z_a varies with the values of current at different points along the antenna. For a resonant half-wave dipole, Z_a is approximately 72 Ω at the center, as shown in Fig. 25-3. At the ends, Z_a is several thousand ohms. Intermediate points have intermediate values. Furthermore, the value of Z_a at the center is higher than 72 Ω for frequencies off resonance. These values are ac impedances

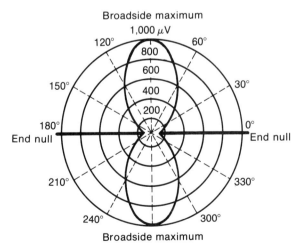

FIGURE 25-2 POLAR DIRECTIVITY PATTERN IN HORIZONTAL PLANE FOR HALF-WAVE DIPOLE ANTENNA.

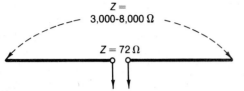

FIGURE 25-3 IMPEDANCE OF HALF-WAVE DIPOLE ANTENNA.

corresponding to an E/I ratio, which cannot be measured with an ohmmeter.

Antenna bandwidth. The half-wave antenna is equivalent to a resonant circuit with resistance and reactance. Therefore, the antenna can be considered as having a value of Q, which determines its bandwidth. Larger diameter for the antenna conductors decreases the reactance, allowing lower Q and wider frequency response. For this reason, metal tubing of $1/4$ to $1/2$ in. diameter is used for the receiving antenna.

Antenna gain. This is a term used to express the increase in signal for one antenna over a standard antenna, usually a half-wave dipole having the same polarization. Antenna gain is generally used in connection with directional antenna systems with multiple elements and is measured in the optimum direction. The gain is usually stated in decibels. An antenna with a gain of 3 dB, as an example, has a power gain of 2 or a voltage gain of 1.4. Double the voltage is a gain of 6 dB. Typical curves for gain are shown in Fig. 25-17b for a VHF antenna and Fig. 25-22b for a UHF antenna.

Front-to-back ratio. This indicates the amount of signal the antenna receives from the front compared with signal received from the back. As an example, if the antenna intercepts 1,000 μV of signal from a transmitter in front, but only 500 μV for signal of the same frequency arriving from the back, the front-to-back ratio is 2 times in voltage, or 6 dB.

Summary of desirable antenna characteristics. Multiple dipole elements can provide many types of combinations. In general, more dipoles intercept more signal, but the antenna characteristics determine how much useful signal is delivered by the transmission line to the receiver. The antenna should have high gain to provide enough signal for a good signal-to-noise ratio, especially in fringe areas. However, high gain is generally associated with narrow bandwidth, which may limit reception to one channel. An antenna impedance of 72 to 300 Ω is desirable for matching to transmission line. Very important is a good front-to-back ratio to prevent ghosts in the picture caused by receiving the same signal from two different directions. In addition, a narrow forward lobe for the antenna directivity pattern helps reduce ghosts. Such an antenna with a sharp directional pattern generally has high gain and narrow bandwidth.

25-3 How Multipath Antenna Signals Produce Ghosts

As shown in Fig. 25-4, a ghost is a duplicate image, a little to the side of the original. Usually, ghosts are caused by multiple antenna signals from the same station which are received over paths of different lengths. Referring to Fig. 25-5, the receiving antenna at point C receives signal from the transmitting antenna at A by two separate paths. The path ABC for signal reflected from the building at B has a length of $6 + 3 = 9$ mi. The direct path AC is 7 mi. Thus the path for reflected signal is 2 mi longer in this example. Since the velocity of the radiated signal is 186,000 mi/s and the extra length is 2 mi, the

FIGURE 25-4 GHOST CAUSED BY REFLECTED SIGNAL.

reflected wave is delayed by 2/186,000 s at the receiving antenna, compared with the direct wave. This time delay is approximately 11 μs.

The electron beam scanning the screen of the picture tube requires about 55 μs to scan across one horizontal line. On a picture having a width of 20 in., then, it requires only 5.5 μs to scan 2 in. The reflected signal, delayed in reception by 11 μs, will produce a second image displaced from the original by 4 in. The ghost is displaced to the right, in the direction of scanning, because the reflected signal arrives at the receiving antenna later in time than the direct signal. Any conducting material can produce reflection, as induced current in the conductor reradiates the signal. However, a nonconductor does not produce reflections.

With multiple reflections there may be multiple ghosts. The intensity of the ghost may be nearly as strong as the original image or just noticeable. Any difference in relative intensity is the result of more attenuation of the reflected wave in its longer travel. The ghost may be a positive or negative image, depending on the relative phase between the multipath rf signals. Reflection by a conductor generally shifts the phase and polarization of the reradiated signal.

The interference in the picture may vary from definite ghost images to reflections that are not noticeable as duplicate images because of insufficient time delay but that cause the picture to appear fuzzy. A delay distance of about 50 ft or less can be considered negligible, since the

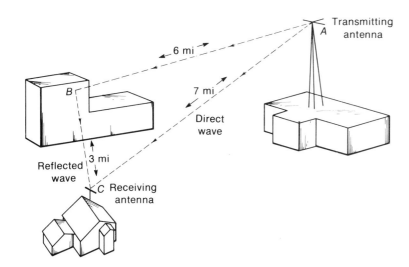

FIGURE 25-5 RECEPTION OF MULTIPATH SIGNALS. THE DISTANCE ABC FOR THE REFLECTED SIGNAL IS 2 MI LONGER THAN THE PATH AC FOR DIRECT SIGNAL.

resultant horizontal displacement on the screen of the picture tube is then considerably less than the width of a picture element.

For the problem of multipath reception, an antenna that has a good front-to-back ratio and a narrow forward lobe with minimum side responses can be rotated horizontally to minimize ghosts. Sometimes changing the antenna location reduces the intensity of the ghost. Once the multipath signals are coupled from the antenna to the receiver, however, it is impossible to eliminate the ghost. The only remedy is in the antenna system.

25-4 Straight Dipole

The antenna illustrated in Figs. 25-1 to 25-3 is a Hertz, doublet, half-wave, or simply straight dipole. Its distribution of current and voltage for half-wave resonance is illustrated in Fig. 25-6. With rf excitation by the radiated electromagnetic field, current is induced in the antenna with the same variations as the applied voltage. The dotted lines for E and I indicate an envelope of peak values for the varying ac signal at different points along the antenna.

The electron flow in the antenna conductor is not instantaneous but travels along the wire in free space with approximately the speed of light. When the electrons reach the end of the wire, the resulting accumulation of charge at the end provides a potential for moving the charge in the opposite direction, reversing the direction of current flow. The resultant current is zero at the ends, with two currents of equal amplitude flowing in opposite directions.

Farther back on the wire, the outgoing and returning currents are not the same, because the charges causing the currents have been supplied to the antenna at different parts of the rf cycle. Maximum current is at the center, where the reflected current from the ends can add to the original current.

The ends of the antenna in free space are points of maximum voltage and zero current. Because of capacitance of the ends, however, the current is normally not zero at the ends but has a definite value. Therefore, the antenna must be foreshortened to give the same current distribution that would be obtained for a half wave in free space.

The voltage and current distribution illustrate why the physical length of an antenna makes it resonant for a particular frequency. When the electron charges in the antenna conductor can travel from the center out to the ends and back to the center in the time of one-half cycle, the current and voltage values are maximum. At other frequencies, partial cancellation of the incident and returning electrons reduces the amount of antenna signal.

Harmonic resonances of a dipole. When the antenna is used to receive several channels, the directional pattern and impedance change at higher frequencies because of the different I and E distribution on the conductor. As an example, a λ/2 dipole cut for channel 4 at 66 to 72 MHz has a length of 3λ/2 for channel 11 at 198 to 204 MHz. The reason is that 198 MHz is the third harmonic of 66 MHz.

Referring to Fig. 25-7a, the antenna has the figure-eight directivity pattern at its fundamental resonant frequency as a half-wave dipole. The impedance at the center is 72 Ω.

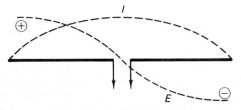

FIGURE 25-6 CURRENT AND VOLTAGE DISTRIBUTION ON HALF-WAVE ANTENNA.

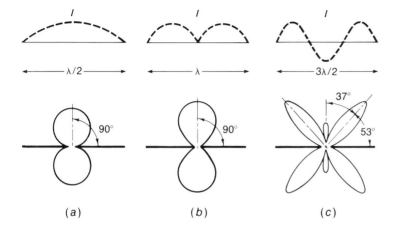

FIGURE 25-7 DIRECTIONAL CHARACTERISTICS OF CENTER-FED DIPOLE ANTENNA AT FUNDAMENTAL AND HARMONIC FREQUENCIES. (a) FIGURE-EIGHT PATTERN AT HALF-WAVE RESONANCE; (b) BROADENED RESPONSE FOR FULL-WAVE OPERATION AT SECOND HARMONIC FREQUENCY; (c) LOBES SPLIT BECAUSE OF HARMONIC OPERATION AT THREE TIMES THE RESONANT FREQUENCY.

Maintaining the same physical length, the antenna in Fig. 25-7b is a full-wave center-fed dipole, at double the frequency of Fig. 25-7a. The dipole still has the figure-eight directivity pattern, but the impedance at the center is now a maximum equal to several thousand ohms.

At three times the fundamental resonant frequency, the same antenna length is three half-waves, as shown in Fig. 25-7c. The center again has minimum impedance, equal to about 100 Ω. However, notice that the directional pattern splits into four major lobes. There is a gain of 1 dB in the direction of maximum reception, but this is 53° off the ends, with little pickup from the broadside direction.

At 4λ and higher harmonics, the response of the dipole splits into more lobes closer to the line of the antenna itself. The conductor then operates as a long-wire antenna, with maximum reception in the direction off the ends. In order to use the broadside response, therefore, a center-fed dipole is limited to a frequency range of 2 to 1 or less.

V dipoles. In order to maintain broadside response over a wider frequency range, the dipole can be angled in the form of a V. For instance, assume the two poles in Fig. 25-7c are moved in 53° to line up with the two forward lobes shown. Then both top lobes would overlap in the forward direction. The result, therefore, is to maintain the broadside response when the antenna operates at harmonics of its fundamental half-wave resonant frequency. The smaller the angle of the V in the dipole, the higher is the frequency at which the dipole maintains its forward response. Note that the V dipole also has a better front-to-back ratio, as the back lobes spread farther apart. However, "V-ing" the dipole for better response at high frequencies lowers its gain for low frequencies because then the antenna extends less than a half wavelength in the space across the ends of the poles.

Capacitive loading of an antenna. The distribution of antenna current and its directional pattern can be altered by inserting series inductance or shunt capacitance. This *antenna loading* makes the antenna electrically longer. The reason is that it takes more time for the current to reach the ends of the antenna. For VHF antennas, a technique used sometimes is to mount short metal rods as wings on the dipole, or to use circular disks. This method is equivalent to adding more shunt capacitance, as more time is required to charge the open ends of the antenna.

25-5 Folded Dipole

As shown in Fig. 25-8, the folded dipole is constructed of two half-wave rods joined at the ends, with one rod open at the center where it connects to the transmission line. The spacing between the rods is small compared with a half wavelength. This half-wave folded dipole has the same directional characteristics as the half-wave straight dipole, with the same amount of signal pickup. However, the antenna resistance of the folded dipole is approximately 300 Ω, which is convenient for matching to 300-Ω transmission line.

The center of the closed section of the half-wave folded dipole is a point of minimum voltage, allowing direct mounting at this point to a grounded metal mast without shorting the signal voltage. Another difference from the straight dipole is that, with operation at twice the fundamental half-wave resonant frequency, the full-wave folded dipole does not have the figure-eight polar pattern, receiving little signal from the broadside direction.

The resistance of the half-wave folded dipole at the center where it connects to the transmission line is higher than the value for a straight dipole because only part of the total antenna current flows in the open section. As a result, the folded dipole antenna resistance

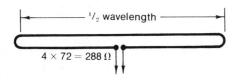

FIGURE 25-8 FOLDED DIPOLE ANTENNA.

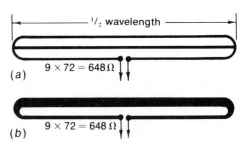

FIGURE 25-9 HIGH-IMPEDANCE FOLDED-DIPOLE ANTENNAS: (a) WITH ADDED CONDUCTOR ACROSS CENTER; (b) WITH DOUBLE DIAMETER FOR THE CLOSED CONDUCTOR SECTION.

equals 72 Ω multiplied by the square of the ratio of the total diameter of all conductor sections to the diameter of the open section. As an example, in Fig. 25-8, with the same diameter for the two conductor sections, the total diameter is twice the diameter of the open section. Therefore, the antenna impedance is 4 × 72, or 288 Ω, which is generally considered as 300 Ω.

In applications where a higher resistance is desired for the folded dipole, the diameter of the closed conductor section is increased. In Fig. 25-9a, an additional closed conductor of the same diameter is added, making the total conductor diameter triple the diameter of the open section, to provide an antenna resistance of 9 × 72, or 648 Ω. The same impedance transformation can be obtained by increasing the diameter of the closed conductor, as shown in Fig. 25-9b, instead of adding sections.

25-6 Broad-Band Dipoles

A *thick dipole* antenna, which has a cross-sectional dimension approximately 0.1λ or greater, can provide more uniform response over a wider band of frequencies, compared with a *thin dipole* conductor having negligible diameter.

Figure 25-10 shows several types of thick broad-band dipoles. These can be constructed of wire conductors, metal sheets, wire screening, or metallic foil. The increased thickness has the following three effects on the dipole characteristics:

1. The antenna resistance decreases and the reactance is lowered even more, resulting in an antenna with lower Q. The antenna has more uniform impedance values over a wide band of frequencies.
2. The broadside response of the directional pattern is maintained over a wider frequency range, before splitting into multiple lobes.
3. More foreshortening is needed with increasing thickness to provide physical lengths equivalent to the electrical wavelengths.

Referring to Fig. 25-10, the antennas in *a* and *b* are operated as half-wave dipoles. When the conductors are separated by 0.1 or less they can be considered as one uniform antenna of wider cross section. The centers are joined in Fig. 25-10a and the ends can also be joined, as in *b*, since the end points of the two antennas are at the same potential.

The triangular dipole in Fig. 25-10c is operated as a full-wave antenna, however, with a gain of approximately 3 dB. The included angle shown provides an antenna resistance of approximately 300 Ω at the center, for full-wave resonance. Smaller angles result in higher impedance. The overall physical length can be foreshortened about 10 percent. The triangular dipole, often called a *bowtie antenna,* is commonly used as a broad-band receiving antenna to cover all the UHF channels.

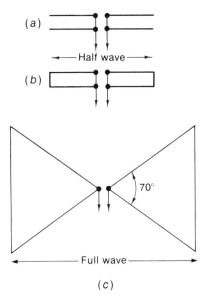

FIGURE 25-10 BROAD-BAND DIPOLE ANTENNAS. *(a)* HALF-WAVE DIPOLE WITH DOUBLE CONDUCTORS OPEN AT THE ENDS; *(b)* SAME DIPOLE, BUT WITH CONDUCTORS SHORTED AT ENDS; *(c)* FULL-WAVE TRIANGULAR DIPOLE. THIS IS A "BOWTIE" ANTENNA.

25-7 Long-Wire Antennas

Compared with a half-wave dipole, a long-wire antenna, which is several wavelengths, has the advantages of increased signal pickup and sharper directivity. As the antenna wire is made longer in terms of the number of half waves, the directional pattern changes because of the current distribution, increasing the directivity along the line of the antenna wire itself.

V antenna. When two long wires are combined in the form of the horizontal V antenna shown in Fig. 25-11, the major lobes of the directional pattern for each wire reinforce along the line bisecting the angle between the two wires. Therefore, the V antenna receives best along the line of the bisector. The greater the electrical length of the conductor legs, the smaller the included angle of the V should be for maximum antenna gain.

592 BASIC TELEVISION

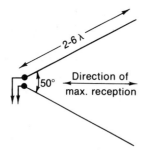

FIGURE 25-11 LONG-WIRE V ANTENNA.

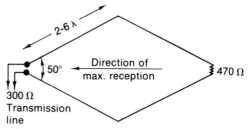

FIGURE 25-12 LONG-WIRE RHOMBIC ANTENNA.

Rhombic antenna. A more efficient arrangement is the rhombic antenna shown in Fig. 25-12, which consists of two horizontal V sections. To make it unidirectional, the rhombic antenna can be terminated with a resistor of 470 Ω, for an approximate match to 300-Ω transmission line. Both the V and rhombic antennas are mounted horizontally for horizontal polarization as television antennas. Each leg should be at least two wavelengths at the lowest operating frequency, the gain and directivity of the antenna increasing with the length. The angle of 50° is a compromise value suitable for leg lengths of 2 to 6λ. Longer legs should have a smaller angle.[1]

With each leg of four wavelengths, the V antenna has a gain of 7 dB, while the rhombic antenna gain is 10 dB, approximately. At ultrahigh frequencies these lengths are practicable, to provide a high-gain antenna for the UHF television band. The rhombic can provide a uniform value of antenna impedance over a total frequency range of 3 to 1, with high gain and sharp directivity.

[1]Details of the construction of long-wire antennas are described in the A.R.R.L. Antenna Handbook, published by the American Radio Relay League, Newington, Conn.

25-8 Parasitic Arrays

When current flows in the receiving antenna it radiates part of the intercepted signal, as in a transmitting antenna. If a conductor approximately one-half wave long is placed parallel to the half-wave dipole antenna but not connected to it, as illustrated in Fig. 25-13, the free wire will intercept some of the signal radiated by the antenna. This signal is reradiated by the free wire to combine with the original antenna current. As a result, part of the intercepted signal lost by radiation in the receiving antenna is recovered by using the free wire, providing increased gain and directivity. The free wire is called a *parasitic element* because it is not connected to the

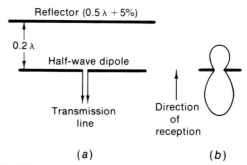

FIGURE 25-13 DIPOLE ANTENNA WITH PARASITIC REFLECTOR IN BACK.

dipole antenna. A parasitic element placed behind the antenna is a *reflector;* a parasitic element in front of the antenna is a *director.* The dipole antenna itself, to which the transmission line is connected, is the *driven element.* This can be either a straight dipole or a folded dipole.

A dipole antenna with one or more parasitic elements is a *parasitic array.* This is the most common type of television receiving antenna because it is simple to construct, can be oriented easily, provides enough gain for average signal strengths, and increases the directivity compared with dipole alone. The main directional effect of the parasitic element is to reduce the strength of signals received from the rear of the antenna, making its response unidirectional. Therefore, an antenna with a parasitic element is useful for reducing the strength of multipath reflected signals arriving from directions behind the antenna, to eliminate ghosts in the picture.

Dipole and reflector. Referring to Fig. 25-13, the reflector is usually placed approximately 0.2λ, or a little less than a quarter wave, behind the dipole, to reinforce signals arriving from the front. Also, the reflector is about 5 percent longer than the dipole, or a little more than a half wave. The antenna response depends on the spacing between the elements, and the tuning of the parasitic, which can be adjusted by changing its length. Closer spacing lowers the antenna impedance and narrows the frequency response. For the typical arrangement shown in Fig. 25-13 the dipole and reflector have a voltage gain of approximately 5 dB. The antenna impedance is about one-half the impedance of the antenna itself, resulting in 150 Ω for the folded dipole and reflector and 36 Ω for the straight dipole and reflector. The front-to-back ratio is approximately 3 dB, resulting in about 1.4 times more signal voltage from the front than from the back.

It is important to note that the parasitic element is effective only within the frequency range for which it is approximately tuned to half-wave resonance. Furthermore, the increase in gain and directivity with a reflector cuts off sharply at frequencies lower than the resonant frequency. For this reason, a dipole with reflector is usually cut for the lowest frequency in the range to be covered. The reflector can then be effective up to a frequency about 30 percent higher than the resonant frequency.

The response of the dipole with reflector can be explained as follows. Signal from the front produces current in the reflector 90° later than in the dipole. The 90° lag results because of the reflector spacing and its extra length. Current in the reflector reradiates signal to the dipole, which arrives an additional 90° later. Meanwhile, the signal at the dipole has varied through 180° of its cycle. The reradiated signal from the reflector, which has been delayed 180° also, then combines with the antenna current in the dipole to provide more signal for the transmission line. This increase in signal provides a forward gain up to 6 dB, or double the signal voltage of a dipole alone. For signal arriving from the back, however, the reradiated signal from the reflector is just 90° out of phase with antenna current on the dipole. The combined signal from the back, therefore, is less than the combined signal from the front.

Dipole with director. Referring to the parasitic director element in Fig. 25-14, it is placed 0.12λ in front of the dipole and is about 4 percent shorter than the driven element. The gain and directivity of the dipole with director drop off

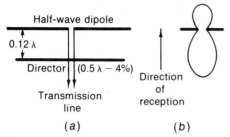

FIGURE 25-14 DIPOLE ANTENNA WITH PARASITIC DIRECTOR IN FRONT.

sharply at frequencies higher than the resonant frequency, which is opposite to the operation of the dipole and reflector off resonance. Practically all television antennas with one parasitic element use a reflector, instead of a director, because the director needs closer spacing for the same gain and front-to-back ratio, which reduces the antenna resistance and narrows the frequency response. For an antenna operating over a relatively narrow frequency, however, directors are combined with a dipole and reflector.

A dipole with one reflector and one director, having the same spacings as for a single parasitic element, provides about 7 dB gain, while the antenna resistance is approximately one-eighth the value of the driven element by itself.

Yagi antenna. A dipole with one reflector and two or more directors, as illustrated in Fig. 25-15, is called a *Yagi antenna*. This is a compact high-gain array, with a sharp forward broadside lobe and narrow bandwidth, often used in low-signal areas to cover one television channel or several adjacent channels. The gain of the Yagi antenna with three parasitic elements is about 10 dB, with a front-to-back ratio of approximately 15 dB. A high-impedance folded dipole is generally used for the driven element so that the reduced value of antenna resistance with the parasitic elements can be about 150 to 300 Ω. More directors can be added but generally only one reflector is used because additional reflectors do not improve the response.

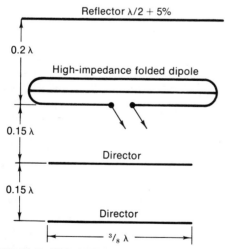

FIGURE 25-15 YAGI ANTENNA ARRAY, WITH DIPOLE, REFLECTOR, AND TWO DIRECTORS.

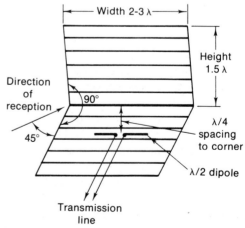

FIGURE 25-16 DIPOLE WITH CORNER REFLECTOR FOR UHF BAND.

Dipole with corner reflector. The antenna in Fig. 25-16 has a reflector constructed as a corner conducting sheet behind the half-wave dipole. The corner reflector can be either a solid metal sheet or a grid consisting of wires or wire screening, provided that the spacing between grid wires is 0.1λ or less. The dipole antenna, insulated from the parasitic reflector, is mounted along the line bisecting the 90° corner. Maximum signal is received from the front along this line. The construction of this antenna is practical for the UHF band because of the small size at these frequencies.

25-9 Multiband Antennas

All the television broadcast channels can be considered in three bands: 54 to 88 MHz for the low-band VHF channels, 174 to 216 MHz for the high-band VHF channels, and 470 to 890 MHz for the UHF channels. The main problem in using one dipole for both VHF bands is maintaining the broadside response, as the directional pattern of a low-band dipole splits into side lobes at the third and fourth harmonic frequencies in the 174- to 216-MHz band. A high-band dipole cut for a half wavelength in the 174- to 216-MHz band is not suitable for the 54- to 88-MHz band because of insufficient signal pickup at the lower frequencies.

As a result, antennas for both VHF bands generally use either separate dipoles for each band or a dipole for the 54- to 88-MHz band modified to provide broadside unidirectional response in the 174- to 216-MHz band also. For the UHF band, a VHF antenna can operate as a long-wire antenna or be used as a reflector sheet behind a broad-band dipole added in front. Also, an array of very short UHF dipoles can be mounted at the front of the VHF antenna.

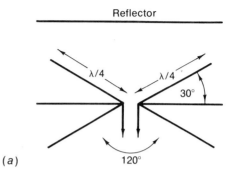

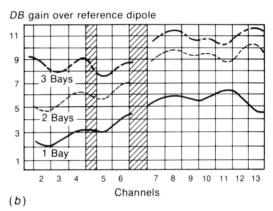

FIGURE 25-17 VHF CONICAL DIPOLE WITH REFLECTOR. (a) SPACING OF ELEMENTS; (b) RESPONSE CURVES FOR CHANNELS 2 TO 13.

It should be noted that higher gain is necessary for a UHF antenna to provide the same signal strength as on the VHF channels because the reference dipole is shorter at ultrahigh frequencies. Furthermore, the UHF band is relatively narrow, as the ratio of 890 MHz to 470 MHz is less than 2:1.

Conical dipole. The VHF dual-band antenna illustrated in Fig. 25-17 is generally called a *conical* or *fan* dipole. This antenna consists of two half-wave dipoles inclined about 30° from the horizontal plane, similar to a section of a

FIGURE 25-18 IN-LINE ANTENNA WITH TWO FOLDED DIPOLES AND REFLECTOR FOR VHF CHANNELS 2 TO 13.

cone, and usually a horizontal dipole in the middle. All the dipoles are tilted inward toward the wavefront of the arriving signal at an angle approximately 30° from the broadside direction, resulting in a total included angle of 120° between the two conical sections. Smaller values of included angle reduce the amount of signal intercepted at low frequencies, as the distance across the front is decreased. Either a straight reflector or a conical reflector can be used behind the conical dipole with approximately the same results.

Cut for a half wavelength at channel 2, the conical dipole with reflector is commonly used as a receiving antenna to cover both VHF bands. The antenna resistance is about 150 Ω. For the 54- to 88-MHz band, the antenna is a conical-type half-wave dipole with a parasitic reflector, providing the desired unidirectional broadside response. For the 174- to 216-MHz band, the tilting of the dipole rods shifts the direction of the split lobes produced at the third and fourth harmonic frequencies, so that they combine to produce a main forward lobe in the broadside direction.

In-line antenna. As shown in Fig. 25-18, this antenna consists of a half-wave folded dipole with reflector for the 54- to 88-MHz band, in line behind the shorter half-wave folded dipole for the 174- to 216-MHz band. The distance between the two folded dipoles is approximately one-quarter wavelength at the high-band dipole frequency. This is the length of line connecting the short dipole to the long dipole, where the transmission line to the receiver is connected. For the low-band channels, the long folded dipole with reflector supplies antenna signal to the transmission line, as the short dipole has little pickup at these low frequencies. For the high-band channels, the short folded dipole supplies signal to the transmission line, with the long folded dipole operating as a reflector. The directivity pattern of the in-line antenna is relatively uniform on all VHF channels, with a unidirectional broadside response. The average antenna gain is approximately 2 dB on the low-band channels and 5 dB on the high-band channels. The antenna resistance is about 150 Ω.

VHF and UHF antenna. Figure 25-19 shows a VHF conical dipole combined with a UHF bowtie

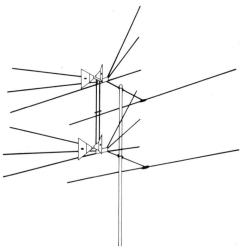

FIGURE 25-19 VHF CONICAL ANTENNA WITH UHF BOWTIE. TWO BAYS STACKED VERTICALLY.

ANTENNAS AND TRANSMISSION LINES 597

FIGURE 25-20 U/V SPLITTER TO SEPARATE VHF AND UHF ANTENNA SIGNALS FOR RECEIVER. LENGTH IS 2-5 IN.

antenna, in two bays, to cover all the television channels. For channels 2 to 6, the conical dipole has a parasitic reflector. For channels 7 to 13, the conical dipoles form a V antenna. On the 470- to 890-MHz band, the conical elements form a reflector for the bowtie antenna. This covers the UHF channels 14 to 83, with an average antenna gain of 6 dB. One downlead from the center of the array is used. At the receiver, the UHF and VHF signals to the antenna terminals on the receiver are separated by a U/V splitter (Fig. 25-20).

Log-periodic antennas. The basic construction for a seven-element array is illustrated in Fig. 25-21. The longest dipole is at the back, and each adjacent element is shorter by a fixed ratio,

typically 0.9. As an example, with the back dipole 108 in. for channel 2 at 54 MHz, the next length is 90 percent, or $0.9 \times 108 = 97.2$ in. Then the next dipole is $0.9 \times 97.2 = 87.5$ in. Also, the distance between dipoles becomes shorter by a constant factor, which is typically 35 percent of quarter-wave spacing. As a result, the resonant frequencies for the dipoles overlap to cover the desired frequency band. All the dipoles are active elements, without parasitic reflectors or directors. The active dipoles are interconnected by the crossed phasing harness, which transposes the signal by 180°.

When the longest dipole is cut for channel 2, the array will cover the low-band VHF channels from 54 to 88 MHz as antenna resonance moves toward the shorter elements at the front. For the high-band VHF channels from 176 to 216 MHz, however, the elements operate as $3\lambda/2$ dipoles. They are angled in as a V, to line up with the split lobes in the directional response for third-harmonic resonance.

When the longest dipole is cut for channel 14 in the UHF band of 470 to 890 MHz, the array can cover all the UHF channels. The V angle is not necessary in the UHF array because this frequency range is less than 2:1.

Figure 25-22 shows a commercial antenna using log-periodic arrays. The larger unit is the VHF antenna for channels 2 to 13. The shorter array at the front is the UHF antenna. Both are connected to one transmission line for the downlead. At the receiver, a U/V splitter connects the signal to the separate VHF and UHF antenna terminals.

25-10 Stacked Arrays

In order to increase the antenna gain and directivity, two or more antennas of the same type can be mounted close to each other and connected to a common transmission line. The individual

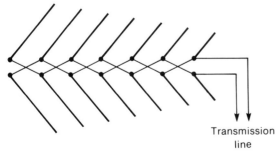

FIGURE 25-21 CONSTRUCTION OF LOG-PERIODIC ANTENNA. DIPOLES ARE DRAWN TO SCALE TO ILLUSTRATE SHORTER DIPOLES AND CLOSER SPACING FROM BACK TO FRONT.

598 BASIC TELEVISION

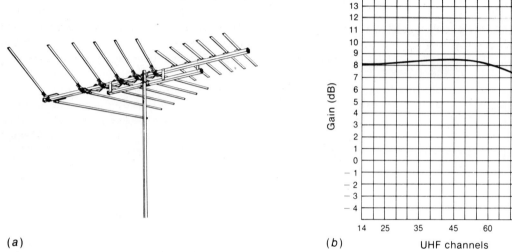

(a)

(b) UHF channels

FIGURE 25-22 (a) LOG-PERIODIC VHF ANTENNA WITH UHF DIPOLES AT FRONT. LENGTH OF CROSSARM OR BOOM IS 10 FT. (b) RESPONSE CURVE OF UHF CHANNELS.

antennas are called *bays* and the combined unit is a *stacked array*. Mounting the antennas one above the other, as illustrated in Fig. 25-23, is vertical stacking. This can be considered a *broadside array,* as the antenna receives or transmits best in the direction broadside to the stacking.

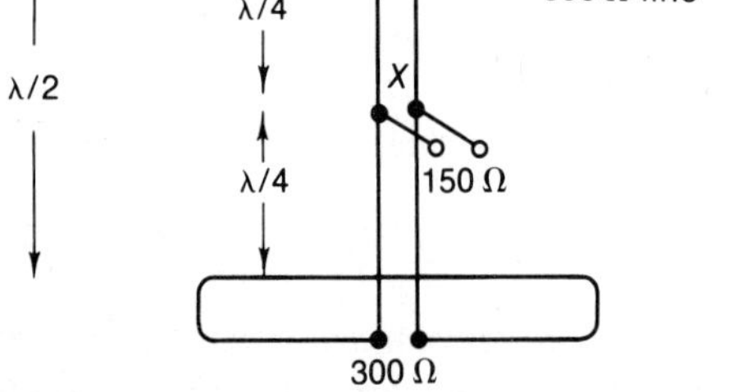

In general, stacking can provide an additional gain up to 3 dB for each antenna bay added, since the larger the area of the array, the greater the amount of signal that can be intercepted. To utilize the signal, however, the individual bays must be phased correctly with respect to the common junction for the transmission line, so that the antenna signals can add to provide the required antenna gain with the desired directional response. Rods used for interconnecting the bays in the stacked array are called *phasing rods*. In phasing separate antennas to a common transmission line, the following points should be noted:

1. A difference of one-quarter wavelength in the distance the antenna signal must travel is equivalent to a phase-angle difference of 90° between the two antenna signals.

FIGURE 25-23 TWO ANTENNA BAYS STACKED VERTICALLY IN BROADSIDE ARRAY. REFLECTORS OMITTED FOR CLARITY.

2. A difference of one-half wavelength is equivalent to a phase-angle difference of 180°.
3. Reversing the connections is equivalent to a phase-angle difference of 180°. With twin lead, a half twist of the line reverses the connections.

Vertical stacking. This is often done in weak signal areas to increase gain and sharpen directivity in the vertical plane, which reduces pickup of external noise from sources usually below the antenna. Figure 25-23 illustrates vertical stacking and how the antenna bays are usually connected to the common transmission line. Folded dipoles are shown, without reflectors for simplicity, but the same principles apply for any other antenna. The half-wave spacing generally used between stacked antenna bays is convenient for interconnecting them with two quarter-wave sections. Because of the symmetry, the antenna signals from both bays travel the same distance to point X and are in phase at the common junction for the transmission line to the receiver. The impedance here is one-half the impedance of either section, as the two lines are in parallel. With more than two antennas in the array, they can be grouped in adjacent pairs, and the pairs are then interconnected to the common transmission line. The horizontal directivity pattern of the vertically stacked array has the same broadside response as the individual antennas, but with more gain.

Undirectional end-fire array. In Fig. 25-24, two antennas are stacked one behind the other a quarter wavelength apart. They are connected to the common transmission line at point X to form an array that has the unidirectional response shown, without any parasitic reflector or director. The arrangement is an end-fire array because it has maximum response through the line of the array. This double V is often used as a VHF antenna. Each V is a dipole, cut for the low-band channels but angled in for broadside response on the high-band channels.

The end-fire directivity results from making the currents in the individual antennas out of phase with each other. With 90° phasing, the response is unidirectional off the end farthest from the transmission line. The antenna to which the transmission line connects is then back of the array. When receiving from the front, antenna 1 intercepts the signal a quarter cycle sooner than antenna 2. However, the quarter-wave line connecting the two antennas delivers this signal at point X in the same phase as the signal intercepted by antenna 2. The array provides a gain of 3 dB, therefore, for signal arriving from the front.

Signal from the back is intercepted by antenna 1 a quarter cycle later than antenna 2. With the additional quarter wavelength of the connecting line, the signal delivered by antenna 1 arrives at point X 180° out of phase with the signal from antenna 2. The two signals cancel, therefore, resulting in minimum reception from the back.

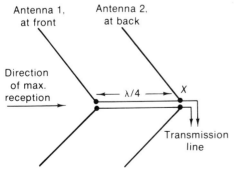

FIGURE 25-24 TWO V ANTENNAS MOUNTED ONE BEHIND THE OTHER IN HORIZONTAL END-FIRE ARRAY. NO REFLECTORS OR DIRECTORS ARE NEEDED.

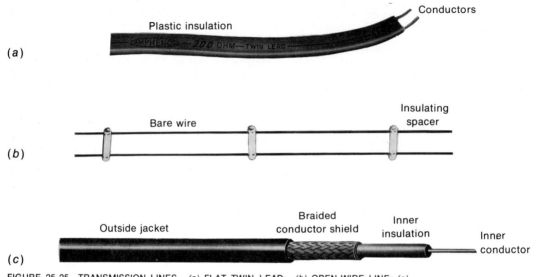

FIGURE 25-25 TRANSMISSION LINES. (a) FLAT TWIN LEAD; (b) OPEN-WIRE LINE; (c) COAXIAL CABLE.

25-11 Transmission Lines

The main types are shown in Fig. 25-25. The transmission line is just a conductor that has the function of delivering the antenna signal to the receiver. Three important requirements of the line are: (1) minimum losses, (2) no reflections of signal on the line, and (3) no stray pickup of signal by the line itself. To prevent signal pickup, the line should be balanced or shielded or both.

A line is balanced when each of the two conductors has the same capacitance to ground. The balance corresponds to the dipole antenna itself, which has balanced signals of opposite phase in the arms. A balanced line is connected to the opposite ends of a center-tapped antenna input transformer. Then any in-phase signal currents caused by stray pickup by the line can be canceled because of the balanced receiver input circuit. However, the antenna signal from the dipole has opposite phases in the two sides of the line. Then they reinforce each other in the antenna input transformer.

A shielded line is completely enclosed with a metal sheath or braid that is grounded to serve as a shield for the inner conductor. The shield prevents stray signal from inducing current in the center conductor. Usually, the shield is grounded to the receiver chassis. With only one inner conductor, the line is unbalanced. However, a balancing transformer (balun) can be used. When two inner conductors are used within the shield, the line is both balanced and shielded.

The parallel-wire line in Fig. 25-25a made in the form of a plastic ribbon is generally called *twin lead*. The type in Fig. 25-25b is open-wire line. These are balanced but not shielded lines.

The coaxial line in Fig. 25-25c is shielded but unbalanced. Shielded lines generally have more capacitance and greater losses. This is shown in Table 25-1, which lists the characteristic impedance and attenuation for the common types of line.

Line losses. The attenuation of the line is caused mainly by I^2R losses in the ac resistance of the wire conductor. These losses reduce the amount of antenna signal delivered by the line to the receiver. The longer the line and the higher the frequency, the greater the attenuation.

For long runs, the line can have appreciable losses. As an example, 200 ft of RG-59U coaxial cable has about 6 dB attenuation for 100 MHz, which means a loss of one-half the signal. The losses are even greater in the UHF band of frequencies. At 500 MHz, the RG-59U cable has an attenuation of 8.3 dB per 100 ft, compared with 3.7 dB at 100 MHz. Note that RG-11U has less losses than RG-59U.

The characteristic impedance (Z_0) of the line depends on the size of the wire conductors and their spacing. Uniform spacing of the conductors is what makes a transmission line with a characteristic Z_0. Wider spacing results in higher Z_0. In practical terms, Z_0 is just the impedance needed to terminate the line to prevent reflections from the end. Two important factors about the characteristic impedance are: (1) Z_0 is an ac value that cannot be measured with an ohmmeter, and (2) Z_0 is a characteristic that is the same for a line of any length. A 1-ft length and a 100-ft length of the same type of line have the same Z_0. However, the longer line has more attenuation losses.

Flat twin lead. This is the type of line generally used for television receivers because it has low losses, is available with 300-Ω Z_0, costs less, and is flexible for ease of handling. The plastic is a dielectric such as polyethylene, with the parallel conductors embedded in the ribbon. For 300-Ω Z_0, the spacing is typically $3/8$ in. between conductors of gage number 20 to 22. The 75- and 150-Ω twin lead have closer spacing.

Since television receivers have a balanced input impedance of 300 Ω, the twin lead of 300-Ω Z_0 is convenient for matching the transmission line to the receiver. With the receiver end matched to prevent reflections, an impedance match at the antenna end may not be necessary.

There are different qualities of twin lead with the same 300-Ω Z_0. Stronger line is preferable for outside use because it lasts longer. Also, thicker conductors, with a lower gage number, have less losses.

The attenuation of flat twin lead increases when the line is wet, especially for old line with

TABLE 25-1 TRANSMISSION LINES

TYPE	CHARACTERISTIC IMPEDANCE $Z_0 \Omega$	ATTENUATION AT 100 MHz, dB/100 FT	CAPACITANCE PER FOOT, pF	VELOCITY FACTOR V†
Flat twin lead	300*	Dry, 1.2; Wet, 7.3	6	0.8
Open-wire line	300–600	0.2	1	0.98
Coaxial, RG-11U, diameter 0.4 in.	75	1.9	20	0.66
Coaxial, RG-59U, diameter 0.24 in.	73	3.7	21	0.66

*Flat twin lead is also available in Z_0 of 75 Ω and 150 Ω.
†V is ratio of velocity in dielectric to velocity in free space.

tiny cracks that absorb water. Ultraviolet light from the sun causes stresses in the plastic which produce almost invisible cracks that collect moisture and dirt, increasing the line losses.

Foam twin lead. This type has the two conductors embedded in polyethylene foam, for low losses, especially for the UHF band of frequencies. The twin line is encased in a strong plastic jacket for protection against the weather.

Shielded twin lead. This type is encased like the foam line, but a shield of solid aluminum foil inside the jacket surrounds the twin lead. It may also have a drain wire connected to the receiver chassis to remove static charges.

Coaxial cable. As illustrated in Fig. 25-25c, this line consists of a center conductor in a dielectric that is completely enclosed by a metallic shield. This may be tubing or a flexible braid, of copper or aluminum. A plastic jacket is molded over the entire line as a protective coating. This line is practically immune to any stray pickup because the outer conductor acts as a grounded shield. In effect, the inner conductor is one side of the line and the conductor shield is the other side.

Coaxial cable is used in noisy locations, or where stray pickup of interfering signals by the line is a problem, and where multiple lines must be run close to each other. In cable distribution systems, coaxial cable is usually a necessity, in spite of its losses. The types generally used for television are RG-11U, with an outside diameter of 0.4 in. for Z_0 of 75 Ω, and RG-59U, with a diameter of 0.24 in. for Z_0 of 73 Ω.

Foam coaxial cable. This type is also available with foam dielectric. This reduces the attenuation about 15 percent at 100 MHz and 25 percent at 500 MHz. Also, the velocity factor V in Table 25-1 is 0.81 with foam dielectric.

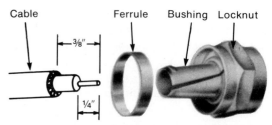

FIGURE 25-26 F-TYPE CONNECTOR FOR COAXIAL CABLE.

F-connector. See Fig. 25-26. This type is often used with coaxial cable. The center conductor of the cable itself is the connecting pin, which eliminates soldering on the male plug. A soft ferrule ring is crimped around the shield to hold the cable. Usually, a locking nut is screwed down to make a tight connection to the female fitting.

25-12 Characteristic Impedance

When the transmission line has a length comparable with a wavelength of the signal frequency carried by the line, the line has important properties other than its resistance. The small amount of inductance of the conductor and the small capacitance between conductors are distributed over the entire length of the line. The result is a distributed inductance and capacitance which can make the line a reactive load, equivalent to lumped reactance in an ordinary circuit. Furthermore, the uniform spacing between conductors provides constant inductance and capacitance per unit length. Therefore, the characteristic impedance can be defined in electrical terms as

$$Z_0 = \sqrt{\frac{L}{C}} \quad \Omega \qquad (25\text{-}3)$$

Z_0 is in ohms, L the inductance per unit length in henrys, and C the shunt capacitance per unit length in farads. As an example, assume a line with 0.54 μH inductance per foot for both conductors and 6 pF capacitance. Then

$$Z_0 = \sqrt{\frac{0.54 \times 10^{-6}}{6 \times 10^{-12}}} = \sqrt{0.09 \times 10^6} = 0.3 \times 10^3$$
$$Z_0 = 300 \ \Omega$$

The closer the conductor spacing, the greater the capacitance and the smaller the Z_0 of the line.

Physical characteristics. In terms of physical construction, Z_0 depends on the conductors and their spacing. For parallel-conductor line with air dielectric,

$$Z_0 = 276 \log \frac{s}{r} \quad \Omega \qquad (25\text{-}4)$$

where s is the distance between centers and r is the radius of each conductor, with s and r both in the same units. As an example, for open-wire line, No. 12 gage wire has a radius of 0.04 in. With 1-in. spacing, the ratio of s/r equals 1/0.04, or 25. The log of 25 is 1.398. Therefore, Z_0 is 276 × 1.398, which equals 386 Ω.

For air-insulated coaxial line,

$$Z_0 = 138 \log \frac{d_o}{d_i} \quad \Omega \qquad (25\text{-}5)$$

where d_o is the diameter of the outside conductor and d_i the diameter of the inside conductor. As an example, No. 18 wire for the inside conductor has a diameter of 0.08 in. If the diameter of the outside conductor is 1/4 in. the ratio of d_o/d_i then is 0.250/0.08, or 3.125. The log of 3.125 is 0.4949. Therefore, Z_0 equals 138 × 0.4949, or 68 Ω.

Resonant and nonresonant lines. When the line has a resistive load equal to Z_0 connected across the end, all the energy traveling down the line will be used in the load. Maximum power transfer results, and no energy is reflected back into the line. This termination of the line in its characteristic impedance makes it effectively an infinitely long line. The reason is that the line has no discontinuity at the load when the load equals Z_0. The effect is the same as though the line had no end. Such a line terminated in its Z_0 is a nonresonant or *flat line* because there are no reflections. Then the length of the line is not critical.

When the transmission line has a termination not equal to Z_0, the current will be reflected from the end. The result is standing waves of E and I, just as on an antenna. This effect makes the line resonant. Its length is critical because a connection at different points on the line has different values of signal.

The ratio of the voltage at a maximum point to the voltage at a minimum point is defined as the voltage standing-wave ratio, abbreviated VSWR. For a nonresonant line terminated in Z_0, the VSWR is 1. For a resonant line not terminated in Z_0, the VSWR is more than 1. The greater the mismatch of the load to the Z_0 of the line, the higher the VSWR of the line. The extreme case of a resonant line with a high VSWR is either a short or open at the end.

As a practical application of a nonresonant line, assume 300-Ω twin lead is connected to the 300-Ω antenna input terminals on the receiver. The line can be cut at any length because there are no standing waves. However, more length still produces more I^2R losses. Also, any extra length should not be coiled because of the inductive effect. If you add hand capacitance to a nonresonant line by touching it, there will be either little effect or less signal. If you touch a resonant line, the result can be much more or much less signal, but only at particular points on the line for different frequencies.

25-13 Transmission-Line Sections as Resonant Circuits

When the transmission line is not terminated in its characteristic impedance, the values of current and voltage change along the line, the magnitudes varying with a wave motion that is the same as for an antenna. Therefore, the impedance for different points on the line varies from maximum at the point of highest voltage on the line to minimum at the point of highest current, as the impedance at any point equals the ratio of voltage to current. This is illustrated in Fig. 25-27 showing transmission lines being used as resonant circuits. Since the action of the line in such an application depends on the existence of reflections, the lines are not terminated in their characteristic impedance but are either shorted or open at the end in order to produce the maximum standing-wave ratio and the highest Q for the equivalent resonant circuit.

In analyzing the action of the transmission-line sections, it should be noted that an open end must be a point of maximum voltage, minimum current, and maximum impedance. For the opposite case, a shorted end must be a point of maximum current, minimum voltage, and minimum impedance. For each length equal to a quarter wave back from the end of the line the voltage and current values are reversed from maximum to minimum or vice versa. Intermediate values of impedance are obtained for points along the line between the maximum and minimum points.

In Fig. 25-27 it is shown that a quarter-wave section shorted at the end is equivalent to a parallel-tuned circuit at the generator side because there is a high impedance across these terminals at the resonant frequency. For a length shorter than a quarter wave the line is equivalent to an inductance.

The open quarter-wave section provides a very low impedance at the generator side of the line. A length less than a quarter wave makes the line appear as a capacitance.

The half-wave sections, however, repeat the impedance at the end of the line to furnish

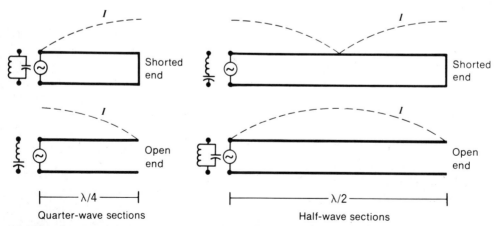

FIGURE 25-27 TRANSMISSION-LINE SECTIONS AS RESONANT CIRCUITS.

TABLE 25-2 RESONANT LINE SECTIONS

LENGTH	TERMINATION	INPUT IMPEDANCE	PHASE SHIFT
Quarter wave	Shorted	Open circuit	90°
Quarter wave	Open	Short circuit	90°
Half wave	Shorted	Short circuit	180°
Half wave	Open	Open circuit	180°

the same impedance at the generator side. A half wave corresponds to a half-cycle of signal, though, with a phase reversal of the voltage and current. The main features of $\lambda/4$ and $\lambda/2$ sections are summarized in Table 25-2.

Transmission-line sections are often called *stubs*. These can be used for

1. Impedance matching
2. An equivalent series resonant circuit to short an interfering rf signal
3. Phasing signals correctly in antennas

For phasing sections, a quarter wave produces a 90° change in phase angle between the signal at the input and output ends. A half-wave section shifts the phase by 180°.

To reduce interference, an open $\lambda/4$ stub at the interfering signal frequency can be used. One side is connected across the antenna input terminals on the receiver, while the open end produces a short at the receiver input one-quarter wave back. The same results can be obtained with a half-wave stub shorted at the end.

25-14 Impedance Matching

This means making the impedance of the load equal to the impedance of the generator or source, for maximum power transfer. In the receiver antenna system, the transmission line is the load for the antenna, while the receiver input circuit is the load for the line. A mismatch at either end causes a loss of signal. In addition, reflections can be produced on the line when all the energy is not absorbed at the load. However, only one end need be connected to its characteristic impedance to eliminate standing waves on the line, since there cannot be any reflections from the matched end.

General practice is to use 300-Ω twin lead to match the receiver input. This impedance is designed to be approximately constant for all channels. As a result, the impedance match is maintained throughout the television band for maximum transfer of signal from the line to the receiver and to eliminate reflections in the line.

Matching the impedance of a multiband antenna to the Z_0 of the transmission line is usually not critical. The reason is that the antenna impedance may vary over a wide range for different channels, as the electrical length of the antenna changes for different frequencies. An impedance mismatch of 2.5 to 1 results in a 1-dB loss of signal.

Quarter-wave matching section. When a quarter-wave section of transmission line with a characteristic impedance Z_0 is neither shorted nor open at the end but has an impedance Z_1 connected across one end, the impedance at the other end Z_2 is

$$Z_2 = \frac{Z_0^2}{Z_1} \quad \text{or} \quad Z_0 = \sqrt{Z_1 Z_2} \qquad (25\text{-}6)$$

Z_0 is the geometric mean of Z_1 and Z_2. Therefore, if a quarter-wave section of line having an impedance equal to $\sqrt{Z_1 Z_2}$ is used to couple two unequal impedances Z_1 and Z_2, the section will provide an impedance match at both ends.

Referring to Fig. 25-28, a 35-Ω antenna is used with 300-Ω line, and the matching section has an impedance equal to $\sqrt{35 \times 300}$, or approximately 100 Ω. This equivalent quarter-wave matching transformer is often called a *Q section*.

The length of the quarter-wave section is calculated for the desired frequency from the formula

$$\frac{\lambda}{4}(\text{ft}) = V \times \frac{246}{f(\text{MHz})} \tag{25-7}$$

V is the *velocity factor*, which depends on the velocity of propagation along the line. This is less than the speed of light because of the reduced velocity in solid dielectric materials. Values of the velocity factor *V* are listed in Table 25-1 for different types of line. As an example, a quarter-wave matching stub at 67 MHz using twin lead would have a length of 2.9 ft.

The quarter-wave matching section has the advantage of producing an impedance match with very little attenuation of the signal, but the section can provide a match only for frequencies at which it is approximately resonant. However, the stub functions in the same way when its length is $3/4\lambda$ or any odd multiple of a quarter wave. Therefore, it can be cut to serve for both the low- and high-frequency VHF television bands.

Balancing unit. Two Q sections can be combined to make the balancing and impedance-transforming unit illustrated in Fig. 25-29. This is called a *balun,* for matching between balanced and unbalanced impedances, or an *elevator transformer*. At one end, the two 150-Ω quarter-wave sections in Fig. 25-29a are connected in parallel, resulting in 75-Ω impedance across points A and B. Either A or B can be grounded to provide an unbalanced impedance at the ungrounded point with respect to ground. At the other end, the two 150-Ω quarter-wave sections are connected in series to provide 300-Ω impedance between points C and D. The quarter wavelength of line isolates the ground point from C or D, allowing a balanced impedance with respect to ground.

Either side of the balun can be used for input, with the opposite side for output. Useful applications are matching 72-Ω coaxial line to 300-Ω receiver input or, in the reverse direction, 300-Ω twin lead to 75-Ω unbalanced input.

The equivalent transformer in Fig. 25-29b shows how bifilar windings are used to simulate the 150-Ω transmission-line sections. This small transformer is inside the balun unit in Fig 25-29c. The windings are usually on a ferrite core to increase the inductance, which makes the line electrically longer. The 75-Ω side has a jack to take coaxial cable, while the 300-Ω side has twin-lead terminals.

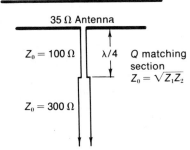

FIGURE 25-28 QUARTER-WAVE SECTION MATCHING ANTENNA TO TRANSMISSION LINE.

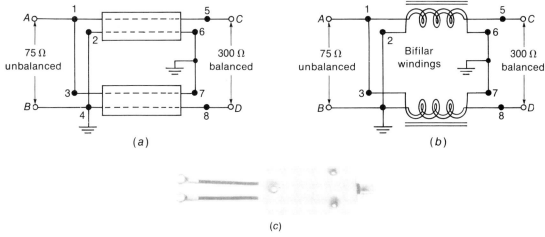

FIGURE 25-29 BALANCING UNIT (BALUN) TO MATCH BETWEEN 75-Ω UNBALANCED AND 300-Ω BALANCED Z. (a) WITH λ/4 MATCHING SECTIONS; (b) EQUIVALENT TRANSFORMER; (c) TYPICAL UNIT. LENGTH IS 3 IN.

25-15 Antenna Installation

A typical installation is illustrated in Fig. 25-30 for a private house in a suburban area. A mounting with chimney straps is shown, but a roof mount or wall mount can be used instead. Mount the antenna near a line of sight, broadside to the transmitter, and as high as possible. Increased height results in more antenna signal and less interference. The antenna should be at least 6 ft away from other antennas and large metal objects.

A suburban location 25 to 35 mi from the transmitter can be considered an average installation with one antenna array. Farther away, larger antennas and two or more bays stacked vertically may be necessary for sufficient antenna signal, especially on the UHF band. With multiple bays and antennas having a long cross

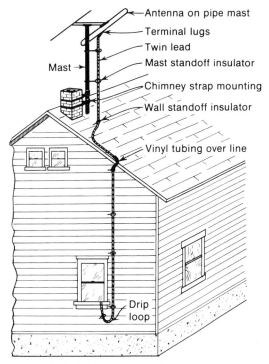

FIGURE 25-30 AN ANTENNA INSTALLATION.

arm, it probably will be necessary to use a roof mount with guy wires to hold the antenna mast steady in the wind. An antenna is usually installed in good weather, but remember, it must be able to hold up in bad weather.

Locations in crowded city areas close to the transmitter but surrounded by tall buildings can have very weak signal with severe reflections. Changing the antenna placement only a few feet horizontally or vertically can make a big difference in signal. The reason is that standing waves of signal are set up in areas where there are large conductors nearby, especially between buildings.

Antenna orientation. Before being clamped tight in its mounting, the antenna is rotated to the position that results in the best picture and sound. For suburban locations and fringe areas, the best orientation is usually broadside to the transmitter location. In city areas, however, the antenna may receive more signal over a reflected path from a high building nearby.

Antenna rotator. This is a motor which turns the antenna mast, operated by a remote control at the receiver. The rotator is very useful when channels must be received from different directions. One example is a town between two big cities that have most of the television stations. In addition, in any installation a rotator can be helpful in finding the best antenna orientation. A special cable is used, with conductors for the antenna signal and power to the rotator.

Selecting the transmission line. In most cases, 300-Ω twin lead is used. For long runs, heavy-duty foam twin lead is preferable. The open-wire line may be necessary for the extreme case of a long run that must have minimum losses.

When there is a problem of pickup of noise and interference by the line, coaxial cable is generally used. In locations where the transmission line must be run down a corner of an apartment house parallel to many other lines, for instance, shielded line is preferable in order to minimize pickup of interference from the other lines. Shielded coaxial cable is the type used in cable distribution systems for multiple receivers.

The transmission-line run. It should be as short as possible. Excess line is cut off in order to minimize attenuation of the signal. Do not coil extra line because the inductance acts as an rf choke to reduce the signal. Twin lead is generally twisted about one turn per foot to improve the electrical balance and strengthen the line mechanically so that it does not flap too much in the wind.

Grounding. To prevent accumulation of static charge on the mast, it can be grounded by connecting a heavy ground wire to a cold-water pipe or to a metal stake driven into the earth. In some areas, local regulations require grounding the mast.

Lightning arrestor. When the antenna is a high point in the area, a lightning arrestor should be installed on the transmission line. The lightning arrestor provides a high-resistance discharge path that prevents static charge from accumulating on the antenna; and, if lightning should strike the antenna, the arrestor is a short circuit to ground that protects the receiver. Indoors, the lightning arrestor is usually mounted by a grounding strap on a cold-water pipe. For use outdoors, a weatherproof type of arrestor can be installed on a grounded mast.

Impedance-matching applications. It is best to use transmission line having a characteristic impedance which is the same as the input impedance of the receiver, since then a match at one end of the line is approximately maintained for all channels. This generally means using 300-Ω line. The impedance match can be checked by noting the effect of added capacitance when you touch the line. When your hand capacitance at any position on the line decreases the picture strength, this indicates the impedance match is correct.

At the antenna, it is preferable not to match to a higher value of line impedance when the antenna is used for several channels and is cut for the lowest frequency. The antenna impedance then increases with frequency and the mismatch allows a broader frequency response in terms of antenna signal delivered at the receiver. If the antenna is used for only one channel, however, more signal can be obtained by matching the antenna to the transmission line, using a quarter-wave matching section. When 72-Ω coaxial cable is used in order to have a shielded transmission line, it can be matched to 300 Ω by means of a balun.

Indoor antennas. These can be used in strong-signal locations without ghosts. Most common is a simple dipole assembly with telescoping rods of adjustable length. The rods are extended to full length for the low-band VHF channels. Shortening the rods provides half-wave resonance for the high-band channels. A convenient indoor antenna for just two or three channels is a folded dipole formed from 300-Ω twin lead, as shown in Fig. 25-31.

The location of an indoor antenna is usually determined by trial and error to find the spot that provides the most signal. The best position is not always horizontal because the antenna pickup is generally reflected signal from nearby conductors, which is usually not horizontally polarized.

Another possibility is a power-line antenna, where one antenna input terminal is connected to the high side of the ac power line through an rf coupling capacitor of about 100 pF. The signal return path is to earth ground, through stray capacitance. However, the antenna signal is usually weak, depending on peaks and nodes for standing waves of rf signal on the power line.

Built-in antennas. With portable receivers, a telescoping dipole assembly is often mounted on the cabinet. When a single pole is used for a *monopole antenna,* it operates as a grounded quarter-wave type, connected to the receiver input with a balun. For UHF reception a single-turn loop antenna is used. The diameter is about 8 in., making the loop conductor several wavelengths for the UHF channel frequencies.

If the built-in antenna is used, the location and orientation of the receiver cabinet can make a big difference in the amount of antenna signal. When it is not being used, the built-in antenna should be disconnected from the receiver input terminals, using only the external antenna.

Attic antennas. In a private house, an outdoor type of antenna can usually be mounted in the attic with good results. Since wood and brick are insulators, they have very little effect on the antenna signal. However, sheets of aluminum insulation can reduce the signal level.

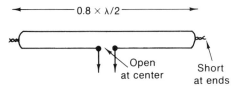

FIGURE 25-31 FOLDED DIPOLE ANTENNA MADE WITH 300-Ω TWIN LEAD.

25-16 Multiple Installations

When multiple receiver outlets from a common antenna system are needed, there are three main requirements:

1. The amount of antenna signal available for each receiver must be at least several hundred microvolts for a good picture without excessive snow. In addition, the signal from the antenna distribution system must be much greater than the amount of signal picked up directly by the transmission line or receiver chassis, without the antenna. Otherwise, there can be ghosts in the picture caused by duplicate signals.
2. Isolation should be provided between receivers to attenuate the local oscillator signal, which can produce beat-frequency interference patterns consisting of diagonal bars in other receivers on the common distribution line.
3. The impedance of the transmission line should be matched at the receiver end. This is important with a long line, to prevent ghosts caused by reflections on the line, or when the impedance match is necessary for maximum signal to provide a satisfactory picture.

Considering these requirements, it may be useful to note that several receivers can simply be connected in parallel to one transmission line, with adequate results if there is enough antenna signal, local oscillator interference is no problem, and the transmission line is less than about 50 ft long.

Multiset coupler. As shown in Fig. 25-32, this is a small, inexpensive unit designed to provide multiple 300-Ω output terminals for two to four

FIGURE 25-32 TWO-SET COUPLER. LENGTH IS 2½ IN. (THE FINNEY CO)

receivers, with 300-Ω input for the line from the antenna. Since no amplifier is included, a strong antenna signal is necessary because the coupler attenuates the distributed signals for each set.

Amplified distribution systems. When there is not enough antenna signal to operate several receivers, or when many receivers must be fed from a single antenna system, as in apartment houses, hotels, and motels, a master antenna distribution system using rf amplifiers is necessary. Furthermore, if individual antennas must be used for different channels, the amplifiers provide a means of mixing the antenna signals for common distribution. Signal for the FM broadcast band of 88 to 108 MHz is usually provided also to feed FM receivers. Coaxial cable is generally used for long distribution lines, as it is shielded against noise and direct pickup of signal.

More details of cable distribution systems are described in the next chapter. However, Fig. 25-33 illustrates a relatively simple amplified system with twin lead for distribution to four receivers. These lines should not be run close together. Also, any branch not used must be ter-

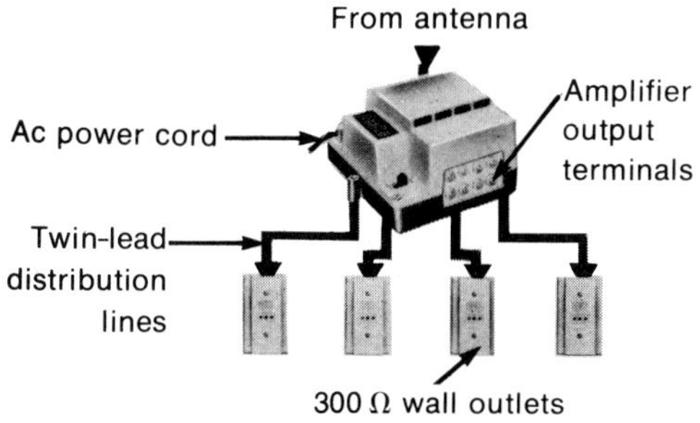

FIGURE 25-33 AMPLIFIED DISTRIBUTION SYSTEM FOR FOUR OUTLETS. (JERROLD ELECTRONICS CORPORATION)

minated with a 300-Ω resistor. Otherwise, oscillations in the amplifier or reflections in a line can be coupled to all the other lines, causing interference patterns in the receivers. This type of distribution system is convenient for installation in a private home, with the antenna and amplifier in the attic.

25-17 Troubles in the Antenna System

The trouble symptom of weak picture with excessive snow, on either some channels or all channels, is often caused by insufficient antenna signal. Also, ghosts in the picture can be eliminated only by providing antenna signal without any reflections.

When the transmission line is open, the line serves as an antenna. The signal may actually be stronger than normal on some channels but very weak on most channels. If the signal is better with only one side of the line connected to the receiver input, this effect shows the line is open.

The transmission line can be checked for continuity by an ohmmeter. Remember that the ohmmeter cannot indicate characteristic impedance but only dc resistance. With a folded dipole antenna connected to the line, it should have practically zero resistance at the receiver end to indicate continuity. With other antenna types, the line can be temporarily shorted at one end to check continuity at the opposite end.

If the transmission line is intermittently open, it will produce flashing in the picture, especially in windy weather. Flashing can also be caused by static discharge of the antenna when there is no lightning arrestor. If the signal is weak only in rainy weather, this usually means worn insulation on the line, which absorbs too much water, causing excessive losses for the antenna signal input to the receiver.

For the UHF channels, vibration of the antenna may produce fading of the picture, because of the very short wavelengths. Clean, tight contacts at the antenna are especially important for ultrahigh frequencies to avoid intermittent connections and loss of signal caused by leakage across the terminals.

SUMMARY

1. The length of a half-wave dipole equals $462/f$, where f is in MHz to calculate the half wavelength in feet. The antenna is horizontal for signals with horizontal polarization. The impedance at the center of a straight dipole is 72 Ω. The characteristics of a half-wave folded dipole are similar to the straight dipole but the impedance at the center is 300 Ω.

2. The polar directivity pattern of a half-wave dipole is a figure eight, showing maximum reception broadside and minimum off the end of the antenna. At harmonic frequencies, however, the broadside response decreases and reception increases off the ends. For this reason, dipoles are often angled in a V to improve the broadside response for higher frequencies.
3. Front-to-back ratio compares how much more signal the antenna receives from the front than the back. A ratio of 2:1 equals 6 dB.
4. Antenna gain indicates how much more signal an antenna array can receive, compared with a half-wave resonant dipole. A voltage gain of 2 equals 6 dB.
5. Ghosts in the picture are generally caused by reflected signal received by the antenna, in addition to direct signal from the transmitter. A directional antenna, with a narrow forward lobe, can eliminate ghosts.
6. A parasitic reflector mounted behind the dipole increases the antenna gain and front-to-back ratio, to minimize ghost problems. See Fig. 25-13 for length and spacing. However, a parasitic director is mounted in front of the dipole. See Fig. 25-14 for length and spacing.
7. A Yagi antenna is a parasitic array with one reflector and two or more directors. Its features are high gain and good front-to-back ratio with a narrow forward lobe, but relatively narrow bandwidth.
8. Antennas can be stacked vertically one above the other to increase the broadside response of the array. See the broadside array in Fig. 25-23.
9. Antennas can be stacked horizontally one behind the other to increase the response through the line of the array. See the double-V end-fire array shown in Fig. 25-24.
10. The main types of transmission line are flat twin lead and coaxial cable. The coaxial line is shielded but unbalanced. Flat twin lead is most common because it is economical, it has low losses, and its 300-Ω impedance matches the receiver input impedance.
11. The characteristic impedance $Z_0 = \sqrt{L/C}$, where L is the inductance and C the capacitance per unit length of line. Z_0 is independent of length.
12. Insufficient antenna signal results in a weak, snowy picture. Then interference in the picture is more noticeable and synchronization is poor because of the weak signal.

Self-Examination (Answers at back of book)

Choose (a), (b), (c), or (d).

1. The length of a half-wave dipole at 100 MHz is: (a) 4 m; (b) 4.62 ft; (c) 100 ft; (d) 100 m.
2. To eliminate ghosts in the picture : (a) use a longer transmission line; (b) con-

nect a booster; (c) change the antenna orientation or location; (d) twist the transmission line.
3. In Fig. 25-2, the antenna signal for a transmitter 60° off the ends equals: (a) 100 μV; (b) 300 μV; (c) 600 μV; (d) 800 V.
4. A Yagi antenna features: (a) stacked V dipoles; (b) two or more driven elements; (c) high forward gain; (d) very large bandwidth.
5. Which of the following is false for a dipole with reflector? (a) It should be cut for the lowest channel. (b) Gain is increased in the forward direction. (c) Front-to-back ratio is increased. (d) The line should be connected to the reflector.
6. Which of the following is true? (a) Angling a dipole in a V improves the response for low frequencies. (b) A folded dipole cannot be angled in a V. (c) The impedance of a folded dipole is 300 Ω. (d) The impedance of a straight dipole is 300 Ω.
7. Which of the following applies to 300-Ω twin lead? (a) Low losses, shielded and unbalanced; (b) high losses, unshielded and unbalanced; (c) low losses, shielded and balanced; (d) low losses, unshielded and balanced.
8. A nonresonant or flat line has no reflections regardless of length because it: (a) is usually very short; (b) is usually very long; (c) has low value of Z_0; (d) is terminated in Z_0.
9. Which of the following is true? (a) A reflector is in back of the antenna. (b) A director is in back of the antenna. (c) A reflector is connected to the line. (d) A director is connected to the line.
10. Which of the following is true? (a) Characteristic impedance can be measured with an ohmmeter. (b) Twin lead has lower line losses when wet. (c) An open twin-lead line can serve as an antenna. (d) Shielded line picks up more signal than unshielded line.

Essay Questions

1. Define the following terms: antenna gain, front-to-back ratio, parasitic element, harmonic antenna, broadside array, end-fire array.
2. Show a straight dipole antenna with length cut for 54 MHz. Also for 174 MHz. Do the same for folded dipole antennas.
3. What is the cause of a ghost in the picture? Why is the ghost usually to the right of the main image?
4. Show the polar directivity pattern of a half-wave dipole alone, and with a reflector. List two differences between the patterns.
5. A dipole is cut for half-wave resonance at 54 MHz. Show its polar directivity pattern at 54 MHz and at its third harmonic frequency of 162 MHz.
6. Which is more desirable to minimize ghosts: narrow bandwidth or a narrow forward lobe for the antenna response? Explain why.

7. List the velocity factors for twin lead, coaxial cable, and open-wire line. Why is the velocity factor for twin lead less than for open-wire line?
8. A ghost caused by multipath reception is 2 in. to the right of the main image, on a raster 20 in. wide. What is the difference in length between the direct and reflected signal paths?
9. Explain briefly how multiple ghost images can be caused by a long transmission line. What is a minimum line length that will produce a noticeable ghost?
10. Explain briefly how a ghost could be produced to the left of the main image. Assume the ghost image varies in strength when a person walks near the receiver.
11. Give three methods of reducing ghosts caused by multipath antenna signals.
12. Give two advantages of dipole with reflector, compared with a dipole alone.
13. Give two differences between a parasitic reflector and parasitic director.
14. Give two advantages and one disadvantage of a Yagi antenna.
15. Draw a folded dipole with an impedance of approximately 1,200 Ω at the center.
16. Describe briefly five types of television receiving antennas, giving one important feature of each.
17. Give two features of an LPV antenna.
18. Describe briefly how the directional pattern of a dipole cut for 54 MHz changes from channel 2 to channel 6 to channel 13.
19. Explain briefly why a long-wire V antenna receives best along its center line.
20. Explain briefly why an end-fire double-V array does not need a reflector.
21. Give one example where coaxial line would be preferable to twin lead for a line length of 100 ft to the external antenna.
22. Why is the transmission line terminated in its characteristic impedance?
23. Compare Z_0 and one additional feature for flat twin lead and RG-59U coaxial cable.
24. Distinguish between a resonant line and a flat (nonresonant) line. Why is a resonant line critical in length, while a flat line is not? How do they compare in VSWR?
25. Describe briefly five main parts of an antenna installation job.
26. What are two problems to consider when connecting multiple receivers to a common antenna?
27. Why does a weak picture with snow indicate the trouble may be insufficient antenna signal? Give one other cause of this trouble symptom.
28. Give two indications of an open transmission line.

Problems (Answers to selected problems at back of book)

1. Referring to Fig. 25-2, tabulate the values of antenna signal at 90°, 120°, 150°, and 180°.

2. Redraw Fig. 25-2 for an antenna with 6 dB gain, compared with the antenna shown, and front-to-back ratio of 6 dB.
3. Calculate the length of a half-wave dipole with reflector and give spacing, in feet, for 54 MHz.
4. Calculate the lengths for a straight dipole with reflector cut for 54 MHz and a folded dipole with reflector cut for 174 MHz. Show the two antennas with separate lines to a double-pole double-throw knife switch.
5. Calculate the length in feet of a half-wave folded dipole made of twin lead and cut for 98 MHz.
6. Show a $\lambda/4$ Q section made of twin lead, matching 300-Ω line to a 72-Ω antenna. Calculate length and Z_0 for the Q section.
7. Calculate length in feet of a folded dipole for the center frequency of channel 11.
8. (a) Draw a Yagi antenna for channel 11, showing lengths and spacings for the folded dipole, reflector, and two directors. (b) Show two of these Yagi antennas stacked vertically, with two alternate methods of phasing the antennas to a common transmission line.
9. What is the length in feet for each leg of a rhombic antenna 2λ long at 470 MHz?
10. (a) Draw a double-V end-fire array, showing length and spacing for 69 MHz. (b) Show two of these double-V antennas stacked vertically, with two alternate methods of phasing the antennas to a common transmission line.
11. Using Eq. (25-5) for air-insulated coaxial line, what is the Z_0 for $d_o = 0.64$ in. and $d_i = 0.2$ in.?
12. Calculate Z_0 for the following examples of parallel-conductor line with air insulation: (a) $L = 0.225$ μH/ft, $C = 1$ pF/ft. (b) Conductors of 0.04 in. radius with 0.6-in. spacing. (c) No. 18 gage wire conductors spaced 0.5 in. apart.

26 Cable Distribution Systems

The requirements of cable distribution systems are illustrated by the system in Fig. 26-1, which delivers antenna signal for the VHF and UHF channels to 24 television receivers. Note the symmetry, with two identical trunk lines of 12 sets each. Sufficient signal is provided by the head end, which consists of an amplifier fed by the master antenna. A two-way splitter divides the amplified output for trunk lines A and B. In the trunk line, four-way multitaps are used. Each taps off part of the signal to wall outlets for a group of four receivers. This layout would be for a dealer's showroom, to operate 24 sets from one antenna. Each receiver should have a signal of approximately 1 to 5 mV. All the distribution lines use coaxial cable with a nominal impedance of 75 Ω. The amplifier gain and the distribution losses are generally expressed in decibel (dB) units because these values can be simply added or subtracted. More details are explained in the following topics:

- 26-1 Head-end equipment
- 26-2 Distribution of the signal
- 26-3 Types of distribution losses
- 26-4 Calculation of total losses
- 26-5 System with multitaps
- 26-6 System with single taps at wall outlets
- 26-7 Decibel (dB) units
- 26-8 Decibel conversion charts

26-1 Head-End Equipment

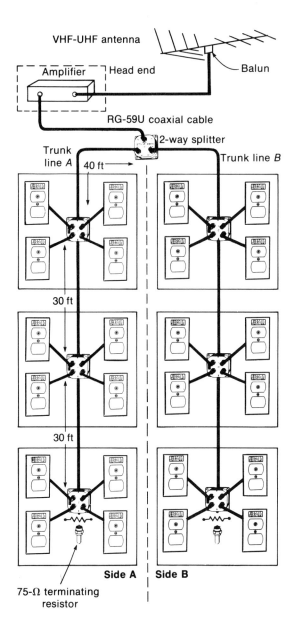

FIGURE 26-1 CABLE DISTRIBUTION SYSTEM FOR 24 RECEIVERS WITH FOUR-WAY MULTITAPS FOR GROUPS OF FOUR WALL OUTLETS. SEE TEXT FOR EXPLANATION.

This consists of the antenna and amplifier to feed signal into the main line for the splitter to the trunk lines. Note the following possibilities for the antenna:

1. Multiband antenna for the VHF channels
2. Multiband antenna for the UHF channels
3. All-channel antenna for the VHF and UHF channels
4. Separate antennas for each channel, using a high-gain Yagi array with a narrow forward lobe to reduce ghosts

When all stations are in the same direction, an all-channel antenna can generally be used. The amplifier provides the required gain. When UHF channels are to be received, they can be converted to unused VHF channels in order to reduce line losses for the distribution system.

The function of the amplifier is to provide the amount of gain needed to make up for the distribution losses. Then each receiver has the signal it would have as a single installation. The signal required for each receiver is at least 1 mV for no snow in the picture, but less than 5 mV to prevent overload distortion. The antenna signal can be determined with a field-strength meter (Fig. 26-2). Or, a battery-operated portable receiver can be connected temporarily to the antenna in order to judge the picture.

VHF amplifier. This amplifies all the VHF channels (2 to 13), with a typical gain of 45 dB.

UHF amplifier. This amplifies all the UHF channels (14 to 83), with a typical gain of 45 dB. More gain will usually be needed for the UHF channels because of greater line losses.

618 BASIC TELEVISION

FIGURE 26-2 FIELD-STRENGTH METER. WIDTH IS 9 IN. (BLONDER-TONGUE LABORATORIES INC.)

All-channel VHF-UHF amplifier. This amplifies the VHF and UHF channel frequencies. A typical unit is shown in Fig. 26-3, with a gain of 40 dB. The input can be one line from a combination antenna or separate lines from VHF and UHF antennas.

Mast-mounted preamplifier. With weak antenna signal, an amplifier may be needed before the main line run from the antenna. Otherwise, the signal-to-noise ratio will be too low for the main amplifier, causing snow in the picture. This factor is especially important if the antenna is far from the main amplifier. Remember that a line run of 100 ft could cut the antenna signal voltage by one-half, or 6 dB, depending on the frequency and type of line. A typical gain for the preamplifier is 28 dB. However, the main requirement of the preamplifier is low noise.

Single-channel amplifier. This amplifies the 6-MHz band for any one VHF channel, with a typical gain of 60 dB.

UHF-VHF channel converter. This can be used to convert any one UHF channel to an unused VHF channel. A typical conversion gain is 6 dB. The reason for converting is the lower line losses for the VHF channel frequencies.

Mixer-amplifier. This can be used to combine the signal from several antennas, or from single-channel amplifiers, for amplified output to the trunk splitter. A typical gain is 30 dB.

Passive antenna mixer. This can combine several antenna signals, but not for adjacent channels. It is passive, without amplification. A typical loss is −2 dB.

FIGURE 26-3 ALL-CHANNEL AMPLIFIER. (JERROLD ELECTRONICS CORPORATION)

Traps and filters. These passive elements attenuate signals that may cause overload in the amplifier and interference in the picture. They are available for the FM band of 88 to 108 MHz or for an interfering adjacent channel, usually at the sound carrier frequency.

26-2 Distribution of the Signal

In general, a splitter just divides the signal from the main cable to the individual trunk lines. A multitap or single tapoff takes a small part of the signal from the trunk line for individual receivers.

The tapoff has isolating resistors in order to separate each receiver from the other receivers on the common trunk line. All these components have fittings for coaxial cable, generally with F-type connectors.

Splitters. A two-way splitter divides the signal for two trunk lines; a four-way splitter divides for four trunks. Each division of the signal into two paths means one-half power into each trunk. This loss is −3 dB, which is 50 percent of the power or 70 percent of the input signal voltage. A four-way split has double these losses, or −6 dB. Then the result is one-fourth power and 50 percent signal voltage in each trunk line.

A typical splitter is shown in Fig. 26-4. The splitter is also considered a *coupler*. Resistance may be included to isolate the output connections from each other and to match the split to the transmission line. When there is less isolation for one outlet, it is a *directional coupler*.

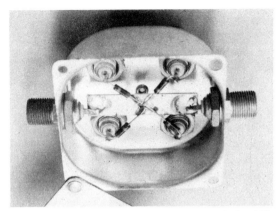

FIGURE 26-5 FOUR-WAY MULTITAP WITH COVER OFF. WIDTH IS 4 IN. NOTE ISOLATION RESISTORS.

FIGURE 26-4 TWO-WAY SPLITTER. WIDTH IS 3 IN. *(JERROLD ELECTRONICS CORPORATION)*

Tapoffs. A four-way tapoff with isolating resistance in each branch is shown in Fig. 26-5. Each tap can be considered as a series voltage divider. As an example, when 1,125 Ω is in series with 75 Ω, the total equals 1,200 Ω. The 75 Ω of the cable is then $1/16$ of 1,200 Ω. Therefore, the tapoff provides $75/1{,}200 = 1/16$ of the trunk signal to a 75-Ω cable. Typically, the fraction of signal tapped off is $1/4$ to $1/16$. The corresponding losses are −12 to −24 dB.

Less signal is usually tapped off at the start of the trunk to allow more signal for receivers at the end of the line. It should be noted that the signal voltage on the trunk is not reduced by the tapoffs because they are parallel paths. Insertion loss reduces the trunk signal.

Wall outlet. This fits into the standard box for an electrical outlet, just to provide a convenient fitting for connecting the distribution line to the television receiver. The wall outlet can be simply a feedthrough connection, without any isolation losses. However, in a layout for single taps the wall outlet itself serves a single tapoff on the trunk, in addition to a feedthrough connection to continue the line to the next tapoff.

Cable termination for the last tap. In general, a tap has connections for the tapoff, plus the input and output lines. The in and out connections are only a continuation of the trunk line. However, the last tap has no place to go on the output side. This end must be terminated with 75 Ω to match the impedance of the line. In Fig. 26-1, each of the trunk lines is terminated with a plug that has the required 75-Ω resistance.

In any distribution system, all the lines must be terminated. An open line is a big source of trouble because it has reflections. These can be coupled into the entire distribution system, causing interference, overload, and ghost problems.

Balun units. A balun matching transformer is generally used to connect the coaxial cable to the antenna. In addition, a balun may be needed at each receiver to connect the 75-Ω cable to the 300-Ω antenna input terminals. However, most sets now have input connections for either coaxial cable or twin lead.

26-3 Types of Distribution Losses

These include the isolation loss at each tapoff, insertion losses in the distribution components, and the attenuation of the cable.

Isolation loss. This is the attenuation caused by isolation resistance. Any tapoff has series resistance to separate each receiver from the trunk line. In the forward direction, from the line to the receiver, the isolation has two effects: (1) matching the impedance of the taps to the impedance of the transmission line, and (2) reducing the amount of signal coupled to the receiver. In the backward direction, from the receiver to the line, the isolation prevents each receiver from affecting the other receivers on the trunk line. Common problems are oscillator radiation from the receiver and different input impedances. At the receiver end, the impedance may depend on whether the set is on or off and what channel it is operating on.

The isolation loss is present at any tapoff because of the necessary series resistance. This loss applies to multitaps, single taps, and wall-outlet tapoffs. Typical losses are -12 to -24 dB, corresponding to voltage ratios of $1/4$ to $1/16$.

Insertion loss. This includes attenuation of the signal caused by either (1) splitting or (2) I^2R loss in the feedthrough connections. Typically, a two-way splitter has -3 dB loss for the split, plus a feedthrough loss of -0.5 dB for VHF and -1 dB for UHF. The total insertion loss of the splitter, then, is -3.5 dB for VHF or -4 dB for UHF. These insertion losses are twice as great for a four-way splitter.

Cable loss. This is the cumulative effect of I^2R losses in a long line. It can be considered as the insertion loss of the cable. Using low-loss, foam coaxial cable similar to RG-59U, the attenuation can be considered to be 3.2 dB per 100 ft for VHF channel 13 and 6.4 dB for UHF channel 70.

Summary of distribution losses. These are listed in Table 26-1. In general, the isolation loss refers to a tapoff. The insertion losses apply to all distribution components. It should be noted that the isolation loss for a tapoff would normally be counted only once. However, the insertion losses must be added because of their cumulative effect on signal distributed throughout the entire system.

26-4 Calculation of Total Losses

The distribution losses are added to find out how much amplifier gain is needed to make up for attenuation of the signal. Note the following points about the general procedure:

1. Take the worst possible case to be sure to have enough signal. Calculate losses for the highest channel to be received. Use the longest line run, if they are not all the same.
2. It is only necessary to calculate the losses

TABLE 26-1 TYPES OF DISTRIBUTION LOSSES

COMPONENT	INSERTION LOSS	ISOLATION LOSS	NOTES
Foam cable RG-59U	−3.2 dB/100 ft	None	At 216 MHz
	−6.4 dB/100 ft	None	At 812 MHz
RG-11U	−2.0 dB/100 ft	None	At 216 MHz
	−4.5 dB/100 ft	None	At 812 MHz
Splitter, two-way	−3.5 dB for VHF	None	Insertion loss includes −3 dB for split
	−4.0 dB for UHF	None	
Four-way	−7.0 dB for VHF	None	Insertion loss includes −6 dB for split
	−7.7 dB for UHF	None	
Tapoff, single	−0.75 dB for VHF	−12 to −24 dB	Wall outlet or pressure tap on cable
	−1.5 dB for UHF		
Four-way	−0.75 dB for VHF	−12 to 24 dB	Insertion loss for each tap
	−1.5 dB for UHF		
Wall outlet, tapoff	0 dB	0 dB	Feed through to receiver
	0.6 dB	−18 to −30 dB	Feed through to trunk
Balun unit	−0.8 dB for VHF	0 dB	Match between 75 Ω and 300 Ω
	−1.3 dB for UHF		
UHF/VHF splitter	−3.5 dB for VHF	None	Includes −3 dB for split
	−4.0 dB for UHF		

for one trunk line. If there is signal for the longest trunk, all the lines will have enough signal.

The procedure involves supplying at least 1-mV signal for the last receiver on the trunk line.

The distribution losses can be grouped in three categories: cable loss, insertion loss, and isolation loss. Insertion losses are added throughout the trunk-line distribution. The isolation loss is just at the tapoff for each receiver. Specific examples will be calculated now for the distribution with four-way tapoffs in Fig. 26-1 and the commonly used method with single tapoffs, which is illustrated in Fig. 26-6.

26-5 System with Multitaps

Refer to the distribution in Fig. 26-1. The cable length for each trunk is 30 ft + 30 ft + 40 ft from the two-way splitter. No length is added for the wall outlets because we assume they are close to the multitaps. The total cable length, then, is 100 ft. We can use foam RG-59U, which has a loss of −6.4 dB for the UHF channels. All the losses here are for a VHF-UHF system.

The insertion loss for the splitter can be taken as −4 dB for UHF. It includes −3 dB for the two-way split and −1 dB for the feedthrough loss. The splitter may have isolation between the outputs, but this is not considered as a series tapoff loss.

The four-way multitap has an insertion loss of −1.5 dB for UHF. This loss is for the whole unit, from the input to output connections. However, the loss of the last unit on the trunk is not added because this signal does not feed through. In Fig. 26-1, therefore, the total insertion loss of the multitaps is 2 × −1.5 = −3 dB for each trunk line.

Each tap can be assumed to have an average isolation loss of −18 dB. This value is counted only once because the loss is in the line to each receiver. Tabulating the losses:

Attenuation of 100 ft of cable	= − 6.4 dB
Insertion loss for two-way splitter	= − 4.0 dB
Insertion loss for two multitaps	= − 3.0 dB
Isolation loss of each tapoff	= − 18.0 dB
Total loss of trunk line	= − 31.4 dB

This distribution loss can be made up by a VHF-UHF amplifier with a gain of 30 to 40 dB. The 40 dB corresponds to a voltage gain of 100. Typically, the amplifier can take antenna signal of 2 mV and supply output of 200 mV for the splitter. This output signal is divided by one-half for approximately 100 mV to the two trunk lines. Then there is enough signal on both trunks for all the tapoffs on the line.

26-6 System with Single Taps at Wall Outlets

Refer to the distribution in Fig. 26-6. The losses in this layout are calculated for each of the four trunks. Each line has a length of 70 + 30 = 100 ft. The calculations are for VHF only, assuming that any UHF channels will be converted to VHF. There are seven tapoffs, plus one termination, for each trunk. The tabulation is:

Attenuation of 100 ft of foam cable, RG-59U, at VHF	= − 3.2 dB
Insertion loss for four-way splitter	= − 7.0 dB
Insertion loss for seven tapoffs (7 × 0.75)	= − 5.25 dB
Isolation loss of each tapoff	= − 18.0 dB
Total loss of trunk line	= − 33.45 dB

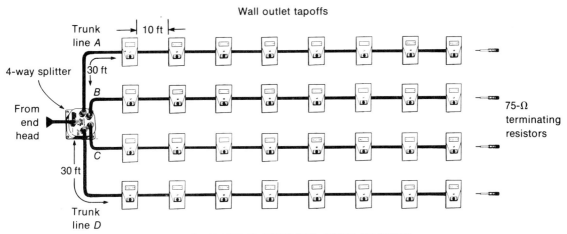

FIGURE 26-6 CABLE DISTRIBUTION SYSTEM FOR 32 RECEIVERS, USING FOUR-WAY SPLITTER AND SINGLE TAPOFFS.

This distribution loss of approximately −34 dB can be made up by an amplifier that has a voltage gain of 50 or 34 dB. With 2 mV into the amplifier, the output signal is 100 mV. This is divided into a little less than 25 mV for each of the four trunks. Then each trunk has enough signal for the tapoffs (Fig. 26-7).

The tapoffs near the splitter should have minimum insertion loss to feed through enough signal for the end of the line. Also, these tapoffs at the start of the cable could have a little more isolation loss than the average, taken as −18 dB, while those near the end of the cable would have less isolation loss. Remember that the signal level becomes smaller at the end of the trunk because of cable attenuation and insertion losses.

The differences in isolation losses could progress in steps of −2 dB, which is the approximate attenuation between two tapoffs. This is done to equalize the signal strength for all receivers on the trunk line. Adjustable tapoffs are available to vary the isolation loss from −10 to −20 dB, corresponding to a tapoff of one-third to one tenth of the signal on the cable.

26-7 Decibel (dB) Units

The decibel (dB) unit is a logarithmic ratio that is convenient in calculating gains and losses for power, voltage, or current. The reason is that dB values are added or subtracted, not multiplied or divided. The logarithmic factor also compresses a wide range of values into smaller numbers. Originally, the bel unit was defined for audio work as $\log(P_2/P_1)$. The logarithm is to base 10. P_2/P_1 is the ratio of two values of power. Either IF or rf signals can be compared.

The logarithmic unit generally used is the decibel, equal to one-tenth of a bel. The value of 1 dB indicates a gain or loss of approximately 26 percent in power or 12 percent in voltage. This much change in audio signal is just perceptible to the ear.

Feed through

FIGURE 26-7 COMBINATION WALL OUTLET AND SINGLE TAPOFF. (JERROLD ELECTRONICS CORPORATION)

Power ratios. The formula for calculating the dB comparison between two values of power is

$$dB = 10 \times \log\left(\frac{P_2}{P_1}\right) \qquad (26\text{-}1)$$

To calculate dB values with this formula, the following method can be used:

1. Find the numerical ratio of the two powers. Be sure to use the same units for both. For instance, if P_2 is 1 W and P_1 equals 1 mW, this ratio is 1,000 mW/1 mW = 1,000.
2. Always let P_2 in the numerator be the larger number. Then the ratio must be 1 or more. This eliminates the problem of working with fractions less than 1, which have a negative logarithm.
3. Find the logarithm of the power ratio. Here, 1,000, or 10^3, has the logarithm 3.
4. Multiply the logarithm by 10 to calculate the answer in decibels. Here we have $10 \times 3 = 30$ dB.

This answer is +30 dB for a gain from 1 mW to 1,000 mW. For a decrease from 1,000 mW to 1 mW the loss would be −30 dB. The dB value is the same for an equal gain or loss. Also, the power ratio is independent of the impedances that develop the powers.

Voltage ratios. Since power is E^2/R or I^2R, the decibel formula can also be expressed in terms of voltage and current. The dB formula is often used for voltages as:

$$dB = 20 \times \log\left(\frac{E_2}{E_1}\right) \qquad (26\text{-}2)$$

The two voltages E_2 and E_1 must be across the same value of impedance. If they are not, multiply Eq. (26-2) by the correction factor $\sqrt{Z_1/Z_2}$.

For example, for a gain from 1 mV to 1,000 mV across the same value of Z, the dB voltage gain is

$$dB = 20 \times \log\left(\frac{E_2}{E_1}\right)$$
$$dB = 20 \times \log(1{,}000) = 20 \times 3 = 60$$

The 60-dB gain for a voltage ratio of 1,000 is double the dB gain for a power ratio of 1,000 because the voltage formula has the multiplying factor of 20 instead of 10. The reason is that power is proportional to the square of voltage. In logarithms, squaring a number corresponds to doubling its logarithm. Therefore, a voltage ratio always equals twice the decibels of the same numerical power ratio.

Reference levels. In order to make the dB units more useful, standard values are used for comparison. Then only one value of P or E compared with the reference can be specified by a decibel value. There are several references in common use, but you can tell which is being used by the abbreviation:

dB = 6 mW across 500 Ω as the reference.
dBm = 1 mW across 600 Ω as the reference.
dBmV = 1 mV across 75 Ω. This reference is used for cable television because we are interested in the amount of signal voltage for coaxial cable with a nominal impedance of 75 Ω. As a formula, the dBmV reference can be stated:

$$\text{dBmV} = 20 \times \log\left(\frac{E_2}{1\text{ mV}}\right) \quad (26\text{-}3)$$

For an rf signal voltage of 100 mV, as an example, this voltage ratio is 100 or 10^2 compared with the reference of 1 mV. The log of 100 is 2. Then

$$\text{dBmV} = 20 \times \log\left(\frac{100\text{ mV}}{1\text{ mV}}\right)$$
$$\text{dBmV} = 20 \times 2 = 40$$

When using reference levels, it is important to note that 0 dB indicates the value equal to the reference. The reason is that the log of 1 happens to be zero. As an example, assume a voltage of 1 mV as the value compared with the dBmV reference:

$$\text{dBmV} = 20 \log \frac{1\text{ mV}}{1\text{ mV}}$$
$$= 20 \times \log 1$$
$$= 20 \times 0 = 0$$

In all cases of decibel comparisons, the 0 dB means equal values are being compared, with a ratio of unity.

Adding and subtracting dB values. Adding the logarithms of numbers corresponds to multiplying the numbers. Also, subtracting logarithms corresponds to dividing numbers. Therefore, dB values in cascade are added or subtracted. For instance, two amplifiers in cascade, each with a dB gain of 13, provide a total gain of 26 dB. As another example, when signal voltage at a level of 20 dBmV is fed into a line with an attenuation of −9 dBmV, the signal level at the end of the line is 20 − 9 = 11 dBmV.

Common decibel values. The ratios listed in Table 26-2 are worth memorizing because they can be used as short-cuts in dB calculations. Although shown as +dB for gain, the same ratios inverted apply for −dB with losses.

Consider some voltage ratios for dBmV with the reference of 1 mV. Double this value is

TABLE 26-2 COMMON DECIBEL VALUES

dB	0	1	3	6	10	12	20	40
Voltage ratio	1	1.12	1.4	2	3.16	4	10	100
Power ratio	1	1.26	2	4	10	16	100	10,000

6 dBmV, or 2 mV. One-half the reference voltage is −6 dBmV, or 0.5 mV. Ten times the reference is 20 dBmV or 10 mV, because the log of 10 is 1. Then 20 × 1 = 20 dBmV. Also, 100 mV is 40 dBmV because the log of 100 is 2. Then 20 × 2 = 40 dBmV.

Doubling the voltage ratio always adds 6 dB. Not only is 6 dB two times the voltage, but 12 dB is four times, 18 dB is eight times, 24 dB is 16 times, and 48 dB is 256 times the voltage, as examples. Each addition of 6 dB means doubling of the voltage. It should also be noted that 10 dBmV is approximately three times the voltage. See Table 26-2.

A useful technique is dividing the ratio into factors that are easier to convert to decibels and then adding the dB values. For instance, a voltage ratio of 200 is the same as 100 × 2. The factor of 100 equals 40 dB, while double the voltage is 6 dB. Therefore, the factors of 100 × 2 yield the values of 40 dB + 6 dB = 46 dB for the voltage ratio of 200. With the 1 mV reference, this equals 200 mV for 46 dBmV.

26-8 Decibel Conversion Charts

Tables 26-3 and 26-4 list the dB values for different voltage ratios, to eliminate the need to use the decibel formula with logarithms. Table 26-3 is in a general form for comparison of two voltages without any reference. However, the values in Table 26-4 are specifically for the dBmV reference of 1,000 μV or 1 mV.

Note that starting with 6 dB for a voltage

TABLE 26-3 DECIBELS (dB) AND VOLTAGE RATIOS*

dB	VOLTAGE RATIO	dB	VOLTAGE RATIO	dB	VOLTAGE RATIO
1	1.12	13	4.48	25	18
2	1.26	14	5	26	20
3	1.4	15	5.6	27	22
4	1.6	16	6.32	28	25
5	1.78	17	7	29	28
6	2.0	18	8	30	32
7	2.24	19	9	31	36
8	2.52	20	10	32	40
9	2.8	21	11.2	34	50
10	3.16	22	12.6	36	64
11	3.56	23	14	38	80
12	4	24	16	40	100

*Calculated as dB = 20 log(V_2/V_1) with both voltages across the same impedance.

ratio of 2, each increase of 6 dB shows a doubled voltage ratio. These values are in boldface type. Any doubled voltage ratio is 6 dB higher.

The last four values in the table are in multiples of 2 dB in order to complete the list with 40 dB for a voltage ratio of 100. For a larger voltage ratio, divide it into factors and add the dB values for the factors. As an example, a voltage ratio of 10,000 is 100 × 100. This corresponds to 40 dB + 40 dB = 80 dB.

For the opposite conversion with more than 40 dB, separate it into terms that can be added to give the total dB value, and then multiply the corresponding voltage ratios. As an example, 46 dB = 40 dB + 6 dB. The voltage ratios are 100 for 40 dB and 2 for 6 dB. Then multiply 100 × 2 = 200 to get the voltage ratio for 46 dB.

In Table 26-4, note that 1 mV = 1,000 μV, which is the reference value for 0 dBmV. Values less than 1,000 μV correspond to −dBmV; voltages more than 1,000 μV are +dBmV. The voltage ratios are approximate values, having been rounded off to fit the pattern of 6 dB more for double voltage and 6 dB less for one-half voltage.

TABLE 26-4 MICROVOLTS AND dBmV*

1,000 μV OR LESS		1,000 μV OR MORE	
dBmV	E_2, μV	dBmV	E_2, μV
0	1,000	0	1,000
−1	890	1	1,120
−2	790	2	1,260
−3	700	3	1,410
−4	630	4	1,580
−5	560	5	1,780
−6	**500**	**6**	**2,000**
−7	445	7	2,240
−8	395	8	2,520
−9	350	9	2,820
−10	315	10	3,160
−11	280	11	3,560
−12	**250**	**12**	**4,000**

*Calculated as dBmV = 20 log(E_2/1 mV) across 75-Ω impedance.

SUMMARY

The following definitions, in alphabetical order, summarize the terminology of cable distribution systems.

Amplifier. A device which increases signal level by 20 to 55 dB. Types include VHF, UHF, combined UHF-VHF, and single-channel amplifiers.
Attenuator. A device which reduces signal level.
Balun. A device which matches 300-Ω balanced and 75-Ω unbalanced impedances. Also called an *elevator transformer.*
Booster. A small preamplifier for antenna signal.
Cable equalizer. A device which compensates for tilt in cable attenuation from channel 2 to channel 13, for very long lines.

Cable losses. Attenuation of the distribution line.

CATV. Cable television. This abbreviation is generally used for systems that serve many receivers in individual homes.

CCTV. Closed-circuit television. Distribution by cable to selected receivers, but usually for video signal rather than antenna rf signal.

Characteristic impedance (Z_0). A property of a cable independent of length. Z_0 is nominally 75 Ω for coaxial cable and 300 Ω for twin lead.

Converter. A device which converts a UHF channel to a VHF channel for lower losses in cable distribution.

dB. Abbreviation for decibel. The dBmV unit is compared with the reference of 1 mV.

Elevator transformer. Same as balun, to match between 75-Ω unbalanced and 300-Ω balanced impedances.

ETV. Educational television for schools, often using a cable distribution system.

Filter. A device which passes the desired band but attenuates other frequencies.

FSM. Abbreviation for field-strength meter, a device which measures signal at the antenna location.

Head end. A section which includes the antenna, amplifier, and any additional equipment needed to feed signal into the cable distribution system.

Impedance matching. Terminating a line in its characteristic impedance.

Insertion loss. Attenuation in the series path of the signal. This loss is cumulative in the trunk line.

Isolation. Separation between signal paths, usually by series resistance in each of the parallel branches. This loss is mainly in the tapoff for each receiver.

MATV. Master antenna television. Feeding multiple receivers from a master antenna, as in motels and apartment buildings.

Mixer. A device which combines signals from several antennas.

Pad. A fixed attenuation, usually with resistors.

Passive component. A device which has no amplification—no tubes or transistors.

Preamplifier. A low-noise amplifier for boosting the antenna signal before the main amplifier.

Splitter. A device which divides signal from one line into two or more lines.

Standard signal. For cable television, 1 mV across 75 Ω for each receiver. This level is 0 dBmV.

Tapoff. A voltage divider to take part of the cable signal to each receiver.

Terminating plug. A 75-Ω resistor to terminate the end of a trunk cable in its characteristic impedance.

Tilt. Unequal cable attenuation for different channels.

Trap. A device which attenuates a specific frequency.

Trunk. A line from the splitter to tapoffs for the receivers.

Wall outlet. A connection between the cable and the receiver without isolation resistance.

Wall tapoff. A connection between the cable and the receiver with isolation resistance. It also feeds signal through to the next tapoff.

Self-Examination   (Answers at back of book)

1. Typical gain of an amplifier for the VHF channels is: (a) 0 dB; (b) −5 dB; (c) 2 dB; (d) 40 dB.
2. Which of the following lines has the greatest losses: (a) twin lead; (b) open-wire; (c) RG-59U coaxial cable; (d) foam RG-11U coaxial cable.
3. The standard signal voltage used as the reference for the dBmV unit is: (a) 6 mV in 500 Ω; (b) 1 mV in 75 Ω; (c) 1 mW in 600 Ω; (d) 6 mW in 1 Ω.
4. The loss in a 2-for-1 split is approximately: (a) 3 dB; (b) 8 dB; (c) one-third voltage; (d) one-quarter power.
5. A tapoff with an isolation loss of −20 dB takes the following fraction of signal voltage: (a) one-half; (b) one-fourth; (c) one-tenth; (d) one-hundredth.
6. The number of trunk lines in Fig. 26-1 is: (a) 1; (b) 2; (c) 3; (d) 4.
7. The number of wall-outlet tapoffs on each trunk line in Fig. 26-6 is: (a) 4; (b) 6; (c) 7; (d) 8.
8. The terminating resistance at the end of a trunk line is: (a) 75 Ω; (b) 100 Ω; (c) 150 Ω; (d) 300 Ω.
9. The number of insertion losses added for the wall-outlet tapoffs in Fig. 26-6 equals: (a) 7; (b) 8; (c) 16; (d) 31.
10. The minimum signal required for each receiver in the distribution system is approximately: (a) 0 mV; (b) 1 mV; (c) 6 mV; (d) 10,000 μV.

Essay Questions

1. Give three types of amplifiers for the antenna signal.
2. What is the difference between a split and a tapoff?
3. Why is coaxial cable used for distribution systems?
4. What is the advantage of converting a UHF channel to a VHF channel for the distribution system?

5. What is the function of a mast-mounted preamplifier?
6. Give the function of the following: balun, U/V splitter, FM trap, and terminating plug.
7. What is the difference between a four-way splitter and a four-way multitap?
8. For an average isolation loss for a tapoff on the trunk line, give the dB value and the corresponding fraction of signal voltage tapped off the line.
9. List the three main types of losses in a distribution system.
10. Show the layout for 40 receivers and calculate typical losses for the VHF television channels. Use a four-way splitter with four trunk lines and single wall-outlet tapoffs, as in Fig. 26-6.

Problems (Answers to selected problems at back of book)

1. Give the dB values for the following voltage ratios: (a) 2:1; (b) 3:1; (c) 10:1; (d) 20:1; (e) 100:1.
2. Give the voltage ratios for the following dB values: (a) 6 dB; (b) 10 dB; (c) 20 dB; (d) 26 dB; (e) 40 dB.
3. Give the mV values for the following dBmV levels: (a) 0 dBmV; (b) 6 dBmV; (c) 10 dBmV; (d) 20 dBmV; (e) 40 dBmV; (f) 46 dBmV.
4. Give the dBmV levels for the following signal voltages: (a) 1 mV; (b) 2 mV; (c) 3 mV; (d) 10 mV; (e) 100 mV; (f) 200 mV.
5. Calculate the net dB gain for -4 dB loss, 50 dB gain, -7 dB loss, and -24 dB loss.
6. Calculate the signal level for 80 dBmV.
7. Calculate the dBmV level for a signal of 2 V.
8. An antenna signal of 100 μV has a gain of 20 dB, loss of -6 dB, gain of 40 dB, and loss of -34 dB. Calculate the output signal level in millivolts.

27 The FM Sound Signal

The sound associated with the picture is transmitted in the same 6-MHz channel but on a separate rf carrier as a frequency-modulated signal. In frequency modulation (FM), the instantaneous frequency of the modulated rf carrier is made to vary with the amount of audio modulating *voltage*. In amplitude modulation (AM), the audio modulating voltage varies the carrier amplitude. FM has several advantages because the amplitude of the modulated rf carrier can be kept constant, while the audio information is in the frequency variations. The main benefit of FM is its freedom from noise and interference. However, AM is used for broadcasting the picture signal because multipath reception of FM picture signals would produce severe distortion, instead of ghosts.

It should be noted that the principles of FM for the television sound signal apply in the same way for the FM radio broadcast band of 88 to 108 MHz. In television, the carrier frequency is shifted ±25 kHz for the maximum amount of audio voltage. In the FM radio broadcast band, the maximum frequency swing is ±75 kHz. The range of audio modulation frequencies is 50 to 15,000 Hz for both services. Furthermore, the principles of FM apply to video tape recording, repeater stations, and other uses where the benefits of an FM signal are desired. The details of frequency modulation are explained in the following topics:

27-1 Frequency changes in an FM signal
27-2 Audio modulation in an FM signal
27-3 Definitions of FM terms
27-4 Preemphasis and deemphasis
27-5 Advantages and disadvantages of FM
27-6 Reactance tube
27-7 Receiver requirements for the FM sound signal
27-8 Slope detection of an FM signal

27-9 Discriminator operation
27-10 Center-tuned discriminator
27-11 The limiter
27-12 Ratio detector
27-13 Quadrature-grid FM detector
27-14 Sound IF alignment
27-15 The audio amplifier section
27-16 Complete circuits for the associated sound signal
27-17 Intercarrier buzz
27-18 Multiplexed stereo sound

27-1 Frequency Changes in an FM Signal

The idea of frequency modulation is illustrated in Fig. 27-1. This circuit is called a *wobbulator*, as the frequency is wobbled around the center value of 100 kHz by the motor-driven capacitor C_V. The inductance L_1 and fixed capacitance C_1 form the tuned circuit for the rf oscillator, resonant at 100 kHz. Also in the resonant circuit is the variable C_V. Its shaft is driven by a motor which rotates the plates in and out of mesh to vary the oscillator frequency. The result is frequency modulation of the rf output from the oscillator.

Generally, C_V is made to rotate at the power-line frequency of 60 Hz. A capacitive semiconductor diode can be used instead of the motor-driven capacitor, however, to modulate the oscillator with the 60-Hz power-line voltage. In either case, the rf output frequency is shifted periodically, at the rate of 60 Hz, to produce an FM signal. This method is used in sweep generators, where the rf signal sweeps through a range of frequencies at the repetition rate of 60 Hz.

Referring to Fig. 27-1, assume that the oscillator is at 100 kHz with C_V halfway in mesh. Then the frequency varies above and below this center value of 100 kHz as the capacitor is driven in and out of mesh. With C_V completely in mesh, the added capacitance in the tuned circuit is maximum and the output from the oscillator is at its lowest frequency. When C_V is all the way out of mesh, the oscillator is at its highest frequency. For values between the two extremes, the oscillator frequency varies continuously between its highest and lowest values around the center frequency of 100 kHz. If the time for one complete revolution of C_V is taken as $1/60$ s, the rf output of the oscillator will appear as in Fig. 27-2. The amplitude remains the same at all times but the frequency is changing continuously, with a swing of ± 20 KHz.

The frequency swing can be made almost any amount and has no relation to the repetition rate. For the same 60-Hz repetition rate, the carrier frequency could be varied by ± 30 kHz or

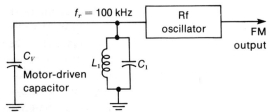

FIGURE 27-1 AN EXAMPLE OF PRODUCING FREQUENCY MODULATION. THE SHAFT OF THE VARIABLE AIR CAPACITOR C_V ROTATES IN AND OUT OF MESH TO CHANGE THE OSCILLATOR FREQUENCY.

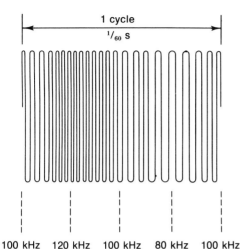

FIGURE 27-2 THE FM SIGNAL PRODUCED BY THE FREQUENCY MODULATION IN FIG. 27-1. THE AMPLITUDE IS CONSTANT BUT THE INSTANTANEOUS FREQUENCY IS CONTINUOUSLY CHANGING.

±50 kHz, depending on the amount of capacitance variation. Or, the same frequency swing of ±20 kHz could be obtained at a rate slower or faster than 60 Hz by changing the speed of the shaft rotation.

27-2 Audio Modulation in an FM Signal

Figure 27-3 illustrates four examples of audio voltage producing frequency modulation of an rf carrier. An audio modulating amplitude of 10 V is assumed to produce a frequency change of 20 kHz with a 100-kHz carrier. Also, we assume linear modulation, so that one-half the audio voltage produces one-half this frequency swing, or twice the audio voltage would double the frequency swing. Notice that in all cases the amount of audio voltage decides the amount of frequency swing. The audio frequency is in the repetition rate of the frequency swings.

In Fig. 27-3a, the output of the transmitted signal is 100 kHz when the audio modulating voltage is at its zero value because this is the carrier frequency for no modulation. With modulation, the transmitted frequency continuously varies between the values of 100 ± 20 kHz. If the frequency increases for positive values of modulating voltage, it will decrease for negative values. For the positive half-cycles of audio, the instantaneous frequency increases from 100 kHz to the maximum value of 120 kHz. The frequencies are between 100 and 120 kHz for values of audio voltage between 0 and +10 V. During the negative half-cycles the output frequency varies between 100 and 80 kHz as the audio voltage varies between 0 and −10 V.

For the voltage shown in Fig. 27-3b, the amount of frequency change is the same 20 kHz because the audio amplitude is again 10 V. However, the rate at which the transmitted signal goes through its complete frequency swings is now 2,000 Hz because of the 2,000-Hz audio modulating frequency. Note that the carrier modulated by 2,000-Hz audio voltage goes through two complete cycles of frequency swing while the carrier modulated by 1,000 Hz goes through one complete cycle. A complete cycle of frequency swing is from the center frequency up to maximum, down to the middle frequency, decreasing to the lowest frequency value, and returning to center frequency.

For the audio modulating voltage shown in Fig. 27-3c, the maximum frequency change is only 10 kHz instead of 20 kHz, because the peak value of the audio is 5 V instead of 10 V. The rate at which the output signal swings about its center frequency is 1,000 Hz. For the modulating voltage of Fig. 27-3d, the maximum frequency change is still 10 kHz for 5 V audio. However, the repetition rate is 2,000 complete swings per second, which is the audio frequency.

The amount of frequency change in the

634 BASIC TELEVISION

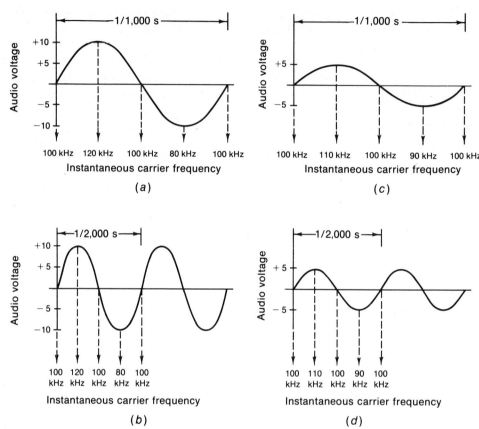

FIGURE 27-3 HOW THE AMOUNT OF FREQUENCY CHANGE IN THE RF CARRIER DEPENDS ON THE AMOUNT OF VOLTAGE IN THE AUDIO MODULATION. (a) AUDIO OF 10 V AND 1,000 Hz; (b) AUDIO OF THE SAME 10 V BUT 2,000 Hz; (c) AUDIO REDUCED ONE-HALF TO 5 V, WITH FREQUENCY OF 1,000 Hz; (d) AUDIO OF 5 V AND FREQUENCY OF 2,000 Hz.

TABLE 27-1 COMPARISON OF FM AND AM SIGNALS

FM	AM
Carrier amplitude is constant	Carrier amplitude varies with modulation
Carrier frequency varies with modulation	Carrier frequency is constant
Modulating-voltage *amplitude* determines rf carrier frequency	Modulating-voltage *amplitude* determines rf carrier amplitude
Modulating frequency is rate of frequency changes in the rf carrier wave	Modulating frequency is rate of amplitude changes in the rf carrier wave

transmitted carrier varies with the amplitude of the audio and should not be confused with the audio frequency. The frequency of the audio modulating voltage is the rate at which the carrier goes through its frequency swings. This determines the pitch of the sound as it is reproduced at the receiver. The amount of audio voltage determines the amount of frequency swing. This determines the intensity or loudness of the sound reproduced at the receiver. These characteristics of FM are summarized in Table 27-1.

27-3 Definitions of FM Terms

The following characteristics can help in explaining how an FM system operates.

Center frequency. This is the frequency of the transmitted rf carrier without modulation, or the output frequency at the time when the modulating signal voltage is at its zero value. The center frequency is also called *rest frequency.*

Frequency deviation. The frequency deviation is the change from the center frequency. As an example, an audio modulating voltage having a peak value of 5 V might produce the frequency change of 10 kHz at 2 V, and 50 kHz for its peak value of 5 V. The peak deviation in this case is 50 kHz. The amount of frequency deviation depends on the amplitude of the audio modulating voltage.

Frequency swing. With equal amounts of change above and below center, the frequency swing is twice the deviation. As an example, when the audio modulating voltage has a peak amplitude on either its positive or negative half-cycle of 5 V to produce a frequency deviation of 50 kHz, the frequency swing is ±50 kHz, or a total of 100 kHz.

Percent modulation. This is the ratio of actual frequency swing to the amount defined as 100 percent modulation, expressed as a percentage. For commercial FM broadcast stations, ±75 kHz is defined by the FCC as 100 percent modulation. For the aural or sound transmitter of commercial television broadcast stations, 100 percent modulation is defined as ±25 kHz. If, for example, the audio modulating voltage for the associated sound signal in television produces a frequency swing of ±15 kHz, the percent modulation is $^{15}/_{25}$, or 60 percent. Less swing is used for the FM sound in television, compared with FM broadcasting, in order to conserve space in the television channel.

The percentage of modulation varies with the intensity of the audio voltage. For weak audio signals, the audio voltage is small, and there is little frequency swing, with a small percent modulation. The audio voltage is greater for the louder signals, and there is more frequency swing, producing a higher percent modulation. The frequency swing produced for the loudest audio signal should be that amount defined as 100 percent modulation.

Phase modulation (PM). In a phase modulator, the phase angle of the rf carrier is shifted in proportion to the amplitude of the audio modulating voltage. The varying phase causes changes in carrier frequency. Therefore, phase modulation results in an equivalent FM signal or *indirect FM.* In many cases, the FM transmitter actually uses a phase modulator in a crystal-controlled oscillator, for excellent stability of the center frequency.

An important characteristic of PM is the fact that the amount of equivalent indirect FM increases with higher audio frequencies. An audio correction filter is used in the modulation, though, to provide the same frequency swings as in direct FM.

27-4 Preemphasis and Deemphasis

The preemphasis refers to boosting the relative amplitudes of the modulating voltage for higher audio frequencies, from 2,000 Hz to approximately 15 kHz. Deemphasis means attenuating these frequencies by the same amount they were boosted. However, the preemphasis is done at the transmitter, while the deemphasis is in the receiver. The purpose is to improve the signal-to-noise ratio for FM reception.

For the transmitter, FCC standards specify that preemphasis shall be employed in accordance with the impedance-frequency characteristic of an *LR* network having a time constant of 75 μs. This characteristic increases the audio modulating voltage for higher frequencies. At the receiver the deemphasis should have the same time constant of 75 μs, but with an opposite characteristic that reduces the amplitude of higher audio frequencies.

A low-pass *RC* filter, as in Fig. 27-4, is generally used for deemphasis in FM receivers, usually in the detector output circuit. The deemphasis network attenuates high audio frequencies fed to the audio amplifier. Since the shunt capacitance *C* has less reactance as the frequency increases, the deemphasized audio voltage has less amplitude for the higher audio frequencies. With a 75-μs time constant, the filter attenuates frequencies at about 1,000 Hz and above. The response is down 3 dB at 2 kHz.

While it would seem that no progress is made if the audio voltage is deemphasized to the same extent that it is preemphasized, a great improvement in signal-to-noise ratio is actually the result. The reason is that the interference is indirect FM produced as phase modulation by noise. When the signal and noise are both reduced by deemphasis, the signal returns to normal while the noise is reduced below normal. This is more effective in FM than in AM because the noise level in FM increases for higher audio frequencies. Finally, the deemphasis has the greatest attenuation for the highest audio frequencies, doing the most good where it is most needed.

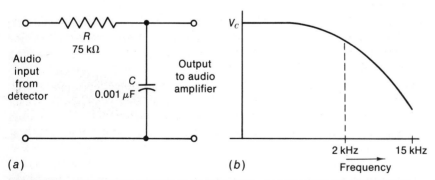

FIGURE 27-4 DEEMPHASIS CIRCUIT IN FM RECEIVER. (a) LOW-PASS *RC* FILTER WITH TIME CONSTANT OF 75 μs; (b) FREQUENCY RESPONSE.

27-5 Advantages and Disadvantages of FM

The greatest advantage of FM is its ability to eliminate from the desired signal the effects of interference. The interference can be a modulated carrier from another FM or AM station, atmospheric or man-made static, or receiver noise. In any case, the effects of the interference on the desired signal can be made negligible in an FM system. This important advantage is an inherent part of the FM system. The reason is that the instantaneous frequency variations of the modulated carrier correspond to the desired signal, allowing interfering amplitude variations to be eliminated.

Transmitter efficiency. Because the carrier amplitude is constant in FM transmission, low-level modulation can be used, with class C operation permissible for maximum efficiency in amplifying the FM signal. These factors allow a smaller, more economical, and efficient transmitter for FM. Furthermore, additions can be made to an FM transmitter just by adding rf power stages to increase the output.

Broadcast channels. Any FM service with wide frequency swings must be in the VHF or UHF band to allow for the greater bandwidth of the modulated signal. The commercial FM broadcast band is 88 to 108 MHz. The assigned channel is 200 kHz wide for each station. For the FM sound in television broadcasting, 50 kHz of the 6-MHz channel is used. The higher broadcast frequencies in the VHF and UHF bands have the disadvantage of reduced transmission distance, compared with the standard AM radio broadcast band of 535 to 1,605 kHz.

Audio frequency range. In general, the use of the VHF band allows a wider audio modulating frequency range. At the higher frequencies, a wider channel to accommodate the resultant side-band frequencies is feasible. This is true of either FM or AM. An AM radio station is limited to approximately 5 to 10 kHz as the highest audio modulating frequency because of the close spacing of assigned carrier frequencies in the standard broadcast band. Given a wide enough channel, though, an AM system can use as wide a modulating frequency range as FM. The picture carrier in television, for example, is amplitude-modulated with video frequencies as high as 4.2 MHz.

The audio modulating frequency range is 50 to 15,000 Hz for commercial FM broadcast stations and the FM sound in television. However, the extent to which this greater audio frequency range is made useful depends on the quality of the audio system in the receiver.

Multipath reception. Since the instantaneous frequency of an FM signal varies with time, the multipath signals at the receiver generally will have different frequencies at any instant. As a result, heterodyning action between the FM multipath signals at the receiver produces interfering beats that continuously change in frequency. The changing beat frequency can produce garbled sound, similar to the effect produced by nonlinear amplitude distortion in an audio amplifier. In picture reproduction, the interfering FM beat produces an interference pattern of bars, with a shimmering effect as the bars continuously change with the beat frequency. This is why AM is preferable to FM for broadcasting the picture signal, since multipath AM signals simply produce multiple ghost images.

FM is generally used for transmitting the picture and sound signals between microwave radio-relay stations. In this service, however, multipath reception is not a problem because directive antennas beam the signal directly from transmitter to receiver.

27-6 Reactance Tube

The circuit in Fig. 27-5 illustrates how a reactance tube can be used to produce FM directly by varying the frequency of the oscillator stage. The oscillator has a tuned *LC* circuit without crystal control. Then the plate-cathode circuit of the reactance tube can be in parallel with the resonant *LC* circuit. As the reactance of the plate-cathode circuit is varied, the oscillator frequency changes.

Definition of reactance. The fundamental characteristic is that the voltage across a reactance be 90° out of phase, or in quadrature, with the current through it. Lagging voltage is capacitive; leading voltage is inductive. It is not necessary to have an actual capacitor or coil. The plate-cathode circuit of the reactance tube has quadrature feedback that makes the plate current i_p 90° out of phase with the plate voltage e_p. This feature makes the plate-cathode circuit reactive. Since the reactance tube is across the *LC* circuit, its resonant frequency depends on the reactance of the plate-cathode circuit. Furthermore, this added reactance depends on the transconductance g_m of the reactance tube. Varying the grid voltage changes the g_m to vary the oscillator frequency.

Circuit arrangement. The tube shown in Fig. 27-5 is convenient because it has two control grids. Grid 3 is used for audio modulating voltage, while grid 1 has the required quadrature feedback voltage. In the plate circuit, L_2 is an rf choke to prevent the rf signal from shorting to ground, while C_2 is a dc blocking capacitor for the B+ voltage. Cathode bias is provided by R_k and C_k. The $R_g C_c$ coupling circuit feeds in the audio modulating voltage.

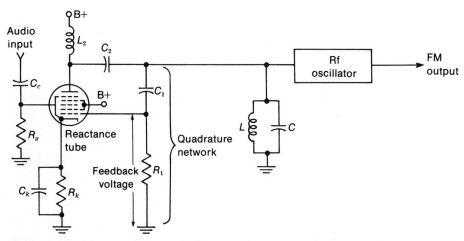

FIGURE 27-5 REACTANCE TUBE FOR PRODUCING FM. ITS PLATE-CATHODE CIRCUIT IS ACROSS THE *LC* TUNED CIRCUIT OF THE OSCILLATOR.

The quadrature network. Note that the R_1C_1 branch is connected from plate to cathode, across the LC tuned circuit. All the plate signal voltage is across the R_1C_1 branch, but only the voltage across R_1 is fed back to grid 1. Most important, this feedback voltage is made to be in quadrature with e_p.

The phase shift of 90° is accomplished by making the reactance of C_1 at least 10 times greater than the resistance of R_1. Then the capacitive current in this branch leads the plate voltage by practically 90°. The voltage across any resistance has the same phase as the current through it. Therefore, the feedback voltage across R_1 is also in quadrature with the plate voltage, 90° leading e_p.

The feedback voltage to grid 1 controls the plate current of the reactance tube. As a result, i_p is 90° leading e_p to make the plate-cathode circuit a capacitive reactance. The tube can also be made an inductive reactance by feeding back quadrature voltage that is 90° lagging e_p.

Functions. A reactance tube can be used two ways. When audio signal voltage varies the grid bias, this is an FM modulator. Then the rf oscillator frequency changes in step with the modulating voltage amplitude.

However, if a dc voltage is applied to shift the grid bias, then the stage is a *reactance control tube* to determine the oscillator frequency. The dc bias can be a control voltage for automatic frequency control (AFC) to hold the oscillator at its correct frequency. In the television receiver, this AFC function can be applied to:

1. The rf local oscillator in the tuner to provide the correct oscillator frequency for each channel. This function is also called automatic fine tuning (AFT).
2. The 3.58-MHz color subcarrier oscillator to maintain the correct phase with respect to the color sync burst. This application is also called automatic phase control (APC).
3. A sine-wave horizontal deflection oscillator to keep the horizontal scanning synchronized. This application is one type of horizontal AFC.

A reactance tube used as a control tube, however, can easily be replaced by a varactor semiconductor diode. Varying the dc voltage on a varactor connected across an LC tuned circuit accomplishes the same result of shifting the resonant frequency.

27-7 Receiver Requirements for the FM Sound Signal

An FM receiver is a superheterodyne like a typical AM receiver. The fact that the signal is frequency-modulated does not affect the heterodyning, as the original frequency swing is maintained around a lower center frequency corresponding to the IF sound carrier. Figure 27-6 illustrates the requirements for the sound section in television receivers. After the desired station has been selected by the tuner, the main functions are IF amplification of the FM signal, AM limiting to remove noise or interference, and detection to recover the audio signal from the frequency swings. The audio amplifier drives the loudspeaker.

In Fig. 27-6, the circuit starts at the left with the intercarrier sound signal of 4.5 MHz. This is the second sound IF carrier frequency in all television receivers, for VHF and UHF channels and for color and monochrome. In color receivers, the 4.5-MHz sound converter is a separate diode detector that beats the 41.25-MHz

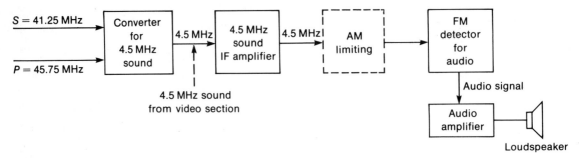

FIGURE 27-6 THE SOUND SECTION OF TELEVISION RECEIVERS. THE AM LIMITING CAN BE IN A SEPARATE LIMITER STAGE OR AN ADDED FUNCTION OF THE FM DETECTOR.

sound IF signal S with the 45.75-MHz picture IF signal P to produce the difference frequency of 4.5 MHz for the intercarrier sound signal.

Both the S and P signals are usually coupled into the 4.5-MHz converter from the last picture IF amplifier stage. In monochrome receivers, the video detector is generally used as the 4.5-MHz sound converter. Then the 4.5-MHz intercarrier sound signal is coupled from either the video detector or the video amplifier to the 4.5-MHz sound IF amplifier, without the need for a separate 4.5-MHz sound converter. This circuit arrangement is indicated by the dotted arrow into the sound IF amplifier. All the stages in Fig. 27-6 are for the sound alone, separate from the picture.

The 4.5-MHz sound IF signal is amplified to a level of several volts for the FM detector. This stage recovers the audio modulation from the frequency swings in the FM signal. Then the detected audio signal is amplified in the audio section to have enough output to drive the loudspeaker.

The FM detector input is tuned to 4.5 MHz, the same as the output circuit of the sound IF amplifier. It is the tuned input circuit of the detector that can recognize the frequency variations in the FM signal.

Furthermore, it is important to realize the separate functions of the sound converter and FM detector stages. The sound converter produces the 4.5-MHz sound IF signal, but this is not the audio signal of 50 to 15,000 Hz needed by the audio amplifier. The 4.5-MHz signal has rf variations corresponding to the audio modulation. The purpose of having the 4.5-MHz signal is to obtain the advantages of intercarrier sound. However, the FM detector is needed to rectify and filter the 4.5-MHz intercarrier sound signal in order to recover the audio modulation.

IF bandwidth. It may be of interest to note that the standard value is 10.7 MHz for the IF center frequency for the FM commercial broadcast band (88 to 108 MHz) in radio receivers. The IF bandwidth required is at least 150 kHz for the maximum frequency swing of ± 75 kHz around 10.7 MHz. For the associated sound signal in television receivers, the required IF bandwidth is 50 kHz, or more, for a maximum frequency swing of ± 25 kHz around 4.5 MHz. In both cases, though, the bandwidth needed for the FM signal is only a small percentage of the center frequency. This factor means that the IF tuned circuits can be aligned for maximum output at the intermediate frequency.

We can compare three examples to see the ratio of IF bandwidth to the center frequency. For ±75 kHz, or 150 kHz, around 10.7 MHz, the ratio is 0.15 MHz/10.7 MHz, which equals 0.014 or 1.4 percent. For ±25 kHz, or 50 kHz, around 4.5 MHz, the ratio is 0.05/4.5, which equals 0.011 or 1.1 percent. For the last example of an AM radio, which is considered a narrowband receiver, an IF bandwidth of ±5 kHz, or 10 kHz, is 2.2 percent of the 455-kHz intermediate frequency.

FM detector. An FM detector rectifies and filters the IF signal to recover the audio modulation, as in an AM detector. In addition, however, the FM detector requires a tuned circuit that has different output voltages for frequency swings above and below center. In effect, the response of the tuned circuit converts the FM signal to amplitude variations, which are then rectified and filtered. The main types of FM detector circuits are the center-tuned discriminator in Fig. 27-9, the ratio detector in Fig. 27-12, and the quadrature-grid detector in Fig. 27-16. They are all tuned to the center frequency.

AM limiting. Probably the main feature of FM is that amplitude limiting can be used to eliminate AM interference. This is possible because limiting the amplitude variations does not distort the frequency variations of the FM signal.

The function of AM rejection in the FM receiver is accomplished either by a separate limiter stage or by using an FM detector that does not respond to amplitude modulation. A limiter is an IF amplifier tuned to center frequency, but with the dc operating voltages reduced so that the stage operates as a saturated amplifier. For specific applications, the discriminator as an FM detector requires a limiter to remove AM from the IF signal input; the ratio detector and quadrature-grid detector do not use a limiter stage because these circuits are insensitive to AM in the FM signal.

27-8 Slope Detection of an FM Signal

The sloping side of the IF response curve provides a varying amplitude response for different signal frequencies, as illustrated in Fig. 27-7. This effect is produced by having the IF center frequency at the side of the curve instead of the middle. Then the different amounts of IF gain for the varying frequencies result in corresponding amplitude variations. With the corresponding AM signal coupled into a diode rectifier, the output is the desired audio signal.

It should be noted that the frequency variations of the IF output are the same as in the input. The frequency variations of the FM signal do not disappear. What happens is that the output signal has amplitude variation added, corresponding to the frequency variations.

The principle of slope detection illustrates two requirements of FM detection: (1) Converting the FM signal into equivalent amplitude variations for the IF signal input to a rectifier, and (2) rectifying the audio variations in IF signal amplitude.

Slope detection is the reason why an FM signal can produce audio output in an AM receiver. In general, slight mistuning shifts the response to put the IF center frequency at the sloping side of the response curve. A specific example for television receivers is slope detection of the associated sound signal in the picture IF section, causing audio signal in the video detector output. In this case, the gain for the IF sound carrier of 41.25 MHz is too high on the sloping side of the picture IF response curve.

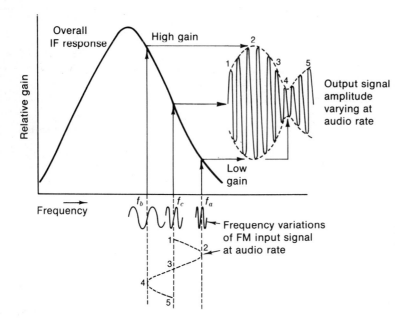

FIGURE 27-7 SLOPE DETECTION OF AN FM SIGNAL. AMPLITUDE VARIATIONS FOR THE OUTPUT SIGNAL AT THE SIDE OF THE RESPONSE CURVE CORRESPOND TO THE FREQUENCY VARIATIONS OF THE INPUT SIGNAL SHOWN AT THE BOTTOM. CENTER FREQUENCY IS f_c; f_a IS ABOVE; f_b IS BELOW.

27-9 Discriminator Operation

As shown in Fig. 27-8a, the discriminator circuit uses two diodes for a balanced FM detector. The operation is similar to slope detection, but with two response curves back to back, one for frequencies above center frequency and the other for frequencies below center. The basic characteristics are:

1. Discriminator input transformer for the two diodes. The tuned transformer proportions the amount of signal for each diode in accordance with the frequency deviation above and below center.
2. Two rectifiers in a balanced circuit. The purpose of the balance is to provide zero output at center frequency. Semiconductor diodes are generally used.
3. The response of the discriminator itself, sep-

arate from the IF amplifier, has the S-shaped curve shown in Fig. 27-8b. This S curve with two opposite polarities represents the result of two balanced diodes and the opposite responses of the discriminator transformer above and below center frequency.

The discriminator in Fig. 27-8 is shown for a center frequency of 10.7 MHz. However, the same results apply for a center frequency of 4.5 MHz, or any other value. The transformer is shown here with three stagger-tuned circuits, in order to illustrate the different responses above and below center frequency. The primary L_p is tuned to the center at 10.7 MHz. The secondary L_1 is tuned 100 kHz, or 0.1 MHz, above at 10.8 MHz. The other secondary L_2 is symmetrical below center, tuned to 10.6 MHz. This frequency separation of ±0.1 MHz gives a total IF bandwidth of 0.2 MHz, or 200 kHz.

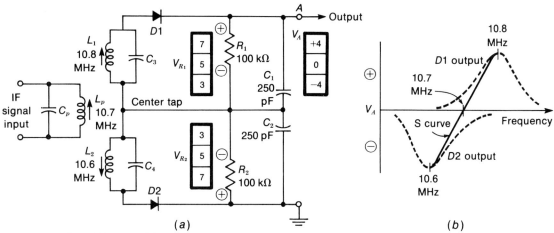

FIGURE 27-8 TRIPLE-TUNED DISCRIMINATOR FOR IF CENTER FREQUENCY OF 10.7 MHz.
(a) CIRCUIT; (b) S RESPONSE CURVE.

Balanced diode detection. Each diode is a rectifier with its own load resistor. When positive signal voltage is applied to the anode of D1, we can consider electron flow in the diode in the direction opposite to the arrow. The electron flow is from cathode to anode, through L_1 and the center-tap connection, returning to the diode cathode. The bypass C_1 across R_1 is a filter to remove IF signal from the audio output, just as in a conventional diode AM detector. Note that the polarity of rectified voltage across R_1 is positive here, at the cathode side of the diode. The amount of rectified voltage depends on the amount of IF signal provided by the tuned circuit with L_1 resonant at 10.8 MHz.

In the same way, when IF signal voltage is applied to D2, it produces rectified output voltage across R_2. This voltage must also be positive at the cathode side of D2, which is the bottom end of R_2.

We can look at the two voltages V_{R1} and V_{R2} as being in series with each other because the output is taken from point A to ground, across both resistors. The polarities of these two voltages are opposing, as the top of R_1 is positive to ground while the top of R_2 is negative to ground. If V_{R1} and V_{R2} were both 5 V, the net output at point A would be 0 V to ground. Therefore, we can consider the two diodes of a discriminator as being in series opposition for the rectified output. The zero-balance point occurs at center frequency, when the discriminator transformer supplies equal amounts of IF signal to the two diodes.

Audio output voltage. Assume now that the amount of signal voltage varies for the two diodes. For example, the signal voltage for D1 increases from 5 to 7 V while the signal for D2 decreases to 3 V. Then D1 will conduct more plate current, producing a larger IR drop across R_1. Also, D2 conducts less current, to produce a smaller voltage drop across R_2.

Now the voltages are unbalanced. The voltage across R_1 rises to 7 V while the voltage across R_2 decreases to 3 V. The net output at point A is the difference between 7 and 3 V. This equals 4 V, positive with respect to ground because the positive voltage is greater.

When the applied signal voltages are reversed, 7 V is applied to D2 and 3 V to D1. Then the voltage drop across R_2 is 7 V, with only 3 V across R_1. The circuit is again unbalanced with an output voltage of 4 V, but this is now negative.

IF input voltage. The frequency swings are converted to corresponding amplitude variations in the signal applied to the two diodes by means of the triple-tuned coupling circuit for the IF signal. Assume first that the instantaneous frequency of the FM signal deviates above center frequency. Then the secondary with L_1 develops a greater signal voltage because the frequency of this tuned circuit is closer to resonance. The secondary voltage developed across L_2, however, is now less than at center frequency because the signal frequency is farther removed from its resonant frequency.

With more signal voltage applied to D1 and less to D2, for a frequency deviation above rest frequency, diode 1 produces more rectified output. A net output voltage of positive polarity results. When the instantaneous frequency of the FM signal deviates below center frequency, more signal voltage is developed across the tuned circuit with L_2 for diode 2, and a negative output voltage is obtained.

With the primary tuned to center frequency, the two secondary circuits alternately provide more or less IF signal voltage for the two diode rectifiers. These amplitude variations correspond to the frequency variations in the FM signal.

Discriminator response curve. The typical S-shaped curve for this balanced detector is shown in Fig. 27-8b. The output at center frequency is zero. For frequencies above center, the output voltage is positive and increases progressively for an increasing swing away from center frequency. Similarly, the output voltage is negative when the input signal is below center frequency. However, the same S-shaped curve with opposite polarity can be obtained by reversing connections to the diodes.

The composite response curve is the net result of the sloping responses of the two individual secondary tuned circuits. They provide outputs of opposite polarity because of the balanced arrangement of the diodes. The positive and negative peaks on the discriminator response curve occur at the resonant frequency for each of the two secondary tuned circuits.

This S-shaped response curve is characteristic of FM detector circuits. With FM signal input, the instantaneous frequency variations swing up and down the linear slope on the S curve to provide amplitude variations corresponding to the audio modulation. You can see the discriminator S curve on an oscilloscope connected to the audio output point A, with a sweep generator supplying the IF signal input.

The triple-tuned discriminator gives good linearity on the S response curve, but it is seldom used because mutual coupling of three tuned circuits makes alignment difficult. Similar results are obtained with a double-tuned transformer for the discriminator in Fig. 27-9 and the ratio detector in Fig. 27-12. In these transformers, both the primary and secondary are tuned to center frequency. However, the phase shift in the secondary varies with the frequency deviation.

27-10 Center-tuned Discriminator

As shown in Fig. 27-9, the input transformer is double-tuned, with primary and secondary resonant at the center frequency of 4.5 MHz. This

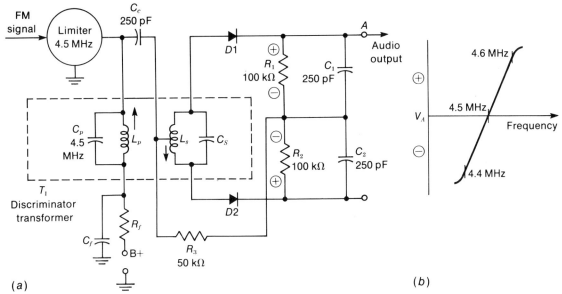

FIGURE 27-9 CENTER-TUNED DISCRIMINATOR FOR IF CENTER FREQUENCY OF 4.5 MHz. (a) CIRCUIT; (b) S RESPONSE CURVE.

circuit is also called a *phase-shift discriminator* because the transformer distributes the IF signal to both diodes in proportion to the phase angle between the primary and secondary voltages. It is also called a *Foster-Seeley discriminator*.

In the input transformer, L_p and L_s couple the IF signal inductively from the limiter stage to the diodes. The main feature of a discriminator transformer, though, is that the IF signal voltage from the primary is also coupled to the center tap of the secondary. The 250-pF coupling capacitor C_c has negligible reactance at the IF signal frequency. The 50-kΩ load resistor R_3 is used to provide a dc return path for diode plate current, while maintaining a load impedance across the primary. An rf choke can be used instead of R_3.

Balanced detection. The two diode rectifiers are balanced in exactly the same way as in the triple-tuned discriminator. When equal signal voltages are applied to the two diodes, the rectified output is the same for the load resistors R_1 and R_2. The net output voltage across the two equal series-opposing voltages is zero. As the applied voltage for each of the diodes alternately increases, while signal for the other decreases, an output voltage of either positive or negative polarity is obtained.

Coupling circuit. As the FM signal varies in frequency, the IF signal voltage applied to the diodes is made to vary in amplitude by means of the coupling arrangement between the limiter stage and the discriminator. Note that signal from the limiter stage is simultaneously coupled to the discriminator in two ways: by induction across the IF transformer and by C_c, which is independent of the mutual induction between L_p and L_s.

646 BASIC TELEVISION

Furthermore, the induced voltage across the secondary is applied to the diode plates in push-pull by means of the center-tap return to cathode. The directly coupled voltage is the primary voltage itself and is applied in parallel to the two diodes. Both diodes are connected to the primary at the same point, which is the center tap. Therefore, the IF signal voltage for the two diodes is the resultant sum of an induced secondary voltage applied in push-pull and the primary voltage applied in parallel to the two diodes.

Quadrature phase. The secondary voltage e_s is 90° out of phase with the primary voltage e_p at center frequency. This phase relation results because the secondary is tuned to resonance. In general, for a transformer where the secondary is tuned, the voltage across the secondary tuned circuit at resonance is 90° out of phase with the voltage across the primary. Furthermore, the phase angle varies around 90° as the applied signal varies above and below the resonant frequency.

Since the primary and secondary of the discriminator transformer are tuned to the IF carrier frequency, e_p and e_s are 90° out of phase at center frequency. Both e_p and e_s are applied at the same time to the diode rectifiers $D1$ and $D2$. The primary signal voltage e_p connected to the center tap is effectively in series with the top half of L_s for signal to $D1$ and the bottom half of L_s for signal to $D2$. The phase relations between e_p and e_s are shown by the phasor diagrams in Fig. 27-10. At center frequency, they are 90° out of phase. As the FM signal swings around center frequency, the phase angle varies above and below 90°.

Phase-angle swings. Figure 27-10a shows the case of resonance with the signal at center frequency. Then e_s and e_p are 90° out of phase. Note that e_s is divided into two equal voltages of opposite polarity, e_{s_1} and e_{s_2}, by the center-tap connection. Therefore, the 90° phase between primary and secondary voltages results in a 90° lagging angle for e_p to e_{s_1} but a 90° leading angle for e_p to e_{s_2}.

The E_1 voltage for $D1$ is the resultant of its two components e_{s_1} and e_p. This combined sig-

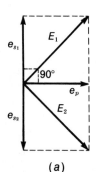
(a)

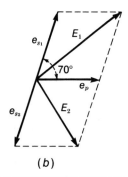
(b)

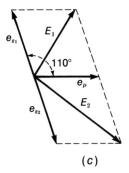
(c)

FIGURE 27-10 PHASE ANGLES BETWEEN PRIMARY VOLTAGE e_p AND SECONDARY VOLTAGE e_s FOR CENTER-TUNED DISCRIMINATOR TRANSFORMER; e_s IS SPLIT INTO TWO OPPOSITE VOLTAGES e_{s_1} and e_{s_2} BY THE CENTER TAP. (a) FOR RESONANCE AT CENTER FREQUENCY; 90° BETWEEN e_p AND e_s; (b) ABOVE RESONANCE; 70° BETWEEN e_p AND e_{s_1}, (c) BELOW RESONANCE; 110° BETWEEN e_p AND e_{s_1}.

nal is the actual voltage applied to the diode for rectification. Similarly, the E_2 voltage for $D2$ is the combined resultant of e_{s2} and e_p. The resultant voltages E_1 and E_2 for the two diodes are equal in this case because of the 90° phase between e_s and e_p. Then e_p is out of phase by the same amount for either e_{s1} or e_{s2}.

With equal amounts of signal voltage E_1 and E_2 applied to the two diodes, the rectified output of each is the same. Then the net voltage from the balanced output circuit at point A equals zero.

In Fig. 27-10b, when the IF signal is above center frequency, e_p is shown 70° out of phase with e_{s1} instead of 90°. The reason for the change in phase angle is that L_s tuned to center frequency is not resonant for the signal frequency above center. As an example, L_s is tuned to 4.5 MHz, but the signal now can be 25 kHz higher. The new phase angle of 70° shown in Fig. 27-10b makes e_p closer to the phase of e_{s1}. As a result, the combined voltage E_1 increases, while E_2 decreases. More signal is applied to $D1$ than to $D2$, and the net output voltage at point A is positive.

In Fig. 27-10c, the phase angle swings in the direction opposite that in b because the signal is below center frequency, instead of above. Now e_p is shown 110° out of phase with e_{s1}. Or, e_p is 70° out of phase with e_{s2}. The result is that e_p is closer to being in phase with e_{s2}. Then E_2 is greater than E_1. $D2$ produces more output than $D1$ now, as the net voltage is negative at point A.

Discriminator response. The S curve in Fig. 27-9b has a bandwidth of ± 100 kHz on the slope, with a center frequency of 4.5 MHz. Although the frequency swing is only ± 50 kHz for television sound, this means that just the linear part of the discriminator response around center frequency is used for FM detection. The linear slope is about one-half the distance between peaks.

Summary of audio detection. How the discriminator response recovers the desired audio voltage from the FM signal can be seen by reviewing the action of the phase-shift discriminator as the transmitted signal is modulated. When the audio modulating voltage is zero, the transmitted rf signal is at center frequency. Tuned in at the receiver, the signal is converted to the intermediate frequency, amplified in the IF section, and coupled to the discriminator for rectification. With the signal at center frequency, the IF transformer coupling the signal to the discriminator is resonant and the secondary voltage is 90° out of phase with the primary voltage. The signal voltages applied to the discriminator diodes then are equal, and the output is zero.

As the transmitter is modulated, the audio modulating voltage produces frequency deviations above and below center frequency. The audio modulation on the rf carrier is then reproduced at the output of the discriminator as a changing dc voltage. This varies in amplitude with the frequency deviation in the FM signal. The rate of these voltage changes is the same as the rate of the frequency swings, which is the audio modulating frequency.

Effect of interfering amplitude modulation. Since the diodes are balanced at center frequency, the net output voltage is zero, regardless of the amplitude of the FM signal. For signal frequencies other than center frequency, however, any variation in amplitude of the IF signal is reproduced in the output of the discriminator. Therefore, a discriminator is preceded by a limiter stage, which has the function of eliminating AM interference in the FM signal coupled to the discriminator.

27-11 The Limiter

How the limiter functions is illustrated by the waveforms in Fig. 27-11a. Note that the signal has a peak amplitude greater than cutoff. The grid-cutoff value depends on the tube and its dc voltages. The amount of grid-leak bias, though, depends on the amount of signal. The grid-leak bias voltage developed will be approximately equal to the peak value of the signal swing. The bias results from grid current, which flows for a very small part of the positive half-cycle at the tip of the positive signal swing. Plate current, however, flows for almost the entire positive half-cycle, as indicated by the shaded area in the illustration.

When the amplitude of the input grid signal increases, a greater negative bias voltage is developed. The grid-cutoff voltage remains the same, however, and the average plate current changes very little. Therefore, the amount of plate-current flow in the limiter stage is approximately constant for all signals with an amplitude large enough to develop grid-leak bias greater than the cutoff voltage.

With a relatively uniform value of average plate current, then, the output voltage across the tuned circuit is constant. Note that the frequency variations in the FM signal are maintained in the output, since the plate-current pulses are produced at the grid signal frequency. The LC tuned plate circuit provides complete sine-wave cycles of the signal.

The limiter stage is also an IF amplifier, with a voltage gain of about 5. Although a vacuum tube amplifier is shown in Fig. 27-11, the limiter stage can use a CE transistor amplifier. Then the RC coupling circuit for signal to the base provides varying reverse bias proportional to signal strength, corresponding to grid-leak bias. A typical RC time constant is 1 to 4 μs.

When the peak amplitude of the signal is too small, the limiting action fails because the stage is then a class A amplifier. Therefore, the

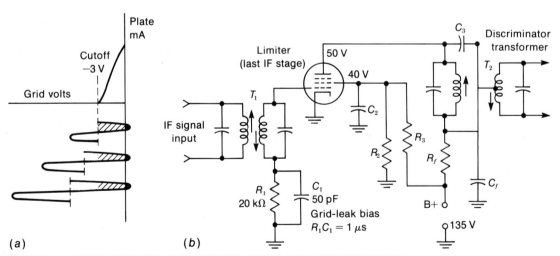

FIGURE 27-11 LIMITER STAGE FOR AN FM RECEIVER. BOTH T_1 AND T_2 ARE TUNED TO THE IF CENTER FREQUENCY. (a) WAVEFORMS FOR DIFFERENT AMPLITUDES OF GRID SIGNAL; (b) CIRCUIT WITH LOW PLATE AND SCREEN VOLTAGES ON PENTODE AMPLIFIER.

preceding stages must have enough gain to saturate the limiter with the smallest useful antenna signal. Assuming 50 μV signal at the antenna, a voltage gain of 60,000 is needed for 3 V at the limiter grid, for good limiting in this example. This is the *limiter-threshold* value for limiting action. Below the limiting threshold, the receiver produces the rushing sound of noise generated in the mixer stage.

27-12 Ratio Detector

This is an FM detector circuit insensitive to amplitude variations in the FM signal. Because no limiter stage is necessary, allowing fewer IF amplifiers, the ratio detector circuit is often used. As shown in Fig. 27-12, the circuit is almost like a discriminator, but for a ratio detector the two diodes are connected in series. Notice that D1 has IF signal applied to its cathode, while D2 has signal at the anode.

Circuit arrangement. In Fig. 27-12, the input coupling transformer T_1 has the same function as in the phase-shift discriminator. Both the primary and secondary tuned circuits are resonant at the IF center frequency. The secondary is center-tapped to produce equal voltages of opposite polarity for the diode rectifiers, while the primary voltage is applied in parallel to both diodes. The transformer is shown in its equivalent form, with L_M indicating the mutual inductance between L_p and L_s.

In the ratio detector, one diode is reversed so that the two half-wave rectifiers are in series for charging the stabilizing voltage source E_3 in the output side of the circuit. Another difference from a discriminator is that audio output is taken from point A at the junction of the two diode capacitors C_1 and C_2, with respect to the center tap on the stabilizing voltage.

Stabilizing voltage. In order to make the ratio detector insensitive to AM interference effects in the audio output, the total voltage E_3, equal to the diode output voltages $E_1 + E_2$, must be stabilized so that it cannot vary at the audio frequency rate. Then audio output is obtained at point A only when the ratio between E_1 and E_2 changes. This is why the circuit is called a

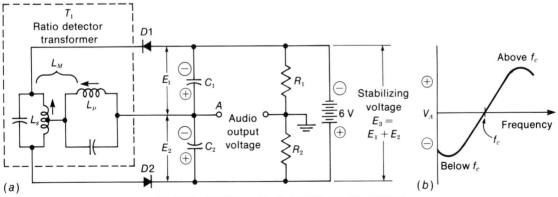

FIGURE 27-12 BASIC RATIO DETECTOR CIRCUIT. T_1 IS TUNED TO IF CENTER FREQUENCY. (a) BALANCED CIRCUIT WITH AC AUDIO OUTPUT AT POINT A; (b) S. RESPONSE CURVE.

ratio detector. The total voltage E_3 remains fixed by the stabilizing voltage source.

Audio output signal. As each diode conducts it produces the rectified output voltage E_1 or E_2 across C_1 or C_2, approximately equal to the peak value of the IF signal applied to each rectifier. At center frequency, the input transformer proportions the IF signal voltage equally for the two diodes, resulting in equal voltages across C_1 and C_2. The output voltage at the audio takeoff point A is zero, therefore, since E_1 and E_2 are equal. Note that these two voltages have opposite polarity at point A with respect to chassis ground.

When the FM signal input is above center frequency, however, we can assume D1 has more IF signal input than D2. Then the rectified diode voltage E_1 is greater than E_2. This makes point A more positive, producing audio output voltage of positive polarity.

Below center frequency, D2 has more IF signal input and E_2 is greater than E_1. The result is audio output voltage of negative polarity at point A. This response is illustrated by the ratio detector S curve in Fig. 27-12, which is essentially the same as the discriminator response curve.

An important characteristic of the ratio detector output circuit can be illustrated by numerical examples. Assume that the frequency deviation above center frequency increases E_1 by 1 V. This makes point A more positive. At the same time, E_2 decreases by 1 V. This makes point A less negative by 1 V, which is the same as making it 1 V more positive. The two rectified diode voltages E_1 and E_2 thus produce the identical voltage change of 1 V in the positive direction at point A. Since the audio signal is taken from point A, the amount of output is the same as though only one diode were supplying audio voltage corresponding to the frequency variations in the FM signal.

As a result, the audio voltage output of a ratio detector is one-half the output of a discriminator, where the audio signal voltages from the two diodes are combined in series with each other at the audio takeoff point. The output in the ratio detector must be taken from the junction of the two diode loads because there is no audio signal voltage across the stabilizing voltage source.

Typical circuit. If a battery were used for the stabilizing voltage, the diodes would operate only with a signal at least great enough to overcome the battery bias on each diode. A large capacitor is used instead, as illustrated by C_3 in the schematic diagram of a typical ratio detector circuit in Fig. 27-13. C_3 charges through the two diodes in series, automatically providing the desired amount of stabilizing voltage for the IF signal level. The 5-μF capacitance of C_3, generally called the *stabilizing capacitor*, is large enough to prevent the stabilizing voltage from varying at the audio frequency rate. A discharge time constant of about 0.1 s is required, as provided by R_1 in series with R_2 across C_3.

The tertiary winding L_t of the ratio detector transformer is used to couple the primary signal voltage in parallel to the two diodes. L_t is wound directly over the primary winding for very close coupling, so that the phase of the primary signal voltage across L_p and L_t is practically the same. The construction of a ratio detector transformer with the tertiary winding is illustrated in Fig. 27-14. This arrangement with the tertiary winding, instead of direct coupling to the secondary center tap, is commonly used with the ratio detector circuit in order to match the high-impedance primary and the relatively low-impedance secondary, which is loaded by the

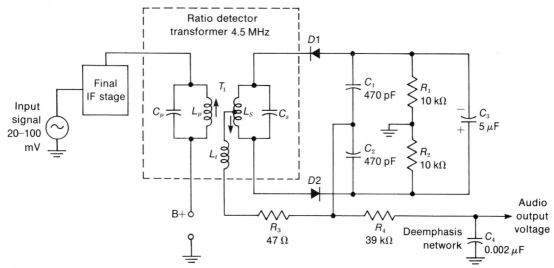

FIGURE 27-13 TYPICAL BALANCED RATIO DETECTOR FOR IF CENTER FREQUENCY OF 4.5 MHz. THE 5-μF ELECTROLYTIC CAPACITOR C_3 PRODUCES THE REQUIRED DC STABILIZING VOLTAGE.

FIGURE 27-14 RATIO DETECTOR TRANSFORMER. HEIGHT IS 2½ IN.

conduction in the diodes. The resistor R_3 in series with L_t limits the peak diode current, to improve the balance of the dynamic input capacitance of the diodes at high signal levels.

Because of the 90° phase relation between the voltages across the secondary and across L_t at resonance, the circuit provides detection of the FM signal in the same manner as the phase-shift discriminator. Both the primary and secondary of the ratio detector transformer are tuned to the IF center frequency, which can be 10.7 MHz in FM broadcast receivers or 4.5 MHz for intercarrier-sound television receivers. The audio output voltage of the ratio detector is taken from the junction of the diode load capacitors C_1 and C_2, in series with the audio deemphasis network, to supply the desired audio signal for the first audio amplifier.

The deemphasis circuit for the detector output consists of R_4 and C_4 with a time constant of 78 μs. Since the high audio frequencies are preemphasized for a better signal-to-noise ratio in transmission, the deemphasis circuit attenuates the high frequencies to provide the original audio modulating signal. C_4 serves as a bypass for high audio frequencies, as illustrated by the deemphasis response curve in Fig. 27-4b.

Single-ended ratio detector. Figure 27-15 illustrates a ratio detector circuit that has an unbalanced output circuit. The input circuit, which is shown in its equivalent form, is the same as in the balanced circuit. However, only the one diode load capacitor C_2 is used in the audio output circuit. Also, the stabilizing voltage across R_1 and C_1 is not center-tapped. It is not necessary to ground the center point of the stabilizing voltage source.

The only effect of the ground connection in Fig. 27-15 is to change the dc level of the audio output signal at A with respect to chassis ground. The same audio signal variations are obtained, but around a dc voltage axis equal to one-half the stabilizing voltage. However, the audio output always has a zero dc level with respect to the midpoint of the stabilizing source.

Only the one diode load capacitor C_2 is necessary in the output circuit to obtain the audio signal because it serves as the load for both diodes. Each diode charges C_2 in proportion to the IF signal input for each rectifier, but in opposite polarity. Therefore, the voltage across C_2 is the same audio output signal as in the balanced ratio detector. The capacitance of C_2 is doubled in the single-ended circuit because it replaces two capacitors that are effectively in parallel for signal voltage in the balanced circuit.

Ratio-detector receiver characteristics. Referring back to Fig. 27-13, note that only 20 to 100 mV signal is applied to the grid of the IF stage driving the ratio detector. With a gain of about 100, the IF output of 2 to 10 V will be enough to drive the diodes. Since no limiter is required, less IF gain is needed, as there is no fixed threshold voltage that must be exceeded. The ratio detector automatically adjusts itself to the IF signal level.

27-13 Quadrature-Grid FM Detector

This circuit uses a pentode combining the functions of FM detector, AM limiter, and first audio amplifier. As shown in Fig. 27-16, the audio sig-

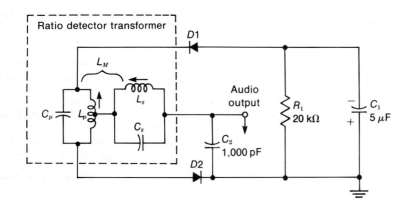

FIGURE 27-15 RATIO DETECTOR WITH SINGLE-ENDED OR UNBALANCED OUTPUT CITCUIT.

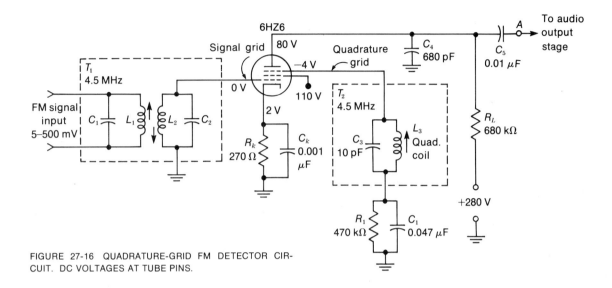

FIGURE 27-16 QUADRATURE-GRID FM DETECTOR CIRCUIT. DC VOLTAGES AT TUBE PINS.

nal output has enough amplitude to drive the audio power output stage.

Either of two types of tubes can be used. One is a gated-beam tube, such as the 6BN6. The other is a sharp-cutoff pentode where grid 3 and grid 1 can be used as dual control grids. Common tubes for this function are the 6DT6 or 6HZ6 pentodes and the 6V10 or 6Z10, combining a sharp-cutoff pentode with a beam-power audio output stage. These tubes are often used with heater ratings of 3 V, 4 V, 5 V, 13 V, and 17 V for series strings. In any case, this detector circuit requires a pentode with two sharp-cutoff control grids to control plate current. Typically, the cutoff is −3 V for grid 1 and −6 V for grid 3.

Referring to Fig. 27-16, the IF FM signal is applied to grid 1, which is the signal grid. Grid 3 also has the signal, but in quadrature phase, by space-charge coupling within the tube. The detected audio output is produced in the plate circuit because of the 90° phase between signals at grid 1 and grid 3.

Quadrature grid. The characteristics of grid 3 are important in the operation of this FM detector circuit. In tubes with multiple grids, when one grid has positive voltage with respect to cathode and the next grid is negative, this grid develops its own space charge corresponding to a virtual cathode. In the 6HZ6 here, grid 3 is at −4 V, while grid 2 is the screen grid at +110 V. Therefore, electrons in the space current to the plate are slowed down enough near grid 3 to form an outer space charge. This is in addition to the normal space charge near the cathode. The outer space charge serves as a source of electrons for the outer electrodes, including grid 3 and the plate.

With signal input voltage to grid 1, the space current is varied. Then the amount of space charge at the virtual cathode near grid 3 also varies. As a result, electrostatic induction by the varying space charge causes signal current to flow in the grid 3 circuit. The resonant circuit of L_3 here provides signal voltage corresponding to the induced current.

The induced current in the grid 3 circuit lags by −90° the signal voltage at grid 1. At resonance, the voltage across the tuned circuit for grid 3 also lags the grid 1 voltage by −90°. This is why grid 3 is called the quadrature grid. The −90° phase applies at center frequency of the FM input. When the signal deviates below and above center frequency, the phase of the quadrature-grid signal varies below and above −90°.

Gating action. In a tube with two control grids, each controls the plate current. Also, either one can cut off the output. Therefore, the grids serve as voltage-controlled gates to turn the output on or off. When the grid voltage allows plate current, the gate is open; when the grid voltage is more negative than cutoff, the gate is closed. Either grid can gate the plate current on or off. When one gate is closed there is no plate current, even if the other gate is open. Therefore, both gates must be on at the same time to have output plate current. This feature is the basis for detecting the FM signal at grid 1 which is gated by the quadrature signal at grid 3 (see Fig. 27-17).

Circuit operation. In Fig. 27-16, the 4.5-MHz sound signal is coupled by T_1 to the signal grid of the FM detector. This signal is also in the quadrature-grid circuit, because of space-charge coupling, but 90° out of phase with the input signal. The quadrature coil L_3 is adjusted to resonate with C_3 at the IF center frequency of 4.5 MHz. Grid-leak bias is provided by R_1C_1.

The phasing between the signals at grid 1 and grid 3 is shown in Fig. 27-17. At center frequency, in a, the two signal voltages are 90° out of phase. Plate current flows only for part of the cycle, when both grids gate i_p on. This time is indicated by the shaded area of the i_p waveform.

Above center frequency, in Fig. 27-17b, e_{g3} lags e_{g1} by more than −90°. Then the two signals

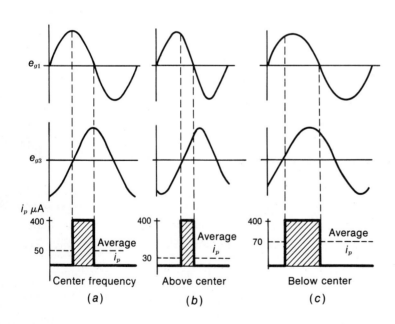

FIGURE 27-17 GATING ACTION OF GRIDS 1 AND 3 ON PLATE-CURRENT OUTPUT OF QUADRATURE-GRID DETECTOR. THE AVERAGE i_p VARIES WITH FM DEVIATION. (a) AT CENTER FREQUENCY e_{g3} LAGS e_{g1} BY 90°; (b) ABOVE CENTER FREQUENCY e_{g3} LAGS MORE; (c) BELOW CENTER FREQUENCY e_{g3} LAGS LESS.

are closer to being of opposite phase. Therefore, the on time for i_p is shorter, resulting in narrower pulses of plate current. Although the amplitude is the same, equal to saturation i_p, the narrower pulses mean a smaller value of average plate current.

Below center frequency, in Fig. 27-17c, e_{g3} lags e_{g1} by less than $-90°$. Then the two signals are closer to being in the same phase, and therefore the on time for i_p is longer, resulting in wider pulses, with a higher average value.

Remember that these phase changes for e_{g3} result from frequency deviations in the FM signal, which follow the audio modulating voltage. Similarly, the variations in average i_p follow the modulation in the FM signal. Although the plate-current pulses are at the IF signal frequency, the average plate current varies with the audio signal.

In the plate circuit, the integrating capacitor C_4 bypasses the IF variations, allowing the audio output voltage to be developed across the plate-load resistor R_L. The audio signal amplitude at the audio takeoff point A is about 14 V peak value, for 25 kHz deviation from the center frequency of 4.5 MHz. This voltage is enough to drive the audio power output stage, without the need for a first audio voltage amplifier.

AM rejection. Undesired amplitude modulation is attenuated because of the following factors: The cathode bias resistor R_k is bypassed for IF but not for audio frequencies. The resultant degeneration of the audio signal reduces the effect of any high-amplitude interference. Also, the IF transformer T_1 is damped by grid current when the signal amplitude at grid 1 exceeds the cathode bias of 2 V. The shunt damping lowers the Q of the tuned transformers, reducing the amount of signal voltage. Finally, grid 3 produces saturation plate current when it is driven positive. This grid has grid-leak bias, which increases with higher signal amplitudes to clamp the positive peak at zero voltage. The result is saturation plate current at a peak value of approximately 0.4 mA for typical signal input. In some circuits a variable cathode bias resistor is used as a *buzz control*, which is adjusted for best limiting.

Locked-oscillator mode. Without any input, or for weak signal, the 6HZ6 detector will oscillate at relatively low amplitude. The circuit is a tuned-grid tuned-plate oscillator, with grid 3 as the oscillator plate supplying feedback to grid 1 through interelectrode capacitance. Then the FM input becomes a synchronizing signal to lock in the oscillator at the signal frequency. Detection occurs because of the gating action of the two control grids in quadrature, as described. However, the locked-oscillator feature is an advantage in rejecting AM with weak signal input. Then, the signal amplitude in the detector is always high enough to provide the required limiting action.

For weak signal input, the oscillations provide the required amplitude of grid voltage, synchronized by the FM input signal. With normal or strong signal, the input voltage itself has enough amplitude for AM limiting. The result is quadrature detection with good AM rejection for both weak and strong signals, in this locked-oscillator, quadrature-grid detector (LOQG) circuit.

27-14 Sound IF Alignment

This consists of two parts: tuning the IF amplifier for maximum output and balancing the AM detector at center frequency. Typical response curves for the 4.5-MHz television sound are shown in Fig. 27-18. In FM broadcast receivers,

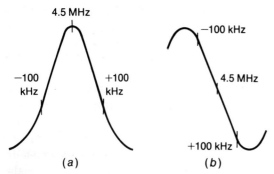

FIGURE 27-18 VISUAL RESPONSE CURVES FOR 4.5-MHz SOUND IF SECTION. (a) MAXIMUM RESPONSE OF IF AMPLIFIERS; (b) S RESPONSE OF QUADRATURE-TUNED CIRCUIT IN FM DETECTOR.

the IF center frequency is 10.7 MHz, with the ±100-kHz markers farther apart for more bandwidth. All these markers are usually available on sweep generators.

Discriminator-limiter combination. The IF amplifiers are aligned for maximum output at the input to the limiter. A visual response curve looks like Fig. 27-18a. However, the IF circuits can also be aligned for maximum grid-leak bias, measured by a dc voltmeter at the limiter grid.

For the discriminator transformer, the S curve can be obtained with an oscilloscope at the audio takeoff point. However, the discriminator can also be balanced with a dc voltmeter, as explained later.

Ratio detector receiver. The IF amplifiers can be aligned for maximum dc output across the stabilizing capacitor with a dc voltmeter. This IF alignment includes the primary of the ratio detector transformer but not the secondary.

A visual IF response curve like Fig. 27-18a can be obtained. However, the stabilizing capacitor must be disconnected temporarily. It bypasses the 60-Hz oscilloscope input signal, which is the modulation frequency of the sweep generator.

The secondary of the ratio detector transformer is aligned by checking the audio output voltage, not the stabilizing voltage. Either an S curve can be obtained at the audio takeoff point, as in Fig. 27-18b, or the ratio detector can be balanced with a dc voltmeter, similar to the discriminator alignment.

Quadrature-grid detector alignment. This circuit is usually aligned with FM input signal from a broadcast station. The procedure is to adjust the quadrature coil, 4.5-MHz IF transformers, and 4.5-MHz sound takeoff trap for best sound, with increasing volume and minimum distortion. In addition, a dc voltmeter to check grid-leak bias on the quadrature grid will read maximum. This voltage can be measured across R_1 in Fig. 27-16.

A specific procedure without any test equipment can be as follows:

1. Tune in a strong channel but attenuate the antenna signal until you hear hiss in the sound. This is receiver noise, indicating weak sound signal below the limiting level.
2. Adjust the quadrature coil for undistorted audio, with minimum hiss and no buzz.
3. The 4.5-MHz IF transformers are adjusted for maximum volume.

Balancing the discriminator. This is essentially tuning the input transformer to center frequency. It can be done without a visual response curve, using just a dc voltmeter and a conventional rf signal generator to supply unmodulated output voltage at the signal frequency. A dc voltmeter is used because the output is a steady voltage when the input signal is not frequency-modulated, since there are no frequency variations to produce variations in the output. The meter should have an impedance of 20,000 Ω/V or higher to avoid detuning the discriminator.

With the generator supplying signal at the IF center frequency, and the dc voltmeter connected across the audio output terminals:

1. Tune the primary of the discriminator transformer for maximum output.
2. Tune the secondary of the discriminator transformer for a sharp drop to zero.

It should be possible to produce either a positive or a negative output voltage when adjusting the secondary. Therefore, it should be tuned for zero indication at the balance point where the output voltage starts to swing from one polarity to the other. A gradual decrease to zero is the wrong indication, as this only means the circuit is being detuned away from center frequency.

When the signal generator frequency is varied manually above and below center frequency, the dc output voltage should vary from zero at center frequency to a maximum value at both sides of center frequency, with opposite polarities below and above center. The response should be symmetrical about center frequency, with equal output voltages produced for the same amount of frequency change below or above center frequency. The two points of maximum output voltage, corresponding to the two peaks on the discriminator response curve, should have the required frequency separation.

The actual polarity of the output voltage for a frequency deviation above or below center does not matter. It is only required that the output voltages be of opposite polarity for frequency deviations above and below center frequency.

Balancing the ratio detector. With the generator supplying signal at the IF center frequency:

1. Tune the primary of the ratio detector transformer for maximum output on the dc voltmeter across the stabilizing capacitor.
2. Move the dc voltmeter to the audio takeoff point. Then tune the secondary for zero balance.

When the ratio detector has a single-ended output circuit, however, the audio output has a dc level equal to one-half the stabilizing voltage. Therefore, usual practice is to insert two balancing resistors temporarily, as shown in Fig. 27-19. This converts the single-ended arrangement to a balanced output circuit, so that the ratio detector secondary can be aligned for balance at zero with the dc voltmeter.

27-15 The Audio Amplifier Section

In receivers with vacuum tubes there is generally a quadrature-grid detector driving one audio output stage. This is a transformer-coupled to a 3.2-Ω loudspeaker (Fig. 27-20). With solid-state circuits, though, the ratio detector or discriminator is generally used. These FM detector circuits require an audio amplifier to drive the

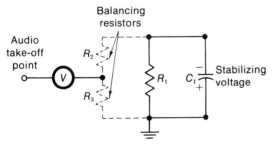

FIGURE 27-19 TEMPORARY RESISTOR CONNECTIONS TO BALANCE AUDIO TAKEOFF POINT TO GROUND IN A SINGLE-ENDED RATIO DETECTOR.

658 BASIC TELEVISION

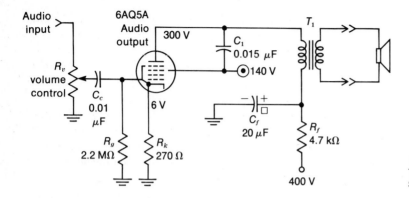

FIGURE 27-20 AUDIO OUTPUT TUBE TRANSFORMER-COUPLED TO LOUDSPEAKER.

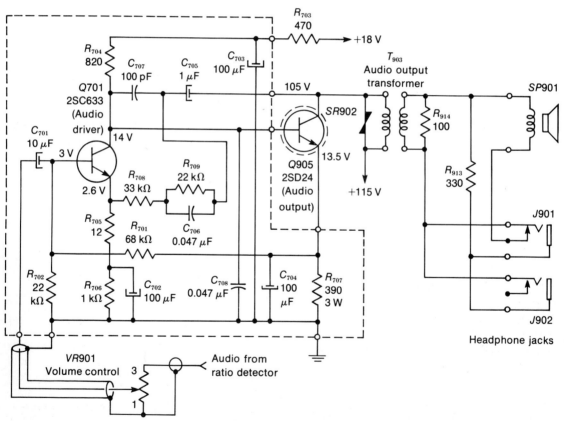

FIGURE 27-21 TRANSISTORIZED AUDIO AMPLIFIER. R VALUES IN OHMS. (SONY CHASSIS KV 1210U)

THE FM SOUND SIGNAL 659

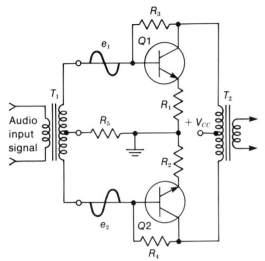

FIGURE 27-22 PUSH-PULL AUDIO AMPLIFIER WITH TRANSISTORS TO DRIVE LOUDSPEAKER.

power output stage (Fig. 27-21). In addition, transistorized audio output stages often use the push-pull circuits illustrated in Figs. 27-22 and 27-23. The loudspeaker impedance is generally 16 to 32 Ω for transistor circuits.

The volume control varies the amount of signal into the audio amplifier. When a tone control is used, it varies the treble response. Either bypassing for high audio frequencies is varied or the amount of negative feedback is adjusted.

Audio output stage. In Fig. 27-20 the audio input at the volume control is obtained from the plate circuit of a quadrature-grid detector. C_c and R_g provide the RC coupling. Cathode bias is produced by the 270-Ω R_k. This resistor is not bypassed, in order to degenerate the input signal. The degeneration, or negative feedback, reduces gain but also reduces distortion. Amplified output from the plate is coupled to the loudspeaker by the audio output transformer T_1. Its function is to match the low impedance of 3.2 Ω for the loudspeaker to the 10-kΩ output impedance of the plate circuit. C_1 between the plate and screen grid is a bypass for high audio frequencies. It compensates for the resonant peak of the output transformer.

In the plate circuit, R_f and C_f form a decoupling filter to isolate the transformer primary from the B+ supply. The 20-μF value for C_f means it is an electrolytic capacitor. Usually this capacitor is one section of a filter capacitor can, as indicated by the square marking. When the cathode resistor R_k is bypassed, the C_k is an electrolytic capacitor of 100 to 200 μF, also as one section of a filter capacitor can.

Transistorized audio amplifier. In Fig. 27-21, transistor Q701 is a CE (common-emitter) stage. It is the first audio amplifier driving the CE power output stage Q905. The audio input from a ratio detector is coupled by C_{701} to the base of Q701. Shielded cable is used to prevent pickup of hum. The audio output from Q905 is transformer-coupled by T_{903} to the 16-Ω

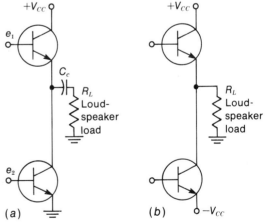

FIGURE 27-23 SERIES OUTPUT CONNECTIONS FOR PUSH-PULL AUDIO AMPLIFIER TO ELIMINATE OUTPUT TRANSFORMER. (a) ONE COLLECTOR SUPPLY V_{CC}; (b) SYMMETRICAL VOLTAGE SUPPLIES OF OPPOSITE POLARITY.

660 BASIC TELEVISION

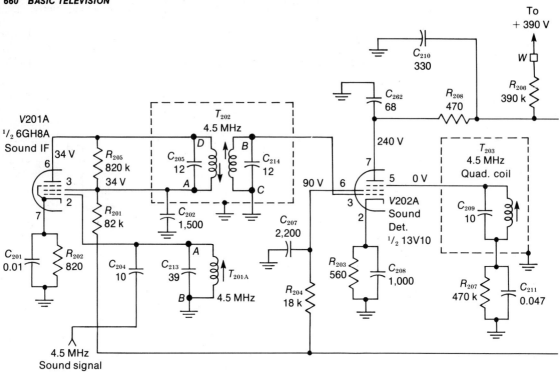

FIGURE 27-24 COMPLETE SOUND SECTION USING TUBES, IN MONOCHROME RECEIVER. R IN OHMS, C ABOVE 1 IN pF. (RCA CHASSIS KCS 183)

loudspeaker and the headphone jacks. The varistor $SR902$ across the primary protects the power output transistor $Q905$ from voltage breakdown if the secondary load should open when the volume is at a high level.

The collector load for $Q701$ is the 820-Ω R_{704}. In the emitter circuit, R_{705} and R_{706} produce the emitter voltage of 2.6 V. R_{705} is not bypassed in order to degenerate the input signal, which reduces distortion and raises the input impedance. The base bias of 3 V on $Q701$ is taken from the emitter of $Q905$, through the 68-kΩ series-dropping resistor R_{701}.

The collector output circuit of $Q701$ is directly coupled to the base of $Q905$ to provide signal drive for the audio output stage. Negative feedback to reduce distortion is provided from the collector of $Q905$ to the base of $Q701$. The feedback line includes C_{705}, R_{709} with its shunt C_{706}, and R_{708} in series. The 0.047-μF C_{708} at the base of $Q905$ is a bypass for high audio frequencies to reduce the treble response. The emitter voltage of 13.5 V on $Q905$ is produced by emitter current through R_{707}.

Push-pull audio output circuits. The basic arrangement is illustrated in Fig. 27-22 with two transistors. Both can be either NPN or PNP. Also, two vacuum tube amplifiers can be used. The push-pull circuit can operate Class B or Class AB, as each stage supplies output signal on alternate half-cycles. The result is more power output with greater efficiency, compared with class A operation. A center-tapped output transformer T_2 is shown in Fig. 27-22 to drive the loudspeaker load. Also, the input transformer T_1

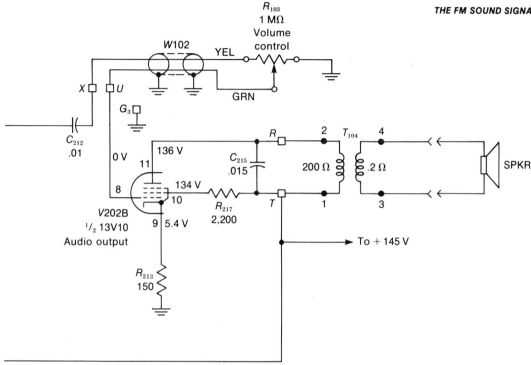

splits the phase of the audio signals e_1 and e_2 driving the push-pull stages.

There are more possibilities for push-pull circuits with transistors than with tubes. First, the center-tapped output transformer T_2 can be eliminated because power transistors provide the low impedance needed to drive the loudspeaker directly. As shown in the series output circuit in Fig. 27-23, the output is taken from the collector-emitter junction of the two transistors in series for dc supply voltage. In Fig. 27-23a, one dc supply voltage is used for both transistors. This circuit uses C_c to block the dc voltage from the loudspeaker load R_L. In Fig. 27-23b, equal voltage supplies of opposite polarity are used on the collector of one transistor and the emitter of the other transistor. Then the dc voltage is zero at the output, allowing a direct connection to the loudspeaker without C_c.

Both circuits in Fig. 27-23 still need a phase splitter for push-pull input. To eliminate this requirement, the push-pull amplifier can use *complementary symmetry,* with one stage NPN and the other PNP. Or, both output stages can be the same type while the driver stages use opposite transistors, in a *quasi-complementary amplifier.*

27-16 Complete Circuits for the Associated Sound Signal

A complete sound section with vacuum tube amplifiers is shown in Fig. 27-24. This is a monochrome receiver, with the 4.5-MHz sound IF signal produced in the output of the video detector. Then the 10-pF C_{204} couples the sound signal to the control grid of $V201A$. This pentode is a tuned IF amplifier, resonant at 4.5 MHz. The transformer T_{201A} is part of the 4.5-MHz sound takeoff trap T_{201B} in the video detector output, not shown in this diagram. The plate transformer T_{202} of the sound IF amplifier is also tuned to 4.5 MHz, driving the quadrature-grid detector V202A. Both the detector and audio output stages use the dual-pentode tube 13V10. The quadrature-grid detector circuit is essentially the same as that illustrated in Fig. 27-16. Also, the audio output stage which drives the loudspeaker is similar to Fig. 27-20.

In Fig. 27-25 an integrated circuit is used for the 4.5-MHz sound IF amplifier and quadra-

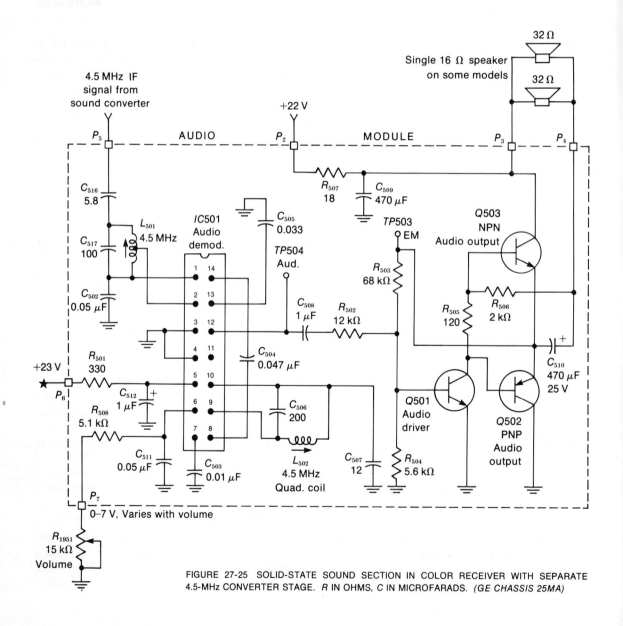

FIGURE 27-25 SOLID-STATE SOUND SECTION IN COLOR RECEIVER WITH SEPARATE 4.5-MHz CONVERTER STAGE. R IN OHMS, C IN MICROFARADS. (GE CHASSIS 25MA)

ture detector, but with discrete transistors for the audio amplifier. This entire sound section is on one small module board. The input is sound IF signal from a separate 4.5-MHz sound converter stage on the picture IF module board. The audio amplifier consists of a PNP driver for the push-pull output circuit with complementary symmetry. One collector supply is used for the NPN and PNP transistors in series for dc voltage.

27-17 Intercarrier Buzz

This is 60-Hz interference in the 4.5-MHz sound signal. The buzz is a result of cross-modulation by the picture signal, especially for white information when the picture carrier is at its minimum level. At the opposite extreme, the picture carrier is high for blanking and sync. The effect of the cross-modulation is AM interference in the sound signal, therefore, at the 60-Hz rate of the vertical blanking pulses. Normally, intercarrier buzz is not audible, though, because of the following factors:

1. The amplitude of the first sound IF carrier of 41.25 MHz is very low, at 5 to 10 percent.
2. Limiting in the 4.5-MHz sound IF circuits reduces the AM interference.
3. Exact balance in the FM detector eliminates AM interference at the center frequency.

Typical troubles that can cause intercarrier buzz are incorrect balance in the FM detector, a defective stabilizing capacitor in a ratio detector, and overload distortion in any of the IF stages common to picture and sound. The overload can be caused by a weak amplifier or incorrect bias, often because of AGC trouble.

27-18 Multiplexed Stereo Sound

Since stereophonic broadcasting with left (L) and right (R) audio signals is used in the FM broadcast band, similar methods are being considered to multiplex two audio signals for television sound. The problem is a little different, though, because of the smaller deviation of 25 kHz, compared with 75 kHz. Also, the horizontal deflection current at 15,750 Hz can cause interference in television receivers.

The general idea of multiplexing two signals on one subcarrier is illustrated by color television, where two color video signals are combined on the 3.58-MHz subcarrier. One suggested method for stereo sound in television uses the following standards: The $L+R$ signal, which is compatible with monaural audio, uses the same modulating frequencies of 50 to 15,000 Hz with the same frequency deviation of ± 25 kHz. The $L-R$ component, which is the stereo component, is on a suppressed subcarrier of 23.625 kHz. This frequency is $1.5 \times 15,750$ Hz. The pilot frequency signal is 39.375 kHz for the subcarrier regenerator in the receiver. This pilot frequency is $2.5 \times 15,750$ Hz. The audio modulating frequency range of the $L-R$ signal is 50 to 7,875 Hz, which is enough to give the extra dimension of stereo sound.

SUMMARY

1. In FM, the frequency of the transmitted carrier varies with the amplitude of the audio modulating voltage. Center or rest frequency is the carrier frequency with zero modulation voltage. Deviation is the frequency change from center frequency. Swing is the total deviation above and below center frequency.

2. The FCC defines 100 percent modulation as ±25 kHz swing for the FM sound in television broadcasting and ±75 kHz for commercial FM broadcast stations in the 88- to 108-MHz band.
3. A reactance-tube circuit provides either X_L or X_c that can be varied by its control-grid voltage. The reactance tube then varies the frequency of a tuned oscillator.
4. Preemphasis in FM transmission boosts the audio amplitude and the resulting carrier deviation to improve the signal-to-noise ratio for higher audio frequencies. At the receiver, deemphasis after the FM signal is detected reduces the amplitude of higher audio frequencies to provide the original audio frequency response. The specified time constant is 75 μs.
5. For slope detection of an FM signal, the IF center frequency is at the side of the IF response curve, to produce amplitude variations corresponding to the frequency variations. A diode detector can then recover those variations corresponding to the audio modulating signal. Slope detection is the reason why an AM receiver can receive FM signals.
6. A discriminator is a balanced FM detector using two diodes. The tuned input transformer changes the frequency variations of the IF signal to amplitude variations that are detected to produce audio output signal. The response curve of a discriminator is an S curve, consisting of two IF response curves of opposite polarity (see Fig. 27-9).
7. In the center-tuned or phase-shift discriminator, both the primary and the secondary are tuned to center frequency.
8. A limiter stage is a saturated class C amplifier to reduce amplitude variations in the IF signal.
9. The ratio detector uses a center-tuned input transformer with two diodes in series to charge the stabilizing capacitor (see Fig. 27-12). No limiter stage is necessary, as the ratio detector rejects AM interference.
10. In the quadrature-grid detector, the input IF transformer and quadrature coil are tuned to center frequency (see Fig. 27-16). No limiter stage is necessary, as the detector rejects AM interference. The amplified audio output in the plate circuit is enough to drive the audio output stage.
11. The quadrature coil is aligned with weak signal from a station for minimum hiss in the sound and maximum undistorted audio output.
12. Push-pull audio output stages can use PNP and NPN transistors for complementary symmetry.
13. In intercarrier sound receivers, the difference frequency of 4.5 MHz is produced as a beat in the sound converter stage, with the modulation of the FM sound signal. The sound IF amplifier and FM detector are always aligned at the center frequency of 4.5 MHz.
14. Intercarrier buzz is produced by 60-Hz blanking pulses from the picture signal in the 4.5-MHz intercarrier sound signal. The buzz can be reduced by AM rejection circuits and exact alignment of the FM detector. In addition, the sound IF response must be low in the common IF amplifier, compared with picture signal.

Self-Examination (Answers at back of book)

Part A Fill in the missing answer.

1. The assigned carrier frequency of an FM broadcast station is 96.3 MHz. Its FM signal has a rest frequency or center frequency of _____ MHz.
2. If 4-V peak audio amplitude changes the rf signal frequency from 200 to 210 kHz, the frequency deviation is _____ kHz.
3. The total swing, then, for 8-V peak-to-peak audio is _____ kHz.
4. For 100 percent modulation of the associated sound signal, the total frequency swing is _____ kHz.
5. If 2-kHz deviation is produced by 2-V audio, then modulating voltage of 4V will produce a deviation of _____ kHz.
6. If the audio modulation in Question 5 has the same amplitude but double the frequency, the deviation will be _____ kHz.
7. The term quadrature indicates a phase angle of _____ degrees.
8. The time constant for preemphasis is _____ μs.
9. The dc output of a balanced discriminator at center frequency is _____ V.
10. The FM detector for the intercarrier sound signal is tuned to _____ MHz.

Part B Answer True or False.

1. In FM, the loudest sounds produce maximum frequency deviation.
2. In FM, the audio modulating frequency determines the rate of frequency swing.
3. A 2-MHz carrier swings ±100 kHz. Therefore, the instantaneous carrier frequency varies between 1.9 and 2.1 MHz.
4. A frequency swing of ±5 kHz corresponds to 5 percent modulation for the television sound signal.
5. Deemphasis of the detected signal provides the most attenuation for the highest audio frequencies.
6. A separate diode is generally used in color receivers for the 4.5-MHz sound converter.
7. In a reactance-tube circuit, the grid feedback voltage is 90° out of phase with the plate voltage.
8. AM rejection circuits in the FM receiver reduce noise and interference.
9. Push-pull amplifiers with transistors can eliminate the need for center-tapped transformers.
10. In a limiter stage, a typical time constant for grid-leak bias is 0.2 s.
11. Slope detection enables an AM receiver to detect an FM signal.
12. In an intercarrier sound receiver with a ratio detector, the primary and secondary are aligned at 4.5 MHz.
13. In the center-tuned discriminator, the primary and secondary voltages are 90° out of phase at center frequency.

14. In the quadrature-grid detector, the signal grid and quadrature grid are 90° out of phase at center frequency.
15. The audio output voltage of a quadrature-grid detector is usually 2 to 5 μV.
16. In a quadrature-grid detector, the audio output voltage is taken from the plate circuit.
17. The IF stages in a ratio-detector receiver can be aligned for maximum stabilizing voltage.
18. If the sound IF response is too high in the common IF amplifier, the result can be sound in the picture and buzz in the sound.
19. The sound takeoff trap is always tuned to 4.5 MHz in intercarrier sound receivers.
20. The S response curve of a discriminator shows two opposite polarities of detected output voltage.

Essay Questions

1. Compare how the carrier wave varies with audio modulation for AM and FM.
2. Define center frequency, deviation, swing, and percent modulation.
3. What characteristic of the FM signal determines loudness of the reproduced audio signal?
4. Define phase modulation. Why does PM produce equivalent FM?
5. Give two reasons for the improved noise reduction in an FM system.
6. Give two advantages of FM over AM and one disadvantage of FM, with both in the VHF band.
7. Why is deemphasis necessary in the FM receiver?
8. What are two important differences between an FM receiver and an AM receiver?
9. Give numerical values for the beat frequencies in the output of the mixer, showing that the amount of frequency swing of the 41.25-MHz IF signal in the receiver is the same as the FM sound signal transmitted in channel 2 at 54 to 60 MHz.
10. Compare how the 4.5-MHz IF signal is obtained in monochrome and color receivers.
11. Give three requirements for detecting an FM signal.
12. What is meant by slope detection?
13. Why is AM rejection an essential requirement of an FM receiver?
14. Draw the schematic diagram of a phase-shift discriminator circuit for an IF center frequency of 4.5 MHz and briefly describe how the circuit operates. Label the audio takeoff point.
15. Draw the schematic diagram of a balanced ratio circuit for an IF center frequency of 4.5 MHz. Indicate the audio takeoff point and the stabilizing voltage.

16. In the quadrature-grid detector circuit, where is the IF signal applied and where is the audio takeoff point?
17. Give two advantages of the quadrature-grid detector, compared with the discriminator.
18. Describe how to align the quadrature-grid detector at 4.5 MHz, using the transmitted sound signal.
19. Give two reasons why television receivers use intercarrier sound. Give one disadvantage of intercarrier sound.
20. When aligning by the voltmeter method, what is the required indication when tuning the secondary in a phase-shift discriminator and in a ratio detector? Where is the meter connected in both cases?
21. Show the connections of the dc voltmeter and balancing resistors for aligning the secondary in a single-ended ratio detector.
22. In a sound IF circuit consisting of two IF stages and a ratio detector, describe how to align the entire section by: (a) the dc voltmeter method with a signal generator, and (b) the visual response curve method. Show the IF and detector curves, with marker frequencies for an IF center frequency of 4.5 MHz.
23. In a circuit consisting of two IF stages, limiter, and center-tuned discriminator, describe how to align the entire section by: (a) the dc voltmeter method with a signal generator, and (b) the visual response curve method. Show curves and marker frequencies for an IF center frequency of 10.7 MHz.
24. Describe how to align the 4.5-MHz sound IF section in Fig. 27-24.
25. In Fig. 27-21, give the functions of the following components: Q701, Q905, C_{701}, C_{708}, R_{707}, and T_{903}.
26. In Fig. 27-21, (a) how is the base voltage of 3 V obtained for Q701? (b) How is the base voltage of 14 V obtained for Q905?
27. In Fig. 27-24, give the functions of the following components: R_{202}, R_{201}, C_{202}, R_{207}, T_{203}, C_{210}, C_{212}, C_{215}, and T_{104}.
28. In Fig. 27-25, give the functions for L_{501}, L_{502}, and IC501.

Problems (Answers to selected problems at back of book)

1. An audio modulating voltage of 3 V at 1,000 Hz produces a frequency deviation of 15 kHz. Assuming linear modulation, calculate the deviation for the following examples of audio: (a) 1 V at 1,000 Hz; (b) 3 V at 1,000 Hz; (c) 6 V at 500 Hz; (d) 9 V at 100 Hz; (e) 0.3 V at 10 Hz.
2. Repeat Prob. 1 for a modulation system in which 3-V audio produces 10-kHz deviation.

3. For the RC deemphasis circuit in Fig. 27-4, calculate the exact frequency at which X_c equals R. What is the attenuation in dB at this frequency, compared with the response at 100 Hz?
4. A deemphasis circuit has R of 75,000 Ω and C of 0.001 μF. Consider this a series ac circuit with 10-V rms audio signal applied. (a) Make a table listing the values of X_c, I, and V_c at 100 Hz, 500 Hz, 1 kHz, 5 kHz, 10 kHz and 15 kHz. (b) Draw a graph showing the V_c values plotted against frequency.
5. In Fig. 27-21, determine the emitter current of Q905.
6. Give the function of each stage, with input and output signals, for the block diagram in Fig. 27-26 below, for the associated sound.

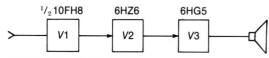

FIGURE 27-26 FOR PROBLEM 6.

28 Receiver Servicing

The work required may include antenna installation, replacing the picture tube, adjusting the setup controls, and troubleshooting defective circuits. Alignment is seldom necessary. Most troubles are caused by a defective tube, capacitor, resistor, or semiconductor device. Since tubes are the most common cause of trouble, they should be checked first, preferably by substituting a new tube. Also, circuits with high values of current or voltage are more likely to have trouble. These include the power supplies and deflection circuits, especially the horizontal output stage. The general idea of localizing the troubles to a section for raster, picture, color, or sound is described in Chap. 19. Here, we can analyze specific methods for finding the defective component. In addition, complete receiver circuits are included to show how all the sections fit together. The topics described here are:

- 28-1 Troubleshooting techniques
- 28-2 Test instruments
- 28-3 Dc voltage measurements
- 28-4 Oscilloscope measurements
- 28-5 Alignment curves
- 28-6 Signal injection
- 28-7 Color-bar generators
- 28-8 Schematic diagram of monochrome receiver with tubes
- 28-9 Schematic diagram of solid-state color receiver
- 28-10 Transistor troubles

28-1 Troubleshooting Techniques

We assume the trouble has been localized to one stage or section. Remember that the receiver was probably working fine until the trouble happened. Therefore, check for simple problems first. Substitute a new tube and other plug-in components that can be replaced easily. Check for good connections at the pins of a module board or an interconnecting cable. Power circuits often have a fuse, which should be tested.

Then voltage and resistance measurements can be made, using a multimeter (Fig. 28-1). Normal dc voltages, measured to chassis ground, are on the manufacturer's schematic diagram. In the signal circuits, the dc voltages are given for no ac signal input in order to eliminate the effects of AGC bias.

Dc voltage tests. Always check the dc supply voltages. Then measure the electrode voltages at the pins for tubes, transistors, and IC units. Normal cathode bias voltage on a tube generally means that the plate and screen-grid currents are correct. Similarly, normal emitter bias or source bias on a transistor amplifier indicates normal operation. Although the stage is usually amplifying ac signal, this function cannot be accomplished without the dc operating voltages.

A dc voltage that is too high indicates an open circuit. Look for an open resistor or coil. Also, the tube or transistor may be off, creating an open circuit.

A dc voltage that is too low indicates a short circuit. Look for a shorted capacitor. Also, the tube or transistor may be conducting too much current because of incorrect bias.

Resistance tests. With the power off, resistors can be checked for an open, which reads infinity

FIGURE 28-1 VOLT-OHM-MILLIAMETER OR VOM. ACCURACY IS 2 PERCENT FOR DC MEASUREMENTS. *(TRIPLETT CORPORATION)*

on the ohmmeter. An open coil winding or open fuse also has infinite resistance. Check capacitors for a short circuit, which reads zero on the ohmmeter. In a circuit that has B+ voltage, a shorted bypass or coupling capacitor changes the dc voltage distribution. However, an open capacitor does not affect the dc voltages.

For resistance checks in transistor circuits, be sure that the ohmmeter does not give a false indication by biasing a junction on. You can open any parallel paths or use reverse polarity for the ohmmeter leads.

Resistors generally do not short but can become open because of age or excessive current. When an open resistor has a bypass capacitor, look for a short in the capacitor. This allows too much current in the resistor, burning it open. A resistor also can be partially open, with too high a resistance. Metal-film resistors often open where the leads connect to the film.

Effect of an open circuit. Not only does an open circuit have infinite resistance, but its effect can be checked with a dc voltmeter. The voltage is too high at one side of an open circuit and zero at the other side. This voltage check can be used to test a resistor, coil, or fuse for an open.

Ac voltage tests. An oscilloscope is used for checking ac video signal and deflection voltages. Normal waveshapes with peak-to-peak amplitudes are shown in the manufacturer's service notes. In the color circuits, the waveforms are shown with signal from a color-bar generator.

Open capacitor. This possibility can be checked by temporarily bridging the suspected capacitor with a good capacitor in parallel. Capacitor substitution boxes for this purpose are available with a wide range of capacitance values. With an ohmmeter, an open capacitor does not show charging action.

Tube-socket adaptor. As shown in Fig. 28-2, this unit is an extension about 2 in. high. You plug the adaptor into the tube socket and the tube into the adaptor. Each pin on the adaptor has a tab for connecting the meter or oscilloscope. This way you can make connections to the tube pins from the top of the chassis.

Replacing components. Keep the same positions for connecting leads and ground returns. This *lead dress* is often critical for high frequen-

FIGURE 28-2 TUBE-SOCKET ADAPTOR. HEIGHT IS 2 IN.

cies because of feedback and with high voltages to prevent corona and arcing. When replacing a resistor, a higher wattage rating than the original can be used. Also, a higher dc voltage rating is permissible for capacitors, except for electrolytic capacitors, which need the specified forming voltage. Do not replace mica or ceramic capacitors with tubular capacitors, because their inductance can affect the coupling or bypassing for high frequencies.

On printed-wiring boards, an individual capacitor or resistor usually can be removed easily by clipping the leads close to the component. Then the new unit is just soldered to the old leads without disturbing the eyelet connections on the board. However, desoldering tools are available to suck out the solder at the eyelet. Just be careful not to apply excessive heat to the printed wiring.

Product safety. Many parts in television receivers have special safety-related characteristics to protect against arcing, breakdown, x-rays, or fire hazard. These features may not be evident just from visual inspection. Such special components are often identified by the manufacturer by a shaded area on the schematic diagram and notes on the parts list. When replacement is necessary, the new components should have the required safety characteristics.

28-2 Test Instruments

The main requirement is a volt-ohmmeter. The multimeter actually makes the test that tells which component is defective. An oscilloscope is used to observe the video, sync, and color signals, to check waveshapes in the deflection circuits, and for visual response curves. More test instruments and a brief description of their functions are listed in the following sections.

Multimeter. This unit can be a volt-ohm-milliammeter (VOM) or vacuum tube voltmeter (VTVM). However, the VTVM now is generally solid-state and may be battery-operated. This feature makes it portable, without the need for plugging into the ac power line. The VOM is a portable meter without any internal amplifier. When the meter does not operate from the power line, the black ground lead of the meter can be connected to any point in the receiver. Otherwise, the black lead must be at chassis ground or the meter readings may be incorrect.

The advantage of the VTVM type of dc voltmeter is its high input resistance of 11 MΩ on all ranges. A VOM generally has a sensitivity of 20,000 Ω/V. On the 100-V range, as an example, the input resistance is 100×20 kΩ = 2 MΩ. This value is really high enough to prevent meter loading in typical transistor circuits.

The dc voltmeter should have a low range of 1 to 3 V, full-scale, for transistor circuits. Then the meter can read bias voltages from base to emitter. Typical values are 0.2 to 0.7 V.

Oscilloscopes. The CRT in the oscilloscope shows a picture of the measured voltage. In order to observe color signals, response above 3.58 MHz is necessary for the vertical amplifier in the oscilloscope. Most units for this purpose have response up to 10 MHz.

Remember that 3.58 MHz is the frequency of the chroma signal before the demodulators. However, after detection, the color video signals have the bandwidth of 0 to 0.5 MHz. Most general purpose oscilloscopes can easily show the waveforms of $B-Y$, $R-Y$, $G-Y$, B, R, and G video signals.

For the internal horizontal deflection, or time base, some oscilloscopes have triggered sweep. This method provides better synchronization of the oscilloscope trace, compared with a free-running time base. In addition, the triggered sync allows the time base to be calibrated.

As an additional feature, the oscilloscope can be a dual-trace type. This CRT shows two waveforms, one above the other, synchronized in time for comparisons of phase angle or waveform distortion.

It should be noted, however, that simpler oscilloscopes can perform the basic functions in testing television receivers. For alignment, the oscilloscope needs response only for 60 Hz, as the waveform is a detected signal. In video signal waveforms, the main requirement is to see the peak-to-peak amplitude, with vertical or horizontal sync. In the deflection circuits, the waveforms have the frequencies of 60 Hz for vertical scanning and 15,750 Hz for horizontal scanning.

Many oscilloscopes have calibrated vertical deflection. Then you can read the peak-to-peak amplitude of the trace.

A low-capacitance probe (LCP) is generally needed to avoid distorting the waveshapes because of detuning. The LCP reduces the signal to the oscilloscope to $1/_{10}$ of the input to the probe.

It should be noted that in transistor circuits, the oscilloscope waveshapes at the collector are better than those at the base. The reason is that the base-emitter junction has a nonlinear E/I characteristic.

Dot-bar generator. This unit supplies black-and-white dots and crosshatch pattern for convergence adjustments, plus 3.58-MHz chroma signal for color bars. The chroma signal is used for checking the color circuits in the receiver.

Sweep generator. This unit is used to obtain visual response curves with an oscilloscope. In addition to the FM signal output, or sweep frequencies, the generator usually has built-in markers for the following alignment curves: rf, IF, 4.5-MHz sound, and 3.58-MHz color. A typical sweep generator is shown in Fig. 23-22 in the section on IF alignment.

Audio sine-wave or square-wave generator. This unit provides signal up to about 200 kHz for testing audio and video amplifiers. The square waves are useful for checking phase response.

Transistor tester. This unit tests for opens, shorts, leakage, and transconductance with both junction transistors and field-effect transistors.

Transistor curve tracer. This unit is used with an oscilloscope to provide a trace of the transistor characteristic curve for testing the transistor.

Dc power supply. This small unit provides B+ voltage for operating solid-state circuits, without the need for the power supply in the receiver.

FIGURE 28-3 TEST JIG WITH COLOR PICTURE TUBE. *(MAGNAVOX CO.)*

Picture tube tester-rejuvenator. Besides testing emission from the cathode of the CRT, the rejuvenator can sometimes boost cathode emission, burn out internal shorts, or weld connections.

Picture tube test jig. As shown in Fig. 28-3, this unit has a color picture tube, with all its associated circuits, to operate the receiver chassis on a test bench in the service shop.

28-3 DC Voltage Measurements

Signal voltage in an amplifier is actually a pulsating or fluctuating dc voltage. The reason is that the plate or collector needs the B+ voltage to provide voltage gain in the stage. A dc voltmeter reads the average dc level. The oscilloscope shows the ac signal variations. If the dc voltages are far from normal, the ac signal will usually not be correct. This factor is why dc voltage measurements are useful in localizing troubles in the ac signal.

In Fig. 28-4a, the fluctuating dc voltage represents signal at the plate or collector of a video amplifier. The variations in dc voltage result from signal variations in current through the load resistor. We can assume a B+ supply voltage of 140 V for this stage to allow for a signal swing between the peaks of 120 and 30 V. The average axis, shown at 50 V, is not through the middle of the voltage waveform because the signal is close to white for most of the cycle. Since a dc voltmeter responds only to the average level of the signal, the meter at the plate or collector reads 50 V, the average dc value. This dc component of 50 V is shown in Fig. 28-4b. The ac component of the signal consists of the variations around the average dc axis, as shown in Fig. 28-4c. In this example, the ac signal variations are between 120 and 30 V, for a peak-to-peak value of 90 V.

It should be noted that an ac-coupled oscilloscope shows just the signal variations in Fig. 28-4c, around the zero axis at the center of the screen. However, a dc-coupled oscilloscope shows the dc level also, as in a.

Zero voltage at the plate or collector. Assuming the stage has a circuit for B+ voltage, this trouble means the supply path is open. Look for an open in the load resistor or decoupling resistor. With a transformer in the B+ supply line, check for an open winding.

Full B+ voltage at the plate or collector. This trouble means the supply path is not open but there is no voltage drop by the plate or collector current. Assuming a resistance of 1 kΩ or more, either for the load or for isolation, no voltage drop means zero current. Typical causes are (1) a defective tube or transistor, (2) an open cathode or emitter circuit, (3) zero screen-grid voltage, (4) excessive negative bias at the control grid, and (5) zero forward bias at the base.

When a stage has a resistance plate load, full B+ at the plate or collector shows that the supply voltage is connected but there is no IR drop. Without plate or collector current, there cannot be any output signal. Stages with a

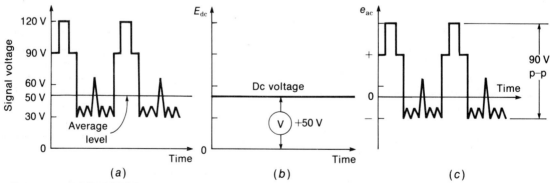

FIGURE 28-4 FLUCTUATING DC VOLTAGE AT PLATE OR COLLECTOR OF VIDEO AMPLIFIER. (a) AC SIGNAL COMBINED WITH AVERAGE DC LEVEL. (b) DC LEVEL OF 50 V MEASURED WITH DC VOLTMETER. (c) AC SIGNAL OF 90 V p-p, MEASURED WITH OSCILLOSCOPE.

resistance load include the video amplifier, sync amplifier, first audio amplifier, and multivibrator deflection oscillators.

It should be noted that in tuned stages, such as the rf amplifier, picture IF amplifier, 4.5-MHz sound IF amplifier, and 3.58-MHz chroma amplifier, there may not be any resistance to drop the supply voltage. Then full B+ voltage at the plate or collector is normal.

Zero screen-grid voltage. This trouble usually results when a shorted bypass capacitor causes excessive current through the screen-grid resistor, which then burns open. In most tubes designed for positive screen-grid voltage, without this accelerating potential there is no plate current and no output signal.

Cathode bias. An open cathode circuit opens the return path for plate current. The result is zero plate current and full B+ voltage at the plate. It is important to note that when you read the cathode voltage, the indication is not zero because the meter completes the cathode circuit. Since the meter resistance is much higher than the normal cathode resistance, the voltage reading is higher than normal instead of zero. Actually the dc voltmeter across the open cathode circuit reads the grid cutoff voltage of the tube.

Emitter bias. Like a cathode, the emitter circuit completes the path for collector current. An open emitter, therefore, results in zero output current and full B+ voltage at the collector.

Dc-coupled stages. In this case, high dc voltages at the input circuit of the tube or transistor are normal because there is no coupling capacitor to block the B+ voltage. As an example for a tube, the control grid may be at 105 V, but the cathode at 108 V results in a grid-cathode bias of $105 - 108 = -3$ V. Similarly, with the base at 30.6 V and the emitter at 30 V, the V_{BE} is 0.6 V for forward bias on an NPN transistor.

Grid-leak bias. For tubes, the negative dc voltage at the control grid indicates the amount of ac signal rectified by the grid-cathode circuit. Oscillator stages generally use grid-leak bias, produced by the feedback. The presence of the bias indicates the oscillator is operating; no bias means there is no feedback and the oscillator is not operating.

In an amplifier that uses grid-leak bias, this dc voltage indicates the amount of ac signal input. The horizontal output stage usually has grid-leak bias of -40 to -70 V, produced by drive from the horizontal oscillator. Stages that generally use grid-leak bias are listed in Table 28-1, with typical values. These voltages are preferably measured with a VTVM for high impedance.

Detected signal voltage. The dc voltmeter can also be used to measure rectified ac signal voltage out of a detector. Typical values are 1 to 3 V. For instance, rectified voltage out of the video detector shows IF signal input.

28-4 Oscilloscope Measurements

The oscilloscope is used for measuring peak-to-peak ac voltages. Besides showing the voltage waveform, the oscilloscope is necessary for measurements on a signal that is not a sine wave. Common uses in troubleshooting include checking video signal, chroma signal, sync pulses, and vertical or horizontal deflection voltages.

A typical oscilloscope is shown in Fig. 28-5, with the accessory probes in Fig. 28-6. Its

TABLE 28-1 TYPICAL GRID-LEAK BIAS VOLTAGES FOR TUBES

STAGE	BIAS, DC VOLTS	NOTES
Horizontal amplifier	−40 to −70	Shows drive from horizontal oscillator
Sync amplifier	−30	Receiver must have signal input
Horizontal MV	−20	At V2 grid
Vertical MV	−75	At V1 grid
Mixer	−2	Produced by oscillator injection voltage
Local oscillator	−3	May be less on high channels
Color oscillator	−3	Without burst input
Quadrature-grid detector	−5	Checks 4.5-MHz sound signal

vertical amplifier is calibrated so that peak-to-peak voltage can be read directly. The cable for signal input to the vertical amplifier is used with one of the following: low-capacitance probe (LCP), direct probe, or demodulator probe.

Direct probe. This is just an extension of the shielded lead, without any isolating resistance. Shielding is needed to prevent pickup of interfering signals. However, the shielded cable has a capacitance of about 60 pF. Also, the input capacitance of the oscilloscope is usually 10 pF. The total C is 70 pF. Therefore, the direct probe can be used only in circuits where this capacitance has little effect on the circuit to which the lead is connected. This usually means audio frequency circuits. In visual alignment, the direct probe can be used for detected sweep signal, as this frequency is 60 Hz.

Low-capacitance probe. This probe has an internal series R to isolate the capacitance of the cable and oscilloscope. Usually, the probe R is 9 times the internal R of the oscilloscope. Then the voltage divider provides one-tenth of the probe input to the oscilloscope. This means you must multiply the oscilloscope reading by 10 for the actual signal input at the probe. As an example, when the oscilloscope trace measures 8 V p-p with the LCP, then the actual signal voltage in the receiver circuit is 80 V p-p.

The low-capacitance probe must be used in high-frequency or high-resistance circuits. Also, the LCP is needed to observe the correct waveform of nonsinusoidal voltages. These factors mean that the LCP is used for video, sync, and deflection voltages, especially since there is usually enough signal to allow for 1:10 voltage division.

Demodulator probe. This probe has an internal crystal diode rectifier. Therefore, the probe can be used to detect the envelope of an amplitude-modulated IF or rf signal. Also, the probe rectifies the sweep-generator signal, if this is necessary for visual response curves in the video, chroma, or IF circuits. The polarity for the dc output voltage of the detector probe is usually negative. About 0.5 V or more signal is necessary for the diode to operate as a rectifier.

Current measurements. Although the oscilloscope is a voltage-operated device, current waveforms can be observed and peak-to-peak

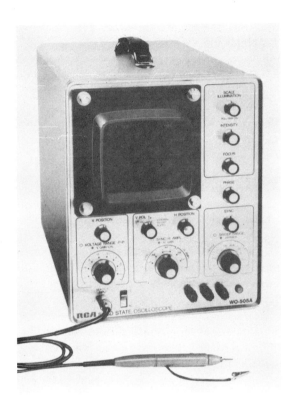

FIGURE 28-5 TYPICAL 5-IN. OSCILLOSCOPE WITH DIRECT/LOW-CAPACITANCE PROBE. *(RCA)*

measurements made the same way. This is done by inserting a known resistance in series with the circuit in such a manner that the current to be measured flows through the resistor. The voltage across this resistor is then coupled to the oscilloscope so that the waveshape can be observed and its amplitude measured as usual. With the voltage measured, I can be calculated as E/R. The current amplitudes are in peak-to-peak values, like the voltage amplitudes. As an example, suppose sawtooth deflection voltage across a 10-Ω R measures 6 V p-p. The current I then is $^6/_{10}$, equal to 0.6 A or 600 mA p-p.

The current waveshape is identical with the observed voltage waveshape. Current and voltage are in phase with each other in a resistance, and there is no change in waveshape.

The resistance to be inserted temporarily for the current measurement should be as small as possible so that it will not change the current in the circuit appreciably. In high-current circuits such as the deflection coils for transistors, a resistance of 1 Ω is high enough.

Frequency comparisons. In order to check frequencies with the oscilloscope, the unknown

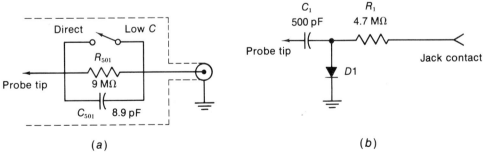

FIGURE 28-6 SCHEMATIC DIAGRAMS OF OSCILLOSCOPE PROBES. *(a)* DIRECT AND LOW-CAPACITANCE. *(b)* DEMODULATOR. *(RCA)*

signal voltage is compared with a voltage source of known frequency. Since we are usually interested in the scanning frequencies of 60 and 15,750 Hz, which are audio frequencies, an audio signal generator with an accurate frequency calibration is used. The usual procedure is as follows:

1. Connect the voltage of unknown frequency to the vertical input of the oscilloscope. Adjust the internal sweep frequency for a trace of two cycles on the screen.
2. Disconnect the voltage and connect in its place the output of the audio oscillator.
3. Adjust the audio oscillator frequency for two cycles on the screen. Do not change the oscilloscope frequency controls.
4. The frequency of the audio oscillator is now equal to the frequency of the unknown voltage previously connected, since they both produce a trace of two cycles with the same sweep frequency.

The setting of the gain controls does not matter in the frequency comparison, and they can be adjusted for a pattern of convenient size. The results can be very accurate when no sync is used for the oscilloscope sweep and the signal generator frequency is correct.

This same procedure can be used to check the accuracy of the signal generator at 60 and 15,750 Hz. For 60 Hz, compare the generator frequency with the ac power line. For 15,750 Hz, compare the generator frequency with the output of the horizontal deflection oscillator in a receiver where the picture is in horizontal sync, for monochrome. On a color broadcast, the horizontal line frequency is 15,734 Hz.

Phase-angle measurements. The oscilloscope can also be used to compare the phases of two sine-wave voltages of the same frequency. One sine-wave voltage is applied to the vertical input of the oscilloscope. At the same time, the other sine-wave voltage is applied to the horizontal input. The oscilloscope internal sweep is not used. Both the vertical and horizontal voltages should produce the same amplitude in the oscilloscope trace.

The combined trace of two sine-wave voltages will look like one of the patterns in Fig. 28-7. The phase angle θ can then be determined by calculating the ratio of the lengths B and A as shown, which gives the sine of the angle. For a

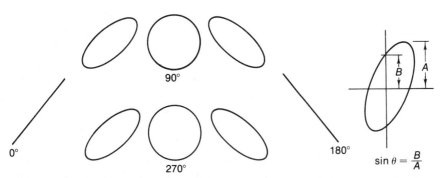

FIGURE 28-7 LISSAJOUS PATTERNS COMPARING PHASE BETWEEN TWO SINE WAVES OF THE SAME FREQUENCY. THE PHASE ANGLE IS θ.

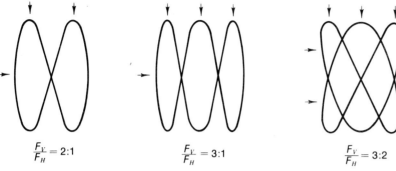

FIGURE 28-8 LISSAJOUS PATTERNS COMPARING FREQUENCIES OF TWO SINE WAVES. F_V IS FREQUENCY OF SIGNAL TO OSCILLOSCOPE VERTICAL INPUT; F_H IS FREQUENCY OF HORIZONTAL INPUT. ARROWS INDICATE CLOSED LOOPS COUNTED FOR THE FREQUENCY RATIO.

circle, B/A equals 1 for the sine, and the angle is 90°. When B is one-half A, the sine equals 0.5 and θ is 30°. Remember, though, that phase-angle measurements apply only for sine waves.

Lissajous patterns. The phase-angle patterns actually show a 1:1 frequency ratio. For higher ratios, the patterns in Fig. 28-8 are produced. Count the loops across either the top or bottom of the trace for the value of F_V, and count the loops at either side for F_H. The frequency ratio is then equal to F_V/F_H. Be sure to count only closed loops: open loops should not be counted at all.

As an example, assume the horizontal input is 60-Hz power-line voltage as a reference frequency. The vertical input is signal from an audio oscillator; the calibration of the oscillator's frequency dial is to be checked. For a circle or line pattern, where F_V/F_H is 1, the audio signal generator frequency is 60 Hz. With a ratio of 2 for F_V/F_H, the vertical input frequency is 120 Hz. Furthermore, frequencies that are not exact multiples can be compared using this method. For the case of F_V/F_H equal to $3/2$, in this example the vertical input is $3/2 \times 60 = 90$ Hz.

How to reproduce the picture on the oscilloscope screen. The cathode-ray tube in the oscilloscope can be used to see the picture on a green screen, if this is desired in checking a receiver. Note the following requirements:

1. The video signal output from the receiver is connected to the Z axis or intensity-modulation terminal of the oscilloscope. This terminal couples the signal to the control grid of the cathode-ray tube.
2. Sawtooth deflection voltage at 60 Hz from the receiver is connected to the vertical amplifier of the oscilloscope.
3. Sawtooth deflection voltage at 15,750 Hz is coupled to the horizontal amplifier of the oscilloscope, or the oscilloscope internal sweep can be used. With oscilloscope sweep, synchronization of the picture is difficult. With deflection voltage from the receiver, use an isolation resistor to prevent detuning of the horizontal oscillator.

The oscilloscope vertical and horizontal amplifiers then provide a scanning raster on the screen. Use the oscilloscope gain controls to adjust for convenient height and width. The video signal may produce a positive or negative picture, depending on polarity. A minimum of about 25 V p-p video signal is necessary.

Vectorscope. This test instrument combines a color-bar generator with an oscilloscope to show the phase angles of the hues. However, a separate generator can be used with a conventional oscilloscope to produce the vectorscope pattern. The output signals of the $R-Y$ and $B-Y$ demodulators are fed to the vertical and horizontal amplifiers of the oscilloscope. The trace is a circle with spokes or petals, as in Fig. 28-9. Each petal is a vector in the rosette pattern, corresponding to the phase angle of one of the color bars.

28-5 Alignment Curves

Figure 28-10 summarizes the visual response curves for different sections of the receiver, including chroma. The curves are shown upward on the oscilloscope screen for positive polarity. However, the correct curve may be down from the zero baseline, with negative input voltage to the oscilloscope. Be sure to have the curve in the polarity specified by the manufacturer's service notes. The wrong polarity can result from oscillations or overload distortion.

Which side of the curve corresponds to picture signal or sound signal is determined by the marker frequencies. The sweep frequencies may increase toward either the right or the left, depending on the generator. Some generators

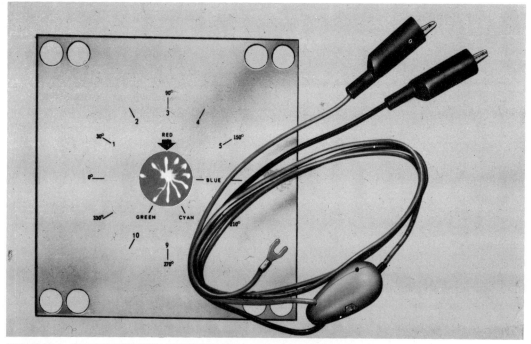

FIGURE 28-9 THE VECTORSCOPE TECHNIQUE WITH ACCESSORY PROBE AND TYPICAL ROSETTE PATTERN ON SCREEN, FOR OSCILLOSCOPE IN FIG. 28-5. *(RCA)*

have a frequency-reversing switch. Notice that the picture carrier with its video modulating frequencies is at one side of the curves, while the sound and chroma subcarrier frequencies are on the opposite side. The sound and chroma frequencies are very close to each other.

For correct curves, the operating conditions of the circuits must not be changed by the test equipment. Use typical bias voltages and signal levels. Do not connect the generator directly to any circuit being tested unless a matching pad is used.

The following checks can be made to be sure the curve you see is the actual receiver response:

1. Reduce the signal input to make sure that the curve just becomes smaller, without changing its shape.
2. A curve that is perfectly flat probably is produced by overload distortion because of too much signal in either the receiver or oscilloscope.
3. The curve should disappear when you turn off the signal generator. If the curve remains, the circuit is oscillating.
4. Reduce the marker amplitude to zero to see if the curve remains the same with or without the marker. If you see two markers that move in opposite directions when the marker frequency is varied, this usually means the curves were not phased correctly before the blanking was turned on.

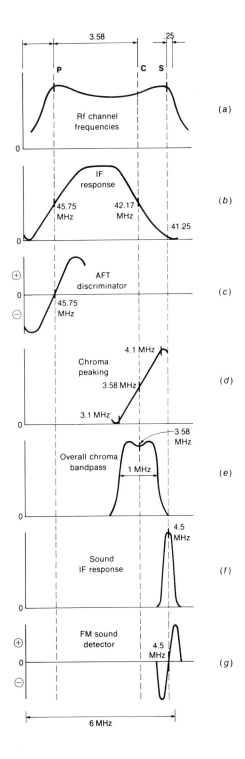

FIGURE 28-10 VISUAL RESPONSE CURVES FOR DIFFERENT SECTIONS OF TELEVISION RECEIVER. (a) RF AMPLIFIER; (b) PICTURE IF AMPLIFIER; (c) DISCRIMINATOR FOR AUTOMATIC FINE TUNING; (d) PEAKING FOR INPUT TO CHROMA BANDPASS AMPLIFIER; (e) OVERALL 3.58-MHz BANDPASS, INCLUDING CHROMA PEAKING AND IF AMPLIFIER; (f) SOUND IF AMPLIFIER; (g) FM SOUND DETECTOR.

682 BASIC TELEVISION

5. When the marker is too broad, it can be made narrower by connecting a 0.01-μF capacitor across the vertical input of the oscilloscope.
6. If the curve changes its shape when you touch the chassis, check the ground connections.
7. With weak signal input, the oscilloscope trace has "grass," which corresponds to snow in the picture, produced by receiver noise.

To summarize the requirements, you should have a clean curve without grass and with enough amplitude to see the markers clearly, and, most important, the curve should change only when you change the alignment adjustments.

28-6 Signal Injection

In this method, your own test signal is used in the receiver circuits to see the effect on the screen of the picture tube. Figure 28-11 illustrates how a signal generator can be connected to check the video, IF, and rf sections. Individual stages also can be tested. Working back from the picture tube to the antenna input, you can localize a stage that does not pass the signal to produce bars on the screen. Note that this method is the opposite of using the signal from the station and checking waveforms with the oscilloscope. Now your generator supplies the signal and the picture tube shows the output. It is not necessary to use a sweep generator, as only one frequency is used for each test.

Generator signals. The following frequencies are used for the different sections:

1. Audio signal for the video section. This can be taken from an audio signal generator or the 400-Hz audio modulation on an rf signal generator.
2. IF signal at about 43 MHz for the picture IF section. This signal must have audio modulation, usually at 400 Hz, to produce bars on the screen.
3. Rf signal for any of the channel frequencies also with audio modulation.

Audio signal for the video section can be injected at the test point in the output circuit of the video detector. The IF signal input can be

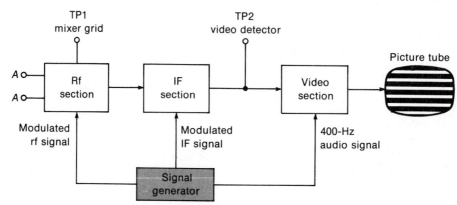

FIGURE 28-11 USING A GENERATOR TO INJECT TEST SIGNAL INTO RECEIVER CIRCUITS.

connected to the test point at the mixer grid, on the rf tuner. Finally, the modulated rf signal is connected to the antenna input terminals. Be sure the receiver is set for the channel frequency of the generator. The audio amplitude is generally 1 to 20 V. However, the rf output of the generator is usually less than 1 V.

Injecting 60-Hz deflection voltage. With tubes, the 60-Hz heater voltage can be connected to the grid of the vertical amplifier to check operation of the output stage when there is no vertical scanning. Some vertical deflection on the screen shows the amplifier is operating.

Patching methods. This technique uses signal from a normal receiver to test a defective receiver. A few possibilities are "borrowing" CRT anode voltage, horizontal drive for the output stage, vertical deflection voltage, and audio signal. Also, the IF output cable from the tuner can be plugged into another receiver. Extension cables can be used. Also, a jumper wire may be needed for a common chassis ground between the two receivers. Use an isolation transformer for a chassis that is not isolated from the power line.

28-7 Color-Bar Generators

A typical unit is shown in Fig. 28-12, with a similar generator in Fig. 11-3. These are larger units that have multiple test signals in addition to color bars. Smaller dot-bar generators are available to provide just bars, dots, and crosshatch

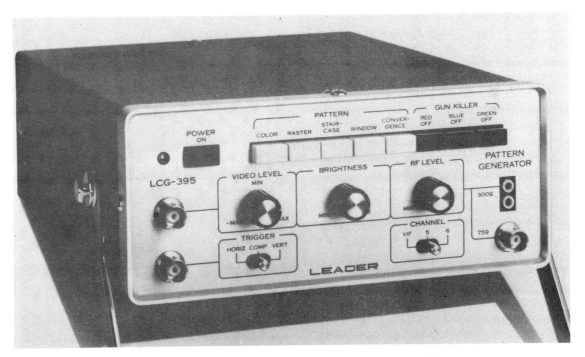

FIGURE 28-12 TYPICAL DOT-BAR GENERATOR. NOTE PATTERNS AVAILABLE FROM TOP ROW OF SWITCHES. *(LEADER INSTRUMENTS CORP.)*

684 BASIC TELEVISION

patterns. All these patterns are in black-and-white for setup adjustments on a color picture tube, except the color-bar pattern for testing the chroma circuits. The rainbow of color bars and their phase angles are shown in color plates XI and XII.

In addition to producing color bars, the generator in Fig. 28-10 has the following functions:

Dot-crosshatch pattern. In this pattern, the white dots and crosshatch are combined. This pattern is used for convergence adjustments on a color picture tube.

Staircase signal. This provides shades of white and gray to check black-and-white tracking.

Raster. A white, noise-free raster is provided for purity adjustments. If the receiver does not have a normal-service switch, the generator can be used to provide the raster without a picture.

Window signal. The white rectangle about one-half the screen area provides low-frequency video signal. Furthermore, the window is exactly in the center for centering adjustments.

Trigger signals. These are vertical and horizontal synchronizing pulses for external sync on the oscilloscope when observing waveforms produced by the generator.

Gun-killer switches. These can be used to cut off the red, green, or blue guns in the color picture tube without adjusting the screen-grid controls in most vacuum tube receivers using a delta-gun CRT.

Generator connections. Signal output is available from 300-Ω terminals for twin lead to the antenna terminals on the receiver. Most genera-

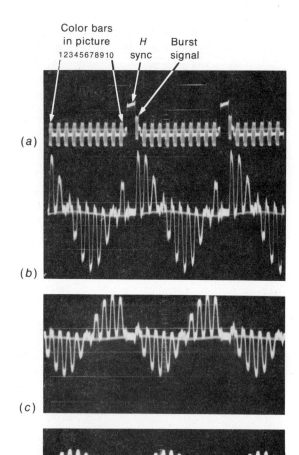

FIGURE 28-13 VIDEO SIGNAL WAVEFORMS WITH COLOR-BAR SIGNAL. RELATIVE AMPLITUDES ARE ALL TO THE SAME SCALE. (a) VIDEO DETECTOR OUTPUT; 1 V p-p FOR TRANSISTOR CIRCUITS. (b) DEMODULATOR OUTPUT FOR $-(B - Y)$ PHASE. MAXIMUM AMPLITUDE AT BARS 1 AND 6 FOR $B - Y$ AXIS AND MINIMUM AT BARS 3 AND 9 FOR $R - Y$ AXIS. (c) DEMODULATOR OUTPUT FOR $-(R - Y)$. MAXIMUM AMPLITUDE AT BARS 3 AND 9 FOR $R - Y$ AXIS AND MINIMUM AT BARS 1 AND 6 FOR $B - Y$ AXIS. (d) OUTPUT FOR $-(G - Y)$. MAXIMUM AMPLITUDE AT BARS 4 AND 10 FOR $G - Y$ AXIS AND MINIMUM AT BARS 1 AND 7 FOR AXIS 90° AWAY.

TABLE 28-2 COLOR BARS IN KEYED RAINBOW PATTERN

ANGLE*	BAR	HUE	NOTES
0°	None	Yellow-green	Used as burst signal. Bar is blanked
30°	1	Yellow-orange	First bar at left
60°	2	Orange	Phase of I
90°	3	Red	Phase of $R-Y$
120°	4	Magenta	Phase of $-(G-Y)$
150°	5	Red-blue	Phase of Q
180°	6	Blue	Phase of $B-Y$
210°	7	Green-blue	Complement of first bar
240°	8	Cyan	Phase of $-I$
270°	9	Blue-green	Phase of $-(R-Y)$
300°	10	Green	Phase of $G-Y$
330°	None	Yellow-green	Phase of $-Q$. Bar is blanked. H sync interval

*Phase angles are clockwise from burst.

tors supply modulated rf output on channel 3, 4, or 5. The receiver must be switched to the channel of the generator. The rf output level is 1 to 10 mV. In addition, IF output can be taken from a 75-Ω jack for coaxial cable to the IF section of the receiver. Finally, the video output can be connected directly to the video amplifier. The video level is 1 to 3 V, in either positive or negative sync polarity.

The gated rainbow. The color bars and their phase angles in 30° steps are listed in Table 28-2. The generator has an oscillator operating exactly 15,750 Hz below the color oscillator at 3.579545 MHz in the receiver. For every horizontal scanning line, therefore, the phase of the test signal shifts 360° with respect to the receiver oscillator. Furthermore, the generator has gating pulses to cut off the signal output every 30°. As a result, the generator output is divided into 30° intervals of bars with different hues. Each bar has a specific hue, 30° different from the next bar, with a black line between the bars for the time the signal is gated off.

Dividing 360° by 30° results in 12 intervals. However, only 10 bars are visible. This specific number of bars can be used to see if there is horizontal overscan in the receiver. The two missing bars are blanked out by horizontal blanking in the receiver. In the generator, a horizontal sync pulse is inserted at a time that would be the twelfth bar. The generator output that would be the first bar is used as a color sync burst signal, also during horizontal blanking time. You can see this timing of signals in Fig. 28-13. The oscilloscope waveforms in the manufacturer's service notes for the color section of the receiver are generally with input signal from a gated-rainbow generator.

Oscilloscope waveforms for color bars. The photos in Fig. 28-13 are taken with a dual-trace, triggered oscilloscope to show the relative timing of the test signals from a color-bar generator. In Fig. 28-13a, at the top is the input signal of the generator, after the video detector. This shows 3.58-MHz chroma input before the color demodulation. Note the timing of the horizontal sync pulse, burst, and 10 color bars.

Figure 28-13b shows the color video output of the $B - Y$ demodulator. Here there is no 3.58-MHz chroma but a geartooth of detected color video signal with a spoke for each bar. The demodulator output is maximum positive for the first bar. This hue is close to $-(B - Y)$ phase or burst phase, as listed in Table 28-2. The output is maximum negative for the sixth bar, which is $B - Y$ phase. This demodulator output is at a minimum or null at bars 3 and 9 for the $R - Y$ axis, in quadrature phase with the $B - Y$ axis.

Figure 28-13c shows the color video output of the $R - Y$ demodulator. Here the demodulated voltage is maximum at bars 3 and 9 for phase angles on the $R - Y$ axis. The null is at bars 1 and 6 for the axis of $B - Y$.

Figure 28-13d shows the output for $G - Y$ color video signal. Here the voltage is maximum at bars 4 and 10 for phase angles on the $G - Y$ axis. For hues 90° away, the resulting null is at bars 1 and 7.

NTSC color-bar generator. This test instrument produces bars for red, green, and blue, with their complementary colors, at 100 percent saturation. The synchronizing signals and blanking pulses at 15,734 Hz, with the color burst, are inserted exactly as in television broadcasting. Its function is testing broadcast equipment. However, the gated-rainbow generator, with hues 30° apart, is generally used for receiver servicing.

28-8 Schematic Diagram of Monochrome Receiver with Tubes

See Fig. 28-15 on foldout page. Note that the top and left borders have numbers and letters to mark zones for finding the components. For instance, the notation at the top "to +390 V (3C)" indicates a connection to boosted B+ of 390 V at the bottom right in zone 3C. In addition, the pictorial diagram in Fig. 28-14 shows a road map of the components on the printed-wiring board PW200. The numbers in circles or balloons for oscilloscope waveshapes 1 to 14 match these points on the schematic diagram. As an example, waveshape 10 for video detector output of 2 V p-p is taken at circle 10, which is also test point (TP) 3.

In the power supply at the bottom of the diagram, a full-wave bridge rectifier is used for B+ of 145 V. The power transformer T101 provides isolation from the ac line. All the heaters are in series, but they have direct current from the bridge rectifier. The diode CR1 blocks current in the heater line when the filter capacitors discharge. A 5-A fuse is in the primary of T101, and a 2.5-Ω fusible resistor protects the B+ line

In the signal circuits, starting at zone 7F, the IF cable from the tuner feeds the first IF amplifier. There are two IF stages driving the video detector CR201. The one video amplifier, using the pentode section of an 11LQ8 tube, supplies 80-V p-p signal for the grid of the picture tube. This signal is shown in oscilloscope waveform 1.

The contrast control in the plate circuit varies the amount of video signal. The brightness control varies the cathode bias on the picture tube, using the screen-grid voltage from the horizontal output tube in zone 5D.

The 4.5-MHz intercarrier sound signal is taken from the video detector output to the sound section at the top of the diagram. The sound circuits include one 4.5-MHz amplifier, a quadrature-grid detector, and the audio output stage for the loudspeaker.

The sync separator is the triode $V21B$ in zone 6F. This section uses the transistor $Q201$ as a noise gate in the cathode return of the sync separator.

The keyed AGC stage uses the triode $V205A$. Horizontal flyback pulses for the plate are coupled in at terminal Z on the board, from C_{109} in the damper circuit. Video signal input is dc-coupled to the control grid of the AGC triode from the video output stage. AGC bias is supplied directly to the first IF amplifier and through terminal B to the rf tuner.

The vertical deflection circuit in zone 4E uses the 21LR8 triode-pentode for a plate-coupled MV circuit. The vertical sawtooth capacitor is C_{238}, with a value of 0.033 μF. The vertical output transformer $T103$ is actually an autotransformer, with the bottom section connected to the plug for the deflection yoke. The secondary winding supplies feedback for the MV. It also provides vertical blanking pulses to the control grid of the picture tube.

The horizontal deflection oscillator in zone 7D uses the 8FQ7 twin-triode in a cathode-coupled MV circuit. The sync discriminator with diodes $SR201$ and $SR202$ forms the horizontal AFC circuit. The sawtooth capacitor is the 180-pF capacitor C_{258}, in series with R_{265} for trapezoidal driving voltage to the grid of the horizontal amplifier. The width control is a series of taps for screen-grid voltage on the 22JR6 output tube.

In zone 4D, $T102$ is the horizontal output transformer. Terminal 3 goes to the plate of the horizontal amplifier, terminal 2 goes to the 17CT3 damper diode, and terminal 1 supplies current to the horizontal scanning coils through the yoke plug. The boosted B+ of 390 V is available at terminal 4, with $C107$ used as the boost capacitor. The high-voltage winding at the top of $T102$ supplies ac input to the 1B3/1G3, which produces anode voltage of 20 kV for the picture tube. In addition, the auxiliary winding at terminals 5 and 6 provides horizontal blanking pulses to the CRT screen grid.

28-9 Schematic Diagram of Solid-State Color Receiver

See Fig. 28-16 on foldout pate with oscilloscope waveforms. Starting at the top left in the diagram, the IF cable supplies signal from the VHF tuner. The IF amplifiers are integrated circuits in $IC101$. The 14 pins are numbered on the inside of the IC symbol, with dc voltages on the outside. Where two voltages are shown, the top value is with signal.

$IC101$ also includes the AGC circuits. At pin 6, the IF bias is taken out for adjustment by R_{122}. The rf bias adjustment is at pin 13. Also, the AGC bias from pin 12 is amplified by $Q100$ to provide enough AGC action for the rf amplifier on strong signals.

A separate sound converter is used, in the path with $Q103$. This video IF amplifier is not for video signal. It amplifies the IF signal to feed the 4.5-MHz sound detector $Y103$. In addition, the IF signal from $Q103$ is coupled to the AFC module for automatic fine tuning.

The 4.5-MHz intercarrier sound signal from Y103 is amplified and detected in the integrated circuit IC301. The quadrature coil is L_{301}. Q301 is the audio output stage driving the loudspeaker.

In the video circuits, Y102 is the video detector. The video preamplifier includes the transistors Q104, Q105, and Q106. The video driver stage Q109 supplies Y signal to the RGB drive section for matrixing with the color video signal. In the video preamplifier circuits, the delay line for Y signal is between Q108 and Q107. The chroma takeoff for 3.58-MHz signal is from the emitter of Q104.

In the chroma section, Q501, Q502, and Q503 make up the 3.58-MHz bandpass amplifier. The $B - Y$ and $R - Y$ demodulators, with the $G - Y$ adder, are in the integrated circuit IC501. Then the blue, red, and green amplifiers are used to provide enough color video signal to drive the three control grids of the picture tube. The Y signal is injected into the emitter of each of these stages. which are Q600, Q604, and Q606.

The three cathodes of the CRT return to ground through the drive controls for red, green, and blue, with the common beam-current limiter stage Q113. Each drive control varies the amount of degeneration for the input signal at the control grid. Q113 limits the beam current by controlling the cathode bias on the CRT.

The burst separator is Q504 at the bottom of the diagram. Signal input is from the emitter of the chroma amplifier Q502. The separated burst rings the crystal at 3.579545 MHz. This output is amplified by Q505, Q506, and Q507. Then the subcarrier cw signal is coupled by T502 to IC501 at pins 6 and 7, for the $B - Y$ and $R - Y$ demodulators.

The automatic chroma control (ACC) uses Q653 and Q652 to control the gain of Q501 in the chroma section. Input signal proportional to burst amplitude is obtained from the crystal X501 in the subcarrier regenerator circuit. The color killer stage is Q650. Input is obtained from Q507 in the subcarrier generator circuit. The output of the color killer stage can cut off the chroma amplifier through Q652 in the ACC section.

In the vertical deflection circuits, the 0.47-μF capacitor C_{268} at the collector of Q262 is the sawtooth capacitor. Q263 and Q264 are driver stages for the pair of vertical output transistors Q267 and Q268. This output stage is a push-pull circuit using complementary symmetry with NPN and PNP transistors.

In the horizontal deflection circuits, Q203 is a sine-wave oscillator. Its frequency is corrected by the reactance-control stage Q202. The sync discriminator for horizontal AFC stage consists of Y201 and Y202. The driver stage Q205 feeds the horizontal output stage Q206 at the bottom of the diagram. The diode Y206 is the damper. The timer stage Q204 provides the correct pulse width in the waveform for driving the power output transistor.

The high-voltage rectifier is Y210, in the

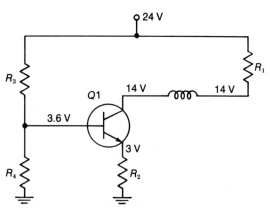

FIGURE 28-17 NORMAL DC VOLTAGES ON NPN SILICON TRANSISTOR FOR CE AMPLIFIER WITH R IN COLLECTOR CIRCUIT.

output of the horizontal output transformer T204. The focus voltage is obtained by tapping down on the anode supply for the CRT.

The low-voltage power supply is at the bottom left on the diagram. Y400 is a half-wave rectifier for most of the supply voltages. However, the scan-rectified output of 22 V is produced by Y402 from an auxiliary winding at terminal 8 on the horizontal output transformer. A series-regulator circuit for this 22-V output keeps the supply voltage constant for the transistors in the signal circuits. Note that in the automatic degaussing circuit R404 is a positive-temperature-coefficient resistor (P.T.C.R.). The CRT heater winding of 6.3 V is wired for instant-on operation.

28-10 Transistor Troubles

In most cases, a defective transistor has an open lead internally or a shorted junction. These troubles usually are the result of overheating or excessive voltage, even temporarily. The transistor also can develop excessive leakage, which reduces gain and makes the transistor noisy. Opens and shorts can be checked for with a multimeter.

Ohmmeter tests. Avoid the $R \times 1$ scale because it often has too much current for low-power junctions. To check a semiconductor diode out of the circuit, just measure the forward and reverse resistances by reversing the ohmmeter leads. In the forward direction the resistance should be low, usually about 100 Ω or less. In the reverse direction, the resistance is much higher. For silicon junctions, the reverse resistance is practically infinite because there is so little leakage current. If the resistance is very low in both directions, the junction must be shorted. If there is infinite resistance in both directions, the diode is open.

For a PNP or NPN transistor, consider it as two diode junctions. One is from base to emitter and the other from base to collector. To check the emitter junction, measure the forward and reverse resistances to the base. Also, measure the forward and reverse resistances from the base to the collector to check this junction.

For an SCR, the resistance from gate to cathode should be checked like a diode junction. However, from the anode to the cathode or gate, the resistance is 10 MΩ or more. The reason is that for either polarity of the ohmmeter at the anode, one of the internal junctions is reverse-biased.

With an FET, the gate electrode of the junction type can be tested as a diode with the source. However, with an insulated gate, the resistance to the source or drain is practically infinite in either polarity. For both types of FET, the resistance from drain to source is the R of the channel. This value is usually in kilohms, but check against a good FET of the same type.

Transistor voltage tests. In the circuit, the main question is usually whether the transistor is conducting or not. We assume a class A stage, where the amplifier conducts with or without signal. Consider the CE amplifier in Fig. 28-17 with normal dc operating voltages. The collector voltage is 14 V because of the 10-V drop across R_1 from the 24-V supply. This IR drop is produced by collector current. Note that there is no voltage drop across L_1 because of its low dc resistance.

The emitter voltage is 3 V because of emitter current through R_2. The R_3R_4 voltage divider provides the base voltage of 3.6 V. The net forward bias here is $3.6 - 3 = 0.6$ V. This forward bias from base to emitter allows collector current to flow.

When the collector circuit has resistance, the current can be checked by measuring the voltage drop. With a VOM, measure directly across the resistance. Then $I_C = V/R$. Typical values of I_C are 1 to 10 mA for small-signal transistors rated at 250 mW or less.

An important test is to measure the forward bias. With a VOM, measure directly from base to emitter. A typical bias is 0.6 V for silicon transistors. Any voltage above 0.7 V, approximately, indicates the junction is open.

A quick test of whether the transistor is operating can be done by temporarily shorting the base to emitter. Without forward bias, the transistor is cut off. Normal operation is indicated when the temporary short changes the output. However, be careful not to short the collector voltage to the base.

SUMMARY

1. Common troubles in receivers include defective tubes, capacitors, resistors, and semiconductor devices.
2. Resistors generally open rather than short. The open circuit can change the dc voltages. Coils and transformers also can open.
3. Capacitors can be either open or shorted. A shorted capacitor affects the dc voltages, but an open capacitor does not.
4. The main items of test equipment for receiver servicing are a multimeter, oscilloscope, and color-bar generator, with a sweep generator for visual alignment if desired.
5. Typical alignment curves for the television receiver are shown in Fig. 28-10.
6. The dc voltmeter is used to measure electrode voltages on tubes and transistors to localize a defective component. Normal dc voltages are shown on the manufacturer's schematic diagram.
7. The ohmmeter is used to check for an open resistor or shorted capacitor. An open capacitor does not show charging action.
8. Transistors and diodes can be checked out of the circuit with an ohmmeter. Test the PN junction with forward and reverse voltages by reversing the ohmmeter leads. A very low reading with both polarities indicates a short; a very high reading in both directions indicates an open.
9. The oscilloscope is used for ac voltage measurements in the video, color, sync, and deflection circuits. Typical peak-to-peak values are shown by the waveforms for receiver circuits in Figs. 28-15 and 28-16.
10. The three oscilloscope probes are used as follows: (a) Low-capacitance probe (LCP) for most measurements. Multiply the reading by 10 to allow for the voltage division by the LCP. (b) Direct probe for visual alignment curves where the sweep generator signal is demodulated. (c) Demodulator probe for IF, rf, and chroma circuits before demodulation.

Self-Examination (Answers at back of book)

Match the items in the column at the right with those at the left.

1. Infinite ohms
2. Zero ohms
3. Low-capacitance probe
4. V_{BE} of zero
5. V_{BE} of 0.6 V
6. Zero collector current
7. 2 V p-p video signal
8. −50 V grid-leak bias
9. Hues every 30°
10. Horizontal bars on screen
11. One pair of bars on screen
12. No raster and no sound

(a) Open collector circuit
(b) Cutoff bias
(c) Shorted capacitor
(d) Gated rainbow generator
(e) Open resistor
(f) 1:10 voltage divider
(g) Open in series heater string
(h) Forward bias
(i) 60-Hz hum
(j) Audio signal in video amplifier
(k) Horizontal output tube
(l) Detector output

Essay Questions

1. Give the uses for five types of test equipment.
2. Give two comparisons between a VOM and a VTVM.
3. Compare two types of oscilloscopes.
4. How is a vectorscope used?
5. Give the uses of the three probes for an oscilloscope.
6. Describe five ways to use a dot-bar generator.
7. How can you check an open fuse with a voltmeter while power is on?
8. How much is the resistance of a voltmeter on the 3-V range if it has a sensitivity of 20,000 Ω/V? Why is the resistance the same for any reading on this range?
9. What is the internal sweep frequency of the oscilloscope required to see two cycles of the following waveshapes: (a) vertical oscillator output; (b) video signal with vertical sync; (c) horizontal oscillator output; (d) video signal with horizontal sync and color burst?
10. Which oscilloscope probe would you use at the following points: (a) grid of horizontal amplifier; (b) video signal at grid or cathode of picture tube; (c) video detector output for IF alignment; (d) output of bandpass amplifier for chroma alignment?
11. Compare the methods of taking off the 4.5-MHz intercarrier sound signal for the monochrome receiver in Fig. 28-15 and the color receiver in Fig. 28-16.
12. How is the chroma signal obtained for the 3.58-MHz bandpass amplifier in Fig. 28-16?

692 BASIC TELEVISION

13. Refer to the monochrome receiver in Fig. 28-15. (a) List four test points and their uses. (b) Label the two fuses and give their functions. (c) Describe the setup adjustments for height and width. (d) Give the functions for T102 in zone 4D; SR201, SR202, and L_{207} in zone 7D; T103 in zone 3E; T201A in zone 7G.
14. Refer to the color receiver in Fig. 28-16. (a) List eight sections in the circuit diagram and the function of each. (b) In which circuit is the 4-A fuse? (c) Describe the circuits for drive and screen-grid adjustments on the red, green, and blue guns in the picture tube. (d) Give the functions for T204 in the horizontal section; T201 in the audio section; X501 in the color section; Y201 and Y202 in the sync section; Y102 in the video section. (e) Give the functions for transistors Q503, Q507, Q600, Q604, Q606, Q504, and Q652.

Problems (Answers to selected problems at back of book)

1. Refer to the CE amplifier in Fig. 28-18 with R_L in the collector circuit. Calculate: (a) Normal V_C; (b) Normal V_{BE}; (c) V_C with R_E open.
2. Refer to the CE amplifier in Fig. 28-19 with the tuned LC circuit as the collector load. (a) How much is the normal V_C? (b) Normal V_{BE}? (c) How much will a voltmeter read at the emitter to ground with R_E open?

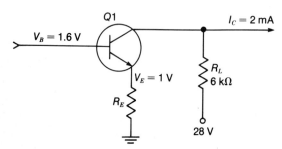

FIGURE 28-18   FOR PROBLEM 1.

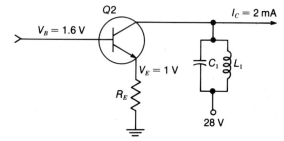

FIGURE 28-19   FOR PROBLEM 2.

A Television Channel Frequencies

CHANNEL NUMBER	FREQUENCY BAND, MHz	PICTURE CARRIER FREQUENCY, MHz	SOUND CARRIER FREQUENCY, MHz	CHANNEL NUMBER	FREQUENCY BAND, MHz	PICTURE CARRIER FREQUENCY, MHz	SOUND CARRIER FREQUENCY, MHz
2	54–60	55.25	59.75	43	644–650	645.25	649.75
3	60–66	61.25	65.75	44	650–656	651.25	655.75
4	66–72	67.25	71.75	45	656–662	657.25	661.75
5	76–82	77.25	81.75	46	662–668	663.25	667.75
6	82–88	83.25	87.75	47	668–674	669.25	673.75
7	174–180	175.25	179.75	48	674–680	675.25	679.75
8	180–186	181.25	185.75	49	680–686	681.25	685.75
9	186–192	187.25	191.75	50	686–692	687.25	691.75
10	192–198	193.25	197.75	51	692–698	693.25	697.75
11	198–204	199.25	203.75	52	698–704	699.25	703.75
12	204–210	205.25	209.75	53	704–710	705.25	709.75
13	210–216	211.25	215.75	54	710–716	711.25	715.75
14	470–476	471.25	475.75	55	716–722	717.25	721.75
15	476–482	477.25	481.75	56	722–728	723.25	727.75
16	482–488	483.25	487.75	57	728–734	729.25	733.75
17	488–494	489.25	493.75	58	734–740	735.25	739.75
18	494–500	495.25	499.75	59	740–746	741.25	745.75
19	500–506	501.25	505.75	60	746–752	747.25	751.75
20	506–512	507.25	511.75	61	752–758	753.25	757.75
21	512–518	513.25	517.75	62	758–764	759.25	763.75
22	518–524	519.25	523.75	63	764–770	765.25	769.75
23	524–530	525.25	529.75	64	770–776	771.25	775.75
24	530–536	531.25	535.75	65	776–782	777.25	781.75
25	536–542	537.25	541.75	66	782–788	783.25	787.75
26	542–548	543.25	547.75	67	788–794	789.25	793.75
27	548–554	549.25	553.75	68	794–800	795.25	799.75
28	554–560	555.25	559.75	69	800–806	801.25	805.75
29	560–566	561.25	565.75	70	806–812	807.25	811.75
30	566–572	567.25	571.75	71	812–818	813.25	817.75
31	572–578	573.25	577.75	72	818–824	819.25	823.75
32	578–584	579.25	583.75	73	824–830	825.25	829.75
33	584–590	585.25	589.75	74	830–836	831.25	835.75
34	590–596	591.25	595.75	75	836–842	837.25	841.75
35	596–602	597.25	601.75	76	842–848	843.25	847.75
36	602–608	603.25	607.75	77	848–854	849.25	853.75
37	608–614	609.25	613.75	78	854–860	855.25	859.75
38	614–620	615.25	619.75	79	860–866	861.25	865.75
39	620–626	621.25	625.75	80	866–872	867.25	871.75
40	626–632	627.25	631.75	81	872–878	873.25	877.75
41	632–638	633.25	637.75	82	878–884	879.25	883.75
42	638–644	639.25	643.75	83	884–890	885.25	889.75

B FCC Frequency Allocations

FCC FREQUENCY ALLOCATIONS FROM 30 kHz to 300,000 MHz

BAND	ALLOCATION	REMARKS
30–535 kHz	Includes maritime communications and navigation	500 kHz is international distress frequency
535–1,605 kHz	Standard radio broadcast band	AM broadcasting
1,605 kHz–30 MHz	Includes amateur radio and international short-wave broadcast	Amateur bands 3.5–4.0 MHz and 28–29.7 MHz
30–50 MHz	Government and nongovernment, fixed and mobile	Includes police, fire, forestry, highway, and railroad services
50–54 MHz	Amateur radio	6-m band
54–72 MHz	Television broadcast channels 2 to 4	Also fixed and mobile services
72–76 MHz	Government and nongovernment services	Aeronautical marker beacon on 75 MHz
76–88 MHz	Television broadcast channels 5 and 6	Also fixed and mobile services
88–108 MHz	FM broadcast	Also available for facsimile broadcast; 88–92-MHz educational FM broadcast
108–122 MHz	Aeronautical navigation	Localizers, radio range, and air traffic control
122–174 MHz	Government and nongovernment, fixed and mobile, amateur	144–148-MHz amateur band
174–216 MHz	Television broadcast channels 7 to 13	Also fixed and mobile services
216–470 MHz	Amateur, government and non-government, fixed and mobile, aeronautical navigation	Radio altimeter, glide path, and meteorological equipment; civil aviation 225–400 MHz
470–890 MHz	Television broadcasting	UHF television broadcast channels 14 to 83; translator stations in channels 70 to 83
890–3,000 MHz	Aeronautical radio navigation, amateur, studio-transmitter relay, government and nongovernment, fixed and mobile	Radar bands 1,300–1,600 MHz; educational television 2,500–2,690 MHz; microwave ovens at 2,450 MHz
3,000–30,000 MHz	Government and nongovernment, fixed and mobile, amateur, radio navigation	Superhigh frequencies (SHF); radio relay; Intelsat satellites
30,000–300,000 MHz	Experimental, government, amateur	Extremely high frequencies (EHF)

Radio-Frequency Bands

The main groups of radio frequencies and their wavelengths are:

- VLF = Very low frequencies, 3 to 30 kHz; wavelengths 100 to 10 km
- LF = Low frequencies, 30 to 300 kHz; wavelengths 10 to 1 km
- MF = Medium frequencies, 0.3 to 3 MHz; wavelengths 1,000 to 100 m
- HF = High frequencies, 3 to 30 MHz; wavelengths 100 to 10 m
- VHF = Very high frequencies, 30 to 300 MHz; wavelengths 10 to 1 m
- UHF = Ultrahigh frequencies, 300 to 3,000 MHz; wavelengths 1,000 to 100 mm
- SHF = Superhigh frequencies, 3 to 30 GHz; wavelengths 100 to 10 mm
- EHF = Extra-high frequencies, 30 to 300 GHz; wavelengths 10 to 1 mm

Note that wavelengths are shorter for higher frequencies. Also, 1 gigahertz (GHz) = 1,000 MHz.

Microwaves have wavelengths of 1 m down to 1 mm (300 GHz). The spectrum of light rays starts at frequencies of 300 GHz and up, with infrared radiation having wavelengths from 1 mm to 10 μm.

Radio Services and Television Interference

Principal services that can cause interference in television receivers are in the following list. The interference is generally at a harmonic frequency or image frequency. All frequencies are in MHz.

Amateur radio. These bands include 1.8 to 2, 3.5 to 4, 7 to 7.3, 14 to 14.25, 21 to 21.45, 28 to 29.7, 50 to 54, 144 to 148, 220 to 225, 420 to 450.

Industrial, scientific, and medical. This band is 13.36 to 14, with diathermy equipment at 13.56. The old frequencies were 27.12 and 40.66 to 40.70.

Citizens radio. These bands include: 26.48 to 28, class D with 23 channels; 220 to 225, class E (proposed); 450 to 470, class A with 16 channels.

FM broadcast. This band is 88 to 108, with 100 channels spaced 0.2 MHz apart.

Public safety (police, fire, etc.). These frequencies are around 33, 37, 39, 42, 44, 154, 156, and 158.

C Television Systems Around the World

Many countries use television standards that are not the same as in the United States. These systems are listed in Table C-1, along with our own FCC standards. The field rate of 50 Hz for vertical scanning is used where this is the ac power-line frequency, instead of 60 Hz as in this country. Note that the combination of 625 lines per frame and 25 frames per second, as used in Western Europe, results in the line-scanning frequency of 15,625 Hz, which is very close to the 15,750 Hz in our standards. The scanning frequencies for color television are not listed, but they are very close to the monochrome standards, for compatibility. In all cases, odd-line interlace is used with two fields per frame, the aspect ratio is 4:3, and AM is used for the picture carrier with multiplexing of the color subcarrier.

TABLE C-1 PRINCIPAL TELEVISION SYSTEMS

	NORTH AND SOUTH AMERICA; INCLUDES U.S., CANADA, MEXICO, AND JAPAN	WESTERN EUROPE; INCLUDES GERMANY, ITALY, AND SPAIN	ENGLAND*	FRANCE†	U.S.S.R.
Lines per frame	525	625	625	625	625
Frames per second	30	25	25	25	25
Field frequency, Hz	60	50	50	50	50
Line frequency, Hz	15,750	15,625	15,625	15,625	15,625
Video bandwidth, MHz	4.2	5 or 6	5.5	6	6
Channel width, MHz	6	7 or 8	8	8	8
Video modulation	Negative	Negative	Negative	Positive	Negative
Sound signal	FM	FM	FM	AM	FM
Color system	NTSC	PAL	PAL	SECAM	SECAM
Color subcarrier, MHz	3.58	4.43	4.43	4.43	4.43

*England also uses 405-line system in 5-MHz channel.
†France also uses 819-line system in 14-MHz channel.

D RC Time Constant

Time Constant and Transient Response

In capacitive and inductive circuits, the voltage or current that results with a sudden change of applied voltage is called the *transient response*. This application is important for the nonsinusoidal waveshapes used in sawtooth deflection, sync pulse circuits, and square-wave testing.

RC charge. Figure D-1a shows a circuit for charging C through the series R by means of the 100-V battery. Note that the battery and switch are used here to provide a square-wave pulse of applied voltage. The leading edge of such a pulse means that maximum voltage is applied. Furthermore, the voltage is maintained for the time of the pulse width. This input voltage corresponds to closing the switch in the simplified circuit. The trailing edge of the pulse is equivalent to removing the applied voltage by opening the switch.

When the switch is closed, current flows in the direction shown to charge C. Since the voltage across C is proportional to its charge Q, a small voltage v_C appears across the capacitor. Remember that $Q = CV$ for a capacitor; or the charge develops a potential difference V equal to Q/C. If the charging continues until C is fully charged, v_C will equal the battery voltage. When the voltage across C equals the applied voltage, the charging current drops to zero.

Initially, when the switch is closed the amount of current flowing is maximum, since at that time there is no voltage across C to oppose the applied voltage. Therefore, C charges most rapidly at the beginning of the charging period. As C becomes charged, the capacitor voltage opposing the battery becomes greater and the net applied voltage for driving current through the circuit becomes smaller. Plotting the capacitor voltage v_C against time as it charges, the RC charge curve of Fig. D-1a is obtained. The capacitor voltage increases at a decreasing rate, as shown by the exponential curve, because the charging current decreases as the amount of charge on the capacitor increases.

The voltage across the resistor v_R is always equal to $i \times R$. This voltage has the same waveshape as the current because the resistance is constant, making the voltage vary directly with the current. When the capacitor is completely charged, the current is zero and the voltage across the resistor is also zero.

At any instant the voltage drops around the circuit must equal the applied voltage. Or, v_R must be equal to $E - v_C$. As an example, with 100 V applied, at the time when C is charged to 63 V, then v_R is 37 V.

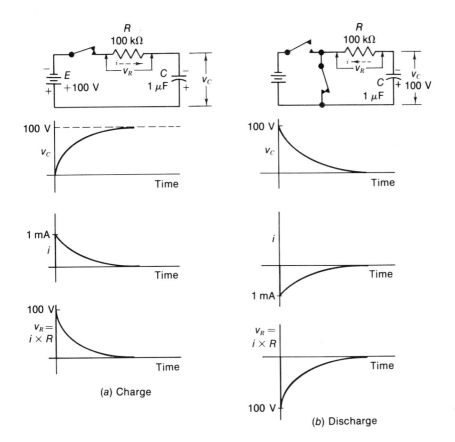

FIGURE D-1

RC discharge. Figure D-1b shows a circuit for discharging C through the same R. Let the voltage across C be 100 V at the instant the switch is closed for discharge. As electrons from the negative plate of the capacitor flow around the circuit to the positive plate, C loses its charge and v_C decreases exponentially as shown. The voltage decreases most rapidly at the beginning of the discharge because v_C, now acting as the applied voltage, then has its highest value and can drive maximum discharge current around the circuit.

The discharge current has its peak value of v_C/R at the first instant of discharge. Its direction is opposite to the current flow on charge because the capacitor is now acting as the generator and is connected on the opposite side of the series resistance R. As C discharges, its current decreases because of the declining value of v_C. After C is completely discharged the current is zero. The voltage across R is equal to $i \times R$ and has the same waveshape as the current.

RC time constant. The series R limits the amount of current. Therefore, higher resistance results in a longer time for charge or discharge. Larger values of C also require a longer time. A convenient measure of the charge or discharge time of the circuit, therefore, is the RC product. The time constant is defined as

$R \times C = t$

where C = capacitance in farads, R = resistance in ohms in series with the charge or discharge current, and t is the time constant in seconds. As an example, with 1-μF C in series with 0.1-mΩ R, the time constant equals 0.1 s.

The time constant states the time required for C to charge to 63 percent of the applied voltage. For discharge, the time constant states the time required for C to discharge 63 percent, or down to 37 percent of its original voltage. In our example, C charges 63 percent in 0.1 s. If the applied voltage is 100 V, after 0.1 s of charge v_C will be 63 V. If the capacitor discharges from 100 V, then v_C will be 37 V after 0.1 s. The capacitor is practically completely charged to the applied voltage after a time equal to five time constants. On discharge, the capacitor voltage is practically down to zero after five time constants. See the universal charge and discharge curves in Fig. D-3.

Note the following facts about charge or discharge of the capacitor. When the applied voltage is more than v_C the capacitor will charge. The capacitor will continue to charge as long as the applied voltage is maintained and is greater than v_C. The rate of charging is determined by the RC time constant. When v_C equals the applied voltage, however, the capacitor cannot take on any more charge, regardless of the time constant. Furthermore, if the applied voltage decreases, the capacitor will discharge. The capacitor will discharge as long as its voltage is greater than the applied voltage. When the capacitor voltage discharges down to zero it cannot discharge any more.

In summary, the capacitor charges as long as the applied voltage is greater than v_C. Similarly the capacitor discharges as long as v_C is able to produce discharge current. Note that if C is discharging and the applied voltage changes to become greater than v_C, the capacitor will stop discharging and start charging.

Also, it should be noted that the 63 percent change of voltage in RC time refers to 63 percent of the net voltage available for producing charging current on charge, or discharging current on discharge. As an example, if 100 V is applied to charge a capacitor that already has 20 V across it, the capacitor voltage will increase by 63 percent of 80 V, in one RC time, adding 50.4 V. The 50.4 V added to the original 20 V produces 70.4 V across the capacitor.

L/R time constant. When voltage is applied to an inductive circuit, the current cannot attain its steady-state value instantaneously but must build up exponentially because of the self-induced voltage across the inductance. This rise of inductive current corresponds to the rise of voltage across a capacitor (see Fig. D-2). The transient waveshapes are exactly the same, with current in the inductive circuit substituted for capacitor voltage.

Since the voltage across a resistor is $i \times R$, the v_R voltage with inductive current corresponds to the v_C voltage. This parallel between inductive current and capacitor voltage applies to either a current rise or decay, corresponding to the capacitor charge or discharge.

The time constant for an inductive circuit is given by the formula

$\dfrac{L}{R} = t$

where L is in henrys, R in ohms, and t in seconds. This time constant indicates the same 63 percent change as for the RC time constant, but

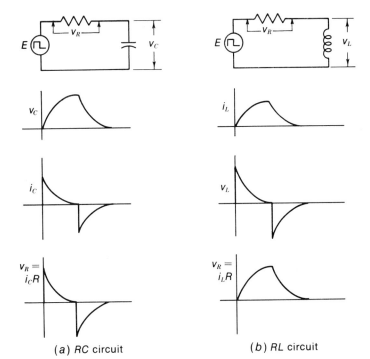

FIGURE D-2 (a) RC circuit (b) RL circuit

for current in the RL circuit. Also, the transient response is completed in five time constants. However, note that higher resistance provides a shorter time constant for the current rise or decay in an inductive circuit.

Universal time-constant graph. With the curves in Fig. D-3 you can determine voltage and current values for any amount of time. The rising curve a shows how the voltage builds up across C as it charges in an RC circuit; the same curve applies to the current increasing in the inductance for an RL circuit. The decreasing curve b shows how the capacitor voltage declines as C discharges in an RC circuit or the decay of current in an inductance.

Note that the horizontal axis is in units of time constants, rather than absolute time. The time constant is the RC product for capacitive circuits but equals L/R for inductive circuits. For example, suppose that the time constant of a capacitive circuit is 5 µs. Therefore, one RC unit corresponds to 5 µs, two RC units equals 10 µs, three RC units equals 15 µs, etc. To find how much voltage is across the capacitor after 10 µs of charging, take the value on graph a corresponding to two time constants, or 86 percent of the applied charging voltage.

The point where curves a and b intersect shows that a 50 percent change is accomplished in 0.7 time constant. These values are listed in Table D-1. The curves can be considered linear within the first 20 percent of change.

In Fig. D-3 the entire RC charge curve actually adds 63 percent of the net charging voltage for each increment of RC time, although it

TABLE D-1 CHARGE OR DISCHARGE PERCENTAGES

FACTOR	% CHANGE
0.2 time constant	20
0.5 time constant	40
0.7 time constant	50
1 time constant	63
2 time constants	86
3 time constants	96
4 time constants	98
5 time constants	99

may not appear so. In the second interval of RC time, for instance, v_C adds 63 percent of the net charging voltage which is 0.37 E. Then 0.63 × 0.37 equals 0.23, which is added to 0.63 to give 0.86 or 86 percent.

Long and short time constants. A long time constant is at least five times longer than the period of the applied voltage. Then the capacitor cannot take on any appreciable charge before the applied voltage drops to start the discharge. Also, there is little discharge before charging voltage is applied again. In an RC circuit with a long time constant, therefore, the applied voltage is mainly across R, with little voltage across C. An RC coupling circuit is an example. Note that a long time constant corresponds to little capacitive reactance, compared with R, as both require a relatively large value of capacitance.

A short time constant is one-fifth, or less, of the period of applied voltage. This combination permits C to become completely charged before it discharges. Then the applied voltage is on long enough to charge the capacitor for five time constants or more. Also, the capacitor can discharge completely before it is charged again by the applied voltage. In an RC circuit with a

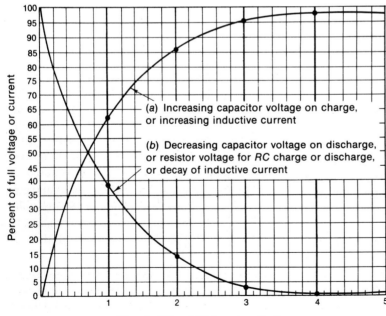

FIGURE D-3

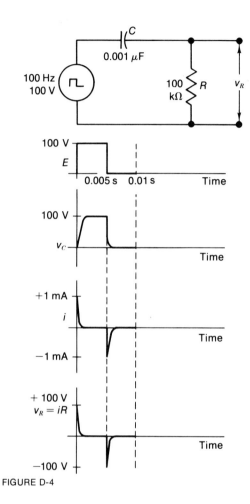

FIGURE D-4

the leading and trailing edges of the square-wave input. Such sharp pulses are useful for timing applications. In general, differentiated output for pulses corresponds to a high-pass filter for sine waves. In both cases, the high-frequency components of the input are in the output circuit.

In any circuit with nonsinusoidal input, the differentiated output has sharp changes in amplitude. This applies to i and v_R in an RC circuit but not to v_C, as the voltage across the capacitor cannot change instantaneously because of its stored charge. In an RL circuit, v_L can provide differentiated output. The current cannot change instantaneously in an inductance because of its magnetic flux.

The instantaneous values of v_L in an RL circuit can be calculated as $L\,di/dt$, where di/dt is the time rate of change of the current. Similarly, in an RC circuit the instantaneous value of i_C is equal to $C\,dv/dt$, where dv/dt is the time rate of change of the capacitor voltage. In making these calculations, use the basic units for all the factors.

short time constant, therefore, the applied voltage is mainly across C. An example of a short time constant is shown for the differentiating circuit in Fig. D-4.

Differentiated output. In an RC circuit with a short time constant, the voltage across R is called differentiated output because its amplitude can change instantaneously, in either polarity. In Fig. D-4, the voltage across R consists of positive and negative spikes corresponding to

Integrated output. In general, an integrated waveform does not have instantaneous changes in amplitude and cannot reverse in polarity. Therefore, v_C in an RC circuit can provide integrated output. In comparison with RC circuits for sine waves, the integrated output corresponds to the action of a low-pass filter. In this case, a shunt capacitor can be considered a bypass for the high-frequency components of the input. For RL circuits, i_L and V_R can provide integrated output, since the current cannot change instantaneously.

Nonsinusoidal waveshapes. For sine-wave circuits, the capacitive or inductive reactance determines the phase angle. With applied voltage that is not a sine wave, there can be changes

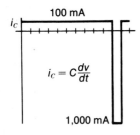
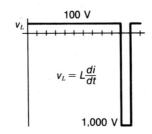
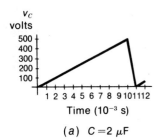
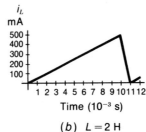

(a) $C = 2\ \mu F$ (b) $L = 2\ H$ FIGURE D-5

of waveshape, in either voltage or current, instead of the changes in phase angle that apply only for sine waves. Instead of reactance, the time constant of the RC or RL circuit must be considered.

Some examples of changes in waveshapes by RC and RL circuits are illustrated in Fig. D-5. The sawtooth capacitor voltage in a results from charging C slowly through a high R in a long time constant circuit, and discharging C fast through a small R in a short time constant circuit. The capacitor charge and discharge current has the rectangular waveshape because of the uniform rate of change for the sawtooth voltage. The values of i_C are calculated as $C\ dv/dt$ for a 2-μF capacitor. Similarly, in b the sawtooth current through an inductance corresponds to rectangular voltage. Here, the values of v_L are calculated as $L\ di/dt$ for a 2-H inductance.

Bibliography

Television

Bierman, H., and M. Bierman. *Color Television Principles and Servicing,* Hayden Book Company Inc., Rochelle Park, N.J., 1973.

Fink, D.G.: *Television Engineering Handbook,* McGraw-Hill Book Company, New York, 1957.

International Radio Consultive Committee: *Text of Plenary Assembly, Volume V, Part 2,* International Telecommunications Union, Geneva, Switzerland, 1970.

Kiver, M.S.: *Color Television Fundamentals,* McGraw-Hill Book Company, New York, 1964.

Levy, A., and M. Frankel: *Television Servicing,* McGraw-Hill Book Company, New York, 1959.

Markus, J.: *Television and Radio Repairing,* McGraw-Hill Book Company, New York, 1961.

Middleton, R.G.: *Transistor Color-TV Servicing Guide,* Howard W. Sams & Co., Inc., Indianapolis, 1971.

National Television System Committee: *Color Television Standards,* McGraw-Hill Book Company, New York, 1961.

Reed, C.R.G.: *Principles of Color Television Systems,* Sir Isaac Pitman & Sons, Ltd., London, 1969.

Simons, K.: *Technical Handbook for CATV Systems,* Jerrold Electronics Corporation, Philadelphia, 1968.

Tinnell, R.W.: *Television Symptom Diagnosis,* Howard W. Sams & Co., Inc., Indianapolis, 1973.

Towers, T.D.: *Transistor Television Receivers,* Hayden Book Company Inc., Rochelle Park, N.J., 1963.

Zbar, P., and P. Orne: *Basic Television Text-Lab Manual,* McGraw-Hill Book Company, New York, 1971.

Transistors

Cutler, P.: *Solid-State Device Theory,* McGraw-Hill Book Company, New York, 1972.

GE Transistor Manual, General Electric Company, Syracuse, N.Y.

Hibberd, R.G.: *Integrated Circuits,* McGraw-Hill Book Company, New York, 1969.

Kiver, M.S.: *Transistors,* 3d ed., McGraw-Hill Book Company, New York, 1962.

RCA Transistor, Thyristor and Diode Manual, RCA Electronic Components, Harrison, N.J.

Schure, A.: *Basic Transistors,* Hayden Book Company Inc., Rochelle Park, N.J..

Semiconductor Cross-Reference Guide, Motorola Semiconductors, Phoenix, Ariz.

Sowa, W.A.: *Active Devices for Electronics,* Rinehart Press, a division of Holt, Rinehart and Winston, Inc., Corte Madera, Calif., 1971.

Tocci, R.J.: *Fundamentals of Electronic Devices,* Charles E. Merrill Books, Inc., Columbus, Ohio, 1970.

Tomer, R.B.: *Semiconductor Handbook,* Howard W. Sams & Co., Inc., Indianapolis, 1968.

Audio-Visual Aids

Education Department, Hayden Book Company, Rochelle Park, N.J.

McGraw-Hill Films, McGraw-Hill Book Company, New York.

National Aeronautics and Space Administration, Washington, D.C.

National Audiovisual Center, General Services Administration, Washington, D.C.

Answers to Self-Examinations

Chapter 1

1. T
2. F
3. T
4. F
5. T
6. T
7. T
8. T
9. T
10. T

Chapter 2

1. 30
2. 525
3. 2
4. $262\frac{1}{2}$
5. 15,750
6. 15,750
7. 60
8. contrast
9. camera
10. picture
11. 6
12. AM
13. FM
14. 60 to 66
15. 4.5
16. sync
17. blanking
18. zero
19. 3.58
20. chrominance

Chapter 3

1. T
2. T
3. T
4. T
5. F
6. T
7. T
8. T
9. T
10. T
11. F
12. T
13. T
14. T
15. F
16. T
17. T
18. T
19. T
20. F

Chapter 4

1. b
2. a
3. d
4. a
5. b
6. c
7. d
8. a
9. b
10. d

Chapter 5

1. T
2. T
3. F
4. T
5. F
6. F
7. F
8. T
9. T
10. T
11. T
12. T
13. F
14. T

15. T
16. T
17. T
18. T
19. T
20. F

Chapter 6

1. b
2. b
3. d
4. a
5. c
6. d
7. b
8. a
9. c
10. c

Chapter 7

Part A

1. d
2. e
3. a
4. b
5. c

Part B

1. f
2. a
3. b
4. c
5. g

6. e
7. h
8. d

Part C

1. b
2. c
3. d
4. e
5. a

Part D

1. T
2. T
3. T
4. F
5. T
6. T
7. F
8. T
9. T
10. T

Chapter 8

1. c
2. a
3. d
4. a
5. d
6. a
7. d
8. b
9. b
10. c
11. c
12. a

Chapter 9

1. T
2. T
3. T
4. F
5. T
6. F
7. T
8. T
9. T
10. T
11. T
12. T
13. T
14. T
15. F
16. T
17. T
18. T
19. T
20. F

Chapter 10

1. T
2. F
3. T
4. T
5. F
6. T
7. T
8. T
9. T
10. T
11. T
12. T

ANSWERS TO SELF-EXAMINATIONS 711

13. F	13. f	3. T
14. T	14. t	4. F
15. T	15. m	5. T
16. T	16. h	6. T
17. T	17. j	7. T
18. T	18. d	8. T
19. T	19. q	9. T
20. T	20. l	10. T

Chapter 11

1. F
2. T
3. T
4. F
5. T
6. F
7. F
8. T
9. T
10. T

Chapter 12

1. k
2. n
3. o
4. i
5. s
6. r
7. p
8. c
9. b
10. e
11. a
12. g

Chapter 13

1. T
2. F
3. T
4. T
5. T
6. T
7. T
8. T
9. T
10. F
11. T
12. T
13. T
14. T
15. T
16. T
17. T
18. T
19. F
20. T

Chapter 14

1. T
2. F

Chapter 15

1. T
2. T
3. T
4. T
5. T
6. F
7. T
8. T
9. F
10. T
11. F
12. T

Chapter 16

1. T
2. T
3. T
4. T
5. T
6. T
7. T
8. F
9. T

10. T
11. T
12. T
13. T
14. T
15. T
16. F
17. T
18. T
19. T
20. F

Chapter 17

1. b
2. a
3. d
4. c
5. c
6. b
7. a
8. b
9. d
10. a

Chapter 18

1. T
2. T
3. T
4. T
5. T
6. F
7. T
8. F
9. T
10. T

Chapter 19

1. j
2. o
3. k
4. e
5. a
6. b
7. c
8. f
9. g
10. p
11. i
12. m
13. d
14. l
15. n
16. h

Chapter 20

Part A

1. T
2. F
3. T
4. T
5. F
6. F
7. T
8. T
9. T
10. T
11. T
12. T
13. T
14. T
15. T
16. F
17. F
18. T
19. T
20. T

Part B

1. 0.15
2. 0.00045
3. 0.003
4. 250
5. 1,000
6. 15,750
7. 16.4
8. 28
9. 30
10. 300

Chapter 21

1. a
2. d
3. a
4. b
5. a
6. c
7. c
8. b
9. b
10. d

Chapter 22

1. d
2. c
3. b
4. a
5. b

6. c
7. c
8. b
9. d
10. d
11. c
12. d

Chapter 23

1. T
2. T
3. F
4. F
5. T
6. T
7. T
8. T
9. T
10. T
11. T
12. T
13. F
14. T
15. T
16. T
17. T
18. T
19. F
20. T
21. T
22. T
23. T
24. T
25. T
26. T
27. T
28. T
29. T
30. T

Chapter 24

1. d
2. c
3. a
4. d
5. a
6. b
7. a
8. d
9. b
10. c
11. b
12. d

Chapter 25

1. b
2. c
3. c
4. c
5. d
6. c
7. d
8. d
9. a
10. c

Chapter 26

1. d
2. c
3. b
4. a
5. c
6. b
7. d
8. a
9. a
10. b

Chapter 27

Part A

1. 96.3
2. 10
3. 20
4. 50
5. 4
6. 4
7. 90°
8. 75
9. zero
10. 4.5

Part B

1. T
2. T
3. T
4. F
5. T
6. T
7. T
8. T
9. T
10. F
11. T
12. T
13. T
14. T

15. F
16. T
17. T
18. T
19. T
20. T

Chapter 28

1. e
2. c
3. f
4. b
5. h
6. a
7. l
8. k
9. d
10. j
11. i
12. g

Answers to Selected Problems

Chapter 1

No problems

Chapter 2

2. (a) 31.75 μs
4. 0.25 μs

Chapter 3

1. 200 mV
2. 24 mV
3. 3 μA

Chapter 4

2. 5 and 667 μs
3. (a) 63.5 μs
   (b) $1/60$ s
4. (a) 60 Hz
   (b) 15,750 Hz
   (c) 18.8 kHz

Chapter 5

1. 2.82 MHz
3. (a) 1,000 μs
   (b) 960 μs
   (c) 5.2 kHz

Chapter 6

1. (a) 55.0, 55.5, and 55.25 MHz
   (b) 80.25 and 77.25 MHz
   (c) 470.75, 471.75, and 471.25 MHz
   (d) 475.25 and 471.25 MHz
2. (d) 65 to 70 percent
4. (a) 4.5 MHz
   (b) 1.5 MHz
   (c) 6 MHz
   (d) 6 MHz
5. (a) 3.58 MHz
   (b) 920 kHz

Chapter 7

1. (a) 129 MHz
   (b) 221 MHz
   (c) 257 MHz
   (d) 517 MHz
   (e) 931 MHz

2. (a) 4.5 MHz
   (b) 4.5 MHz
   (c) 4.5 MHz
   (d) 3.58 MHz
   (e) 3.58 MHz

Chapter 8

1. 0.11, 0.30, 0.59, 0.89, and 1.0
3. (a) $C = 0.5$
   (b) Magenta hue
4. 0.68 V peak-to-peak and 0.34 average level

Chapter 9

1. (a) 107 MHz
   (b) 61.25, 64.83, 65.75 MHz
   (c) 45.75, 42.17, 41.25 MHz
   (d) 3.58 MHz
2. (a) 300 V
   (b) 100 V, positive

Chapter 10

1. Turn deflection yoke
2. Centering magnet rings
4. (a) Weak emission in picture tube
   (b) Weak high-voltage rectifier
5. 0.2, 0.3, 0.35, and 0.55 mA

Chapter 11

1. Deflection yoke
2. Screen controls
3. TB pincushion
4. Blue lateral magnet

Chapter 12

1. (a) 50 Ω
   (c) 18 W
3. 292 V
4. 4.5 kV

Chapter 13

1. (a) 1.6 MHz
2. (a) 3.2 Hz
3. 4 kHz
4. (a) 0.025 µs
   (b) 2,500 µs
5. 120 µH

Chapter 14

1. 10 V
3. (a) 150 V
   (b) 150 V
   (c) 60 V
4. 22 V for white; 73 V for black

Chapter 15

1. (a) 0.2 s
   (b) 0.2 s
2. 20 kΩ
3. 2,000
4. 66 dB

Chapter 16

1. 50 μs
3. (a) 0.1 s
   (b) 0.07 s
   (c) 0.1 s
   (d) 0.5 s
4. 360 Hz

Chapter 17

No problems

Chapter 18

No problems

Chapter 19

1. (a) $P = 55.25$ MHz, $S = 59.75$ MHz
       $P = 61.25$ MHz, $S = 65.75$ MHz
       $P = 67.25$ MHz, $S = 71.75$ MHz
   (b) $P = 51.75$ MHz, $S = 47.25$ MHz
       $P = 45.75$ MHz, $S = 41.25$ MHz
       $P = 39.75$ MHz, $S = 35.25$ MHz
3. (a) 1 pair
   (b) 7 pairs
4. (a) 59 pairs
   (b) 222 pairs
5. 0.75 in.

Chapter 20

1. (a) 10 V
   (b) 40 V
   (c) 50 V
   (d) 63 V
   (e) 86 V
   (f) 99 V
3. (a) 73 V
   (b) 27 V
5. (a) 100 μA
   (b) 100 V
   (c) 50 μA
   (d) 50 V
7. (a) +200 V
   (b) −200 V
   (c) −2,000 V

Chapter 21

1. 40 V
2. (a) 3.6 mA
   (b) 0.14 A
3. 15 Ω
4. 990 Ω

Chapter 22

1. (a) 5,000 µs
   (b) 79
   (c) Yes
2. −18 V, −36 V, and −72 V
3. (a) 100 mA
   (b) 20 W
4. 30 W
5. 40 µs

Chapter 23

1. (a) 4,000
   (b) 4,000
   (c) 800
3. 13 V
5. 1.37 µH

Chapter 24

1. 101 MHz and 257 MHz for VHF
2. $f = 107$ MHz; $L = 0.44$ µH
3. 0.028 ft
4. $X_C = 20.4$ MΩ for 60 Hz

Chapter 25

1. 1,000 µV, 600 µV, 200 µV, and 0
3. Dipole = 8.6 ft
   Reflector = 9.0 ft
   Spacing = 3.44 ft
5. 3.77 ft
7. 2.33 ft
9. 4 ft
11. 70 Ω

Chapter 26

1. Values in Problem 2
2. Values in Problem 1
3. Values in Problem 4
4. Values in Problem 3
5. 15-dB gain
6. 10,000 mV or 10 V
7. 66 dBmV
8. 10 mV

Chapter 27

1. (a) 5 kHz
   (b) 15 kHz
   (c) 30 kHz
   (d) 45 kHz
3. 3 dB down at 2.133 kHz
5. 34.6 mA

Chapter 28

1. (a) 16 V
   (c) 28 V
2. (a) 28 V
   (c) 1.2 V approx.

Index

ABL (automatic brightness limiter) circuit, 177, 399–401
Absorption trap, 528
ACC (automatic color control) circuit, 176, 397
Additive colors, 142
ADG (automatic degaussing) circuit, 177, 219
Adjacent channels, 95, 527, 417–418
AFC (automatic frequency control):
   horizontal, 340–347
   tuner, 561–564
AFPC (automatic frequency and phase control) circuit, 177, 370–376
   alignment, 375–376
AFT (automatic fine tuning) circuit, 176, 561–564
   alignment, 563
Afterglow, 205
AGC (automatic gain control) circuits, 109, 309–328
   amplified, 311, 322
   delayed, 311, 317
   keyed, 314–317
   level controls, 114–115, 323–324
   troubles with, 324–326
Airplane flutter, 312
Alignment curves, 680–682
Alpha, transistor, 123
Aluminized screen, 191
AM (amplitude modulation), defined, 89–90
Amateur radio bands, 697

Amplified AGC, 311, 322
Antenna:
   bandwidth, 586
   directivity patterns, 585
   distribution systems, 616–630
   front-to-back ratio, 586
   gain, 586
   impedance, 585–586
   loading, 589–590
   orientation, 608
   resonant length, 583–584
   stacked, 597–600
   troubles in, 611
   types of, 582–615
Antiringing network, 492
Aperture mask, 193–194
Aquadag coating, 186
Aspect ratio, 25
Associated sound signal, 26, 107, 111–112, 631–668
Astable MV, 439
ATC (automatic tint control) circuit, 176, 398–399
Audio circuits, 657–662
   push-pull, 660–661
Automatic brightness limiter (ABL) circuit, 177, 399–401
Automatic color circuits, 176–177, 394–404
   brightness limiter, 177, 399–401
   chroma control, 395–397

Automatic color circuits:
   color control, 176, 397
   color killer, 395–397
   degaussing, 177, 219
   fine tuning, 176, 561–564
   frequency and phase control, 177, 370–376
   one-button tuning, 177, 401–403
   tint control, 176, 398–399
Automatic color control (ACC) circuit, 176, 397
Automatic degaussing (ADG) circuit, 177, 219
Automatic fine tuning (AFT) circuit, 176, 561–564
Automatic frequency and phase control (AFPC) circuit, 177, 370–376
Automatic frequency control (*see* AFC)
Automatic gain control (AGC), 109, 309–328
Automatic tint control (ATC) circuit, 176, 398–399
Autotransformer, 228–229
   horizontal, 493–495
   vertical, 461–462
Avalanche diode, 125

B+ supply line, 230, 250–251
Back porch, 68–69
Balun, 606–607, 620, 628
Bandpass amplifier, 172–173, 365–368
Barkhausen oscillations, 507
Barrel distortion, 57
Bars in picture, 411–413
Base, transistor, 119
Baseband, defined, 92
Beam-bender magnet, 189
Beat:
   4.5 MHz, 286–287, 414
   920 kHz, 413
Bend in picture, 411
Beta, transistor, 123
Bias clamp circuit, 318
Bifilar IF coil, 522
Bipolar transistor, 119

Bistable MV, 439
Black setup, 72
Blanker stage, 387–388
Blanking:
   amplifier, 387–388
   CRT, 360
   frequencies, 60–61
   horizontal, 67–69
   vertical, 69–71
Blocking oscillator:
   circuit, 428–434
   controls, 434–436
   frequency, 429
   synchronization, 436–438
   transistor, 433–434
   waveshapes, 430–432
Blooming, 205, 254
Blue lateral magnet, 195, 211
Bonger unit, 576
Boosted B+, 479, 486–489
Booster:
   antenna, 627
   CRT, 205
   station, 99
Box-office TV, 9
Breathing, CRT, 205
Bridge rectifier, 238–239
Bridged-T trap, 528
Brightener, CRT, 205
Brightness:
   control, 115, 200–201
   defined, 23
   limiter, 399–401
   troubles with, 409
   varying, 297–298
Broadcasting, television, 2–3, 44–45
Broadside antenna array, 597–599
Burn, screen, 205
Burst:
   amplifier, 174, 369–370
   color sync, 148, 155
   separator, 174, 369–370

Buzz in sound, 112, 114, 663
B—Y:
   amplifier, 384–385
   axis, 377
   demodulator, 376–382
   signal, 150, 153–157

C (chrominance, chroma, or color) signal (see Chroma signal)
Cable television (CATV):
   channels, 569–570
   distribution systems, 616–630
   uses of, 6–7
Camera chain, 44–45
Camera tubes, 31–49
   image orthicon, 36–37
   plumbicon, 40–41
   silicon, 41–42
   solid-state, 43
   spectraflex, 43–44
   vidicon, 37–40
Capristor, 564
Cascade voltage doubler, 240
Cascode amplifier, 554–555
Cathode drive, CRT, 202
CATV (cable television), 6–7, 628
CB (common-base) circuit, 120–122
CC (common-collector) circuit, 120–122
CCTV (closed-circuit television), 7–8, 628
CE (common-emitter) circuit, 120–122
Center frequency, 635
Center-tuned discriminator, 645–648
Centering adjustments, 188
Channels, television, 4–5, 93–97, 694
Characteristic impedance, 602–604, 628
Cheater cord, 236
Chroma:
   amplifier, 172–173, 365–369
   defined, 144–145, 148
   demodulators, 376–382
   signal, 22–23, 144–145

Chrominance (see Chroma)
Circuit breakers, 255–257
CK (color killer) circuit, 174–175, 395–397
Clamping, 299
Closed-circuit television (CCTV), 38
Cochannel interference, 417–418
   stations, 95
Collector, transistor, 119
Color addition, 141–144
Color amplifier, 365–368
Color-bar generators, 213, 673, 683–686
Color circuits, 385–393
   automatic, 394–404
   troubles in, 388–390
Color controls, 176
Color killer (CK) circuit, 174–175, 395–397
Color oscillator, 173, 371–375
Color picture tubes:
   convergence, 195, 212–216, 222–223
   degaussing, 218–219
   drive adjustments, 169, 218
   gray scale, 218, 223
   pincushion correction, 219–222
   purity, 195, 211–212, 222
   screen adjustments, 217–218
   setup adjustments, 210–215
   shield, 210–211
   types of, 193–200
Color receivers, 167–184
   circuits, 358–404
   picture tubes, 193–200, 210–225
   troubles in, 180–182
Color subcarrier, 161
Color television:
   broadcasting, 5–6, 22–23, 145–149
   systems, 168–169, 688–689
Color temperature, 218
Colorplexed video signal, 157–158
Common-base (CB) circuit, 120–122
Common-collector (CC) circuit, 120–122
Common-emitter (CE) circuit, 120–122
Compactron tube, 116

Compatible color, 23, 145
Complementary colors, 143–144
Complementary symmetry, 470–471, 661
Composite video signal, 65–87
Conical dipole, 595–596
Continuing spot, 205
Contrast, 23–24, 264
    control, 266–268
    preset, 401
Convergence:
    dynamic, 195, 213–216, 222–223
    procedure for, 222–223
    static, 195, 213–215, 222
    waveshapes, 216–217
    yoke, 169, 213–214
Conversion gain, IF, 556
Converter:
    frequency, 105, 556–557
    sound, 111, 639–640
Corner reflector, 595
Corona, 254–255
Coupler, antenna, 610, 619
Crossover point, CRT, 192
Curve tracer, transistor, 673

Damper diodes, 116–117, 482–485
Damping:
    horizontal, 482–485
    IF, 518, 520–521
    vertical, 459
Darlington pair, 122
dB (decibel) units, 623–627
Dc component, 78–80, 196–208, 298–300
Dc coupling, 301–302, 305–306, 389–390, 675
Dc insertion, 299, 303–305
Dc reinsertion, 299, 303–305
Dc restorer, 299, 303–305
Decibel (dB) units, 623–627
Decoding, color, 140–141, 149–150
Deemphasis, 636

Deflection amplifiers:
    horizontal, 477–512
    vertical, 457–476
Deflection angle:
    CRT, 187
    yoke, 491–493
Deflection oscillators:
    blocking, 428–434
    multivibrators, 438–447
    synchronizing, 436–438
    triggering, 436–438
    wrong frequency, 450–451
Deflection yoke:
    adjustment of, 187–188
    troubles in, 473
    types of, 491–493
Degaussing:
    automatic, 218–219
    coil, 219
Delay line, 170–172, 362–363
Delayed AGC, 311, 317
Delta gun, 197
Demodulator probe, 676–677
Demodulators, color, 173–174, 376–382
Depletion, FET, 123
Desaturated colors, 159
Desoldering, 130
Detail, in picture, 24–25, 268–269, 272
Detector, video, 283–287, 675
Detent, tuner, 550
Deviation, frequency, 635
Diac, 125
Diagonal lines, in picture, 415–417, 451
Diathermy interference, 416, 697
Differentiator, RC, 705
DIP package, 127
Dipole antennas:
    conical, 595–596
    folded, 590
    straight, 588–589
    V-type, 588
Directional coupler, 619

Director with dipole, 593–594
Discriminators:
    FM, 642–652
        alignment of, 655–658
        center-tuned, 645–648
        Foster-Seeley, 645–648
        phase-shift, 645–648
        triple-tuned, 642–645
    sync, 341–345
        push-pull, 342–343
        single-ended, 343–345
Doping, transistor, 118–119
Dot-bar generator, 213, 673, 683–686
Double superheterodyne, 111
Double-tuned stages, 519–522
Drain, transistor, 122–123
Drive controls, receiver, 169, 218
Dud, CRT, 205
Dynamic convergence, 195, 213–216

Eccles-Jordan trigger, 439
Efficiency diode, 117, 478
EHF (extra-high frequencies), 697
EIA (Electronic Industries Association), 14
Electron beam:
    deflection, 192–193
    focus, 191–192
Electronic Industries Association (EIA), 14
Electronic tuning, 568–573
Element, picture, 17–18, 75–77
Elevator transformer, 606–607, 628
Emitter, transistor, 119
Emitter-follower, 121–123
Encoding color, 140–141, 145–149
End-fire antenna, 599
Enhancement, transistor, 123
Envelope, carrier, 91–93
Equalizer, cable, 627
Equalizing pulses, 59–60, 336
Extra-high frequencies (EHF), 697

F-connector, 602
f-number, 46
Facsimile, 10
FCC (Federal Communications Commission) allocations, 696–697
Feedthrough capacitors, 552–553
Ferrite beads, 128
FET (field-effect transistors), 122–123, 689
Field, television, 20–21
Field-effect transistors (FET), 122–123, 689
Field-strength meter, 617–618
Film camera, 44–45
Filters:
    active, 245
    capacitor input, 244–245
    choke input, 244–245
    ripple, 245
Fine tuning control, 114, 548, 560–561
Fishtailing effect, 58
Flicker, 20, 56
Flyback:
    defined, 19
    high voltage, 110, 252–255, 479, 489–490
    transformer, 407, 478
Flying-spot scanner, 35–36
FM broadcast band, 697
FM detectors, 641–657
    alignment, 655–658
    discriminators, 642–652
    quadrature-grid, 652–655
    ratio detector, 649–652
    S-response curve, 644
FM receivers, 639–641, 661–663
Focal length, 46
Focus:
    electrostatic, 191–192
    high-voltage, 192
    low-voltage, 192
    picture tube, 188
Fold-over:
    horizontal, 508
    vertical, 472

Folded dipole, 590
Forward AGC, 313–314
Forward voltage, 119
Foster-Seeley discriminator, 645–648
Frame, television, 18–21
Frequency dividers, 446
Frequency interlace, 162
Frequency modulation, 631–668
 advantages of, 637
 center frequency, 635
 deemphasis, 636
 detection of, 641–657
 deviation of, 635
 disadvantages, 637
 discriminators, 642–652
 limiting, 641, 648–649
 multipath reception, 637–638
 PM, 635–636
 preemphasis, 636
 quadrature-grid detector, 652–655
 ratio detector, 649–652
 reactance tube, 638–639
 receivers, 639–641, 661–663
 swing, 636
Fringing, color, 196, 212
Front end, 105–106, 546–581
Front porch, 68–69
Front-to-back ratio, 586
Full-wave rectifiers, 230–231, 237–239
Fuses, 132, 255–257
 testing of, 256
Fusible resistors, 257

Gain, antenna, 586
Gamma, 80–81
Gate, FET, 122–123
Gated AGC, 314–317, 350–351
Gated rainbow, 684–686
Germanium, 117–118
Ghosts, 414–415, 586–588
Grainy picture, 540

Gray scale, 77–78
 tracking, 218, 223
Grid-leak bias, 479, 497, 499, 675–676
Grounded-grid amplifier, 554
G–Y signal, 150, 154, 156–157, 384–386

Hammerhead pattern, 83, 352–353
Harmonic tuning, horizontal, 481
Hash filter, 236
Head-end equipment, 617–618
Heat sink, 124
Heater circuits:
 parallel, 241–243
 series, 243–244, 250–251
Height control, 434–435
Herringbone effect, 416
HEW (U.S. Department of Health, Education, and Welfare), 133, 203
High voltage:
 flyback, 110, 252–255, 489–490
 focusing, 253
 limit control, 503
 troubles, 507–508
Hold control:
 horizontal, 115, 331, 354
 vertical, 115, 330, 353, 435
Hole:
 charge, 119
 current, 119
Horizon distance, 97–98
Horizontal AFC, 340–347
 discriminator, 341–345
 filter, 345–346
 hunting, 346
 phasing, 351–357
Horizontal amplifier, 479–482
 protective bias, 499–500
 SCR, 502–503
 transistors, 500–502
Horizontal deflection circuits, 477–512
 analysis, 495–497

Horizontal deflection circuits:
    controls, 490–492
    functions, 478–479
    output transformer, 493–495
    ringing, 482–485
    troubles with, 503–508
    yoke, 491–493
H.O.T. (horizontal output transformer), 478, 493–495
Hot carrier diode, 127
Hue:
    defined, 25, 144
    phase angles, 155–157
    preset, 403
Hum:
    in power supply, 251–252
    in sync, 354
    troubles with, 419–420
Hunting, in AFC, 346
Hybrid receivers, 105

I signal, 146–148, 152–153
IC (integrated circuit) units, 126–127
Iconoscope, 36
IF amplifiers:
    alignment, 532–539
    double-tuned, 519–522
    frequencies, 514–515
    link coupling, 520–521
    neutralizing, 522–524
    response curve, 524–527
    single-tuned, 517–519
    stagger-tuned, 519
    troubles with, 539–541
    wave traps, 527–529
IF noise, 550
IGFET (insulated gate field-effect transistors), 123
Ignition noise, 418–419
Image dissector, 35
Image frequencies, 551

Image orthicon (I.O.), 36–37
Implosion, 203
Industrial radio band, 697
In-line antenna, 596
In-line guns, 197–200
Insertion loss, 620–623, 628
Instant-on power supply, 110, 235–236
Insulated gate field-effect transistors (IGFET), 123
Integrated circuit (IC) units, 126–127
Integrated output, RC, 705
Integrated thyristor and rectifier (ITR), 503
Integrator, vertical, 334–337
Intensity modulation, 201
Intercarrier buzz, 26, 663
Intercarrier sound, 26, 107–108, 111–112, 639–640
Interference, 413–420
    adjacent channel, 527
    cochannel, 417–418
    diagonal line, 415–417
    diathermy, 416
    frequencies, 697
    ghosts, 414–415
    herringbone effect, 416
    hum bars, 419–420
    ignition noise, 418–419
    sound bars, 413–414
    venetian-blind effect, 417–418
    windshield-wiper effect, 417–418
Interlace, frequency, 162
Interlaced scanning, 52–56, 354
Internal blanking, 464–465
Inversion of sidebands, 111, 525
I.O. (image orthicon), 36–37
Ion-trap magnet, 189
Isolation loss, 620–623, 628
ITR (integrated thyristor and rectifier), 503

Jitter, vertical, 471–472
Junction transistors, 117–129

Keyed AGC, 314–317
Keyer, burst, 174, 369–370
Keystone distortion, 57, 473
Killer, color, 174–175, 395–397

Lag, vidicon, 39
LCP (low-capacitance probe), 676–677
LED (light-emitting diode), 127
Lenses, camera, 34
Light splitter, camera, 35
Light units, 46–47
Lightning arrestor, 608
Limiter, AM, 641, 648–649
Line of sight, 97–99
Line-connected supply, 229
Line crawl, 56
Line-isolated supply, 229
Line pairing, 58
Linearity:
   horizontal, 490–491
   vertical, 462–464
Link alignment, IF, 538
Link coupling, 520–521
Lissajous patterns, 679
Local oscillator, 557–561
Log-periodic antenna, 597
Low-capacitance probe (LCP), 676–677
Low-voltage power supply, 109–110, 226–260
   heater circuits, 241–244
LPV (log-periodic V) antenna, 597
L/R time constant, 702
Luminance:
   amplifier, 287–288
   defined, 145
   signal, 5, 24, 145, 151–152

Marconi antenna, 583
Master antenna television (MATV), 628
Matrix, 146, 150–152, 180, 382–384
MATV (master antenna television), 628
Memory circuit, 576–577
Memory fine tuning, 560–561
Metal-oxide semiconductor field-effect transistor (MOSFET), 123
Microvolts per meter, 584–585
Microwaves, 697
Midband channels, 569
Miller effect, 277
   integrator, 469–470
Mixer stage, 556–557
Modulation hum, 420
Modules, 129
Moiré effect, 58
Monopole antenna, 609
Monostable MV, 439
MOSFET (metal-oxide semiconductor field-effect transistor), 123
Motion pictures, 19–20
Multimeters, 670–675
Multipath antenna signals, 586–588
Multiplexing, color, 6, 145
Multitaps, 622
Multivibrator (MV) circuits, 438–447
   astable, 439
   cathode-coupled, 441–445
   collector-coupled, 446–447
   frequency, 441
   monostable, 439
   plate-coupled, 439–441
   synchronizing, 445–446
   vertical deflection, 465–467
MV [see Multivibrator (MV) circuits]

National Television Systems Committee (see NTSC)
Negative transmission, 89
Neutralization:
   IF, 522–524
   rf, 553–554
Noise:
   in picture, 550–551, 578

Noise:
   snow, 550–551, 578
   in sync, 337–340
Noise gate, 318, 338, 349–350
Noise inverter, 318, 338
Noise switch, 340
Normal-service switch, 169–170
Novar base, 115
NPN transistors, 119–124
NTSC (National Television Systems Committee):
   defined, 14
   generator, 685–686
   system, 162–163, 698–699

OBT (one-button tuning), 177, 401–404
Octave, 270
Odd-line interlace, 52–56
Offset carrier, 98
One-button tuning (OBT), 177, 401–404
Orientation, antenna, 608
Oscillator:
   color, 173–174, 371–375
   deflection, 424–456
   tuner, 107, 557–561, 573
     radiation, 551
Oscilloscopes, 672, 675–680
   picture on screen, 679–680
   probes, 676–677
   vectoscope, 680
Overloaded picture, 310, 324, 411

Pairing, line, 58
PAL (Phase Alternation by Line), 162–163, 699
Parasitic antennas, 592–595
   director, 593–594
   reflector, 593
   Yagi, 594
Patching methods, 683
Pay TV, 9

Peak reverse voltage (PRV) rating, 228
Peaking:
   coil, 278
   control, 278–279
Percent modulation:
   AM, 90
   FM, 635
Persistence:
   screen, 190
   vision, 19–20
Phase-angle measurements, 678–679
Phase angles, color, 155–157, 377
Phase distortion, 273–276
Phase modulation (PM), 635–636
Phase-shift discriminator, 645–648
Phasing rods, antenna, 598
Phosphors, screen, 189–190
Photocathode, 32
Photoconductivity, 33
Photoemission, 32
Picture elements, 18–19, 75–77
Picture phone, 9–10
Picture signal:
   modulated, 88–102
   negative transmission, 89
   vestigial sidebands, 89–93
Picture tubes, 185–225
   aluminized screen, 190
   aperture mask, 193–194
   base connections, 205–206
   booster, 204–205
   centering, 188
   color, 193–200, 210–225
   deflection angle, 187–188
   delta guns, 197
   electron beam, 190–191
   focusing, 188, 191–193
   in-line guns, 197–200
   matrix, 180
   monochrome, 185–193
   phosphors, 189–190
   rejuvenator, 673

Picture tubes:
   shadow mask, 193–194
   signal drive, 201–202
   test jig, 673
   tester, 673
   troubles with, 204–205
   type numbers, 186, 206–207
   x-radiation, 203–204
PIN (pincushion) correction, 219–222
Pincushion (PIN):
   correction, 219–222
   distortion, 57, 219
Plumbicon, 40–41
PM (phase modulation), 635–636
PN junction, 119
PNP transistors, 119–124
Polarization, antenna, 584
Power supplies, 226–260
   B+ line, 230, 250–251
   circuit breakers, 255–257
   filter circuits, 244–245
   fuses, 255–257
   heater circuits, 241–244, 250–251
   high-voltage, 252–255
   hum, 251–252
   instant-on, 235–236
   leakage current, 133
   low-voltage, 226–252
   quick-on, 235–236
   rectifier circuits, 231–241
   transformers, 228–229
   troubles with, 250–251
   voltage regulators, 228, 245–248, 251
Power transformers, 228—229
   autotransformer, 228
   isolation, 229
Preamplifier:
   antenna, 618
   video, 361–362
Preemphasis, 636
Preselector, 550
Preset controls, 401–403

Primary colors, 142–143
Printed wiring, 129–130
Probes, oscilloscope, 676–677
Product safety, 672
Projection TV, 8–9
PRV (peak reverse voltage) rating, 228
Pulse-cross display, 353
Pulse repetition rate, 429–430
Purity:
   adjustments, 195, 211–212, 222
   magnet, 169, 211–212
Push-pull transistors, 470–471, 660–661

Q matching section, 605–606
Q (quadrature) signal, 146–147, 153
Quadrature-grid detector, 652–655
Quasi-complementary symmetry, 470–471, 661

Radiation:
   oscillator, 551
   x-ray, 133, 203–204
Rainbow generator, 784–786
Ramp voltage, 427
Raster:
   defined, 50
   troubles with, 405–408
Ratio detector, 649–652
RC circuits, 700–706
RC coupling, 300–301
RC time constant, 700–706
   universal graph, 703–704
Reactance tube, 638–639
Reaction scanning, 479, 485–487
Receiver noise, 550
Rectifier circuits, 227–241, 258
   high-voltage, 253–254
Reflector, with dipole, 593
Regulation, voltage, 228, 245–248, 251
Relaxation oscillator, 430
Remote cutoff tube, 529

Remote tuning, 574–577
Resolution, 24–25, 77
   color, 159—160
Rest frequency, 635
Restorer, dc, 299, 303–305
Retrace, 19, 52
   suppressor, 291, 464–465
Reverse AGC, 313
Reverse voltage, 119
Rf amplifier, 550–556
Rf tuners, 546–581
R,G,B receiver, 179
Rhombic antenna, 592
Ringing:
   color oscillator, 373–374
   in horizontal amplifier, 482–485
   in IF amplifier, 541
   in video amplifier, 411
Rolling picture, 330
R—Y, B—Y receiver, 177
R—Y demodulator, 376–382
R—Y signal, 150, 153–154, 156–157

S response curve, 644
S-shaping capacitor, 449–450
Satellites, 10–11
Saturable reactor, 222
Saturation, color, 25, 144, 149
Sawtooth capacitor, 428
Sawtooth generator, 426–428, 432–434, 443–445
Sawtooth waveform, 51–52, 425–427
Scan rectification, 478
Scanning, 4, 18–19, 50–64
   frequencies, 21, 52
      color, 161–162
Schmidt projection, 8–9
Schmitt trigger, 439
SCR (silicon-controlled rectifier), 124, 689
Screen-grid adjustments, 217–218
SECAM (sequential technique and memory storage), 163, 699

Semiconductors, 117–128
Series peaking, 278
Serrations, 59–60
Servicing:
   product safety, 672
   receiver, 669–692
Setup, black, 72
Shadow area, 99
Shadow mask, 193–194
Sharpness control, 279
SHF (superhigh frequencies), 697
Shunt peaking, 278
Sidebands, 90–93
Signal injection, 682–683
Silicon, 117–118
Silicon-controlled rectifier (SCR), 124, 689
Silicon-diode rectifiers, 227–228
Silicon-diode vidicon, 41–42
Silicon hum bar, 236
Single sideband transmission, 92–93
Size controls, 434–435
Slope detection, 641
Slow-blow fuse, 256
Smear:
   in picture, 411
   tunable, 540
Snivets, 507
Snow, in picture, 410, 578
Solid-state devices, 117–129
Solid-state image sensor, 43
Sound bars, 413–414, 527–528
Sound takeoff, 111, 289–291, 640
Source, transistor, 122–123
Source follower, 123
Spark-gap capacitor, 128, 204
Spectraflex camera tube, 43–44
Split-sound receiver, 112
Splitter, cable TV, 619, 628
Square-wave testing, 276
Stabilizing coil, 338–340
Stacked antennas, 597–600
Stagger-tuning, 518–519

Staircase signal, 77–78, 684
Standards, transmission, 27, 698–699
Standing-wave ratio, 603
Static convergence, 195, 213–215
Stereo sound, TV, 663
Sticking picture, 37
Streaking, in picture, 411
Stubs, transmission line, 605
Subcarrier, color, 145, 147–148
Subscription TV, 9
Subtractive primary colors, 143
Superband channels, 569–570
Superhigh frequencies (SHF), 697
Suppressed subcarrier, 148
Sweep capacitor, 428
Sweep generator, 673, 680–681
Sweep voltage, 427
Swing, frequency, 635
Switching diodes, 572
Sync, color, 148, 155, 174, 370–376
Sync discriminator, horizontal, 340–347
Sync separation, 331–337
Synchroguide, 347
Synchrolock, 347
Synchronizing circuits, 329–357
Synchronizing pulses, 59–61
Synchronous demodulators, 150, 376–382

Tapoff, cable, 619–620, 628
Television:
   applications, 1–15
   history, 13–14
   standards, 27, 688–689
Test equipment, 672–673
Test jig, 673
Test patterns, 77–78
Theatre TV, 8–9
Thermal grease, 124
Thermal runaway, 122
Thermistor, 128

Thyristor, 124–125
Tilt, cable, 628
Time base, 427
Time delay, video, 273–276
Tint:
   automatic, 398–399, 403
   defined, 144, 155–156
Toroidal yoke, 199, 492–493
Transformers:
   horizontal, 493–495
   oscillator, 430
   power, 228–229
   vertical, 461–462
Transients, RC, 700–706
Transistors, 117–129
   circuits, 120–122
   doping, 118–119
   field-effect, 123
   heat sink, 124
   NPN, 119–124
   PNP, 119–124
   tester, 673
   troubles in, 689–690
   type numbers, 123–124
Translator station, 99
Transmission lines, 600–607
   characteristic impedance, 602–604
   coaxial cable, 601–602
   foam types, 602
   losses, 601
   open-wire, 601
   resonant sections, 604–605
   stubs, 605
   twin-lead, 601–602
   velocity factor, 606
Trap circuits:
   4.5 MHz, 288–291
   IF, 527–529
Trapezoidal raster, 57, 408–409, 473
Trapezoidal waveshape, 447–449
Triac, 124

Triggered sync, 436–438
Trinitron, 198–199
Troubles:
  AGC, 324–326
  antenna, 611
  bend in picture, 411
  brightness, 409
  color, 180–182, 388–390, 410
  ghosts, 414–415, 586–588
  height, 407–409
  high voltage, 507–508
  horizontal, 503–508
  IF, 539–541
  interference, 415–419
  localizing, 130–131
  low-voltage supply, 250–251
  multiple, 131–132
  picture tube, 204–205
  smear, 411
  snow, 410
  sync, 353–354
  tuner, 550, 577–578
  vertical, 471–473
    roll, 330
  width, 407–409
Troubleshooting techniques, 670–687
  alignment curves, 680–682
  color bars, 685–686
  grid-leak bias, 675–676
  injecting signal, 682–684
  oscilloscope measurements, 675–680
  patching methods, 683
  transistor troubles, 689–690
  voltage tests, 670–675
Trunk line, 616–617, 622–623, 628
Tunable hum, 420
Tunable smear, 540
Tuner, rf, 105–106, 546–581
Tuner subber unit, 577
Tunnel diode, 127
Turret tuners, 548–549

Twin lead, 601–602

UHF antennas, 596–597
UHF channels, 4–5, 14, 93–97, 547–548, 697
UHF tuner, 107, 547–548, 565–569, 574
Ultrasonic sound waves, 574
Ultraudion oscillator, 558–559
Unipolar transistors, 122
Utilization ratio, 75–76
U/V splitter, 597

V antenna:
  dipole, 595–596, 599
  long-wire, 591
Varactor:
  diode, 126
  tuning, 568–573
Varicap, 570
Varistor, 128
VDR (voltage-dependent resistor), 128
Vectorscope, 680
Velocity factor, 606
Venetian-blind effect, 417–418
Vertical deflection circuits, 457–476
  linearity, 460–464
  Miller integrator, 469–470
  multivibrator, 465–467
  retrace suppressor, 464–465
  transistors, 466–471
  troubles with, 471–473
  waveshapes, 460
Vertical interval test signals (VITS), 82–83
Vertical sync, 334–337
Vestigial sidebands, 93–97, 525–527
VHF channels, 4–5, 14, 93–97, 547–548, 697
VHF tuner, 105–106, 547–550, 564–565
  turret, 548–549
  switch, 548–550
  varactor, 568–573

Video, defined, 2
Video amplifiers, 108–109, 261–295
   circuits, 280–283
   4.5 MHz trap, 289–291
   frequencies, 268–273
   load line, 266–267
   low-frequency compensation, 280
   luminance amplifier, 287–288, 363–365
   peaking, high-frequency, 278–279
   preamplifier, 361–362
   time delay, 273–276
   Y amplifier, 287–288, 363–365
Video detector, 107, 283–287, 675
   dc level, 285
   test point, 286
Video disc recording, 12–13
Video IF amplifiers, 107, 513–545
Video signal, 65–87, 262–265
   amplitudes, 66–67, 262–268
   colorplexed, 157–158
   dc level, 296–308
   distortion, 272–276
   frequencies, 268–273
   functions, 288–289
   polarity, 264–265
Video tape recording (VTR), 11–13, 45
Vidicon, 37–40
Viewing distance, 25
VITS (vertical interval test signals), 82–83
Voltage-dependent resistor (VDR), 128
Voltage doubler, 240–241
Voltage quadrupler, 241–242
Voltage-reference diode, 125
Voltage regulators, 245–248
Voltage tripler, 241–242
Volume control, 115
VTR (video tape recording), 11–13, 45

Wall outlets, cable, 622–623

Wave traps:
   4.5 MHz, 288–291
   IF, 527–529
Wavelength:
   antenna, 583–584
   units, 47, 696
White:
   defined, 144
   level, 66
Wideband amplifiers, 269–270
Width control, 490
Willemite, 189
Window signal, 77–78, 684
Windshield-wiper effect, 417–418
Wobbulator, 632
World television, 10–11
   standards, 688–689

X axis:
   color, 157
   demodulator, 177–179
X rays, 133, 203–204

Y signal, 5, 145, 151–152
   amplitudes, 157–158
   bandwidth, 152
Y video amplifier, 287–288, 363–365
   delay line, 362–363
Yagi antenna, 594
Yoke:
   convergence, 169, 213–214
   deflection, 187–188, 491–493

Z axis:
   color, 157
   modulation, 201
Zener diode, 125–126, 246

DATE DUE

NOV 1 1995

NEW ENGLAND INSTITUTE OF TECHNOLOGY LIBRARY

```
TK6630                              3233
G75        GROB, BERNARD
1975          BASIC
COPY 3     TELEVISION:
           PRINCIPLES AND
```